Polymer Insulation Applied for HVDC Transmission

Boxue Du
Editor

Polymer Insulation Applied for HVDC Transmission

Springer

Editor
Boxue Du
School of Electrical and Information Engineering
Tianjin University
Tianjin, China

ISBN 978-981-15-9730-5 ISBN 978-981-15-9731-2 (eBook)
https://doi.org/10.1007/978-981-15-9731-2

This Springer imprint is published by the registered company Springer Nature Singapore Pte Ltd.
The registered company address is: 152 Beach Road, #21-01/04 Gateway East, Singapore 189721, Singapore

Preface

The integration and reliability of HVDC power transmission and transformation equipment are facing great challenges, considering its rapid development of voltage level and capacity. The insulation system must withstand severe electric field distortion, temperature rise, mechanical stress, etc. Advanced insulation materials with unique comprehensive performance provide effective solutions to the above problems. The dielectric properties, failure mechanisms, and exploring novel insulation materials become the core issues. What comes out from the first overview of the contents in this book covers most of the power equipment in the HVDC system, like polymeric insulated cable, gas-insulated transmission line (GIL), film capacitors, which play critical roles in large-scale power transmission.

There are several important properties that matter a lot in polymer insulation such as interface property, space charge, and treeing. The topics range from electrical tree degradation, nanodielectric fabrication, parameter optimization, and feasibility evaluation of advanced materials for HVDC insulation. However, polymer dielectrics is really an interdisciplinary area concerning chemistry, physics, and mechanics. Many principles and theories are not confirmed or even unknown for both academic researchers and industrialists. To this end, we write this book to provide some valuable researches, hoping to inspire readers with new ideas of polymer properties and new research methods.

This book is a collection of previous researches of many scholars in this field. People with some basic background knowledge may find it with abundant content covering many hot topics in recent years. In chapters "DC Insulation Performance of Crosslinked Polyethylene for HVDC Cables"–"Electric Ageing and Breakdown Phenomenon in Polypropylene under Complex DC Operating Condition," we focused on the main insulation in HVDC cables. Space charge properties, breakdown phenomenon in polyethylene/polypropylene insulations and their modification methods were discussed. In chapters "Conduction and Charge Transport Characteristics of Silicone Rubber Composites with Nonlinear Conductivity"–"Theoretical Model and Suppressing Method of Interface Charge Accumulation in HVDC Cable Accessory: A Review," we provided an interesting study on HVDC cable accessories with multilayer insulation structure which required new dielectric properties different from main insulation system. We also paid attention to traditional equipment like GIS/GIL in chapters "Epoxy Resin Insulating Composites for Vacuum Cast Electrical Insulators of GIS"–"Mechanical Reliability of Large-Scaled Epoxy Insulator in GIS." The charge accumulation and electric field distribution of epoxy resin insulators are studied; several typical methods of electric field homogenization of insulators such as Functional Graded Materials and Surface Molecular Structure modification are introduced and discussed. Besides, the edge of new technology is our first concern. In chapter "Insulation Design of Superconducting Gas Insulated Transmission Line," polymer dielectrics and insulation design of superconducting GIL were involved. Besides, in chapters "Polymer Film Capacitors Applied in HVDC Transmission"–"High Temperature Dielectric Materials for Electrical Energy Storage," polymer insulation of power capacitor and energy storage capacitor is discussed. Some of these technologies have just come into practice in power system for a few years and we are aiming to provide data and experiments support for practice with current research.

Accelerating the Discovery of New Dielectric Properties in Polymer Insulation started for me when I worked as the leader of the high voltage lab of Tianjin University, China. This book includes many aspects of HVDC polymer insulation, some of which are quite different from each other. It is really a team effort to finish this book. It is my pleasure to acknowledge the numerous contributions made by my team, among which the proofreading done by Dr. Jin Li and Mr. Chenlei Han are highly recognized.

Tianjin, China Boxue Du

Contents

Contributors

Chi Chen School of Electrical Engineering, Xi'an University of Technology, Xi'an, China

George Chen University of Southampton, Southampton, UK

Wei Chen State Grid Nanjing Power Supply Company, Nanjing, China

Chuanhui Cheng Electric Power Research Institute, China Southern Power Grid, Guangzhou, China

Sang Cheng State Key Lab of Power Systems, Department of Electrical Engineering, Tsinghua University, Beijing, China

Qingguo Chi Key Laboratory of Engineering Dielectrics and Its Application, Ministry of Education, Harbin University of Science and Technology, Harbin, People's Republic of China

Xiaoyang Cui Institute of Electrical Engineering, Chinese Academy of Sciences, Beijing, China

Zhi-Min Dang Tsinghua University, Beijing, China

Jianan Dong School of Electrical and Information Engineering, Tianjin University, Tianjin, People's Republic of China

B. X. Du Tianjin, China

Boxue Du School of Electrical and Information Engineering, Tianjin University, Nankai, Tianjin, China

Bin Du Hefei University of Technology, Hefei, Anhui, China

Qijun Duan Hebei Provincial Key Laboratory of Power Transmission Equipment Security, North China Electric Power University, Baoding, China

Kailun Fan Tianjin University, Nankai District, Tianjin City, China

Baozhong Han Key Laboratory of Engineering Dielectrics and Its Application, Ministry of Education, Harbin University of Science and Technology, Harbin, China

Chenlei Han School of Electrical and Information Engineering, Tianjin University, Tianjin, China

Chong Han Tianjin University, Tianjin, China

Tao Han Key Laboratory of Smart Grid of Education Ministry, School of Electrical and Information Engineering, Tianjin University, Tianjin, China

Yuanhan He Hefei University of Technology, Hefei, Anhui, China

Zhaohao Hou Tianjin University, Tianjin, China

Fei Kong Beijing International S&T Cooperation Base for Plasma Science and Energy Conversion, Institute of Electrical Engineering, Chinese Academy of Sciences, Beijing, China

Chunyang Li Key Laboratory of Engineering Dielectrics and Its Application, Ministry of Education, Harbin University of Science and Technology, Harbin, China

Jin Li School of Electrical and Information Engineering, Tianjin University, Nankai, Tianjin, China

Qi Li State Key Lab of Power Systems, Department of Electrical Engineering, Tsinghua University, Beijing, China

Wei-Kang Li Global Energy Interconnection Research Institute Corporation, Beijing, China

Wendong Li State Key Laboratory of Electrical Insulation and Power Equipment, Xi'an Jiaotong University, Xi'an, Shaanxi, China

Zhonglei Li School of Electrical Engineering and Automation, Tianjin University, Tianjin, China;
Key Laboratory of Smart Grid of Education Ministry, School of Electrical and Information Engineering, Tianjin University, Tianjin, China

Hucheng Liang Tianjin University, Nankai, Tianjin, China

Chuanjie Lin Tsinghua University, Beijing, China

Haoliang Liu Tianjin University, Nankai District, Tianjin City, China

Fangcheng Lü School of Electrical and Electronic Engineering, North China Electric Power University, Beijing, China

Zhaoyu Ran School of Electrical and Information Engineering, Tianjin University, Tianjin, People's Republic of China

Haoou Ruan School of Electrical and Electronic Engineering, North China Electric Power University, Beijing, China

Tohid Shahsavarian University of Connecticut, Storrs, USA

Tao Shao Beijing International S&T Cooperation Base for Plasma Science and Energy Conversion, Institute of Electrical Engineering, Chinese Academy of Sciences, Beijing, China

Jingang Su State Grid Hebei Electric Power Research Institute, Shijiazhuang, China

Tatsuo Takada Tokyo City University, Tokyo, Japan

Hao Wang State Grid Nanjing Power Supply Company, Nanjing, China

Mingyang Wang School of Electrical and Information Engineering, Tianjin University, Tianjin, People's Republic of China

Shihang Wang Xi'an Jiaotong University, Xi'an, Shaanxi, China

Xia Wang State Key Laboratory of Electrical Insulation and Power Equipment, Xi'an Jiaotong University, Xi'an, China

Xuepei Wang Hefei University of Technology, Hefei, Anhui, China

Zehua Wang School of Electrical and Information Engineering, Tianjin University, Tianjin, China

Liqiang Wei State Grid Hebei Electric Power Research Institute, Shijiazhuang, China

Meng Xiao Tianjin University, Nankai District, Tianjin City, China

Jun Xie Hebei Provincial Key Laboratory of Power Transmission Equipment Security, North China Electric Power University, Baoding, China

Qing Xie School of Electrical and Electronic Engineering, North China Electric Power University, Beijing, China

Jiwen Xing Tianjin University, Nankai District, Tianjin City, China

Honghua Xu State Grid Nanjing Power Supply Company, Nanjing, China

Ranran Xu Tianjin University, Nankai District, Tianjin City, China

Hang Xu Jinan, China

Kerong Yang Hefei University of Technology, Hefei, Anhui, China

Zhuoran Yang State Grid Nanjing Power Supply Company, Nanjing, China

Zihe Yang State Grid Nanjing Power Supply Company, Nanjing, China

Hang Yao School of Electrical and Information Engineering, Tianjin University, Tianjin, China

Kai Yin School of Electrical and Electronic Engineering, North China Electric Power University, Beijing, China

Jun-Wei Zha University of Science and Technology Beijing, Beijing, China

Zhenyu Zhan School of Electrical and Electronic Engineering, North China Electric Power University, Beijing, China

Boya Zhang State Key Laboratory of Electrical Insulation and Power Equipment, Xi'an Jiaotong University, Xi'an, Shaanxi, China

Cheng Zhang Tianjin University, Nankai, Tianjin, China;
Beijing International S&T Cooperation Base for Plasma Science and Energy Conversion, Institute of Electrical Engineering, Chinese Academy of Sciences, Beijing, China

Chengcheng Zhang Key Laboratory of Engineering Dielectrics and Its Application, Ministry of Education, Harbin University of Science and Technology, Harbin, China

Guanjun Zhang State Key Laboratory of Electrical Insulation and Power Equipment, Xi'an Jiaotong University, Xi'an, Shaanxi, China

Guixin Zhang Department of Electrical Engineering, Tsinghua University, Beijing, China

Ling Zhang State Key Lab of Power System, Department of Electrical Engineering, Tsinghua University, Beijing, China

Peng Zhang State Grid Hebei Electric Power Research Institute, Shijiazhuang, China

Song Zhang Hefei University of Technology, Hefei, Anhui, China

Tiandong Zhang Key Laboratory of Engineering Dielectrics and Its Application, Ministry of Education, Harbin University of Science and Technology, Harbin, People's Republic of China

Ying Zhang School of Electrical and Information Engineering, Tianjin University, Nankai District, Tianjin City, China

Hong Zhao Key Laboratory of Engineering Dielectrics and Its Application, Ministry of Education, Harbin University of Science and Technology, Harbin, China

Yushun Zhao Hefei University of Technology, Hefei, Anhui, China

Ming-Sheng Zheng University of Science and Technology Beijing, Beijing, China

Yuanxiang Zhou State Key Lab of Power System, Department of Electrical Engineering, Tsinghua University, Beijing, China

Wenbo Zhu Tianjin University, Tianjin, China

Lewei Zhu Tianjin University of Technology, No. 391 Binshui West Road, Xiqing District, Tianjin, China

Polymer Insulation for HVDC Cables

DC Insulation Performance of Crosslinked Polyethylene for HVDC Cables

Shihang Wang

Abstract Crosslinked polyethylene is one of the main insulation materials in the high-voltage DC cables, which was produced from the low-density polyethylene crosslinked by the decomposed organic peroxide at elevated temperatures. The DC insulation performance of the polyethylene materials determines the insulation reliability of high-voltage DC cables. This chapter reviews the DC insulation performance of polyethylene, including DC electrical conductivity characteristics, space charge characteristics, as well as DC electrical breakdown characteristics. The morphology and physical-chemical defects of the polyethylene materials greatly affect the charge trap characteristics, thereby regulating the DC insulation performance.

1 Introduction

High-voltage-direct-current (HVDC) transmission is likely to have profound effects on electricity transmission strategy throughout the word over the next few decades. There will be a huge demand for HVDC extruded cable systems in the future. The extruded HVDC cables offer significant advantages over traditional paper insulated cable types, such as a higher conductor temperature can be used, lighter moisture barriers can be used, jointing of extruded cables is simpler, increased system lengths and reduced maintenance [1]. Insulation technology is the core technology of HVDC cables. The quality and performance of the insulation determine the reliability of the HVDC cable. Therefore, in-depth research on the DC insulation performance of the HVDC cable insulation materials is needed to lay the foundation for maintaining the insulation state of HVDC cables and developing the next generation high-performance insulation materials. This chapter mainly introduces the DC insulation performance of polyethylene, and is organized as follows. First, the HVDC cable insulation materials, namely polyethylene material, is introduced, including the characteristic of the morphology and the physical-chemical defects. Then, the key DC insulation performance of polyethylene insulation materials is discussed in

S. Wang (✉)
Xi'an Jiaotong University, Xi'an, Shaanxi, China
e-mail: wangshih@mail.xjtu.edu.cn

B. Du (ed.), *Polymer Insulation Applied for HVDC Transmission*,
https://doi.org/10.1007/978-981-15-9731-2_1

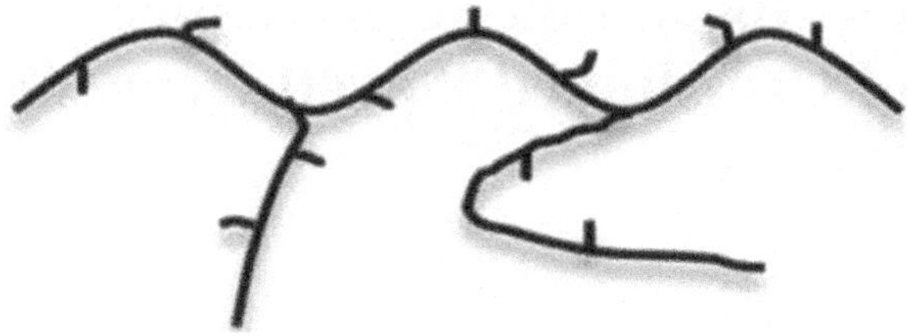

Fig. 1 Molecular chain structure of LDPE

turn, that is, DC electrical conductivity characteristics, space charge characteristics and DC electrical breakdown characteristics. And the relation between DC insulation performance and the morphology as well as the defects are also mentioned.

2 DC Cable Used Polyethylene Materials

Polyethylene is the simplest hydrocarbonic polymer, and is obtained from the polymerization of ethylene. Polyethylene are classified into liner, branched and crosslinked polyethylene. Polyethylene has been used as the insulation materials of high voltage cable for about five decades. The first 225 kV low density polyethylene (LDPE) cable was installed in 1969. Nowadays, crosslinked polyethylene (XLPE) is the most commonly used cable insulation material.

2.1 Low Density Polyethylene

Polyethylene vary in molecular weight and degree of branching. LDPE has the highest degree of branching (see in Fig. 1), which determines its rheological properties and is suitable for the extrusion process during the cable manufacture. Other polyethylene are unbranched high-density polyethylene (HDPE) and short-branched linear low density polyethylene (LLDPE). The rheological properties or crystalline properties of these polyethylenes determine that they are not suitable for cable insulation. However, the blended LDPE/HDPE have been studied to used as the potential recyclable cable insulation materials.

2.2 Crosslinked Polyethylene

XLPE is LDPE crosslinked by an organic peroxide, typically dicumyl peroxide (DCP), at elevated temperatures. There are by-products in the crosslinked reaction, that are methane, acetophenone, and cumyl alcohol (see Fig. 2). Therefore, after

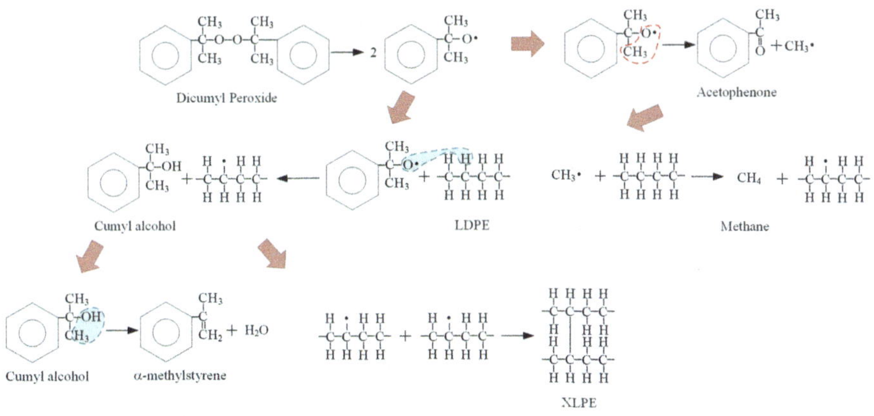

Fig. 2 Crosslinking process and formation of crosslinking by-products

the cross-linking reaction is completed, XLPE needs to undergo a degassing treatment at 60~70 °C to remove as much crosslinking by-products as possible. After crosslinking, XLPE can tolerate a conductor temperature of 90 °C, 105 °C for up to 36 h during emergencies, and 250 °C for a few seconds during short circuit.

It is precisely because of the introduction of the crosslinking agent and the generation of crosslinking by-product that there are more impurities in the XLPE insulation. The different types of crosslinking by-products have large differences in diffusion rate, and the crosslinking by-products cannot be completely removed.

In the cable industry, the crosslinking process (also call vulcanization) occurs at elevated temperature, after the cable insulation is extruded. The temperature of the insulation layer in the vulcanized pipeline can reach 150~280 °C and the crosslinking reaction is fast. However, during the extrusion process (at 120 °C), due to the heat generated by the viscous dissipation, the cable insulation materials melt may locally crosslink prematurely, which is also called scorch. In order to prevent the scorch phenomenon, a small amount of antioxidant is added at the same time as the crosslinking agent is added to the cable insulation materials. Therefore, it can be seen that the crosslinking agent and the antioxidant are the two major additives, and the residual antioxidant and the crosslinking by-product will eventually remain in the cable insulation, which will have a significant impact on the electrical insulation performance.

The structure of the polymer determines its insulating properties. The structure includes molecular chain structure and morphology, as well as defect structure. The entanglement of molecular chains during crystallization will cause physical defects. The introduction of crosslinking agents and antioxidants in crosslinkable cable insulation compounds reduces purity and act as the chemical defects. Components that may become chemical defects in the cable insulation are residual antioxidants, crosslinking by-products, moisture and other impurities. The accurate understanding of the structure, composition and processing of polyethylene insulation materials is the basis for understanding its electrical insulation properties of the cable insulation layer.

2.3 *Morphology of Polyethylene Materials*

Polyethylene consists of spherulites, amorphous regions and interphase between the spherulites and amorphous regions (see Fig. 3) [2]. The spherulite consists of ordered crystalline lamellar ribbons growing radially from the nucleating site. Between these lamellar ribbons will be similar ribbon-like phases of amorphous regions. A lamellar crystal will have a length approximately equal to that of a spherulite radius with a crosssection of ~20 nm in the chain (c-axis) direction and ~100 nm wide [2]. The spherulites of LDPE have diameters of several micrometers or more that 10 micrometers, but after crosslinking, the crystallization ability of crosslinked polyethylene decreases [3]. The spherulite size decreases with the increasing DCP contents, and the morphology changes from consisting of almost spherical banded structures to axialitic nonbanded structures and finally to more or less randomly distributed lamella stacks [3].

The crystallinity is an important parameter for characterizing the morphology of polyethylene insulation materials. It can generally be obtained by X-ray diffraction (XRD) measurement or differential scanning calorimetry (DSC) measurement. The

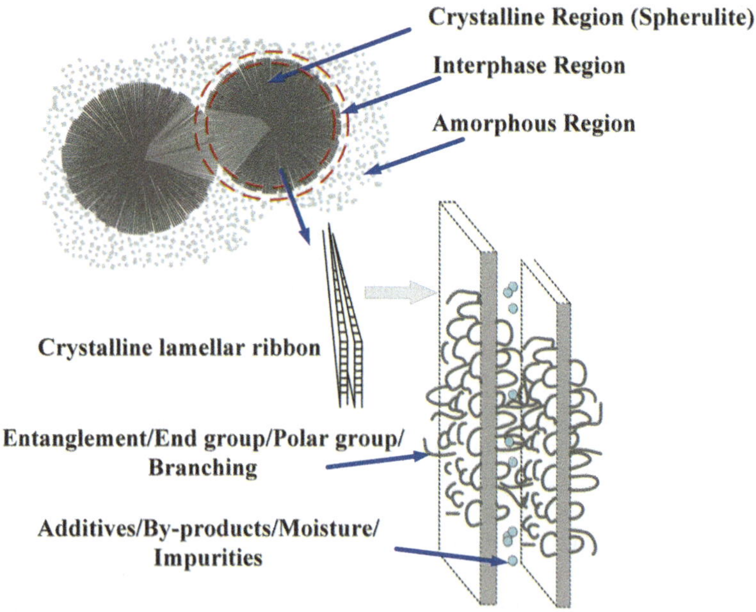

Fig. 3 Morphology of semi-crystalline polyethylene showing spherulites composed of crystalline-amorphous ribbons. (After [2])

crystalline regions in polyethylene insulation materials can produce X-ray diffraction. The peaks at about 21.4° and 23.6° are characteristic of the (110) and (200) lattice planes. A peak at about 20.8° can be obtained through peak separation analysis by Gaussian function to the XRD spectra, which represents the amorphous region. The crystallinity can be caculated from the following relationship:

$$\chi(\%) = \frac{\text{Area 1} + \text{Area 2}}{\text{Area 1} + \text{Area 2} + \text{Area 3}} \times 100$$

where, χ (%) is the crystallinity percentage; Area1 is the area under the principal crystalline peak at 21.4°; Area2 is the area under the secondary crystalline peak at 23.6°; Area3 is the area under the amorphous halo (see Fig. 4). The XRD spectra can also determine the microscopic parameters such as the size of the crystal grains and interplanar crystal spacing.

There are many factors that affect the morphology of polyethylene, such as the molecular weight, the short-branched structure, the long-branched structure, as well as the introduced additives, impurities and the crosslinking degree. The processing temperature and the cooling rates also affect the morphology.

The nucleating agents gives higher crystallinity and smaller spherulites. It is generally not intended to add a nucleating agent to polyethylene for high-voltage cable insulation, but other additives may play the role of nucleating agent, like antioxidants. Antioxidant is an essential additive for cable insulation materials, which helps to prevent premature crosslinking during extrusion process. On the one hand, it reduces the formation of gel impurities during extrusion. On the other hand, it improves the production efficiency of cables and increases the production length of cables. Morphology influences the electrical properties directly, like the electrical conductivity characteristics, space charge characteristics and DC electrical breakdown characteristics.

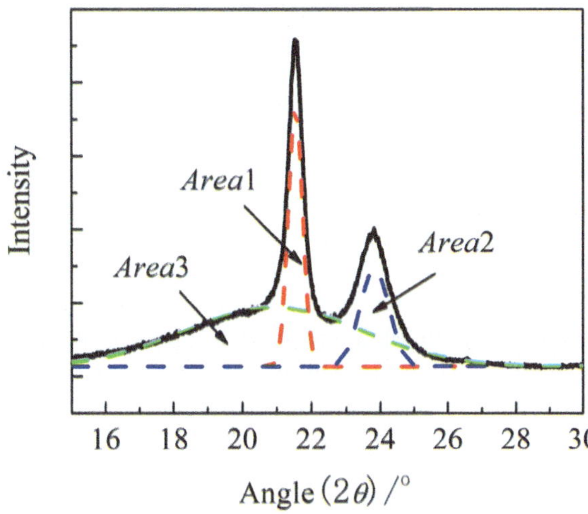

Fig. 4 The XRD spectra of LDPE or XLPE

2.4 Defects of Polyethylene Materials

Defects in the polyethylene insulation materials can be divided into physical defects and chemical defects. It is clear that both physical defects and chemical defects presenting in insulating materials can both trap charges. Physical defects include molecular weight distribution, free volume and vacuoles, various conformation defects in the amorphous regions [4]. Numerical models were widely used to simulate the band structure of alkane molecules, and the results obtained by molecular modeling showed that conformational defects were predicted to produce shallow traps with energies below 0.3 eV [5], with a density of about 10^{20} cm^{-3}, leading to an average distance between traps of 15 Å. The residence time of the electrons in these states was estimated to be between 10^{-11} and 10^{-13} s. The number of charge traps along a conformationally disordered polyethylene chain is approximately equal to the sum of the traps obtained for segments making up the polyethylene chain. Physical traps in polyethylene assist the charge transport process and do no provide sites for long lasting space charge.

Chemical defects include branching, end of chains and chemical irregularities, additives and by-products and so on. It has been noted in many studies that the amorphous interspherulite regions in the polyethylene are preferred trapping locations of charge carriers. On the one hand, residual additives, crosslinking by-products, moisture, impurities and other chemical defects are mainly concentrated in this regions. On the other hand, the amorphous regions also have more physical defects. Defects create charge traps that hinder the migration of electrons, holes, or ions, causing space charge to accumulate.

3 DC Electrical Conductivity Characteristics

3.1 Charge Conduction

3.1.1 Low-Field Conduction and High-Field Conduction

The electrical conductance of the polyethylene under a low-field (below the threshold strength) is mainly ion conductance, for the little charge injected into polyethylene from the electrodes. The charge carriers mainly come from the ionization of additives or impurities and are thermally activated. The mechanism of the ion conductance is hopping conductance. Schematic illustration of the charge transport is shown in the following figure. Figure 5a is in the absence of an electric field and Fig. 5b is with an applied filed *E*. In the absence of an electric field, the probability per unit time for the charge carrier to move a neighboring position is given by:

$$w = \nu_0 \exp\left(-\frac{E_i}{\mathrm{kT}}\right)$$

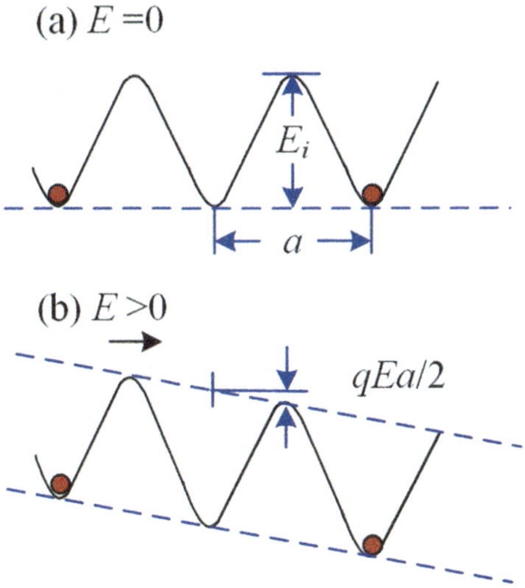

Fig. 5 Schematic diagrams of charge hopping conductance under low-field

ν_0 is the number of attempted escapes per second. E_i is the activation energy which is the activation barrier. Under an applied field, the probability changes because the barrier height is changed by an amount of *qEa/2*. In the direction of the field, the velocity of the positive charge q can be written as:

$$v = a\nu_0\left[\exp\left(-\frac{E_i - \frac{qEa}{2}}{kT}\right) - \exp\left(-\frac{E_i + \frac{qEa}{2}}{kT}\right)\right] = a\nu_0\exp\left(-\frac{E_i}{kT}\right)\left[2\sinh\left(\frac{qEa}{2kT}\right)\right]$$

In the case of low field, $qEa \ll kT$, the last term sin*h*(*qEa*/2*kT*) can be approximately equal to *qEa/*2*kT.* Hence, the mobility can be express as:

$$u_+ = \frac{q\nu_0 a^2}{kT}\exp\left(-\frac{E_i}{\mathrm{kT}}\right)$$

Therefore, only under a low-field that satisfies $qEa \ll kT$, the activation energy of electrical conductivity can be accurately obtained. For example, at 20 °C, 50 °C, 70 °C and 90°C, *kT* are 0.025 eV, 0.028 eV, 0.030 eV and 0.031 eV respectively. Suppose a is 2 nm, and when E is 3 kV/mm, *qEa* equals 0.006 eV which satisfies the above condition. If we use the relatively high electrical field, such as filed above 10 kV/mm, then the *qEa* is above 0.02 eV and the accurate E_i will not be obtained accurately by the Arrhenius plot.

By comparing the low-field conduction at different temperatures of XLPE insulation materials for HVAC cables and HVDC cables, it can be found that there is a significant difference in activation energy (E_i). The E_i obtained under 3 kV/mm are 0.89 eV for *DC material-1* and 1.09 eV for *AC material-1*, and the E_i obtained under 30 kV/mm are 0.6 eV for *DC material-2* and 1.0 eV for *AC material-2* (see

Fig. 6) [6, 7]. The E_i *of DC materials* is significantly smaller than that of *AC materials*. In addition, it can also be seen that after the increase of the test electric field, the E_i obtained is smaller. At the relative low electrical field below the threshold for space charge accumulation, the conduction in XLPE is a hopping mechanism driven by charge carriers, and migration of charge carriers is restricted by the traps in the crystalline regions and amorphous regions. Therefore, it is considered that the conductivity characteristics are associated with trap characteristics.

The thermally stimulated currents (TSC) can be used to obtain the trap densities and the energy levels of the traps in polyethylene materials. There are several distinct current peaks exhibited at different temperatures (see Fig. 7) [8]. The TSC of XLPE, which used as AC material or DC material, has three peaks P_1–P_3, and the TSC of LDPE with previous X-ray irradiation has five TSC peaks C_1–C_5 [8]. The thermal stimulated current can be described by the following equation:

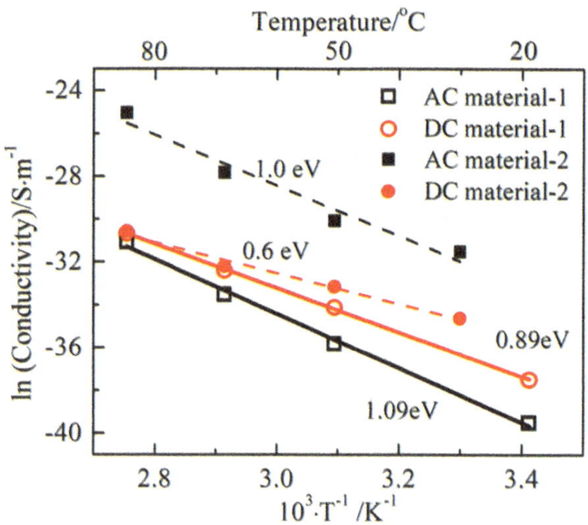

Fig. 6 The DC conductivity characteristics of XLPE materials [6, 7]

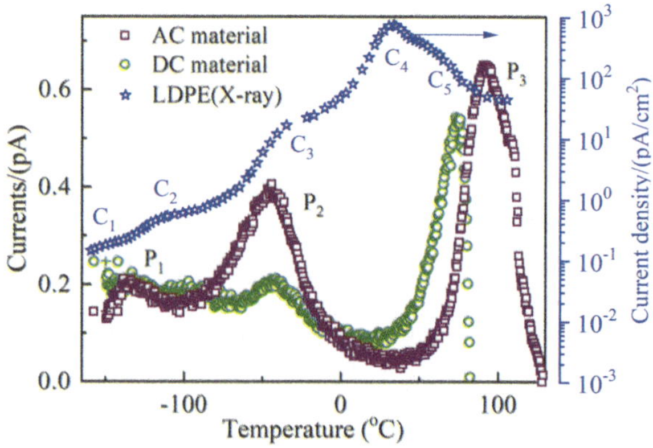

Fig. 7 The thermal stimulated currents of LDPE and XLPE materials (after [6, 8])

$$I(T) = \mathrm{A}\exp\left[-\frac{H}{kT} - \frac{\mathrm{B}}{v}\int_{T_0}^{T}\exp\left(-\frac{H}{kT}\right)\mathrm{d}T\right]$$

where T is the absolute temperature and k is the Boltzmann's constant, H is the energy depth of the traps, v is the heating rate, A and B are coefficients independent of T and H, but dependent on the model one uses to analyze the TSC. H can be calculated through curve fitting using this equation. The calculated energy depths of the deep traps of *AC material* and *DC material*, 1.09 eV and 0.89 eV respectively, are same with the E_i obtained by the DC electrical conductivity at different temperatures. These experimental data further confirmed the correlation between conductivity characteristics and trap characteristics.

The areas of the current peaks have a significant difference for these two samples which indicate the difference amount of trapped charges. In *AC material*, the amounts of the trapped charge corresponding to α and β peaks are both obviously larger than that in *DC material*. Consequently, the trap density of *DC material* is considered to be much lower than that of *AC material*. It can be seen that the difference of the DC conductivities between *AC material* and *DC material* is mainly determined by the trap densities, and *DC material*, with a relatively lower trap density, has a higher DC electrical conductivity. The trap characteristics of different polyethylene materials are significantly different. In general, the trap energy level of XLPE is higher than that of LDPE, because of the additives, by-products and more complex molecular structure. The C_4 peak and C_5 peak increase with density or degree of crystallinity in polyethylene materials. Correspondingly, we can also infer the conductivity characteristics of different polyethylene materials based on their trap characteristics.

It is worth noting that the test time of the conductance current has a great influence on the DC conductivity test results. Reference [9] studied the DC conductivity of LDPE in the temperature range of 20 to 90 °C under electric fields from 4 kV/m to 20 MV/m. For a long time (6 days), the conduction mechanism is dominated by the space charge trapped in the materials. The current does not attain a steady-state value after 29 days at 50 °C and 8 MV/m. It oscillates continuously, the variations are less regular and the conductivity decreases significantly as the sample thickness increases. The activation energy decreases from 0.8 to 0.58 eV when determined from long time measurements. An explanation is proposed considering the constraints imposed by the trapped space charge on charge injection and transport.

The electrical conductance of polyethylene materials under the high-field and extra high-field is complicated. At high fields, the current in polyethylene materials deviates from Ohm's law and rises super-linearly with field. Many papers have studied the high-field conductivity characteristics. Several mechanisms listed in Table 1 have been proposed for high-field conduction in polyethylene.

Table 1 Several mechanisms for high-field conduction in polyethylene (after [8])

Mechanisms	Formula
Electrode-limited type	
Schottky	$J = J_0 \exp\left(\beta_S \sqrt{E}/kT\right)$
Tunnelling	$J = AE^2 \exp(-B/E)$
Bulk-limited type	
Poole-Frenkel	$\sigma = \sigma_0 \exp\left(\beta_{pe} \sqrt{E}/2kT\right)$
Hopping	$J = J_0 \sinh(eE\lambda/2kT)$
Space charge limited conduction (SCLC)	$J = 9\varepsilon_0\varepsilon_r \mu V^2/8d^3$

3.1.2 Morphology and Charge Transport

The results of the density-functional electronic structure calculations reveal that the wave function of the conduction-band edge is of interchain character, as opposed to the intrachain character of all the filled valence-band states. Thus, while a hole added to neutral polyethylene will mainly belong to the polyethylene chain backbone bonds, an added electron in polyethylene will mostly reside between the chains, and far from the existing bonds. Moreover, the added electron state charge centroid is predicted to move further out from the chain backbone towards the low-density interstitial region, if and when the chains are pried apart. This suggests that injected electrons will naturally flow to low-density regions inside real PE, and that the experimentally established propensity of polyethylene to expel electrons out of the bulk, should be directly related to the interchain nature of the conduction states [10]. Therefore, electron transport and hole transport in the polyethylene have different characteristics (see in Fig. 8).

It has been reported that the important role of the interphase region formed between the crystal region and the amorphous region on charge carrier migration, and believed that the interphase region hindering effect on electron migration is much smaller than that on hole migration [12].

3.2 Electrical Conductivity and Electric Field Distribution

The electric field distribution within the insulation under the AC voltage depends on the capacitive field which inversely proportional to the electrical permittivity of the insulation. The electrical permittivity of the HVAC cable insulation changes little within the operating temperature range. So the electric field distribution is less affected by the properties of polyethylene. The highest electric field is located inside the insulating layer near the conductor. But in HVDC cable, the electric field distribution is more complicated and is temperature and time dependent.

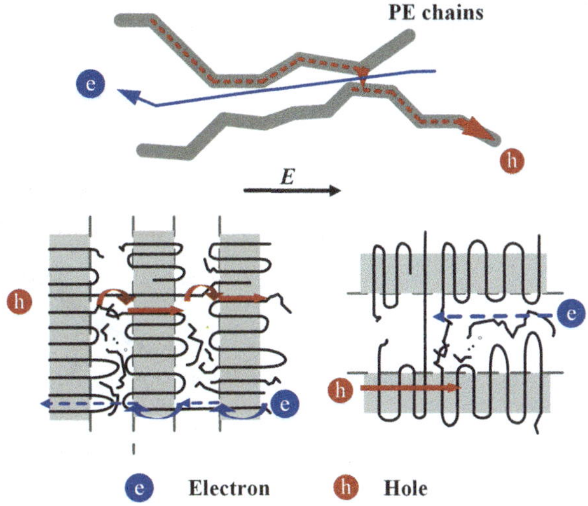

Fig. 8 Schematic diagrams of the electron path, positive hole path and the polyethylene molecular chains (After [2, 11])

The electrical conductivity of polymer dielectric depends both on temperature and electric field. In the temperature range of 20~90 °C, the electrical conductivity of polyethylene can change by 2~3 orders of magnitude. So the highest electric field is located at the outer part of the insulating layer, as outside the insulating layer has a relatively low temperature. Therefore, a smaller increment of electrical conductivity of polyethylene materials in the working temperature range is an advantageous property for the DC insulation materials, which was determined by the activation energy (E_i) of the electrical conductivity. As mentioned above, the activation energy can be obtained from the DC conductivities measured at different temperature under a low-field. The *DC material* has a lower E_i, which is conducive to the uniform distribution of the electric field under the DC field.

When the temperature or electric field strength changes, the electric field at any point in the insulating layer will change. The redistribution of the electric field will also cause the DC conductivity at different locations to change further, so that the distribution of the electric field is further adjusted until it reaches a steady state. In short, the electric field distribution in the insulation layer of a DC cable is determined by the coupling of temperature and electric field. After the DC cable is loaded, the maximum electric field strength of the insulation layer tends to move to the surface of the insulation layer (see in Fig. 9). The greater the change in DC conductivity of polyethylene materials with the increasing electric field, the more favorable the electric field distribution tends to be. Therefore, the parameters controlling the electric field distribution in the polyethylene insulation materials are the activation energy (E_i) of DC conductivity and the electric field dependence coefficient (B), as shown in the following formula.

$$\sigma(T, E) = A \exp\left(-\frac{E_i \cdot q}{kT}\right)\frac{\sinh(B|E|)}{|E|}$$

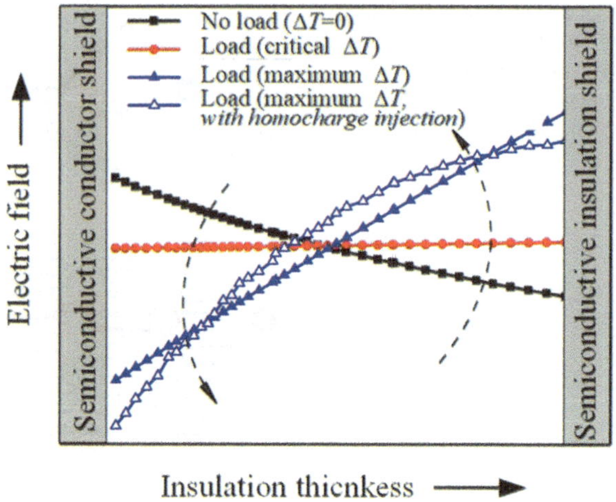

Fig. 9 Schematic diagram of electric field distribution in DC cable insulation layer

Excellent HVDC cable materials should have a lower conductivity activation energy and a higher B [9]. In addition, the presence of space charge also affects the electric field distribution in the DC cable insulation (see in Fig. 9).

4 Space Charge Characteristics

When the polymer insulation material is under a DC electric field, space charge will accumulate more or less, depending on the composition and characteristics of the materials, voltage level, and electrode condition. Space charge strongly affects the electric field distribution over the insulation thickness, which controls the insulation behaviors, both the short-time insulation properties and long-time reliability. The space charge phenomenon is the key difference between DC insulation and AC insulation, and it is also the difficulty in the design of insulation systems of HVDC extruded cables. The space charge characteristics can be understood from the aspects of charge generation and charge transport.

4.1 Charge Generation

As to the source of space charge in the polyethylene insulation of the cable systems, it can be divided into the internal charge and the injected charge. The internal charge is formed due to the presence of ionic dissociable additives, impurities, dislocations, and other chemical defects, and the external charge is mainly injected from the conductor into the insulation.

4.1.1 Internal Charge

The homocharge and the heterocharge are terms that are often used to describe the space charge characteristics. Homocharge refers to the charge with the same polarity to the adjacent electrode while heterocharge refer to the opposite polarity. Homocharges are usually formed from the charge injection or charge extraction near the electrode, which will reduce the electric field in the vicinity of the electrode interface but enhance the bulk electric field. When HVDC voltage is applied across the polymer insulation, small dissociable chemical species may become ionised and drift towards the electrode with opposite polarity where they can trapped, which eventually produce heterocharges.

In XLPE insulation, substances that can affect the internal heterocharge include crosslinking by-products, antioxidants, moisture and other impurities. This type of space charges will increase the electric field at the electrode interface which will make it easier for the charge injection. It has been reported that crosslinking by-products and antioxidants can lead to heterocharge accumulation [6, 13–15]. Moisture also affects the accumulated space charge behavior of LDPE and XLPE very strongly. Moisture even at a very low concentration can also promotes the ionization of crosslinking by-products in XLPE, such as acetophenone, cumylalcohol and α-Methylstyrene [16]. The testing temperature has numerous effects on the space charge generation such as the enhancement of ionic dissociation of polar crosslinked by-products, charge mobility and DC electrical conductivity [17].

XLPE insulation materials used in HVAC cables usually exhibit heterocharge testing in the PEA measurement (see Fig. 10). But, polyethylene insulation materials used in HVDC cables must avoid the presence of heterocharge. From this perspective, insulation materials used for HVDC cables should have a higher purity. At present, the development of cable insulation materials with little or no additives has been studied.

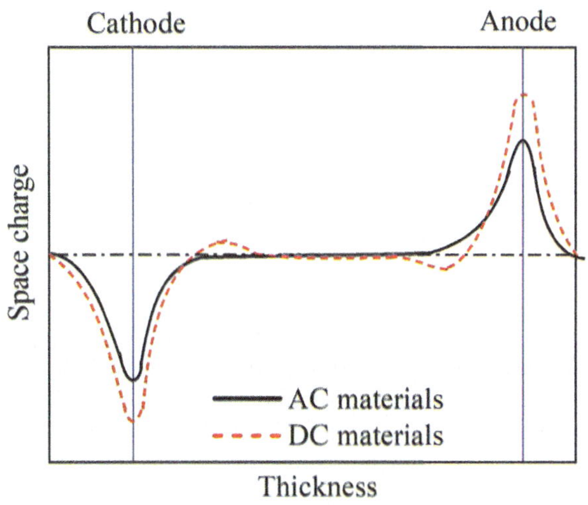

Fig. 10 Schematic diagram of the space charge curves of HVDC and HVAC cable used XLPE materials (After [6, 21])

As reported from [18], it was proposed a viable alternative to peroxide crosslinking that still fulfils demand of the processing advantages and high electrical resistivity of ultra-clean LDPE and commercial XLPE. Click chemistry reactions between two polyethylene copolymers (with epoxy and acrylic acid functional groups) allow the design of a curing process that is additive-free and does not result in the release of any byproducts [18]. For antioxidants, it has been studied to graft them on the molecular chain of HDPE or PP, which can inhibit its migration while playing an antioxidant role [19, 20]. In the future, LDPE or XLPE will still be the key insulation materials for HVDC cables, but the introduction and action of additives will change greatly.

4.1.2 Charge Injection

In HVDC cable, there will be charges injected to the XLPE layer from the conductor shield layer, which always be called homocharge. There are two main mechanisms in the charge injection process, that are the Schottky injection and the Fowler-Nordheim injection. Schottky injection theory is based on the potential barrier height (ΦD) of the metal-insulator interface. The Schottky injection can be expressed as:

$$J = AT^2 \exp\left[-\left(\phi_D - \sqrt{e^3 E/4\pi\varepsilon_0\varepsilon_r}\right)/kT\right]$$

The Fowler-Nordheim effect is also known as the tunnelling effect from bulk metal to other bulk crystalline solids. It is found that electron may tunnel through a potential barrier although it has less energy to overcome the barrier's height. This is due to the particle wave duality characteristic of electron. As the electron wave travel through the barrier, the wave will attenuate reducing its' amplitude but leaving the energy unchanged. The Fowler-Northeim injection can be expressed as:

$$J = AE^2 \exp(-B/E)$$

The charge injection is significantly influenced by the temperature and electric field. There will be more charge injected to the insulation at a higher temperature or under a higher electric field. There is a threshold in the relation between charge injection and electric field. When the applied field is below the threshold for charge injection, the amount of injected charge is small and the insulation system behaves according to Ohm's law. The same amount of injected charge from one electrode is extracted by the other electrode, and thus no net space charge remains trapped and accumulates in the insulation bulk [22, 23].

As soon as the field exceeds the threshold, charge injection from the electrodes increases and the electrodes are not able to extract the increasing amount of injected charge, which finally accumulates in the insulation bulk [24]. The typical current-poling field characteristic of XLPE specimen peeled from the power cable with a thickness of 150 μm is shown in Fig. 11, and a threshold field separating the ohmic and the power law regimes can be defined about 17 kV/mm [23]. The PEA technique

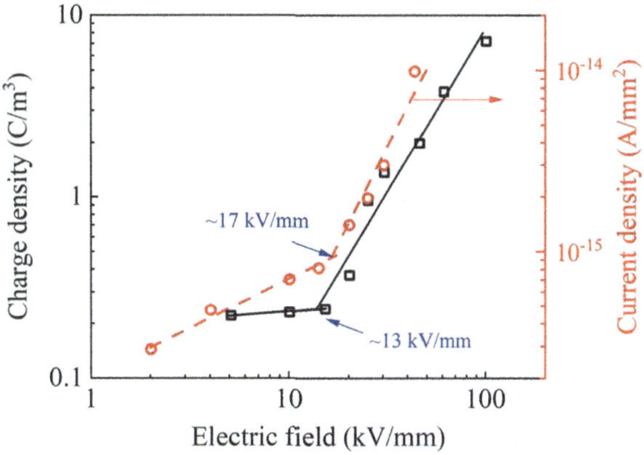

Fig. 11 The current-poling field characteristic and charge-poling field characteristic of XLPE sample peeled from the power cable (After [23])

is also used to test the same sample, and the space charge density accumulated in the sample after a given polarization time was obtained. The charge-poling field characteristic is also shown in Fig. 11 [23], and the threshold field is about 13 kV/mm. The threshold obtained by these two techniques are very close. When the applied field is lower than the threshold field, the typical current-poling field characteristic and the charge-poling field characteristic are different. This is because that the current density increases with the increasing the electric field, but the charges are not captured by the traps and fail to reflect the space charge accumulation.

Generally, as the temperature increases, the threshold field decreases. As shown in Fig. 12, the threshold field of XLPE cable at 70 °C obtained by charge-poling field characteristic (~3 kV/mm) is much lower than that at 25 °C (~8 kV/mm). The threshold field of XLPE cable is much lower than that of XLPE sample, and the threshold field of XLPE sample is much lower than that of LDPE sample. These phenomena are directly related to the multi-level structure of the polyethylene materials.

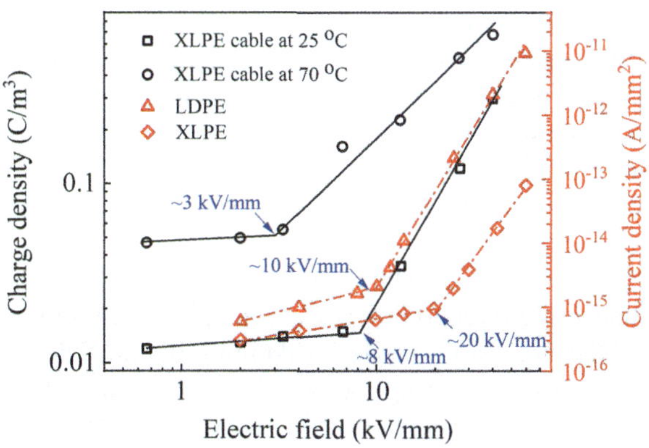

Fig. 12 Threshold characteristics for LDPE, XLPE, and XLPE cable (after [22, 25])

The characteristics of charge injection at the insulation/electrode interface are affected by the electrode materials, the conformation of polyethylene molecular chains, the defects [26] and the contact form of conductor and insulation [27]. The presence of functional groups such as terminal carbonyl groups and conjugated double bonds can reduce the injection barriers for holes and electrons at the polyethylene/electrode interface [28]. So the conditions of the insulation/electrode interface is very important for determining the charge injection. It can be concluded that high purity is a requirement for DC cable insulation materials. High purity means the performance of a higher charge injection threshold field and less internal ionizable impurities and charge traps.

In the research and development of new insulating materials for HVDC cables, the general focus is to suppress the homocharge injection. For example, the introduction of nanoparticles or grafting polar groups to polyethylene matrix brings deep charge traps, thereby suppressing charge injection.

4.2 Charge Transport

In addition to conductivity-controlled electric field distribution, the problem of space charge accumulation under the DC field makes the DC electric field distribution more complicated than the AC electric field, and the design is more difficult. For example, when homocharge injection occurs on the insulation layer of a DC cable, the electric field on both sides of the insulation layer will decrease, while the internal electric field will increase. When the insulating layer accumulates heterocharge, the electric field on both sides of the insulating layer will increase. Therefore, the electric field distribution of DC cable insulation is determined by the complex charge transport characteristics including electrical conduction and space charge characteristics. The space charge problem is a key problem in the DC cables, which limits the application of XLPE and hinders the development of DC cables.

Studying the space charge characteristics at different temperatures is one of the key indicators for evaluating the insulation properties of cable materials. Reference [29] showed that when the XLPE sample is in a temperature gradient environment, it is easier to accumulate heterocharges on the low temperature region, and the amount of accumulated charge increases with the increasing temperature gradient. Reference [30] studied the effect of temperature on the charge trapping characteristics of XLPE, and developed the technique of joint testing of space charge and electrical conductivity.

In addition, the voltage polarity of the HVDC cables is sometimes reversed to change the direction of power energy transmission. At this time, the effect of the homocharge in the cable insulation will cause the local electric field distortion or the breakdown during the polarity reversal or voltage withdrawal. This is because there is a certain depth of charge injection in the polymer, which ranges from a few microns to tens of microns. When the applied voltage is removed or the polarity is reversed, the reverse electric field formed near the electrode will be much higher than the average electric field strength during the working period.

It is reported that the polarity reversing charge was generated in the middle of the insulation and moved towards the appropriate electrodes under the influence of a field in excess of the maximum applied field [31]. These charges will increase the local field strength in the cable sample by 50–70% [31].

5 DC Electrical Breakdown Characteristics

The electrical breakdown characteristics of polyethylene have been correlated with many factors, including morphology, chemical composition, additives, impurities. In general, morphology has the greatest impact on breakdown strength of polyethylene, for the breakdown strength of spherulites is much higher than that of the amorphous regions. Taking polypropylene as an example, the DC breakdown strength of polypropylene spherulites can reach an average value of 540~690 kV/mm, while the DC breakdown strength of the amorphous regions between spherulites is only 120~330 kV/mm [32]. The crystallinity of the polyethylene is positively related to the DC breakdown strength and the greater the crystallinity, the higher the breakdown field strength [33–35]. Besides crystallinity, the spherulite size is found to be another key parameter affecting the DC breakdown strength [34]. It is reported that the internal spherulite size is of 17~20 μm in the low density polyethylene sheet sample with a molecular weight of 2500, and the sample's DC breakdown strength is only 250 kV/mm [32]. But the low density polyethylene sheet sample with a molecular weight of 37,000 contains spherulites with size of only 6~8 μm, and its DC breakdown strength is as high as 430 kV/mm [32]. Other related research has also proved that the large number of small spherulites in polyethylene means higher breakdown field strength. For example, nanoparticles [36] and small molecules [34] can be introduced to enhance the DC breakdown field of polyethylene for the nucleation effect of these fillers.

It is concluded from these results that the initial breakdown channels occur mainly in the interspherulite space of the material. The effect of the spherulite size is accounted for by the interspherulite regions have a lower density and irregular structure. As the spherulite size increases, the density of the polymer within the interspherulite space decreases and a reduction in electrical strength and other properties is observed.

The DC electrical breakdown characteristics of polyethylene are also affected by the sample thickness and test temperature (see Fig. 13 as an example). Because the operating temperature of the polyethylene insulating cables is up to 70~90 °C, studies of breakdown characteristics is generally from room temperature to 90 °C. In general, the maximum values of the breakdown voltage are obtained in the low-temperature region.

The morphology also affects the temperature-dependent characteristics of the DC breakdown strength. Is observed in [6, 39] that the larger the spherulite, the higher

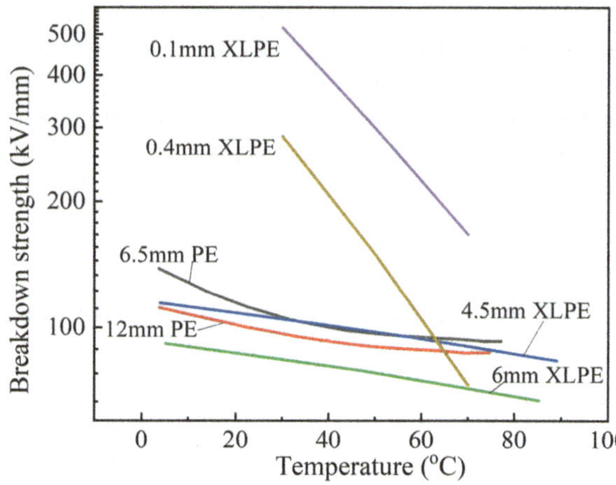

Fig. 13 Effect of sample thickness and test temperature on the DC breakdown strength of HVDC cable used polyethylene insulation (After [37, 38])

DC breakdown strength at relatively high temperature, around 70~90 °C (see Fig. 14 as an example). The breakdown at relatively higher temperatures is tend to be thermo-mechanical breakdown. The large spherulites tend to make the polyethylene have a higher mechanical modulus, and thus can withstand higher electro-mechanical stress, leading to a higher breakdown strength. The author believes that so far, the correlation between polymer morphology and breakdown has not established a correlation mechanism at the microscale, and further research is needed.

The additives and impurities also affect the DC breakdown strength. The distorted electric field caused by the heterocharge often has a significant effect on the long-term insulation characteristics. But as to the short-term breakdown, the significant homocharge injection inhibits the effect of heterocharge. The presence of additives and impurities, like by-products, mainly increases the conductivity of XLPE, and the dissipation energy in the XLPE is induced by the accumulation of space charge, which promote the dielectric breakdown [40]. Although the intrinsic breakdown strength of XLPE is high (about 800 kV/mm under DC voltage), due to various

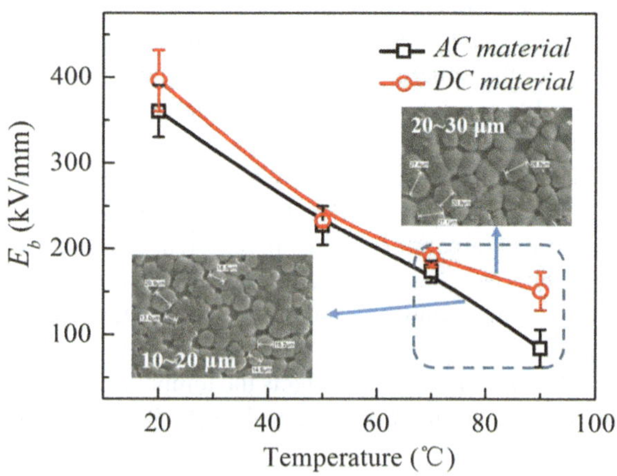

Fig. 14 DC breakdown strength of XLPE insulation materials at higher temperatures (After [6])

factors, the actual DC breakdown strength is much lower than the intrinsic breakdown field strength. It can be seen that there is still a lot of room for optimization of cable insulation materials, regulating the purity and morphology to improve the DC breakdown strength.

6 Conclusion

Polyethylene is a semi-crystalline polymer, and its morphology will regulate the transport path of charges, thereby determining its electrical conductivity characteristics, space charge characteristics, and breakdown characteristics under a DC electric field. Crosslinked polyethylene is made of low-density polyethylene through peroxide cross-linking, and its morphology has changed compared to low-density polyethylene, so the DC insulation performance has changed accordingly. From the perspective of defects, polyethylene materials have physical defects and chemical defects. In particular, due to the introduction of additives in crosslinked polyethylene and the generation of by-products, its chemical defects are more significant compared to low-density polyethylene. Defects will affect the transport process of charge carriers in the form of charge traps with different energy levels. Therefore, it can be seen that the regulation of the morphology and defects of polyethylene materials can effectively regulate their DC insulation performance.

References

1. G. Mazzanti, M. Marzinotto, *Extruded Cables for High Voltage Direct Current Transmission* (Wiley, 2013)
2. J.P. Jones, J.P. Llewellyn, T.J. Lewis, IEEE T. Dielect. El. In. **12**(5), 951–966 (2005)
3. S. Nilsson, T. Hjertberg, A. Smedberg, Eur. Polym. J. **46**(8), 1759–1769 (2010)
4. G. Teyssedre, C. Laurent, IEEE T. Dielect. El. In. **12**(5), 857–875 (2005)
5. M. Meunier, N. Quirke, J. Chem. Phys. **113**(1), 369–376 (2000)
6. S. Wang, P. Chen, H. Li, J. Li, Z. Chen, IEEE T. Dielect. El. In. **24**(3), 1809–1817 (2017)
7. M. Tefferi, L. Chen, S. Nasreen, H. Uehara, R. Ramprasad, Y. Cao, Tailoring polymeric insulation materials for dc cable dielectrics, in *Paper presented at the 2019 IEEE Conference on Electrical Insulation and Dielectric Phenomena (CEIDP)* (2019)
8. M. Ieda, IEEE Trans. Electr. Insul. EI **19**(3), 162–178 (1984)
9. E.R. Neagu, J.N. Marat-Mendes, Appl. Phys. Lett. **82**(12), 1920–1922 (2003)
10. S S, T E, I S Phys. Rev. B **62**(7), 4389–4393 (2000)
11. T.J. Lewis, IEEE T. Dielect. El. In. **9**(5), 717–729 (2002)
12. A.K. Upadhyay, C.C. Reddy, J. Appl. Phys. **122**(6), 64105 (2017)
13. T. Mizutani, *Effects of additives and morphology on space charge in LDPE*. IEEE, pp. 487–492 (2001). https://doi.org/10.1109/iseim.2001.973710
14. B. Vissouvanadin, S. Le Roy, G. Teyssedre, C. Laurent, I. Denizet, M. Mammeri, B. Poisson, *Effect of cross-linking agent on space charge accumulation in XLPE under DC electric stress*. IEEE, pp. 727–730 (2009). https://doi.org/10.1109/ceidp.2009.5377868
15. G. Teyssedre, C. Laurent, A. Campus, U.H. Nilsson, G.C. Montanari, *Antioxidant and its reaction products as charge trapping centres in crosslinked polyethylene*. IEEE, pp. 96–99 (2003) https://doi.org/10.1109/ceidp.2003.1254803
16. F. Aida, S. Wang, M. Fujita, G. Tanimoto, Y. Fujiwara, J. Electrostat. **42**(1), 3–15 (1997)

17. Y.L. Chong, G. Chen, Y.F.F. Ho, IEEE T. Dielect. El. In. **14**(1), 65–76 (2007)
18. M. Mauri, A. Peterson, A. Senol, K. Elamin, A. Gitsas, T. Hjertberg, A. Matic, T. Gkourmpis, O. Prieto, C. Müller, J. Mater. Chem. C **6**(42), 11292–11302 (2018)
19. S. Al-Malaika, S. Riasat, C. Lewucha, in *AIP Conference Proceedings*, vol. 1736(1)
20. M. Yuan, G. Zhang, B. Li, T.C.M. Chung, R. Rajagopalan, M.T. Lanagan, ACS Appl. Mater. Inter. **12**(12), 14154–14164 (2020)
21. T. Yamanaka, S. Maruyama, T. Tanaka, The development of DC ±500 kV XLPE cable in consideration of the space charge accumulation (689–694) (2003) (Reprinted)
22. G.C. Montanari, IEEE T. Dielect. El. In. **7**(3), 309–315 (2000)
23. L.A. Dissado, C. Laurent, G.C. Montanari, P.H.F. Morshuis, IEEE T. Dielect. El. In. **12**(3), 612–620 (2005)
24. D. Fabiani, G.C. Montanari, C. Laurent, G. Teyssedre, P.H.F. Morshuis, R. Bodega, L.A. Dissado, IEEE Electr. Insul. M. **24**(2), 5–14 (2008)
25. S. Delpino, D. Fabiani, G.C. Montanari, C. Laurent, G. Teyssedre, P.H.F. Morshuis, R. Bodega, L.A. Dissado, IEEE Electr. Insul. M. **24**(1), 14–24 (2008)
26. M. El-Shahat, A. Huzayyin, H. Anis, IEEE T. Dielect. El. In. **25**(6), 2178–2182 (2018)
27. J. Andersson, A. Smedberg, R. Persson, U.H. Nilsson, *Influence of sample preparation on space charge behavior of crosslinked polyethylene for HVDC cables*. IEEE, pp. 409–412 (2013). https://doi.org/10.1109/icsd.2013.6619844
28. M. El-Shahat, A. Huzayyin, H. Anis, IEEE T. Dielect. El. In. **26**(2), 642–647 (2019)
29. Z. Lv, J. Cao, X. Wang, H. Wang, K. Wu, L.A. Dissado, IEEE T. Dielect. El. In. **22**(6), 3186–3196 (2016)
30. L. Lan, J. Wu, Y. Yin, X. Li, Z. Li, IEEE T. Dielect. El. In. **21**(4), 1784–1791 (2014)
31. M. Fu, L.A. Dissado, G. Chen, J.C. Fothergill, IEEE T. Dielect. El. In. **15**(3), 851–860 (2008)
32. S.N. Kolesov, IEEE Trans. Electr. Insul. EI **15**(5), 382–388 (1980)
33. S. Nilsson, T. Hjertberg, A. Smedberg, B. Sonerud, J. Appl. Polym. Sci. **121**(6), 3483–3494 (2011)
34. K.M. Seven, J.M. Cogen, T. Person, J.R. Reffner, J.F. Gilchrist, J. Appl. Polym. Sci. **135**(22), 46325 (2018)
35. D. Li, L. Zhou, X. Wang, L. He, X. Yang, Materials **12**(11), 1746 (2019)
36. S. Wang, P. Chen, J. Xiang, J. Li, Study on DC breakdown strength and morphology in XLPE/Al(OH)3 nanocomposites, in *Technical Committee on Dielectrics & Electrical Insulation*, IEEJ, pp. 355–358 (2017). https://doi.org/10.23919/iseim.2017.8088758
37. M. Salah Khalil, IEEE Electr. Insul. M. **13**(6), 35–47 (1997)
38. Q. Kaiyu, S. Pengfei, W. Jiandong, Y. Yi, Proceedings of the CSEE **38**(24), 7121–7130 (2018)
39. S. Wang, S. Yu, J. Xiang, J. Li, S. Li, IEEE T. Dielect. El. In. **27**(2), 482–488 (2020)
40. T. Kanai, H. Kasuga, H. Miyake, Y. Tanaka, Property of DC breakdown in XLPE containing cross-linking by-products, in *Paper presented at the 2018 12th International Conference on the Properties and Applications of Dielectric Materials (ICPADM)* (2018)

Surface Ligand Engineering of Polymer Nanodielectrics for HVDC Cables

Ling Zhang, Xiaoyang Cui, and Yuanxiang Zhou

Abstract The development of insulating materials is a crucial knob for HVDC cables. The space charge effect and positive conductivity-temperature relationship of cable insulating materials render electric field distortion, which is detrimental to the reliability of HVDC cable insulation. To address this issue, polymer nanodielectrics have been used to improve the insulation performance. However, great challenges lie ahead including controllable large-scale fabrication and performance regulation of polymer nanodielectrics, mainly because of the complexity, uncertainty, and variability of the polymer matrix-particle interface. In recent years, surface ligand engineering has emerged as an effective way to tune the particle dispersion by controlling enthalpic and entropic interactions among nanoparticles, ligands, and polymer matrices, so as to introduce tunable functionalities and optimize the performance of polymer nanodielectrics. Herein, we outline recent advances of surface ligand engineering to achieve various exciting properties of polymer nanodielectrics for HVDC cables, leading to inhibited space charge, reduced DC conductivity and temperature sensitivity, improved breakdown strength, inhibited electrical tree aging, and improved thermal conductivity, etc. We aim to summarize surface ligand strategies for advanced polymer nanodielectrics and how they function, point out the knowledge gaps between lab results and industrial applications, and offer suggestions to guide the rational design of more effective and efficient polymer nanodielectrics for future HVDC transmission engineering.

L. Zhang (✉) · Y. Zhou
State Key Lab of Power System, Department of Electrical Engineering, Tsinghua University, Beijing 100084, China
e-mail: zhangling15@tsinghua.edu.cn

Y. Zhou
e-mail: zhou-yx@tsinghua.edu.cn

X. Cui
Institute of Electrical Engineering, Chinese Academy of Sciences, Beijing 100190, China
e-mail: cuixy@mail.iee.ac.cn

B. Du (ed.), *Polymer Insulation Applied for HVDC Transmission*,
https://doi.org/10.1007/978-981-15-9731-2_2

1 Introduction

High voltage direct current (HVDC) transmission play a pivotal role in transporting electrical energy with large capacity and long distance. The next generation of HVDC transmission demands high voltages, good reliability, low cost, and free maintenance. Plastic HVDC cables, owing to the advantages of small size, large transmission capacity, simple and firm structure, etc., are quite suitable for large-scale application in both onshore and submarine flexible DC transmission. Nowadays, developing high-performance plastic HVDC cables has become a crucial knob for the next generation of HVDC transmission [1].

Advanced insulating materials must be developed to meet the great challenge of modern HVDC cable engineering due to the significant difference between the dielectric properties under DC electric field and those under AC electric field. In 1978, the Japanese tried to employ cross-linked polyethylene (XLPE) AC cables to the ±250 kV HVDC transmission system between Hokkaido and Honshu Island. However, insulation breakdown frequently occurred during the trial operation, and it was found that the space charge effect of DC pre-stressing was the main reason for the breakdown voltage drop of XLPE AC cables. At that time, there was no suitable polymer insulating materials for ±250 kV HVDC cables, so that traditional oil-paper insulated cables had to be adopted instead [2].

During the past decades, the space charge effect, including the injection, accumulation, and movement processes, have attracted considerable attention in the field of polymer insulation for HVDC cables. Uneven space charge distribution can be easily formed in the cable insulation under DC electric field. An average charge concentration of 1 $C{\cdot}m^{-3}$ can generate an electric field of 50 $kV{\cdot}mm^{-1}$, which affects the dielectric strength and electrical aging of the cable insulator (as illustrated in Fig. 1), leading to the fracture of polymer chain, formation of electrical branches and discharge paths, and even unexpected breakdown of the cable insulation [3]. Moreover, the cable conductor will be heated under operation conditions, thereby generating a temperature gradient field in the cable insulator with high temperature (T) near the cable conductor and low T near the outside insulation shielding layer. This temperature gradient field will cause a dramatic change of electric conductivity in the cable insulator by the orders of 2–3 magnitudes and the formation of a reverse electric field in the cable insulator, thereby promoting charge injection from the conductor shielding layer on the high T side and accelerating charge accumulation and electric field distortion near the outside insulation shielding layer on the low T side. Such space charge effect seriously threatens the reliability of the cable insulation [4]. Currently, in order to meet the requirement that the maximum electric field should be lower than 20 $kV{\cdot}mm^{-1}$, the working temperature of HVDC cables is restricted to 70 °C, which is lower than that of HVAC cables (90 °C). Figure 1 illustrates the distributions of the temperature gradient field and the internal electric field in the insulator of HVDC cables, as well as the processes of space charge injection, accumulation, migration, and dissipation, in which charge migration is accompanied by recombination, trapping, hopping, and transition processes [5].

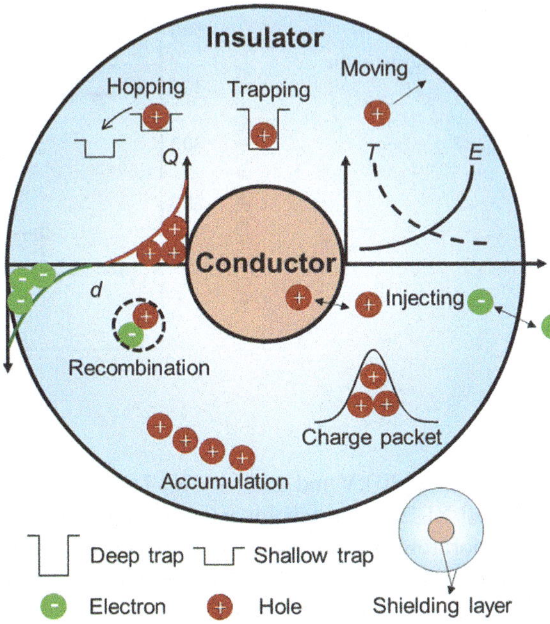

Fig. 1 Space charge evolution processes in the insulator of HVDC cables

The major critical scientific challenge to be addressed is the elucidation of the influence and mechanism of the space charge effect on the dielectric strength, long-term aging, and failure of insulating materials. In recent years, a considerable amount of research effort has focused on the dynamic evolution processes of space charge in polymer dielectrics under multi-physical fields, and the mechanism of insulation aging and failure caused by space charge during the long-term operation.

To meet the requirement of larger capacity and higher reliability for next generation of HVDC cables with increasing transmission voltage levels, advanced cable insulating materials should have the features of less space charge accumulation, low DC conductivity and temperature sensitivity, high dielectric strength, and high thermal conductivity, etc. With the development of flexible DC transmission technology, the voltage source converter makes the system control more flexible, and the polarity reversal of DC voltages no longer occurs.

In 1991 and 2003, the Japanese successively developed ±250 kV and ±500 kV HVDC cables using MgO/XLPE nanocomposites, and ±250 kV HVDC cables were applied to Hokkaido-Honshu DC link line project [6, 7]. In 1999, ABB Company put the world's first ±80 kV plastic HVDC cable line into operation [8]. At the same time, J-Power Company developed ±250 kV HVDC cables with XLPE nanodielectrics [9]. In 2011, South Korea reported the research and development progress of ±250 kV HVDC cables with XLPE nanodielectrics, and passed the type test based on CIGRE TB496 Recommendations in 2014 [10]. The past ten years have witnessed a rapid increase of voltage levels of HVDC transmission in China, catching up with the rest of the world [11], as shown in Fig. 2. At present, the most widely used insulating

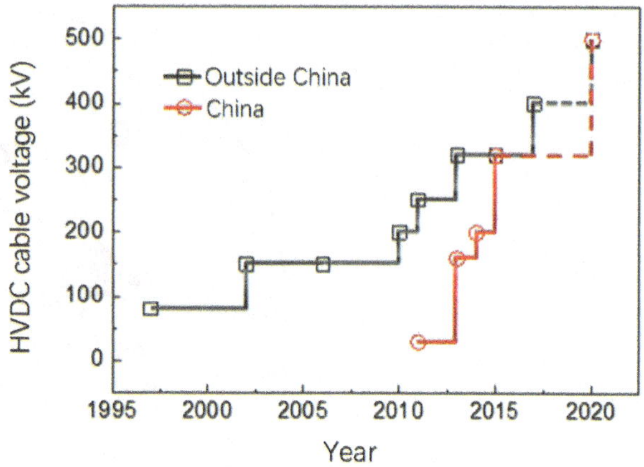

Fig. 2 Development of plastic HVDC cable transmission projects in the world

materials for ±320 kV and ±525 kV HVDC cables are LE4253 DC and LS4258 DCE ultra-clean XLPE materials invented by Nordic Chemical Industry, respectively.

Incorporating nanosized fillers into polymer matrices can result in significant improvement of various dielectric properties, owing to the size and interface effect of the nanocomposites [12]. Polymer nanodielectrics refer to the composite materials formed by homogeneously dispersing nanosized fillers (i.e., nanospheres, tubes, or sheets) into the traditional high-molecule-weight polymer matrix, and its applications in the field of electrical insulation include HV cables and accessories, capacitors, motors, and nonlinear dielectrics, etc. Here is a brief history of nanodielectrics. In 1984, an American patent revealed that adding 5–50 nm SiO_2 or Al_2O_3 particles to a polymer matrix improved the resistance capacity to corona aging [13]. In 1994, Lewis T J put forward the theory of nanodielectrics [14]. In 1998, Segal V used nanoparticles to improve the insulation strength of transformer oil [15]. In 1999, Henk et al. improved the electrical strength of epoxy resin by adding SiO_2 nanoparticles [16]. However, these early studies did not receive much attention at that time. Experimental research since 2002 has shown that nanodielectrics has notable advantages in improving space charge, DC conductivity, breakdown, tree aging, partial discharge, heat conduction, etc., and has become an important and hot research field for developing advanced insulating materials [17]. A key issue in this field is the rational surface modification of inorganic nanoparticles along with fabrication technologies of polymer nanodielectrics [18]. There are several grand challenges to be tackled with: (1) controllable dispersion of inorganic nanoparticles in polymer matrices, (2) systematic elucidation of the complexity of nanodielectric interfaces, (3) cross-scale regulation of micro-meso-macro properties, and (4) efficient extension of the boundary from lab-scale to industry-scale preparation.

2 Surface Ligand Engineering of Polymer Nanocomposites

Nanosized inorganic fillers, such as metal oxide particles, have a surface energy as high as 500–2000 mJ·m^{-2}, whereas the surface energy of polymers is typically 20–50 mJ·m^{-2} [19]. Hence, inorganic nanoparticles are typically immiscible with polymer matrices, leading to nanoparticle agglomerates. To address this issue, surface ligand engineering has been employed to achieve a good and homogeneous nanoparticle dispersion in the polymer matrix.

Surface ligand strategies can be divided into two categories: one is the "grafting to" method, and the other is the "grafting from" method [20], as illustrated in Fig. 3. The former refers to the attachment of free organic small molecules or readily made end-functionalized polymer brushes to the surface of nanoparticles, whereas the latter refers to the direct growth of polymer brushes initiated from nanoparticle surfaces functionalized with an initiator or chain transfer agent [21].

The most commonly used "grafting to" strategy includes silane coupling agent methods, ligand exchange methods, and click chemistry methods. Alcohol groups on silane react with hydroxyl groups, and silane molecules are bonded on the surface of inorganic nanoparticles after dehydration [22]. There are many kinds of silanes, such as ≡SiOR (R is alkyl), hydrosilane ≡ SiH, chlorosilane ≡ SiCl, etc., which are suitable for dispersing nanoparticles, i.e. SiO_2 nanoparticles, into silicone rubber, and the formed Si–O–Si bond keeps the whole system stable. Compared with carboxyl group, phosphate functional group forms a stronger covalent bond with hydroxyl group, which can be utilized for surface ligand exchange [23]. Besides, in the reaction process of click chemistry, ethynyl groups on the surface of nanoparticles are bonded with molecules with azido groups under the catalysis of monovalent copper. The reaction conditions are mild and the conversion efficiency is high [24]. However,

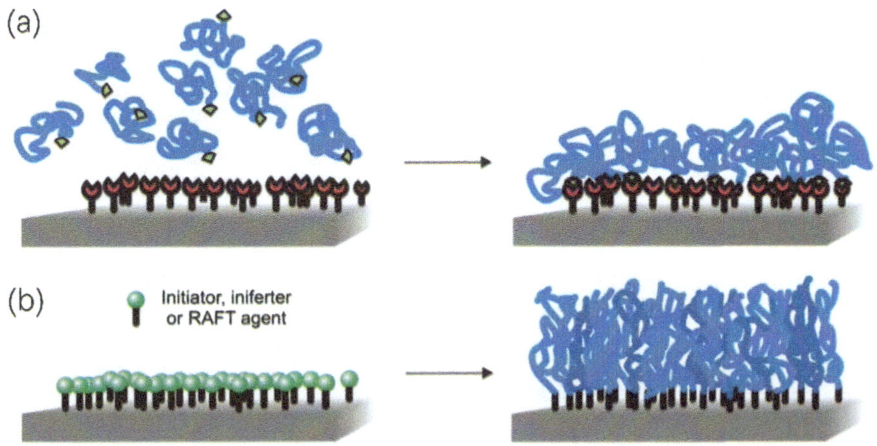

Fig. 3 Surface ligand strategies. **a** Grafting to, **b** Grafting from (Reprinted with permission from Ref. [20]. Copyright 2009 American Chemical Society)

the "grafting to" strategy suffers from a limited grafting density due to the steric repulsions of the readily-made polymer chains.

In contrast to the "grafting to" strategy, the "grafting from" strategy enables deliberate control of composition, molecular weight, architecture, and polydispersity of polymer brushes via controlled radical polymerization. Specific methods include atom transfer radical polymerization (ATRP), nitroxide mediated polymerization (NMP), and reversible addition-fragmentation chain transfer (RAFT) polymerization [25]. A high grafting density can be achieved due to largely mitigated steric hindrance. The molecular weight of polymer brushes can range from 1 to 300 kg·mol^{-1}, and the grafting density can range from 0.01 to 1 ch·nm^{-2} using the "grafting from" strategy. Since 1998, RAFT polymerization method proposed by Chiefary et al. has been widely investigated owing to its wide range of applicable monomers, relatively mild polymerization conditions, narrow distribution of molecular weight, as well as some available industrialized RAFT transfer agents [26]. A typical RAFT Polymerization route is presented in Fig. 4.

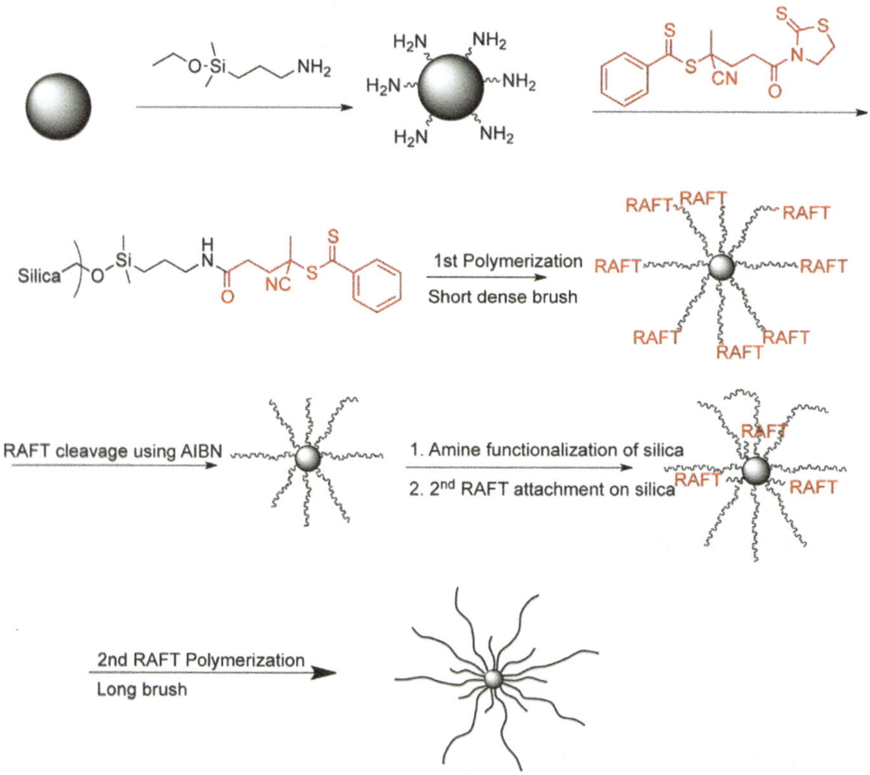

Fig. 4 Sequential RAFT Polymerization for the synthesis of bimodal brush nanoparticles (Reprinted with permission from Ref. [21]. Copyright 2013 American Chemical Society)

The dispersion state of nanoparticle grafted with polymer brushes in the polymer matrix is governed by the balance between the enthalpic attraction of nanoparticles and the entropy related steric hindrance provided by surface ligands. The van der Waals interaction between the nanoparticles primarily determines the enthalpic attraction and induces agglomeration. Grafting polymer brushes on nanoparticles at a low grafting density effectively reduces the enthalpic attraction and enhances the dispersion due to the repulsion of the brush layer. On the other hand, entropic constraints due to the brush–matrix chain interactions significantly influence the inclusion (wetting) and expulsion (dewetting) of the matrix chains from the brush chain layers. The dispersion state can be determined by three parameters—the degree of polymerization of the matrix chains (P), the degree of polymerization of the grafted chains (N), and the grafting density (σ) [27], on the premise that the grafted chains are chemically identical to the matrix chains. At a low grafting density and with grafted chains long enough to form brushes, i.e., $\sigma N^{1/2} < (N/P)^{1/2}$, the free polymer matrix chains can interpenetrate and swell the brush chains, leading to wetting of the brush chains and promoting miscibility between the nanoparticles and the polymer matrix. When the grafting density is very high, the entropic interactions between the brush chains and the matrix chains are dominant, leading to autophobic dewetting of the matrix chains from the brush chains, hence phase separation occurs. Furthermore, it is worth noting that spherical nanoparticles (e.g. SiO_2) grafted with long polymer brushes in chemically identical homopolymer matrix (e.g. polystyrene) can self-assemble into a variety of anisotropic superstructures, as a result of the balance between core–core enthalpic attractions and entropy change due to the redistribution of polymer brush chains. With increasing brush molecular mass and increasing σ, the self-assembled structures evolve from spherical aggregates to sheet-like structures, strings and finally well-dispersed nanoparticles [28].

Nanoparticles grafted with monomodal polymer brushes encounter the difficulty to shield the van der Waals force between nanoparticles and reduce entropy attraction at the same time. Figure 5 describes a roadmap of surface ligand engineering of nanoparticles from simple to complex [29]. The bimodal grafting strategy of polymer brushes provides a feasible solution to the above-mentioned problem, using short chains with a high grafting density to reduce the attraction between nanoparticles, and long chains with a low grafting density to reduce the entropy attraction. On the basis of bimodal brush grafting, small molecule groups with tunable functions could be further grafted on the surface of nanoparticles, forming mixed bimodal grafting and even mixed multimodal integrated grafting strategies to realize the directional decoupling of entropic and enthalpic interactions [30].

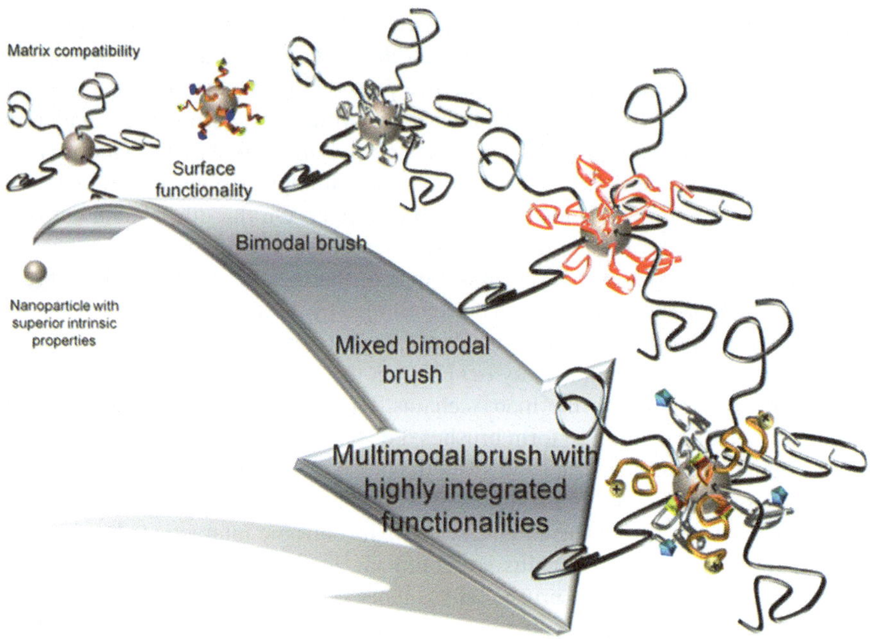

Fig. 5 Surface ligand engineering of nanoparticles: from simple to complex (Reprinted with permission from Ref. [29]. Copyright 2014 American Chemical Society)

3 Surface Ligand Engineering for Advanced Cable Insulating Materials

3.1 Space Charge

Incorporating nanoparticles into polymer matrices will form massive interfacial area in polymer nanodielectrics, introducing new trapping sites, thus affecting the generation, migration, accumulation, and dissipation of space charges. In recent years, research on TiO_2/LDPE, SiO_2/LLDPE, and SiO_2/XLPE nanocomposites has proved that nanoparticle surface grafting can leads to the suppression of space charge effect [31–33].

To reduce van der Waals forces between nanoparticles and improving interface compatibility, small-molecule-weight silane coupling agents, such as vinyl trimethoxy silane, 3-aminopropyltriethoxysilane, and hexamethyldisilazane [34–36], have been employed as a "grafting to" approach, especially on the nanoparticle surface with hydroxyl groups via condensation reaction. It has the advantages of simple process, wide applicability, and low cost. Choosing coupling agents with the same or similar molecular structure as the matrix polymer chains can improve nanoparticle dispersion, resulting in notable suppression of space charge effect [37].

It was found that surface grafting also significantly affects trapping site distribution and charge transport characteristics. For instance, grafting coupling agent onto nanoparticles increases the concentration of deep trapping sites near the shielding layers of cable insulation, capture homocharges, and then reduce the electric field strength between the homocharge layer and the shielding layer, increasing the barrier height of subsequent homocharges injection [38]. Besides charges captured in deep trapping sites near the shielding layers, the number of shallow trapping sites in the bulk of composite materials also increases due to the nanoparticle surface grafting, which promotes the dissipation process of bulk space charge [39].

There are limitations when surface grafting of coupling agents is applied to high-molecule-weight polymer systems, because the grafting parameters should be compatible with the molecular weight of the polymer matrix to achieve better dispersion [40]. Well-dispersed nanoparticles could be obtained by grafting polymer brushes with similar molecular structure to matrix polymer chains, and the aggregation size can be reduced to tens of nanometers [41]. Zhang et al. obtained homogeneous nanoparticle dispersion (about 100 nm) by grafting poly(stearyl methacrylate) (PSMA) brushes onto SiO_2 nanoparticles, thereby introducing more deep trapping sites compared with the non-grafted SiO_2/XLPE nanocomposites, realizing a remarkable suppression effect of space charge even under a strong electric field of −100 kV/mm [42], as shown in Fig. 6.

In order to endow polymer nanodielectrics with stronger trapping ability of charge carriers, it is essential to accurately control the interfacial region between nanoparticles and polymer matrix. Electroactive organic functionalities with small molecule weight are usually introduced into the interfacial region [43], but they tend to aggravate the aggregation of nanoparticles due to their polar groups. In this text, the mixed

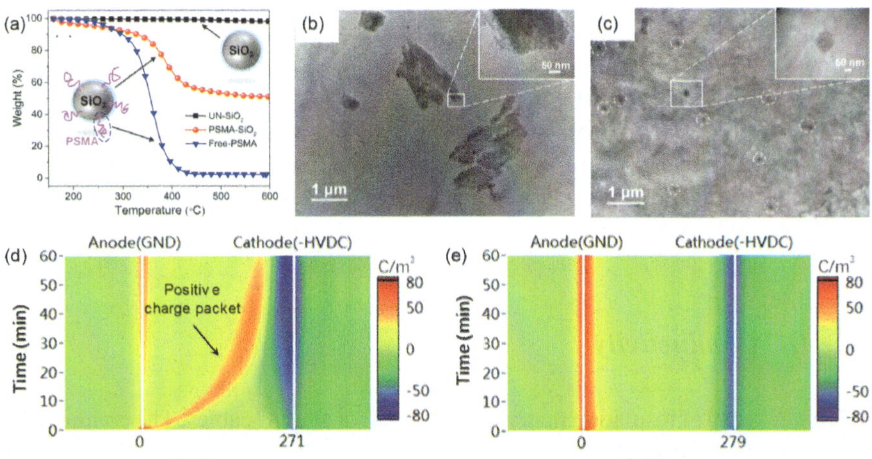

Fig. 6 **a** TGA curves, **b** and **c** are TEM images of unmodified SiO_2/XLPE and PSMA-SiO_2/XLPE; **d** and **e** are space charge patterns of XLPE and PSMA–SiO_2/XLPE under −100 kV/mm (Reprinted with permission from Ref. [42]. Copyright 2017 American Institute of Physics)

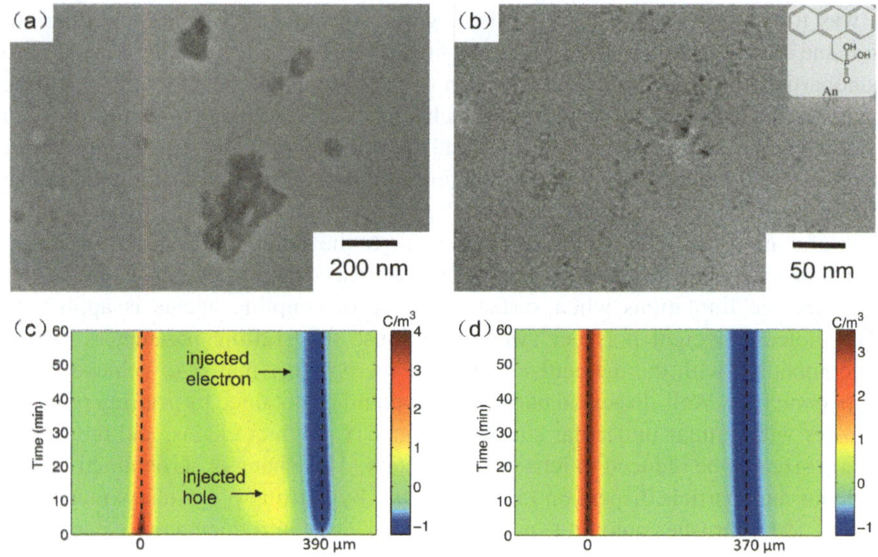

Fig. 7 TEM images of **a** 4 wt.% as-synthesized TiO_2/SIR and **b** An-PDMS-TiO_2/SIR nanocomposites. Space charge profiles of **c** PDMS-TiO_2/SIR and **d** An-PDMS-TiO_2/SIR nanocomposites. (Reprinted and adopted with permission from Ref. [44]. Copyright 2016 American Institute of Physics)

bimodal grafting strategy, i.e., adopting both polymer brushes and organic functionalities, has been employed to improve the incompatibility between polar functional groups and nonpolar polymer matrix. For example, Huang et al. reported the attachment of 9-anthracenemethylphosphonic acid (An) functionalities and polydimethylsiloxane (PDMS) brushes onto TiO_2 nanoparticles, achieving nanosized dispersion in silicone rubber (SIR) matrix and inhibited the injection of space charge under 10 kV/mm electric field [44], as shown in Fig. 7.

Polymer brush grafting technology have been proved to improve the charge transport characteristics of composite materials, however, the fabrication process and the reaction conditions are relatively complex, posing a limit to industry-scale production for HVDC cable application.

3.2 DC Conductivity

DC conductivity of insulating materials is pivotal for HVDC cables. The conductivity of polymer materials increases with the temperature. On the premise of satisfying the multi-layer insulation coordination, it is expected that the insulating materials of HVDC cables have lower DC conductivity and smaller temperature sensitivity.

Generally speaking, the conductivity of inorganic nanoparticles is higher than that of the polymer matrix, and charge carriers could easily migrate through the interfacial region between nanoparticles and the polymer matrix. When the distance between nanoparticles is short, aggregates are easily formed and the adhesion to the matrix material is not tight enough, hence, DC conductivity of nanocomposites will increase. It is reported that DC conductivity of nanocomposites can be reduced via nanoparticle surface grafting [45]. Well-grafted nanoparticles display low entropic interactions that leads to homogeneous dispersion, thus reducing the local conduction current. Pourrahimi et al. improved the dispersion of ZnO nanoparticles and obtained better interfacial adhesion through surface grafting of methyltrimethoxysilane coupling agent, thus achieving a stronger trapping ability of charge carrier [46]. As a result, the conductivity of the ZnO/LDPE nanocomposites decreased by two orders, demonstrating the effectiveness of DC conductivity regulation via the nanoparticle surface grafting strategy.

It is of great importance to improve the temperature stability of DC conductivity of insulating materials for HVDC cable applications. Zhang et al. found that coupling agent grafted SiO_2/XLPE nanocomposites render lower conductivity and smaller temperature coefficient [47]. Zhou et al. also found that the current density and conductivity temperature coefficient of grafted MgO/polypropylene nanocomposites were lower than those of pure polypropylene [48]. Currently, research effort mainly focuses on reducing DC conductivity of nanocomposite materials, while there is little attention being paid to directional control of temperature coefficient of DC conductivity, which demands rational structural design and systematically experimental work.

3.3 Breakdown Strength

Surface grafting changed the physical and chemical characteristics of polymer nanodielectrics, such as nanoparticles dispersion status, crystallization, and glass transition temperature, thereby increasing average trap energy level or the concentration of trapping sites, thus affecting the breakdown strength of polymer nanocomposites. However, when the aggregation size of nanoparticles reaches micron level and acts as insulation defects, the internal electric field will be distorted, resulting in the breakdown strength drop of nanodielectric materials [49]. Luckily, the breakdown strength of polymer nanodielectrics is higher than that of pure polymers due to the presence of well-dispersed nanoparticles [50]. On the other hand, according to the multi-core model, low-density interlayer regions between nanoparticles and the polymer matrix due to less stoichiometrically crosslinked layer result in an increase in the free volume and local conduction current [51]. Hence, electrons are more easily accelerated in the free volume, thus having higher energy to destroy the polymer chains, while the rise of conduction current will increase the risk of thermal breakdown. To address this issue, nanoparticle surface grafting strategy is employed to strengthen the bonding between nanoparticles and the polymer matrix, so that the low-density regions can be reduced and thus the breakdown strength of nanocomposites can be improved [52]. In addition, surface grafting promotes the crystallization of the polymer matrix, which also accounts for the breakdown strength improvement of polymer nanodielectrics [53].

To further inhibit the injection or migration of charges and realize the synchronous improvement of breakdown strength, electroactive functional groups are introduced into the interfacial region of the nanocomposites to form more deep trapping sites. For instance, anthracene functional groups and poly(stearyl methacrylate) (PSMA) polymer brushes were grafted onto SiO_2 nanoparticles. This bimodal surface ligand modification leads to an increase in the DC breakdown strength of polypropylene nanocomposites by 33% [54]. In 2004, Lewis pointed out that the capture of space charge by the electric double layer formed at the interfacial region of nanodielectrics was a key factor affecting the breakdown strength [55]. In 2005, Tanaka et al. put forward the multi-core model, and proposed that the average free path of electrons in the third layer and the coulomb force of electric double layer affect the breakdown strength [56]. In 2010, Li et al. proposed the interaction region model, pointing out that nano-doping after surface grafting could increase the trap energy level and trap density of the interaction region, improve the distribution characteristics of space charge, and reduce the distortion degree of electric field, thus improving the breakdown field strength [57]. Moreover, by the combination of simulation and experimental results, Min et al. explained the important role of trap characteristics and free volume in the interaction area to the improvement of breakdown strength [58]. All the above-mentioned theoretical interface models reveal that charge transport is closely related to the electrical breakdown process. Therefore, it is necessary to systematically establish the intrinsic relationship between the nanoparticle surface grafting, charge transport, and breakdown strength in the interfacial region.

3.4 Electrical Tree Aging

Electrical tree is a pre-breakdown phenomenon that causes failure of insulating materials, threatening the operation safety and reliability of HVDC cables and cable accessories. It is originated from impurities, gas voids, and mechanical defects that are inevitably introduced during the preparation process of polymer insulating materials. The local electric field is distorted within small regions of insulating materials, resulting in partial discharge and electrical tree aging. Figure 8 shows the typical shapes of electrical tree in silicone rubber insulation of cable accessories in a wide range of voltage frequencies from DC to high frequency (130 kHz), where single-branch electrical tree is usually found under DC voltages.

The research history can be traced back to 1936 when Robinson first discovered electrical tree aging phenomena in high voltage cables [59]. In 1972, IEEE Conference on Electrical Insulation and Dielectrics made it clear that electrical tree aging is one of the main causes of solid dielectric insulation deterioration [60]. Therefore,

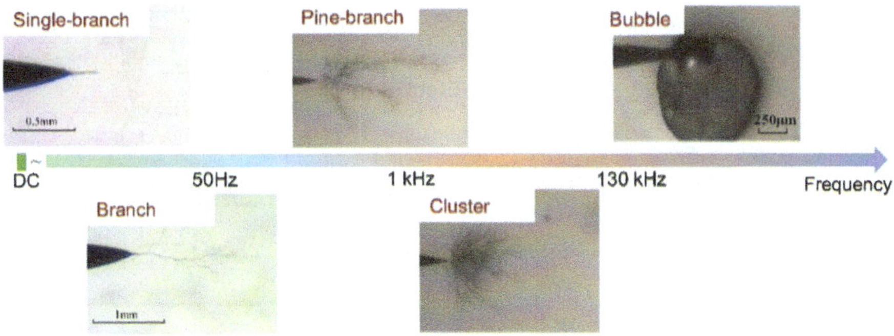

Fig. 8 Typical electrical tree patterns in a wide range from DC to high frequency in silicone rubber insulation of cable accessories

it is of great significance to inhibit electrical tree aging to improve the insulation reliability of electrical equipment.

Voltage stabilizers, i.e. aromatic compounds with low ionization potential and high electron affinity, have been widely used for inhibiting electrical tree aging after massive engineering practice [61]. The inhibiting mechanisms of the voltage stabilizers are as follows: (1) small-molecular-weight voltage stabilizer can migrate to the defect regions of polymer insulation. Because the conductivity of the voltage stabilizer is greater than that of the polymer matrix, the electric field at the defects is weakened; (2) voltage stabilizer can capture high-energy electrons, absorb their energy, and reduce the impact of electrons on polymer chains, thus improving the electrical tree resistance of insulating materials.

In recent years, it has been found that grafting aromatic compound voltage stabilizer onto polymer chains can resist electrical tree aging more effectively. For example, grafting 1-(4-vinyloxy)phenylethenone onto XLPE chains can absorb the energy of high-energy electrons, and render an initiation voltage of electrical tree increase by 25.8% [62]. Moreover, fullerene derivatives with high-electron affinity can also serve as voltage stabilizers to increase the initiation voltage of electrical tree by 26% [63].

Furthermore, electrical tree aging can be effectively inhibited by the regulation of the polymer matrix crystallization via grafted nanoparticles [64]. For instance, in the 3-glycidoxypropyl-methyl-di-ethoxysilane coupling agent grafted SiO_2/epoxy nanocomposites, SiO_2 nanoparticles hindered the developing path of electrical tree branches, resulting in a triple increase in the duration from tree initiation to breakdown [65]. In another case, the KH550 coupling agent grafted montmorillonite/polyethylene nanocomposite achieved uniform and dense crystallization, which hindered the development of electrical tree [66], as shown in Fig. 9.

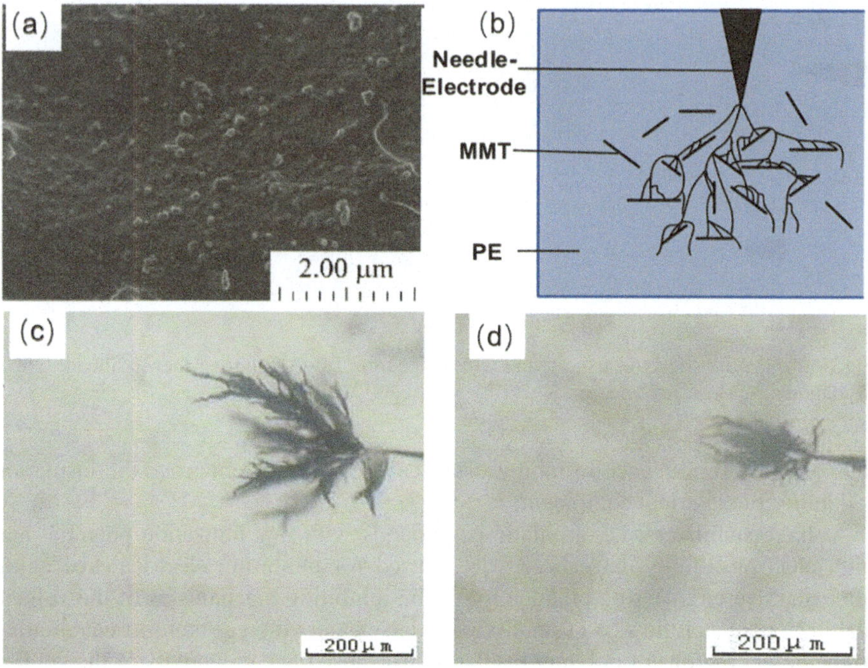

Fig. 9 **a** TEM image of KH550 modified MMT/XLPE, **b** inhibiting effect of MMT on growth path of an electrical tree, **c** and **d** are electrical trees in XLPE and its nanocomposites (Reprinted with permission from Ref. [66]. Copyright 2015 IEEE)

3.5 Partial Discharge

Partial discharge occurs due to voids and cracks in insulating materials under external electric field. In the HVDC cable engineering, partial discharge exerts a continuous accumulation effect on the aging process of solid insulating dielectrics. Hence, it is of great importance to restrain the occurrence of partial discharge.

Incorporating nanoparticles into polymer matrices is widely used to improve the partial discharge resistance, because nanoparticles can reduce partial discharge, and the partial discharge resistance increases with the decrease of particle size. For example, in the KH550 coupling agent grafted MgO/XLPE nanocomposites, the intensity of partial discharge is obviously weakened due to the resistance of inorganic materials to partial discharge corrosion, and the hindrance of well-dispersed nanoparticles to high-energy discharge [67, 68]. In the KH570 grafted boron nitride (BN)/ethylene propylene diene monomer nanocomposites, the initiation voltage of partial discharge increases with the filler loading, for the embedded nanoparticles enhance the barrier property and reduce the net electric field in the air gap [69]. In another case, the partial discharge characteristics of the virgin XLPE, unmodified SiO_2/XLPE, and octylsilane grafted SiO_2/XLPE nanocomposites were compared

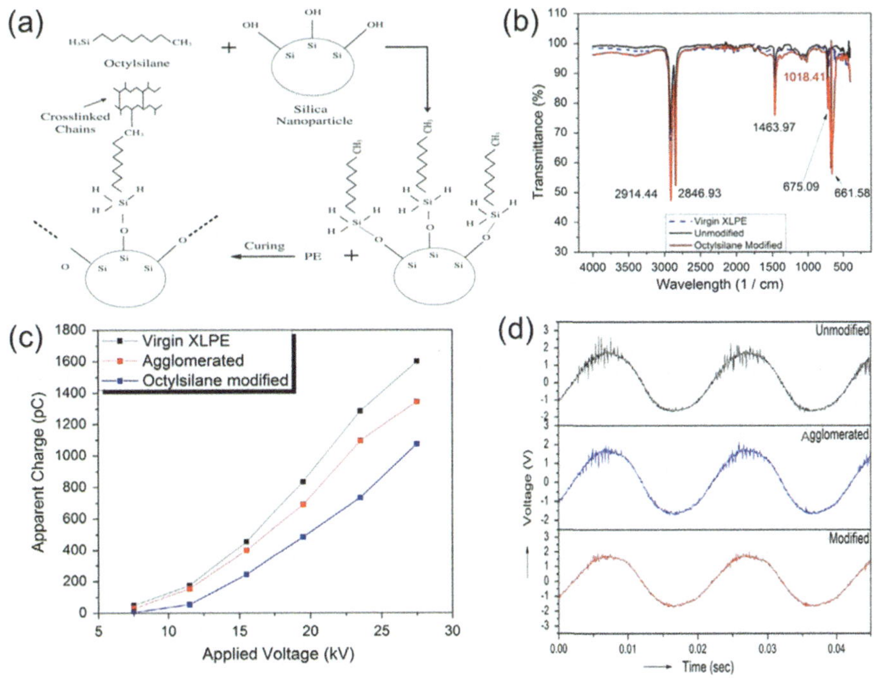

Fig. 10 The effect of modified SiO_2 on partial discharge. **a** Octylsilane-modified silica with polymer chains, **b** FTIR characterization, **c** and **d** are partial discharge characteristics of XLPE and its nanocomposites (Reprinted and adopted with permission from Ref. [70]. Copyright 2017 Springer)

(Fig. 10), it was found that surface grafting renders the lowest partial discharge activity and the highest discharge initiation voltage, demonstrating that nanoparticle surface grafting is effective to suppress partial discharge activities [70].

3.6 *Thermal Conductivity*

Temperature gradient occurs in the cable insulation under the operation conditions. Since most polymers are poor conductor of heat, a larger temperature gradient leads to a greater reverse degree of electric field distribution. Increasing the thermal conductivity of insulating materials can reduce the temperature gradient in cable insulation. The essence of heat transfer is the macroscopic manifestation of the collision and energy transfer of electrons and microscopic particles. Heat transfer in solids is typically conducted through either free electrons or lattice wave function of lattice vibrations known as phonons. However, there are few free electrons in polymer materials, and the macromolecular chains are largely disordered. Thus, complete crystals cannot be formed and the thermal conductivity is very low.

Thermal conductivity of polymer nanodielectrics could be improved by incorporating fillers with much higher thermal conductivity than that of the matrix. The main additives are BN, AlN, carbon nanotubes, SiC, Si_3N_4, Al_2O_3, and ZnO [71, 72]. For polymers, improving the regularity of molecular chains is the key to obtain higher thermal conductivity. Fillers plays a decisive role in the thermal conductivity of polymer composites, while the interface is the main obstacle of heat transfer. When the filler loading is low, particles will be isolated from each other and the polymer composites have low thermal conductivity. With the increase of filler loading, particles come into contact with each other and form a thermally conductive network. When the filler loading further increases, there is little improvement in thermal conductivity due to the percolation effect [73] (Fig. 11). Surface ligand engineering provides a effective toolbox to improve the interfacial compatibility and reduce the interfacial thermal resistance.

Inorganic nanoparticles have been widely used in the preparation of thermally conductive insulating polymer nanocomposites due to their high thermal conductivity and good electrical insulation properties. The thermal conductivity of hyperbranched polyaramid grafted Al_2O_3/epoxy nanocomposites is obviously higher than that of the unmodified group, because the strong covalent bond between nanoparticles and epoxy matrix generates a heat flow network and reduces phonon scattering [74]. The thermal conductivity of KH550 coupling agent grafted AIN/polypropylene nanocomposites increases with the AIN loading, and KH550 grafting forms an organic active monolayer between AIN and polypropylene, which reduces the interfacial thermal

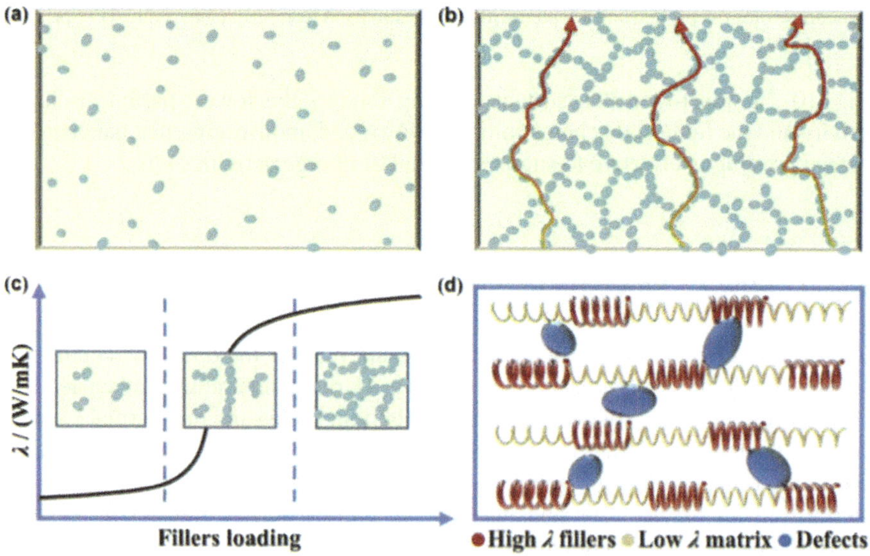

Fig. 11 Schematic diagram of thermal conduction path. **a** "Sea-island" in low filler loadings, **b** thermal conduction paths in high filler loadings, **c** percolation phenomenon, **d** thermoelastic coefficient theory (Reprinted with permission from Ref. [73]. Copyright 2020 Elsevier)

resistance [75]. The thermal conductivity of silane grafted BN/epoxy nanocomposites is improved due to the formation of covalent bridges at the interface, reducing the scattering of phonons at the interface and improving the interfacial thermal resistance [76, 77].

There is a paradox that polymer composites with excellent dielectric properties usually possess low thermal conductivity, while composites with high thermal conductivity usually possess high dielectric constant and high dielectric loss. To further meet the requirements of both high thermal conductivity and excellent insulation, special functional groups can be grafted onto nanoparticles. For example, in the polyglycidyl methacrylate (PGMA) grafted h-BN/epoxy nanocomposites, the thermal conductivity of 3%, 9%, and 15% filler-loading of PGMA-h-BN can be increased by 60%, 203%, and 505%, respectively [78]. Surface grafted h-BN nanosheets are beneficial to enhance the compatibility between fillers and epoxy matrix, reduce the apparent viscosity of composites, thus improving the thermal conductivity. Polyethylene glycol grafted graphene oxide suppresses dielectric loss by the increase of free volume, thus promoting the dispersion of graphene oxide with improve interfacial interaction and thermal conductivity [79]. A remarkable example is the addition of polyhedral oligosilsesquioxane (POSS) functionalized boron nitride nanotubes (BNNTs) into the epoxy matrix, achieving much lower dielectric constant and an effectively reduced coefficient of thermal expansion (CTE). By adding 30 wt% of POSS grafted BNNTs, the CTE is decreased by about 30%. The covalent interactions between the epoxy matrix and the POSS is the key factor for the enhancement of thermal conductivity [80]. In another case, a mixed filler strategy was employed to improve the interfacial condition between the fillers and the matrix [81]. Specifically, h-BN microparticles modified with silane-coupling agent KH550 were added into LDPE matrix, forming a thermal conduction path to improve the thermal conductivity, whereas SiO_2 nanoparticles with the same surface modification were used to improve the charge transport characteristics.

In the next generation of HVDC cable engineering practice, it is urgent to improve thermal conductivity while maintaining good mechanical properties and dielectric properties. Nanoparticle surface ligand engineering has shown great potential in achieving multi-performance improvement.

4 Conclusions and Outlooks

4.1 Conclusions

To sum up, surface ligand engineering promotes the rapid development of advanced polymer nanodielectrics for HVDC cable applications by improving the interfacial compatibility of polymer nanodielectrics, manifesting the inherent properties of nanoparticles and polymer matrices, and introducing tunable functionalities. It is a grand challenge to disperse inorganic nanoparticles in the high-molecular-weight polymer matrices. To address this issue, the mixed bimodal and mixed multimodal grafting strategies on the surface of nanoparticles have been used to broaden the dispersion window by directionally decoupling enthalpy and entropy interactions. Currently, polymer nanodielectrics based on surface ligand grafting have shown a pivotal role and great potential in inhibiting space charge, reducing DC conductivity and temperature sensitivity, improving breakdown strength, inhibiting electrical tree aging, and improving thermal conductivity, etc. However, due to a lack of integrated interface design theory and precise control of fabrication technology, surface ligand modified polymer nanodielectrics have not yet become the mainstream choice for advanced plastic HVDC cables. Moreover, the long-term operation characteristics, multi-factor influencing mechanism, and multi-performance coordinated control methods of surface ligand modified polymer nanodielectrics need to be further studied.

4.2 Outlooks

While exciting progress has been made in this field, there is still a lack of precise and facile surface ligand fabrication technologies, and the accurate compostion-structure–property relationship between surface ligand modified nanoparticle-matrix interface and the dielectric performance is still at a nascent stage. Future research directions may include:

(1) Rational design and precise control of mixed multimodal brush grafting methods, which offers excellent polymer-matrix compatibility and integrated functionalities, thus may fulfill the needs for multi-performance improvement.
(2) With the rapid development of microscopic characterization technology, an interface model considering multi-level structure regulation is expected to be established and subsequently simulation should be carried out, which provides important insight and guidance for the searching of advanced insulating materials.
(3) The physical and chemical structure of the nanoparticle surface grafts is more fragile than that of the matrix polymer chains, and the long-term stability is controversial. Long-term aging starts from bond fracture and dissociation of molecules, which affects the chemical environment of the interfacial region and macroscopic dielectric properties. It is necessary to carry out quantitative experimental on the long-term characteristics of the interfacial structure state.
(4) To push the boundary from lab-scale to industry-scale fabrication, it is difficult to ensure the nanoparticle dispersion status and the stability of material properties. In the extrusion process of HVDC cables, the aggregation of nanoparticles will block the filter screen, resulting in the inability of continuous extrusion. Therefore, it is necessary to further study the fabrication and processing of polymer nanodielectrics with mature technology, low cost, and facile safe operation.

References

1. G. Mazzanti, M. Massimo, *Extruded Cables for High Voltage Direct Current Transmission* (Wiley, Hoboken, NJ, 2013)
2. Y. Maekawa, A. Yamaguchi, S. Yoshida et al., Development of DC ±250 kV XLPE cable and factory joints, in *3rd Proceedings International Conference on Insulated Power Cable (Jicable'91)* (Paris, France, 1991), pp. 554–561
3. Y. Li, T. Takada, Progress in space charge measurement of solid insulating materials in Japan. IEEE Electr. Insul. Mag. **10**(5), 16–28 (1994)
4. R.N. Hampton, Some of the considerations for materials operating under high-voltage direct-current stresses. IEEE Electr. Insul. Mag. **24**(1), 5–13 (2008)
5. J.M. Alison, R.M. Hill, Model for bipolar charge transport, trapping and recombination in degassed crosslinked polyethene. J. Phys. D Appl. Phys. **27**(6), 1291–1299 (1994)
6. K. Terashima, M. Hara, H. Suzuki et al., Research and development of 250 kV DC XLPE cables. IEEE Trans. Power Deliv. **13**(1), 7–16 (1998)
7. S. Maruyama, N. Ishii, M. Shimada et al., Development of a 500-kV DC XLPE cable system. Furukawa Rev. **25**, 47–52 (2004)
8. T.L. Hanley, R.P. Burford, R.J. Fleming et al., A general review of polymeric insulation for use in HVDC cables. IEEE Electr. Insul. Mag. **19**(1), 13–24 (2003)
9. T. Katayama, T. Yamazaki, Y. Murata et al., Space charge characteristics in DC-XLPE cable after 400 kV PQ test, in *9th Proceedings International Conference on Insulated Power Cables (Jicable'15)* (Versailles, France, 2015), A7.2
10. S. Lee, T. Lee, E. Jung et al., Development of 250kV HVDC XLPE cable system in Korea, in *CIGRE AORC Technical meeting* (Niigata, Japan, 2014), pp. 334–337
11. S. Xie, M. Fu, Y. Yin et al., Triple jumps of XLPE insulated HVDC cable development in China: from 160 kV, 200 kV to 320 kV, in *9th Proceedings International Conference on Insulated Power Cables (Jicable'15)* (Versailles, France, 2015), A9.6
12. A.C. Balazs, T. Emrick, T.P. Russell, Nanoparticle polymer composites: where two small worlds meet. Science **314**(5802), 1107–1110 (2006)
13. R.D. Johnston, M.Markovitz, Corona resistant insulation: USA, 4760296 A.1988–07–26.
14. T.J. Lewis, Nanometric dielectrics. IEEE Trans. Dielectr. Electr. Insul. **1**(5), 812–825 (1994)
15. V. Segal, A. Hjortsberg, A. Rabinovich et al., AC (60 Hz) and impulse breakdown strength of a colloidal fluid based on transformer oil and magnetite nanoparticles, in*Conference Record on the IEEE International Symposium on Electrical Insulation* (Arlington, VA, USA, 1998), pp. 619–622
16. P.O. Henk, T.W. Kortsen, T. Kvarts, Increasing the electrical discharge endurance of acid anhydride cured DGEBA epoxy resin by dispersion of nanoparticle silica. High Perform. Polym. **11**(3), 281–296 (1999)
17. T. Tanaka, A. Bulinski, J. Castellon et al., Dielectric properties of XLPE/SiO_2 nanocomposites based on CIGRE WG D1.24 cooperative test results. IEEE Trans. Dielectr. Electr. Insul. **18**(5), 1482–1517 (2011)
18. S.T. Milner, Polymer brushes. Science **251**(4996), 905–914 (1991)
19. W. Caseri, Nanocomposites of polymers and metals or semiconductors: historical background and optical properties. Macromol. Rapid Commun. **21**(11), 705–722 (2000)
20. R. Barbey, L. Lavanant, D. Paripovic et al., Polymer brushes via surface-initiated controlled radical polymerization: synthesis, characterization, properties, and applications. Chem. Rev. **109**(11), 5437–5527 (2009)
21. S.K. Kumar, N. Jouault, B. Benicewicz et al., Nanocomposites with polymer grafted nanoparticles. Macromolecules **46**(9), 3199–3214 (2013)
22. Y.K. Huang, Q. Liu, X.D. Zhou et al., Synthesis of silica particles grafted with weil-defined living polymeric chains by combination of RAFT polymerization and coupling reaction. Macromolecules **42**(15), 5509–5517 (2009)
23. Y. Li, P. Tao, A. Viswanath et al., Bimodal surface ligand engineering: the key to tunable nanocomposites. Langmuir **29**(4), 1211–1220 (2013)

24. P. Tao, A. Viswanath, L.S. Schadler et al., Preparation and optical properties of indium tin oxide/epoxy nanocomposites with polyglycidyl methacrylate grafted nanoparticles. ACS Appl. Mater. Interfaces. **3**(9), 3638–3645 (2011)
25. J. Yan, M.R. Bockstaller, K. Matyjaszewski, Brush-modified materials: control of molecular architecture, assembly behavior, properties and applications. Prog. Polym. Sci. **100**, 101180 (2020)
26. J. Chiefari, Y.K. Chong, F. Ercole et al., Living free-radical polymerization by reversible addition-fragmentation chain transfer: the RAFT process. Macromolecules **31**(16), 5559–5562 (1998)
27. P.G. Ferreira, A. Ajdari, L. Leibler, Scaling law for entropic effects at interfaces between grafted layers and polymer melts. Macromolecules **31**(12), 3994–4003 (1998)
28. P. Akcora, H.J. Liu, S.K. Kumar et al., Anisotropic self-assembly of spherical polymer-grafted nanoparticles. Nat. Mater. **8**(4), 354–359 (2009)
29. Y. Li, T.M. Krentz, L. Wang et al., Ligand engineering of polymer nanocomposites: from the simple to the complex. ACS Appl. Mater. Interfaces. **6**(9), 6005–6021 (2014)
30. Y. Huang, L.S. Schadler, On the nature of high field charge transport in reinforced silicone dielectrics: experiment and simulation. J. Appl. Phys. **120**(5), 055101 (2016)
31. D. Ma, T.A. Hugener, R.W. Siegel et al., Influence of nanoparticle surface modification on the electrical behaviour of polyethylene nanocomposites. Nanotechnology **16**(6), 724–731 (2005)
32. X.Y. Huang, P.K. Jiang, Y. Yin, Nanoparticle surface modification induced space charge suppression in linear low density polyethylene. Appl. Phys. Lett. **95**(24), 242905 (2009)
33. L. Zhang, Y.X. Zhou, X.Y. Cui et al., Effect of nanoparticle surface modification on breakdown and space charge behavior of XLPE/SiO_2 nanocomposites. IEEE Trans. Dielectr. Electr. Insul. **21**(4), 1554–1564 (2014)
34. M. Roy, J.K. Nelson, R.K. MacCrone et al., Polymer nanocomposite dielectrics-the role of the interface. IEEE Trans. Dielectr. Electr. Insul. **12**(4), 629–643 (2005)
35. K. Lau, A. Vaughan, G. Chen et al., On the space charge and DC breakdown behavior of polyethylene/silica nanocomposites. IEEE Trans. Dielectr. Electr. Insul. **21**(1), 340–351 (2014)
36. M.Z. Gao, J.M. Yang, H. Zhao et al., Preparation methods of polypropylene/nano-silica/styrene-ethylene-butylene-styrene composite and its effect on electrical properties. Polymers **11**(5), 797 (2019)
37. S.J. He, J.B. Hu, C. Zhang et al., Performance improvement in nano-alumina filled silicone rubber composites by using vinyl tri-methoxysilane. Polym. Testing **67**, 295–301 (2018)
38. B. Han, X. Wang, Z. Sun et al., Space charge suppression induced by deep traps in polyethylene/zeolite nanocomposite. Appl. Phys. Lett. **102**(1), 345–349 (2013)
39. S.J. Wang, J.W. Zha, Y.H. Wu et al., Preparation, microstructure and properties of polyethylene/alumina nanocomposites for HVDC insulation. IEEE Trans. Dielectr. Electr. Insul. **22**(6), 3350–3356 (2015)
40. S. Boyes, A. Granville, M. Baum et al., Polymer brushes-surface immobilized polymers. Surf. Sci. **570**(1), 1–12 (2004)
41. Y. Huang, T.M. Krentz, J.K. Nelson et al., Bimodal brush functionalized TiO_2/silicone nanocomposites with improved dielectric properties, in *IEEE Electrical Insulation Conference* (Seattle, USA, 2015), pp. 325–328
42. L. Zhang, M.M. Khani, T.M. Krentz et al., Suppression of space charge in crosslinked polyethylene filled with poly(stearyl methacrylate)-grafted SiO_2 nanoparticles. Appl. Phys. Lett. **110**(13), 132903 (2017)
43. Y.H. Gao, X.Y. Huang, D.M. Min et al., Recyclable dielectric polymer nanocomposites with voltage stabilizer interface: toward new generation of high voltage direct current cable insulation. ACS Sustain. Chem. Eng. **7**(1), 513–525 (2018)
44. Y.H. Huang, K. Wu, M. Bell et al., The effects of nanoparticles and organic additives with controlled dispersion on dielectric properties of polymers: charge trapping and impact excitation. J. Appl. Phys. **120**(5), 55102 (2016)
45. Y. Murakami, M. Nemoto, S. Okuzumi et al., DC conduction and electrical breakdown of MgO/LDPE nanocomposite. IEEE Trans. Dielectr. Electr. Insul. **15**(1), 33–39 (2008)

46. A.M. Pourrahimi, T.A. Hoang, D. Liu et al., Highly Efficient interfaces in nanocomposites based on polyethylene and ZnO nano/hierarchical particles: a novel approach toward ultralow electrical conductivity insulations. Adv. Mater. **28**(39), 8651–8657 (2016)
47. L. Zhang, Y.X. Zhou, M. Huang et al., Effect of nanoparticle surface modification on charge transport characteristics in XLPE/SiO_2 nanocomposites. IEEE Trans. Dielectr. Electr. Insul. **21**(2), 424–433 (2014)
48. Y. Zhou, C. Yuan, C.Y. Li et al., Temperature dependent electrical properties of thermoplastic polypropylene nanocomposites for HVDC cable insulation. IEEE Trans. Dielectr. Electr. Insul. **26**(5), 1596–1604 (2019)
49. J.C.Fothergill, J.K. Nelson, M. Fu, Dielectric properties of epoxy nanocomposites containing TiO_2, Al_2O_3 and ZnO fillers, in *Annual Report Conference on Electrical Insulation and Dielectric Phenomena* (Boulder, USA, 2004), pp. 406–409
50. S.T. Ju, H. Zhang, M.J. Chen et al., Improved electrical insulating properties of LDPE based nanocomposite: effect of surface modification of magnesia nanoparticles. Compos. A Appl. Sci. Manuf. **66**, 183–192 (2014)
51. T. Tanaka, Dielectric Nanocomposites with insulating properties. IEEE Trans. Dielectr. Electr. Insul. **12**(5), 914–928 (2005)
52. F. Tian, J. Yao, P. Li et al., Stepwise electric field induced charging current and its correlation with space charge formation in LDPE/ZnO nanocomposite. IEEE Trans. Dielectr. Electr. Insul. **22**(2), 1232–1239 (2015)
53. G. Yu, Y.J. Cheng, X.H. Zhang, The dielectric properties improvement of cable insulation layer by different morphology nanoparticles doping into LDPE. Coatings **9**(3), 204 (2019)
54. T. Krentz, M.M. Khani, M. Bell et al., Morphologically dependent alternating-current and direct-current breakdown strength in silica–polypropylene nanocomposites. J. Appl. Polym. Sci. **134**(1), 44347 (2017)
55. T.J. Lewis, Interfaces are the dominant feature of dielectrics at the nanometric level. IEEE Trans. Dielectr. Electr. Insul. **11**(5), 739–753 (2004)
56. T. Tanaka, M. Kozako, N. Fuse et al., Proposal of a multi-core model for polymer nanocomposite dielectrics. IEEE Trans. Dielectr. Electr. Insul. **12**(4), 669–681 (2005)
57. S.T. Li, G. Yin, G. Chen et al., Short-term breakdown and long-term failure in nanodielectrics: a review. IEEE Trans. Dielectr. Electr. Insul. **17**(5), 1523–1535 (2010)
58. D.M. Min, H.Z. Cui, W.W. Wang et al., The coupling effect of interfacial traps and molecular motion on the electrical breakdown in polyethylene nanocomposites. Compos. Sci. Technol. **184**(10), 107873 (2019)
59. D.M. Robinson, *Dielectric Phenomenon in High Voltage Cables* (Chapman and Hall, London, 1936)
60. M. Ieda, M. Nawata, A consideration of treeing in polymers, in *Annual Conference on Electrical Insulation and Dielectric Phenomena*, Buck Hills Falls, PA, USA (1972), pp. 143–50
61. V. Englund, R. Huuva, S.M. Gubanski et al., High efficiency voltage stabilizers for XLPE cable insulation. Polym. Degrad. Stab. **94**(5), 823–833 (2009)
62. W. Dong, X. Wang, B. Tian et al., Use of grafted voltage stabilizer to enhance dielectric strength of cross-linked polyethylene. Polymers **11**(1), 176 (2019)
63. M. Jarvid, A. Johansson, R. Kroon et al., A new application area for fullerenes: voltage stabilizers for power cable insulation. Adv. Mater. **27**(5), 897–902 (2015)
64. F.N. Musa, N. Bashir, M.H. Ahmad et al., Electrical treeing in high voltage insulations: a review on nanocomposite insulating materials and their processing techniques. J. Optoelectron Adv. Mater. **17**(3–4), 462–476 (2015)
65. T. Iizuka, T. Tanaka, Effects of nano silica filler size on treeing breakdown lifetime of epoxy nanocomposites, in *Proceedings of the 9th International Conference on Properties and Applications of Dielectric Materials* (Harbin, China, 2009), pp. 733–736
66. X.H. Chi, J.G. Gao, X.H. Zhang, Electrical tree propagating characteristics of polyethylene/nano-montmorillonite composites. IEEE Trans. Dielectr. Electr. Insul. **22**(3), 1530–1536 (2015)

67. X. Zhu, J.D. Wu, Y.L. Wang et al., Characteristics of partial discharge and AC electrical tree in XLPE and MgO/XLPE nanocomposites. IEEE Trans. Dielectr. Electr. Insul. **27**(2), 450–458 (2020)
68. T. Tanaka, Y. Mizutani, T. Imai et al., Digest report of investigation committee on polymer nanocomposites and their applications as dielectrics and electrical insulation, in *Proceedings of International Symposium on Electrical Insulating* (Mie, Japan, 2008), pp. 4664460
69. M.T. Nazir, B.T. Phung, A. Sahoo et al., Surface discharge behaviours, dielectric and mechanical properties of EPDM based nanocomposites containing nano-BN. Applied Nanoscience **9**(8), 1981–1989 (2019)
70. P.A. Sharad, K.S. Kumar, Application of surface-modified XLPE nanocomposites for electrical insulation-partial discharge and morphological study. Nanocomposites **3**(1), 30–41 (2017)
71. S. Zhang, X.Y. Cao, Y.M. Ma et al., The effects of particle size and content on the thermal conductivity and mechanical properties of Al_2O_3/high density polyethylene (HDPE) composites. Express Polym. Lett. **5**(7), 581–590 (2011)
72. W.Y. Zhou, S.H. Qi, H.D. Li et al., Study on insulating thermal conductive BN/HDPE composites. Thermochim. Acta **452**(1), 36–42 (2007)
73. Y.Q. Guo, K.P. Ruan, X.T. Shi et al., Factors affecting thermal conductivities of the polymers and polymer composites: a review. Compos. Sci. Technol. **193**, 108134 (2020)
74. J.H. Yu, X.Y. Huang, L.C. Wang et al., Preparation of hyperbranched aromatic polyamide grafted nanoparticles for thermal properties reinforcement of epoxy composites. Polym. Chem. **2**(6), 1380–1388 (2011)
75. J.W. Gu, W.X. Chen, A.J. Ma et al., Thermal conductivity polypropylene/aluminium nitride composites. Adv. Mater. Res. **194–196**, 1577–1580 (2011)
76. Y.S. Xu, D.D.L. Chung, Increasing the thermal conductivity of boron nitride and aluminum nitride particle epoxy-matrix composites by particle surface treatments. Compos. Interfaces **7**(4), 243–256 (2000)
77. X.Y. Huang, P.K. Jiang, T. Tanaka, A Review of dielectric polymer composites with high thermal conductivity. IEEE Electr. Insul. Mag. **27**(4), 8–16 (2011)
78. Y.L. Jiang, X.J. Shi, Y.Z. Feng et al., Enhanced thermal conductivity and ideal dielectric properties of epoxy composites containing polymer modified hexagonal boron nitride. Compos. A Appl. Sci. Manuf. **107**, 657–664 (2018)
79. W.Q. Wu, J.Y. Tu, H.R. Li et al., Suppressed dielectric loss and enhanced thermal conductivity in poly(vinylidene fluoride) nanocomposites using polyethylene glycol-grafted graphene oxide. J. Mater. Sci. Mater. Electron. **31**(1), 807–813 (2020)
80. X.Y. Huang, C.Y. Zhi, P.K. Jiang et al., Polyhedral oligosilsesquioxane-modified boron nitride nanotube based epoxy nanocomposites: an ideal dielectric material with high thermal conductivity. Adv. Func. Mater. **23**(14), 1824–1831 (2013)
81. Y. Gao, B.B. Xu, X.F. Wang et al., Charge transport in low density polyethylene based micro/nano-composite with improved thermal conductivity. J. Phys. D Appl. Phys. **52**(28), 285302 (2019)

Voltage Stabilizer and Its Effects on Polymer's DC Insulation Performance

Chunyang Li, Chengcheng Zhang, Hong Zhao, and Baozhong Han

Abstract Adding voltage stabilizer is a promising method to develop high-performance polymer insulating materials for HVDC cables. In this chapter, voltage stabilizers are classified into three categories according to the different mechanisms of inhibiting electrical tree initiation, namely inhibiting partial discharge, capturing high-energy electrons and inhibiting macromolecular degradation. The hot research topics and relevant research progresses of voltage stabilizer as well as the influences of voltage stabilizers on the DC insulation performances of polymer insulating materials are summarized. The satisfactory progress has been made by relevant researches on alkylated voltage stabilizer, quantum chemical calculation and the combined use of voltage stabilizer and nanoparticles, which are the hot research topics of voltage stabilizer recently. The voltage stabilizers exhibit excellent effects on improving the DC breakdown strength and DC electrical tree resistance of the polymeric insulations including polyethylene and polypropylene. The proper use of voltage stabilizers also has positive effects on inhibiting space charge accumulation in polymeric insulation. However, the mechanism of voltage stabilizers' effects on the charge transport behaviors in polymeric insulation and how to use voltage stabilizer to regulate the charge transport need further study.

C. Li (✉) · C. Zhang · H. Zhao · B. Han
Key Laboratory of Engineering Dielectrics and Its Application, Ministry of Education, Harbin University of Science and Technology, No. 52 Xuefu Road, Heilongjiang Province, Harbin, China
e-mail: lichunyang_hust@163.com

C. Zhang
e-mail: cczh0111@126.com

H. Zhao
e-mail: hongzhao@hrbust.edu.cn

B. Han
e-mail: hbzhlj@163.com

B. Du (ed.), *Polymer Insulation Applied for HVDC Transmission*,
https://doi.org/10.1007/978-981-15-9731-2_3

1 Introduction

Crosslinked polyethylene (XLPE) is the most widely used polymer insulating material for high voltage direct current (HVDC) power cables due to its excellent electrical performance, mechanical performance and low price. However, under DC electric field, the electric field distortion caused by the space charge accumulation and stress inversion in XLPE insulation would lead to the two typical irreversible electrically induced damages of electrical tree and breakdown, which seriously threaten the reliable operation and limit the voltage grade of power cable.

To improve the performance of XLPE insulation, several methods have been proposed. Most of the improvements have been made by reducing impurities or voids generated in the crosslinking process and making the insulating material very clean [1]. Although this method indeed improved the quality of polymer materials to a certain extent, it is not always feasible to produce a cleaner material in an economical or practical manner today, as the cleanliness of the XLPE has already reached an extremely advanced level.

Therefore, to improve the voltage grade and operation reliability of power cables, more efforts should be focused on the modification of XLPE material to further improve the electrical tree resistance and breakdown strength of XLPE. Voltage stabilizer is one of the special additives used firstly in polyethylene (PE) or XLPE insulating materials for power cables. It can date back to the 1960s and has outstandingly contributed to improve the long-term operation stability and voltage grade of PE or XLPE insulated power cables. As an anti-aging agent, the voltage stabilizer is mainly used to inhibit the electrical aging especially the electrical treeing of insulating materials under high electric field.

Since most of the researches on voltage stabilizer sprang up in the period far before the polymer insulated HVDC power cables were widely applied, the voltage stabilizers reported and applied were mostly used in alternating current (AC) cable insulating materials. Up to now, the studies and applications of voltage stabilizers still focus on the influences of voltage stabilizer on the AC electrical tree resistance of XLPE. Regarding to the effects of a voltage stabilizer on the DC electrical properties, especially space charge characteristics, very little is known.

However, it does not mean that the application of voltage stabilizers is only limited to the polymeric insulation for high voltage alternating current (HVAC) cables. Since the dielectric strength of a polymer insulating material has a certain limit, as the operating voltage of extruded polymer insulated DC cables further rising, only using a single modification method may not be able to obtain a satisfactory insulating material. The unique advantage of voltage stabilizer is that its molecule can absorb and dissipate the energy of high-energy electrons [2], which can significantly improve the breakdown strength and electrical tree resistance of polymeric insulation and is beneficial to further improve the voltage grade and service life of polymer insulated cables. For the purpose of exploring the mechanism of voltage stabilizers, some researchers have already found the positive effects of voltage stabilizers on the DC insulation properties of polymers by experiments.

In addition, voltage stabilizers are mostly derived from various additives used in plastic industry. The cost of the raw materials is low, and almost no additional manufacture process or equipment is further required when a voltage stabilizer is introduced into the insulating material [3]. Moreover, the rise and development of new material modification technologies and research methods since the new century provide a lot of opportunities for the development of voltage stabilizers, and the recyclable polymeric insulation materials applied in HVDC cables also have an urgent demand for highly enhanced dielectric strength. Therefore, the application of voltage stabilizer to improve DC insulation performances of polymer materials has become a research hotspot recently.

Further systematic researches on the influences of voltage stabilizer on the DC insulation performances of polymers is not only favorable to expanding the application of voltage stabilizer but also helpful to promote the development of polymer insulated HVDC cables. In this chapter, voltage stabilizers are divided into three categories according to their different mechanisms of inhibiting electrical tree initiation, and the hot research topics and relevant research progresses of voltage stabilizer as well as the influences of voltage stabilizers on the DC insulation performances of polymer insulating materials are summarized.

2 Categories, Mechanisms and Hot Research Topics of Voltage Stabilizers

2.1 Categories and Mechanisms of Voltage Stabilizers

The main function of voltage stabilizer is to improve the electrical tree resistance of polymeric insulation, therefore, the selection and research of voltage stabilizer are usually based on the inhibition mechanism of electrical tree initiation. With the development of the understanding of electrical tree initiation mechanism, the variety of voltage stabilizer gradually becomes more abundant and the research on the mechanism of voltage stabilizer is also gradually deepening. The mechanisms of various voltage stabilizers suppressing the electrical tree initiation are schematically summarized as shown in Fig. 1.

According to the different mechanisms of inhibiting the electrical tree initiation, voltage stabilizers can be divided into three categories: voltage stabilizers that inhibit partial discharge, voltage stabilizers that capture high-energy electrons and voltage stabilizers that inhibit macromolecular degradation. As shown in Fig. 1, the three categories of voltage stabilizers respectively interfere with the different processes of the electrical tree initiation, thus improving the electrical tree resistance. This section will introduce the mechanisms of voltage stabilizers by category.

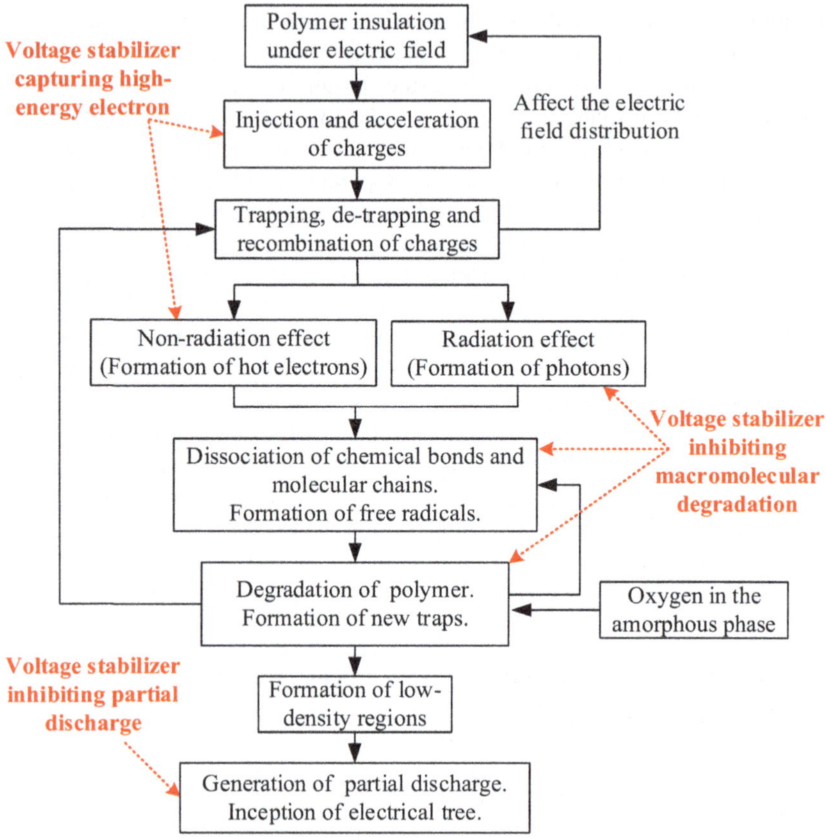

Fig. 1 Schematic diagram of electrical tree initiation mechanism and functions of voltage stabilizers

2.1.1 Inhibiting Partial Discharge

At the early age after the discovery of electrical tree phenomenon, it was considered that the defects in the insulation and the resultant partial discharges were essential for the electrical tree initiation. At that time, the cable manufacture equipment was relatively backward, and the purity of insulation was unsecured. So, there were many defects in the insulation construction of power cables, such as impurities and voids in the insulation layer, and surface imperfections of the screen layer, etc. The defects may lead to a divergent electric field in the insulation and then partial discharge occurs. The partial discharges accelerate the deterioration of the surrounding polymeric insulation, and generate heat and pressure, which facilitates the gradual expansion of defects and the further development of electrical tree [4]. The mechanisms of voltage stabilizers emerged in this period were mostly improving the partial discharge resistance of polymeric insulation, or restraining the occurrence of partial discharge by homogenizing the electric field distribution at the defects. For example, it was reported that by addition of several quaternary ammonium salts or organometallic compounds could increase the corona aging life of various polymer insulating materials including PE [5, 6], and adding dodecanol could inhibit the growth of both the electrical tree and water tree in PE insulation because the additive could migrate into the discharge channels of electrical trees to homogenize the adjacent electric field distribution [7].

The mechanism that electrical tree originates from partial discharges was widely accepted under the manufacture and experimental conditions at that time. The voltage stabilizers that inhibit partial discharge were also practically applied and had helped to improve the long-term reliability and operating voltage of PE insulated cables. Thereafter, with the improvement of purification technology of insulating material, the application of super smooth semiconducting screen material and the industrialized application of triple extrusion technology, the defects in the cable insulation reduced greatly and the operation reliability and service life of PE and XLPE insulation improved significantly [4, 8]. Accordingly, the application of voltage stabilizers to inhibit partial discharge were gradually abandoned. In the meantime, researchers gradually found that the mechanism that electrical tree originates from partial discharge was incomplete.

2.1.2 Capturing High-Energy Electron

With the development of experimental methods, it was found by further researches on the mechanism of electrical tree initiation that the partial discharge is not the primary cause of electrical tree. The electrons injected from electrodes under electric field may obtain enough kinetic energy through acceleration and attack the polymer molecular chains to yield cation radicals by ionizing the saturated aliphatic macromolecules [2, 9]. The free radical degradation process of the polymer induced by the cation radicals may gradually generate the low-density region and even the void. Partial discharge eventually occurs in the void and gives rise to the subsequent inception and growth of a visible electrical tree [2].

The addition of voltage stabilizers which are mostly aromatic compounds can capture the high-energy electrons and attenuate the energy of electrons in polymeric insulation, through which the additive can keep the polymer matrix less harmed under the condition of high electric field and eventually improve the electrical tree resistance of the insulation. The mechanism is schematically shown in Fig. 2. The energy required for excitation or ionization of most aromatic compounds is below the energy required to ionize the saturated aliphatic macromolecules. Under high electric field, the aromatic compounds will be preferentially attacked by the electrons due to the higher electron affinity, and then turn into excited state or be ionized and yield secondary electrons with low energy which are unable to destroy the macromolecular chain any more. The excited state molecules generated by the excitation of aromatic compounds will release the gained energy in a relatively harmless way through luminescence or vibration and be restored to the initial state [10–12]. While the cation

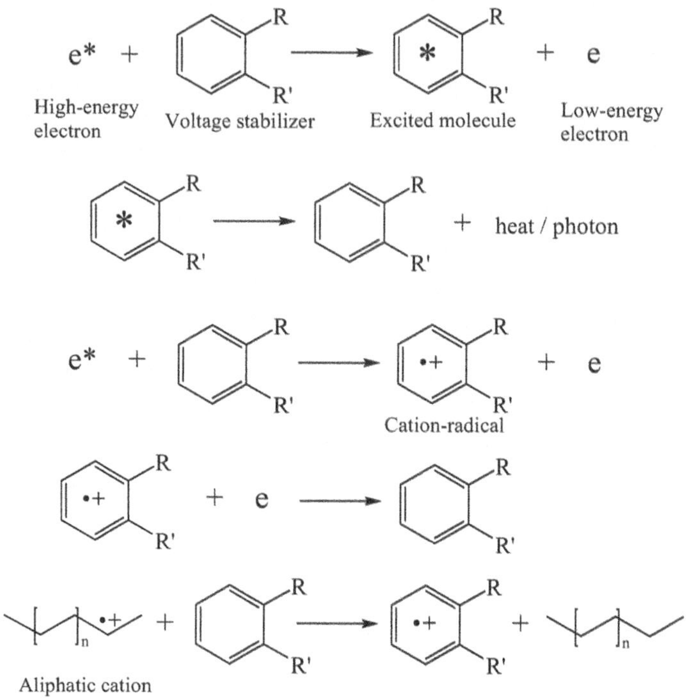

Fig.2 Mechanism of voltage stabilizer that captures high-energy electron

radicals generated by ionization will be restored to the original voltage stabilizer through the recombination with other electrons [2].

Moreover, if the aliphatic cation inevitably appears due to the ionization under a divergent electric field, since the aromatic cation of voltage stabilizer possesses the much stronger ability of π-electron delocalization than that of the aliphatic cation, the voltage stabilizers can also restore the aliphatic cation radical through transforming the aliphatic cation to relatively stable aromatic cation [12]. Therefore, any ionizations of the macromolecules of the dielectric can be quenched by direct electron transfer from a nearby voltage stabilizer molecule, thus preventing the free radical chain reactions from leading to polymer degradation. As a result, the electric field induced deterioration of polymeric insulation is fundamentally restrained by the addition of voltage stabilizers that capture high-energy electrons hence an improved dielectric strength of polymer insulating material.

Voltage stabilizer that can capture high-energy electrons acts on the charge injection and acceleration which are the fundamental process of electric field induced damage. Therefore, this type of voltage stabilizer can improve not only the resistance of the polymeric insulation to electrical tree under extremely distorted electric field but also the short-time breakdown strength of the material under uniform electric field [13–15]. In addition, it also has a good inhibitory effect on the partial discharge

[16, 17]. According to the mechanism as shown in Fig. 2, the voltage stabilizer that captures high-energy electrons is theoretically an inconsumable additive. Therefore, the researches on this category of voltage stabilizers have been the most active since its birth.

2.1.3 Inhibiting Macromolecular Degradation

According to the hot electron theory and photodegradation theory which explain the mechanism of electrical tree initiation, the electrons could hardly be directly accelerated to gain enough kinetic energy by the electric field to cause collision ionization due to the existence of a large number of charge traps in the polymer [18–20]. It was argued that the energy released in the trapping and recombination processes of the injected charge carriers is the main reason that causes degradation of polymeric insulation. In making a transition from an upper to a lower energy state due to trapping or recombination, an energy equal to the energy difference between the two states will be evolved either radiatively or nonradiatively [18]. Through the radiation effect and the nonradiation effect, hot electrons (high-energy electrons) and ultraviolet photons can be generated respectively, which will induce free radical reactions and then lead to the degradation of polymer molecules to yield the low-density region [19, 20].

According to the more complete understanding of electrical tree initiation mechanism, it can be inferred that any additives that can interfere with the free radical reactions of polymeric insulation can be used as voltage stabilizers to suppress electrical tree. Bamji et al. [21] found that different kinds of light stabilizers including ultraviolet absorbent, hindered amine light stabilizer, and excited state quencher can all prolong the electrical tree inception time of PE. Among the light stabilizers, hindered amine light stabilizers which are radical scavengers can eliminate the macromolecular free radicals generated under high electric field [22]. The excited state quenchers can transfer the excited state macromolecule to the ground state through energy transfer to avoid the generation of macromolecular free radicals. The ultraviolet absorbent can directly absorb ultraviolet photons to avoid the excitement of macromolecules. Different light stabilizers can severally prevent the different processes of photodegradation, and then inhibit the degradation of macromolecules and the initiation of electrical trees [23].

The antioxidant as an essential additive of polymer insulating materials can inhibit the chain oxidation reaction in polymer and also the electrical tree initiation. Sekii et al. [24, 25] studied the inhibiting effects of single antioxidant and synergistic antioxidants on AC electrical tree inception voltage of XLPE. As shown in Fig. 3 [25], both types of antioxidants can improve the electrical tree inception voltage, and the phenolic antioxidants show better effect than the sulfur-containing antioxidant. The combination use of the phenol antioxidant and sulfur-containing antioxidant can yield the synergistic effect and improve the electrical tree inception voltage by 85% at most. The mechanism is shown in Fig. 4 [25]. The phenolic antioxidants as the primary antioxidant can transform the peroxy radical ROO· into the hydro-peroxide

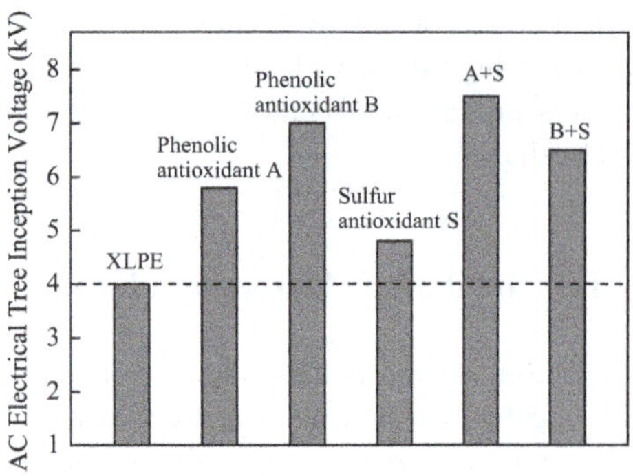

Fig. 3 Effect of antioxidants on the AC electrical tree inception voltage of XLPE [25]

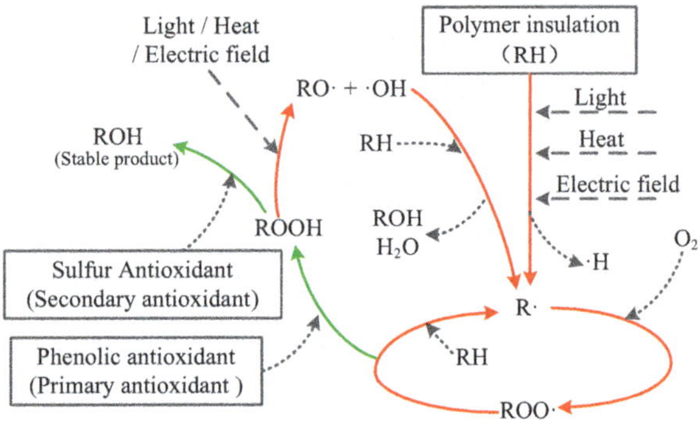

Fig. 4 Chain oxidation reaction and mechanisms of antioxidants [25]

ROOH by donating hydrogen and become stabile radical at the same time. And the sulfur-containing antioxidant as the secondary antioxidant can decompose the hydro-peroxide into the stable product ROH. The two antioxidants severally prevent different processes of the chain oxidation reaction and then suppress the degradation of polymers under high electric field [25].

The researches above show that the anti-aging additives such as antioxidants and light stabilizers can also play the role of voltage stabilizer, which offer an important tip for the development of polymer insulating materials designed for use in electrical apparatus working at high electric field. When choosing the anti-aging additives used in polymer insulating materials, one should not only pay attention to the inhibition effects on the thermal-oxidative aging or UV-irradiation aging, but also to the inhibition effects on the electrical trees. The proper combination of the primary antioxidant and the secondary antioxidant can yield double function of inhibiting both thermal-oxidative aging and electrical treeing.

2.2 *Hot Research Topics of Voltage Stabilizers*

2.2.1 Alkylation of Voltage Stabilizer

If the voltage stabilizer have poor compatibility with polymer matrix, it is easy to migrate outside and precipitate on the surface of polymeric insulation under high operating temperature and electric field during the long-term service of cables, which would gradually lead to the vanishment of its effect on improving the dielectric strength of polymeric insulation. In order to overcome the problem, most researchers improved the molecular structure of voltage stabilizer by introducing long alkyl chains to expand the molecular structure of voltage stabilizer and enhance the interaction of voltage stabilizer molecule with polymer molecules.

Ferrocene is verified to be a highly effective voltage stabilizer. However, ferrocene has a poor compatibility with PE insulation which hinders its practical application in cable insulation. To solve the problem, Gao et al. [26] synthesized a ferrocene derivative voltage stabilizer containing long alkyl side chain. Experimental results show that the electrical tree inception voltage of PE is enhanced by 61% when the content of ferrocene derivative is 1%. The ferrocene derivative with long alkyl side chain exhibits improved compatibility with PE and thermal stability, and the electrical tree inception voltage of the sample is higher than that of pure PE sample by 72.4% after thermally treated at 80 °C for 30 days.

The patent [27] proposed that by adding benzophenone substituted with at least one group selected from alkyl, arylalkyl, and alkylaryl as voltage stabilizer can increase the lifetime and dielectric strength of the polyolefin-based insulating material. The migration of the additives were tested through characterization of the additive concentration after the molded samples were severally subjected to aging at room temperature and 90 °C. The results indicate that the benzophenone substituted with alkyl or alkoxy containing more than eight carbon atoms exhibits a higher stability than the unsubstituted benzophenone in the insulting material over time due to the reduced migration towards the external surface of the polymer.

Englund et al. [28, 29] found that the voltage stabilizers derived from the two basic molecular structures as shown in Table 1 can increase the electrical tree inception voltage of XLPE by more than 70%. In order to improve the compatibility of voltage stabilizer with XLPE, long alkyl side chains were introduced into the basic molecular structure of voltage stabilizer during the synthesis process. Some representative molecular structures are also shown in Table 1 [30–33]. The experimental results show that the attachment of side chains to the voltage stabilizers has the benefits of decreasing the vapor pressure and melting temperature compared to the unmodified compound and the voltage stabilizers with long side chains display significantly less weight loss at high temperature and negligible influence on the gel content of XLPE. From a processing perspective, it can be concluded that the compatibility of voltage stabilizer with PE is improved because of the modified thermal properties. However, the efficiency of voltage stabilizers improving the electrical tree resistance would be reduced if the alkyl side chain exceeds a certain length [33].

Table 1 Basic molecular structures, examples and voltage stabilizers with long alkyl chain

Basic molecular structure	Example	Voltage stabilizer with long alkyl chain
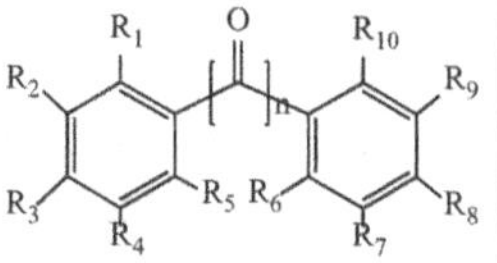 Structure I	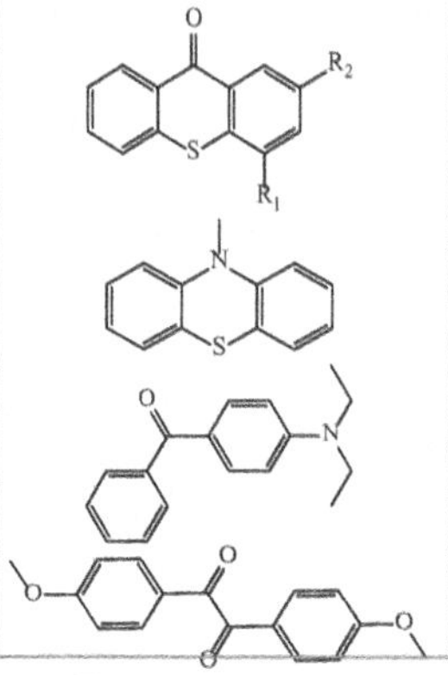	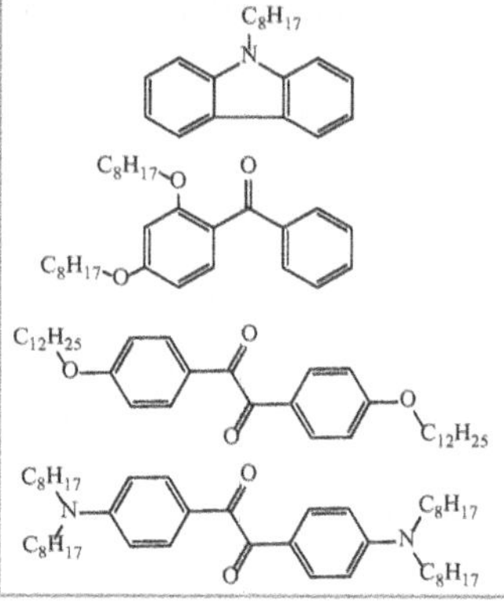
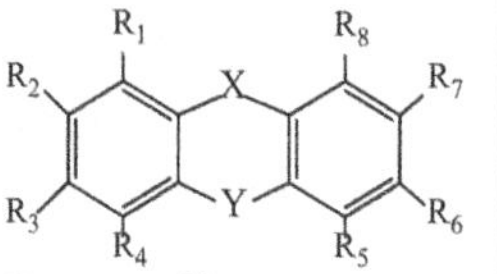 Structure II		

As far as the present research progress is concerned, alkylation of voltage stabilizer is the most widely used method to solve the problem of voltage stabilizer's poor compatibility with polymer matrix. However, the effect of alkylation on the efficiency of voltage stabilizer has not been clearly concluded at present, and how to achieve a low-migration voltage stabilizer without interfering with its efficiency is worth further study.

2.2.2 Investigation via Quantum Chemical Calculation

In recent years, quantum chemical calculation is extensively applied in engineering and many valuable results have emerged in several fields of research. Through the quantum chemical calculation, one can investigate the electronic energy level, density of states, ionization potential, electron affinity and three-dimensional electrostatic potential distribution of molecules [34, 35]. Voltage stabilizers are very likely to undergo transition of energy state or chemical reaction under high electric field and these changes are all accompanied by the change of electron energy. The breakdown, aging and electrical treeing behavior of polymers are all closely related to high-energy electrons. Therefore, the quantum chemical calculation can help to select promising voltage stabilizers as well as provide foundation for deeper physico-chemical understanding of the mechanism of voltage stabilizers.

In the author's research group, Zhang et al. [10–12, 34] presented theoretical researches on the mechanism of acetophenone and its analogues as voltage stabilizers. It is indicated that the intramolecular keto-enol tautomerism and valence bond isomerization in acetophenone molecule play an important role for its high efficiency in enhancing the dielectric strength of PE. The energy of high-energy electron can be consumed through the repeating isomerization reaction and transformed to relatively harmless energy to weaken the impact of high-energy electrons on polymer molecules. In addition, the researchers also provided suggestions for designing the molecular structure of aromatic ketone voltage stabilizers. The proper introduction of carbonyl and benzene ring to form conjugated molecular structure is beneficial for increasing the electron affinity and reducing the HOMO–LUMO energy gap and ionization potential of the molecule to significantly improve the efficiency of voltage stabilizer.

In Ref. [36], it was proposed that a high-efficiency voltage stabilizer should possess both a high electron affinity and a low ionization potential by comparing the relationship between the efficiencies of eight voltage stabilizers and their electron affinities and ionization potentials. Jarvid et al. [37] further investigated the relationship between the efficiencies of voltage stabilizers and the quantum chemical properties of the molecules by the combination of theoretical calculation and electrical tree inception experiments. The ionization potential, electron affinity, E_{HOMO} and E_{LUMO} of a set of previously reported voltage stabilizers were obtained through quantum chemical calculation. The electrical tree inception experiments of materials containing thirteen voltage stabilizers with divergent molecular structures were carried out. The parameter φ (unit: kg·mol^{-1}) was defined to identify the efficiency of voltage stabilizers, as shown in the formula 1.

$$\varphi = \frac{E_{63} - E_{63}^{XLPE}}{E_{63}^{XLPE}} \times \frac{1}{c} \tag{1}$$

where E_{63} and $E_{63}{}^{XLPE}$ are the scale parameters from the Weibull distribution fitted to the electrical tree inception electric field of the sample containing voltage stabilizers and pure XLPE sample, respectively (unit: kV/mm); c is the molar concentration of voltage stabilizers (unit: mol·kg^{-1}). The relationship between the efficiency of voltage stabilizers and the ionization potential, electron affinity and E_{HOMO}-E_{LUMO} of molecules is shown in Fig. 5 [37]. It can be seen that the efficiency of voltage stabilizers is largely unrelated to the ionization potential, and has a weak correlation with E_{HOMO}-E_{LUMO} and a significant correlation with the electron affinity. It is concluded that a high-efficiency voltage stabilizer should possess firstly a high electron affinity and secondly a narrow HOMO–LUMO energy gap. Accordingly, the researchers proposed that the candidate range of voltage stabilizers should be extended to organic semiconductor materials with high electron affinity. It is confirmed by the subsequent experiments that the two fullerenes of C_{60} and PCBM with high electron affinities show excellent performance as voltage stabilizers, especially the PCBM exhibits better effect at a low molar concentration. If evaluated by the voltage stabilizer efficiency parameter of φ, the PCBM is the most efficient among voltage stabilizers reported so far [38].

Su et al. [35] observed the inhibition effect of antioxidants on the electrical treeing of EPDM, and then investigated the voltage stabilization mechanism of combined use of pentaerythritol tetrakis(3-(3,5-di-tert-butyl-4-hydroxyphenyl) propionate) and tris(2,4-ditert-butylphenyl) phosphite (denoted as AOA and AOD) as synergistic antioxidants through quantum chemical calculation based on density functional

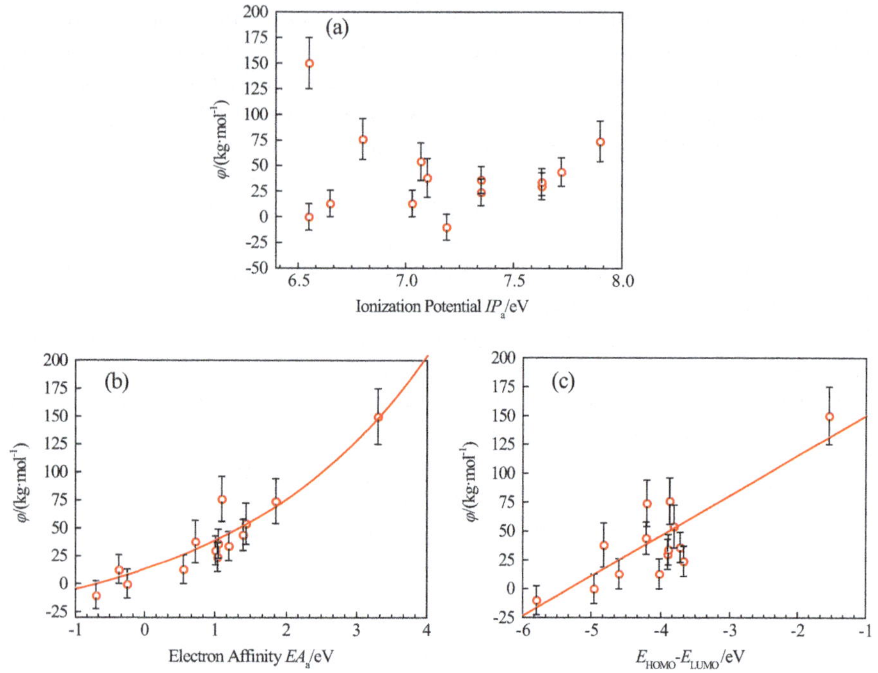

Fig. 5 Relationship between voltage stabilizer efficiency and molecular quantum chemistry characteristics [37]: **a** ionization potential, **b** electron affinity and **c** E_{HOMO}-E_{LUMO}

theory (DFT). The theoretical calculation indicates that when AOA and AOD are simultaneously filled into EPDM, the multistep charge-trapping properties are more efficient compared to either antioxidant used alone. Moreover, it is demonstrated that the electronic structures of synergistic antioxidant byproducts could facilitate the realization of long-term charge modulation and the synergistic antioxidants could act as high-efficiency and high-versatility voltage stabilizers with multistep and multi-scale charge-trapping to highly efficiently modulate electrical degradation in polymer dielectrics.

With the help of quantum chemical calculation, the experimental quantity is greatly reduced and the research efficiency of voltage stabilizer is significantly improved. Moreover, the action mechanism of voltage stabilizer can be explored, and the effective functional groups in voltage stabilizer molecules can be identified, which is instructive for the synthesis of novel alkylated voltage stabilizer.

2.2.3 Combined Use of Voltage Stabilizer and Nanoparticles

At present, many important theoretical achievements have been obtained in the research of nanocomposites. However, the nanoparticles are easy to agglomerate in the polymer due to the high surface energy, which severely affects the performance of nanocomposites. To realize the uniform dispersion of nanoparticles, it is the most feasible way to modify the surface of nanoparticles to reduce the surface energy and increase the compatibility with polymers [39]. This also provides a new approach for the research of voltage stabilizers. If the voltage stabilizer is used as a surface modifier of nanoparticles through interaction of chemical bond or Van der Waals force, it can not only reduce the surface energy of nanoparticles, but also inhibit the migration and precipitation of voltage stabilizer itself, and it is possible to obtain the synergistic effect of voltage stabilizer and nanoparticle.

Yamano et al. [40] first reported the combined use of voltage stabilizer and nanoparticle in low density polyethylene (LDPE) and revealed the synergistic effect of Al_2O_3 nanoparticles and phthalocyanine voltage stabilizers on the inhibition of electrical trees. It is found that the phthalocyanine voltage stabilizer can increase the electrical tree inception voltage of LDPE and the Al_2O_3 nanoparticles can reduce the growth rate of electrical trees. While the voltage stabilizer and nanoparticles are simultaneously mixed into LDPE, the extra inhibition effect on the electrical tree growth is obtained which is more than 3 times higher than the nanocomposite without voltage stabilizer. The results of AFM observation suggest that the mixed addition of phthalocyanine voltage stabilizers improves the dispersion of Al_2O_3 nanoparticles, increases the interface area between nanoparticles and LDPE matrix and then enhances the inhibition effect of nanoparticles on the growth of electrical trees. The mechanism of the synergistic effect is speculated that the voltage stabilizer molecules and nanoparticles attract with each other by Van der Waals fore, which can restrain the mutual attraction and agglomeration of nanoparticles.

Andersson et al. [41] conducted a comprehensive research on the mixed effect of voltage stabilizer and nanoparticles. The 0.03 wt% thioxanthone voltage stabilizer and 3 wt% Al_2O_3 nanoparticles were mixed into LDPE in different ways as shown in Fig. 6 [41]. The results show that the electrical tree inception electric field of LDPE blended physically with both voltage stabilizers and nanoparticles is higher than that of LDPE containing only nanoparticles. However, when the nanoparticles with the surface grafted by voltage stabilizers are mixed into LDPE, the electrical tree resistance of LDPE is significantly impaired. The reason is inferred that grafting voltage stabilizer onto the surface of Al_2O_3 nanoparticles would weaken the charge trapping effect of nanoparticles.

As far as the current research results are concerned, the influence of the combination of voltage stabilizers and inorganic nanoparticles is inconsistent. The reason may be due to the diverse mechanisms and characteristics of different nanoparticles or voltage stabilizers. Therefore, relevant research should be conducted based on fully understanding the respective mechanism of voltage stabilizer and inorganic nanoparticle. If the complementarity of the two mechanisms can be realized, it would be a promising approach to develop high-performance polymer insulating material. A successful combined use of voltage stabilizer and nanoparticle will be expounded in the Sect. 3.2, which exhibits the synergistic effect of nanoparticle and voltage stabilizer by functionalizing silica nanoparticle with voltage stabilizer.

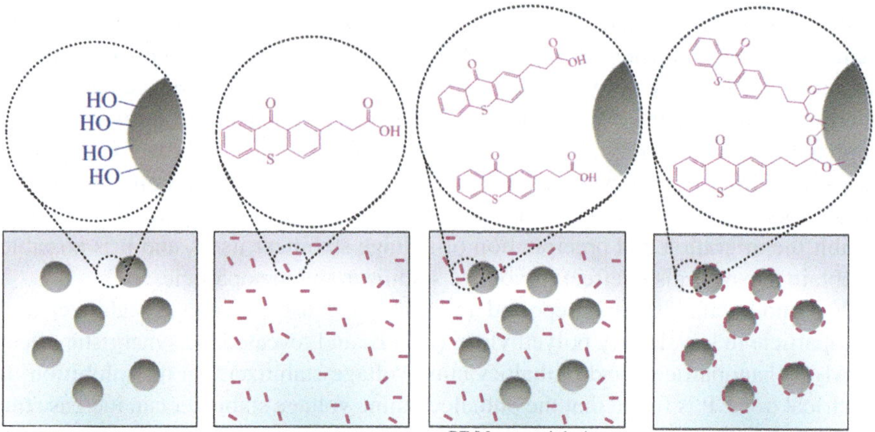

Fig. 6 Schematic illustration of different ways of mixing nanoparticle and voltage stabilizer [41]

3 Effects of Voltage Stabilizer on Polymer's DC Insulation Performance

Due to the different electric field distribution rules, the studied electrical properties of HVDC cable insulation are different from those of HVAC cable. As a polymer insulating material for HVDC cables, people mainly focus on the properties such as space charge accumulation, DC breakdown strength and conductivity which changes with temperature and electric field. In addition, electrical treeing under DC voltage is significantly different from that under AC voltage in mechanism. Under a constant DC voltage, it is difficult for an electrical tree to occur due to the shielding effect of space charge on the defect. But when there is an instantaneous change of the external voltage such as shorted or superimposed with a reverse polarity transient voltage, the trapped space charges may instantly be released and burst into the electrode, which can cause scissions of polymer chains and eventually yield a DC electrical tree [42]. The grounded DC electrical tree experiment which is carried out under a periodically grounded DC high voltage is usually used to investigate the electrical treeing behavior of polymeric insulation under DC field [43, 44].

3.1 Influences of Voltage Stabilizer and Its Analogues

In the long history of voltage stabilizer, most attention has been paid to the influences of voltage stabilizer on the dielectric constant, dielectric loss, AC breakdown and AC electrical tree, while little attention is paid to the insulation performance of polymer materials under DC electric field. Nevertheless, the influences of voltage stabilizer and its analogues on the DC insulation performance of polymeric insulation can partly be acquired with insufficient reports.

Yamano et al. [14] added six azo dyes with different molecular structures to LDPE and tested the influence of additives on DC breakdown strength of LDPE in the temperature ranging from −35 °C to 30 °C. It is inferred that electron avalanche plays a leading role in the breakdown process in the tested temperature range. The experimental results indicate that the azo additives increase the breakdown strength by up to 50% and also reduce the conductivity of LDPE, the mechanism is that the azo dyes can capture charge carriers, and consume the energy through the excitation of molecule itself. Besides, it is found that the efficiency of voltage stabilizers depends upon the type of radical connected to the dye molecule, and the higher breakdown strength and resistivity can be obtained when the electron-accepting radical is connected to the azo dye molecule.

Yamano [15] also studied the effects of a series of polycyclic compounds (naphthalene, anthracene, tetracene, pentacene, 9,10-dibromoanthracene, 9-nitroanthracene and 1-aminoanthracene) on the DC breakdown strength and impulse breakdown strength of LDPE. It is found that the additives have similar enhancement effect on DC breakdown strength and impulse breakdown strength of LDPE. Among the polycyclic compounds without radicals, anthracene is the most effective and can increase the breakdown strength of LDPE by 1.5 times. After two nitro or a bromine radical are connected to anthracene molecules, the breakdown strength increment is further improved. The results of thermal stimulated depolarization current indicate that when the electron-accepting radical is connected to the voltage stabilizer molecule, the deeper traps are introduced into PE, which are beneficial to the further improvement of breakdown strength.

As mentioned above, antioxidants can also act as voltage stabilizers. Researchers have found that antioxidants also have a positive effect on the DC insulation performance of polymer insulating materials. Sekii et al. [44] studied the effects of two antioxidants on the inception of grounded DC electrical trees of XLPE. As shown in Fig. 7 [44], the experimental results indicate that the phenolic antioxidants have no obvious effect on grounded DC electrical tree inception voltage, while the sulfur-containing antioxidant slightly increases grounded DC electrical tree inception

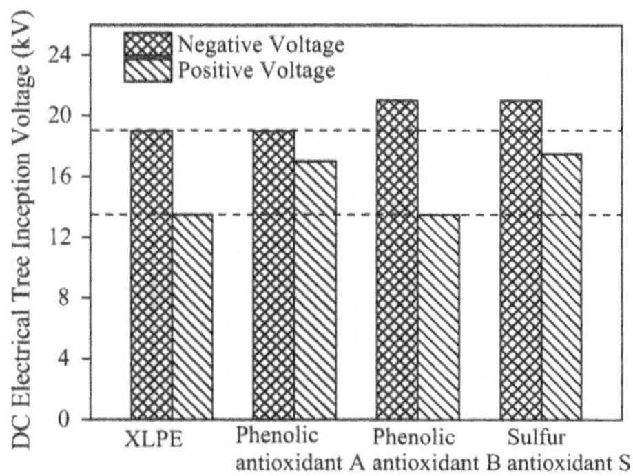

Fig. 7 Effect of antioxidants on the grounded DC electrical tree inception voltage of XLPE [44]

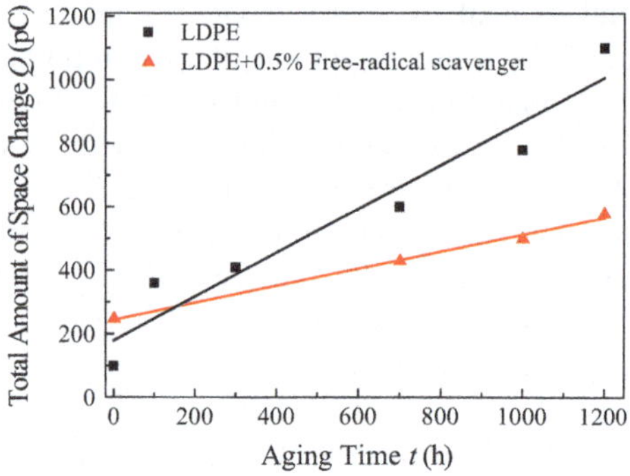

Fig. 8 Effect of voltage stabilizer on the space charge accumulation amount in LDPE samples before and after electrical aging [45]

voltage of materials under voltage of both polarities. The DC conductivity measurement results show that the sulfur-containing antioxidant can reduce the conductivity of XLPE. It is speculated that sulfur-containing antioxidant may not only stabilize the free radical, but also yield charge traps, and then suppress grounded DC electrical trees.

As for the influence of voltage stabilizer on space charge accumulation characteristics, Yin et al. [45] studied the space charge accumulation characteristics of samples for neat LDPE and LDPE containing free radical scavenger voltage stabilizer before and after aging under 50 kV/mm power frequency electric field. As shown in Fig. 8 [45], the total space charge amounts in both material increase linearly with the extension of aging time. The space charge amount of the sample containing voltage stabilizer is slightly larger than that of the neat LDPE sample before aging, while the space charge increases with the aging time at a lower rate, which leads to less space charge accumulation in the sample of LDPE containing voltage stabilizer after a certain period of aging time. It is inferred that the voltage stabilizer can inhibit the electrical aging of LDPE thus reducing the space charge traps caused by aging and hence a reduced accumulation amount of space charge.

Acetophenone, one of the voltage stabilizers discovered in the early stage, also exists in HVDC cable insulation as a crosslinking byproduct. Therefore, its influence on space charge in XLPE is a concern. Some studies have shown that acetophenone can significantly increase the hetero charge accumulation, the conductivity and space charge migration rate in PE [46, 47]. In addition, as a necessary additive in the XLPE insulation for HVDC cable, the effect of antioxidant on space charge behavior also draws much attention. Montanari et al. [48] studied the effect of antioxidants on the space charge characteristics of LDPE, it is found that the space charge accumulation is suppressed and the threshold field of space charge injection is slightly increased in the sample containing antioxidant, it is inferred that antioxidants can increase the depth or density of traps in LDPE. Sekii et al. [49, 50] found that phenolic and sulfur-containing antioxidants do not generate hetero space charges in PE insulation by themselves, but yield hetero charges near the anode in the crosslinked samples. It is inferred that the charge may originate from the interaction of crosslinking agents, crosslinking byproducts and antioxidants.

3.2 Voltage Stabilizers for HVDC Cable Insulation

Exploitation of insulating materials for HVDC cables based on nanocomposites has been the most popular technical route for the past two decades. But in recent years, researchers encountered the technological problem in ensuring the uniform dispersion of nanoparticles during the real manufacture of cable insulation and exposed the potential imperfection of performance degradation of nanocompsites during long-term service [51]. Meanwhile, the advantages of chemical modification and the use of voltage stabilizer to improve the DC insulation performance of polymer materials are drawing more attention [52]. It was not until the recent three years that researchers started to search for voltage stabilizers suitable for polymeric insulation used in HVDC cables and systematically investigate the influences of voltage stabilizer on DC insulation properties of polymer materials.

Du et al. [53] selected 4,4′-difluorobenzophenone, 4,4′-dihydroxybenzophenone, and 4,4′-bis(dimethyl amino)benzyl as voltage stabilizers (denoted as A, B, and C, respectively) and investigated their effects on the DC conductivity, breakdown strength and space charge accumulation of XLPE insulation with the additive content of 0.5 wt%. The results indicate that though the influence of voltage stabilizers on DC conductivity is not obvious (the values of DC conductivity of all the XLPE samples are at the same order of magnitude), a higher breakdown strength is found in the voltage stabilizer modified XLPE samples as shown in Fig. 9 [53].

The space charge distribution in Fig. 10 [53] reveals that the effect of voltage stabilizer A on space charge suppression is not obvious. For the samples XLPE-B and XLPE-C, less space charges are injected into the bulk of the samples, with the space charges mainly accumulated near the anode. As a result, almost no electric field distortion occurs in the bulk of the XLPE samples, that is, the voltage stabilizers B

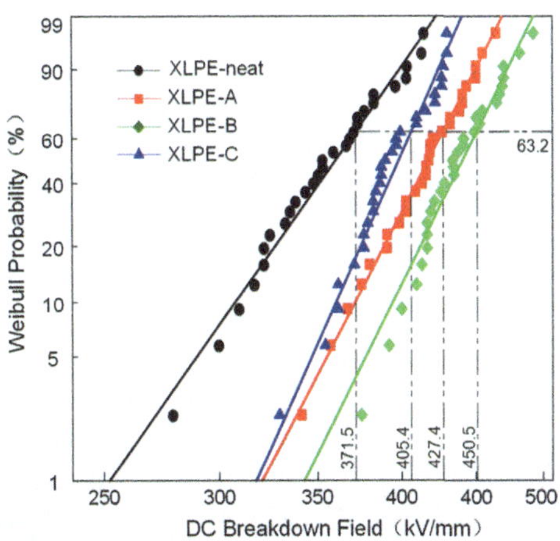

Fig. 9 Relationship between the Weibull probability and the DC breakdown field of the different voltage stabilizer modified XLPEs [53]. Copyright Clearance Center's RightsLink® License Number: 4854060375827

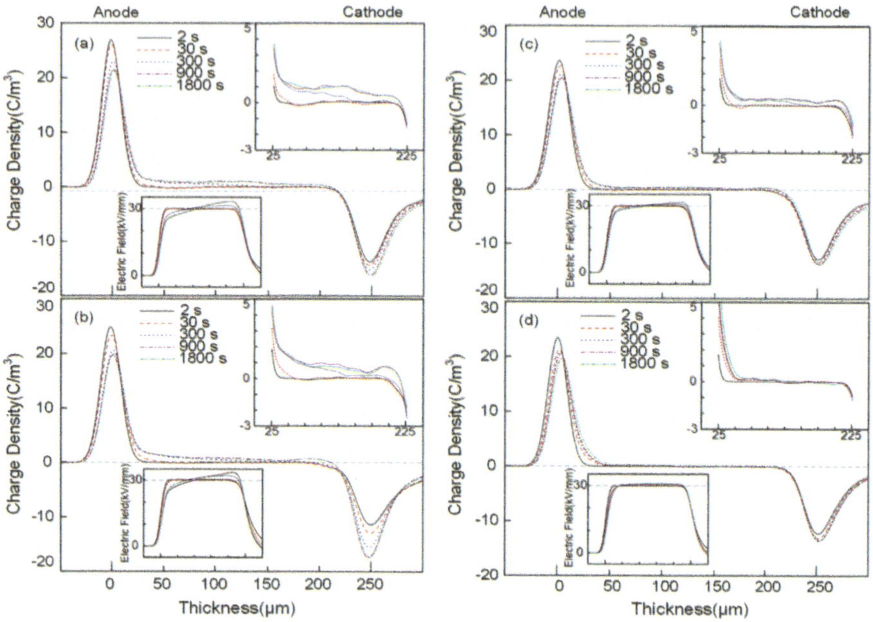

Fig. 10 Space charge characteristics and electric field distortions in voltage stabilizer modified XLPE samples as a function of polarization time under -30 kV/mm [53]: **a** XLPE-neat; **b** XLPE-A; **c** XLPE-B; **d** XLPE-C. Copyright Clearance Center's RightsLink® License Number: 4854060375827

and C show great ability to suppress space charge in XLPE insulation. Moreover, as shown in Fig. 11 [53], the XLPE-B and XLPE-C samples show a better ability to promote charge dissipation compared with the neat XLPE and XLPE-A samples.

Combining the experimental results of the space charge characteristics, the apparent trap-controlled carrier mobility and the trap level, the mechanism of voltage

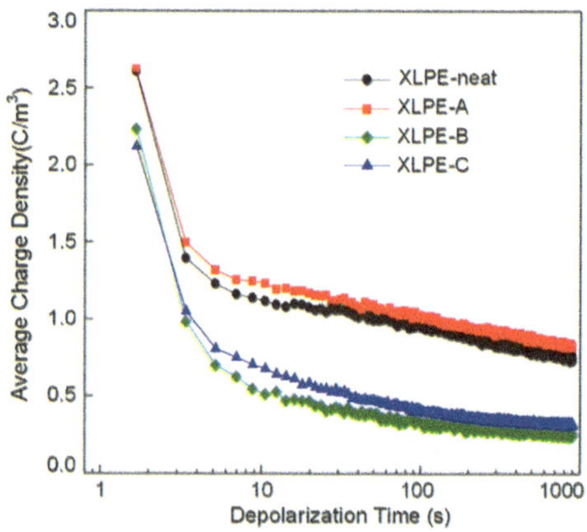

Fig. 11 Relationships between average charge density and depolarization of different voltage stabilizer modified XLPEs [53]. Copyright Clearance Center's RightsLink® License Number: 4854060375827

stabilizer to restrain space charge was inferred. A suitable voltage stabilizer, such as the voltage stabilizers B and C for XLPE in this experiment, could absorb the energy of injected energetic charges and release energy in a harmless manner, leaving low-energy charges behind. On the one hand, it is difficult for the low-energy charges to inject into the bulk of samples, thereby causing reduced space charge injection and accumulation inside the samples in the process of polarization. On the other hand, most of the injected low-energy charges are easily trapped by shallow traps, resulting in a faster electron de-trapping rate and a higher apparent carrier mobility of the voltage stabilizer modified XLPE samples in the process of depolarization.

Chen et al. [54] investigated the effect of different weight percent of 3-aminobenzoic acid voltage stabilizer on the insulation properties of XLPE. The electrical measurements involved with DC insulting performance included the DC step-by-step breakdown, space charge, DC conductivity and surface potential decay. The results indicate that the addition of 1wt% voltage stabilizer can significantly increase the DC breakdown strength and voltage endurance life of XLPE. Moreover, the samples containing 1wt% 3-aminobenzoic acid exhibit negligible space charge accumulation and the least electric field distortion inside the insulation bulk, and also exhibit lower DC conductivity by one order of magnitude comparing to the neat XLPE samples. The optimum performance of the XLPE with 1wt% 3-aminobenzoic acid voltage stabilizer is related to the deeper charge traps and higher trap density introduced by the additive which are characterized by the surface potential decay measurement.

XLPE, as we know, is a thermoset plastic and cannot be recycled and reused once the cables were retired, which will lead to high cost of reutilization and environmental problems. Therefore, the currently widely used XLPE insulated HVDC cables cannot fulfill the environmental sustainability and a high performance recyclable thermoplastic material for HVDC cable insulation is highly desirable [52]. Some efforts have been paid to develop thermoplastic material for HVDC cable insulation by using voltage stabilizer.

In order to develop PP based insulating material for HVDC cable, Huang et al. [55] synthesized a voltage stabilizer (BVA) functionalized silica nanoparticle (as shown in Fig. 12 [55]), blended it into isotactic polypropylene (PP) with different additive contents and investigated the doping effects on the DC insulation performances of PP.

After applying a DC electric field of 50 kV/mm for 1.5 h, the samples were short-circuited and the average volume space charge densities during decay behavior of space charges were calculated. As shown in Fig. 13a [55], the charge density shows the order of PP > PP/SiO_2 > PP/SiO_2 -sh > PP/SiO_2 -sh-BVA. In order to investigate whether the nanocomposite can still perform well under the cable operation environment, the space charge behavior of the nanocomposites films was tested under 20 kV/mm DC electric field at 80 °C, which is close to the operation condition of a real cable. After applying a 20 kV/mm DC electric field at 80 °C for 1.5 h, the average volume space charge densities in the samples during charge decay behavior exhibit the same order under the test condition of room temperature as shown in Fig. 13b [55]. The results indicate that the surface modification shows positive effect

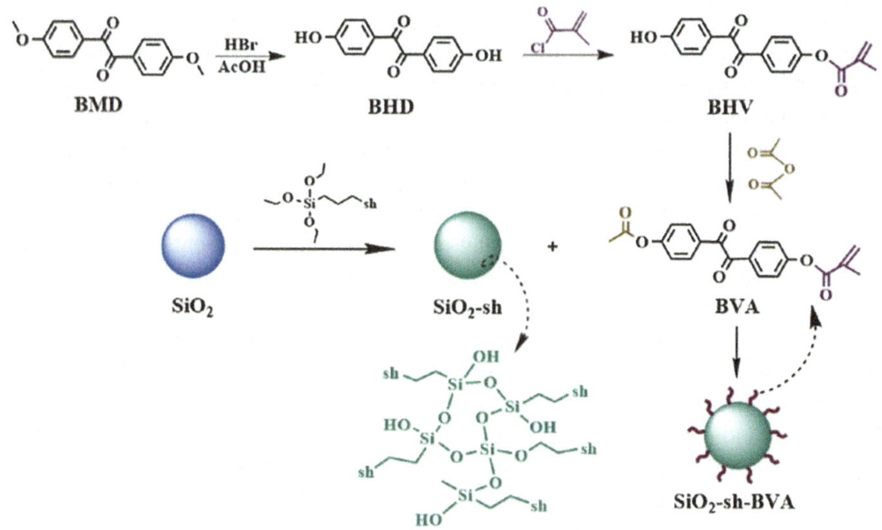

Fig. 12 Scheme for the preparation of BVA and SiO_2 -sh-BVA nanoparticles [55]. Reprinted with permission from reference [55]. Copyright (2019) American Chemical Society

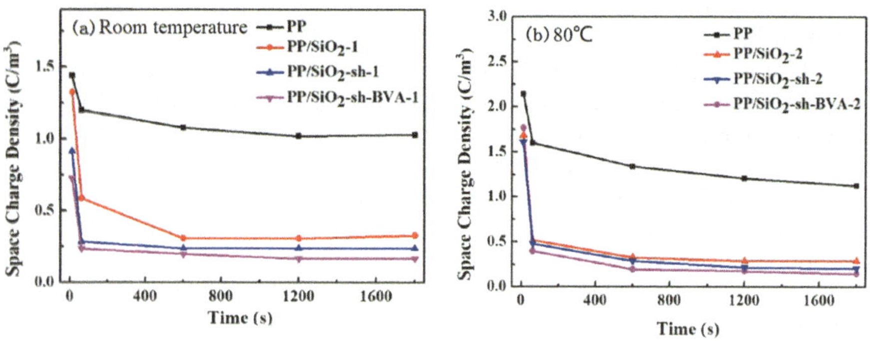

Fig. 13 Time dependent average volume space charge densities of PP and its nanocomposite with different fillers during the depolarization process at room temperaure (**a**) and 80 °C (**b**) [55]. Adapted with permission from reference [55]. Copyright (2019) American Chemical Society

on suppressing the space charge accumulation in the nanocomposites, and the voltage stabilizer functionalized nanoparticle shows the highest capability in suppressing the space charge injection and accumulation no matter at high temperature or room temperature.

The electric field dependent leakage current densities of PP and the PP nanocomposites were also measured. The result indicates that the SiO_2-sh-BVA nanoparticles show the most apparent suppressing effect on the leakage current in PP under high electric field. Since the leakage currents mainly originate from the charge injection from electrodes, the suppressed leakage current in PP/SiO_2 -sh-BVA means lower charge injection, which is consistent with the aforementioned space charge behavior of the nanocomposites.

The surface potential decay behavior of PP and the nanocomposites were measured to characterize the trap energy and trap density. As shown in Fig. 14 [55], The results indicate that both the deep trap energy and the deep trap density are significantly decreased in the nanocomposites of SiO_2-sh-BVA. Deep traps are usually introduced in bonded layer (the first layer) and bound layer (the second layer), and shallow traps are generated in loose layers (the third layer) according to the multicore model proposed by Tanaka. Therefore, it can be concluded that the voltage stabilizer functionalized SiO_2 can tailor the third layer in the interface between nanoparticles and PP. Namely, the voltage stabilizer functionalized SiO_2 nanoparticles have a strong interaction with the macromolecular matrix, resulting in significantly increased volume fractions of loose layers.

The characteristic breakdown strength of the different samples is presented in Fig. 15 [55]. Compared with pure PP, the nanocomposites with 1 wt % unmodified SiO_2 exhibit slightly lower breakdown strength, while the nanocomposites with 1

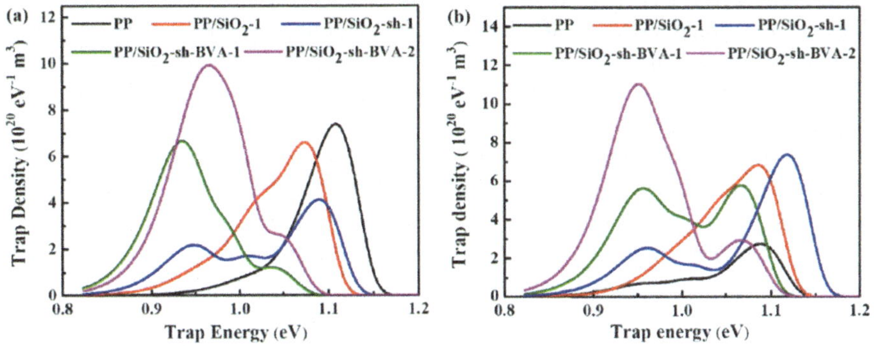

Fig. 14 Trap energy versus trap density of PP, PP/SiO_2-1, PP/SiO_2-sh-1, PP/SiO_2-sh-BVA-1, and PP/SiO_2-sh-BVA-2 after charging by positive corona (a) and negative corona (b) [55]. Reprinted with permission from reference [55]. Copyright (2019) American Chemical Society

Fig. 15 Characteristic breakdown strength of PP and PP nanocomposites [55]

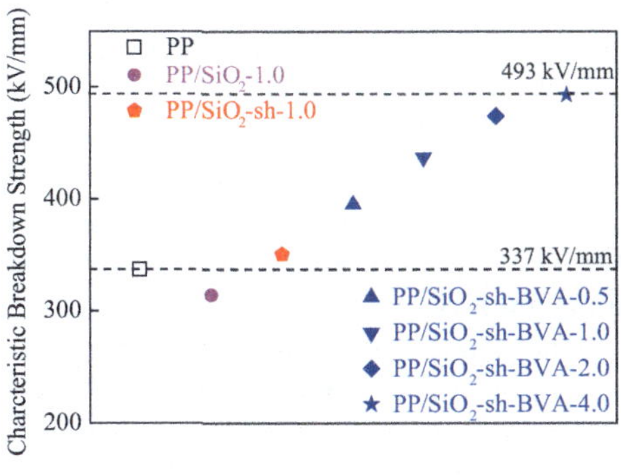

wt% silane modified SiO_2 exhibit slightly higher breakdown strength. In the case of nanocomposites with voltage stabilizer modified SiO_2, each nanocomposite shows apparently enhanced breakdown strength, and the breakdown strength increases with the increase of SiO_2-sh-BVA. At a SiO_2-sh-BVA concentration of 4 wt%, the characteristic breakdown increased to 493 kV/mm from 337 kV/mm of the pure PP, resulting in an enhancement of 46%. The enhanced breakdown strength of the PP/SiO_2-sh-BVA nanocomposites should be mainly determined by improved interfacial compatibility, the introduction of shallow traps, and the high electron affinity and low ionizing potential of the voltage stabilizer. However, the enhancement can realize only under the condition that the nanoparticles have excellent interfacial compatibility with the PP matrix. Despite the significantly enhanced electrical property of the PP/SiO_2-sh-BVA nanocomposites, further investigation is needed to improve their flexibility while maintaining their excellent electrical property.

Chen et al. [56] used LDPE blend containing 10 wt% high density polyethylene (HDPE) as polymer matrix and investigated the effects of four voltage stabilizers, m-aminophenylboric acid, 2-methoxy-5-pyridineboric acid, m-aminobenzoic acid and 4-dimethylaminobenzoic acid, on the insulation properties of the LDPE/HDPE blends. The effects of 1 wt% voltage stabilizers on the insulation properties of blends were studied based on AC electrical tree inception experiments, space charge measurements and surface potential decay experiments. The results show that when the voltage stabilizers are added into the blends, the tree inception voltages are increased, in particular the blend with m-aminobenzoic acid shows the highest increase of 41% compared to the neat LDPE/HDPE. It is also found that space charge accumulations of the blends are apparently suppressed by adding voltage stabilizers and the blend with the m-aminobenzoic acid has the least space charge accumulation. The estimated trap energy levels and trap densities demonstrate that the voltage stabilizers increase the trap density and decrease the trap depth, which can facilitate the de-trapping of charge and release the captured charge faster, thus effectively suppressing the accumulation of the space charge in the blends.

Most of the above studies mainly focus on the effects of voltage stabilizer on the DC insulation performances of polymer at room temperature. However, the actual cable insulation works at elevated temperature, and the insulation performances at high temperature play a decisive role in the practical application of the material.

Through a large number of quantum chemical calculations and experimental studies, the author [57] selected out a kind of graftable aromatic ketone voltage stabilizer (AKVS) which can be grafted onto XLPE macromolecules in the thermo-crosslinking process through free radical addition reaction on its ethenyl group. And then the DC insulation properties of unmodified XLPE (XLPE) and XLPE grafted with aromatic ketone compounds (XLPE-AK) were investigated by testing the DC breakdown strength, conductivity, and space charge of the samples at different temperatures.

The changes of characteristic breakdown strengths of the two materials over temperature are shown in Fig. 16. The error bars in the figure represent the 95% confidence interval of the Weibull distribution for characteristic breakdown strength. It can be seen that the breakdown strength of XLPE-AK material at 20 °C, 50 °C,

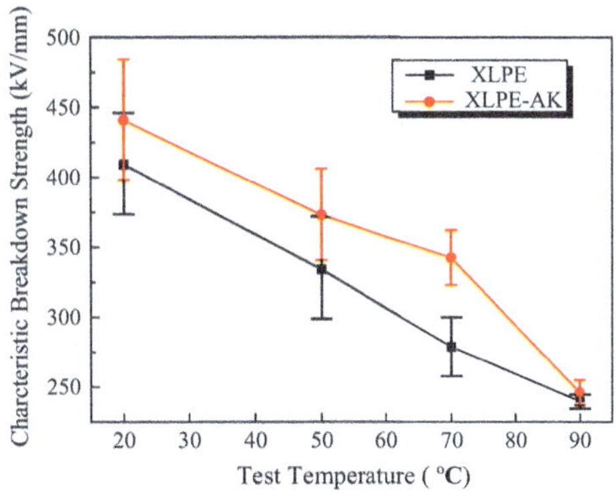

Fig. 16 Dependence of characteristic breakdown strength on temperature

70 °C and 90 °C is 8.0%, 12.0%, 23.1%, and 3.0% higher than that of XLPE, respectively. Compared to unmodified XLPE, XLPE-AK exhibits superior DC breakdown characteristics at various test temperatures due to the graft of voltage stabilizer.

The curves of the conductivity of two XLPE materials with the temperature and electric field are shown in Fig. 17. On the whole, compared with XLPE, the conductivity of the voltage stabilizer modified XLPE decreases under room temperature and high electric field, while it increases under high temperature and low electric field, the differences are indicated by the arrows.

In order to avoid the electric field distortion of stress inversion phenomenon in the insulating layer when the cable is fully loaded, the HVDC cable insulation design usually requires that the dependence of conductivity of the insulating material on temperature is as small as possible. From this perspective, the conductivity of XLPE-AK is more dependent on temperature, which is a disadvantage in practical application. The influence of the conductivity difference between the two materials on the electric field distribution of HVDC cables was compared using finite element simulation according to the typical structure of the ±525 kV HVDC cable. The electric field distributions in the insulating layers severally using the two materials as insulation were calculated through the coupling simulation of electric and thermal

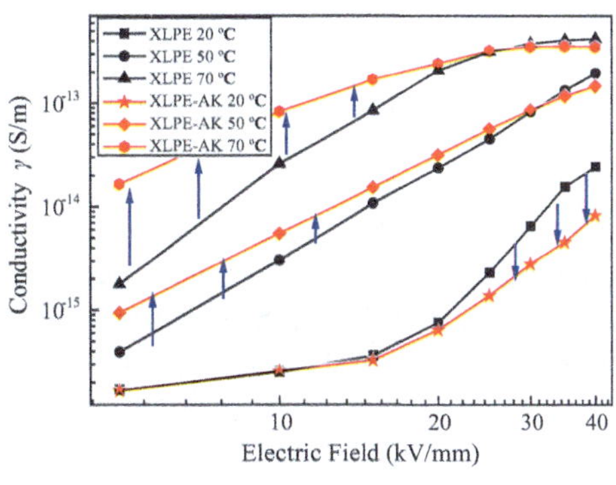

Fig. 17 γ–E curves of two materials at different temperatures

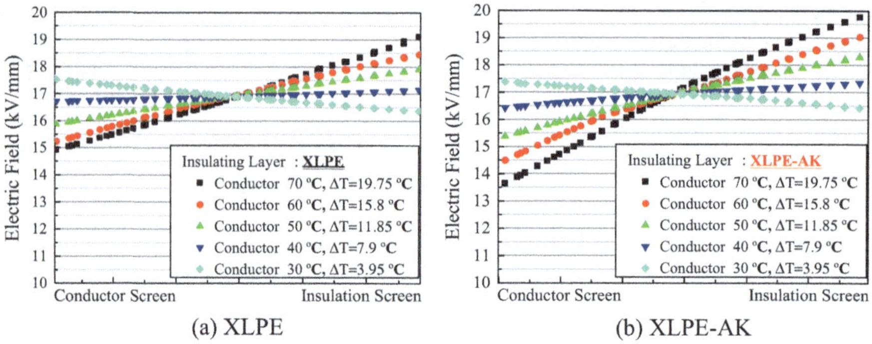

Fig. 18 Electric field distributions of HVDC cables with different materials as insulating layer

fields. The results are shown in Fig. 18. When the temperature of conductor exceeds 40 °C, stress inversion occurs in the insulating layers. Although the stress inversion in XLPE-AK insulating layer is relatively more obvious, the maximum electric field is merely increased by less than 1 kV/mm. The negative influence of the voltage stabilizer on the electric field distribution of the insulating layer is acceptable since there is a more significant improvement in the breakdown strength of the XLPE-AK material at elevated temperature.

Figure 19 shows the space charge distribution curves of XLPE and XLPE-AK samples during depolarization after stressed at the selected three test temperatures. The accumulated charge densities in the bulk of both the XLPE and XLPE-AK samples increase significantly with elevated temperature. The maximum charge density of XLPE sample at room temperature is similar with that of XLPE-AK sample, while the maximum charge densities of XLPE-AK samples at 50 °C and 70 °C are less than those of XLPE samples. The average charge densities are calculated and their decay over time at different temperatures for the two materials are shown in Fig. 20. It is indicated that in both the XLPE and XLPE-AK samples, the average space charge densities in the initial seconds of short circuit as well as the charge decay rates become higher with the temperature rising. From the results in Figs. 19 and 20, it can be inferred that the XLPE-AK sample exhibits similar space charge accumulation amount and decay characteristics at room temperature with XLPE sample, while exhibits less space charge accumulation, more uniform space charge distribution and higher space charge dissipation rate at elevated temperatures than XLPE sample.

As shown in Fig. 21, the apparent trap-controlled mobility of XLPE-AK is lower than that of XLPE at room temperature, while higher than that of XLPE at elevated temperature. It can be inferred that the grafting of AKVS molecules reduces the depth and density of the inherent deep traps in XLPE matrix, and simultaneously introduce a large number of shallow traps. Based on the calculated apparent trap-controlled mobility of two materials at different temperatures, the schematic diagrams of trap distribution of XLPE and XLPE-AK materials are inferred as shown in Fig. 22. The

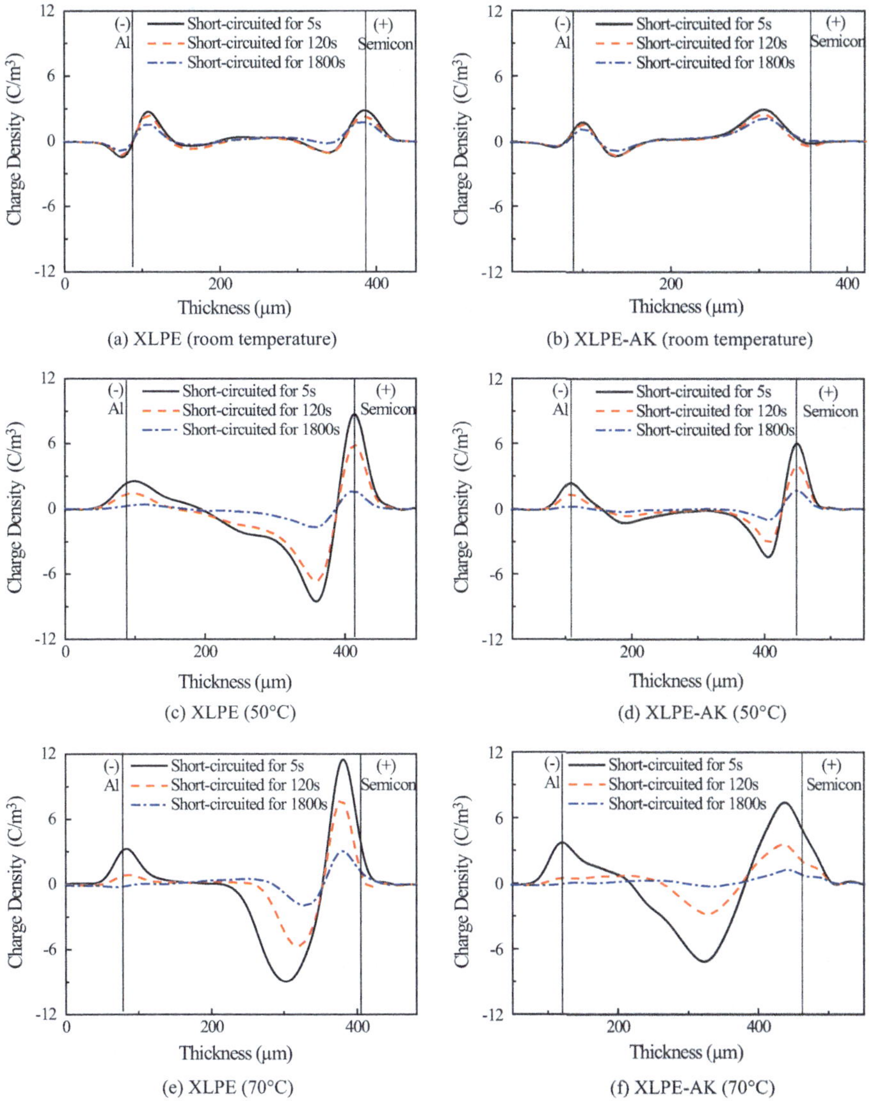

Fig. 19 Space charge distribution in samples during depolarization process

AKVS molecules bonded to the PE molecular chains by grafting reaction can help to form a uniform, dense and long-lasting distribution of the additive in the polymer and change the original microstructures and trap distribution in XLPE. Although the reduction of deep traps and the increase of shallow traps in XLPE-AK materials may result in easier injection of charge carriers, the uniformly and densely distributed shallow traps are capable to promote the migration, evacuation and extraction of the charges, which may facilitate a more uniform charge distribution and less charge

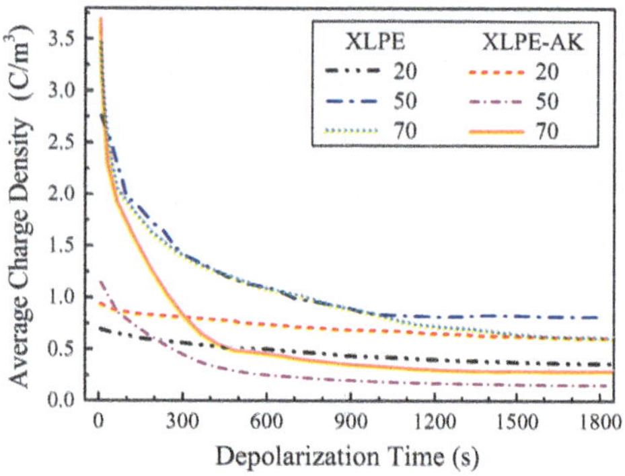

Fig. 20 Decay curves of average space charge density as depolarization time

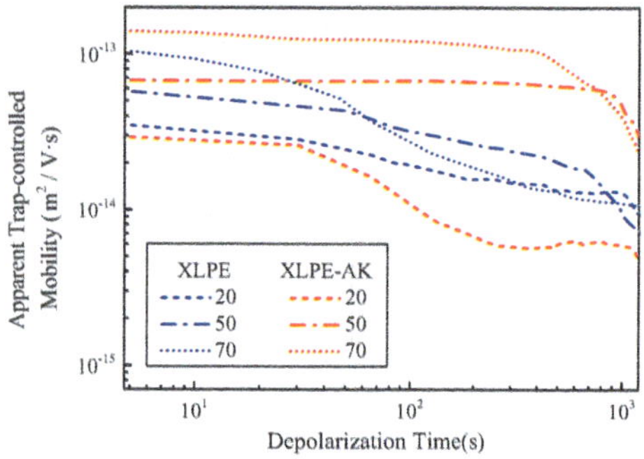

Fig. 21 Apparent trap-controlled mobility of two materials at different temperatures

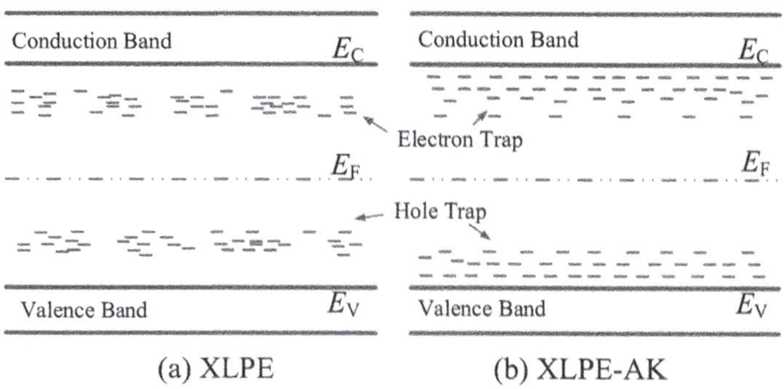

Fig. 22 Schematic diagrams of the trap distribution of two materials

accumulation as well as promoting the decay and dissipation of space charges in the material.

The electrical treeing phenomenon under DC voltage is closely related to the space charge accumulation and dissipation. It is unclear whether the voltage stabilizer that captures high-energy electrons can suppress the DC electrical tree. Therefore, the authors also investigated the influences of voltage stabilizer grafting on the growth

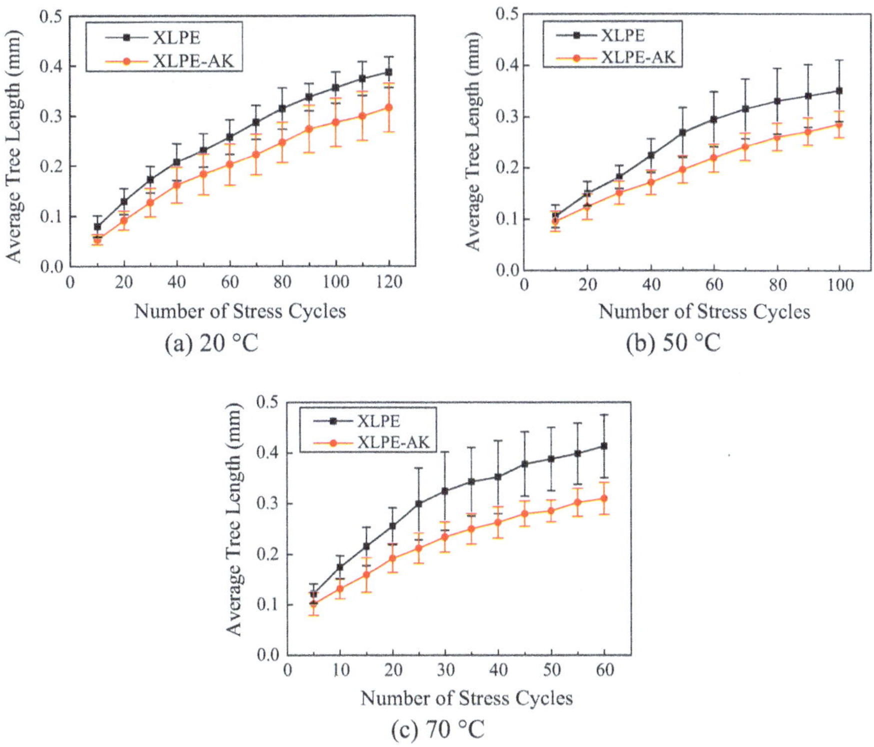

Fig. 23 Dependence of the grounded DC tree length on the stress period amount under −20 kV at different temperatures

rate of grounded DC electrical tree in XLPE. The dependences of grounded DC electrical tree length on the number of stress cycles at different temperatures are shown in Fig. 23. One stress cycle herein consists of prestressing under −20 kV DC voltage for 3 min and subsequent 3 s of short circuit process.

As shown in Fig. 23, AKVS grafting can significantly inhibit the growth of grounded DC trees of XLPE at different temperatures. Grounded DC electrical tree is considered initiated by the damage caused by the rapid de-trapping and the subsequent collision with polymer chains of the trapped charges under a shorted DC voltage. It is reported that the grounded DC electrical tree can be inhibited by nanoparticles due to the restrained charge injection and migration as well as the decreased accumulation amount of space charges [43]. Though the grafted voltage stabilizer cannot effectively restrain the injection and migration of space charges as confirmed by the space charge measurement, the destructive effect of charge de-trapping process can be weakened by the voltage stabilizer through capturing high-energy electrons, thus the growth of grounded DC electrical trees at different temperatures is significantly suppressed.

Based on the results of above studies, it can be seen that the voltage stabilizers can improve the DC insulation performances of breakdown strength and DC electrical tree resistance of polymer insulating materials including PE and PP. It can be concluded that proper use of voltage stabilizers has considerable prospects to promote the research and development of XLPE materials and recyclable materials for HVDC cable insulation. Nevertheless, there still remains some problems to be solved. Firstly, it is generally accepted that voltage stabilizer can improve the breakdown strength and electrical tree resistance of polymer insulating material, whereas the influences of voltage stabilizer and its analogues on the space charge behavior of polymeric insulation are very complicated. The voltage stabilizer may give rise to hetero space charge accumulation by dissociation, and may also suppress the homo charge accumulation or yield homo charge accumulation by affecting the trap distribution in the polymer. In short, the mechanisms of voltage stabilizers influencing the trap distribution and charge transport in polymeric insulation are still unclear, and the attempts to use voltage stabilizer to improve the DC dielectric strength of insulating materials and simultaneously suppress the space charge accumulation needs further study. Secondly, more attention should be paid to the effects of voltage stabilizer on the DC insulation performances at elevated temperature. Last but not least, whether the migration, volatilization and dissociation of voltage stabilizer and the interaction of voltage stabilizer with other additives have negative influences on the long-term DC insulation performances of polymeric insulation requires verification by sufficient experimental data.

4 Conclusion

(1) Voltage stabilizers are widely used to improve the AC electrical tree resistance of PE and XLPE. According to the different mechanisms of inhibiting electrical tree initiation, voltage stabilizers can be classified into three categories: voltage stabilizers that inhibit partial discharge, voltage stabilizers that capture high-energy electrons, and voltage stabilizers that inhibit macromolecular degradation.

(2) Alkylation of voltage stabilizer is the most widely used method to solve the problem of voltage stabilizer's poor compatibility with polymer matrix; Quantum chemical calculation is playing an increasingly important role in the research of voltage stabilizer, which can help to select and synthesize promising voltage stabilizers as well as provide foundation for deeper physico-chemical understanding of the mechanism of voltage stabilizers; The proper combined use of voltage stabilizer and nanoparticles can yield the synergistic effect on improving the dielectric strength of polymeric insulation, which is a promising approach to develop high-performance polymer insulating material.

(3) The proper use of voltage stabilizer can improve the DC insulation performances of polymer insulating materials including PE and PP, which offers a promising

way to develop XLPE materials and recyclable materials for HVDC cable insulation. However, more attention should be paid to the effects of voltage stabilizer on the DC insulation performances at elevated temperature and more efforts should be made to reveal the mechanisms of voltage stabilizer influencing the trap distribution and charge transport in polymeric insulation. Besides, the influences of the migration, volatilization and dissociation of voltage stabilizer and the interaction of voltage stabilizer with other additives on the long-term DC insulation performances of polymeric insulation also require further studies.

References

1. J.O. Bostrom, E. Marsden, R.N. Hampton, U. Nilsson, Electrical stress enhancement of contaminants in XLPE insulation used for power cables. IEEE Electr. Insul. Mag. **19**, 6–12 (2003)
2. A.C. Ashcraft, R.M. Eichhorn, R.G. Shaw, Laboratory studies of treeing in solid dielectrics and voltage stabilization of polyethylene, in *IEEE International Symposium on Electrical Insulation* (1978), pp. 213–218
3. H.J. Davis, Electrical insulation composition based on polyolefin containing dye additives as voltage stabilizers. U.S. Patent 4,216,101 (1980)
4. R.M. Eichhorn, Treeing in solid extruded electrical insulation. IEEE Trans. Electr. Insul. **12**, 2–18 (1976)
5. D.E. Maloney, Quaternary ammonium salt containing polyolefin covered electrical conductor. U.S. Patent 3,499,791 (1970)
6. J.J. McKeown, Insulated electrical conductors having corona resistant polymeric insulation containing organo metallic compounds. U.S. Patent 3,577,346 (1971)
7. E.J. McMahon, A tree growth inhibiting insulation for power cable. IEEE Trans. Electr. Insul. **16**, 304–318 (1981)
8. G. Teyssedre, C. Laurent, Advances in high-field insulating polymeric materials over the past 50 years. IEEE Electr. Insul. Mag. **25**, 26–36 (2013)
9. T. Tanaka, Effects of charge injection and extraction on tree initiation in polyethylene. IEEE Trans. Power Appar. Syst. **97**, 1749–1759 (1978)
10. H. Zhang, Y. Shang, H. Zhao, B. Han, Z. Li, Mechanisms on electrical breakdown strength increment of polyethylene by acetophenone and its analogues addition: a theoretical study. J. Mol. Model. **19**, 4477–4485 (2013)
11. H. Zhang, Y. Shang, X. Wang, H. Zhao, B. Han, Z. Li, Mechanisms on electrical breakdown strength increment of polyethylene by aromatic carbonyl compounds addition: a theoretical study. J. Mol. Model. **19**, 5429–5438 (2013)
12. H. Zhang, H. Zhao, X. Wang, Y. Shang, B. Han, Z. Li, Theoretical study on the mechanisms of polyethylene electrical breakdown strength increment by the addition of voltage stabilizers. J. Mol. Model. **20**, 2211–3281 (2014)
13. L.J. Heidt, Solid dielectric polyolefin compositions containing various voltage stabilizers. U.S. Patent 3,522,183 (1970)
14. Y. Yamano, H. Endoh, Increase in breakdown strength of pe film by additives of azocompounds. IEEE Trans. Dielectr. Electr. Insul. **5**, 270–275 (1998)
15. Y. Yamano, Roles of polycyclic compounds in increasing breakdown strength of LDPE film. IEEE Trans. Dielectr. Electr. Insul. **13**, 773–781 (2006)
16. Y. Yamano, M. Okada, Reduction of PD in a void by additives of azobenzoic compound in HDPE insulating material. IEEE Trans. Dielectr. Electr. Insul. **8**, 889–896 (2001)

17. Y. Yamano, Y. Fukuchi, Decrease in AC partial discharges on insulating materials by use of dye additives. IEEE Trans. Dielectr. Electr. Insul. **3**, 425–431 (1996)
18. K.C. Kao, New theory of electrical discharge and breakdown in low-mobility condensed insulators. J. Appl. Phys. **55**, 752–755 (1984)
19. C. Lurent, C. Mayoux, S. Noel, Mechanisms of electroluminescence during aging of polyethylene. J. Appl. Phys. **58**, 4346–4353 (1985)
20. S.S. Bamji, Luminescence and space charge phenomena in polymeric dielectrics, in *IEEE Annual Report Conference on Electrical Insulation Dielectric Phenomena* (2008), pp. 1–12
21. S.S. Bamji, A.T. Bulinski, J. Densley, Electrical tree suppression in high-voltage polymeric insulation. U.S. Patent 4,870,121 (1989)
22. Z. Li, Y. Yin, X. Wang, D.M. Tu, K.C. Kao, Formation and inhibition of free radicals in electrically stressed and aged insulating polymers. J. Appl. Polym. Sci. **89**, 3416–3425 (2003)
23. W.L. Hawkins, *Polymer Degradation and Stabilization* (Springer, Berlin, 1984), pp. 74–86
24. K. Hirota, Y. Kanemitu, N. Kamada, Y. Sekii, The influence of antioxidants on electrical tree generation in XLPE. Electr. Eng. Jpn. **135**, 154–159 (2001)
25. Y. Sekii, D. Tanaka, M. Saito, N. Chizuwa, K. Karasawa, Effects of antioxidants on the initiation and growth of electrical trees in XLPE, in *IEEE Annual Report Conference on Electrical Insulation and Dielectric Phenomena* (2003), pp. 661–665
26. L.Y. Gao, D.M. Tu, Y. Liu, B. Qiu, G.L. Huang, L.H. Wang, Effect of ferrocene derivatives on the dielectric properties of polyethylene, in *International Conference on Properties and Application of Dielectric Materials* (1991), pp. 796–799.
27. L. Martinotto, F. Peruzzotti, M.D. Brenna, Cable, in particular for transport or distribution of electrical energy and insulating composition. U.S. Patent 6,696,154 B2 (2004)
28. T. Hjertberg, V. Englund, Polyolefin composition for medium/high/extra high voltage cables with improved electrical breakdown strength. U.S. Patent 8,519,037 B2 (2013)
29. T. Hjertberg, V. Englund, Polyolefin composition for medium/high/extra high voltage cables comprising benzophenone-type voltage stabiliser. U.S. Patent 8,765,843 B2 (2014)
30. V. Englund, R. Huuva, S.M. Gubanski, T. Hjertberg, High efficiency voltage stabilizers for XLPE cable insulation. Polym. Degrad. Stab. **94**, 823–833 (2009)
31. V. Englund, R. Huuva, S.M. Gubanski, T. Hjertberg, Synthesis and efficiency of voltage stabilizers for XLPE cable insulation. IEEE Trans. Dielectr. Electr. Insul. **16**, 1455–1461 (2009)
32. H. Wutzel, M. Jarvid, J.M. Bjuggren, A. Johansson, V. Englund, S. Gubanski, M.R. Andersson, Thioxanthone derivatives as stabilizers against electrical breakdown in cross-linked polyethylene for high voltage cable applications. Polym. Degrad. Stab. **112**, 63–69 (2015)
33. M. Jarvid, A. Johansson, J.M. Bjuggren, H. Wutzel, V. Englund, S. Gubanski, C. Müller, M.R. Andersson, Tailored side-chain architecture of benzil voltage stabilizers for enhanced dielectric strength of cross-linked polyethylene. J. Polym. Sci. Part B Polym. Phys. **52**, 1047–1054 (2014)
34. H. Zhang, Y. Shang, H. Zhao, B. Han, Z. Li, Study of the effect of valence bond isomerizations on electrical breakdown by adding acetophenone to polyethylene as voltage stabilizers. Comput. Theor. Chem. **1062**, 99–104 (2015)
35. J. Su, B. Du, T. Han, Z. Li, M. Xiao, J. Li, Multistep and multiscale electron trapping for high-efficiency modulation of electrical degradation in polymer dielectrics. J. Phys. Chem. C **123**, 7045–7053 (2019)
36. S. Kisin, J.D. Doelder, R.F. Eaton, P.J. Caronia, Quantum mechanical criteria for choosing appropriate voltage stabilization additives for polyethylene. Polym. Degrad. Stab. **94**, 171–175 (2009)
37. M. Jarvid, A. Johansson, V. Englund, A. Lundin, S. Gubanski, C. Müller, M.R. Andersson, High electron affinity a guiding criterion for voltage stabilizer design. J. Mater. Chem. A **3**, 7273–7286 (2015)
38. M. Jarvid, A. Johansson, R. Kroon, J.M. Bjuggren, H. Wutzel, V. Englund, S. Gubanski, M.R. Andersson, A new application area for fullerenes: voltage stabilizers for power cable insulation. Adv. Mater. **27**, 897–902 (2015)
39. K.Y. Lau, A.S. Vaughan, G. Chen, Nanodielectrics: opportunities and challenges. IEEE Electr. Insul. Mag. **31**, 45–54 (2015)

40. Y. Yamano, M. Iizuka, Improvement of electrical tree resistance of LDPE by mixed addition of nanoparticles and phthalocyanine. IEEE Trans. Dielectr. Electr. Insul. **18**, 329–337 (2011)
41. M. Andersson, *Polyethylene Blends, a Material Concept for Future HVDC Cable Insulation*. Chalmers University of Technology (2017)
42. M. Ieda, Nawata, DC treeing breakdown associated with space charge formation in polyethylene. IEEE Trans. Electr. Insul **12**, 19–25 (1977)
43. Y. Wang, G. Li ,Y. Yin, The effect of nano-MgO addition on grounded DC tree in cross-linked polyethylene, in *2015 IEEE 11th International Conference on the Properties and Applications of Dielectric Materials*, pp. 285–288 (2015)
44. Y. Sekii, H. Kawanami, M. Saito, K. Sugi, I. Komatsu, DC tree and grounded DC tree in XLPE, in *IEEE Annual Report Conference on Electrical Insulation and Dielectric Phenomena*, pp. 523–526 (2005)
45. Y. Yin, D. Xiao, D. Tu, An application of space charge in valuing the electric ageing degree of insulating polymer. Proc. CSEE **22**, 43–48 (2002)
46. R. Minami, N. Hirai, Y. Ohki, M. Okashita, T. Maeno, Effects of liquid chemicals on space charge evolution in low-density polyethylene, in *IEEE International Symposium on Electrical Insulating Materials* (2001), pp. 87–90
47. N. Hussin, G. Chen, Analysis of space charge formation in LDPE in the presence of crosslinking byproducts. IEEE Trans. Dielectr. Electr. Insul. **19**, 126–133 (2012)
48. G.C. Montanari, C. Laurent, G. Teyssedre, A. Campus, U.H. Nilsson, From LDPE to XLPE: investigating the change of electrical properties. Part I: space charge, conduction and lifetime. IEEE Trans. Dielectr. Electr. Insul. **12**, 438–446 (2005)
49. Y. Sekii, T. Maeno, Generation and dissipation of negative heterocharges in XLPE and EPR. IEEE Trans. Dielectr. Electr. Insul. **16**, 668–675 (2009)
50. H. Suzuki, Y. Sekii, K. Noguchi, K. Shimura, Influence of antioxidants on heterocharge generation in polymeric dielectrics. Electr. Eng. Jpn. **172**, 1–9 (2010)
51. W. Lei, K. Wu, Y. Wang, Y. Cheng, X. Zheng, L.A. Dissado, S.J. Dodd, N.M. Chalashkanov, J.C. Fothergill, C. Zhang, W. Li, Are nano-composites really better DC insulators? A study using silica nanoparticles in XLPE. IEEE Trans. Dielectr. Electr. Insul. **24**, 2268–2270 (2017)
52. J.W. Zha, Y.H. Wu, S.J. Wang, D.H. Wu, H.D. Yan, Z.M. Dang, Improvement of space charge suppression of polypropylene for potential application in HVDC cables. IEEE Trans. Dielectr. Electr. Insul. **23**, 2337–2343 (2016)
53. B.X. Du, C. Han, J. Li, Z. Li, Effect of voltage stabilizers on the space charge behavior of XLPE for HVDC cable application. IEEE Trans. Dielectr. Electr. Insul. **26**, 34–42 (2019)
54. X. Chen, A. Paramane, H. Liu, J. Tie, Y. Tanaka, Enhancement of service life and electrical insulation properties of polymeric cables with the optimum content of aromatic voltage stabilizer. Polym. Eng. Sci. **4**, 1–15 (2020)
55. Y. Gao, X. Huang, D. Min, S. Li, P. Jiang, Recyclable dielectric polymer nanocomposites with voltage stabilizer interface: toward new generation of high voltage direct current cable insulation. ACS Sustain. Chem. Eng. **7**, 513–525 (2019)
56. X. Chen, L. Yu, C. Dai, A. Paramane, Y. Tanaka, Enhancement of insulating properties of polyethylene blends by delocalization type voltage stabilizers. IEEE Trans. Dielectr. Electr. Insul. **26**, 2041–2049 (2019)
57. C. Li, H. Zhao, B. Han, H. Zhang, C. Zhang, Y. Ai, Effect of voltage stabilizer on the DC insulation properties of XLPE. Proc. CSEE **23**, 7071–7079 (2018)

Polypropylene Insulation Materials for HVDC Cables

Jun-Wei Zha, Ming-Sheng Zheng, Wei-Kang Li, George Chen, and Zhi-Min Dang

Abstract High-voltage direct current (HVDC) transmission plays an important role in the development of sustainable transmission networks and the conversion of energy systems. As an important medium in the transmission system, HVDC cables have become a hot research topic. Flexible DC transmission using HVDC plastic cables is the mainstream direction advocated by the international power grid. Therefore, the demand for high-performance HVDC plastic cables is increasing. Aiming at the development process of thermoplastic environmental polypropylene insulation materials for HVDC cables, the development status and technical bottlenecks of HVDC cable insulation materials were summarized. This chapter discussed the research progress of polypropylene insulation materials from the perspective of the basic structure of polypropylene, intrinsic modification, and its nanocomposite, etc. The necessity and urgency of research and development of insulation materials for HVDC cables were clarified. Finally, the future development direction of HVDC cables was summarized and prospected.

1 Introduction

High-voltage direct current (HVDC) transmission has unique advantages in long-distance large-capacity power transmission and grid interconnection. Moreover, the

The original version of this chapter was revised: The affiliations of the authors have been updated. The corrections to this chapter are available at https://doi.org/10.1007/978-981-15-9731-2_27

J.-W. Zha (✉) · M.-S. Zheng
University of Science and Technology Beijing, Beijing, China
e-mail: zhajw@ustb.edu.cn

W.-K. Li
Global Energy Interconnection Research Institute Corporation, Beijing, China

G. Chen
University of Southampton, Southampton, UK

Z.-M. Dang
Tsinghua University, Beijing, China

B. Du (ed.), *Polymer Insulation Applied for HVDC Transmission*,
https://doi.org/10.1007/978-981-15-9731-2_4

power regulation of HVDC transmission system is fast and flexible. HVDC is the main trend of future power grid development. It is suitable for power supply to islands, capacity expansion of urban load centers, grid connection of wind power, etc. especially for the development of urban direct current (DC) distribution systems, high-voltage DC transmission is essential. Therefore, it has received more and more attention. There are three types of HVDC cables: oil-filled cables, oil-impregnated paper cables, and plastic extruded cables. Plastic extruded HVDC cables have an indispensable position in wind power grid connection, island power supply, and cross-sea long distance transmission. The critical issues of HVDC cable insulation materials have been further studied to promote the application of HVDC cables in the power transmission with large-capacity, long-distance, and complex environments [1–5].

Cross-linked polyethylene (XLPE) has been used for many years as an insulating material for HVDC transmission systems. XLPE is the most popular material owing to its superior electrical and mechanical properties in HVDC cables. However, the traditional XLPE is gradually becoming a thorny issue as it is a thermoset material that is not easily recyclable, which has the following problems: cross-linking process equipment takes up a lot of space, low production efficiency, large energy consumption, large equipment investment and energy loss. In addition to the shortcomings of the production process and the material itself, recently the world has paid more and more attention to environmental issues. XLPE is a material that cannot be degraded and is difficult to reuse. In general, common XLPE cables are treated by incineration, which leads to environmental pollution and waste material disposal [6–10]. Therefore, the use of new non-cross-linked cable insulation materials is of great significance, and it is more important to try to find new materials to solve this problem.

Polypropylene (PP), as a thermoplastic, is cheap and has high mechanical strength, good processability, heat resistance and electrical insulation properties [11–17]. PP ranks third among the five general plastics, second only to PE and PVC, and is currently one of the fastest growing plastics with the fastest output growth. The research and development of PP cable materials have obvious economic and social benefits [18, 19]. However, the research on PP insulation for wire and cable is still in its infancy. PP has excellent electrical insulation properties, and its electrical properties are basically not affected by changes in ambient humidity and electric field frequency. It has excellent arc resistance and good heat resistance, and can be used for a long time at above 100 °C. It is also outstanding in chemical resistance. Furthermore, PP shows good recycling properties. Therefore, PP can be regarded as a potent eco-friendly material in advanced HVDC cables. Compared with XLPE materials, PP has a high operating temperature and a simple cable drawing process. The most important thing is that PP materials have the function of being recyclable compared with XLPE. For XLPE's non-degradable and recyclable, PP materials meet the requirements of low-carbon environmental protection materials proposed today. Hosier et al. conducted a lot of research on recyclable isotactic PP cable materials [20–22]. They studied six kinds of propylene copolymers or homopolymers provided by Sigma-Aldrich Chem. They found that although all materials meet the requirements of cable materials in certain indicators, but no material can have room temperature toughness, high temperature mechanical properties and excellent electrical properties. Further research found that blending isotactic polypropylene and ethylene propylene copolymer can obtain the best results. The group has also studied five kinds of ethylene propylene copolymers produced by the Nordic Chemical Industry and found that with the increase of ethylene content, the low temperature brittleness of isotactic polypropylene has been improved very well. Among them, copolymers with a mass fraction of 10% ethylene are expected to become recycled DC cable insulation material. Yoshino et al. have done a lot of research on the application of syndiotactic polypropylene (s-PP) in the insulation of recyclable cables. It was pointed out that compared with isotactic polypropylene (i-PP), s-PP can form smaller spherulites, have lower crystallinity, and have good thermal stability and electrical properties. It can operate at 90 °C without cross-linking [23–27]. At present, scientists have used a metallocene complex as a catalyst to produce a new type of s-PP. It exhibits several relatively good electrical insulation. The test report shows that s-PP as the main insulation cable has a life span of >30 years, which meets the requirements for cable use. At the same time, its insulation strength is still at a high level before the end of its life, which meets the engineering requirements. Compared with the cable just put into operation, its insulation strength decreased only slightly. In addition, it shows that the water tree aging phenomenon in s-PP cables is better suppressed compared with the currently popular XLPE cable. At the 2015 JICABLES conference, Presman announced the successful preparation of 320 kV polypropylene high-voltage DC cables, taking the lead in becoming an international cable enterprise using polypropylene to prepare HVDC cables. Ordinary polypropylene is difficult to meet the requirements of cable due to its high modulus and insufficient flexibility. Therefore, it is usually necessary to modify polypropylene, mainly including two methods: block copolymerization and doped rubber. He et al. adopted the method of POE toughening PP to effectively reduce the PP modulus, while the POE/PP composites have excellent electrical insulation properties [28]. In addition, Jiang et al. found that block polypropylene also effectively reduces PP modulus. The space charge injection is significantly reduced, and the breakdown strength is higher than low-density polyethylene (LDPE) [29].

2 Issues of Space Charge in HVDC Cables

Space charge is a crucial issue in the research of HVDC cable insulation materials. During the operation of the DC plastic cable, space charge is particularly likely to accumulate in the insulating material of its insulating layer due to the electric field. For example, the commonly used XLPE materials have good insulating properties due to low carrier mobility and high trap concentration. However, it is precisely because of these characteristics that space charges accumulate inside the XLPE under DC voltage [30]. Due to the accumulation, movement and dissipation of space charge, the local electric field inside the materials will be distorted, which will affect the conductance, breakdown and aging characteristics of insulating materials [31–34]. Therefore, reducing and eliminating space charge in insulating materials is the key to development of HVDC plastic cables. It is also an important frontier direction for the study of dielectric theory, and it is of great significance to promote the development of polymer insulation aging and breakdown theory.

2.1 The Key Issue in HVDC Cables—Space Charge

For HVDC cable insulation materials, the generation, transfer, and dissipation of space charge will change the electric field distribution of the material, which ultimately affects the insulation performance of the material. Therefore, how to suppress the accumulation of space charge in insulating materials has become an important scientific issue in this field. The space charge is mainly caused by the migratable charge injected into the electrode, the trapped carriers, or the ionization of organic or inorganic impurities. These will migrate within the material and then superimpose at some point to form a charge packet [35, 36]. The formation process of space charge is shown in Fig. 1.

The formation and accumulation of space charge redistributes the electric field in the material, reflecting the change in trap density from another angle. The whole process involves two important aspects: "into the trap" and "out of trap". The rate of "into the trap" and "out of trap" affects charge injection, carrier mobility, etc., directly affects the electric field distribution of the polymer material, and eventually affects the breakdown characteristics of the cable. Therefore, by detecting the space charge in the material, not only can the establishment and attenuation process of the space charge inside the material be understood from the microscopic level, but also the electrical breakdown and aging characteristics of the material can be shown

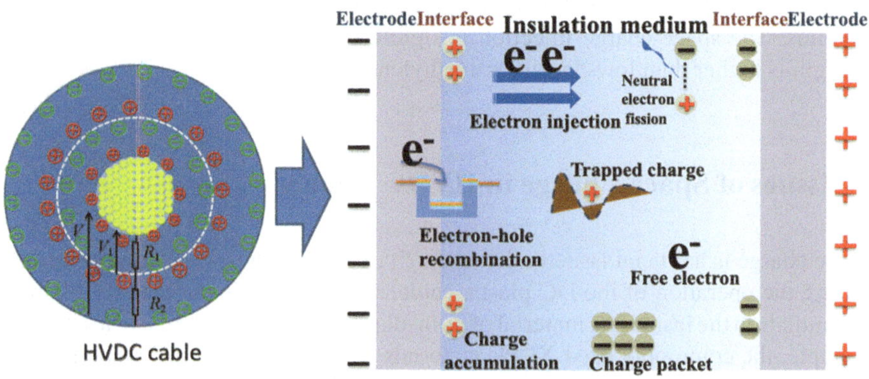

Fig. 1 Schematic diagram of formation mechanism and motion law of space charge in HVDC cable insulation materials

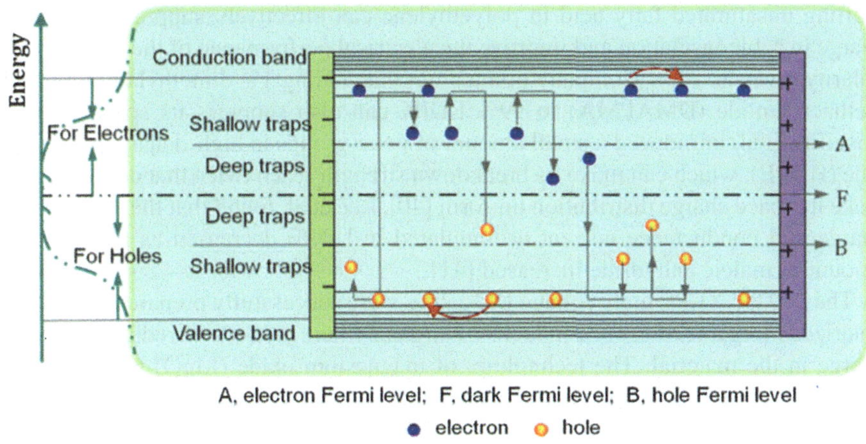

Fig. 2 Schematic diagram of the energy band and dynamic process of the space charge in the insulation layers. *Source* Zha and Dang 2019 [39]. Reproduced with permission of AIP Publishing

from the macroscopic level. It is of great significance to study the comprehensive insulation properties of materials [37, 38].

Under a high operating electric field, a large amount of charge is injected into the insulation layer of the cable. Therefore, the formation of space charge is usually accompanied by the injection of charge. The energy band and movement process of the space charge in the insulating layer between the two electrodes are shown in Fig. 2. The electrons and holes in the conduction and valence bands can be captured by shallow traps. The electrons captured by the shallow trap can be thermally activated into the conduction band, and then they can be recombined at the capture center or captured by the deep trap. Holes trapped in shallow traps can receive electrons, and then they can also recombine or be trapped by deep traps [39].

2.2 *Space Charge Suppression Method*

At present, three methods for suppressing space charge in the insulation of high-voltage DC cables have been proposed, and some results have been achieved. ① grafting modification with special functional groups; ② Adding inorganic nano-oxide; ③ Blending modification. Studies have shown that the introduction of polar groups in the polyethylene (PE) body will generate charge traps, which will help to suppress the accumulation of space charge in the insulating medium. The introduction of polar groups such as carbonyl group, hydroxyl group, nitro group, cyano group or aromatic ring in the copolymer of ethylene and α-olefin can effectively suppress the space charge in the polymer matrix and increase its resistivity. The maleic anhydride grafted onto XLPE introduces a charge trap to suppress space charge well. The grafting unsaturated fatty acid in polyethylene can effectively suppress the space charge in cable insulation and improve the electrical performance of the cable under polarity reversal. ABB Company pointed out that adding 1% dimethylaminopropyl methacrylamide (DMAPMA) to 99% LDPE can also suppress its space charge. Yoshifuji et al. introduced a small amount of polar groups in high-density polyethylene (HDPE), which can make its breakdown strength reach twice that of XLPE and make its space charge distribution uniform [40]. Lee et al. found that the heteropolar charge and conductance current accumulated in LDPE decreased as the grafting amount of maleic anhydride increased [41].

The 250 kV XLPE high-voltage DC cables were successfully prepared by adding polarized inorganic fillers or conductive inorganic fillers to XLPE to reduce the space charge in the material. The technology of magnesium oxide (MgO) nanoparticles filled with cross-linked polyethylene was proposed and then a 500 kV high-voltage DC plastic cable was successfully developed, waiting for commercial application [42, 43]. More scholars have also conducted a lot of research work on the modification of polyolefin insulation materials for cables. Wu et al. found that after adding different amounts of SiO_2 nanoparticles into LDPE, the average bulk charge density inside the composite dielectric can be effectively suppressed, and the average decay rate of its charge decreases with increasing content [44]. Suh et al. studied the space charge behavior of ionic compounds doped with PE, and found that methacrylic ionic compounds can promote the uniform distribution of space charges, thereby improving the dielectric breakdown strength [45]. Tu et al. found that the proper ratio of ethylene-vinyl acetate copolymer (EVA) and PE blended can significantly reduce the injected charge, eliminate the frictional charging phenomenon, and reduce the deep traps in PE [46]. Dang et al. found that trisorbitol with different mass fractions as a nucleating agent was added to PE to change its crystalline form. They used the crystal morphology theory to explain the correspondence between the space charge distribution under a DC electric field and the formation of a water tree under an alternating current (AC) electric field, and found the optimal mass fraction of trisorbitol for suppressing space charge in PE [47].

2.3 Space Charge Suppression Mechanism

In order to gain a deeper understanding of the behavior of space charge, researchers have carried out in-depth research on its mechanism. Based on the previous research theory, T. Tanaka et al. proposed a multi-layer interface model of nanoparticles in polyethylene in 2005 [48]. The model is based on the surface modification of nanoparticles. The commonly used surface modifier is a silane coupling agent. Taking nanoparticles as the center, the interface from inside to outside is divided into three layers in sequence: bonding layer, binding layer, and loose layer. The three-layer interface together constitutes the electric double layer of the polymer, which becomes an important theoretical basis for explaining the suppression of space charge accumulation in nanometers. A schematic diagram based on the multi-core model is shown

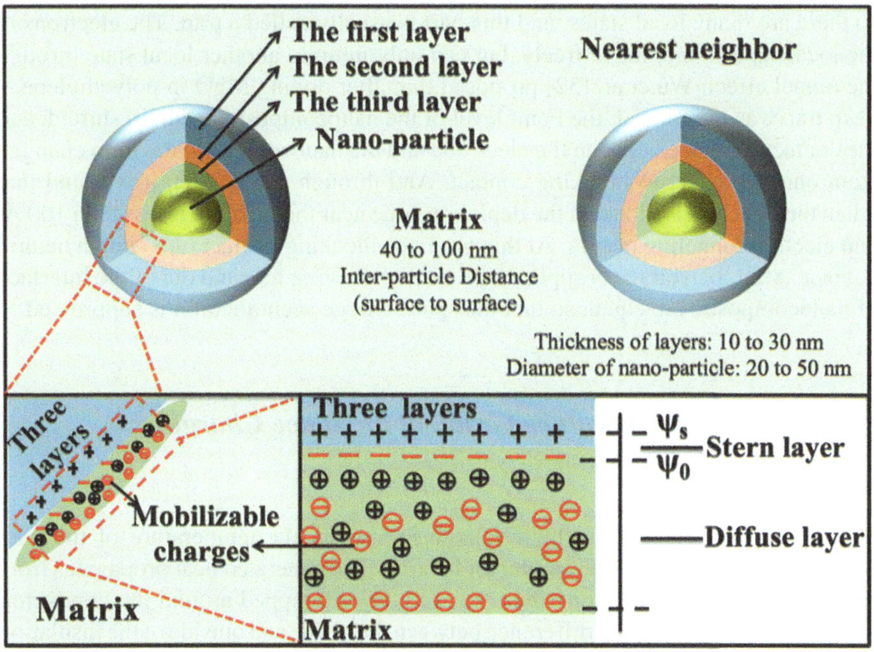

Fig. 3 Schematic diagram of the multi-core model

in Fig. 3. The theory is that when carriers are captured by the nanoshell, redistribution of positive and negative charges occurs in the Gouy-Chapman diffuse layer, forming an electric double layer. The electric double layer is an independent, uniform and stable electric potential, which does not generate a local electric field to the entire material.

After the deep trap theory was proposed, the researchers used an ionic nanoparticle doped with polyethylene [49] and found that when the nanoparticles are evenly dispersed, the deep trap formed at the interface of the nanoparticles can effectively capture carriers and form an independent electric field. The local effective electric field is avoided, and the injection barrier is increased, thereby suppressing space charge accumulation. In addition, T. Takada et al. [50] proposed a deep trap model formed by MgO, and calculated that its trap energy level is between 1.5–5 eV. Other studies have also shown that MgO can not only effectively suppress space charge accumulation, but also increase volume resistivity, reduce carrier migration, improve breakdown strength, and suppress electrical tree growth [51].

Nanoparticles change the energy band of the polymer. By changing the interface contact type between the electrode and the medium, the mechanism of inhibiting space charge accumulation is studied. There are crystalline regions and amorphous regions in polyolefins, and the molecular segments in the crystalline regions are repeatedly arranged in order, and the atoms are combined with each other. In the amorphous region, the molecular chains are arranged disorderly and there is no periodicity, so there are many local states, and this part is usually called a trap. The electrons in the local state cannot move freely, but can only jump to another local state through the tunnel effect. Wu et al. [52] proposed that after doping MgO in polyethylene, a deep trap was introduced, the Femi level of the nanocomposite medium shifted, and the contact interface between the electrode and the nanocomposite medium changed from ohmic contact to blocking contact. And through calculation, it is found that when the interface is blocked, the depletion zone near the medium is less than 100 Å, and electron tunneling occurs. At this time, the blocking contact turns into a neutral contact. After the voltage is applied, the carriers flowing into and out of the interface of nanocomposite are equal, so that the space charge accumulation is suppressed.

2.4 Effect of Temperature Gradient on Space Charge in HVDC Cables

During the operation of the high-voltage DC cable, the temperature of the core conductor will rise due to the conductor heating. The generated heat propagates from the inside to the outside along the insulating layer wrapped around the conductor, which causes a temperature difference between the inside and outside of the insulation layer. Therefore, the cable insulation layer is in a temperature gradient field with high inside and low outside rather than a constant temperature field. Chen et al. studied the space charge distribution of degassed and non-degassed XLPE power cables at room temperature and temperature difference. It was found that the temperature gradient reduced the heteropolar charge in the non-degassed XLPE cable [30]. Montanari et al. studied the distribution law of space charge in model cables and MV-level cables under temperature gradient [53]. The study shows that for model cables, the effect of temperature gradient effect on space charge accumulation only exists below the injection material insulation threshold. For MV grade cables, the temperature gradient effect can cause the electric field in the insulation to reverse, that is, the maximum electric field strength appears outside the cable insulation. However, the conductor temperature during full-load operation of the cable can reach 70°C, and the temperature difference between the inside and outside of the insulation may exceed 50 °C.

In addition, with the increase of transmission voltage level, space charge accumulation and electric field distribution in polymer insulation under high electric field strength and high temperature gradient need to be studied. To this end, Wu et al. conducted a systematic study on the space charge phenomenon of sheet low-density polyethylene samples under different temperature gradients. It was found that the temperature gradient will promote the accumulation of a large amount of heteropolar charges on the low-temperature side of the LDPE, and the greater the temperature gradient, the greater the amount of heteropolar charges, thereby making the electric field distortion on the low-temperature side more severe [54]. However, the mechanism of the effect of temperature gradient on space charge is not perfect.

2.5 *Effect of Crystal Structure on Space Charge in HVDC Cables*

There are both crystalline and amorphous regions in the polymer. This difference has caused differences in electrical properties. The molecular chain structure is ordered and periodic, forming crystalline regions. The arrangement of molecular chains is disordered, forming a disordered region, called the amorphous region. Studies have shown that the electrical properties of the amorphous region are significantly lower than those of the crystal region. The spherulites in polyethylene have a significant effect on the insulation properties of cables. The crystal form is significantly affected by the processing conditions, especially the cooling rate. Zhou et al. studied the changes in the size of spherulites of polyethylene at different cooling rates and their space charge behavior in high fields. The results show that the slower the cooling rate, the larger the spherulite. The sample with higher crystallinity has a larger space charge packet amplitude and the lower the charge packet migration rate [55]. It was also found that the smaller the spherulites in polyethylene, the more obvious the effect of suppressing space charge. The crystal size, structure, and processing conditions in polyethylene have an important impact on the preparation of high-performance DC cables.

Polypropylene usually has three common crystalline forms including monoclinic (α), trigonal (β) and orthorhombic (γ). The most stable one is the α crystal form [56]. The PP produced by ordinary processing is mainly monoclinic. The size of the spherulites is large, and the grain boundaries are obvious. Due to the crystal structure of polypropylene, it exhibits high elastic modulus, surface hardness and other excellent mechanical properties. The disadvantage is that stress concentration is easy to occur at the grain boundary, resulting in poor toughness. These disadvantages limit its application in HVDC cables. The β-crystalline polypropylene belongs to the hexagonal crystal system, and its spherulite size is smaller than that of the α-crystalline, which has high impact strength and good toughness. In the next generation of polypropylene cables, the application prospect is broad.

3 Polypropylene and Their Modification

Polypropylene is more thermally stable than polyethylene and can be used in a non-crosslinked state. It is used to prepare HVDC cable insulation materials and has great application prospects [21]. Among three common crystal form, α crystal forms are widely found in ordinary polypropylene. They have high mechanical strength, belong to the monoclinic crystal system, and they also have large spherulites. More importantly, the electrical properties of the α crystal form have been extensively studied, and there are many investigations on its insulation properties [57, 58]. The β crystal form needs to be induced by a special process and a doping nucleating agent, and it is in a hexagonal crystal system. The spherulite is smaller, the toughness is good, and the strength is lower than the α crystal form. Moreover, there are currently very few reports on the electrical properties of β-crystalline polypropylene. Because its toughness is better than α-crystalline polypropylene, its application in the future non-cross-linked, environmentally friendly new cable has greater application prospects. Therefore, it is of great significance to carry out research on the preparation and electrical insulation properties of β-crystalline polypropylene.

3.1 Morphology and Crystalline-Phase-Dependent

The different molecular structures of PP is illustrated as shown in Fig. 4. Isotactic polypropylene (i-PP) can be prepared by the polymerization of propylene monomers. Due to its excellent electrical insulation properties, high melting point and thermal stability, and good isotacticity and crystallinity, it is a promising material that is expected to replace XLPE. Importantly, i-PP is an environmentally friendly insulating material that can be easily recycled. It is worth noting that the copolymerization of ethylene and propylene molecules can be achieved by adding ethylene to improve the mechanical properties of PP, especially the impact properties. This will improve the reliability and availability of PP for HVDC cables. By adding vinyl groups to the PP molecular chain, block polypropylene (b-PP) and random polypropylene (r-PP) can be synthesized by controlling the proportion of ethylene and polymerization conditions.

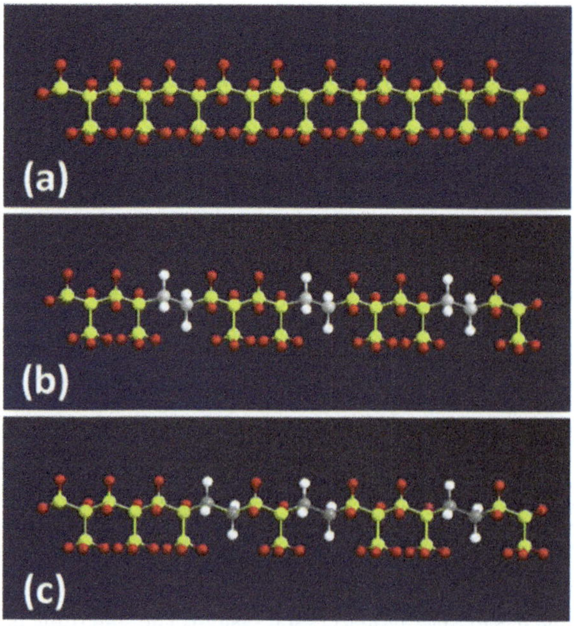

Fig. 4 Schematic illustration of the molecular structure of **a** i-PP, **b** b-PP, and **c** r-PP. *Source* Zha and Dang 2016 [59]. Reproduced with permission of AIP Publishing

Among the three crystalline phases of PP, it has been proved that the α-phase ratio present in i-PP is the largest. The β phase can be achieved through the introduction of β nucleating agents or during the crystallization process with a gradient temperature rise [60]. However, due to the low stability of the γ phase, it is considered that it rarely occurs. During the copolymerization process, the β- and γ-phases may be present in b-PP and r-PP. Zha et al. [59] present a systematic study of the influence of morphology and crystallization on the electrical insulation properties of polypropylene has been conducted. It was concluded that the formation of β and γ phases can induce deep traps, which can capture carriers and significantly affect the migration of carriers, thereby improving insulation performance.

3.2 *Modification of Polypropylene*

Polypropylene is a thermoplastic material that is recyclable, cheap, and easier to process. Therefore, polypropylene is an ideal recyclable cable insulation material. Compared with XLPE, it has more excellent electrical insulation performance, and has a higher using temperature and good heat resistance. More importantly, it is an environmentally friendly material. However, the key problem is that its impact performance is poor, its modulus is large, and it becomes brittle at low temperatures. Not suitable for direct use as a cable. Related scholars doped different types of elastomers, such as ethylene propylene diene monomer (EPDM), polyolefin elastomer (POE) and styrene-(ethylene-co-butene)-styrene triblock copolymer (SEBS), etc., to improve the above shortcomings of polypropylene [61–64]. Among them, SEBS has good compatibility and similar melting point with PP, so it is an ideal modified material. The addition of SEBS forms a sea-island structure in the PP matrix, which can significantly improve the flexibility of the PP matrix. Zha et al. adopted chemical modification on PP with polar functional groups by grafting with maleic anhydride (MAH). Compared with pure PP, MAH induces smaller spherulites, and the β crystal form disappears due to the rearrangement of molecular chains [65]. At the same time, with the increase of MAH grafting content, PP-g-MAH has much better performance in suppressing space charge injection and aggregation than PP.

Many researchers have proved that the addition of β-nucleating agent (β-NA) can greatly improve the mechanical properties of polypropylene due to the transformation of the crystal structure [66–70]. Due to its special crystal structure, it is believed that β crystals have better mechanical properties than α crystals [71, 72]. Therefore, β nucleating agents are usually added to PP. Compared with the modification in the polymerization process, adding β-NA is a simpler and more practical method. β-NA may act as a heterogeneous nucleating agent in b-PP, thereby increasing the crystallization temperature. The observation of crystal morphology showed that the size of spherulites became smaller, which indicated that the morphology of b-PP changed after adding β-NA. Due to the introduction of deep traps, the introduction of β-NA can significantly inhibit the accumulation of space charge, which is consistent with the results of trap level distribution [73].

4 Polypropylene Nanocomposites

The significant suppression of the space charge of PP composite material is of great significance for the development of HVDC transmission cables. In recent years, it has been found that nanoparticles have excellent properties for improving polymer materials due to their quantum size effect and large specific surface area. Since T. J. Lewis proposed the concept of nanodielectrics in 1994 [74], scholars have conducted extensive research on the performance improvement and mechanism of polymer insulating materials after the addition of nanoparticles. Depending on the particle size, shape and doping amount of the nanoparticles, the polymer matrix exhibits different nanodielectric behaviors. Therefore, various levels of space charge suppression, dielectric properties, thermal properties, and enhanced mechanical properties can be achieved by introducing nanoparticles into the polymer matrix.

Polymer nanocomposites combined with inorganic fillers of different nanometer sizes such as MgO, ZnO, Al_2O_3, SiO_2, etc. can be used as HVDC cable materials. These fillers effectively suppress the accumulation of space charge, significantly improve the insulation and mechanical properties, and induce high thermal stability to the above manufacturing process [75–78]. The interface between the particles and the matrix helps trap the carriers in the interface trap between the nanomaterial and the polymer during the transition process, and ultimately limits the migration of electrons and holes. Tanaka proposed a multi-core model and interface model to reveal the mechanism of this behavior [48]. Extensive research has been conducted to improve interface compatibility. In their study, the synthesis of highly efficient interfacial nanoparticles incorporated into a polymer matrix is disclosed to suppress space charge.

4.1 Zero-Dimension (0D) Fillers

Researchers have conducted extensive research on the performance improvement and mechanism of polymer insulating materials after adding nanoparticles, mainly focusing on the space charge distribution, accumulation, attenuation characteristics, breakdown strength, electrical conductivity, and aging mechanism. Due to the different particle size, shape, doping amount of the nanoparticles, and the different polymer matrix, the effects exhibited are different in terms of suppressing space charge, improving mechanical properties, etc.

He et al. introduced polyolefin elastomers into PP, which improved the mechanical properties of PP while reducing the accumulation of space charge. Their further research shows that the introduction of MgO nanoparticles can improve the electrical properties of PP/POE blends, especially the accumulation of space charge [79]. Zha et al. adopted the chemical modification of MgO with KH550 to obtain good dispersibility in the PP/SEBS blends [76]. They also manifested that compared with

PP/SEBS blends, the addition of MgO effectively suppressed the space charge accumulation and improved its breakdown strength due to a certain number of deep traps caused by loading 0.5 phr MgO. In addition, they also prepared PP/SEBS composites filled with surface-modified ZnO (mZnO) nanoparticles by melt blending. The ZnO filler modified by silane coupling agent (KH550) can produce good adhesion to the polymer matrix. Moreover, the incorporation of m-ZnO has little effect on the melting and crystallization process and dielectric properties of PP/SEBS. When the mZnO content is 0.5%, the composite material exhibits higher breakdown strength, lower electrical conductivity, and significant suppression of space charge [80]. In order to further improve the electrical insulation properties of block polypropylene (b-PP), 1 wt% Al_2O_3 nanoparticles were added to b-PP by melt blending process. It can be found that the trap density of the b-PP/Al_2O_3 composites was 3.2 times greater than that of unfilled b-PP, thereby improving the electrical properties, especially the space charge suppression in the b-PP/Al_2O_3 composites [81].

4.2 One-Dimension (1D) Fillers

Wu et al. designed the arrangement of β crystals in isotonic PP by doping one-dimensional N,N'-dicyclohexyl terephthalamide (DCTH) act as a nucleating agent, as shown in Fig. 5 [68]. The doping of DCTH (0.1 wt%) has a great influence on the formation of β crystallite clusters, which makes the proportion of β crystallites (βc) in iPP significantly increase from 0% to 83.2%. Compared with the dispersed α crystallites, the growth of β crystals effectively suppressed the accumulation of space charge. This is because DCTH, which is a microcrystalline nucleus, has an electron affinity, which can prevent charge carriers from passing through the amorphous region of iPP. In addition, the crystallites grown on the surface of the DCTH have a wide bending gap to limit the transfer of charge carriers, thereby improving the breakdown strength and reducing the conduction current.

For comparison, the effects of multi-structure ZnO (i.e., ZnO nanoparticles (ZnOp), nanowires (ZnOw) and teraneedle (ZnOt) like ZnO) on space charge suppression and electric field distortion were systematically and comparatively studied as shown in Fig. 6. The breakdown strength and conductive current characteristics of PP/ZnO nanocomposites were also investigated [82]. By adding ZnO nanowires, the nanocomposites significantly suppress space charge accumulation, while the addition of ZnO nanoparticles and needle-like whiskers could not effectively suppress the space charge accumulation.

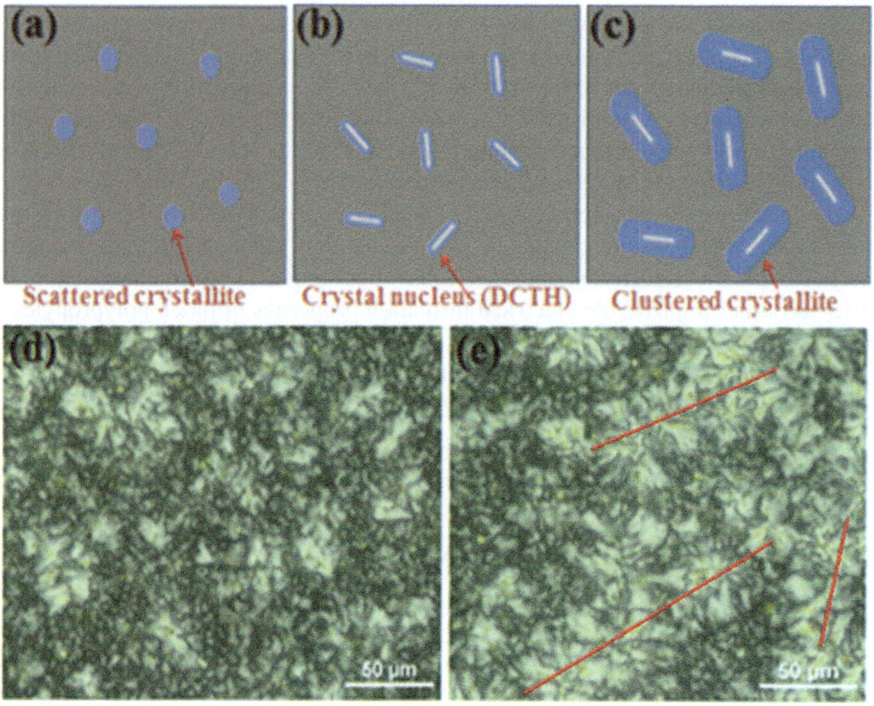

Fig. 5 Schematic illustration of formation of crystal morphology in iPP **a** scattered crystallite, **b** DCTH act as β-nucleating agent and **c** process of clustered β-crystal formation; Polarized optical microscopy images of **d** scattered a and **e** clustered β crystal. *Source* Zha and Dang 2015 [68]. Reproduced with permission of AIP Publishing

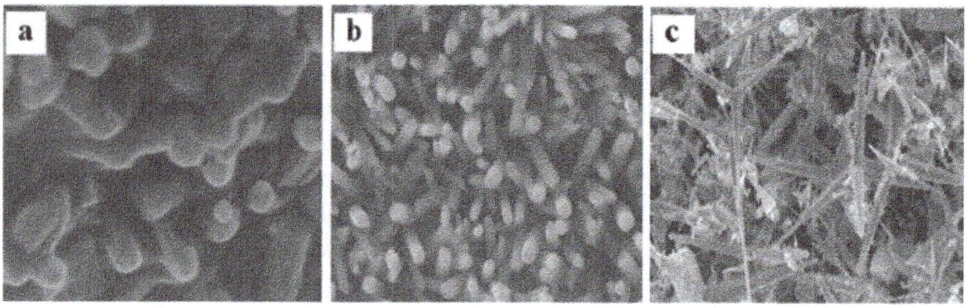

Fig. 6 SEM images of different structure ZnO **a** ZnO-p, **b** ZnO-w, **c** ZnO-t. *Source* Zha and Dang 2017 [82]. Reproduced with permission of IOP science

4.3 Two-Dimension (2D) Fillers

Up to now, though the electrical insulating properties of PP based materials have been extensively studied, few studies have focused on the PP composites with two-dimensional fillers, such as graphene oxide (GO). According to reports, the addition of GO has a strong effect on suppressing the pocket charge in low density polyethylene (low concentration 0.05 wt%) [83]. Zha et al. designed a two-dimensional functional GO (f-GO), and the chemical reaction principle of surface modification is shown in Fig. 7. They found that the introduction of functionalized GO (f-GO) can effectively suppress the accumulation of space charge in PP/SEBS composites [84]. This is because that as the interface traps in the modified PP/GO increase, the number and energy of free charges decrease, which will effectively increase the short-term breakdown strength. In addition, due to the ionization of impurities under high electric fields, the addition of f-GO resulted in a slight increase in conductivity. The conclusion indicates that the incorporation of two-dimensional fillers provides great potential for changing the insulation properties of PP-based composite materials used in HVDC cables.

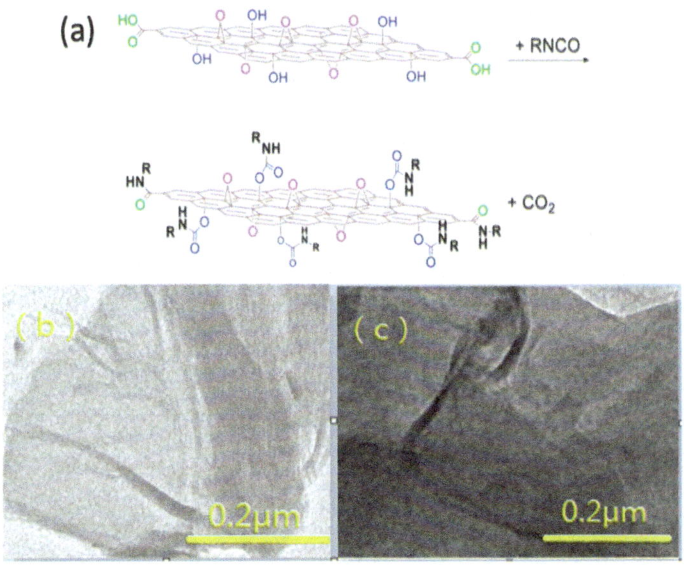

Fig. 7 **a** modification of GO and TEM images of **b** GO and **c** f-GO. *Source* Zha and Dang 2018 [84]. Reproduced with permission of IEEE

4.4 Other Structure Fillers

Many studies have confirmed that the introduction of nanoparticles could introduces deep traps. so that interface traps will capture carriers during the transition process. Therefore, the space charge accumulation is effectively suppressed. Many scholars have done a lot of work to improve interface compatibility. Cheng et al. designed C–MgO nanoparticles with abundant surface defects containing a highly effective interface through surface carbonization. And a ternary nanocomposites were prepared by melt blending with PP and SEBS. As shown in Fig. 8, the space charge is significantly suppressed [85]. In addition, the DC breakdown strength is increased to 304 kV/mm, which may be due to the introduction of deep traps due to the introduction of nanoparticles with high surface vacancy defects. A very small amount of MgO with foam nanostructures synthesized through a designed freeze-drying process was introduced to suppress the accumulation of space charge and further improve the electrical properties of PP/SEBS insulation [39]. The foamed MgO filler almost shows a porous structure and is dispersed in the PP matrix in the form of dendritic distribution, which is very different from the traditional nanoparticles.

In order to avoid filler agglomeration caused by excessive doping, ZnO composite filler (co–ZnO) was synthesized by solvothermal method [86]. ZnO nanoparticles successfully attached to ZnO nanowires, forming a core-shell structure. It has been manifested that doping co–ZnO fillers into polymer insulating materials can reduce the accumulation of space charge due to the synergy between ZnO nanofibers and nanoparticles and polymer matrix. This synergistic effect can effectively inhibit the movement and recombination of the carrier. This is attributed to the introduction of co–ZnO filler, which caused a certain number of traps at the interface between the filler and the matrix and the interface between the ZnO nanofibers and nanoparticles. The deep traps can trap charge carriers, leading to the formation of a local electric field, which increases the barrier for charge carrier injection, and thus exhibits excellent space charge suppression effects.

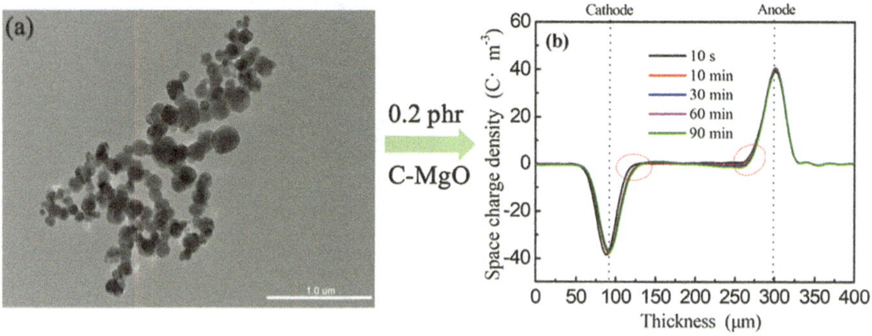

Fig. 8 **a** TEM images of C-MgO nanoparticles and **b** Space charge distribution in the PP/SEBS/C-MgO nanocomposites (0.2 phr C-MgO) under the applied electric field of 60 kV/mm. *Source* Zha and Dang 2019 [85]. Reproduced with permission of AIP Publishing

In summary, a large number of studies have indicated that the addition of nanoparticles such as SiO_2, MgO, ZnO, Al_2O_3, etc. can significantly inhibit the space charge accumulation in the polymer composites. However, the nanoparticles are prone to agglomeration, which can impair the electrical properties of polymer insulating materials. Moreover, it cannot guarantee continuous and stable production during mass production. Compared with nano-doping, blending can ensure the uniformity and stability of the material. Low-density polyethylene is a polyolefin material with a similar structure to PP, and the electrical properties of PP/LDPE composites are rarely studied. Yan et al. found that when the LDPE content is 40 wt% in PP, it can effectively improve the space charge suppression and increase the breakdown strength [87]. The trap level density obviously affects the space charge distribution. Compared with the traps in pure PP, the interface between PP and LDPE can form shallow traps because the shallow traps in pure PP are mainly located in the interface area. Through these shallow traps, the space charge migration becomes easier. The addition of LDPE increases the trap level density in PP, which can change the space charge distribution in the PP composites.

5 Conclusion

At present, inorganic nanofillers doped with PP can inhibit space charge injection and accumulation under different electric fields, which has been verified by a large number of experiments. Scientists have no consensus on the explanation of the mechanism. There are two main theories: One is the trap level theory, which is divided into deep traps and shallow traps. It is mainly supported by the core-shell model and Schottky injection barrier. The second is the interface contact theory between electrode and material, which is divided into barrier contact and ohmic contact. The main theoretical basis is work function and energy band theory. For high-voltage DC cable insulation materials, comprehensive consideration should be given to the interface compatibility of the nanofiller itself, the charge accumulation characteristics in the material, the effect of the nanofiller on the PP structure, the relationship between charge transport and the nanofiller, and the breakdown law of the high-voltage electric field. It is necessary to further explore the mechanism of space charge suppression, and finally establish a theoretical model combining the interface injection characteristics and the nanomaterial structure.

There are few reports on the mechanism of chemical modification to suppress space charge. For chemical modification, the essence is to reconstruct new molecular chains to improve the performance of materials in terms of resistance to breakdown and blocking charge transport. However, this technology requires a large cost investment and deep synthesis experience, and has been monopolized by the incumbent polymer manufacturers. The polyolefin industry is constantly improving, hoping to make breakthroughs in insulating materials in the future.

In view of the gap between China and developed countries in high-voltage and UHVDC cables, advanced impurity filtering and detection technology can be introduced to prepare ultra-clean domestic PP raw materials. Combined with nanocomposite technology, we will focus on breaking through the technical bottleneck of domestic HVDC cables and develop domestic HVDC cables with China's independent intellectual property rights, thus breaking the monopoly of developed countries on the domestic market. The use of domestically produced ultra-clean PP raw materials to prepare high-performance HVDC cables can greatly reduce the production cost of HVDC cables under the premise of satisfying the power transmission performance of the power grid. It will further promote the industrial upgrade of domestic ultra-clean PP raw materials and improve the performance of domestic HVDC cables. At the same time, advanced test equipment should also be invested in research and development, including space charge test equipment and conductance test equipment for three-dimensional full-size cables, to provide technical guarantee for the early realization of high-performance HVDC cables.

References

1. G. Mazzanti, M. Marzinotto, *Extruded Cables for High-Voltage Direct-Current Transmission: Advances in Research and Development*. Wiley-IEEE Press (2013)
2. M. Hao, Y. Zhou, G. Chen et al., IEEE Trans. Dielectr. Electr. Insul. **22**(1), 72–80 (2015)
3. Y.N. Wang, G.D. Li, J.D. Wu, Y. Yin, IEEE Trans. Dielectr. Electr. Insul. **23**, 3704–3711 (2016)
4. M.S. Khalil., IEEE Electr. Insul. Magaz. **13**, 35–47 (1997)
5. A.M. Pourrahimi, T.A. Hoang, D.M. Liu, L.K. H. Pallon, S. Gubanski, R.T. Olsson, U.W. Gedde, M.S. Hedenqvist. Adv. Materi. **28**, 8651–8657 (2016)
6. F. Rogti, M. Ferhat, Appl. Phys. Lett. **104**, 031605 (2014)
7. B.X. Du, J. Li, IEEE Trans. Dielectr. Electr. Insul. **23**, 1190–1197 (2016)
8. W.C. Du, W. Zhong, Y. J. Lin, L. Shen, Q.G. Du, Eur. Polymer J. **40**, 1987–1995 (2004)
9. Y. Zhou, J. Hu, B. Dang, J.L. He, RSC Adv. **6**, 48720–48727 (2016)
10. J.W. Zha, Y.H. Wu, S.J. Wang, D.H. Wu, H.D. Yan, Z.M. Dang, IEEE Trans. Dielectr. Electr. Insul. **23**, 2337–2343 (2016)
11. H. Janet, T.R. Jow, IEEE Trans. Dielectr. Electr. Insul. **19**(3), 990–995 (2012)
12. E. Izci, N. Bowler. 10th IEEE International Conference on Solid Dielectrics. Potsdam, Germany, 1–4 (2010)
13. K. Yoshino, A. Ueda, T. Demura, Y. Miyasgita, K. Kurahashi, Y. Matsuda. Proceedings of the 7th International Conference on Properties and Applications of Dielectric Materials, Nagoya, Japan, pp. 175–178 (2003)
14. J.W. Choi, J.H. Cheon, H.J. Kim et al., IEEE Trans. Appl. Supercond. **20**(3), 1280–1283 (2010)
15. B. Dang, J.L. He, J. Hu et al., J. Nanomater. **2015**(1), 1–9 (2015)
16. B. Dang, Q. Li, Y. Zhou et al., Compos. Sci. Technol. **153**, 103–110 (2017)
17. T.M. Krentz, M.M. Khani, M. Bell et al., J. Appl. Polym. Sci. **44347**, 1–10 (2017)
18. I.N. Hidayah, M. Mariatti, H. Ismail et al., Plastics. Rubber Compos. **44**(7), 259–265 (2015)
19. X. Zheng, G. Chen, IEEE Trans. Dielectr. Electr. Insul. **15**(3), 800–807 (2008)
20. I.L. Hosier, S. Reaud, A.S. Vaughan et al., *International Symposium on Electrical Insulation* (Canada, Vancouver, 2008), pp. 502–505
21. I.L. Hosier, A.S. Vaughan, S.G. Swingler, J. Materi. Sci. **46**(11), 4058–4070 (2011)
22. I.L. Hosier, L. Cozzarini, A.S. Vaughan et al., J. Phys: Conf. Ser. **183**(1), 12015 (2009)
23. D.W. Kim, K. Yoshino, T. Inoue et al., Jpn. J. Appl. Phys. **38**(6), 3580–3584 (1999)

24. D.W. Kim, K. Yoshino, J. Phys. D Appl. Phys. **33**(4), 464–471 (2000)
25. K. Kurahashi, Y. Matsuda, A. Ueda et al., *IEEE/PES Transmission and Distribution Conference and Exposition* (Japan, Yokohama, 2002), pp. 1278–1283
26. K. Yoshino, T. Demura, M. Kawahigashi et al., Electr. Eng. Jap **146**(1), 18–26 (2004)
27. K. Kurahashi, Y. Matsuda, Y. Miyashita et al., Electr. Eng. Jap **155**(3), 1–8 (2006)
28. Y. Zhou, J.L. He, J. Hu, X.Y. Huang, P.K. Jiang, IEEE Trans. Dielectr. Electr. Insul. **22**(2), 673–681 (2015)
29. X. He, J. Shi, L. Wu, P. Jiang, B. Liu, Compos. Sci. Technol. **161**, 115–123 (2018)
30. M. Fu, L.A. Dissado, G. Chen et al., IEEE Trans. Dielectr. Electr. Insul. **15**(3), 851–860 (2008)
31. Y.W. Zhang, J. Lewiner, C. Alquie et al., IEEE Trans. Dielectr. Electr. Insul. **3**(6), 778–783 (1996)
32. N. Hozumi, H. Suzuki, T. Okamoto et al., IEEE Trans. Dielectr. Electr. Insul. **1**(6), 1068–1076 (1994)
33. R. Bartnikas, IEEE Trans. Dielectr. Electr. Insul. **4**(5), 544–557 (1997)
34. S. Boggs, IEEE Electr. Insul. Mag. **20**(4), 22–27 (2004)
35. R.S. Le, P. Segur, G. Teyssedre et al., J. Phys. D Appl. Phys. **37**, 298–305 (2004)
36. G. Chen, J. Zhao. J. Phys. D:Appl. Phys. **44**, 212001 (2011)
37. M. Teruyoshi, IEEE Trans. Dielectr. Electr. Insul. **1**(5), 923–933 (1994)
38. G. Chen, T.Y.G. Tay, A.E. Davies, IEEE Trans. Dielectr. Electr. Insul. **8**(6), 867–873 (2011)
39. J.W. Zha, Y. Wang, S.J. Wang, M.S. Zheng, X. Bian, Z.M. Dang, Appl. Phys. Lett. **114**, 252902-1-5 (2019)
40. N. Yoshifuji, T. Niwa, T. Takahashi et al., IEEE Trans. Power Delivery **7**(3), 1053–1059 (1992)
41. S.H. Lee, J.K. Park, J.H. Han et al., IEEE Trans. Dielectr. Electr. Insul. **2**(6), 1132–1139 (1995)
42. T. Takada, Y. Hayase, Y. Tanaka, IEEE Trans. Dielectr. Electr. Insul. **15**(1), 152–160 (2008)
43. Y. Hayase, H. Aoyama, Y. Tanaka et al., 8th International Conference on Properties and Applications of Dielectric Materials, Bali, pp. 159–162 (2006)
44. J. Wu, W. Liu, Y. Zheng et al., IEEE International Conference on Solid Dielectrics, Potsdam, Germany, pp. 1–4 (2010)
45. K.S. Suh, D. Damon, J. Tanaka, IEEE Trans. Dielectr. Electr. Insul. **2**(1), 1–11 (1995)
46. X. Li, W. Jin, D. Tu, K. Muratu, T. Takada, IEEE Int. Conf. Propert. Appl. Dielectr. Mater. **1**, 311–314 (1997)
47. Z. Dang, J. Kang, D. Tu, IEEE Int. Conf. Propert. Appl. Dielectr. Mater. **1**, 562–566 (2000)
48. T. Tanaka, M. Kozako, N. Fuse, Y. Ohki, IEEE Trans. Dielectr. Electr. Insul. **12**(4), 669–679 (2005)
49. B. Han, X. Wang, Z. Sun, J. Yang, Q. Lei, Appl. Phys. Lett. **102**, 142903–142904 (2013)
50. T. Takada, Y. Hayase, Y. Tanaka, T. Okamoto, IEEE Trans. Dielectr. Electr. Insul. **15**(1), 152–160 (2008)
51. Y. Murakami, M. Nemoto, M. Nagao et al., IEEE Trans. Dielectr. Electr. Insul. **15**(1), 33–38 (2008)
52. K. Wu, Z. Lv, X. Wang, D. Tu, L.A. Dissado, IEEE Conference on Electrical Insulation and Dielectric Phenomena, pp. 299–302 (2012)
53. D. Fabiani, G.C. Montanari, C. Laurent et al., IEEE Electr. Insul. Mag. **24**(2), 5–14 (2008)
54. X. Chen, X. Wang, K. Wu et al., IEEE Trans. Dielectr. Electr. Insul. **17**(6), 1796–1805 (2010)
55. Y. Zhou, Y. Wang, Z. Markus, N. Wang, Q. Sun, X. Liang, Z. Guan, Jpn. J. Appl. Phys. **50**, 017101 (2011)
56. B. Lotz, J.C. Wittmann, A.J. Lovinger, Polymer **37**, 4979–4992 (1996)
57. K. Yoshino, T. Demura, M. Kawahigashi, Y. Miyashita, K. Kurahashi, Y. Matsuda, Electr. Eng. Jap **146**(1), 872–879 (2004)
58. D. Shi, J.H. Yang, Z.H. Yao, Y. Wang, H.L. Huang, W. Jing, J.H. Yin, G. Cost, Polymer **42**, 5549–5557 (2001)
59. J.W. Zha, H.D. Yan, W.K. Li, Z.M. Dang, Appl. Phys. Lett. **109**, 222902-1-5 (2016)
60. F. Luo, C. Geng, K. Wang, H. Deng, F. Chen, Q. Fu, B. Na, Macromolecules **42**, 9325 (2009)
61. A. Vanderwal, J.J. Mulder, Polymer **39**(26), 6781–6787 (1998)
62. X. Hong, G. Nie, Z. Lin, J. Rong, Polymer-Plast. Technol. Eng. **51**(10), 971–976 (2012)

63. O.M. Jazani, A. Arefazar, S.H. Jafari et al., Polymer-Plast. Technol. Eng. **52**(2), 206–212 (2013)
64. A.K. Gipta, B.K. Ratnam, K.R. Srinirasan, J. Apply Polym. Science **46**(2), 281–293 (1992)
65. J.W. Zha, Y.H. Wu, S.J. Wang, D.H. Wu, H.D. Yan, Z.M. Dang, IEEE Trans. Dielectr. Electr. Insul. **23**, 2337–2343 (2016)
66. Y. C. Li, F.Q. Lin, X.Q. Cui, W.H. Guo, J. Thermoplast. Comp. **29**, 381 (2016)
67. C.Q. Jiang, S.C. Zhao, Z. Xin, J. Thermoplast. Compos. **28**, 610 (2015)
68. Y.H. Wu, J.W. Zha, W.K. Li, S.J. Wang, Z.M. Dang, Appl. Phys. Lett. **107**, 112901 (2015)
69. Y. Gao, J. Li, G. Chen, T. Han, B. Du, IEEE Trans. Dielectr. Electr. Insul. **27**, 947–955 (2020)
70. F. Luo, C.L. Xu, K. Wang, H. Deng, F. Chen, Q. Fu, Polymer **53**, 1783 (2012)
71. J. Varga, J. Macromolecul. Sci. Part B **41**, 1121–1171 (2002)
72. A. Menyhárd, G. Faludi, J. Varga, J. Therm. Anal. Calorim. **93**, 937–945 (2008)
73. C. Zhang, J.W. Zha, H.D. Yan, W.K. Li, Z.M. Dang, Appl. Phys. Lett. **112**, 091902 (1–5) (2018)
74. T.J. Lewis, IEEE Trans. Dielectr. Electr. Insul. **1**, 812 (1994)
75. L.K.H. Pallon, R.T. Olsson, D. Liu et al., J. Mater. Chem. A **3**, 7523 (2015)
76. J.W. Zha, Y. Wang, W.K. Li, S.J. Wang, Z.M. Dang, IEEE Trans. Dielectr. Electr. Insul. **24**, 1457 (2017)
77. S.P. Fillery, H. Koerner, L. Drummy, E. Dunkerley, M.F. Durstock, D.F. Schmidt, R.A. Vaia, A.C.S. Appl, Mater. Interfaces **4**, 1388 (2012)
78. I.N. Hidayah, M. Mariatti, H. Ismail, M. Kamarol. Plast., Rubber Compos. **44**, 259 (2015)
79. Y. Zhou, J.L. He, J. Hu, B. Dang, J. Appl. Polym. Sci. **133**, 42863 (2016)
80. J.W. Zha, J.F. Wang, S.J. Wang, Q. Qin, Z.M. Dang, IEEE Trans. Dielectr. Electr. Insul. **25**, 3469–3476 (2018)
81. C. Zhang, J.W. Zha, H.D. Yan, W.K. Li, Y.Q. Wen, Z.M. Dang, IEEE Trans. Dielectr. Electr. Insul. **25**, 1293–1299 (2018)
82. J.W. Zha, Q. Cheng, H.D. Yan, W.K. Li, Z.M. Dang. J. Phys. D: Appl. Phys. **50**, 305301-1-5 (2017)
83. Y.J. Kim, S.T. Ha, G.J. Lee, J.H. Nam, I.H. Ryu, S.H. Nam, C.M. Park, I. In, J. Kim, C.J. Han. J. Nanosci, Nanotechnol **13**, 3464–3469 (2013)
84. J.W. Zha, H.D. Yan, W.K. Li, Z.M. Dang, IEEE Trans. Dielectr. Electr. Insul. **25**, 1088–1094 (2018)
85. Q. Cheng, J. W. Zha, J.T. Zhai, D.L. Zhang, X. Bian, G. Chen, Z.M. Dang, Appl. Phys. Lett. **115**, 102904-1-5 (2019)
86. J.W. Zha, Q.Q. Qin, Z.M. Dang, IEEE Trans. Dielectr. Electr. Insul. **26**, 868–875 (2019)
87. H.D. Yan, C. Zhang, W.K. Li, J.W. Zha, Polym. Compos. **41**, 780–787 (2020)

The Insulating Properties of Polypropylene Blends Modified by ULDPE and Graphene for HVDC Cables

Zhaohao Hou, Boxue Du, Ranran Xu, Jin Li, and Zhonglei Li

Abstract HVDC cable transmission plays an important role in urban center power supply, large-scale utilization of distributed energy, island and cross sea power transmission. Thermoplastic polypropylene (PP) material has excellent heat resistance and insulation performance, and can be recycled. However, it is not suitable to apply PP directly to the cable insulation for its low temperature impact and brittleness. In this chapter, polypropylene was blended with vinyl elastomer ULDPE to improve its mechanical toughness. The optimal proportion of PP/ULDPE blends was determined by the melting and crystallization behavior, mechanical tensile properties, trap distribution characteristics, space charge, conductivity and breakdown properties of the blends. Then, PP/ULDPE blends with different amount of nano graphene were prepared. The space charge, electrical conductivity and breakdown strength of PP and its nano graphene modified composites at different temperatures were measured and analyzed. Based on the isothermal discharge current method, the trap distribution characteristics of PP/ULDPE/graphene nanocomposites were obtained to study the relationship between space charge, electrical conductivity, breakdown strength and nano graphenen.

1 Introduction

In recent years, thermoplastic polypropylene (PP) materials have caused widespread concern for its excellent heat resistance and insulation performance [1, 2]. PP can have high mechanical strength without cross-linking treatment, and can be recycled for environmental protection, which meets the development demand of environment-friendly cable insulation. At the same time, the charge injection threshold of iPP in DC field is high, which is beneficial to restrain the space charge injection and accumulation in cable insulation [3]. Although the insulation performance of iPP

Z. Hou (✉) · B. Du · R. Xu · J. Li · Z. Li
Tianjin University, No. 92 Weijin Road, Nankai District, Tianjin 300072, China
e-mail: houzhaohao@163.com

B. Du (ed.), *Polymer Insulation Applied for HVDC Transmission*,
https://doi.org/10.1007/978-981-15-9731-2_5

is excellent, the mechanical performance of iPP has the disadvantages of poor low-temperature impact performance and brittleness. In the process of cable production, laying and installation, when using the iPP insulation materials alone, there is the risk of cable insulation cracking when bending stress, so the iPP cannot be used directly for high-voltage DC cable insulation.

One of the most important methods to improve the mechanical toughness of PP insulation is to blend PP matrix and elastomer [4]. The toughening effect of PP blend is closely related to the type, content and dispersion of elastomer and the interaction with PP. It has been found that the ultra-low-density polyethylene (ULDPE), which is polymerized by metallocene, has a high homogeneous short chain branching distribution and can produce elasticity and flexibility. At the same time, its linear main chain keeps good rigidity, which can effectively improve the mechanical impact properties and toughness of polypropylene [5]. Moreover, its high melting point (about 120 °C) makes it have a wide temperature range. Therefore, on the basis of improving the mechanical properties of PP, blending ULDPE can keep the better thermal properties of PP blends, which is conducive to the operation of PP insulated cable at a higher working temperature. Using the elastomer ULDPE with homogeneous short chain branched distribution and higher melting point has the advantages of improving the mechanical properties and maintaining good thermal properties of polypropylene. However, it is necessary to further study the mechanical properties, thermal properties and electrical properties of PP/ULDPE blends and determine the optimal parameters.

A lot of traps can be introduced in the interface between polymer and nanoparticles [6]. The electric properties of PP composites, such as space charge and breakdown strength, were improved [3]. However, common nano particles such as MMT, ZnO, SiO_2, TiO_2 are difficult to achieve a specific surface area of 1000 m^2/g. Generally, a large amount of nano addition is needed to improve the electrical properties of the composite. Although the dispersion of nanoparticles can be improved to a certain extent by surface treatment [7], the large amount of additives aggravates the agglomeration of nanoparticles. They are not easy to disperse, and even introduces impurity defects, which threatens the service life and safe operation of cables.

Due to graphene's advantages of high mechanical strength, good flexibility, excellent optical performance, good conductivity and thermal conductivity, many researchers have made extensive exploration in the application fields of graphene in infrared detectors, solar cells, light-emitting diodes, quantum devices, etc. [8]. Graphene, as a kind of nano filler with only atomic layer thickness, has a unique two-dimensional structure and a huge specific surface area (theoretical surface area can reach 2630 m^2/g). In theory, it can interact with polymer matrix more than traditional nano filler. Therefore, it is speculated that the addition of a small amount of graphene (far less than the addition of traditional nano particles) can bring a large number of nano filler polymer interface areas, at the same time, it can avoid the agglomeration problem when the nano amount is large, which is conducive to the further development of the potential of nano dielectrics [9]. At present, in the field of insulating medium research, previous research mainly focused on the electrical, thermal and mechanical properties of polymer/graphene composite near the percolation threshold, and the nano addition amount was about 0.5–5wt%. However, the effect of nano graphene on the electrical properties of PP composite is still very little, and the mechanism is not clear.

In this chapter, ultra low density polyethylene (ULDPE) is selected to modify the toughness of polypropylene, and the influence of ULDPE on the insulation property of polypropylene is focused to obtain the optimal blending ratio parameters. On the basis of the optimal blending ratio, PP/ULDPE/graphene nanocomposites were further prepared, and the space charge, breakdown and conductivity characteristics of the composites were studied. In addition, the conductor will bear different load current and heat during the operation of the cable, which will lead to the temperature change of the insulation layer, and the temperature effect will affect the space charge, conductivity, breakdown and other insulation characteristics. Therefore, the influence of temperature on the insulation properties of PP blends, especially the mechanism of nano graphene on the insulation properties of PP composites under different temperature was studied.

2 The Properties of PP/ULDPE Blends

2.1 PP/ULDPE Blends

In this section, polypropylene and ULDPE elastomer with different mass ratio are melt blended, so that polypropylene and elastomer can be heated above the melting temperature and fully mixed. PP was produced by Sinopec Yangzi Petrochemical Co., Ltd., model F401, isotactic polypropylene, and its isotactic degree is about 96%, melting temperature is about 165 °C, density is 0.9 g/cm^3. Vinyl elastomer (ULDPE) was produced by Dow Chemical Company, model Attane™ 4203, ethylene-1-octene copolymer, and its melting temperature is about 115 °C, density is 0.905 g/cm^3, solution flow rate is 0.8 g/10 min.

Through choosing the flat die with different slotting depth, we can get the samples with different thickness for different experiments. Among them, the thickness of the specimen for morphology observation is about 300 μm, the thickness of the specimen for thermal, mechanical, electrical conductivity and space charge test is about 200 μm, and the thickness of the specimen for breakdown strength test is about 50 μm. For the convenience of description, PP, PU5, PU15 and PU30 are used to express pure PP, blends with elastomer content of 5wt%, 15wt% and 30wt%, respectively.

2.2 Morphology Observation of PP/ULDPE Blends

The mechanical toughening effect of elastomer on polypropylene is not only related to the content of elastomer, but also influenced by the dispersion of elastomer in polypropylene. In this section, SEM was used to observe the brittle section of PP/ULDPE blends, to observe the dispersion of elastomer in PP matrix, and to analyze its effect on the mechanical properties of PP/ULDPE blends. The test results are shown in Fig. 1. Figure 1a is pure PP, and Fig. 1b–d are PU5, PU15 and PU30 blends.

It can be seen from Fig. 1a that the cross section of pure PP is relatively flat and smooth with lamellar structure, which is a typical brittle fracture feature. The lamellar structure decreased with the addition of elastomer, and with the increase of elastomer content, the irregularity of cross section increased. As shown in Fig. 1b, for the PU5 sample, the particle size of ULDPE is relatively uniform and distributed in the polypropylene matrix as a ball. The ULDPE in PU15 is basically distributed as microsphere in the polypropylene matrix, the particle size is relatively uniform, the number of particles is large, and it is evenly inserted in the PP; the cross section of the sample is locally flat, the overall cross section is uneven, with tear marks, and the two-phase interface tends to be fuzzy. When the content of elastomer increased to 30wt%, some regions began to show ULDPE continuous strip distribution; the interface between polypropylene and elastomer was clear, and the dispersion was poor. The dispersion of different content of elastomer in PP matrix is quite different. Under the same blending condition, the elastomer is easier to be cut into smaller particles when the content is small, and the elastomer is not easy to be divided into smaller particles when the content is high, resulting in continuous distribution. According to the SEM test results, when the content of ULDPE elastomer is less

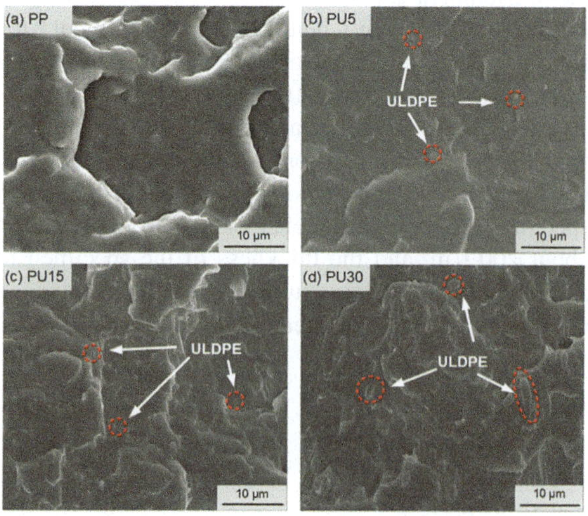

Fig. 1 SEM of brittle section of PP/ULDPE blends

than 15wt%, the dispersion is more uniform, and the interface between ULDPE and PP is fuzzy, which shows that they have good compatibility.

2.3 Melting and Crystallization Behavior of PP/ULDPE Blends

Polypropylene is a semi crystalline polymer, and its crystallization characteristics are closely related to its mechanical properties, trap distribution, charge transport characteristics, breakdown strength and other electrical properties. By studying the melting and crystallization behavior of polypropylene and its blends with elastomer, it is of great significance to analyze its mechanical and electrical properties. In this chapter, differential scanning calorimetry (DSC) was used to test the melting and crystallization behavior of polypropylene and its blends with elastomer. By analyzing the crystallization and melting curves, the important parameters such as crystallization, melting temperature and crystallinity of the blends can be obtained. Figures 2 and 3 are melting and crystallizing behavior of PP and PP/ULDPE blends, respectively.

As can be seen from Fig. 2, there is a single melting endothermic peak in pure polypropylene, with a peak temperature of 163.2 °C. The new melting peak appeared in the blends with the addition of the elastomer ULDPE, and the endothermic peak introduced by the elastomer was about 111 °C; the melting peak temperature of PP and the elastomer ULDPE in the blends increased first and then decreased with the increase of the elastomer content. With the increase of elastomer content, the area under the melting endothermic peak of ULDPE and that under the melting endothermic peak of PP decreased. It can be seen from Fig. 3 that there is a single crystallization exothermic peak in pure polypropylene with a peak temperature of

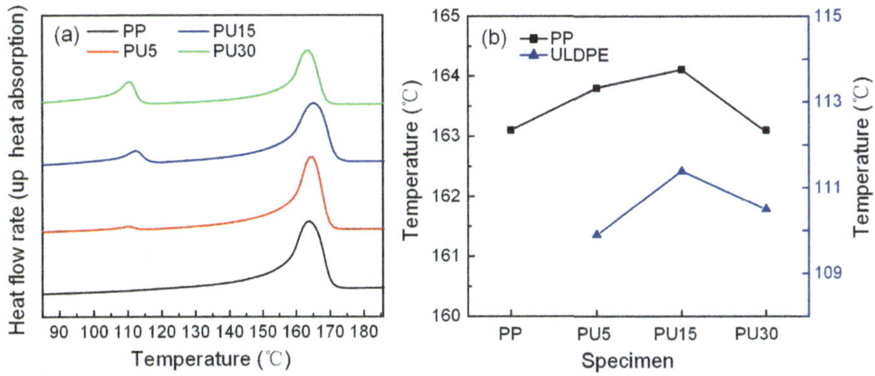

Fig. 2 Melting curve and melting temperature of PP/ULDPE blends

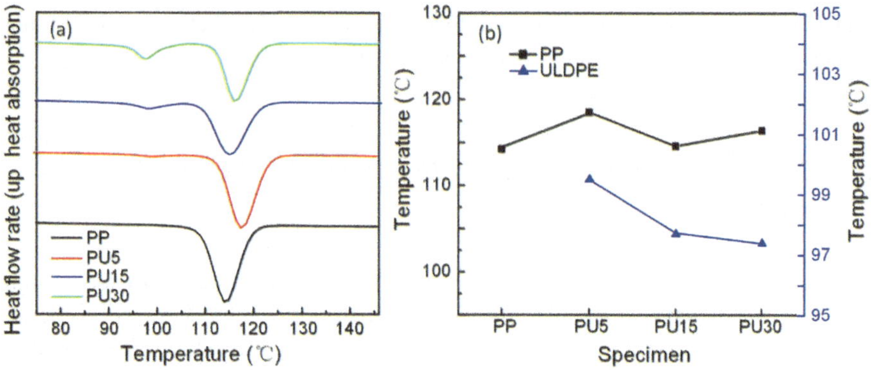

Fig. 3 Crystallizing curve and crystallization temperature of PP/ULDPE blends

114 °C. With the increase of the content of the elastomer in the blends, the crystallization peak temperature of the blends decreased slightly, and the crystallization temperature of PP increased first and then decreased.

The introduction of the elastomer changed the melting temperature and crystallization temperature of the two phases in the blend in a small range. The change of the microstructure of the two phases of PP and ULDPE in the blend can be further analyzed through the crystallinity. The calculation results of the crystallinity of blends with different elastomer ratios are shown in Fig. 4.

It can be seen from Fig. 4 that with the increase of ULDPE in the blend, the crystallinity of polypropylene tends to decrease, while that of elastomer increases. The addition of ULDPE hinders the growth of polypropylene crystal. Although ULDPE

Fig. 4 Crystallinity of polypropylene and elastomer in different blends

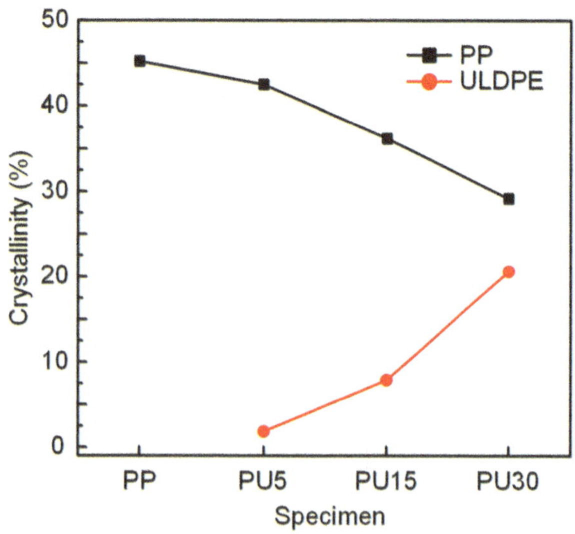

cannot penetrate into the crystal lattice of polypropylene, it destroys the regularity of the molecular chain of polypropylene. This reduces the ordered arrangement of the molecular chain into the crystal lattice, destroys the growth of polypropylene crystal, and reduces the proportion of the crystalline area of polypropylene. At the same time, due to the increase of the proportion of elastomer and the decrease of uniform dispersion effect, the situation of continuous distribution of local ULDPE appears, which makes the proportion of crystallizable part in elastomer increase. In general, the introduction of the elastomer ULDPE does not change the melting point of polypropylene, and the melting point of the introduced ULDPE is also around 110 °C. The higher melting point of the blend is of great significance to the demand of high-voltage and large capacity DC cables for high-temperature operation.

2.4 Mechanical Properties of PP/ULDPE Blends

The mechanical property is an important index to evaluate the development of cable insulation. The mechanical toughness of polypropylene can be improved by using elastomer. In order to find a suitable formula for PP/ULDPE blends with good mechanical properties, the tensile tests of pure PP and PP/ULDPE blends with different elastomer contents were carried out. The stress-strain curve of polymer material in tensile test can reflect the resistance of material to deformation and energy absorption.

The dynamic stress-strain curve in the process of drawing is shown in Fig. 5. Figure 6 shows the calculated values of elongation at break and tensile strength of polypropylene and its blends. As shown in Figs. 5 and 6, with the increase of

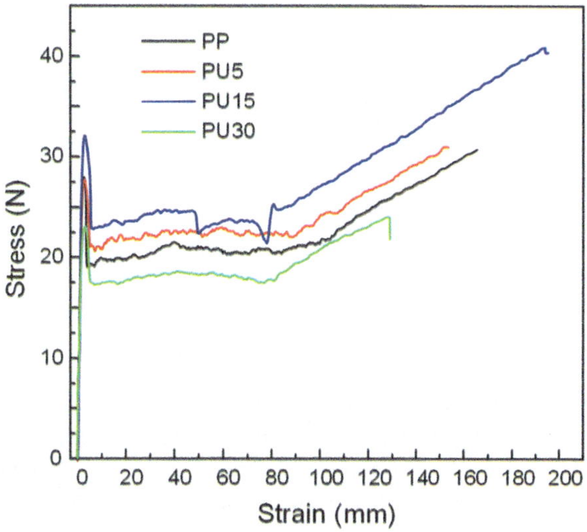

Fig. 5 Stress strain curve of pure PP and PP/ULDPE blends

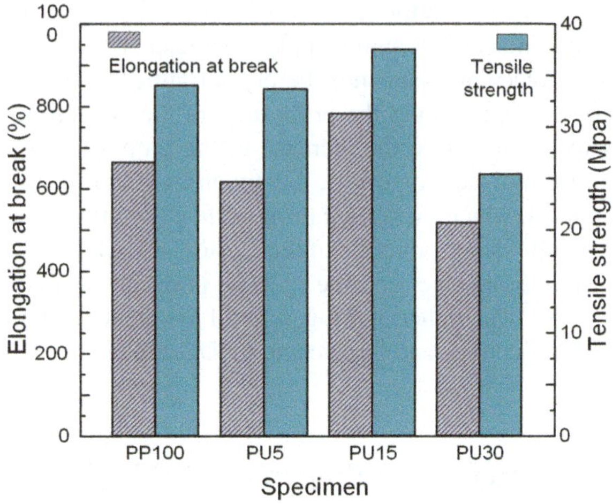

Fig. 6 Elongation at break and tensile strength of pure PP and PP/ULDPE blends

ULDPE content, the tensile strength of PP/ULDPE blends first increased slightly and then decreased. Compared with pure PP, the elongation at break of PP/ULDPE blends decreased slightly when the content of ULDPE was lower (5wt%) and higher (30wt%). When the content of ULDPE was 15 wt%, the elongation at break and tensile strength reached the maximum.

The properties of blends are closely related to processing, crystallization, flexibility of molecular chain, morphology of blends and other factors [10]. As shown by the SEM results in Fig. 1, the particle size of ULDPE as a dispersed phase increases with the increase of the amount of ULDPE added. Polypropylene is a semi crystalline polymer, in which crystalline phase (mostly spherulite) and amorphous phase coexist. From the DSC results in Figs. 2 and 3, it can be seen that the crystallinity of PP in the blend decreases with the increase of elastomer content. According to the craze theory, the average particle size of the dispersed phase should be controlled so that the blends of the two-phase system can achieve better mechanical properties. If the particle size of the elastomer in the matrix is too small, it will be buried in the craze, so that the elastomer cannot be branched and play the role of toughening; if the particle size of the elastomer is too large, it cannot meet the acceleration required for craze expansion in the plastic matrix to achieve toughening. The elongation at break of the blend may increase first and then decrease with the increase of the elastomer content. With the increase of ULDPE content in the blends, the tensile strength of PP/ULDPE blends first increased and then decreased. This is mainly due to the addition of a certain amount of elastomer ULDPE, which destroys the high crystallinity of polypropylene, reduces its crystallinity, and refines the spherulite structure of PP. This is conducive to the uniform distribution and relaxation of stress, thus improving the impact strength of the blend. When the content of elastomer is high, the uneven degree of dispersion of the elastomer increases, and the particle size of the elastomer becomes larger and continuous phase appears Distribution, resulting in a decrease in tensile strength.

2.5 *Electrical Properties of PP/ULDPE Blends*

2.5.1 Dielectric Properties of the Blends

Most of the polymers have the problem of thermodynamic incompatibility, so the morphology of polymer blends is mostly polyphase structure. Dielectric properties are not the primary concern of DC cable insulation, but can reflect the microstructure and properties of polymer insulation to a certain extent, such as the compatibility of blends. The dielectric constant and dielectric loss tangent of PP and PP/ULDPE blends are shown in Fig. 7.

There is no polar group on the molecular chain of polypropylene, and the polarization mode is mainly electronic polarization, which belongs to non-polar polymer. As shown in Fig. 7a, the dielectric constant of polypropylene is around 2.1. With the increase of elastomer content, the dielectric constant and dielectric loss of the blends increased. When the content of ULDPE is 30wt%, the dielectric loss increases obviously, which corresponds to the results of SEM and DSC. As a dispersed phase, the smaller the particle size of elastomer, the more evenly dispersed in the continuous phase of polypropylene, the better the adhesion and compatibility between the two phases, and the smaller the dielectric constant and dielectric loss of the blend system. When the content of elastomer is less than 15wt%, the dielectric constant and

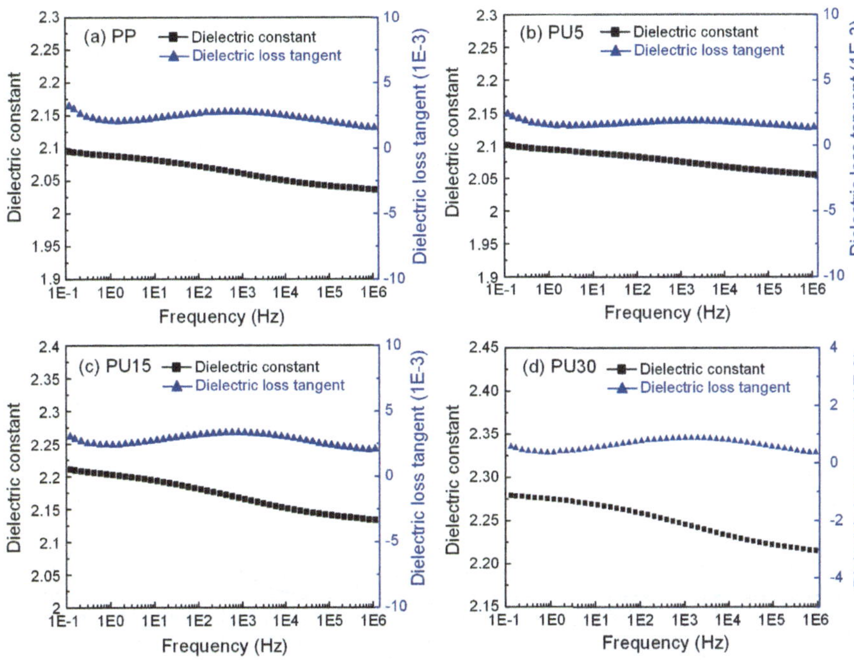

Fig. 7 Dielectric constant and dielectric loss tangent of PP and PP/ULDPE blends

dielectric loss of the blends increase slightly, which indicates that the compatibility between PP and elastomer ULDPE is good.

2.5.2 Trap Distribution Characteristics of PP/ULDPE Blends

The molecular branch, the interface between amorphous region and crystal region, the structural defects and impurities in insulating medium may introduce the local state energy level, i.e. trap, which has binding effect on electrons and holes. The trapping of electrons or holes in solid insulating materials will reduce carrier mobility and conductivity, and the trapped charge is easy to accumulate to form space charge, which will affect the distribution of electric field, breakdown strength, aging and other characteristics in insulation. The charge transport process in dielectrics is accompanied by charge debonding and trapping, which is closely related to the characteristics of traps in materials, such as trap energy level and trap density. After adding elastomer to PP, the internal structure of PP was changed, and the trap distribution of PP blends was changed.

The distribution curve of trap characteristics of polypropylene and its blends is shown in Fig. 8. Figure 8a shows the isotherm discharge current curve, and Fig. 8b shows the trap energy level distribution of polypropylene and its blends.

It can be seen from Fig. 8b that the trap energy level measured by the isothermal discharge current method is narrow. In this section, it varies from 0.75 eV to 1.1 eV, but the curve of experimental results can still reflect a certain trend of change. In this paper, the traps below 1.0 eV are called shallow traps, while the traps above 1.0 eV are called deep traps. Impurities, end groups and branched chains exist in the interface between crystalline phase and amorphous phase in semi crystalline polymers such as polypropylene, which is considered as one of the main sources of traps [11]. It can be seen from the figure that there are two different trap energy levels in pure polypropylene and its blends. The trap energy levels of the two peaks are about

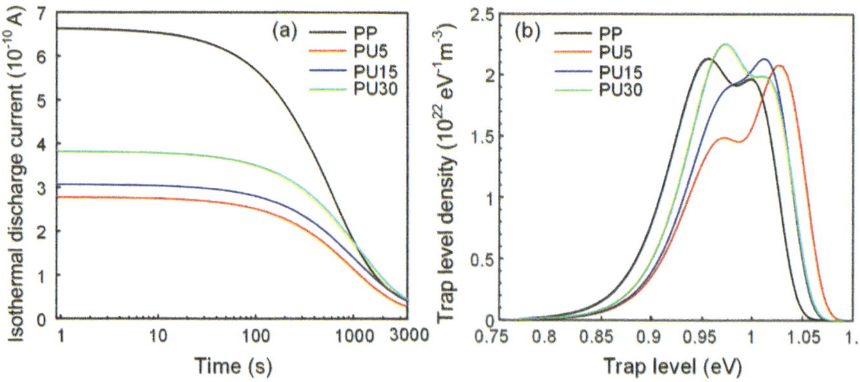

Fig. 8 **a** isothermal discharge current curve and **b** trap distribution characteristics of polypropylene and its blends

0.95 eV and 1.02 eV respectively. Due to a large number of crystalline and amorphous interfaces in polypropylene, the higher crystallinity makes the free volume between spherulites smaller, and has stronger blocking effect on charge transport, which makes a large number of traps exist in polypropylene samples. After the introduction of elastomer, the peak energy levels of deep trap and shallow trap in the blends are higher than that of pure polypropylene. In the blends, the peak value of deep trap energy level of PU15 is slightly lower than that of PU5, but the energy density of deep trap is higher than that of PU5; when the content of ULDPE is 30wt%, the shallow trap energy level and density of PU30 are higher than that of PU5 and PU15, but the deep trap is less. Compared with pure polypropylene, it can be seen from the morphology and DSC test results of the blends in Sects. 2.1 and 2.2 above that with the increase of elastomer content, the crystallinity of PP in the blends decreases. This is because the introduction of elastomer into the blends destroys the original structure of the internal crystalline and amorphous regions of PP, and the introduction of PP/ULDPE interface changes the trap characteristics of the blends. When the elastomer content is low, such as PU5 and PU15, the dispersion, particle uniformity and compatibility with polypropylene are good. At this time, the trap formed at the interface of the two phases has a strong binding effect on the charge; when the elastomer content is 30wt%, there is a non-uniform continuous phase distribution in the elastomer, which reduces the degree of tight binding between polypropylene and elastomer, and forms more shallow traps with weak charge binding.

2.5.3 Space Charge Characteristics of PP/ULDPE Blends

Space charge characteristics of insulating materials are important parameters in insulation design of HVDC cables. At present, the pulse electro acoustic (PEA) method is one of the main methods for nondestructive measurement of space charge distribution in solid media, which has the characteristics of high spatial resolution and simple operation method. Therefore, PEA method is used to measure the space charge characteristics of different polypropylene composites.

The test results of charge distribution in polypropylene and its blends during polarization and depolarization are shown in Figs. 9 and 10 respectively.

The test results of space charge distribution in polypropylene and its blends during polarization are shown in Fig. 9. It can be seen from Fig. 9a that with the increase of pressure time, a small amount of heteropolar space charge appears near the two electrodes in the polypropylene sample, and the amount of space charge injected into the material is small. It can be seen from Fig. 9b that, compared with pure polypropylene, there is almost no heteropolar charge accumulation on both sides of PU5 sample after 3600 s of pressurization, but there is a small amount of homopolar space charge injection, and a small amount of negative space charge is accumulated inside the sample. It can be seen from Fig. 9c that for sample PU15, more negative space charge is injected into the sample from the cathode, and less space charge is injected into the anode side. With the increase of pressurization time, the negative charge is transferred to the sample, and the space charge accumulation is increasing.

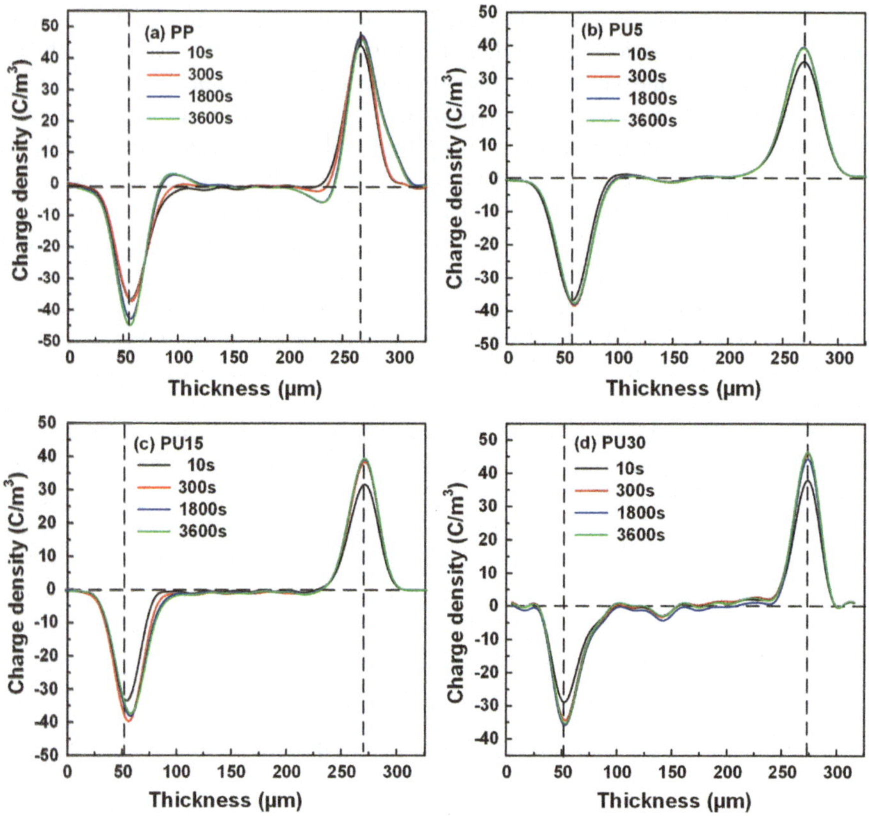

Fig. 9 Space charge distribution of PP and its blends during polarization

It can be seen from Fig. 9d that a large number of negative charges are injected and accumulated on the cathode side of sample PU30, and more space charges of the same polarity are also injected on the anode side. With the passage of time, the injected negative charge continuously migrates to the inside of the sample, and a large amount of negative space charge accumulates in the sample.

Figure 10 shows the test results of space charge distribution in polypropylene and its blends during depolarization (short circuit). As shown in Fig. 10a, the space charge in pure PP sample cannot be released instantly at the beginning of short circuit operation, and a few charges may accumulate near the electrode in the sample, and decrease with time, but dissipate slowly. It can be seen from Fig. 10b that a small amount of negative space charge accumulated in PU5 at the initial stage of short circuit and dissipated rapidly with depolarization. It can be seen from Fig. 10c that after 3600 s polarization, the space charge accumulated in sample PU15 is mainly negative polarity, which is distributed near the cathode and inside the sample. After depolarization for 600 s, there is still a small amount of charge in the sample which is not completely dissipated. It can be seen from Fig. 10d that the space charge

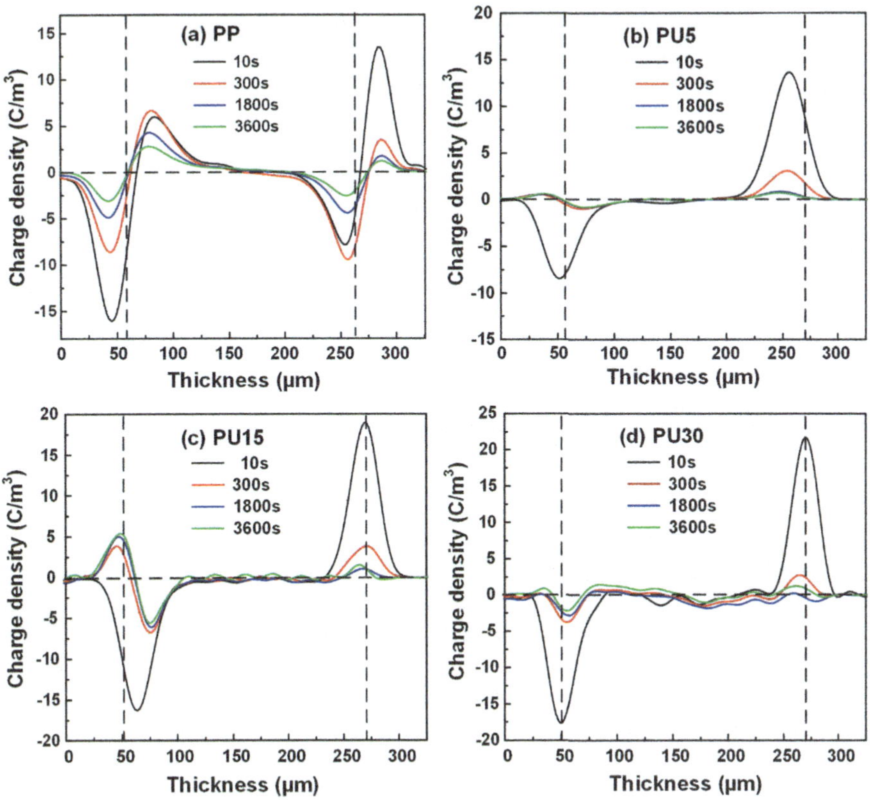

Fig. 10 Space charge distribution of PP and its blends during depolarization

accumulated in the sample PU30 is mainly negative, and the negative space charge near the cathode keeps decreasing with the depolarization time; depolarization 600 At s, the positive space charge appears near the cathode, which may be due to the fact that the space charge accumulated in the sample is difficult to dissipate when trapped by the deep trap, and the positive charge of different polarity is induced by the cathode when it is short circuited.

From the qualitative analysis of the above experimental results, it can be seen that different elastomer contents in polypropylene blends have a great influence on the space charge injection and dissipation characteristics of the blends in polarization and depolarization process. Different samples are polarized for 10 s, 300 s, 1800 s, 3600 s and depolarized for 10 s, 30 s, 300 s, 600 s. The average space charge density changes with time are as shown in Fig. 11a, b, respectively.

It can be seen from Fig. 11a that at the same polarization time point, the average space charge density of blends is higher than that of pure polypropylene samples, and the average space charge density of blends increases with the increase of elastomer content. For pure polypropylene samples, the average space charge density is less, and

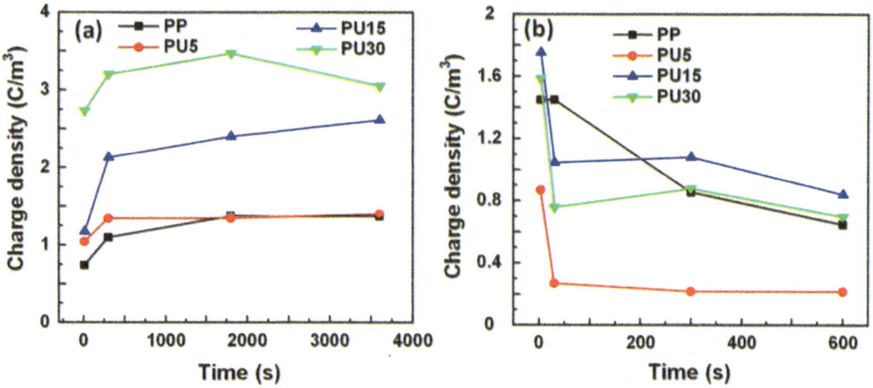

Fig. 11 Average space charge density in the sample during polarization and depolarization

it is almost in equilibrium at 1800 s. Compared with pure polypropylene, the average charge density of PU5 increases rapidly and is larger than that of polypropylene in the first 300 s of polarization. After 300 s of polarization, the dynamic equilibrium state of PU5 is basically the same as that of pure polypropylene. The average charge density of the blends PU15 and PU30 in the same time is higher than that of PU5 and PP, and the charge density of PU30 is higher. Because the thickness of the samples used in the space charge test is basically the same, it shows that more space charges are accumulated in PU30 samples in the same time.

Figure 11b shows the change of average charge density of different samples during depolarization. It can be seen from the figure that the average charge density of all samples decreases rapidly in the first 30 s with the passage of short circuit time. Although the change of space charge density from the beginning of the short circuit to the first 10 s cannot be shown, the trend of the change of average charge density of different samples and the binding effect on the charge can be preliminarily judged from the figure. It can be seen from Fig. 11a that the charge density of blends after 3600 s polarization is greater than or equal to that of pure polypropylene, and the average charge density of blends decreases rapidly within 30 s of initial depolarization, but it shows that the binding capacity of blends to charge is lower than that of pure polypropylene, or the charge mobility in blends is greater. Although the charge density of the blends is higher than that of pure polypropylene after 3600 s polarization, the charge density of the blends is lower or slightly higher than that of pure polypropylene after 600 s depolarization due to the faster charge transfer rate in the blends.

2.5.4 Conductivity of PP/ULDPE Blends

Under DC electric field, the conductivity of cable insulation is an important factor affecting the distribution of electric field in insulation, which is closely related to

Joule heat, loss and insulation temperature generated by insulation during cable operation; in addition, the conductivity characteristic of cable insulation material is the primary factor to be considered in DC cable design and the key material control parameter in DC cable national standard.

The conductivity of polypropylene and its blends are shown in Fig. 12.

It can be seen from the results that the introduction of elastomer makes the conductivity of PP blends larger than that of pure PP. Compared with pure polypropylene, when the elastomer content of the blend is lower than 15wt%, the conductivity of the blend decreases slightly; when the elastomer content is 30wt%, the conductivity of the blend increases to about 7 times of pure polypropylene. Polypropylene is a kind of semi crystalline polymer with a high degree of crystallinity. There are a large number of crystal and amorphous interfaces inside. Impurities, end groups and branched chains at the interface can form traps and bound the movement of charge. After the introduction of elastomer, the original crystal structure of polypropylene was destroyed, and the internal trap characteristics were changed. The trap energy level and density were decreased, the binding effect on the charge was weakened, the carrier mobility was increased, and the conductivity was increased. Compared with the case of higher elastomer content, when the particle size of the elastomer in the blend is smaller and the dispersion is more uniform, the combination of the two phases is closer and the free travel is shorter. At this time, the interface composed of elastomer and polypropylene can form traps, which can capture the charge and restrict the transport of carriers, making the conductivity smaller. When the content of elastomer is high, due to the uneven dispersion of elastomer, the part of elastomer is distributed in continuous phase. The interface between polypropylene and elastomer is loose and the free travel is longer. The charge is more easily transferred at the interface of two phases, which makes the conductivity larger. Therefore, the

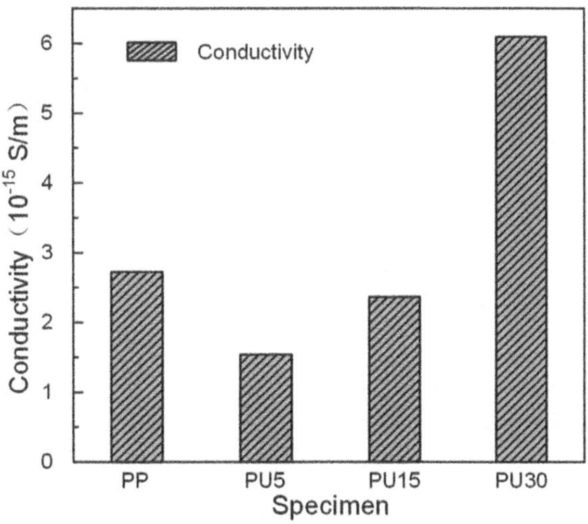

Fig. 12 Conductivity of PP and PP/ULDPE blends

content of elastomer should be in a certain range to obtain better particle size and dispersion, which is helpful to reduce the conductivity of the blends.

2.5.5 Breakdown Strength of PP/ULDPE Blends

The breakdown strength is an important parameter to evaluate the insulation performance of DC cable. In order to accurately evaluate the breakdown resistance of PP composite insulation materials and fully expose the defects of the insulation materials, the cylindrical electrode with larger contact area with the sample is selected in this chapter for DC breakdown strength test. Based on the above experimental steps and data processing methods, the statistical results of breakdown strength of polypropylene and its blends are shown in Fig. 13.

It can be seen from Fig. 13 that the breakdown strength of pure polypropylene insulation material is relatively high. After the introduction of the elastomer ULDPE, the breakdown strength of the blend decreases with the increase of the elastomer content, and the higher the elastomer content, the more obvious the breakdown strength decreases. The change of the free volume in the specimen may also lead to the decrease of the breakdown strength. From the results of DSC test, the addition of elastomer lead the decrease of crystallinity, which will cause the increase of free volume, and further increase the free volume of electrons. Hence, more energy can be obtained in the process of electronic movement. The damage of polymer molecular chain will be more serious when it is impacted by electrons, thus reducing the breakdown strength. It can be seen from the results of space charge test that more space charge is accumulated in the blends than that in the pure polypropylene samples, and the electric field distortion becomes serious, which can also lead to the decrease of breakdown strength. In addition, according to the trap characteristics of

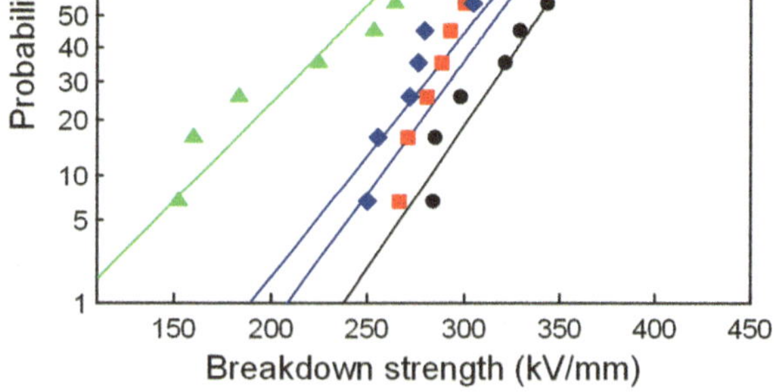

Fig. 13 Breakdown strength distribution of PP and PP/ULDPE blends

polypropylene and its blends, when the elastomer content is low, the trap energy level and depth of polypropylene are smaller than that of the blends, which have strong binding effect on carrier migration, and the conductivity is smaller. While the deep trap energy level and density of the blends decrease with the increase of the elastomer content, and the binding effect on the charge is weaker, the conductivity is larger, and the breakdown strength is lower. Therefore, when the content of elastomer is small, the decrease of breakdown property is small.

3 The Insulation Properties of Nano Graphene Modified PP/ULDPE Blends

For the high voltage direct current polypropylene cable insulation material, on the basis of improving its mechanical toughness, we should focus on the insulation performance of the composite material. From the previous experimental results and analysis, it can be seen that the PP blended with 15wt% elastomer ULDPE has excellent comprehensive properties and its mechanical toughness has been effectively improved. But the insulation performance of the blend has also declined to a certain extent, which still needs further modification. Compared with traditional nano MgO, ZnO and other fillers, nano-graphene has high electron affinity, large specific surface area and prominent nano surface and interface effects, and has great potential in improving the insulation performance of nano-sized dielectric. In addition, the conductor will bear different load current and heat during the operation of the cable, which will lead to the temperature change of the insulation layer, and the temperature effect will affect the space charge, conductivity, breakdown strength of the polymer. Therefore, the influence of temperature on the insulation properties of PP blends, especially the mechanism of nano graphene on the insulation properties of PP composites under different temperatures, needs to be further studied.

In this section, the PP/ULDPE/graphene nanocomposites are prepared. The dispersion of nano graphene in polymer matrix was observed, and the melting and crystallization behavior of the nanocomposites was analyzed by DSC. The influence of different content of nano graphene on the space charge, conductivity and breakdown strength of PP/ULDPE/graphene nanocomposites at 30, 60 and 90 °C were studied. The trap distribution characteristics of composites based on IDC method are analyzed.

3.1 Dispersion of Nano Graphene in PP/ULDPE Blends

The polypropylene PP and elastomer ULDPE materials used in this section are the same as those in the previous section. The nano graphene used is produced by Suzhou carbon rich graphene Technology Co., Ltd. The graphene monolayer (prepared by

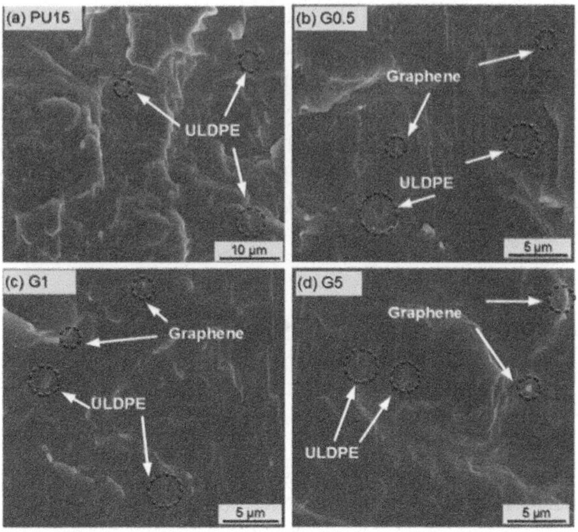

Fig. 14 SEM of PP/ULDPE/graphene nanocomposites

physical stripping method) has a purity of more than 99wt%, a thickness of about 1 nm, a diameter of about 0.2–10 um, a number of 1–2 layers and a specific surface area of 1000–1217 m^2/g. PP/ULDPE/graphene nanocomposites were prepared by blending PU15 with nano graphene content of 0.005wt%, 0.01wt% and 0.05wt% respectively, which are called G0.5, G1 and G5 respectively.

The properties of nano dielectric materials are closely related to the dispersion state of nano particles in the matrix. In this paper, the brittle section of PP/ULDPE/graphene nanocomposites is observed by scanning electron microscope to observe the dispersion of elastomer and nano graphene in the polypropylene matrix. The equipment and test method adopted are the same as those in the previous section, and the test results are shown in Fig. 14. From Fig. 14a, i.e. SEM of the cross-section of the blend PU15, it can be seen that ULDPE, as an elastomer, is dispersed in the polypropylene matrix in an irregular microsphere shape, with relatively uniform particle size. The cross-section of the sample is locally flat, the overall cross-section is uneven, and the two-phase interface tends to be fuzzy. Although the amount of nano graphene added is small, the basic distribution of nano particles can be seen from the section SEM of Fig. 14b–d, that is, PP/ULDPE/graphene nanocomposites G0.5, G1 and G5. In the samples containing graphene, the shape of ULDPE is spherical or elliptical, and the distribution of graphene particles is tiny. It can be seen from Fig. 14b, c that when the amount of nano graphene added to the blend is low, the nano particles are more evenly dispersed and there is almost no serious agglomeration. When the amount of nano graphene is increased to 0.05wt%, that is, sample G5, the agglomeration of some nano graphene can be observed in the matrix.

3.2 Space Charge Characteristics of PP/ULDPE/Graphene Nanocomposites

In order to study the effect of different amount and temperature on the space charge characteristics of PP/ULDPE/graphene nanocomposites, PEA measuring equipment and method are used in this section. The polarization electric field of all samples is 60 kV/mm, the polarization time is 3600 s, and the short-circuit depolarization time is 600 s. In this section, the color map of a certain color is used to represent the distribution change of space charge in the nanocomposites. The polarity and amount of charge corresponding to different colors are shown in the legend. The vertical axis of each graph represents the thickness of the sample, the horizontal axis represents the time, and the white dotted line represents the interface position between the electrode and the sample. The change of space charge with polarization time can be used to characterize the transport process of charge injection and transfer. The space charge distribution test results of different samples during polarization and depolarization at 30 °C, 60 °C and 90 °C are shown in Figs. 15 and 16.

The space charge distribution of different samples in the polarization process at different temperatures is shown in Fig. 15. It can be seen from Fig. 15a-1 that after polypropylene is polarized at 30 °C for 3600 s, only a small amount of heteropolar charge appears near the electrode of the sample, which may be caused by polarization or ionization of a small amount of impurities in the sample under high electric field. As shown in figure (a-2), when the test temperature rises to 60 °C, the homopolar charge injected from the two electrodes increases, and a small amount of negative charge accumulates inside the sample. When the temperature reaches 90° C, as shown in figure (a-3), the homopolar charge injected from the two electrodes intensifies, and a significant space charge injection occurs 15 min after polarization; then the homopolar charge injected slowly drops and reaches a relatively stable state, and still maintains a relatively high charge injection amount at the end of the test. Figures 15b-1–b-3 show the space charge changes of pu15 during polarization at 30, 60 and 90 °C. Compared with PP at the same test temperature, the space charge behavior of pu15 shows different characteristics. As shown in figure (b-1), a small amount of negative charge is injected into the sample from the cathode at 30 °C and increases slowly over time. For PU15 at 60 °C, a small amount of heteropolar charge was accumulated near the two electrodes. As shown in figure (b-3), when the test temperature reaches 90 °C, the negative charge accumulated near the anode in the sample is dominant, and only a small amount of positive charge accumulated near the cathode in the sample.

As shown in Fig. 15c-1–c-3, the space charge distribution of the PP/ULDPE blend with 0.005wt% nano graphene modified nanocomposite G0.5 at different temperatures is shown. It can be seen from the figure that compared with PU15, the space charge accumulation at three temperatures of G0.5 has been significantly improved, among which the same polarity space charge injected by two electrodes at 30 °C, the different polarity charge accumulation near the cathode in the sample at 60 °C, especially the different polarity charge accumulation near the anode in the sample at 90 °C has the most significant change. Figures 15d-1–d-3 show the space charge

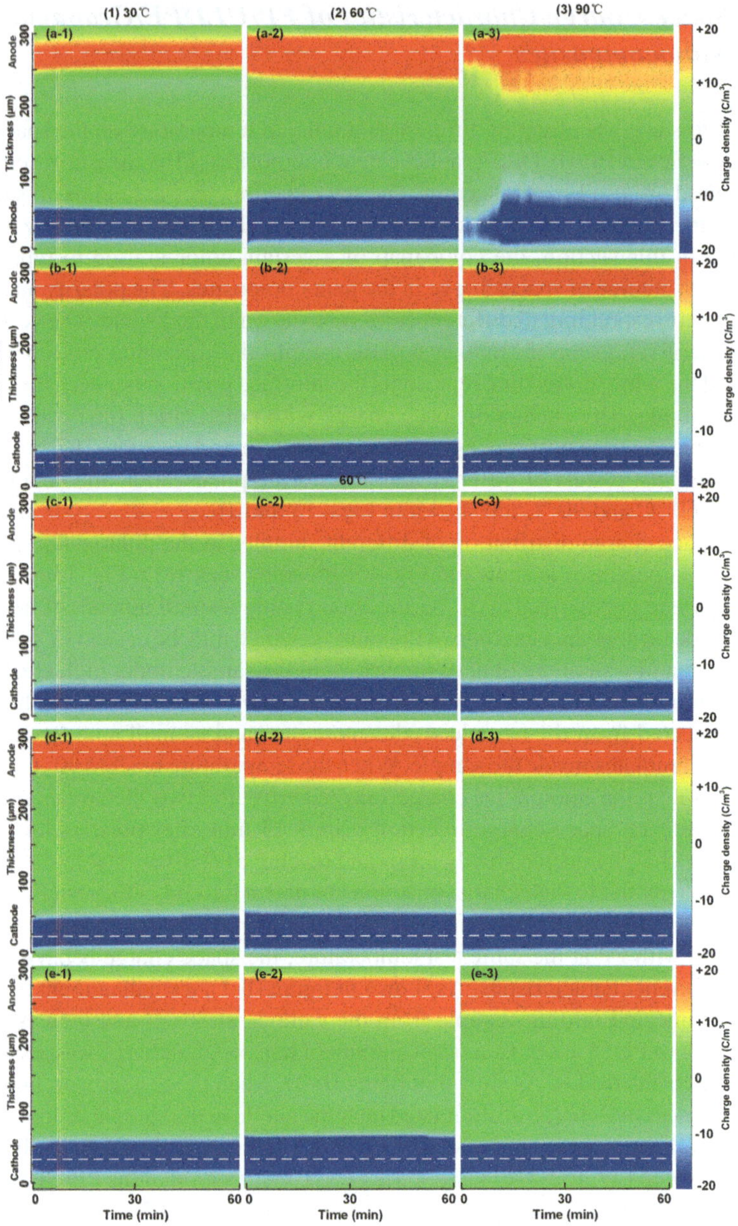

Fig. 15 Space charge change of different samples during polarization

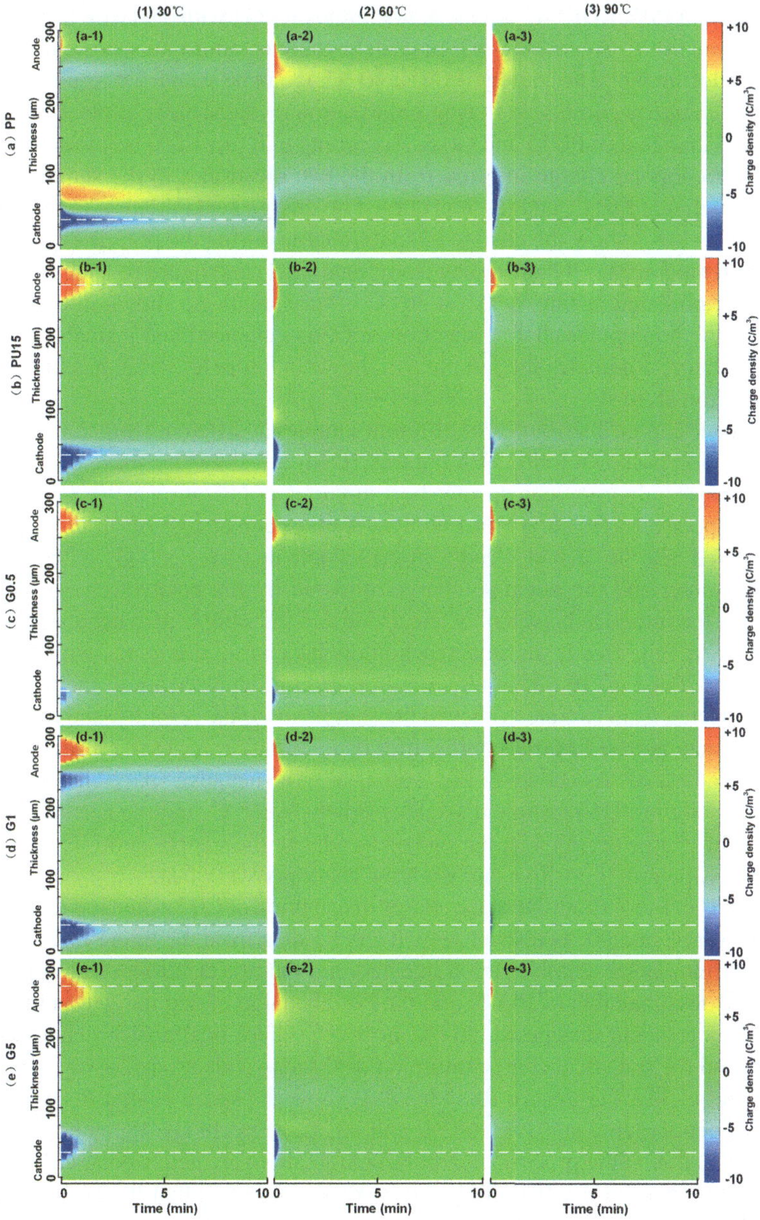

Fig. 16 Space charge change of different samples during depolarization

distribution of PP/ULDPE/graphene nanocomposites G1 with 0.01wt% nano content at 30, 60 and 90 °C. According to figure (d-1), there is almost no space charge accumulation during the polarization of G1 at 30 °C. It can be seen from figure (d-2) that the space charge accumulated during polarization of the sample at 60° C is significantly less than that of PU15 and G0.5. According to figure (d-3) below at 90 °C, the depth of homopolar charge injected from the two electrodes in the sample is shallower than that at 60 °C, and there is almost no charge accumulation in the sample. Figures 15e-1–e-3 show the change of space charge during the polarization of G5 at different temperatures. It can be seen from figure (e-1) and figure (e-1) that the space charge accumulated in sample G5 at 30 °C is very small. As shown in figure (e-2), at 60 °C, the homopolar charge injection in G5 is enhanced and a small amount of negative charge dominates the sample. As shown in figure (e-3), G5 has little charge accumulation at 90 °C.

The space charge distribution of different samples at different temperatures during depolarization is shown in Fig. 16. Figures 16a-1–a-3 show the space charge distribution of pure polypropylene during short-circuit depolarization at 30 °C, 60 °C and 90 °C, respectively. As shown in Fig. 16a-1, the space charge of different polarity accumulated near the two electrodes in the sample decreases slowly with the depolarization time, and the negative charge attracted by the positive charge near the cathode at the interface between the electrode and the sample also decreases slowly. As shown in Fig. 16a-2, the injected homopolar charge dissipates with the short circuit time. For Fig. 16a-3, in 60 s of short-circuit depolarization, the homopolar charge injected from two electrodes in the sample dissipates rapidly, and the amount of charge accumulated in the sample after depolarization for 600 s is less than that at 30 °C and 60 °C. As shown in Fig. 16b-1–b-3, the negative space charge injected by the cathode in PU15 sample dissipates slowly at 30° C, and the accumulation of heteropolar charge occurs near the two electrodes in the sample at 60 °C and 90 °C, and the dissipation of positive and negative charges is very fast.

Figure 16c-1–c-3 show the space charge distribution of G0.5 when it is depolarized at 30 °C, 60 °C and 90 °C respectively. It can be seen from the figure that the initial charge of homopolar depolarization injected into the two electrodes in sample G0.5 at 30 °C is less than that of PU15, and the initial charge of depolarization in sample at 60 °C is less and dissipated slowly, at 90 °C, there is almost no charge in the sample, and the statistical charge initiation near the two electrodes is very small and dissipates quickly. Figures 16 (d-1) to (d-3) show the space charge distribution of G1 when depolarizing at 30 °C, 60 °C and 90 °C, respectively. It can be seen from Figure (d-1) that in the process of depolarization, there are heteropolar charges near the two electrodes in the sample. It can be seen from (d-2) that the positive charge injected into the anode of the sample dissipates slowly with the depolarization time, the charge near the cathode dissipates rapidly, and there is no charge accumulation in the middle of the sample. It can be seen from (d-3) that the charge accumulation at both ends of the sample is very small at the beginning of depolarization, and dissipates rapidly with depolarization time. As shown in Fig. 16e-1–e-3, the initial depolarization charge at both ends of G5 sample is more than G1 at 30 °C, and the dissipation is faster than G1. As shown in Fig. 16e-2, the charge dissipates rapidly

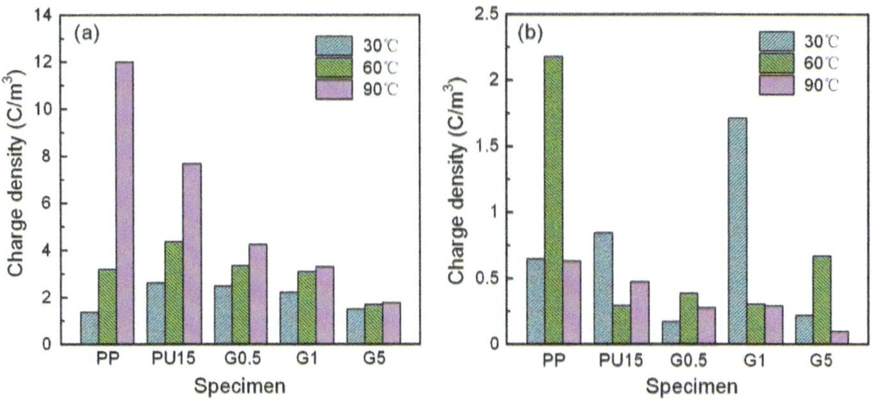

Fig. 17 Average charge density of different samples after **a** 3600 s polarization and **b** 600 s depolarization at 30, 60 and 90 °C

in the sample during depolarization at 60 °C It can be seen from Fig. 16e-3 that at 90 °C, the charge accumulation at both ends of the sample is less at the initial stage of depolarization than that at the same temperature G1, and there is almost no charge in the sample after depolarization for 10 min.

It can be seen from the above experimental results that the space charge characteristics of polypropylene and its blends are quite different at different temperatures. When nano graphene is introduced, the different content of nano particles has a great influence on the space charge injection and dissipation characteristics of PP/ULDPE blends in the process of polarization and depolarization; at the same time, temperature plays an important role in the evolution of space charge characteristics of PP/ULDPE/graphene nanocomposites. The calculation results of the average charge density of different samples after polarizing for 3600 s and depolarizing for 600 s at 30, 60 and 90° C are shown in Fig. 17.

A s shown in Fig. 17a, the average charge density of all samples increases with the increase of temperature after 3600 s polarization, and the average charge density of samples at 30 °C and 60 °C increases first and then decreases at the same temperature, but decreases at 90 °C. Among them, the average charge density of PU15 is lower than that of pure PP at 90 °C, and the charge density of pu15 is higher than that of pure PP at the same temperature. The average charge density of PP/ULDPE/graphene nanocomposites decreased with the increase of nano addition at the same temperature. It can be seen from Fig. 17b that the average charge density of different samples changes greatly after being poled for 600 s at three different temperatures. The average charge density of PP/ULDPE blends was higher than that of pure PP at 30 °C, but it decreased first, then increased and then decreased with the increase of nano graphene content. At 60 °C, the average charge density of PP was higher than that of other samples, and the charge density of the blends modified by nano graphene was higher than that of pure PP/ULDPE blends. Under the condition of depolarization at 90 °C, the charge density of different samples decreased.

3.3 Conductivity Characteristics of PP/ULDPE/Graphene Nanocomposites

The electric field distribution inside the insulation of HVDC cable is related to its conductivity, and it is greatly affected by the temperature. In this section, the conductivity characteristics of the samples at 30, 60 and 90 °C are tested. Figure 18 shows the relationship between the conductivity of polypropylene and its PP/ULDPE/graphene nanocomposites at 30, 60 and 90 °C.

It can be seen from Fig. 18 that the conductivity of all samples increases with the increase of temperature, and the change trend of the conductivity of samples from PP to G5 is basically the same at each same test temperature, among which the conductivity of PP/ULDPE blend PU15 is greater than that of pure PP, the conductivity of sample G1 is the smallest, and the conductivity of sample G5 is the largest. From the test results at different temperatures, it can be seen that the change range of the conductivity increase from 60 °C to 90 °C is larger than that from 30 °C to 60 °C, which shows that the high temperature has a greater impact on the conductivity

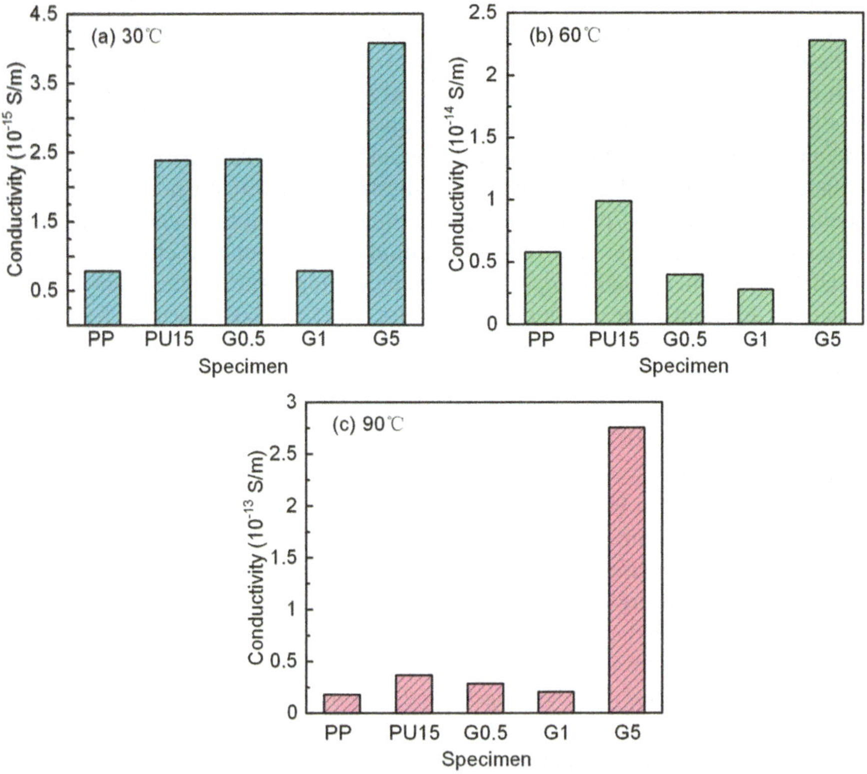

Fig. 18 Conductivity of polypropylene and its PP/ULDPE/graphene nanocomposites at **a** 30 °C, **b** 60 °C and **c** 90 °C

of the material. The conductivity of PP blend PU15 can be reduced by introducing 0.005wt% nano graphene, and the conductivity of G1 will be further decreased when the amount of nano particles is further increased to 0.01wt%. When the amount of nano particles is increased to 0.05wt%, the conductivity of PP/ULDPE/graphene nanocomposites G5 increased several times, and increased significantly with the increase of temperature. This indicated that the conductivity of PP blends could be reduced when the amount of nano graphene was added in a certain range, but the high amount of nano graphene would increase the conductivity of the blends.

3.4 Breakdown Strength of PP/ULDPE/Graphene Nanocomposites

The breakdown strength is an important parameter of the insulation performance of HVDC cables. In order to further study the DC breakdown characteristics of PP/ULDPE/graphene nanocomposites with different nano graphene additions and temperatures, the Weibull distribution of DC breakdown strength of PP/ULDPE/graphene nanocomposites at different temperatures is shown in Fig. 19.

It can be seen from Fig. 19 that with the increase of temperature, the breakdown strength of each sample shows a downward trend; at the same test temperature, the breakdown strength of sample G1 is higher than that of other samples, while the breakdown strength of sample G5 is the lowest. As shown in Fig. 19a, the breakdown strength of pure polypropylene is between 275 kV/mm and 375 kV/mm, and the breakdown strength of the blend PU15 decreases after the introduction of elastomer. When 0.005wt% nano graphene is added, the breakdown strength of PU15 increases to slightly lower than that of pure polypropylene. When the amount of nano graphene is increased to 0.01wt%, the breakdown strength of G1 was significantly improved. While the breakdown strength of G5 decreased significantly when the nano addition was further increased to 0.05wt%. The above experimental results show that the breakdown properties of the blends can be improved by adding a certain amount of nano graphene at different temperatures, but the dominant role of temperature is more obvious at high temperature.

3.5 Trap Characteristics of PP/ULDPE/Graphene Composite

According to the experimental results of space charge, conductivity and breakdown strength, adding a certain amount of nano graphene particles can effectively inhibit the space charge accumulation in PP/ULDPE blends, improve the electric field distribution in the samples, and improve the resistivity and breakdown strength of the blends. The difficulty and time of the trapped charge debonding are closely related to the trap energy level and density. The larger the trap energy level and trap density

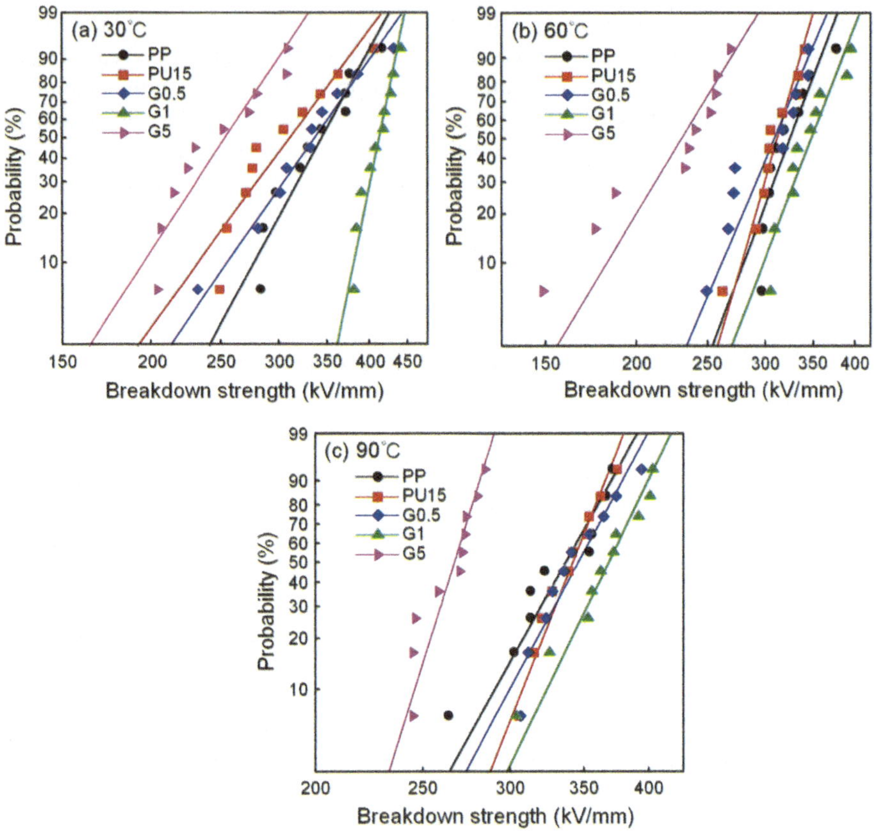

Fig. 19 DC breakdown strength of PP and PP/ULDPE/graphene nanocomposites at **a** 30 °C, **b** 60 °C and **c** 90 °C

is, the larger trap barrier the charge needs to overcome and take a long time to debond, and then become free charge again. It can be seen that the transport process of the charge injected by the electrode and the ion charge produced by the ionization of impurities and polar groups in the polymer under the action of strong field is closely related to the trap characteristics in the dielectric, and then the space charge, breakdown and conductivity characteristics of the nano dielectric [12].

Therefore, the influence of nano graphene on the internal trap distribution characteristics of PP/ULDPE blends is studied based on the IDC method. Figure 20a shows the isotherm discharge current curve of PP/ULDPE blend and its composite with different amount of nano graphene. Figure 20b shows the trap distribution characteristics of PP/ULDPE blend and its composite modified with different content of nano graphene.

It can be seen from Fig. 20a that the initial discharge current of pure polypropylene is the highest, indicating that a large amount of charge debonding takes part in carrier migration at the moment of discharge after 3000 s polarization. Because

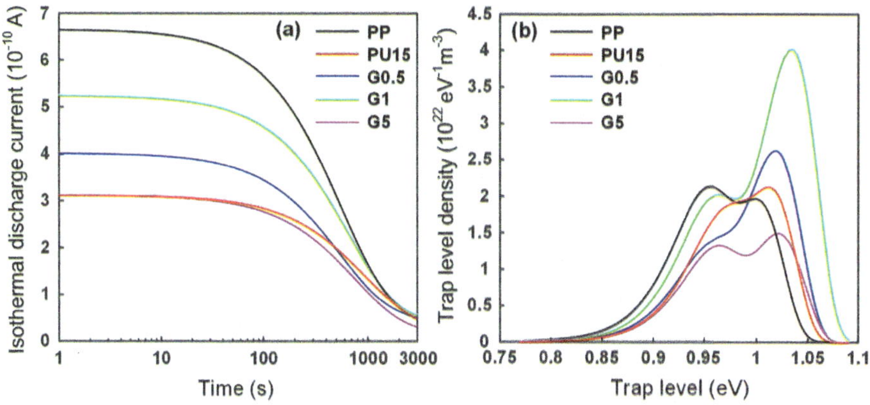

Fig. 20 a Isothermal discharge current curve and **b** Trap distribution characteristics of PP/ULDPE/graphene nanocomposites

the introduction of elastomer changes the original internal structure of polypropylene and forms different interfaces, the starting current value of PU15 is lower than that of pure polypropylene. The initial discharge current of PP/ULDPE/graphene nanocomposites did not change after adding 0.005wt% nano graphene particles, but the decreasing trend was different after discharge for about 100 s. When the content of nano graphene particles increased to 0.01wt%, the initial discharge current of the samples increased; when the nano content further increased to 0.05wt%, the initial value of discharge current of the sample decreased significantly. The results showed that a small amount of nano graphene changed the trap distribution in PU15.

It can be seen from Fig. 20b that the trap energy levels of all samples are mainly distributed between 0.8 and 1.1 eV. Due to the introduction of elastomer into the blend, the original internal crystalline and amorphous structure of polypropylene was destroyed, the crystallinity of PP in the blend decreased, and the PP/ULDPE interface was introduced, which changed the trap characteristics of PP, making the trap level and density of PU15 higher than that of pure PP. The trap level and density of PP blend PU15 increased obviously after adding nano graphene particles. With the increase of the mass fraction of nanoparticles from 0.005wt% to 0.01wt%, the trap level and density of the sample gradually increase; with the further increase of the nano content to 0.05wt%, the trap density and depth of the sample begin to decrease. A large number of deep traps were introduced into PP/ULDPE/graphene nanocomposites with low amount of nano particles, which enhanced the binding ability of the sample to the charge and inhibited the charge migration and dissipation in the matrix. While PP/ULDPE/graphene nanocomposites with high nano content had small trap depth and density, and the charge was easy to dissipate.

3.6 Mechanism of Nano Graphene Regulating Electrical Properties of PP/ULDPE Blends

Combined with the space charge distribution of LDPE/MgO nanocomposites under different electric fields and temperatures, it has been found that the deep traps generated by the induced dipoles of nano MgO particles can capture the carriers with positive and negative polarity, and inhibit the transport of carriers injected from the electrode [13]. Because the surface electron of graphene can move freely in the whole plane structure, under the action of electric field, the free electron will move along the direction of electric field to one end of graphene, and the other side of graphene will be positive, at this time, a dipole with strong force can be produced. It has been found that when an external electric field is applied, nano graphene polarizes, producing a dipole, and increases with the increase of the intensity of the external field [14].

In addition, according to Lewis [15] and Nelson [16] proposed the dielectric double-layer interface model. It can be seen that the surface of nanoparticles will gather charge under electric field. At the same time, due to the action of electric field polarization, the heteropolar charge will be generated outside the nanoparticles to form a shielding layer. Under the action of Coulomb force, the charge in the polymer matrix will migrate and diffuse to form a dispersion layer around the nanoparticles. The interface region produced by the dispersion layer has an important influence on the charge transport in the material, and then affects the electrical properties of the composite. Therefore, under the action of electric field, the strong local heteropolar potential distribution produced by the induced dipole of nano graphene has a strong binding effect on the positive and negative charge transfer near the nanoparticles, which may be an important reason for the introduction of deep traps in the interface area of nano graphene.

Therefore, when the amount of nano graphene is low, such as 0.005wt% in this chapter, a large number of well dispersed interfaces can be formed in PP/ULDPE blends. At the same time, due to the electric field polarization characteristics of nano graphene, it can produce strong attraction and binding effect on the carriers in the composite, and introduce deep traps. With the increase of the content of nanoparticles, the number of interface area between nanoparticles and matrix also increases, which makes the trap level and density gradually increase. That is to say, when the amount added in this paper is 0.01wt%, the introduction of a large number of traps can greatly improve the ability of charge transport regulation [17–19], which makes the space charge injection of PP/ULDPE/graphene composite reduce, the conductivity decreases, and the breakdown strength increases. When the dielectric double layers of adjacent particles overlap, the overlapped parts will form conductive paths under the action of external electric field. The conductivity of the dielectric bilayer is generally much larger than that of the polymer matrix. Therefore, when the nano content is further increased to 0.05wt%, the ability of the interaction area between graphene and polymer matrix to control the electric charge decreases due to the phenomenon of partial agglomeration and uneven dispersion of nano graphene in the blends. In addition, the dielectric double layers between adjacent particles of graphene with higher amount of addition begin to overlap, and even form a conductive path under the effect of external electric field, which leads to the increase of conductivity of the composite and the decreas of breakdown strength [20, 21].

Therefore, adding 0.01wt% nano graphene can improve the electrical properties of PP/ULDPE blends at different temperatures. However, due to the small amount of nano addition, the proportion of temperature effect increases and the nano modification effect has a downward trend with the increase of temperature.

4 Conclusion

In this chapter, the microstructure, melting and crystallization behavior, mechanical properties and electrical properties of PP/ULDPE blends with different elastomer content and PP/ULDPE/graphene nanocomposites with different nano graphene addition were studied. The main conclusions of this chapter are as follows:

Based on the mechanical properties of vinyl elastomer ULDPE modified polypropylene, it is found that the dispersion and particle size in polypropylene matrix are more uniform when the content of elastomer is low, which is conducive to two-phase compatibility, and the mechanical properties and thermal properties of the blends are better at this time. Two phase interfaces were introduced into the blends by elastomer. The deep traps of the blends become larger, the space charge accumulates a little, the conductivity decreases, and the breakdown strength decreases.

PP/ULDPE/graphene nanocomposites were prepared by melt blending. It was found that with the increase of temperature, homopolar charge injection increased, heteropolar charge accumulation occurred in the blends, breakdown strength and conductivity of PP and its blends decreased. Nano graphene with planar structure has a large specific surface area, which leads to a trap between graphene and polymer matrix. Nano doping can introduce a large number of deep traps into the composite, which makes the space charge accumulation, conductivity and breakdown strength of the composite smaller.

References

1. I.L. Hosier, A.S. Vaughan, S.G. Swingler, An investigation of the potential of polypropylene and its blends for use in recyclable high voltage cable insulation systems. J. Mater. Sci. **46**(11), 4058–4070 (2011)
2. K. Kurahashi, Y. Matsuda, A. Ueda et al., The application of novel polypropylene to the insulation of electric power cable. IEEE/PES Transmission and Distribution Conference and Exhibition, Yokohama, Japan: IEEE, pp. 1278–1283 (2002)
3. B.X. Du, H. Xu, J. Li et al., Space charge behaviors of PP/POE/ZnO nanocomposites for HVDC cables. IEEE Trans. Dielectr. Electr. Insul. **23**(5), 3165–3174 (2016)
4. B.X. Du, H. Xu, J. Li, Effects of mechanical stretching on space charge behaviors of PP/POE blend for HVDC cables. IEEE Trans. Dielectr. Electr. Insul. **24**(3), 1438–1445 (2017)

5. A.R. Khare, S.P. Westphal, M.T.K. Ling et al., Thermal and dynamic mechanical analysis on metallocene ULDPE/PP blends to optimize impact properties. Thermochim. Acta **357–358**, 155–160 (2000)
6. T.J. Lewis, Nanometric dielectrics. IEEE Trans. Dielectr. Electr. Insul. **1**(5), 812–825 (1994)
7. X. Huang, P. Jiang, Y. Yin, Nanoparticle surface modification induced space charge suppression in linear low density polyethylene. Appl. Phys. Lett. **95**(24) (2009)
8. D. Akinwande, L. Tao, Q. Yu et al., Large-Area Graphene Electrodes: Using CVD to facilitate applications in commercial touchscreens, flexible nanoelectronics, and neural interfaces. IEEE Nanatechnol. Mag. **9**(3), 6–14 (2015)
9. T. Kuilla, S. Bhadra, D. Yao et al., Recent advances in graphene based polymer composites. Prog. Polym. Sci. **35**(11), 1350–1375 (2010)
10. W.Y. Chiu, S.J. Fang, Mechanical properties and morphology of crosslinked PP/PE blends and PP/PE/propylene–ethylene copolymer blends. J. Appl. Polym. Sci. **30**(4), 1473–1489 (1985)
11. Z. Ye-Wen, L. Ji-Xiao, Z. Fei-Hu et al., Formation and distribution of space-charge in cross-linked polyethylene. Chin. Phys. Lett. **19**(8), 1191–1194 (2002)
12. G. Chen, J. Zhao, S. Li et al., Origin of thickness dependent dc electrical breakdown in dielectrics. Appl. Phys. Lett. **100**(22), 222904 (2012)
13. T. Maezawa, J. Taima, Y. Hayase et al., Space charge formation in LDPE/MgO Nano-composite under high electric field at high temperature[C]: IEEE, pp. 271–274 (2007)
14. E.J.G. Santos, E. Kaxiras, Electric-field dependence of the effective dielectric constant in graphene. Nano Lett. **13**(3), 898–902 (2013)
15. T.J. Lewis, Interfaces are the dominant feature of dielectrics at the nanometric level. IEEE Trans. Dielectr. Electr. Insul. **11**(5), 739–753 (2004)
16. M. Roy, J.K. Nelson, R.K. MacCrone et al., Polymer nanocomposite dielectrics-the role of the interface. IEEE Trans. Dielectr. Electr. Insul. **12**(4), 629–643 (2005)
17. S. Li, N. Zhao, Y. Nie et al., Space charge characteristics of LDPE nanocomposite/LDPE insulation system. IEEE Trans. Dielectr. Electr. Insul. **22**(1), 92–100 (2015)
18. L. Zhang, Y. Zhou, J. Tian et al., Experiment and simulation of space charge suppression in LDPE/MgO nanocomposite under external DC electric field. J. Electrostat. **72**(4), 252–260 (2014)
19. F. Tian, Q. Lei, X. Wang et al., Effect of deep trapping states on space charge suppression in polyethylene/ZnO nanocomposite. Appl. Phys. Lett. **99**(14), 142903 (2011)
20. T. Tanaka, Dielectric nanocomposites with insulating properties. IEEE Trans. Dielectr. Electr. Insul. **12**(5), 914–928 (2005)
21. S. Li, G. Yin, S. Bai et al., A new potential barrier model in epoxy resin nanodielectrics. IEEE Trans. Dielectr. Electr. Insul. **18**(5), 1535–1543 (2011)

Effect of Mechanical Stress on Space Charge Behaviors of PP Insulation Materials

Hang Xu, B. X. Du, and Zhonglei Li

Abstract Polypropylene (PP) blended with polyolefin elastomer (POE) has been considered as a potential choice to replace cross-linked polyethylene (XLPE) as HVDC cable insulating material, which shows excellent performance in electrical, mechanical and thermal properties. The space charge accumulation, formed under dc electric field in PP/POE blend, would distort the local electric field, and furtherly lead to partial discharge or premature breakdown. During the processes of fabrication, installation and operation, the cable insulating material may be exposed to mechanical stress. In order to study the influence of mechanical stretching on space charge behaviors, PP/POE blends were prepared and stretched to different ratios of 1, 1.1, 1.2, 1.3 and 1.4, and the space charge behaviors during polarization and depolarization processes were measured and analyzed by PEA method. Besides, the effect of mechanical stretching on surface and internal morphology were analyzed by SEM test, and the relationship between the surface and internal morphology and space charge characteristics is obtained in this chapter.

1 Introduction

The XLPE is widely applied as insulating materials in high voltage cables. However, as thermosetting material, XLPE is difficult to recycle at the end of lifetime [1–5]. The recyclable blend of PP and polyolefin elastomer (POE) has attracted much attention. It not only modifies the brittleness and stiffness of pure PP, but also shows many outstanding performances in electrical and thermal properties. In addition, it

H. Xu (✉)
No. 336, Nanxinzhuang West Road, Jinan, China
e-mail: cse_xuh@ujn.edu.cn

B. X. Du · Z. Li
Building 26, No. 92 Wejin Road, Nankai District, Tianjin, China
e-mail: duboxue@tju.edu.cn

Z. Li
e-mail: lizhonglei@tju.edu.cn

B. Du (ed.), *Polymer Insulation Applied for HVDC Transmission*,
https://doi.org/10.1007/978-981-15-9731-2_6

can be manufactured without the crosslinking step, reducing much time and energy consumption. With these unique advantages, the PP/POE blend has been considered as a potential HVDC cable insulating material to replace XLPE in the future [6–9].

It is generally recognized that one of the most important problems in HVDC cables is the understanding and suppressing of space charge in insulating materials, which can increase the local electric field in the insulating materials, leading to partial discharge, faster degradation and even premature failure [10–13]. In semi-crystalline polymers like polyethylene and polypropylene, the space charge behaviors are closely related to the morphologies of crystalline and amorphous regions. Wang et al. found that the uniform distribution of space charge in low-density polyethylene (LDPE) was greatly related to the crystallinity and spherulite size [14]. Li et al. annealed PP at different temperatures to form specimens with different morphologies, and found that the formation of imperfect spherulites promoted proportional space charge distribution [15]. Mizutani et al. reported that LDPE with a lower density or higher amorphous fraction showed a higher space charge mobility, and they inferred it was caused by the difference in density or crystallinity [16].

The insulation material of extruded cable has to face the challenge of mechanical stretching in some cases [17–19]. For example, during the cooling process of fabrication, mechanical strain can be generated by the different cooling rates of inner and outer insulation due to temperature gradients. Besides, the cables may be bent when installed. During the operation of cables, thermomechanical strain can be created by the different thermal expansion between conductor and the insulation material. In addition, the electric field can also induce mechanical stress in polymers. Such mechanical stretching can result in complex microstructure changes in the semi-crystalline polymers [20–22]. For example, the amorphous regions may be elongated and the crystalline lamella may be broken, leading to significant influence on the electrical performance of polymers. David et al. investigated the effect of internal mechanical strain on electrical tree characteristics of polyethylene, and found that with increasing the value of residual stresses, the tree inception time decreased and the length of trees increased [23]. Mita et al. reported the dc and ac breakdown strength of LDPE decreased with increasing elongation ratio up to 30% and then increased until the elongation ratio reached to 160%, while both of them decreased monotonically in HDPE [24]. However, little work has been done about the influence of mechanical stretching on space charges. A thorough understanding of space charge behaviors in stretched PP/POE blend is necessary for the development and safe operation of HVDC cables.

This chapter investigates the correlation between space charge behavior and the mechanical stress. PP/POE blends were prepared and stretched to different ratios of 1, 1.1, 1.2, 1.3 and 1.4, and the space charge behaviors during polarization and depolarization processes were measured and analyzed by PEA method. Besides, the effect of mechanical stretching on surface and internal morphology were analyzed by SEM test. The results indicate a larger number of space charges accumulate in stretched specimens after polarization, and they decay faster than those in original specimens during depolarization process. In addition, the charge mobility show a dependence on elongation ratio, which is related to the morphology change caused by mechanical stretching. It is concluded that the space charge accumulation and transport are obviously influenced by the external mechanical stretching.

2 Experimental Arrangement

2.1 Specimen Preparation

The PP was supplied by SINOPEC Yangzi petrochemical Co., Ltd., China, the isotacticity of which is about 97%. POE (8150) is a kind of ethylene-octene copolymer with an ethylene comonomer content of 75%, which was supplied by DuPont Dow.

For the PP/POE blend, the preparation process is as the following. Firstly, POE and PP were mixed uniformly in a twin-roll mill set with a rotation speed of 50 rpm for 10 min at the temperature of 190 °C. Then the mixture was molded in a stainless steel mold at the temperature of 190 °C for 10 min and cooled to room temperature under a pressure of 15 MPa. The weight ratio of POE are 20% and 40%, which is respectively designated as POE20 and POE40 in this chapter.

Specimens in dumbbell shape were uniaxially stretched by a tensile machine at room temperature in air at a rate of 5 mm/min. The schematic diagram of tensile machine is shown in Fig. 1. The ΔL represents the displacement of the speciment in the stretching process. Elongation ratio λ is defined as the ratio of final stretched length to initial length. After being drawn, specimens were left clamped between

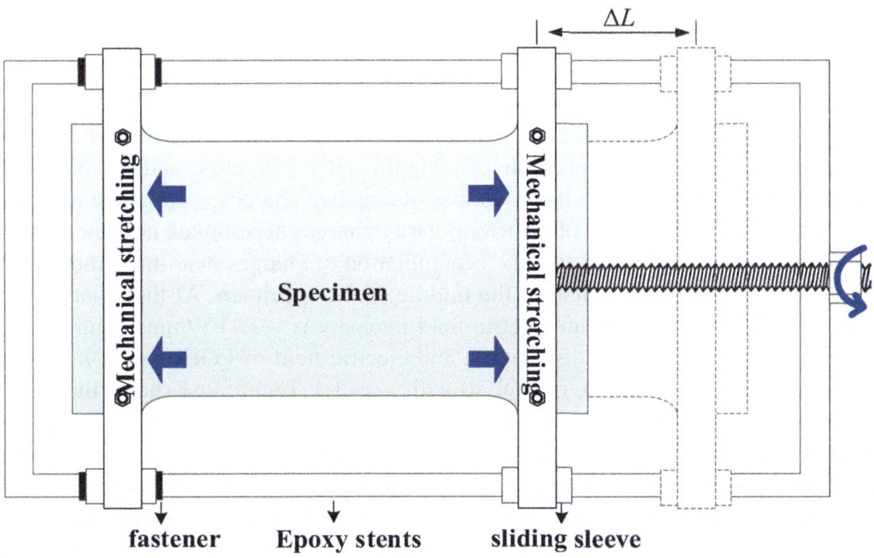

Fig. 1 Schematic diagram of tensile machine

grips of the tensile machine for about 6 h. Different specimens with elongation ratio λ of 1, 1.1, 1.2, 1.3 and 1.4 were prepared, respectively.

2.2 *Space Charge Measurement*

The space charge behavior of each specimen was measured by PEA method at room temperature [25]. Each specimen was polarized under a DC voltage with an electric field of 70 kV/mm for 60 min, and then the voltage-off test was continued for 10 min. The space charge behavior of each group was confirmed by repeating the tests for five times.

3 Effects of Mechanical Stretching on Space Charge Behaviors of PP/POE Blend

3.1 *Effect of Mechanical Stretching on Space Charge Characteristics During Polarization Process*

The morphology of PP/POE blend can be effected by the mechanical stretching, thus leading to the change of the space charge characteristics.

Figure 2 shows the space charge distribution and electric field distribution of the POE20 with different elongation ratios during the polarization process under an applied electric field of 70 kV/mm. From the space charge distribution of POE20 in Fig. 2a1, it can be seen that there is obvious negative charge injection near the cathode, which results in a slight electric field distortion inside the specimen as shown in Fig. 2b1. Figure 2a2 shows the space charge distribution of POE20 with $\lambda = 1.1$. The number of space charges significantly increases, and the depth of injected charges is higher. A large number of negative charges accumulate near the cathode and a small number of positive polarity charges accumulate near the anode. It can be seen from Fig. 2b2 that the accumulation of charges near the cathode and anode distorts the electric field in the middle of the specimen. At the polarization time of 3600 s, the maximum electric field intensity is −77 kV/mm. Figure 2 a3, b3 shows the space charge distribution and electric field of POE20 with $\lambda = 1.2$, which are similar with those of POE20 with $\lambda = 1.1$. Figure 2a4 shows the space charge distribution of POE20 with $\lambda = 1.4$ during the polarization process, and the number of space charges is significantly higher than that of other specimens. A large number of negative space charge is injected near the cathode, and the charge density decreases with the increase of the injection depth. Only a small number of positive polarity charges can be observed near the anode. As can be seen from Fig. 2b4, the maximum electric field distortion of the specimen appears near the anode. At the

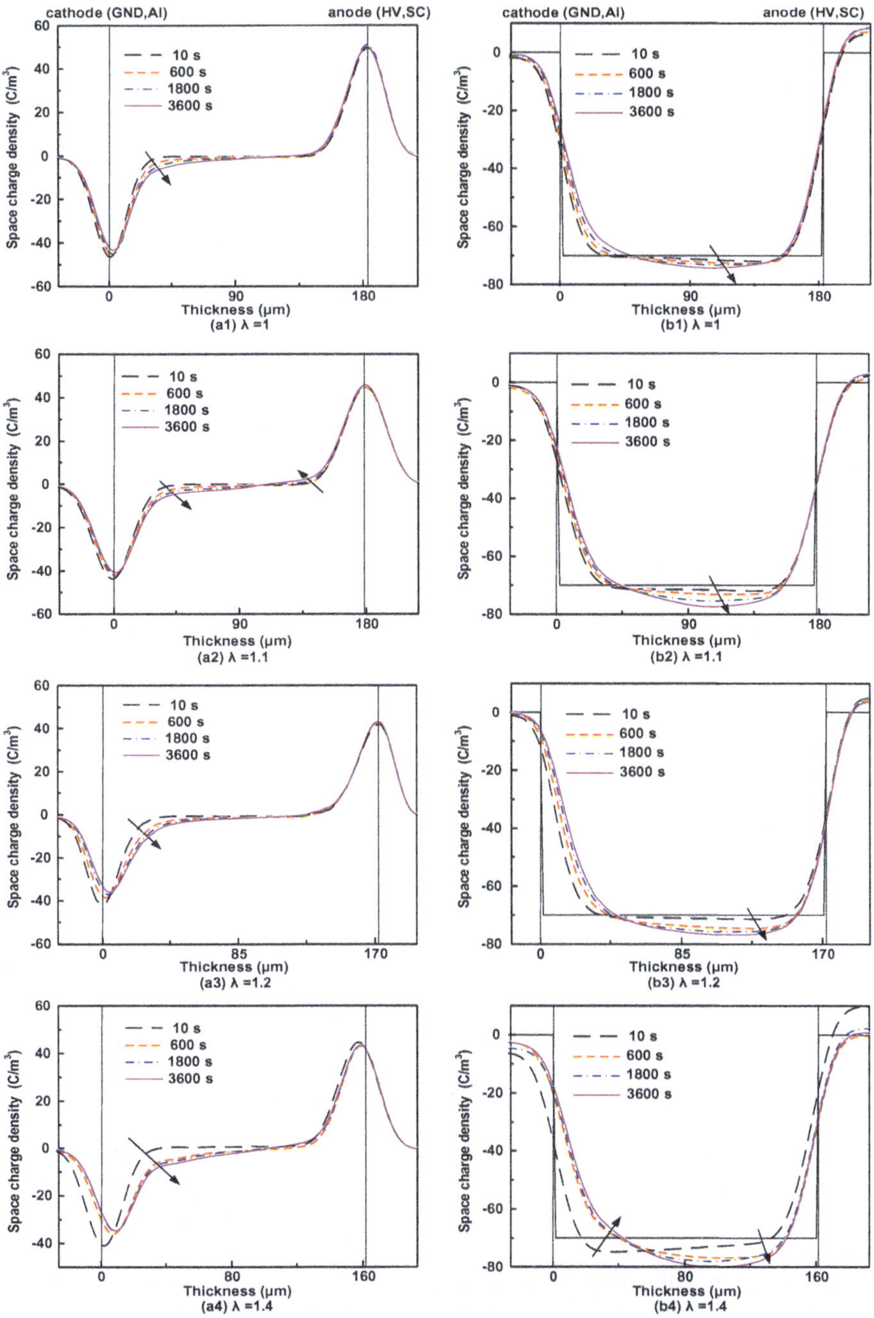

Fig. 2 Space charge behaviors and electric field of POE20 with different elongation ratio during the polarization process. **a** space charge behaviors **b** electric field

polarization of 3600 s, the maximum electric field intensity inside the specimen is about −80 kV/mm.

Figure 3 shows the space charge distribution and electric field distribution of POE40 with different elongation ratios under the electric field of 70 kV/mm. For POE40 specimens without mechanical stretching, negative charges accumulate near the cathode, and the number of injected charges increase with polarization time, but almost no space charges are observed near the anode. The maximum electric field distortion appears near the anode, as can be seen from Fig. 3b1. At the polarization time of 3600 s, the maximum electric field intensity is −75 kV/mm. Figure 3a2 shows the space charge distribution of POE40 with $\lambda = 1.1$ during the polarization process. A number of negative charges are observed in the specimen, the depth of which is higher than that of untreated POE40. Negative charges accumulation is observed near the anode. The electric field distortion near the anode is relatively serious as can be seen from Fig. 3b2. At the polarization time of 3600 s, the maximum electric field distortion intensity is −76 kV/mm. Figure 3a3 shows the space charge distribution of POE40 with $\lambda = 1.2$, in which significant negative charges injection can be observed. The number and depth of space charges increase continuously with the polarization time. As the elongation ratio increases from 1 to 1.2, the number of space charge increases, as wll as the depth of space charge. At the polarization time of 3600 s, the maximum electric field in POE40 with $\lambda = 1.2$ is −79 kV/mm, as shown in Fig. 3b3. Figure 3a4 shows the space charge distribution of the POE40 with $\lambda = 1.4$, where negative charge accumulation can also be observed. As the elongation ratio increases from 1.2 to 1.4, the number of space charge decreases. According to Fig. 3b4, the maximum electric field in POE40 with $\lambda = 1.2$ is −76 kV/mm at the polarization time of 3600 s.

For further analysis, the total amount of space charges in the bulk was calculated to obtain the relationship between space charge accumulation and the elongation ratio. The following equations were used:

$$Q_{total}(t) = \int_0^L |\rho(x,t)| S dx \quad (1)$$

$$Q'_{total}(t) = Q_{total}(t)\frac{L_0}{L} \quad (2)$$

where $\rho(x, t)$ is the measured charge density in depth x (distance from the cathode) at polarization time t, S is the area of electrode and L is the thickness of specimen. Considering the difference in specimen thickness, the total amount of space charges obtained were normalized to value of the same specimen thickness of $L_0 = 180\ \mu m$ according to Eq. 2.

Figure 4 shows the total amount of space charge in PP/POE blends with different elongation ratios during the polarization process. It can be seen that the total amount of space charges accumulating in the specimen increases continuously with polarization time, besides, the total amount of space charge accumulated in the specimen with mechanical stretching was more than that of specimen without mechanical stretching.

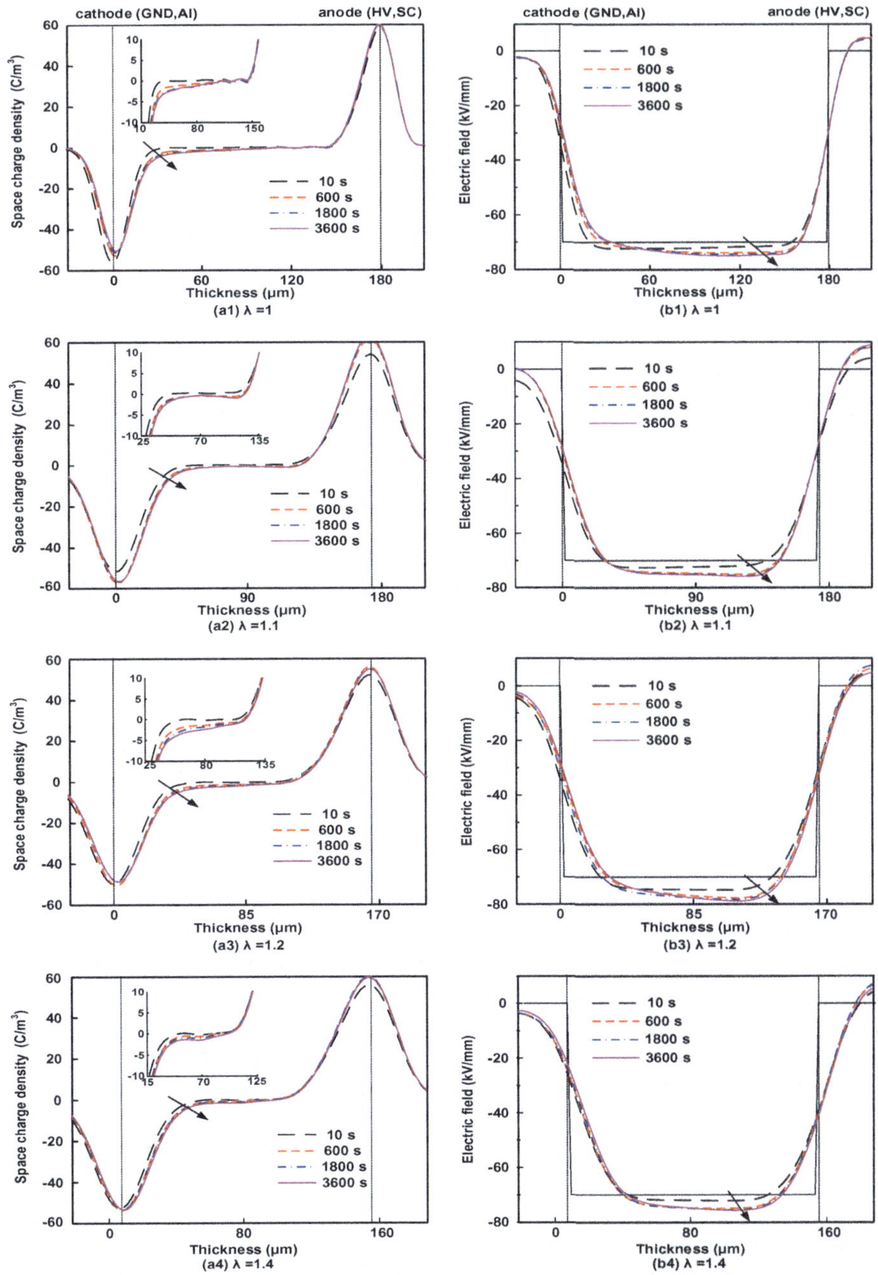

Fig. 3 space charge distribution and electric field distribution of POE40 with different elongation ratio. **a** Space charge distribution **b** electric field distribution.

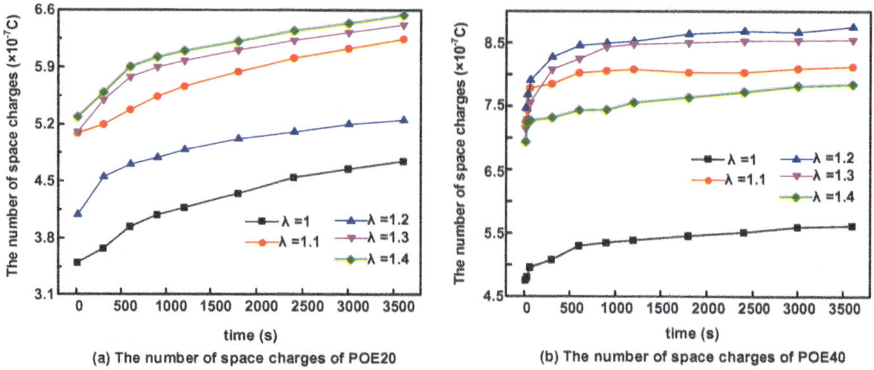

Fig. 4 Relationship between the total charge amount and the poling time. Copyright Clearance Center RightsLink® License Number: 4856981008241

As the elongation ratio increases, the total space charge amount of POE20 and POE40 specimens show different trends.

Figure 4a shows the total space charge amount of POE20 blends with different elongation ratios. The total space charge amount of POE20 λ = 1 is the smallest compared with other POE20 blends. As the elongation ratio increases from 1 to 1.1, the total space charge increases. With the increase of elongation ratio from 1.1 to 1.2, the total space charge amount decreases. With the increase of elongation ratio from 1.2 to 1.4, the total space charge amount continues to increase. The total space charge amount of POE20 λ = 1.4 is the highest, reaching about 650 nC at the polarization time of 3600 s.

Figure 4b shows the total space charge amount of POE40 blends with different elongation ratios. The total space charge amount of POE40 without stretching is the smallest. The total space charge amount increases as the elongation ratio increases from 1 to 1.2. With the increase of elongation ratio from 1.2 to 1.4, the total space charge amount decreases. The total space charge amount of POE40 λ = 1.2 is the highest, reaching about 870 nC at the polarization time of 3600 s.

The electric field distortion is closely related to the accumulation of space charge in specimen bulk, which is a great concern in the operation of HVDC cables. The electric field along the specimen was obtained from the space charge distribution:

$$E(x) = \frac{1}{\varepsilon_0 \varepsilon_r} \int_0^x \rho(x) dx \tag{3}$$

where *E(x)* is the electric field in depth *x*, $\rho(x)$ is the measured charge density,ε_0 is the vacuum permittivity and ε_r is the relative permittivity of the specimen. Thus the distortion factor of the electric field ΔE can be obtained by the following equation:

$$\Delta E = \frac{E\text{max} - E_{av}}{E_{av}} \tag{4}$$

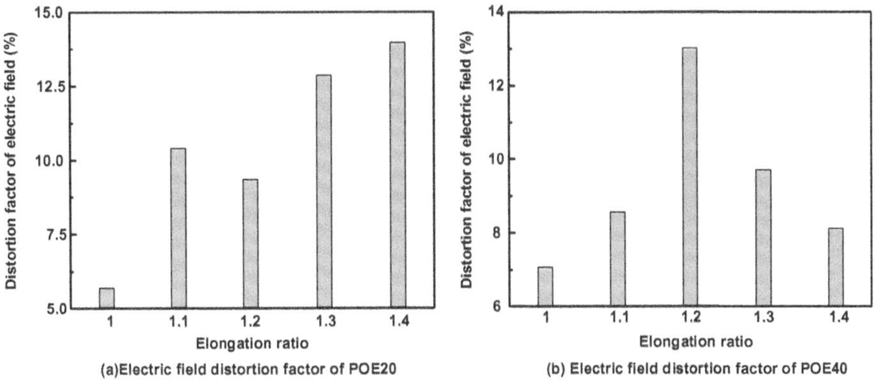

Fig. 5 Distortion factor of the electric field after 1 h of DC polarization. Copyright Clearance Center RightsLink® License Number: 4856981008241

where E_{max} is the highest value of the electric field in specimen after DC polarization for 60 min, E_{av} is the ideal electric field without space charges.

Comparison of the distortion factor of the electric field at 3600 s for each specimen is shown in Fig. 5. PP/POE blends without mechanical stretching have less internal space charge and less electric field distortion factor. Mechanical stretching increases the distortion factor of electric field. For POE20 blends, the distortion factor of the electric field increases with increasing the elongation ratio except the POE20 with $\lambda = 1.2$. For the POE40 blends, when the elongation ratio is lower than 1.2, the distortion factor increases with increasing the elongation ratio. When the elongation ratio is higher than 1.2, the distortion factor decreases with increasing the elongation ratio. As the elongation ratio increases, the total space charge amount and the distortion factor of the electric field in the POE20 and POE40 specimens show different trends, which may be related to the differences in the mechanical toughness of the POE20 and POE40 specimens.

3.2 *Effect of Mechanical Stretching on Space Charge Characteristics During Depolarization Process*

Figure 6 shows the space charge distribution of POE20 blends during depolarization process. Figure 6a–c show that, the space charge dissipates with the lapse of depolarization time, and the space charge is little when the depolarization time is 600 s. For the POE20 with $\lambda = 1.4$, at the depolarization time of 600 s, there is still a large number of space charge. During the depolarization process, space charge in shallow traps dissipates rapidly, and space charge in deep traps dissipates slowly. The above results suggest that the large number of space charges in the POE20 with $\lambda = 1.4$ are trapped in deep traps. Figure 7 shows the space charge distribution of POE40

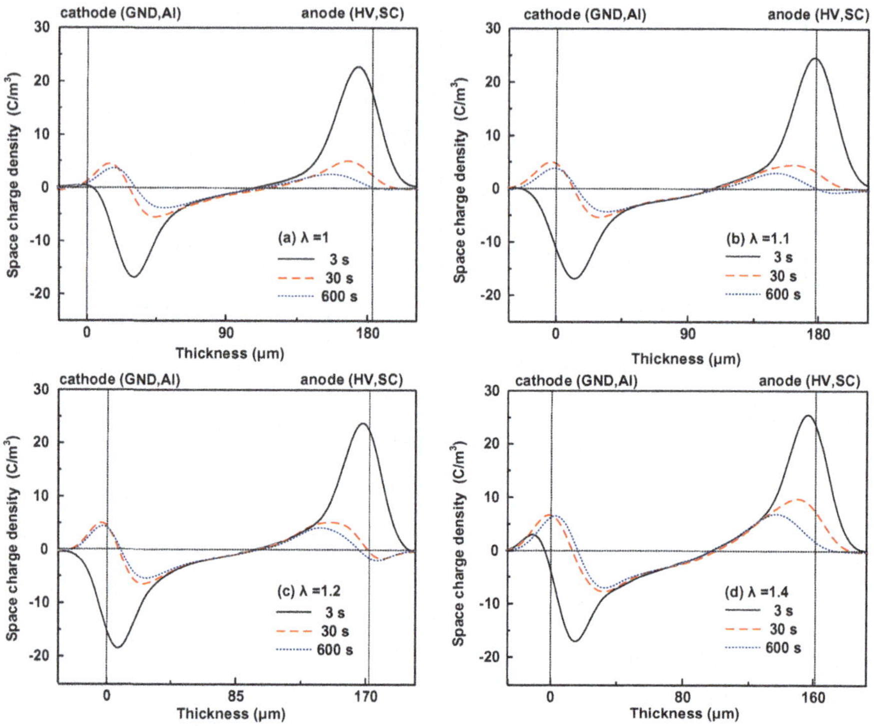

Fig. 6 Space charge distribution in POE20 blends during the depolarizaiton process

blends during depolarization process. For POE40 with $\lambda = 1.1$ and $\lambda = 1.2$, the space charges dissipate rapidly. When the depolarization time is 600 s, there is almost no space charge accumulation, only a small number of space charges accumulate near the anode and cathode. For POE40 with $\lambda = 1$ and $\lambda = 1.4$, the space charges dissipate slowly. During the depolarization process, only the space charge density near the anode and cathode decreases gradually, and the space charge distribution in the middle shows no change. When the depolarization time is 600 s, there is much space charge accumulating in the specimens.In order to quantify the space charge decay characteristics of different specimens, the mean volume charge density at time *t* of depolarization, *q(t)*, is calculated according to Eq. (5).

$$q(t) = \frac{1}{L}\int_0^L |\rho(x, t)|dx \tag{5}$$

Figure 8 shows the relationship between space charge density and depolarization time. The initial space charge density of POE20 without mechanical stretching is the lowest. The POE20 with $\lambda = 1.1$ has a higher initial space charge density, but its space charge density is close to that of the POE20 without mechanical stretching at the depolarization time of 600 s, indicating that the space charge dissipation rate

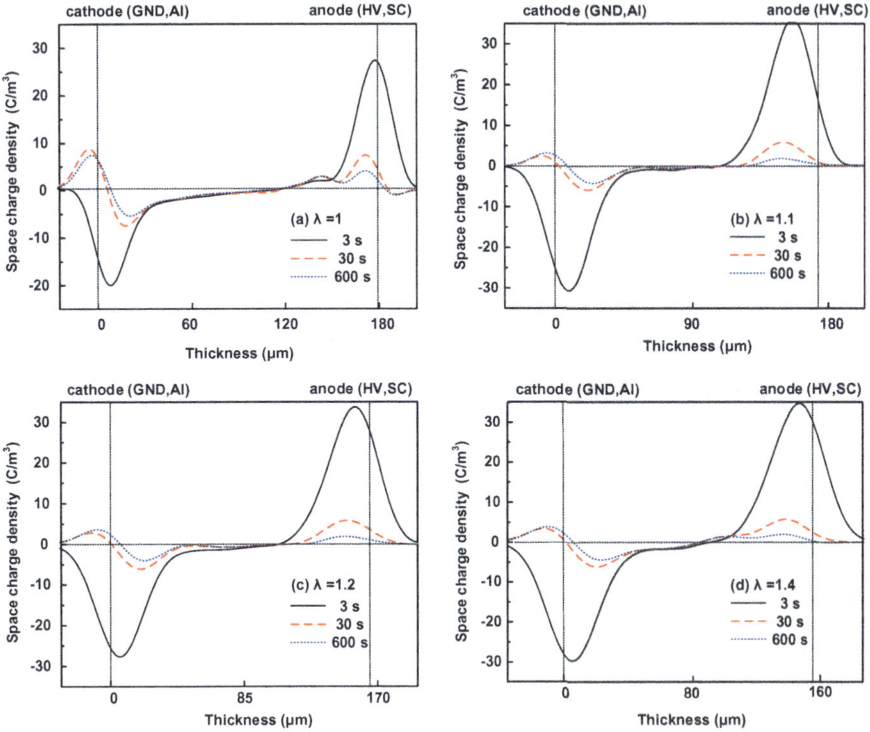

Fig. 7 Space charge distribution in POE40 blends during the depolarizaiton process. Copyright Clearance Center RightsLink® License Number: 4856981008241

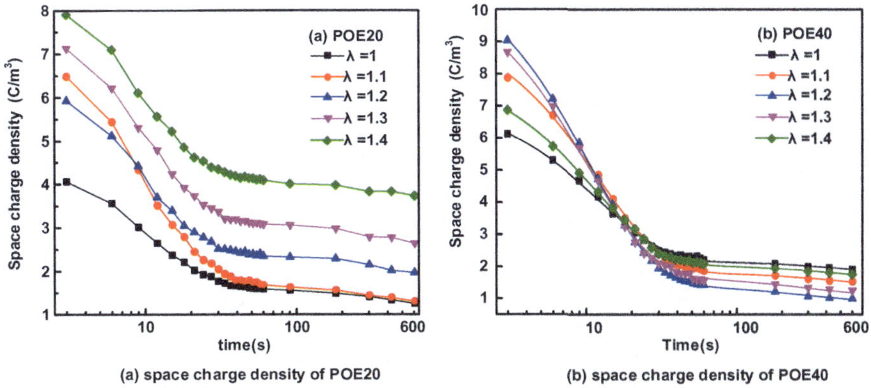

Fig. 8 The relationship between space charge density and depolarization time. Copyright Clearance Center RightsLink® License Number: 4856981008241

of the POE20 with λ = 1.1 is faster, and slight mechanical stretching can facilitate the charge de-trapping process. As the elongation rate increases from 1.2 to 1.4, the initial space charge density increases, so does the space charge density at the end of depolarization process, and the space charge density attenuation curve are relatively gentle, indicating that a large number of deep traps exit in the POE20 blends with large elongation ratio, making it difficult for the space charges to dissipate.

Figure 8b shows the space charge density attenuation curves of POE40 blends. It is shown that the space charge dissipation rate of POE40 with λ = 1.2 is the fastest, with the space charge density dropping from about 9 C/m^3 to about 1 C/m^3 over 600 s depolarization time. The space charge dissipation rate of POE40 with λ = 1 is the slowest, with the space charge density dropping from about 6 C/m^3 to about 2 C/m^3 over 600 s depolarization time. The space charge dissipation rate of other specimens is in the middle of the two.

3.3 *Effect of Mechanical Stretching on Charge Mobility*

The apparent charge mobility was derived based on the space charge decay characteristics [26]

$$\mu(t) = \frac{2\varepsilon}{q(t)^2} \frac{dq(t)}{dt} \tag{6}$$

where *q(t)* is the mean volume charge density according to Eq. (5). The time-dependent carrier mobility of PP/POE blends with different elongation ratios is shown in Fig. 9.

During the depolarization process, the charges are firstly released from shallow traps and then from deep traps. Therefore, the mobility of charges decreases with the

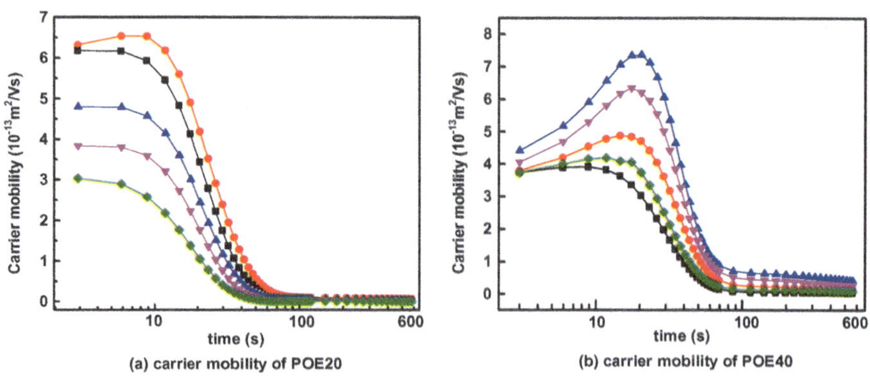

Fig. 9 The relationship between apparent charge mobility and depolarization time. Copyright Clearance Center RightsLink® License Number: 4856981008241

decay time. The stretched specimens generally show higher mobility than original specimen. For the POE20 blends, the apparent carrier mobility of POE20 with $\lambda = 1.1$ specimen is the highest, and the apparent carrier mobility decreases gradually with the increase of elongation ratio. For POE40 blends, it is found that the charge mobility increases with the elongation ratio from 1 to 1.2, and decreases from 1.2 to 1.4. Besides, the mobility curves show an increase in the first 11 s. This may be attributed to the limitation of this method. It is inferred that some charges are recombined during the depolarization process, thus the calculation of charge mobility is influenced [27]. Though this method only provides approximated values of mobility, it is still useful for material characterization.

Comparing the visual carrier mobility of POE20 and POE40, the variation trend was basically the same. With the increase of the elongation ratio, the apparent carrier mobility first increases and then decreases. For POE20 blends, the turning point for apparent carrier mobility appears at an elongation ratio of 1.1, while for POE40 blends, the turning point appears at an elongation ratio of 1.2. Because the POE content of POE20 is lower than that of POE40, the roughness of POE20 is lower than that of POE40. Therefore, POE20 is easier to be damaged by mechanical stretching, which means POE20 blends have much more morphology changes at the same elongation ratio. The morphology changes includes cracks or micropores, and they can introduce new trap energy levels into the specimen, which is the main reason for the decrease of carrier mobility.

4 Effects of Mechanical Stretching on Trap Distributions of PP/POE Blend

When the PP/POE blends are subjected to mechanical stress, its internal morphology and structure undergo complex changes, including the rotation of chain segments and branches, conformational changes of molecular chains, elongation of amorphous region and rearrangement of crystalline region, which lead to the change of trap density and trap depth. The trap distribution of PP/POE blends are investigated by the surface potential decay method.

Figure 10 shows the surface potential attenuation curve of PP/POE blends with different tensile ratios. Different specimens show different surface potential dissipation rates, which means different trap distributions. The trap density *Nt (E)* and trap depth E_t can be calculate from the surface potential attenuation curves:

$$\frac{N_t(E)}{E_t} = \frac{4\varepsilon_0\varepsilon_r}{eL^2k^2T^2\ln(\nu t)}\left|t\frac{dU_s(t)}{dt}\right| \tag{7}$$

where e is the charge of an electron; $U_S(t)$ is the surface potential at dissipation time t.

Figure 11 shows the trap distributions of PP/POE blends with different elongation

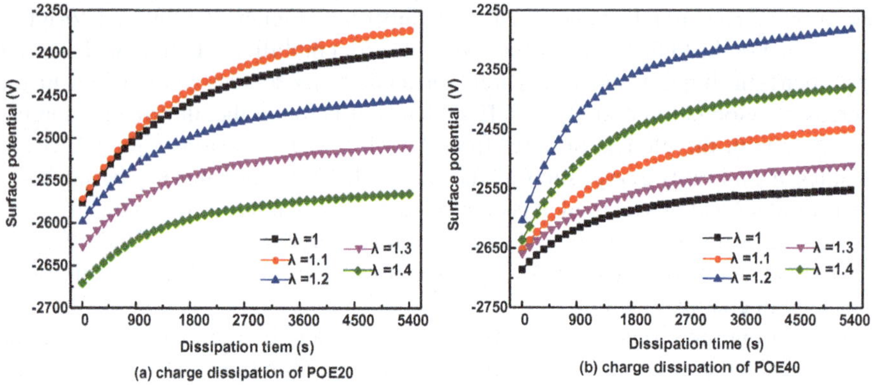

Fig. 10 Relationship between the surface potential and dissipation time

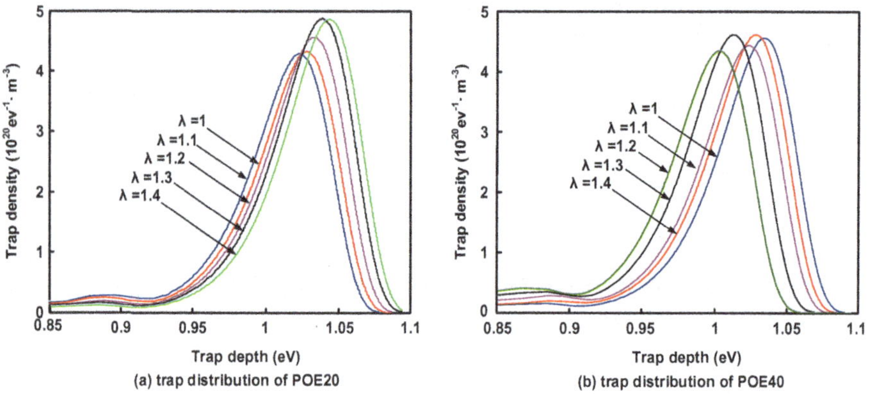

Fig. 11 Relationship between the trap distribution and elongation ratio

ratio. As shown in Fig. 11a. the trap depth of POE20 are mainly distributed between 0.95 and 1.1 eV. Compared with the POE20 without mechanical stretching, the trap depth and density were slightly reduced at the elongation ratio of 1.1. When the elongation ratio is greater than 1.1, the snare depth and density increase significantly. As shown in Fig. 11b, trap depth in the POE40 specimen decreases as the elongation ratio increases from 1 to 1.2. As the elongation ratio increases from 1.2 to 1.4, the trap depth in the specimen begins to increase again, but it is still lower than the trap depth of the POE40 without mechanically stretching.

5 Discussion

5.1 Morphological and Structural Changes of PP/POE Blends Under Mechanical Stretching

All the stretched specimens reveal a completely different space charge behavior from the original specimen, with more space charge accumulation and faster charge decay. The change in microstructure may be responsible for the results. As semicrystalline polymers, PP usually contains crystalline lamella and amorphous region as illustrated in Fig. 12. The crystalline lamellar ribbons are usually in spherulitic arrays, and amorphous regions exist between the lamellar ribbons. In the crystalline lamella, molecular chains are arrayed compactly and orderly, while in the amorphous region, they are arrayed loosely and randomly. The conduction mechanisms of mobile charges are rather different in crystalline lamella and amorphous region, so the space charge behavior is greatly related to the microstructure of polymers.

Under continuous mechanical stretching, some significant changes in the aggregation structure of PP/POE blend can be gradually obtained, and a simplified schematic diagram is shown in Fig. 13. Influenced by relatively weak Van der Waals bonds, amorphous regions would firstly endure the external mechanical stress. Tie molecules, which transverse the amorphous region and connect two neighboring lamellas, would be extended, resulting in extended amorphous regions in the stretched polymer. The crystallites would tilt due to the mechanical stretching. With increasing the mechanical stress further, the tie chains may break up or even be pulled out from the lamella. Besides, the crystalline lamella may be distorted and some blocks can even be pulled out of the crystalline lamellar ribbons due to the highly extension of tie chains [28]. As a result, more defective crystallites with smaller size can be generated during the drawing process, which provides more interfaces of crystalline/amorphous

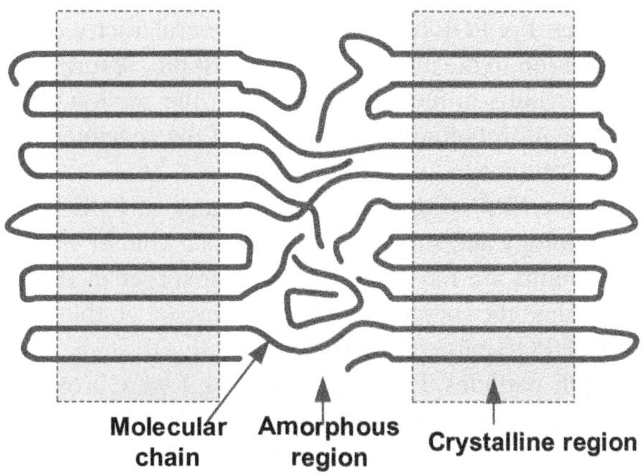

Fig. 12 The internal morphology of PP. Copyright Clearance Center RightsLink® License Number: 4856981008241

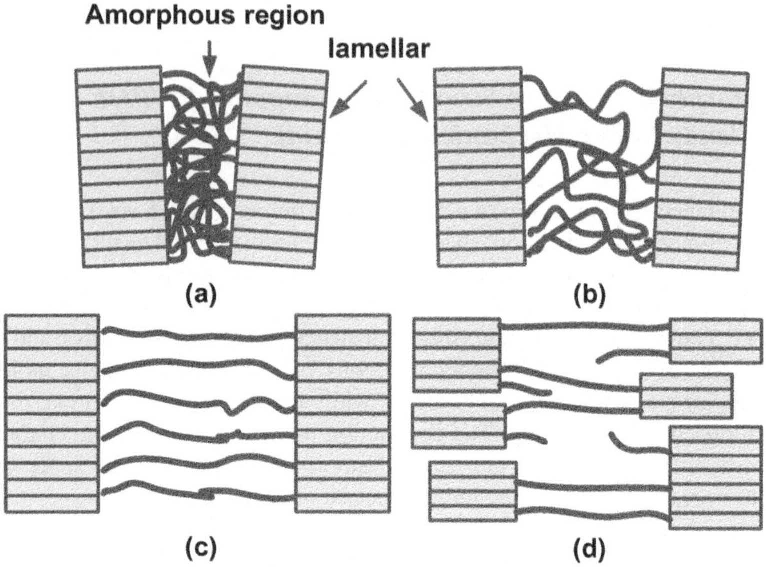

Fig. 13 The change of PP morphology under mechanical stress. Copyright Clearance Center RightsLink® License Number: 4856981008241

phase. The final consequence of successive mechanical stretching is the formation of cracks and fracture.

With the increase of elongation ratio, the internal morphology and structure of PP/POE blends change variously, which can be verified by scanning electron microscope results. The brittle section of POE40 blends with different elongation ratios were observed by SEM, and the results are shown in Fig. 14. The profile of POE40 without mechanical stretching is relatively flat. For POE40 with $\lambda = 1.2$, a wave-like pattern appears in the section, indicating that the internal structure of POE40 blend changes. For POE40 with $\lambda = 1.3$, several micropores can be observed on the section in addition to the more pronounced ripple. A further increase in the elongation ratio to 1.4 results in more micropores in the section. The SEM results further indicate that the morphology and structure of the specimen changed with the increase of the elongation ratio.

In the PP/POE blends, the existence of POE has an important influence on the morphology and structure change. As a kind of elastomer, a large number of molecular chains are loosely arranged in disorder in POE. In the process of mechanical stretching, the curved molecular chains are gradually straightened or even broken up, and the POE particles can deform easily. To study the effect of mechanical stretching on POE particles, POE40 with $\lambda = 1.4$ were broken off in liquid nitrogen, with the fracture surface parallel to the stretching direction, and fractured specimens were etched in heptane under ultrasonic condition for 20 min to observe the phase structure by Scanning Electron Microscope test. Figure 15 shows the dispersion of POE in original PP/POE blend. The dark holes represent POE particles, which were removed

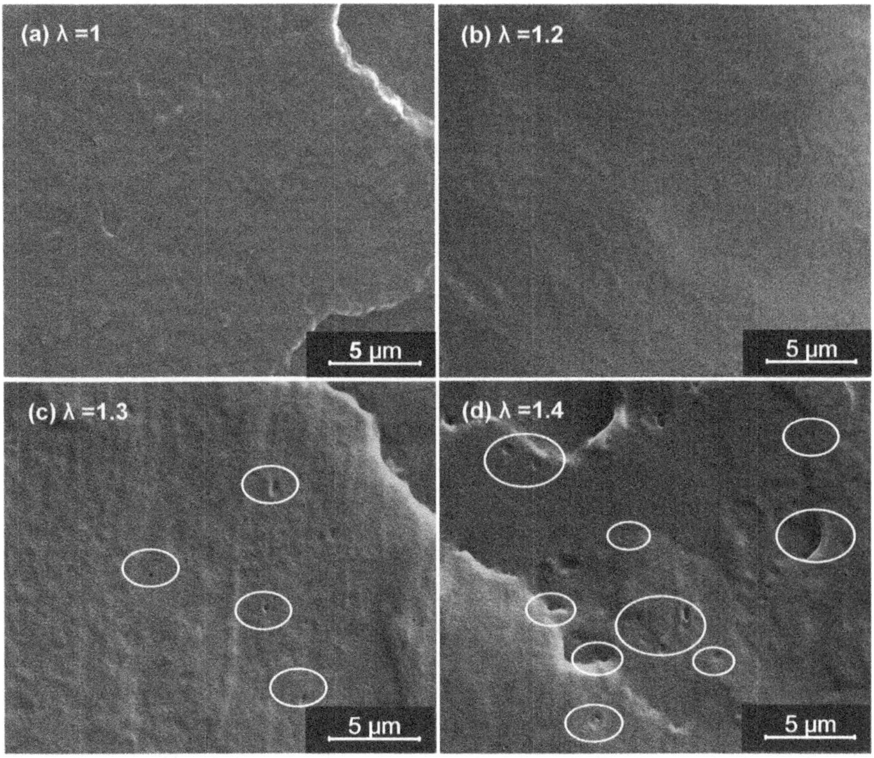

Fig. 14 The effect of mechanical stretching on the section of PP/POE blends

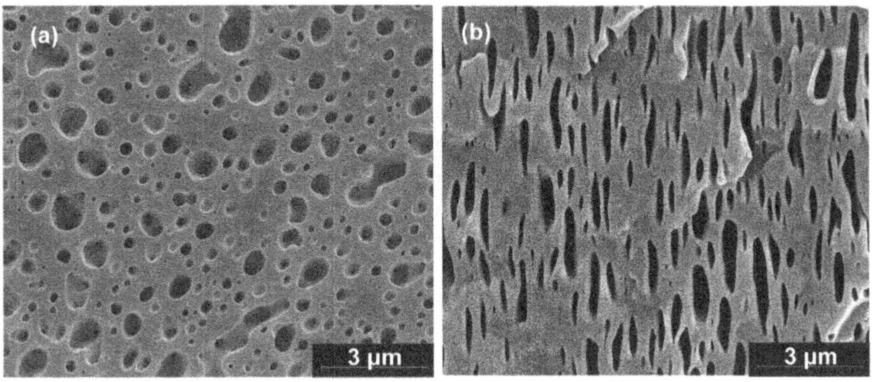

Fig. 15 The effect of mechanical stretching on PP shape.

by heptane etching. It can be observed that POE particles disperse uniformly in the PP matrix with an average diameter of 1 μm. Figure 15b shows the dispersion of POE in stretched blend. Due to the external mechanical stretching, the POE particles are elongated along the stretching direction. These deformed POE particles act as stress concentrators, resulting in shear yielding and/or formation of crazes in the surrounding PP matrix [29]. As a result, the crack propagation resistance and toughness of PP/POE blend can be greatly enhanced. Compared with POE40, the POE content of POE20 is less, therefore, its mechanical toughness is lower than POE40, and mechanical damage is more prone to be formed in POE20.

5.2 The Effect of Mechanical Stretching on Carrier Mobility

The increased charge mobility and decreased trap depth in stretched specimens can be interpreted according to the nature of charge transport. In the specimen bulk, mobile charges consist of three distinct classes, massive molecular ions, electrons and holes, and they show very different transport characteristics. Due to the steric factors, diffusion of massive ions generally take place in free volume paths to avoid the space occupied by the polymer. Saturated hydrocarbons usually have negative electron affinity, PP and POE have only single hydrocarbon bonds in their molecular chains, which are saturated hydrocarbons and have negative electron affinity [30, 31]. The negative electron affinity of the PP and POE allow electrons to transport in the intermolecular spaces between the polymer chains, while hole transport is confined to polymer chains, and long-range transport would require inter-chain hopping. All these charge transports are more likely to take place in the amorphous regions [32].

After stretching, the amorphous region can be extended and the crystalline lamella can break up as shown in Fig. 12. As a result, the density of polymer can be decreased and the free volume may increase as well [33]. For ions and electrons, the increase in free volume can weaken the interaction with chain segments, making it easier for ions and electrons to transport. For holes, the inter-chain hopping would be interrupted by increased free volume. It can therefore be concluded that the acceleration of ions and electrons transport increases the charge mobility more than the inhibition of holes transport decreases it, which may attribute to few positive charges injection during polarization process.

In the process of mechanical stretching, the internal morphology and structure of POE20 and POE40 change variously. In slightly stretched specimens, the amorphous region is elongated, which enlarges the spacing between molecular chains. According to the PEA results, space charge in POE20 and POE40 is mostly negative polar charge, therefore, the apparent carrier mobility of the specimens with small elongation ratios show a significant increase.

With further increase in the elongation ratio, both POE20 and POE40 show significant decreases in the apparent carrier mobility. It is speculated that more microvoids, defective crystallites, interfaces of crystalline/amorphous phase are generated in highly stretched specimens, which can introduce much more trap sites and furtherly decreases the charge mobility [34]. Besides, the binding force between the PP and POE also decreases due to stretching, and more trap sites may be introduced at the interface to inhibit the charge transport and increase the trap depth.

Under continuous mechanical stretching, the POE particles play an important role in preventing crack formation, and the extensive fractures of molecular chains are resisted, which may help to resist the formation of deep trap sites. Compared with POE20, the POE content in POE40 is higher, which can better inhibit the mechanical damage tomolecular chains. Therefore, the POE 40 show a decrease in the apparent carrier mobility at an elongation ratio of 1.3, while POE 20 show a decrease in the apparent carrier mobility at an lower elongation ratio of 1.2.

5.3 The Effect of Mechanical Stretching on the Total Amount of Space Charge

The total amount of space charge of PP/POE blends with relatively low elongation ratio is higher than that of PP/POE blend without mechanical stretching. This may be related to the increase of apparent carrier mobility. It can be seen from the space charge distribution that a large number of negative charges are injected from the cathode and accumulate close to the electrode in all the specimens. For the original specimen, the charge mobility is relatively lower. Therefore, fewer injected charges move into the bulk and most of them accumulate close to cathode, decreasing the effective electric field between the electrodes and the surface of the specimens and thus suppressing further space charge injection. However, in stretched specimens, higher charge mobility leads to more charges migrating into the bulk. Therefore, the inhibitory effect of firstly injected charge on charge injection is greatly migrated, as a consequence, more charges are injected into the stretched specimens bulk [35].

For the POE40 blends, when the elongation ratio increases from 1.2 to 1.4, the apparent carrier mobility decreases, and correspondingly, the number of space charge decreases. However, when the elongation ratio of POE20 specimen increases from 1.1 to 1.4, the apparent carrier mobility decreases, but the total number of space charge in the specimen continues to increase. Obviously, the above explanation is no longer valid, indicating that the space charge accumulation is also affected by other factors.

Compared with the POE40 blends, the POE20 blends show poor toughness and appear more mechanical damage at the same elongation ratio, indicating more and deeper traps. Under the direct current electric field, charges are injected into the specimen, which transport along the direction of the electric field and are captured by the trap. When the traps in the outer layer of the specimen is filled, the free charge will continue to migrate along the direction of the electric field until it is captured by the next trap. The highly mechanical stretching of POE20 specimens contains more traps, thus increasing the capacity for charge per unit volume, and trapping more charges than slightly mechanical stretching POE20 blends.

On the other hand, the surface morphology of the specimen changes after mechanical stretching, which may promote the injection of electrons into the specimen, thus increasing the number of space charge. Figure 16 shows the surface morphology of PP/POE blends observed by SEM. The surface of original PP/POE blend is relatively flat. Some subtle ripples appear on the surface of POE20 with $\lambda = 1.1$. With increasing the elongation ratio, more and more ripples appear on the surface of specimen. When the elongation ratio is 1.4, obvious wavy lines can be observed, and some tiny holes appear. In the process of mechanical stretching, the surface structure of the specimen is destroyed.

Under the combined effect of the above factors, when the elongation ratio of POE20 increases from 1.1 to 1.4, the number of space charge increased significantly. For the POE40 blends, relative high toughness results in less mechanical damage, so the space charge accumulation is mainly affected by the apparent carrier mobility. Therefore, when the elongation ratio of POE40 increases from 1.2 to 1.4, the number of space charge decreases.

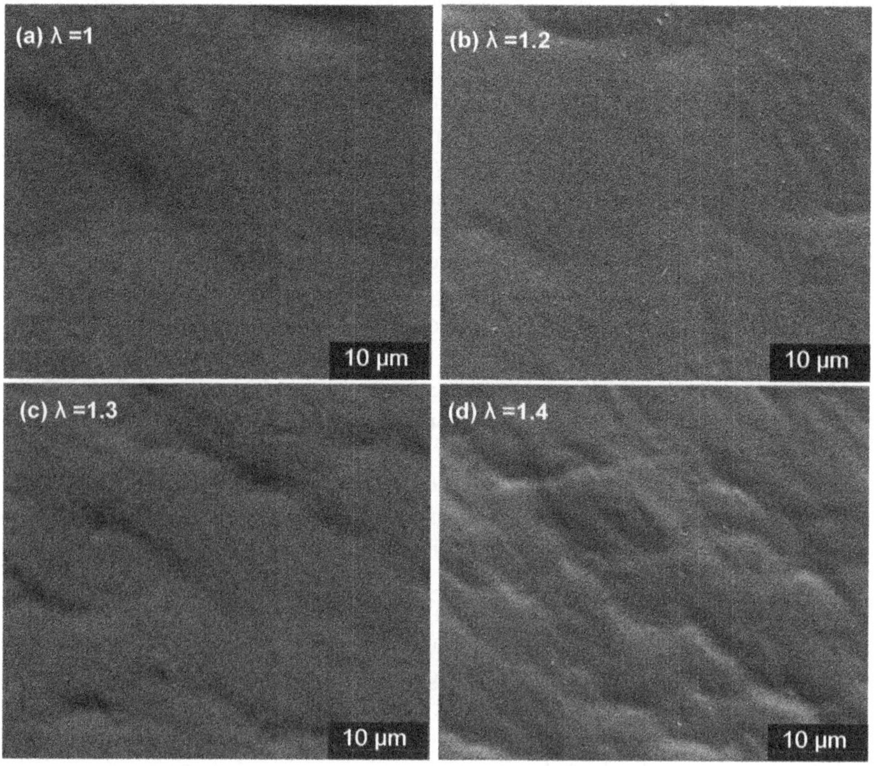

Fig. 16 The effect of mechanical stretching on surface morphology

6 Conclusion

An experimental study of space charge behavior in PP/POE blend with different elongation ratios was conducted. The results reveal that the space charge accumulation and transport are obviously influenced by the external mechanical stretching. The main conclusions can be summarized as follows:

(1) For the POE20 blends, as the elongation ratio increases from 1 to 1.1, the number of space charges and carrier mobility increase, while trap depth and density decrease; as the elongation ratio increases from 1.2 to 1.4, the number of space charges increases, the carrier mobility decreases, and the trap depth and density increase.

(2) For the POE40 blends, as the elongation ratio increases from 1 to 1.2, the number of space charge and carrier mobility increase, trap depth and density decreases; as the elongation ratio increases from 1.2 to 1.4, the number of space charges and carrier mobility decrease, and the trap depth and density increase.

(3) For POE40 and POE20 with low elongation ratio, a higher carrier mobility can promote the migration of space charges to the interior of the specimen, easing the inhibition effect on the electric field between the electrode and the specimen, and resulting in an increase in the total space charge accumulated in the specimen. For POE20 specimens with a larger elongation ratio, the carrier mobility decreases and the space charge accumulation increases. This is due to the serious damage of its internal morphological structure, which introducea large number of deep traps.

References

1. G.C. Montanari, C. Laurent, G. Teyssedre, A. Campus, U.H. Nilsson, From LDPE to XLPE: Investigating the Change of Electrical Properties. Part I. Space Charge, Conduction and Lifetime. IEEE Trans. Dielectr. Electr. Insul. **12**(3), 438–446 (2005)
2. G. Teyssedre, C. Laurent, G.C. Montanari, A. Campus, U.H. Nilsson, From LDPE to XLPE: investigating the change of electrical properties. Part II. Luminescence. IEEE Trans. Dielectr. Electr. Insul. **12**(3), 447–454 (2005)
3. F.N. Lim, R.J. Fleming, R.D. Naybour, Space charge accumulation in power cable XLPE insulation. IEEE Trans. Dielectr. Electr. Insul. **6**(3), 273–281 (1999)
4. C. Green, A. Vaughan, G. Stevens, A. Pye, S. Sutton, T. Geussens, M. Fairhurst, Thermoplastic cable insulation comprising a blend of isotactic polypropylene and a propylene-ethylene copolymer. IEEE Trans. Dielectr. Electr. Insul. **22**(2), 639–648 (2015)
5. M.I. L. Hosier, S. Reaud, A.S. Vaughan, S.G. Swingler, Morphology, thermal, mechanical and electrical properties of propylene-based materials for cable applications. IEEE Int'l. Symp. Electr. Insul.(ISEI), pp. 502–505 (2008)
6. K. Yoshino, T. Demura, M. Kawahigashi, Y. Miyashita, K. Kurahashi, Y. Matsuda, The application of novel polypropylene to the insulation of electric power cable. IEEE Trans. Distrib. Conf. Exhibit., pp. 1278–1283 (2002)

7. A.L.N. Silva, M.I.B. Tavares, D.P. Politano, F.M.B. Coutinho, M.C.G. Rocha, Polymer blends based on polyolefin elastomer and polypropylene. J. Appl. Polym. Sci. **66**(10), 2005–2014 (1997)
8. M.Y. Zhou, J.L. He, J. Hu, X.Y. Huang, P.K. Jiang, Evaluation of polypropylene/polyolefin elastomer blends for potential recyclable hvdc cable insulation applications. IEEE Trans. Dielectr. Electr. Insul. **22**(2), 673–681 (2015)
9. B. Dang, J.L. He, J. Hu, Y. Zhou, Large improvement in trap level and space charge distribution of polypropylene by enhancing the crystalline-amorphous interfaces effect in blends. Polym. Int'l. **48**(2), 206–208 (2016)
10. G.C. Montanari, Bringing an Insulation to Failure: the Role of Space Charge. IEEE Trans. Dielectr. Electr. Insul. **18**(2), 339–364 (2011)
11. T.T.N. Vu, G. Teyssedre, B. Vissouvanadin, S. Le Roy, C. Laurent, Correlating conductivity and space charge measurements in multi- dielectrics under various electrical and thermal stresses. IEEE Trans. Dielectr. Electr. Insul. **22**(1), 117–127 (2015)
12. D. Fabiani, G.C. Montanari, A. Cavallini, G. Mazzanti, Relation between space charge accumulation and partial discharge activity in enameled wires Under PWM-like voltage waveforms. IEEE Trans. Dielec. Electr. Insul. **11**(3), 393–405 (2004)
13. L.A. Dissado, G. Mazzanti, G.C. Montanari, The role of trapped space charges in the electrical aging of insulating materials. IEEE Trans. Dielectr. Electr. Insul. **4**(5), 496–506 (1997)
14. X. Wang, H.Q. He, D.M. Tu, C. Lei, Q.G. Du, Dielectric properties and crystalline morphology of low density Polythylene blended with Metallocene catalyzed polyehtylene. IEEE Trans. Dielectr. Electr. Insul. **15**(2), 319–326 (2008)
15. X. Li, Q.G. Du, J. Kang, D.M. Tu, Influence of microstructure on space charges of polypropylene. J. Polym. Sci. B **40**(4), 365–374 (2002)
16. T. Mizutani, H. Semi, K. Kaneko, Space Charge Behavior in low- density polyethylene. IEEE Trans. Dielectr. Electr. Insul. **7**(4), 503–508 (2000)
17. G. Mazzanti, G.C. Montanari, Elemental strain and trapped space charge in thermoelectrical aging of insulating materials life modeling. IEEE Trans. Dielectr. Electr. Insul. **8**(6), 966–971 (2001)
18. L.A. Dissado, G. Mazzanti, G.C. Montanari, Elemental strain and trapped space charge in thermoelectrical aging of insulating materials part 1: elemental strain under thermo-electrical-mechanical stress. IEEE Trans. Dielectr. Electr. Insul. **8**(6), 959–965 (2002)
19. T. Tanaka, T. Fukuda, S. Suzuki, Water tree formation and lifetime estimation in 3.3 kV and 6.6 kV XLPE and PE power cables. IEEE Trans. Power Appar. Syst. **95**(6), 1892–1900 (1976)
20. J.P. Crine, Influence of electro-mechanical stress on electrical properties of dielectric polymers. IEEE Trans. Dielectr. Electr. Insul. **12**(4), 791–800 (2005)
21. K. Yahagi, Dielectric properties and morphology in polyethylene. IEEE Trans. Electr. Insul. **15**(3), 241–250 (1980)
22. J.Y. Li, F.S. Zhou, D.M. Min, S.T. Li, The enerngy distribution of trapped charges in polymers based on isothermal surface potential decay model. IEEE Trans. Dielectr. Electr. Insul. **22**(3), 1723–1732 (2015)
23. E. David, J.L. Parpal, J.P. Crine, Influence of internal mechanical stress and strain on electrical performance of polyethylene electrical treeing resistance. IEEE Trans. Dielectr. Electr. Insul. **3**(2), 248–257 (1996)
24. S. Mita, K. Yahagi, Effect of elongation on dielectric breakdown strength in polyethylene. Jpn. J. Appl. Phys. **14**(2), 197–201 (1975)
25. Y. Li, M. Yasuda, T. Takada, Pulsed electroacoustic method for measurement of charge accumulation in solid dielectrics. IEEE Trans. Dielectr. Electr. Insul. **1**(2), 188–195 (1994)
26. G. Mazzanti, G.C. Montanari, J.M. Alison, A space-charge based method for the estimation of apparent mobility and trap depth as markers for insulation degradation theoretical basis and experimental validation. IEEE Trans. Dielectr. Electr. Insul. **10**(2), 187–197 (2003)
27. M. Hao, Y. Zhou, G. Chen, G. Wilson, P. Jarman, Space Charge behavior in thick oil-impregnated pressboard under HVDC stresses. IEEE Trans. Dielectr. Electr. Insul. **22**(1), 397–400 (2013)

28. J.P. Jones, J.P. Llewellyn, T.J. Lewis, The Contribution of field- induced morphological change to the electrical aging and breakdown of polyethylene. IEEE Trans. Dielectr. Electr. Insul. **12**(5), 951–966 (2005)
29. Z. Fu, W.L. Dai, H.M. Yu, X.X. Zou, B.Q. Chen, Effect of composition on fracture behavior of polypropylene–wollastonite–polyolefin elastomer system. J. Mater. Sci. **46**(5), 1272–1280 (2011)
30. M. Tahara, Y. Hayase, M. Honjoh, K. Nagasawa, Y. Tanaka, T. Takada, M. Yoshida, Charge accumulation properties in saturated and aromatic hydrocarbons by electron beam irradiation. IEEE Conf. Electr. Insul. Dielectr. Phenom (CEIDP), pp. 165–168 (2008)
31. S. Serra, E. Tosatti, S. Iarlori, S. Scandolo, G. Santoro, Interchain electron states in polyethylene. Phys. Rev. B **62**(62), 4389–4393 (2000)
32. T.J. Lewis, The physico-chemical origins and nature of space charge in insulating solids under electrical stress. IEEE 7th Int'l. Conf. Solid Diel. (ICSD), pp. 223–227 (2001)
33. W. Glenz, N. Morosoff, A. Peterlin, Density of drawn polyethylene. J. Polym. Sci. Polym. Lett. **9**(3), 211–217 (1971)
34. A. Baba, K. Ikezaki, Drawing and annealing effects on thermally stimulated currents in polypropylene films. J. Appl. Phys. **72**, 2057–2059 (1992)
35. J. Zhao, Z. Xu, G. Chen P. Lewin, Effect of field-dependent mobility on current density and dynamics of space charge in polyethylene. IEEE Conf. Electr. Insul. Dielectr. Phenom. (CEIDP), pp. 120–123 (2009)

Space Charge Characteristics of Coaxial Cable Insulation

Chi Chen, Xia Wang, Kai Wu, and Chuanhui Cheng

Abstract The accumulation of space charge inside the insulation of high-voltage direct current (HVDC) cable may lead to serious distortion of the internal electric field distribution and thus extremely influence the long-term reliability. In addition, a temperature gradient effect, caused by the Joule heat originating from the current in the HVDC cable core conductor, forms across the cable insulation when the cable is loaded in service. To date, extensive research about the space charge behaviors in plane samples under a temperature gradient condition has been reported. However, it is still questionable whether the results of plane samples can reflect the charge characteristics in an actual cable directly. Therefore, it is of great significance to measure and research the space charge behaviors in an actual full-size coaxial cable. In this chapter, space charge measurements for full-size coaxial cables are summarized, and the corresponding recovery algorithms are introduced. Besides, the space charge behaviors for 10 and 160 kV cables under different temperature gradients are analyzed, the results show that obvious hetero-charges accumulate near the outer semi-conductor electrode of the cables under temperature gradients, which are injected from the inner conductor electrode and then migrate to the outer semi-conductor one. And the hetero-charges near the outer semi-conductor electrode increase with the temperature gradients.

C. Chen (✉)
School of Electrical Engineering, Xi'an University of Technology, Xi'an, China
e-mail: chenchi@xaut.edu.cn

X. Wang · K. Wu
State Key Laboratory of Electrical Insulation and Power Equipment, Xi'an Jiaotong University, Xi'an, China

C. Cheng
Electric Power Research Institute, China Southern Power Grid, Guangzhou, China

B. Du (ed.), *Polymer Insulation Applied for HVDC Transmission*,
https://doi.org/10.1007/978-981-15-9731-2_7

1 Introduction

Compared with the high-voltage (HV) alternating current (AC) power cables, the HVDC cables have many advantages, such as higher working stress, smaller insulation thickness, lighter weight, smaller outer radius, more load flow, less dielectric and conductor losses, no AC magnetic stress and so on. According to the differences in cable structure and insulation material, HVDC cables include oil-filled cables, oil-impregnated paper cables, inflatable cables, and extruded plastic cables. Among them, cross-linked polyethylene (XLPE) extruded plastic cables have many advantages, such as a larger transmission capacity, a more stable operation situation, and a simpler accessory structure [1]. However, due to the space charge in the cable insulation, more extensive research is still needed before XLPE cables can be put into practical application in HVDC transmission [2].

The insulation of XLPE plastic cables is generally composed of orderly molecular chain crystal phases and disordered molecular chain amorphous phases, there are some defects in the insulation material itself. And at the same time, some residual impurities, such as catalysts and antioxidants, also exist. From the microscopic point of view, the asymmetric distribution of atoms where physical defects and chemical impurities exist will lead to localization of the electron wave function. Thus, these defects and impurities have an ability to attract carriers and become localized states in materials, namely, traps. During the migration process of the carriers, once trapped, it is difficult for them to escape the traps and return to being free charges without sufficient external energy, forming space charge effect.

Space charge can be divided into homo-charge and hetero-charge according to the polarity of the charge and that of its adjacent electrodes [3]. The former is used to describe the charge whose polarity is the same as that of its adjacent electrode, while the latter is used to describe the charge whose polarity is opposite to the adjacent electrode. The accumulation of homo-charge reduces the electrical stress near the interface and increases it inside the medium, while the accumulation of hetero-charge plays an opposite role. Generally, for the XLPE cable insulation, on the one hand, under a lower electrical stress, impurities (such as antioxidants and catalysts) in cable insulation are ionized, and the ionization charges migrate to the opposite polarity electrode and accumulate as hetero-charges. On the other hand, under a higher electrical stress, the injected charges are trapped by the traps near the electrodes and accumulate as homo-charge. It is well known that the accumulation and migration of space charge under DC stress will distort the local field and threaten the long-term reliability of cables in service. Moreover, the space charge distributions are also closely related to many characteristics of the material microstructure, such as dielectric conductance, electrode injection, charge recombination, and trap density variation. Many electrical properties of the material macrostructure, such as conductivity, breakdown, and aging, will also be degraded gradually due to the influence of space charge effect [4].

Besides, when a cable is loaded in service, a temperature gradient field forms across the cable insulation caused by the Joule heat originating from the current in the cable core conductor [5]. This affects the injection and migration characteristics of carrier, leading to a conductivity gradient and then an electric field gradient [6, 7]. The localized field enhancement in a polymeric cable is unpredictable due to space charge accumulation, posing a vital threat to cable insulation in service, especially under applied DC voltage reversal or outage. Therefore, research on space charge accumulation in cable insulation under a temperature gradient is key for HVDC plastic cables [8–13]. In addition, as a temperature gradient exists across the cable insulation, the density of the dielectric will be inhomogeneous, leading to the recovery of the space charge waveform being more complex [12].

This chapter is organized as follows. Firstly, the measurement method for space charge is discussed briefly. Then, the space charge measurements for coaxial cables are introduced. Thirdly, the recovery algorithm for space charge waveform in coaxial cables is reviewed. And lastly, Sect. 5 outlines the space charge behaviors for 10 and 160 kV coaxial cables, and the charge injection, migration and extraction characteristics are also analyzed.

2 Space Charge Measurement Methods

2.1 Development of Space Charge Measurement Technology

Space charge measurement technology provides a convenience for researchers to explore the characteristics of charge transport and trapping in dielectric insulation, and it has received increasing attentions for the measurement and evaluation of dielectric properties. This technology is often combined with other electrical and optical measurement techniques (such as conductance current test and electroluminescence test) to research the dielectric characteristics of insulation materials, such as the conductivity mechanisms, electrical aging characteristics and pre-breakdown characteristics.

The measurement and research of space charge can be traced back to the 1970s, when Hitachi, a Japanese company, first reported a qualitative study of the space charge in an HVDC cable based on the powder image method [14]. However, this technology can only be used for the qualitative research, and the measurement accuracy is not high enough. Later, thermal stimulation methods were used in the research of space charge, including thermally stimulated current (TSC), thermally stimulated surface potential (TSSP), thermoluminescence (TL) and so on. These measurement methods belong to the lossy measurements [15–19]. During the thermal stimulation measurement process, a thermal stimulation source is applied to the measured samples. In response to the external thermal stimulation, the space charge inside the measured samples produces electrical and optical signals, and through these signals the charge information can be obtained indirectly. These thermal stimulation techniques are still widely used until now. However, it is believed that extracting the real charge characteristics from the trap information obtained by these methods is difficult, and it is also difficult to distinguish the physical processes that cause electrical and optical signals during the thermal stimulation (thermionic steering, polarization, charge debonding and charge recombination).

A. Imburgia classified the space charge measurement methods into two categories, destructive and non-destructive methods [20]. And the non-destructive methods can also be divided into the following three categories: thermal, acoustic and optical space charge measurement methods. Presently, the thermal and optical space charge measurement techniques are relatively less used, the most widely used is the acoustic measurement method, which include pressure wave propagation (PWP) space charge measurement methods and pulsed electro-acoustic (PEA) space charge measurement methods [21–23]. According to the generation of the pressure wave pulse, the PWP method can be divided into piezoelectricity induced pressure propagation (PIPP) and laser induced pressure propagation (LIPP). When a pressure wave pulse is induced by an electric pulse and a piezoelectric material, the approach is called the PIPP method. And when it is induced by a laser signal, the approach is called the LIPP method. Both the hardware structures and the pulse signal characteristics for PEA and PWP space charge measurement techniques are different. Therefore, during space charge measurement, it is necessary to make a reasonable choice according to the measurement accuracy [20].

2.2 Application of PEA Space Charge Technology

The PEA space charge measurement technique has been widely used in the past 30 years as a simpler measurement method with a stronger anti-noise ability and a lower cost. The principle of the PEA method is to apply a narrow electric pulse to the measured sample, making the charge of the micro-unit inside the sample produce a small displacement. This small displacement propagates across the sample as a sound wave, the sound wave is captured by a piezoelectric sensor and transformed into a voltage signal, the space charge information of the sample could be obtained from the amplified voltage signal. The complete PEA space charge measurement process has been shown in Fig. 1 [24, 25].

Recently, many studies on space charge characteristics based on the PEA technique have been reported. G. Chen et al researched the influence of different electrode materials on space charge and found that a semi-conductor layer electrode is easier to inject charge than a metal electrode, the injection rate for the semi-conductor is higher [26]. Takada et al found that it is easier to accumulate hetero-charge in non-degassed samples, while vacuum degassed samples tend to accumulate homo-charge [27]. Xia Wang et al found that during voltage polarity reversal, space charge would show a "mirror effect" [28]. With the development of measurement technology and material temperature performance, the PEA space charge measurement equipment suitable for various working conditions has been gradually emerged, and it is widely

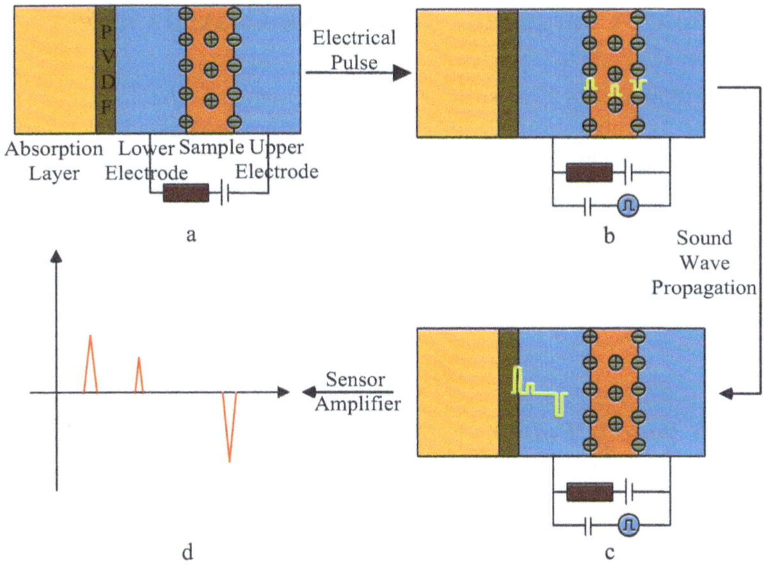

Fig. 1 The PEA space charge measurement process

used in the study of dielectric materials in various extreme temperature conditions and of the charge characteristics in micro areas.

At present, PEA space charge measurement technology can be divided into two types according to the measured sample structure, those are: plane samples and coaxial cable samples. Recently, a few studies on the space charge behaviors in XLPE cables under temperature gradients and DC voltages have been published [12, 13]. Despite recent progress on space charge detection technique in solid insulation, the measurement of space charge in actual coaxial cables with the PEA technique is still challenging (such as the formation of a temperature gradient field and the application of high stresses). In recent years, extensive studies have mainly focused on space charge measurements of plane samples because of the simplicity of the application of DC voltages and pulsed voltages. However, it is still questionable whether the results of a plane sample can reflect the space charge behaviors of an actual cable directly. To accurately assess the space charge characteristics of the DC cable insulation and make a scientific evaluation of the irreversible changes, it is necessary to measure the space charge behaviors for full-size DC cables under the operation conditions. However, due to the thicker insulation layer of the full-size HVDC cable, many technical difficulties should be solved to satisfy the demands of signal resolution and measurement sensitivity. And owing to the specificities of the acoustic propagation and electrical stress distribution in a coaxial cable insulation, the space charge waveform recovery method for a plane sample cannot be directly applied to a coaxial cable.

3 PEA Space Charge Measurement for Coaxial Cables

With the development of PEA space charge measurement technology, space charge measurement for cable sample has also made a great progress. K Fukunaga et al first tried to measure the space charge distribution in a coaxial cable insulation using PEA technology [29]. Until now, PEA space charge measurements for coaxial cables could be classified into the following three categories according to the pulse injection mode: (1) high voltage coupled pulse injection method, (2) outer semi-conductive layer partial stripping injection method and (3) pulse measurement electrode coupling injection method.

3.1 High Voltage Pulse Coupled Injection Method

K Fukunaga et al first measured the space charge distribution in cable insulation using high voltage pulse coupled injection method and M. Fu et al improved the measurement device [29, 30]. Figure 2 shows the schematic diagram of high voltage pulse coupled injection method.

During the measurement process, a high frequency nano second pulse and a HVDC signal are injected into the measured cable's core conductor through a coupling capacitance and a current-limited resistance respectively. This injection method is similar to that for plane PEA space charge measurement devices, and the coupled pulse voltage amplitude depends on the partial voltage value of the coupling capacitance and the cable body capacitance, as formula (1) shows:

$$V_{sa} = \frac{C_0}{C_0 + C_{cable}} V_P \tag{1}$$

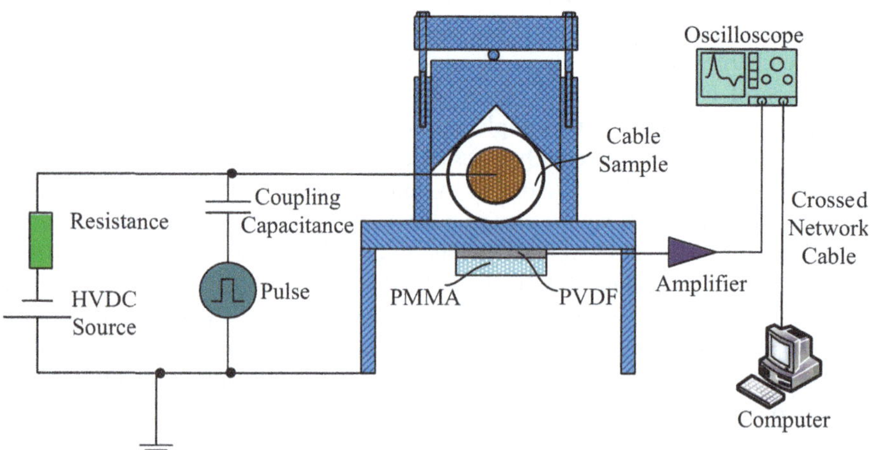

Fig. 2 High voltage pulse coupled injection PEA space charge measurement process [30]

In the formula, C_0 and C_{cable} are the coupling capacitance and the cable body capacitance respectively, V_{sa} and V_{p} are the coupled and injected pulse voltage respectively. However, in this device as mentioned above, the high frequency pulse signal is injected into the cable core. On the one hand, for the pre-identification and type tests of a full-size DC cable based on the CIGRE TB 219 and CIGRE TB 496 standards, the measured cable should be considered as a distributed capacitance when a high-frequency pulse signal is applied. For this high voltage pulse coupled injection method, a serious superposition effect occurs due to the reflection and attenuation during the pulse propagates along the cable, which has a significant effect on the space charge measurement. On the other hand, due to the capacitance coupling of the high frequency pulse, corona discharge may occur at the cable end, which not only influences the measurement but also threatens personal safety.

3.2 Outer Semiconductor Layer Partial Stripping Injection Method

B. Vissouvanadin et al improved the pulse injection method to avoid pulse signal reflection and attenuation, and the measurement schematic diagram of this method is shown in Fig. 3 [32].

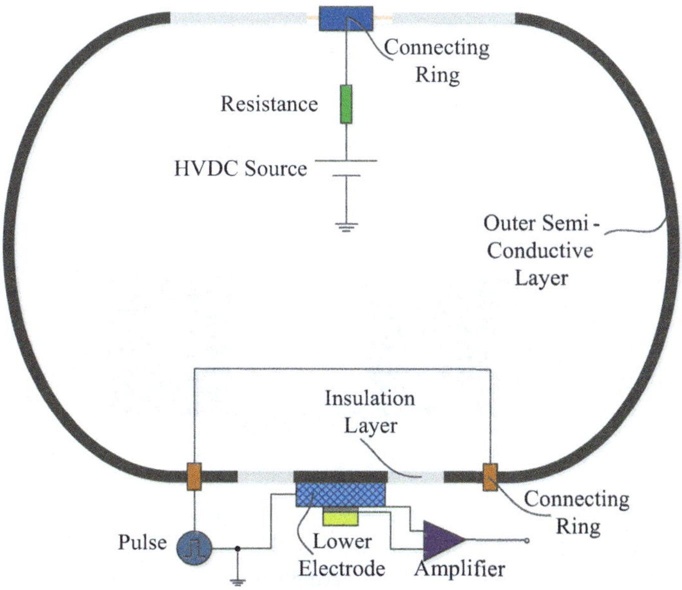

Fig. 3 Space charge measurement for pulse outer semi-conductive layer partial stripping injection [32]

The specific steps of the pulse outer semi-conductor layer partial stripping injection space charge measurement are as follows. First, two sections of the outer semi-conductive layer on both sides of the measurement position are removed. Then, a high frequency nano second pulse is injected into the cable from the outer semi-conductive layer on both sides of the exposed insulation layers through connecting rings, as shown in Fig. 3. This measurement device can be used to measure the space charge distribution of cable insulation with a higher voltage level and a longer length. The voltage amplitude of the coupled pulse depends on the relationship between the equivalent capacitances at the injection point and the other parts of the measured cable, as formula (2) shows.

$$V_{\mathrm{sa}} = \frac{2C}{2C + C_{\mathrm{r}}} V_{\mathrm{P}} \tag{2}$$

In the formula, C is the equivalent capacitance at the cable injection point, and C_{r} is the equivalent capacitance at the other parts. However, it is necessary to peel off some parts of the outer semi-conductor layer of the measured cable samples, which not only destroys the structure of the cable body, but also does not meet the requirements of the actual cable operation with a grounding of the outer semi-conductive layer.

3.3 Pulse Measurement Electrode Injection Method

Hozumi et al improved the cable space charge measurement device by changing the high-frequency nano second pulse injection mode, overcoming the waveform signal attenuation [31]. Therefore, the space charge measurement for long cables could be realized. Figure 4 shows the system diagram of the space charge measurement device

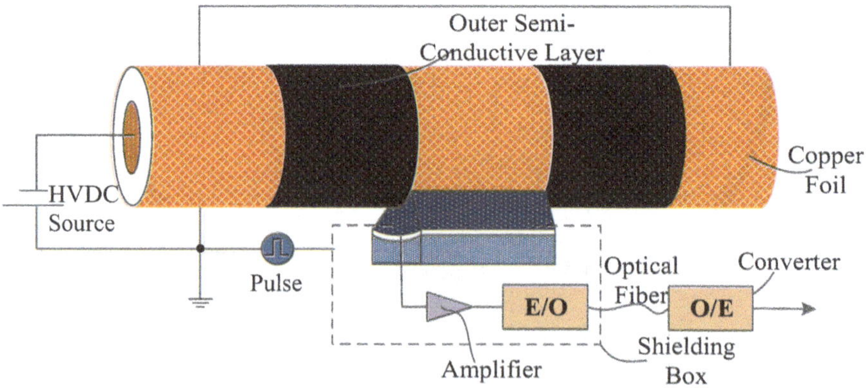

Fig. 4 System diagram of the space charge measurement device for pulse injection from the measurement electrode [31]

for pulse injection from the measurement electrode.

As shown in Fig. 4, only a part of the cable sheath should be stripped to expose the outer semi-conductive layer. Then, a high voltage nano second pulse is applied to the measurement electrode isolated from the ground, and the sheath on both sides of the cable is kept grounded. In addition, as the measurement electrode bears the pulse signal and is isolated from the ground, E/O conversion and optical fiber connection are needed to realize isolated transmission of the measurement signal. This pulse injection mode does not require the destruction of the cable body structure and is expected to be applied to space charge measurement of long cables in actual operation. And as the high-frequency nano-second pulse is injected directly from the measurement electrode, attenuation and distortion during pulse propagation are avoided. However, the separation length and the resistance of the cable outer shield could have significant impacts on the injection amplitude of the nano second pulse and then the measurement system sensitivity. In addition, at present, the maximum conversion frequency of the AD converters, used as the E/O and O/E conversion equipments in this measurement system, is only approximately 1 MHz, while the frequency of the wave signal output by the PEA device is between approximately 50–500 MHz, which leads to a loss of high frequency components of the detected wave signal.

Xia Wang et al developed an improved PEA space charge measurement system for a full-size cable under a temperature gradient condition by designing the signal detection and pulse injection units [33], as shown in Fig. 5. In this improved PEA measurement system, a nano second pulse signal is injected into the measurement electrode

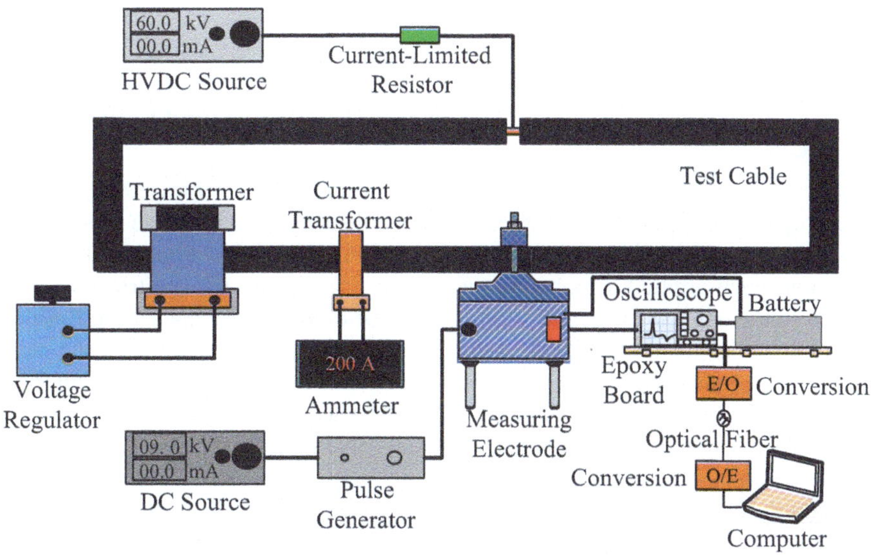

Fig. 5 Schematic diagram of the improved PEA space charge measurement system for coaxial cables under a temperature gradient

insulated from the ground by epoxy columns. Similar to the electrode, an oscilloscope is also placed at a high floating potential. A HVDC voltage signal is applied to the cable core conductor through a current-limited resistor, and the measurement electrode is tightly attached to the outer semi-conductor layer of the cable. During measurement, a operator can operate a computer safely and remotely control the oscilloscope through only a digital optical fiber connected to the oscilloscope.

Through an induced current heating device that consists of a current transformer, a transformer, an ammeter and a voltage regulator, the induced current is generated inside the cable core conductor and flows in a closed loop. The measured cable is passed through the transformer and functioned as the secondary winding. Therefore, the measured cable is heated due to the resistive loss, the temperature of the inner conductor is higher than the outer semi-conductive layer. In this way, a temperature gradient is radially formed along the insulation. A voltage regulator is used to apply a voltage to the transformer primary winding. A current transformer and an ammeter are used to induce and measure the induced current of the measured cable. And a thermocouple sensor is used to measure the temperature of the cable outer semi-conductive layer. This PEA system can not only improve the measurement accuracy but also decrease the costs. Moreover, it has many advantages such as no pulse attenuation or corona discharge, no damage to the outer semi-conductive layer and less bandwidth limitation. Hence, it is suitable for space charge measurement of the long full-size HVDC cables.

4 Recovery Algorithm for the Space Charge Waveform in a Coaxial Cable

4.1 Propagation Principle of a Pressure Wave in a Coaxial Cable

The propagation principle of a pressure wave inside a coaxial cable is shown in Fig. 6. The electrical stress has a non uniform radial distribution along the insulation direction when applied a high external voltage. Therefore, the electrical stress is related to the applied voltage as well as the position inside the insulation. Besides, due to the coaxial structure of the cable, the pressure wave generated by the pulse signal propagates radially along the insulation in a divergent state. As a result, both the divergent pressure wave and non uniform electrical stress together determine the charge distribution behavior inside the coaxial cable insulation [12].

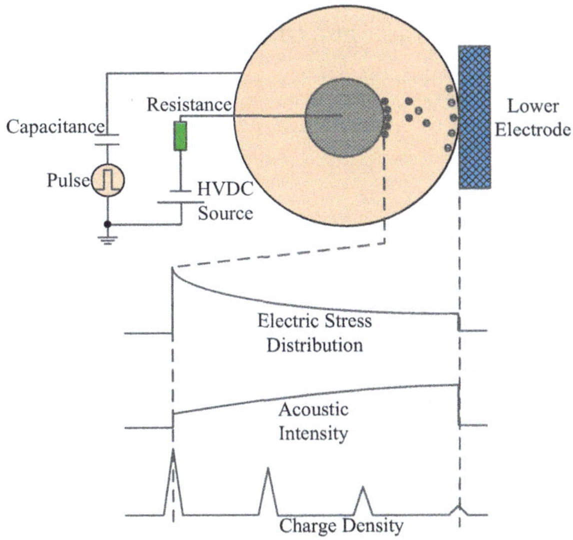

Fig. 6 Schematic diagram of pressure wave propagation and charge distribution in coaxial cable [4]

4.2 Calibration of the Space Charge Distribution in a Coaxial Cable

Once a nano second pulse is applied into a cable, the interaction forces, generated by the charges (internal charge and induced charges on the electrodes) and pulsed electrical stress can be written as follows:

$$f(a,t_c) = \sigma(a)\frac{v_p(t_c)}{a\ln(\frac{b}{a})} + \frac{\varepsilon}{2}[\frac{v_p(t_c)}{a\ln(\frac{b}{a})}]^2 \tag{3}$$

$$f(b,t_c) = \sigma(b)\frac{v_p(t_c)}{b\ln(\frac{b}{a})} + \frac{\varepsilon}{2}[\frac{v_p(t_c)}{b\ln(\frac{b}{a})}]^2 \tag{4}$$

$$\Delta f(r,t_c) = \rho(r)\Delta r\frac{v_p(t_c)}{r\ln(\frac{b}{a})} \tag{5}$$

In these formulas, $f(a,t_c)$ and $f(b,t_c)$ represent the interaction forces generated by the pulse and the charges at the interfaces between the electrodes and the insulation, $\Delta f(r,t_c)$ is generated by the charges in the cable insulation. *a*, *b* and *r* are the radius of the inner conductor, the radius of the outer semi-conductor, and the space charge location in the cable insulation. σ and ρ represent the surface charge density and volume charge density, $v_p(t)$ is the nano second pulse, and ε represent the permittivity of the insulation.

The pressure wave induced by the space charge and pulse can be described as:

$$p(a, t_c) = k_0 k_2 k_3 k_4 \sigma(a) \frac{v_p(t_c - \frac{b-a}{u_{sa}})}{a\ln(\frac{b}{a})} \sqrt{\frac{a}{c}} c \tag{6}$$

$$p(b, t_c) = k_1 k_3 k_4 \sigma(b) \frac{v_p(t_c)}{b\ln(\frac{b}{a})} \sqrt{\frac{b}{c}} c \tag{7}$$

$$\Delta p(r, t_c) = 0.5 k_2 k_3 k_4 \rho(r) \Delta r \frac{v_p(t_c - \tau)}{r\ln(\frac{b}{a})} \sqrt{\frac{r}{c}} c \tag{8}$$

In these formulas, $p(a,t_c)$, $p(b,t_c)$, and $\Delta p(r,t_c)$ represent the pressure waves generated by the pulse interacted with the charges at the inner electrode interface and the inner insulation, at the outer electrode interface and the outer semi-conductive layer, and in the cable insulation layer, t_c is the time when the PVDF piezoelectric sensor receives the pressure wave, and c represent the distance between the cable core and the piezoelectric sensor. k_0 and k_1 are the forward wave coefficients of electrode excitation, which can be defined by the electrodes impedance of the interface and obtained as 0.5 and 0.9 according to the acoustic impedance coefficients of the insulation material and electrode. k_2 and k_3 are the transmission coefficients of pressure waves at the interfaces between the insulation and the outer semi-conductor, and between the outer conductor and piezoelectric sensor, k_4 is the absorption coefficient of the piezoelectric sensor, k_2, k_3 and k_4 can be obtained as 1.8, 0.37 and 1 respectively. τ is the time in which the pressure wave propagates from point r to the measured electrode.

The pressure waves detected by the PVDF piezoelectric sensor can be written as:

$$\begin{aligned} p(c, t_c) &= p(a, t_c) + p(r, t_c) + p(b, t_c) \\ &= \frac{0.5 k_2 c k_3}{\ln(b/a)} [\frac{\sigma(b)}{\sqrt{bc}} v_p(t_c) + u_{sa} \int_a^b \frac{\rho(\tau)}{\sqrt{rc}} v_p(t_c - (b - r)/u_{sa}) dr \\ &\quad + \frac{\sigma(a)}{\sqrt{ac}} v_p(t_c - (b - a)/u_{sa})] \end{aligned} \tag{9}$$

Therefore, the pressure waves in the frequency domain can be obtained as:

$$\begin{aligned} p(c, f) &= p(a, f) + p(r, f) + p(b, f) \\ &= \frac{0.5 k_2 c k_3}{\ln(\frac{b}{a})} (\frac{\sigma(b)}{\sqrt{bc}} + u_{sa} \Delta\tau R_0(f) + \frac{\sigma(a)}{\sqrt{ac}}) V_p(f) \end{aligned} \tag{10}$$

$$u_{sa} \int_a^b \frac{\rho(\tau)}{\sqrt{(b - u_{sa}\tau)c}} v_p(t - \tau) d\tau \Leftrightarrow u_{sa} \Delta\tau R_0(f) V_P(f) \tag{11}$$

The electrical signal detected by the PVDF piezoelectric sensor is as follows:

$$V(f) = S(f)[\frac{\sigma(b)}{u_{sa}\Delta\tau\sqrt{bc}} + R_0(f) + \frac{\sigma(a)}{u_{sa}\Delta\tau\sqrt{ac}} \exp(-j2\pi f \frac{b - a}{u_{sa}})] \tag{12}$$

In the formula, $S(f)$ represents the transfer function of PVDF piezoelectric sensor. For the interface of the lower electrode and the cable insulation, the relationships between the electrical signal and the space charge can be expressed as:

$$V_0(f) = S(f)\frac{\sigma_0(b)}{u_{sa}\Delta\tau\sqrt{bc}} \tag{13}$$

$$\sigma_0(b) = \varepsilon_0\varepsilon_r\frac{V_{dc}}{b\ln(\frac{b}{a})} \tag{14}$$

In these formulas, ε_0, ε_r and V_{dc} represent the vacuum permittivity, relative permittivity, and DC voltage. Through the formulas (13) and (14), the transfer function of the PVDF piezoelectric sensor $S(f)$ can be obtained. So the corrected space charge distribution can be obtained by substituting the transfer function $S(f)$ into formula (12) [12].

$$[\frac{\sigma(b)}{\sqrt{bc}} + u_{sa}\Delta\tau R_0(f) + \frac{\sigma(a)}{\sqrt{ac}}\exp(-j2\pi f\frac{b-a}{u_{sa}})] = \varepsilon_0\varepsilon_r\frac{V_{dc}}{b\ln(b/a)}\frac{V(f)}{V_0(f)} \tag{15}$$

4.3 Recovery of Attenuation and Dispersion of the Space Charge in a Coaxial Cable

In addition, attenuation and dispersion will occur during the pressure wave propagation along the coaxial cable insulation. Similar to the propagation inside the plane samples, the amplitude and pulse width of space charge waveform in a cable are also attenuated and broadened:

$$P(r, t_b + \Delta t) = P(b, t_b)e^{-\alpha(f)(b-r)-j\beta(f)(b-r)} \tag{16}$$

In the formula, $(b\text{-}r)$ represent the distance between the position in the cable insulation and the outer semi-conductor electrode. The attenuation coefficient $\alpha(f)$ and dispersion coefficient $\beta(f)$ satisfy:

$$\alpha(f) = -\frac{1}{b-a}\ln[\frac{P(f,b)}{P(f,a)\sqrt{a/b}}] \tag{17}$$

$$\beta(f) = \frac{1}{b-a}|\phi(f,b) - \phi(f,a)| \tag{18}$$

From the divergent acoustic propagation and the non uniform electrical stress, the acoustic propagation at position a is $(b/a)^{1/2}$ times of that at position b. For the equations from (16) to (18), the effect of the coaxial geometric is eliminated by the factor $(b/a)^{1/2}$. Therefore, the transfer function can be written as [12]:

$$G(f, r) = \frac{P(f, r)}{P(f, a)} = e^{-[\alpha(f)+j\beta(f)](b-r)} \quad (19)$$

5 Space Charge Characteristics of Coaxial Cables

In this chapter, space charge behaviors for the 10 kV XLPE cables under different temperature gradient conditions and different DC electrical stresses were introduced detailed, which were all measured through the PEA coaxial cables space charge measurement system showed in Fig. 5 [33]. During measurements, the amplitude of the applied high voltage pulse is 3 kV and the polarization time is 5 h.

Firstly, space charge behaviors for 10 kV XLPE cables under 10 kV/mm electrical stress and 0° C, 10° C and 20° C temperature gradients were measured, as Fig. 7 shown. Hetero-charges accumulate in the adjacency of the outer semi-conductor electrode and the inner conductor electrode of the cables, and the amount increases with polarization time. In addition, the hetero-charges near the inner conductor are less than those near the outer semi-conductor. Simultaneously, the induced charge peaks of the inner electrodes decrease. Therefore, the accumulation of the hetero-charges in the adjacency of both electrodes could be ascribed to the extraction blocking of the XLPE-SC electrode interfaces. And when applied a constant DC voltage on the cable, the electrical stress of the inner conductor is higher than that of the outer semi-conductor electrode due to the coaxial structure in the absence of space charge. This factor causes the imbalance of charge injection and extraction between the outer semi-conductor and inner conductor electrodes. Therefore, larger numbers of charges are injected from the inner conductor and migrated to the outer semi-conductor electrode, accumulate as hetero-charges there. In addition, the hetero-charges in the adjacency of the outer semi-conductor increase with temperature gradients, while these near the inner conductor show no obvious variety as less hetero-charge accumulates there.

In addition, when XLPE cables used in the traditional current source converter transmission system, there still exist some technical difficulties. This is mainly because during the polarity reversal process, space charge effect not only exists but also has a certain lag effect, which could cause the electric stress distortion in the cable insulation. Therefore, in this chapter, space charge characteristics in the 10 kV XLPE cable during voltage reversal are also measured and analyzed, the temperature differences of the two electrodes are set as 0° C and 20° C respectively. During measurement, 10 kV/mm electrical stress was applied for the measured cables for 5 h, then a quick voltage polarity reversal from 10 to -10 kV/mm was carried out.

Figure 8 shows the space charge behaviors in the 10 kV coaxial cables during voltage reversal under the temperature gradients of 0 and 20° C. Obviously, it could be seen from Fig. 8 that during voltage reversal, the charge peaks of the inner conductor reverse with the applied voltage quickly. However, the charge peaks of the cable outer semi-conductor do not reverse immediately but their amplitudes decrease gradually.

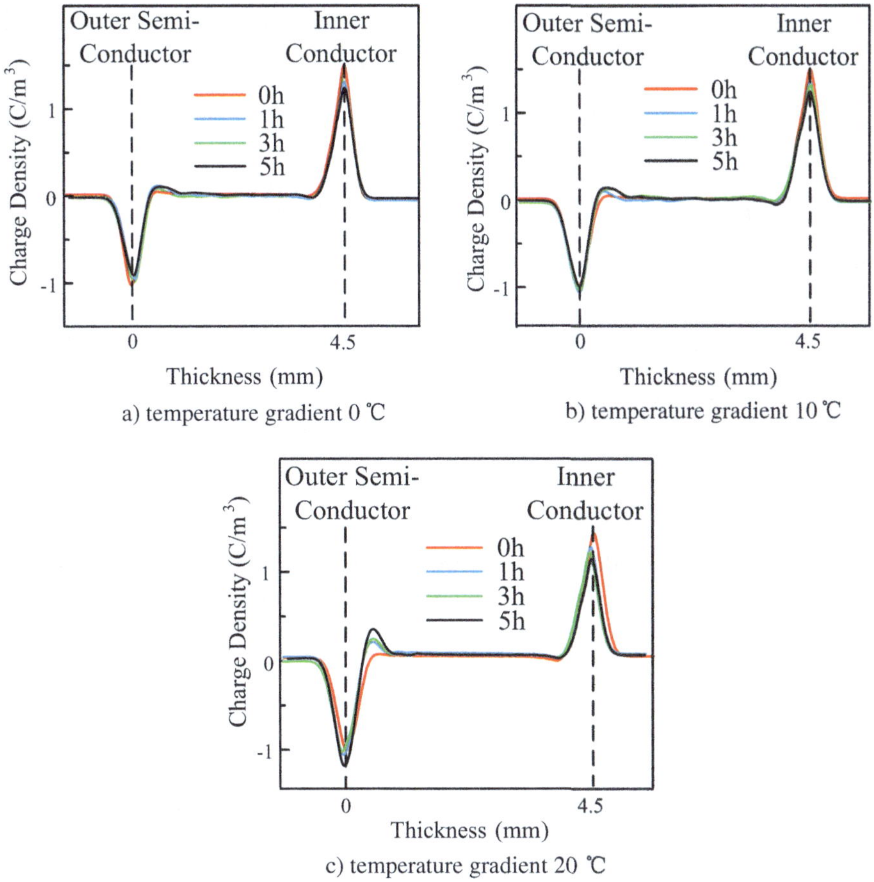

Fig. 7 Space charge waveforms in the 10 kV coaxial XLPE cables under 10 kV/mm electrical stress and different temperature gradients

After approximately 0.5 h, the charge peaks of the outer semi-conductor reverse evidently, and it costs about 3 h reaching another balance state.

When no DC voltage applied, if a point charge q exists at x position of the XLPE cable insulation, the "image charges" on the two electrodes induced by the point charge q are σ_1 and σ_2. So the electrical stresses at the interfacial positions are as follows:

$$E_1 = \frac{\sigma_1}{2\pi a \varepsilon_0 \varepsilon_r}$$
$$E_2 = \frac{\sigma_2}{2\pi b \varepsilon_0 \varepsilon_r} \quad (20)$$

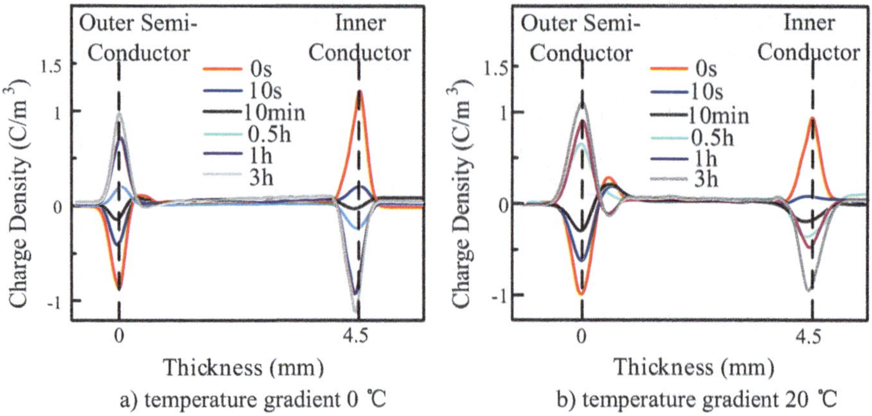

Fig. 8 Space charge behaviors in 10 kV coaxial cables during voltage reversal under different temperature gradients

In the formulas, ε_r indicates the permittivity of XLPE insulation. *a*, *b* indicate the cable inner conductor and outer semi-conductor positions. At *x* position, as the existence of the point charge *q*, the electrical stress would be distorted as follow:

$$E(r) = \frac{\sigma_1}{2\pi r \varepsilon_0 \varepsilon_r}, \ a \le r < x$$
$$E(r) = \frac{\sigma_1 + q}{2\pi r \varepsilon_0 \varepsilon_r}, x \le r \le b \tag{21}$$

When the applied DC voltage is *U*, that is:

$$\int_a^b E(r)\mathrm{d}r = U \tag{22}$$

And the relationship between the charges σ_1, σ_2 and q should also be taken into consideration:

$$\sigma_1 + \sigma_2 + q = 0 \tag{23}$$

From all the equations above, the "image charge" induced on the two electrodes are:

$$\begin{cases} \sigma_1 = \frac{2\pi\varepsilon_0\varepsilon_r U}{\ln(b/a)} - \int_a^b \rho(r)\ln(a/r)\mathrm{d}r \\ \sigma_2 = -\frac{2\pi\varepsilon_0\varepsilon_r U}{\ln(b/a)} - \int_a^b \rho(r)[1 + \ln(r/a)]\mathrm{d}r \end{cases} \tag{24}$$

Here the point charge q is replaced by the distributed charge $\rho(r)$. From Eq. (24), it could be obtained that the charge on the electrodes consist of two parts: accumulative space charge and polarization charge. Figure 7 shows more hetero-charges accumulate near the outer semi-conductor than the opposite electrode under 10 kV/mm. From Fig. 8, during voltage polarity reversal, the inner conductor charge peak reverses quickly while the outer semi-conductor charge peak decreases gradually but without quickly reverse. In the Eq. (24), as more hetero-charges accumulate near the outer semi-conductor, the position x in the equation is near the outer semi-conductor of the XLPE cables. Therefore, $\int_a^b \rho(r)[1+\ln(r/a)]\mathrm{d}r$ may be larger than $\int_a^b \rho(r)\ln(a/r)\mathrm{d}r$ which are all decided by the x position, and as almost no obviously charge accumulates near the inner conductor electrode, $\int_a^b \rho(r)\ln(a/r)\mathrm{d}r$ can be neglected.

During the reversal of the applied voltage, the charge peak near the inner conductor (σ_1) reverses rapidly. And for the charge peak near the outer semi-conductor (σ_2), $\int_a^b \rho(r)[1+\ln(r/a)]\mathrm{d}r$ can not be neglected. Therefore, the polarity of this charge peak depends on the difference of $\frac{2\pi\varepsilon_0\varepsilon_r U}{\ln(b/a)}$ and $\int_a^b \rho(r)[1+\ln(r/a)]\mathrm{d}r$, so the charge peak of the outer semi-conductive layer does not reverse immediately. After the reversal of applied voltage, the voltage polarity on the cable inner conductor changes, the opposite polarity charges inject from the inner conductor and migrate towards the outer semi-conductor, changes the polarity of the difference between $\frac{2\pi\varepsilon_0\varepsilon_r U}{\ln(b/a)}$ and $\int_a^b \rho(r)'[1+\ln(r/a)]\mathrm{d}r$, causes the reverse of the charge peak near the outer semi-conductor electrode.

And to research the space charge characteristics in 10 kV XLPE cables furthermore, charge waveforms under 15 and 20 kV/mm were measured. The space charge waveforms in the 10 kV XLPE cables for steady state under different electrical stresses are collected together and presented in Fig. 9.

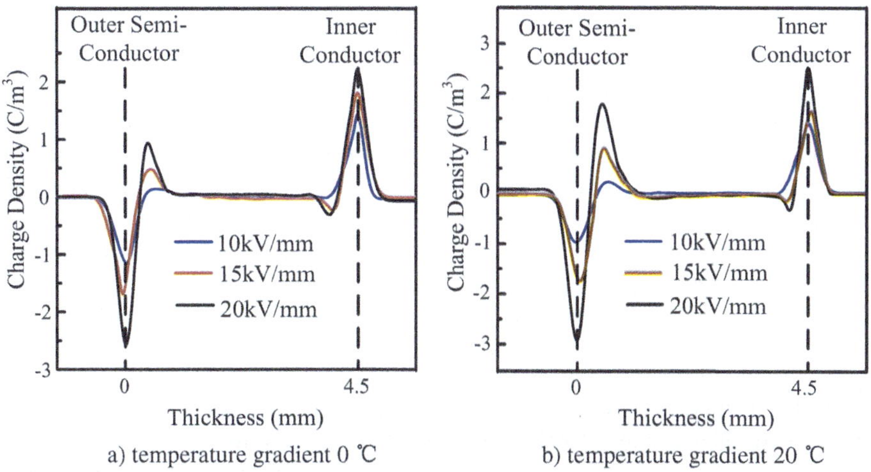

Fig. 9 Space charge waveforms in the 10 kV XLPE cables under different electrical stresses and temperature differences

As shown in Fig. 9, the accumulation of hetero-charge near both sides of the electrodes increases with the applied DC electrical stresses. And compared with the measurement results at room temperature, the amount of hetero-charge accumulates near the inner conductor electrode decreases while those near the outer semi-conductor increases under the temperature gradient conditions. These could be attributed to the influence of the temperature gradient on the charge injection from the electrode, mobility and charge extraction. On the one hand, as the temperature difference between the outer semi-conductor and the inner-conductor, charge injection at the higher temperature electrode (inner conductor electrode) is more powerful than that at the lower temperature electrode (outer semi-conductor electrode), which is equal for the charge extraction rate. On the other, as the temperature dependence of the carrier mobility for XLPE cable insulation, the charge carriers injected from the inner conductor electrode are easier to migrate towards the outer semi-conductor electrode. Therefore, the coaxial structure effect together with the temperature gradient effect promotes the accumulation of the hetero-charge in the adjacency of the outer semi-conductor electrode of the cables.

Besides, the temperature gradient effect enhances the conductivity gradient along the radial direction of cable insulation, causing the accumulation of the hetero-charges. However, to what extent the conductivity gradient has an effect on the hetero-charges' accumulation is questionable. Due to the higher extraction barrier of the PE/electrode interface [34], blocking effect plays a significant role for the charge accumulation characteristics [35]. Previous reports found that compared with space charge behaviors caused by the difference of the charge injection and extraction rates at PE/electrode interfaces, the conductivity gradient effect causes a little effect on charges in PE [34]. The temperature gradient and coaxial structure could enhance this effect.

From the space charge behaviors above, it can be concluded that more hetero-charge accumulation near the outer semi-conductor electrode due to the coaxial structure effect. And under temperature gradient conditions, larger hetero-charges appear near the outer semi-conductor which acts as the lower temperature electrode and the amount of the hetero-charge increases with the temperature gradient. Therefore, both the temperature gradient effect and the coaxial structure effect increase the hetero-charges amounts near the outer semi-conductive layer. And during voltage polarity reversal process, the charge peak at cable inner conductor reverses with voltage quickly. However, the charge peak at the outer semi-conductive layer doesn't reverse quickly but decreases gradually as the influence of residual charges.

In addition, space charge characteristics in 160 kV HVDC cables under a 10 kV/mm electrical stresses and temperature gradients (10, 20, 30 and 40° C) have also been measured and analyzed, the space charge measurement and temperature controlled methods have been reported in ref [13]. Figure 10 shows the space charge behaviors of the measured 160 kV cable at room temperature.

From Fig. 10 it can be seen there is almost no charge injection into the cable insulation, which causes a slight extension of the induced charge peak during poling. And Fig. 11 shows the space charge behaviors in 160 kV DC cables under temperature

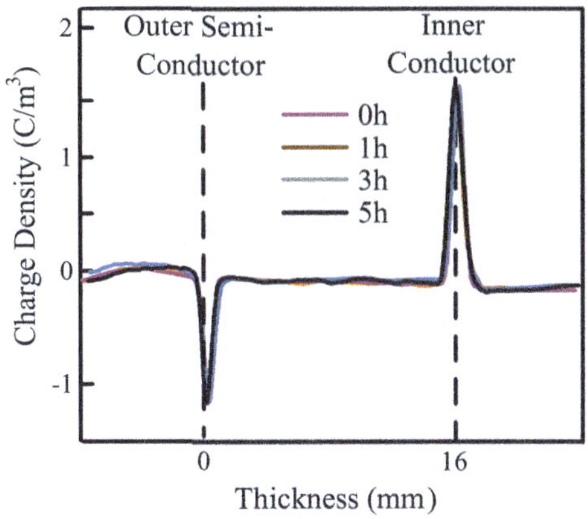

Fig. 10 Space charge waveforms in the 160 kV XLPE DC cables under 10 kV/mm electrical stresses

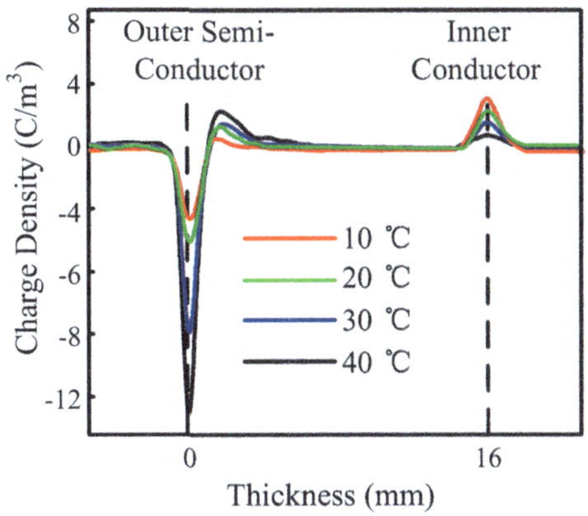

Fig. 11 Space charge waveforms in the 160 kV XLPE DC cables under temperature differences

gradients, it can be concluded from the figure that obvious hetero-charge accumulation appears near the outer semiconductor electrode under a temperature gradient, and the hetero-charge amount increases with the temperature gradient. Simultaneously, the charge peak at the HVDC cable conductor decreases gradually, indicating the hetero-charge comes from the cable inner conductor charge injection and then migrates to the opposite electrode, which shows a similar charge accumulation behavior with the 10 kV cables. Therefore, the outer semi-conductive layer should attract more attention during the cable operation process.

6 Conclusion

During the long-term operation of DC cables, the space charge characteristic serious influences the reliability of the cable insulation. In this section, the PEA space charge measurement methods were introduced detailed. And space charge behaviors for 10 and 160 kV coaxial cables were analyzed and researched respectively. The following conclusions can be obtained:

(1) Due to the coaxial structure effect of the measured cables, the electric stress distribution across the cable insulation layer is uneven, which is higher at the inner conductor than that for the outer semi-conductive layer electrode in the absence of space charge. For the 10 kV coaxial cable, a large number of hetero-charges are injected from the inner conductor electrode, migrate to the outer semi-conductive layer electrode and accumulate there. Compared with the outer semi-conductor, there is less hetero-charge accumulation near the inner conductor. For the measured 160 kV DC cables, at room temperature, there is almost no charge accumulation under a 10 kV/mm electrical stress.
(2) During the polarity reversal test, the charge peak at the inner conductor of 10 kV coaxial cables reverse relatively quickly with the applied voltage reversal, while that at the outer semi-conductive layer decrease gradually with the voltage reversal instead of reversing quickly as the influence of residual charges.
(3) For the 10 kV coaxial cable, compared with the space charge test at room temperature, the amount of hetero-charge accumulated in the outer semi-conductive layer increases, and that near the inner conductive electrode decreases. For the 160 kV DC cables, obvious heterocharge accumulation appears near the outer semi-conductor electrode, and the charge amount increases with the temperature gradient. This is due to the charge injection, migration and extraction rates at the inner conductor electrode are all higher than those at the outer semi-conductive layer electrode caused by the temperature effect. Consequently, the outer semi-conductive layer of cables should be paid more attention for the long-term reliability evaluation to cable safety operations.

References

1. D. Fabiani, G.C. Montanari, C. Laurent et al., Polymeric HVDC cable design and space charge accumulation. Part 1: Insulation/Semicon Interface. IEEE Electr. Insul. Mag. **23**(6), 11–19 (2007)
2. S.S. Bamji, A.T. Bulinski, R.J. Densley, Degradation of polymeric insulation due to photoemission caused by high electric fields. IEEE Trans. Electr. Insul. **24**(1), 91–98 (1989)
3. X. Chen, X. Wang, K. Wu et al., Space charge measurement in LDPE films under temperature gradient and DC stress. IEEE Trans. Dielectr. Electr. Insul. **17**(6), 1796–1805 (2010)
4. T. Muronaka, Y. Tanaka, T. Takada et al. Measurement of space charge distribution in XLPE cable using PEA system with flat electrode. Electrical Insulation and Dielectric Phenomena, Proceedings of Conference on Electrical Insulation and Dielectric Phenomena, pp. 266–269 (1996)
5. K. Wu, X. Chen, X. Wang et al. Effects of temperature gradient on space charge accumulation at applied voltage reversal in LDPE film. 2010 10th IEEE International Conference on Solid Dielectrics, pp. 1–4 (2010)
6. X. Chen, X. Wang, K. Wu et al., Effect of voltage reversal on space charge and transient field in LDPE films under temperature gradient. IEEE Trans. Dielectr. Electr. Insul. **19**(1), 140–149 (2012)

7. Y. Li, M. Yasuda, T. Takada, Pulsed electroacoustic method for measurement of charge accumulation in solid dielectrics. IEEE Trans. Dielectr. Electr. Insul. **1**(2), 188–195 (1994)
8. K. Fukunaga, Innovative PEA space charge measurement systems for industrial applications. IEEE Electr. Insul. Mag. **20**(2), 18–26 (2004)
9. W. Choo, G. Chen, S.G. Swingler, Electric field in polymeric cable due to space charge accumulation under DC and temperature gradient. IEEE Trans. Dielectr. Electr. Insul. **18**(2), 596–606 (2011)
10. W. Wang, D. He, S. Chen et al., Correction of acoustic wave in PEA space charge measurement on cable under temperature gradient. High Voltage Eng. **41**(4), 1084–1089 (2015)
11. H. Wang, K. Wu, Q. Zhu et al., Recovery algorithm for space charge waveform under temperature gradient in PEA method. IEEE Trans. Dielectr. Electr. Insul. **22**(2), 1213–1219 (2015)
12. X. Wang, C. Chen, J. Hao et al., Recovery algorithm for space charge waveform in coaxial cable under temperature gradient. Sens. Mater. **29**(8), 1147–1157 (2017)
13. X. Wang, C. Chen, C. Cheng et al., Space charge characteristics in 160 kV DC XLPE cable under temperature gradient. IEEE Trans. Dielectr. Electr. Insul. **25**(6), 2366–2374 (2018)
14. M.S. Khalil, International research and development trends and problems of HVDC cables with polymeric insulation. IEEE Electr. Insul. Mag. **13**(6), 35–47 (1997)
15. K.S. Suh, J. Tanaka, D. Damon, What is TSC? IEEE Electr. Insul. Mag. **8**(6), 13–20 (1992)
16. T. Nishitani, K. Yoshino, Y. Inuishi, Simultaneous observation of thermally stimulated current and thermoluminescence in polyethylene. Jpn. J. Appl. Phys. **14**(5), 721–722 (1975)
17. Y. Suzuoki, T. Mizutani, M. Ieda, Thermally stimulated currents from carriers trapped in polyethylene. Jpn. J. Appl. Phys. **15**(5), 929–930 (1976)
18. J. Belana, M. Mudarra, J. Calaf et al., TSC study of the polar and free charge peaks of amorphous polymers. IEEE Trans. Electr. Insul. **28**(2), 287–293 (1993)
19. R.E. Collins, Analysis of spatial distribution of charges and dipoles in electrets by a transient heating technique. J. Appl. Phys. **47**(11), 4804–4808 (1976)
20. A. Imburgia, R. Miceli, E.R. Sanseverino et al., Review of space charge measurement systems: acoustic, thermal and optical methods. IEEE Trans. Dielectr. Electr. Insul. **23**(5), 3126–3142 (2016)
21. P. Notingher, S. Agnel, A. Toureille, Thermal step method for space charge measurements under applied dc field. IEEE Trans. Dielectr. Electr. Insul. **8**(6), 985–994 (2001)
22. S. Hole, T. Ditchi, A. Lewiner, Non-destructive methods for space charge distribution measurements: what are the differences? IEEE Trans. Dielectr. Electr. Insul. **10**(4), 670–677 (2003)
23. K. Fukunaga, Progress and prospects in PEA space charge measurement techniques[Feature Article]. IEEE Electr. Insul. Mag. **24**(3), 26–37 (2008)
24. H. Kitajima, Y. Tanaka, T. Takada et al. Measurement of space charge distribution at high temperature using the pulsed electro-acoustic (PEA) method. Seventh International Conference on Dielectric Materials, Measurements and Applications, pp. 8–11 (1996)
25. I.H. Park, Y.H. Kwon, S.H. Nam et al., The effects of space charge on conduction mechanism in polymer dielectric-air gap structure. Ann. Report Conf. Electr. Insul. Dielectr. Phenomena **2000**, 149–152 (2000)
26. G. Chen, T.Y.G. Tay, A.E. Davies et al., Electrodes and charge injection in low-density polyethylene using the pulsed electroacoustic technique. IEEE Trans. Dielectr. Electr. Insul. **8**(6), 867–873 (2001)
27. Y. Li, T. Takada, Experimental observation of charge transport and injection in XLPE at polarity reversal. J. Phys. D Appl. Phys. **25**, 704–716 (1992)
28. X. Wang, D. Tu, Y. Tanaka et al., Space charge in XLPE power cable under dc electrical stress and heat treatment. IEEE Trans. Dielectr. Electr. Insul. **2**(3), 467–474 (1995)
29. K Fukunaga. Measurement of space charge distribution in cable insulation using the pulsed electroacoustic method. Conference. on polymer insulated power cables, 1991
30. M. Fu, G. Chen, A.E. Davies. A modified PEA space charge measuring system for power cables. Proceedings of the 6th International Conference on Properties and Applications of Dielectric Materials, pp. 104–107 (2000)

31. N. Hozumi, T. Okamoto, T. Imajo. "Space charge distribution measurement in a long size XLPE cable using the pulsed electroacoustic method. Conference Record of the 1992 IEEE International Symposium on Electrical Insulation, pp. 294–297 (1992)
32. B. Vissouvanadin, T.T.N. Vu, L. Berquez et al., Deconvolution techniques for space charge recovery using pulsed electroacoustic method in coaxial geometry. IEEE Trans. Dielectr. Electr. Insul. **21**(2), 821–828 (2014)
33. C. Chen, Y. Bu, X. Wang et al., A comparison of space charge behaviors in coaxial cable and film sample under temperature gradient. IEEE Trans. Dielectr. Electr. Insul. **26**(6), 1941–1948 (2019)
34. K Wu, Z Lv, Q Zhu, et al. Space charge formation and conductivity characteristics of PE and oil impregnated paper under a temperature gradient. Proceedings of 2014 International Symposium on Electrical Insulating Materials, pp. 93–96 (2014)
35. D. Fabiani, G.C. Montanari, C. Laurent et al., HVDC Cable Design and Space Charge Accumulation. Part 3: Effect of Temperature Gradient[Feature article]. IEEE Electr. Insul. Mag. **24**(2), 5–14 (2008)

Electric Ageing and Breakdown Phenomenon in Polypropylene Under Complex DC Operating Condition

Lewei Zhu and Boxue Du

Abstract In this chapter, the electric ageing and breakdown phenomenon of polypropylene (PP) under impulse superimposed DC voltage was studied first. Then three different polycyclic compounds were added to PP, namely 4,4′-bis(dimethylamino)benzil, 2-hydroxy-2-phenylacetophenon and 4-phenylbenzophenone. The electrical tree aging experiments were carried out on these samples at 30, 60 and 90 °C under impulse superimposed DC voltage. The effect was compared and the polycyclic compound that inhibited the electrical tree best was selected, and then the addition amount of the polycyclic compound was changed from 0.1 to 0.3 wt.% to get the best inhibition effect. The results show that under impulse superimposed DC voltage, different DC amplitude, pulse amplitude and pulse frequency have an effect on electric ageing and breakdown phenomenon. 2-hydroxy-2-phenylacetophenon can inhibit the growth of electrical tree. However, 4,4′-bis(dimethylamino)benzil and 4-phenylbenzophenone promote electric ageing and breakdown. Adding 0.1 wt.% of 2-hydroxy-2-phenylacetophenon polycyclic compound to PP can effectively inhibit the growth of electrical tree and inhibit electric ageing and breakdown phenomenon. It is a potential environmental friendly type HVDC cable insulation material.

1 Introduction

Electrical tree is an electric aging and breakdown phenomenon of insulating materials. Once it passes through the insulating layer, electric breakdown will occur, which will affect the operation safety of polymer cable [1]. Under the pure DC voltage, it is difficult to trigger the electrical tree, and the voltage of tree initiation is very high [2]. Due to the operation of the isolating switch and the short circuit fault of the

L. Zhu (✉)
Tianjin University of Technology, No. 391 Binshui West Road, Xiqing District, Tianjin, China
e-mail: zhu.lewei@email.tjut.edu.cn

B. Du
Tianjin University, No. 92 Weijin Road, Nankai District, Tianjin, China
e-mail: duboxue@tju.edu.cn

B. Du (ed.), *Polymer Insulation Applied for HVDC Transmission*,
https://doi.org/10.1007/978-981-15-9731-2_8

HVDC cable, operation surge overvoltage often occurs [3, 4]. Electrical trees are easy to be triggered and cause breakdown under impulse voltage [5]. However, the initiation and breakdown of electrical tree under impulse superimposed DC voltage are still unclear. Through the study of the electrical tree growth characteristics in XLPE under impulse voltage, it was found that the growth characteristics of electrical tree under DC and pulse voltages are significantly different. The pulse voltage can cause the formation of electrical tree due to the detrapping of electrical charges [6]. The literature [7] reported the breakdown characteristics of modified high density polyethylene (HDPE) under DC and pulse superimposed voltage. The results showed that the pulse breakdown voltage superimposed on the DC was lower than the DC breakdown voltage alone. The literature [8] reported the electrical tree characteristics of epoxy resins under DC and AC composite voltages. The results showed that the electrical tree growth characteristics had a certain relationship with the polarity of the DC voltage. The literature [9] studied the growth characteristics of cross-linked polyethylene (XLPE) under DC and AC composite voltages. It was found that the amplitude of the AC voltage had a great influence on the growth characteristics of the electrical tree. The literature [10] reported the electrical aging of epoxy resins under positive and negative pulse voltages. The results showed that as the pulse amplitude increases, the lifetime decreases, and the tree length under positive voltage is longer than under negative voltage. The literature [11] reported the electrical treeing of silicone rubber (SiR) at different pulse frequencies. The results show that as the pulse frequency increases, the tree length tends to increase. It can be found that the DC amplitude, DC polarity, pulse frequency and pulse voltage have a certain influence on the electric aging and breakdown phenomenon.

With the emphasis on environmental protection, PP have been proposed to replace XLPE as a HVDC cable insulation [12, 13]. This is due to the heat-setting properties of XLPE, which makes it difficult to recycle after the end of its service life, which will cause environmental pollution. PP itself has high mechanical strength and melting point, so it does not require cross-linking treatment, which can maintain its own thermoplastic properties, so it can be recycled and is an environmentally friendly DC cable insulation material. K. Kurahashi et al. mixed sPP with PE and antioxidants to produce 0.6 and 22 kV AC recyclable cables, tested the AC breakdown field strength and dielectric loss at different temperatures, and found that they can meet the requirements of practical applications [14]. In 2008, HV Cable Solutions manufactured a ±35 kV polypropylene DC cable. In 2015, Italian Prysmian developed a prototype sample of a P-Laser high-voltage DC cable with a voltage rating of 320 kV based on a polypropylene substrate [15]. In 2018, the company prepared samples of high-voltage DC cables with a voltage rating of ±500 kV using polypropylene as the substrate, and ±640 kV cables have passed type tests [16]. In China, in 2018, Shanghai Huapu Cable Co., Ltd. and Shanghai Jiaotong University jointly developed PP insulation materials and production processes for medium voltage cables with a voltage level of ±35 kV and below. PP-based environmentally friendly DC cable insulation materials have shown great application prospects. We have previously studied the electrical tree characteristics of PP under repeated pulse voltages [3, 17]. PP has a lower initial probability and a lower growth rate compared with XLPE. However, at present, there are few studies on the growth characteristics of PP electrical tree under impulse superimposed DC voltage.

The aging of the electrical tree limits the development of cable insulation and the transmission voltage level of the cable [18, 19]. In addition to engineering improvements to the cable structure design and manufacturing process, currently, the methods for suppressing electrical tree in insulation by improving cable insulation materials are: (1) material blend modification; (2) nanoparticle modification; (3) polycyclic compounds modification [20, 21]. The polycyclic compounds can be directly added to the polymer during the production process through mechanical blending, which has the effect of delaying the breakdown, and is one of the current mainstream electrical tree suppression methods [20–23]. The literature [23] studied the inhibitory effect of various polycyclic compounds with an added amount of 0.5 wt.% on LDPE electrical trees, and found that bismuth has the best effect on inhibiting the electrical tree initiation characteristics. Adding polycyclic compounds such as benzene, acetophenone derivatives and benzophenone derivatives to XLPE can also inhibit the growth of electrical trees [24, 25]. However, there are few studies on the inhibitory effect of polycyclic compounds on PP electrical trees under impulse superimposed DC voltage.

2 Experimental Arrangement

2.1 Preparation of PP and PP/Polycyclic Compounds Composites

Using an internal mixer and a hot press molding machine to prepare PP and PP/polycyclic compounds composites. The internal mixer is used to eliminate the stress between the PP particles by mechanical blending, so that the polycyclic compounds and the PP are fully mixed. The hot press forming process helps to obtain the sample shape and thickness required for the experiment.

The specific preparation process of PP/polycyclic compounds composites was as follows:

(1) The experimental materials were placed in an oven at 60 °C for 24 h to remove moisture from raw materials.
(2) Turn on the internal mixer's heating power to heat the mixer, keeping the temperature at about 150 °C. In the experiment, the mixer speed was 50 r/min. Put an appropriate amount of PP into the mixer, mix it for 2–3 min and take it out to achieve the purpose of cleaning the mixer.
(3) The appropriate content of PP pellets was put into the internal mixer at 150 °C with the rotation speed of 50 r/min for 3 min, and then a certain amount of polycyclic compound was added to mixer and was mixed for 10 min.
(4) Put the PP/polycyclic compound composite into a special mold, and then placed the mold between the thermoforming machine.

(5) Melted the composite at 190 °C for 15 min, and then pressed the sample at 190 °C and 15 MPa.
(6) Reduced the pressure to 0 MPa and turned off the heating power. During the cooling process, when the temperature dropped between 120 and 140 °C, the needle electrode was inserted into the special mold.

The above is the preparation process for preparing the PP/polycyclic compounds composites sample. The preparation of the PP sample only needs to be removed in step (4).

The mold size was 150 mm × 15 mm × 3 mm. The distance between the needle tip and the edge of the sample was set to 2 ± 0.1 mm. The diameter and radius of curvature of the needle electrode were 300 μm and 3 μm, respectively.

In order to compare the effects of different polycyclic compounds on the growth characteristics of PP electrical tree, through the above method, PP and PP/polycyclic compounds composites with three different types of polycyclic compounds were prepared respectively. The mass fractions of the polycyclic compounds were all 0.1%. They are: A. 4,4′-bis(dimethylamino)benzil, a kind of benzil. B. 2-Hydroxy-2-phenylacetophenone, a kind of acetophenone derivatives. C. 4-phenylbenzophenone, a kind of benzophenone derivatives. Figure 1 shows the chemical structural formula of them. For the convenience of distinction, as shown in Table 1, PP was used to represent neat PP without polycyclic compound. The PP/4,4′-bis(dimethylamino)benzil composite with the polycyclic compound content of 0.1 wt.% was represented by PP-AN0.1. The PP/2-hydroxy-2-phenylacetophenone composite with the polycyclic compound content of 0.1 wt.% was represented by PP-BN0.1. The PP/4-phenylbenzophenone composite with the polycyclic compound content of 0.1 wt.% was represented by PP-CN0.1. In order to compare the effect of polycyclic compound content on the growth characteristics of electrical tree, 0.2 wt.% and 0.3 wt.% PP/2-hydroxy-2-phenylacetophenone composites were prepared, represented by PP-BN0.2 and PP-BN0.3 respectively.

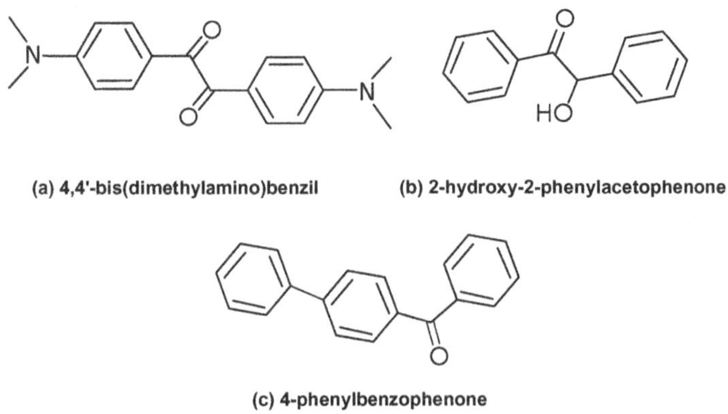

Fig. 1 Chemical structural formula of the chosen polycyclic compounds

Table 1 Specific samples kinds

Sample kind	Polycyclic compounds kind	Content (wt.%)
PP	0	0
PP-AN0.1	4,4′-bis(dimethylamino)benzil	0.1
PP-BN0.1	2-hydroxy-2-phenylacetophenone	0.1
PP-CN0.1	4-phenylbenzophenone	0.1
PP-BN0.2	2-hydroxy-2-phenylacetophenone	0.2
PP-BN0.3	2-hydroxy-2-phenylacetophenone	0.3

2.2 *Electrical Tree Experiment System and Analytical Method*

The experiment platform for room temperature and high temperature environment under complex DC operating condition is shown in Fig. 2. The platform is composed of a sample installation unit and a temperature control system. The sample installation unit includes high-voltage electrode, ground electrode and plastic base. The high-voltage electrode and the ground electrode are respectively attached to the high-pressure side and the grounding side of the PP sample. The impulse superimposed DC voltage was generated by the DC-impulse power source, which was consist of DC power source and impulse power source. The rise time of the single impulse voltage is 50 μs, and the tail time is 60 μs. The plastic base is placed in a heat-resistant glass cylinder, and there is a sufficient amount of dimethyl silicone oil in the cylinder to submerge the high-voltage electrode and the sample. Dimethyl silicone oil can effectively prevent flashover along the surface of the sample during the experiment. In addition, silicone oil can also fill the irregularities on the surface of the sample,

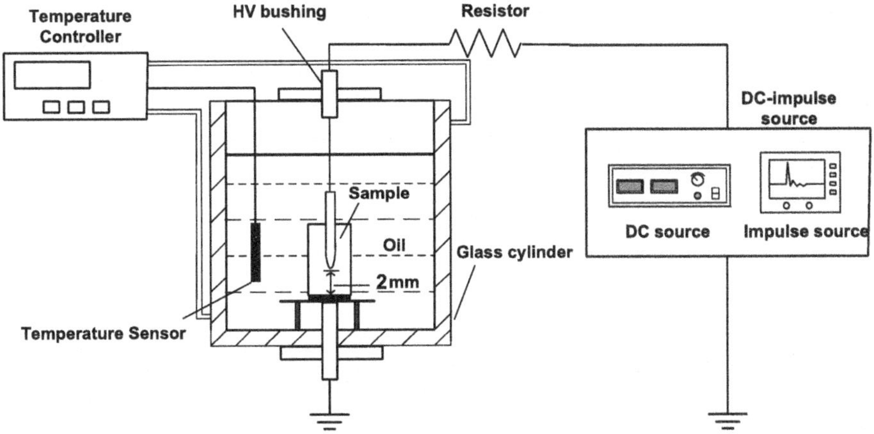

Fig. 2 The experiment platform under complex DC operating condition

increase the transparency of the sample, and improve the clarity of the electrical tree observation. The temperature controller system includes a temperature sensor composed of a temperature controller and a thermocouple. The high temperature environment is provided by the resistance wires on both sides of the heat-resistant glass cylinder, and the maximum working temperature is 200 °C. Immerse the temperature sensor in the oil and place it next to the sample. The thermostat controls the temperature of the sample by opening and closing the electric heating circuit. When the temperature in the box reaches the preset temperature, the heating is automatically stopped and the temperature in the box is maintained near the preset temperature.

The electrical tree imaging system included a computer, a microscope unit, and a cold light source. The microscope unit consisted of an objective lens, an eyepiece and a CCD. The CCD was a high-resolution imaging device. The camera multiplier was 1X, and the highest resolution was 1024 pixels × 768 pixels. The three objective magnifications were 4X, 10X, 40X, and the eyepiece magnification was 10X. The treeing process was recorded by the digital microscope. All experiments were recorded from the electrical tree initiation starting from the needle tip to the electrical tree breakdown with the longest electrical tree channel reaching the ground electrode.

3 Electrical Treeing Initiation and Breakdown Phenomenon Under Impulse Superimposed DC Voltage

3.1 Experimental Method

The sample for electrical tree experiment under impulse superimposed DC voltage is the PP sample prepared in Sect. 2.1. The experiment was conducted on the experimental platform described in Sect. 2.2. The experimental temperature was set to 30 °C. The DC voltage amplitude range was from −25 to +25 kV, and the pulse voltage amplitude range was from −35 to +35 kV. During the experiments, the DC voltage was applied at a rate of 1 kV/s. After 1 min, the impulse voltage was applied at a rate of 1 kV/s, and the impulse superimposed DC voltage was generated. The treeing process is finished when the breakdown occurs, or the treeing time reaches 10 h. In order to reduce the error, each set of experiments was repeated 20 times.

3.2 *Electrical Tree Growth Characteristics Under Impulse Superimposed DC Voltage*

3.2.1 Effects of Impulse Voltage Superimposed Different DC Voltages

The insulation of DC cable bears the rated DC voltage for a long time. When the pulse voltage occurs, the pulse voltage is compounded above the DC voltage, forming the impulse superimposed DC voltage, which is easy to lead to the growth of electrical tree. In different rated voltage and polarity of DC cable, the effect of pulse voltage on the growth of electrical tree is different. In this section, the characteristics of electrical tree growth under different amplitudes of DC and pulse compound voltage are studied. The amplitude of pulse voltage is +25 kV and the frequency is 400 Hz. The electrical tree growth characteristics of ±5, ±15 and ±25 kV DC superimposed impulse voltage are measured.

Figure 3 shows the composite voltage waveform of DC superimposed +25 kV impulse with different amplitudes. Figure 3a is +25 kV impulse superimposed +5, +15 and +25 kV DC voltages, respectively. The three diagrams in Fig. 3a have the same DC voltage polarity as the pulse voltage polarity, and are the same polarity DC-pulse composite voltage. The voltage in Fig. 3b is +25 kV impulse superimposed −5, −15 and −25 kV DC voltages. The three diagrams in Fig. 3b have a DC voltage polarity opposite to the pulse voltage polarity, which is a compound DC voltage of different polarity. It should be noted that this section mainly discusses the electrical tree growth characteristics in the case of DC-pulse composite voltage. Due to the change in DC amplitude and the pulse amplitude does not change, only the magnitude of DC voltage amplitude will be mentioned in subsequent discussions. It is the effect

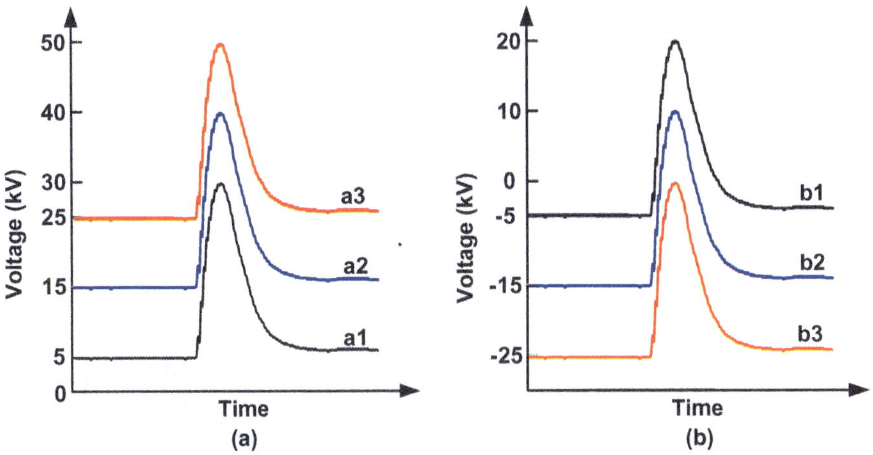

Fig. 3 Diagrammatic waveforms of impulse superimposed different DC voltages

of the combined voltage of this magnitude of the DC voltage and the +25 kV pulse voltage, not the effect of the pure DC voltage.

The typical electrical tree structure of +25 kV impulse superimposed different DC voltages is shown in Fig. 4. When the polarity of the DC voltage is positive, as shown in Fig. 4a1–a3, the structure of the electrical tree is similar at different DC voltage amplitudes, indicating that when the positive DC voltage is combined with the positive pulse voltage, DC voltage amplitude has little effect on the structure of the electrical tree. However, the electrical tree structure changes with the negative DC amplitude. When the DC voltage is −5 and −15 kV, the structure of the electrical tree is similar, but when the DC voltage is increased to −25 kV, the electrical tree channel is obviously darker in color and wider in width, which means that under different magnitudes of negative DC voltage, the electrical tree growth mechanism is different. At the same time, there are more pine branches at −5 and −15 kV DC voltage compared to the electrical tree structure under positive DC voltage. When the DC voltage is applied, a space charge layer is formed around the tip of the needle or electrical tree. Since the injection depth and injection range of electrons are larger than holes, under negative DC voltage, a larger space charge layer will be formed around the electric tree than under positive DC voltage [26]. Therefore, when they are combined with pulse voltages of the same amplitude, under the negative DC voltage, it is easier to produce avalanche discharge in the low-density region, thereby forming more pine branches.

Figure 5 shows the growth characteristics of the electrical tree under different amplitudes of positive DC superimposed +25 kV impulse voltage. Figure 6 shows the electrical tree growth characteristics under different amplitudes of negative DC superimposed +25 kV impulse voltage. As shown in Fig. 5, with the increase of

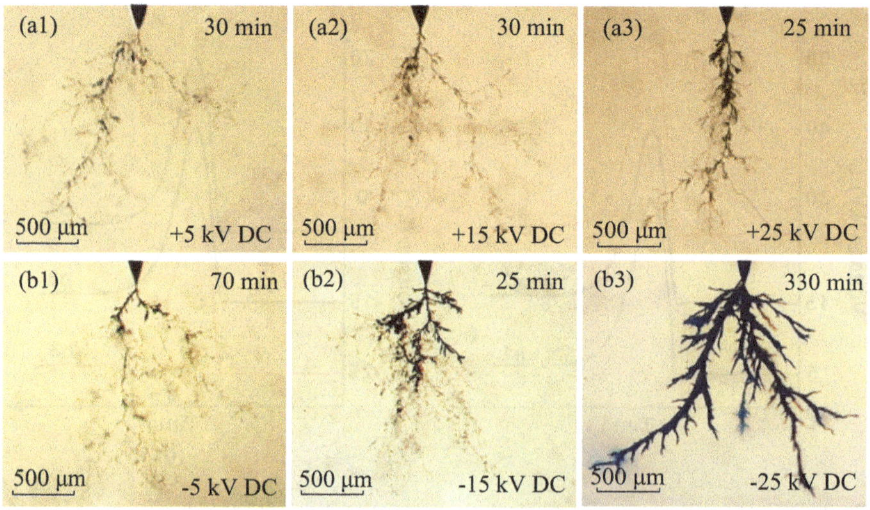

Fig. 4 Typical electrical tree structures under +25 kV impulse superimposed different DC voltages

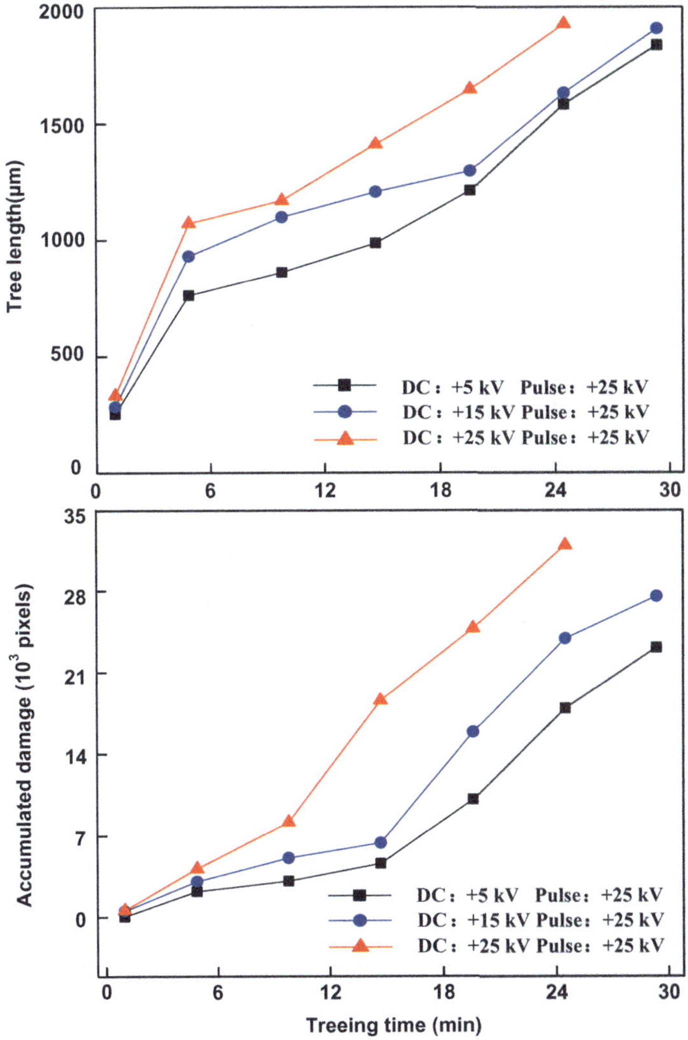

Fig. 5 The electrical tree growth characteristics under different amplitudes of positive DC superimposed +25 kV impulse voltage

positive DC voltage amplitude, the growth rate and accumulation of electrical tree increase. As shown in Fig. 6, with the increase of the negative DC voltage amplitude, the growth rate and cumulative damage of the electrical tree increase first and then decrease. As the magnitude of the negative DC voltage increases, the average value of the applied composite voltage field strength becomes smaller, proving that the growth rate of the electrical tree is not only related to the magnitude of the applied electric field, but also closely related to the movement of different charges under different waveforms. The mechanism will be discussed in detail later.

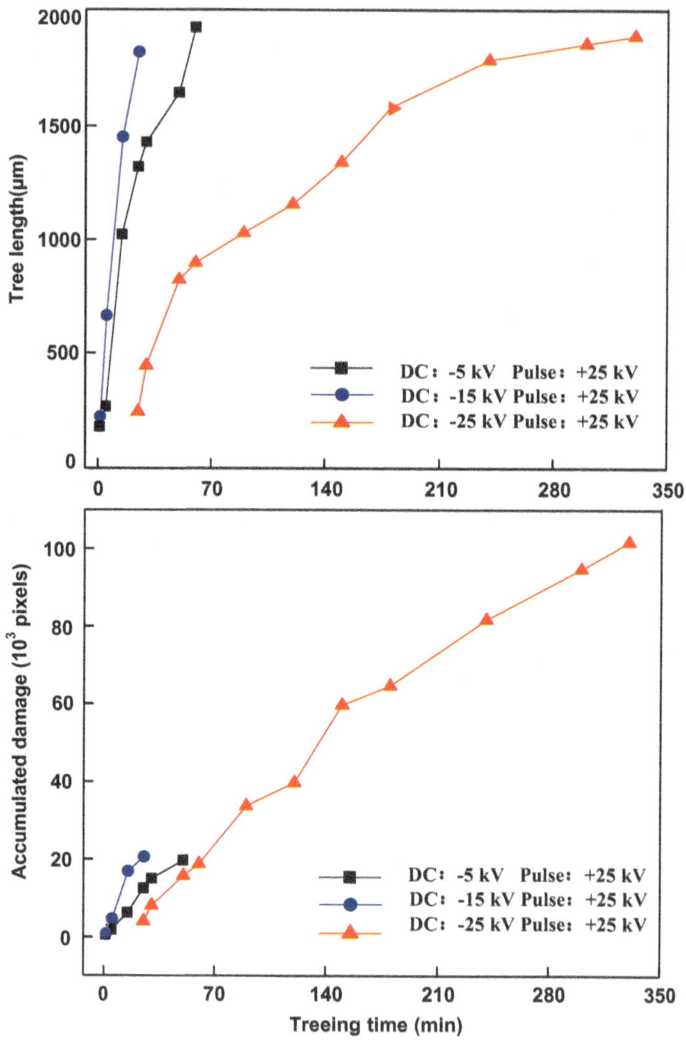

Fig. 6 The electrical tree growth characteristics under different amplitudes of negative DC superimposed +25 kV impulse voltage

Figure 7 shows the breakdown time of the electrical tree under +25 kV impulse superimposed different DC voltages. The electrical tree breakdown time of +5, + 15 and +25 kV DC voltage is 35, 32, 28 min, and the electrical tree breakdown time of −5, −15 and −25 kV DC voltage is 75, 27, 382 min, respectively. As the amplitude of the positive DC voltage increases, the breakdown time of the electrical tree decreases. But with the increase of the negative DC voltage amplitude, the electric tree breakdown time first decreases and then increases. At the same time, the breakdown time of the electric tree at −15 kV DC voltage is less than that

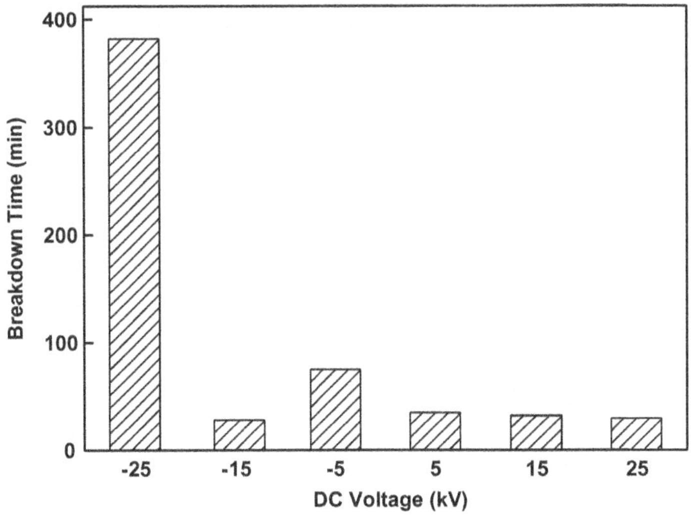

Fig. 7 Breakdown time under +25 kV impulse superimposed different DC voltages

of +5, +15 and +25 kV DC voltage. The breakdown time of −5 kV DC voltage is greater than that of +5, +15, and +25 kV DC voltage. Literature [27] studied the impact of different magnitudes of DC preload voltage on the pulse breakdown voltage of polyethylene. The results showed that when the polarity of the pulse voltage was positive, the breakdown voltage under negative DC voltage preload was lower than that under positive DC voltage. The author attributed the cause of this result to the effect of space charge. As the polarity of the voltage was reversed, the space charges of the same polarity became space charges of different polarity, thereby reducing the breakdown voltage. Literature [28] studied the electric field distribution before and after the polarity change in XLPE. The results showed that the electric field changed from −50 to 130 kV/mm after the polarity change, indicating that the polarity change would cause severe electric field distortion. It can be seen from Fig. 3 that when the DC voltage amplitude is −5 and −15 kV, the composite voltage has a polarity change process. During this process, the distortion of the electric field will cause the breakdown time to be shortened. However, the average value of the applied composite voltage field strength of −5 and −15 kV is less than the average value of the composite voltage field strength of +5, +15 and +25 kV, which will result in a longer breakdown time. The reason why the breakdown time of −5 kV DC voltage is greater than the breakdown time of positive DC voltage, while the breakdown time of −15 kV DC voltage is less than the breakdown time of positive DC voltage depends on which factor dominates. At −5 kV DC voltage, the DC voltage is small, resulting in a small number of injected electrons, so the effect of negative space charge is not obvious. In contrast, when the DC voltage increases to −15 kV, the effect of space charge dominates.

3.2.2 Effect of DC Superimposed Impulse Voltage with Different Frequencies

This section studies the typical electrical tree structure under DC superimposed impulse voltages of different frequencies. The pulse frequencies are 400, 800 and 1200 Hz, respectively. Figure 8 shows the typical electrical tree structure under DC-impulse voltage of different frequencies. The applied voltages in Fig. 8a1–a3 are +5 kV DC and +25 kV pulse composite voltage, which are the same polarity DC pulse composite voltage. The applied voltages in Fig. 8b1–b3 are −15 kV DC and + 25 kV pulse composite voltages, which are composite pulse DC voltages of different polarities. The pressure time is 10 min. As the frequency increases, the electrical tree channel becomes wider and darker.

Figure 9 shows the electrical tree growth characteristics under positive DC superimposed +25 kV impulse voltage of different frequencies. Figure 10 shows the electrical tree growth characteristics under negative DC superimposed +25 kV impulse voltage of different frequencies. The applied voltage in Fig. 9 is the composite voltage of +5 kV DC and +25 kV pulse. The applied voltage in Fig. 10 is a composite voltage of −15 kV DC and +25 kV pulse. With the increase of the pulse frequency in the composite voltage, the growth rate and accumulated damage of the electrical tree increase.

When the electric field in the thin tube of the electric tree exceeds a certain critical value, an avalanche discharge occurs, and the avalanche discharge diffuses along the thin tube. When the electric field strength in the thin tube is lower than the residual electric field strength, the avalanche discharge no longer continues to diffuse. The number of collision electrons produced by an initial electron in an avalanche discharge is N_A [29]:

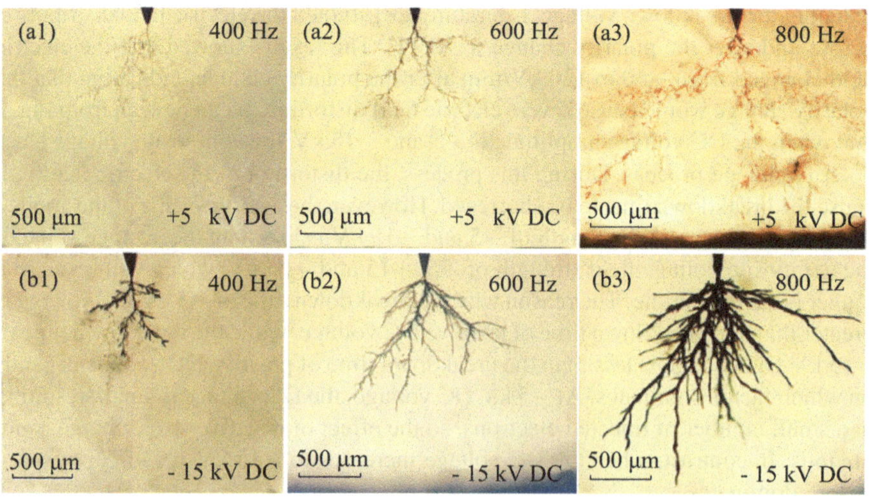

Fig. 8 Typical electrical tree structures under +5 and −15 kV DC superimposed +25 kV impulse voltages with different frequencies

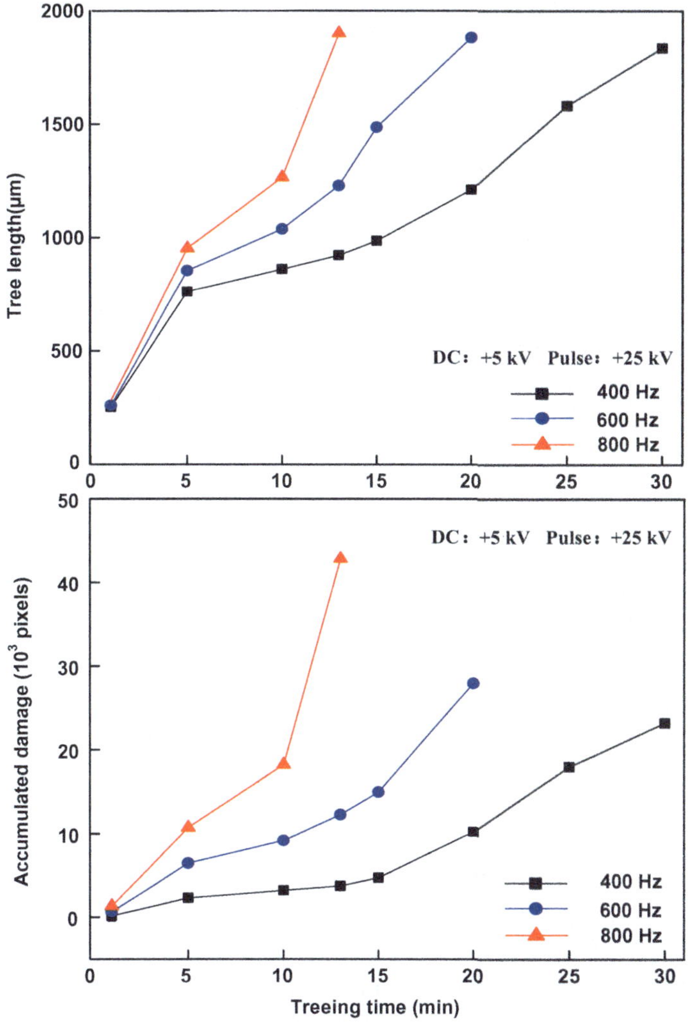

Fig. 9 The electrical tree growth characteristics under positive DC superimposed +25 kV impulse voltage of different frequencies

$$N_A = \exp\left[\frac{L}{\lambda}\exp\left(-\frac{I}{\exp\lambda E}\right)\right] - 1 \tag{1}$$

where L represents the path length of avalanche occurrence; I represents ionization energy; λ represents the average free travel between two collisions.

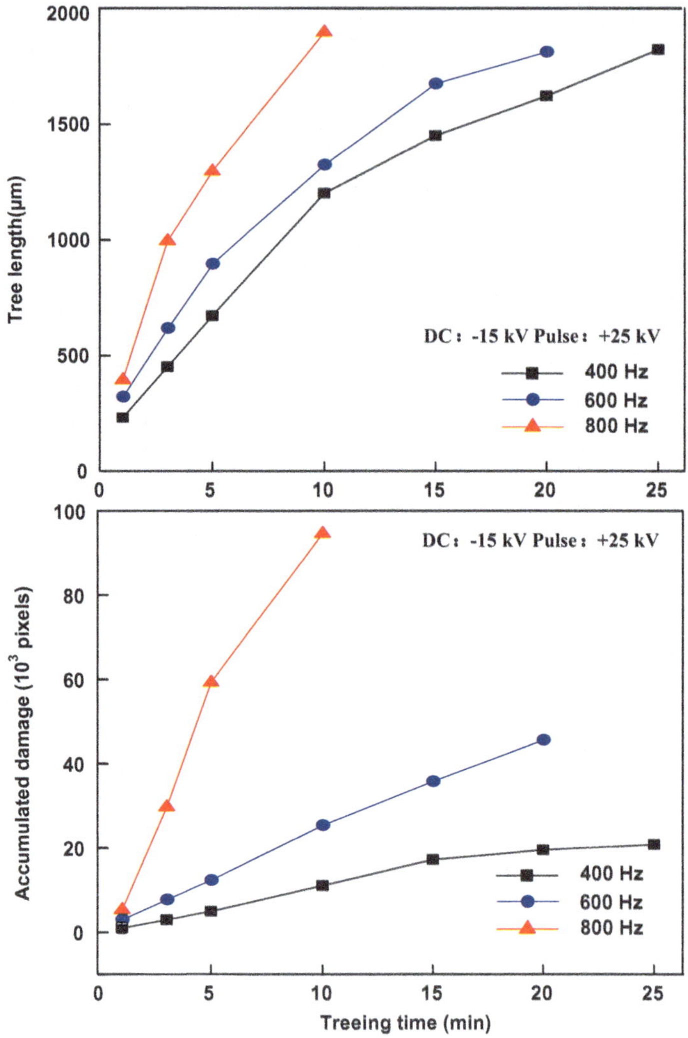

Fig. 10 The electrical tree growth characteristics under negative DC superimposed +25 kV impulse voltage of different frequencies

Then, within a certain period of time in a certain area, the total number of impact ionizations N_t is:

$$N_t = ftbn_0N_A \tag{2}$$

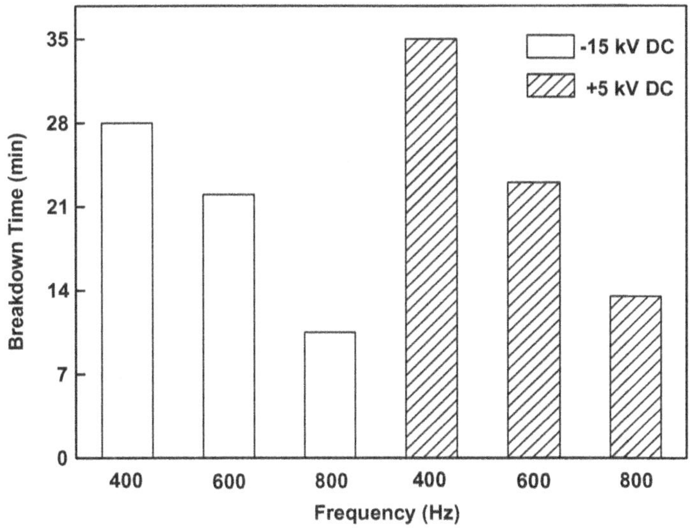

Fig. 11 Breakdown time of −5 and −15 kV DC superimposed +25 kV impulse voltages with different frequencies

where f represents the pulse voltage frequency; t represents unit time; b represents the number of avalanche discharges caused by each pulse voltage; n_0 represents the number of avalanche discharge electrons.

As the pulse frequency increases, the N_t increases and the length of the electrical tree increases. Since N_t is proportional to the area destroyed by avalanche discharge, as the frequency increases, the damage caused by each avalanche discharge is greater during the same time, resulting in a wider electrical tree channel and greater accumulated damage.

Figure 11 shows the breakdown time of the electrical tree under DC superimposed impulse voltages of different frequencies. As the pulse frequency increases, regardless of the magnitude and polarity of the DC voltage, the breakdown time becomes significantly smaller. When the pulse voltage frequency is 600 and 800 Hz respectively, the breakdown time at −15 kV DC and +25 kV pulse voltage is shorter than that at +5 kV DC and +25 kV pulse voltage. This is consistent with the result when the pulse frequency is 400 Hz in Sect. 3.2.1.

3.2.3 Effects of DC Voltage Superimposed Different Impulse Voltages

In this section, the electrical tree growth characteristics under the DC superimposed different impulse voltage were studied. The electrical tree growth characteristics under the ±15, ±25 and ±35 kV pulses superimposed −25 kV DC were measured respectively. Figure 12 is a composite waveform of −25 kV DC superimposed different amplitudes pulses. Figure 12a1–a3 are the composite voltages of

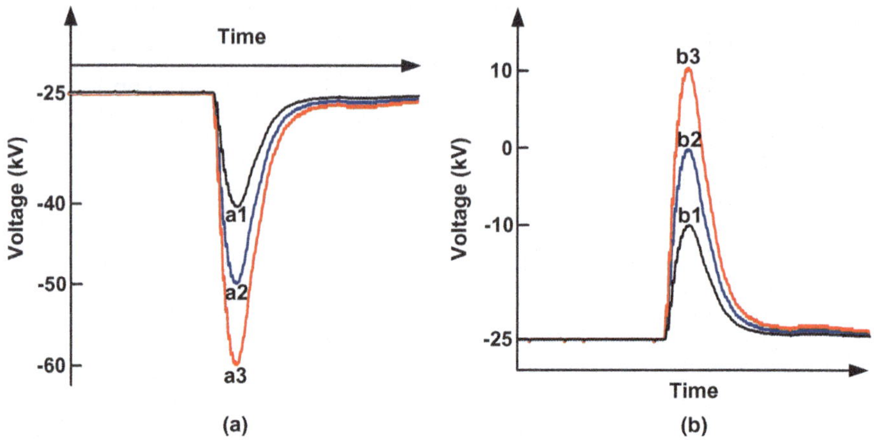

Fig. 12 Diagrammatic waveforms under −25 kV DC superimposed different impulse voltages

−25 kV DC and −15, −25 and −35 kV pulse voltages respectively. The DC voltage polarity of these three graphs is the same as the pulse voltage polarity. It is the composite voltage of the same polarity DC pulse. Figure 12b1–b3 are the composite voltage of −25 kV DC and +15, +25 and +35 kV pulse voltages respectively. It is a composite voltage of DC pulses with different polarities. Since the pulse amplitude in the composite voltage in this section changes but the DC amplitude does not change, for convenience of discussion, only the magnitude of the pulse voltage amplitude will be mentioned to refer to the effect of the combined voltage of the pulse voltage of this amplitude and the DC voltage of −25 kV, rather than the effect of pure pulse voltage.

Figure 13 shows the electrical tree structure under −25 kV DC superimposed different impulse voltages. The pulse amplitude has a great influence on the structure of the electrical tree. As the amplitude of the pulse voltage increases, the depth of charge injection becomes deeper, and the farther the formed charge front is from the needle tip or the end of the main branch of the electrical tree. For electrical trees, it is easier to branch behind the charge front [30]. Therefore, as the voltage amplitude increases, the more lateral branches grow from the main branch, resulting in an increase in the density of electrical branches.

Figure 14 shows the electrical tree growth characteristics under different amplitudes of negative impulse superimposed −25 kV DC voltage. Figure 15 shows the electrical tree growth characteristics under different amplitudes of positive impulse superimposed −25 kV DC voltage. As shown in Figs. 14 and 15, as the pulse voltage increases, the growth rate and accumulated damage increase significantly.

Figure 16 shows the breakdown time under −25 kV DC superimposed different impulse voltages. It can be seen from the figure that under the pulse of +15 kV, the growth rate of the electrical tree is extremely slow, and no breakdown occurs within 10 h. As the pulse amplitude increases, the breakdown time is significantly shortened, indicating that the pulse amplitude has a very large impact on the breakdown time. It

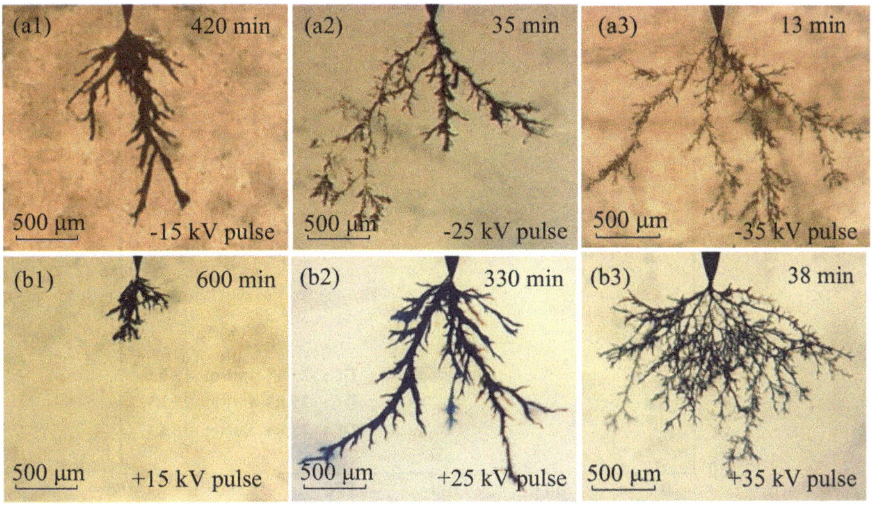

Fig. 13 Typical electrical tree structures under −25 kV DC superimposed different impulse voltages

can be observed that when the pulse voltage polarity is negative, the breakdown time is less than when the pulse voltage polarity is positive. It can be concluded that the composite voltage of DC pulses of the same polarity is more likely to cause electrical tree breakdown than the composite voltage of DC pulses of different polarity. Because the average value of the composite voltage of the negative pulse voltage is greater than that under the positive pulse voltage. According to (1), as the electric field strength E increases, N_A increases, and the amount of avalanche discharge increases. Therefore, voltages of the same polarity are more likely to cause breakdown than voltages of different polarities. It should be noted that for a pulse voltage of +35 kV, there will be a process of voltage polarity change. When the voltage polarity changes from negative to positive, the electric field is severely distorted and the breakdown time is significantly shorter [28]. But the breakdown time of −35 kV pulse is still less than that of +35 kV pulse. The reason for this phenomenon may be that the effect of a larger effective average electric field is greater than the effect of polarity changes.

3.3 Charge Accumulation Process Under Impulse Superimposed DC Voltage

Figure 17 is a schematic diagram of charge transportation process under impulse superimposed different DC voltages. The charge transportation process under positive impulse superimposed positive DC voltages is shown in Fig. 17a. During the application of the positive DC voltage, charges are formed around the needle or

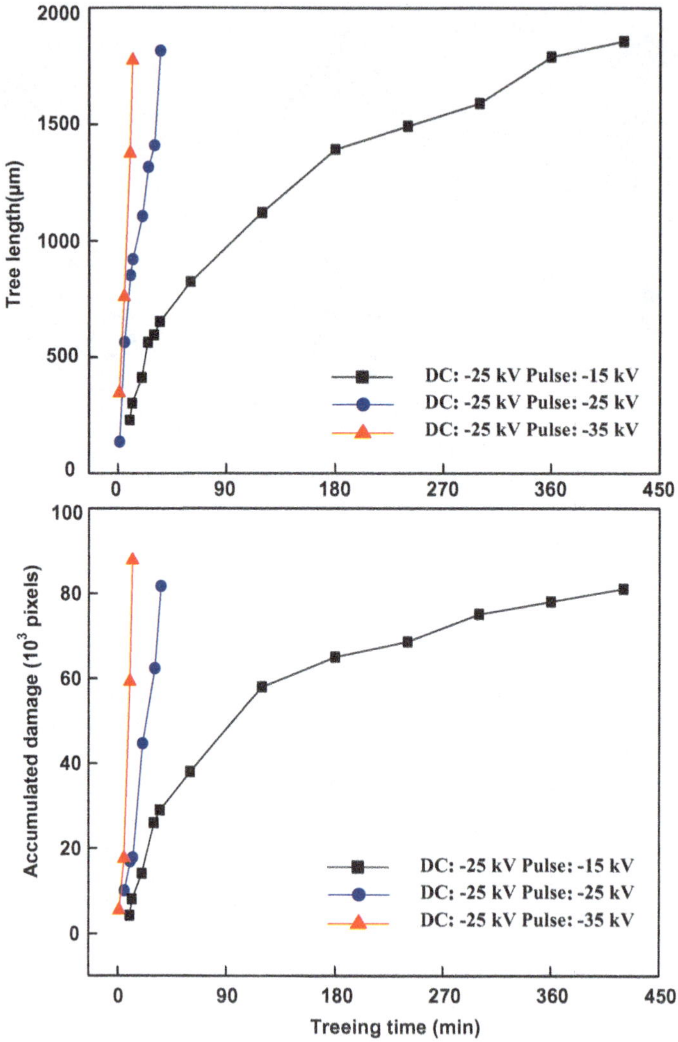

Fig. 14 The electrical tree growth characteristics under different amplitudes of negative impulse superimposed −25 kV DC voltage

the electrical tree branches tips. When the pulse voltage is applied, a large number of charges are instantaneously injected; these charges stimulate the existed trapped charges to detrap and move far away from the needle or the electrical tree branches tips and trap into other traps again. These injected charges themselves are trapped after one or more scattering around the needle tip or the electrical tree branches tips, reducing the electric field strength and electron injection rate [6]. The states of the charges change from high energy to low energy state because of the trapping process.

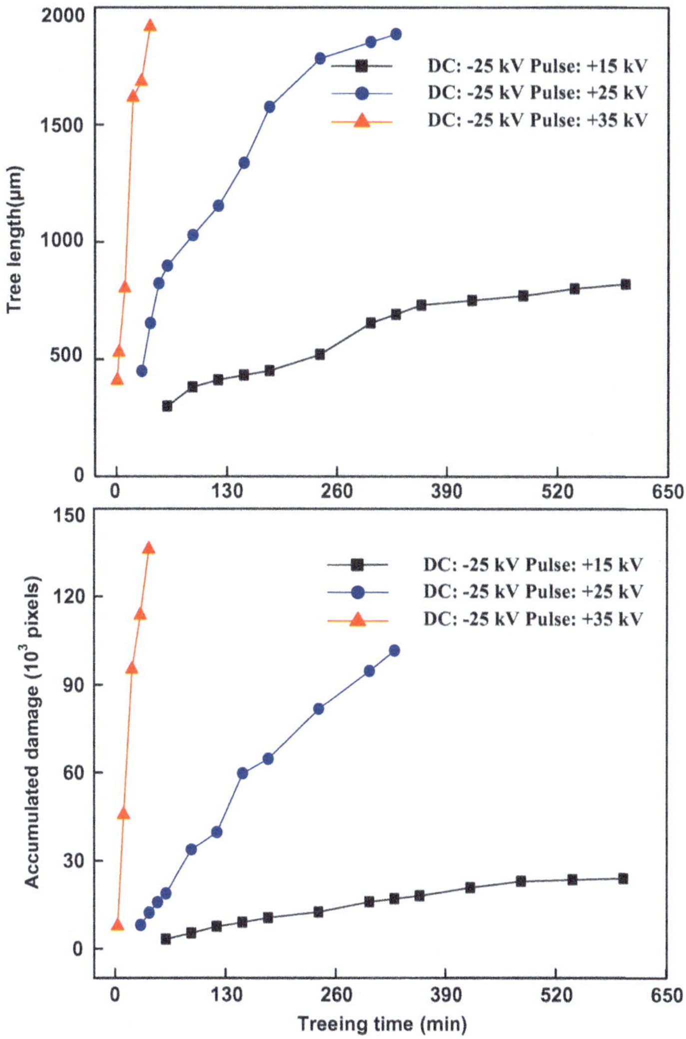

Fig. 15 The electrical tree growth characteristics under different amplitudes of positive impulse superimposed −25 kV DC voltage

The excess energy is passed to other charges, which are converted into hot electrons via the Auger effect. Hot electrons break molecular chains to form low-density areas during this process. Charges collide with each other in low-density areas, releasing energy that breaks more molecular chains [31], accelerating the formation of low-density areas. Charges collide with ionization in low-density regions, releasing energy to destroy more molecular chains, forming micropores. Subsequently, partial discharge occurs in the micropores. Partial discharge will produce local temperature rise and a large amount of gas in a short time, resulting in expansion stress of the

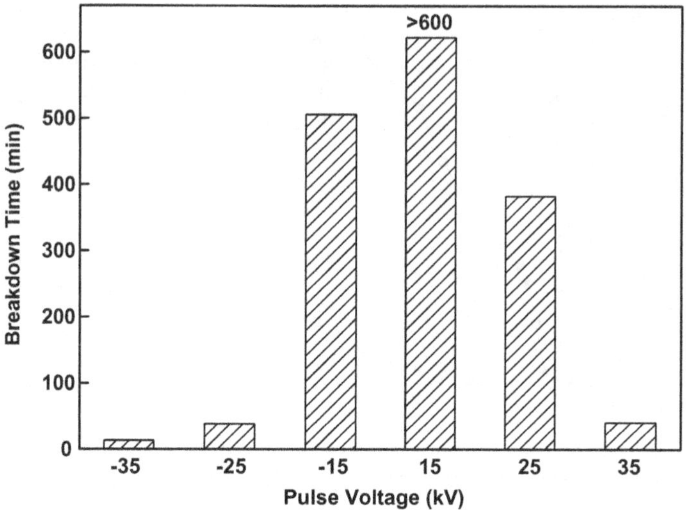

Fig. 16 Breakdown time under −25 kV DC superimposed different impulse voltages

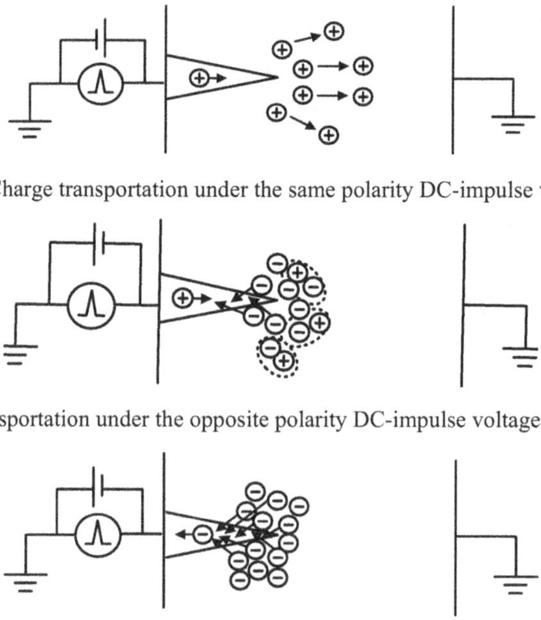

(a) Charge transportation under the same polarity DC-impulse voltages.

(b) Charge transportation under the opposite polarity DC-impulse voltages $|U_{DC}| < |U_{pulse}|$.

(c) Charge transportation under the opposite polarity DC-impulse voltages $|U_{DC}| > |U_{pulse}|$.

Fig. 17 Charge transportation process under different DC amplitudes [35]. Copyright Clearance Center's RightsLink® License Number: 4855741051820

PP material around the micropores. Under this expansion stress and the impact of charged particles, the micropores rapidly expand into the amorphous region with weaker mechanical strength, resulting in the growth of electrical branches. When the DC voltage is negative and the pulse voltage is positive and the absolute value of the DC voltage is smaller than the pulse voltage, the charge transportation process is shown in Fig. 17b (+25 kV impulse superimposed −5 and −15 kV DC voltages, corresponding to Fig. 3b1, b2, respectively). During the application of the negative DC voltage, electrons are accumulated around the needle tip. After the positive pulse voltage is applied, some electrons are extracted, and when the voltage polarity changes from negative to positive, the positive charges are injected. Negative charges recombine with positive charges to produce energy that accelerates the growth of electrical trees [30]. With the voltage polarity changing from negative to positive, the electrons change from homo space charges to hetero space charges, leading to the electric field distortion that causes discharges and the breaking of molecular chains [32]. When the DC voltage is negative and the pulse voltage is positive and the absolute value of the DC voltage is greater than the pulse voltage, the charge transportation process is shown in Fig. 17c (+25 kV impulse superimposed −25 kV DC voltages, corresponding to Fig. 3b3). After the pulse voltage is applied, because of the electric potential reduction at the needle tip, electrons are extracted [33]. Such charge movement causes the breaking of molecular chains, leading to the electrical tree growth [34].

3.4 Analysis of the Correlation Mechanism Between Charge Accumulation and Growth Characteristics of Electrical Tree

The growth process of electrical tree is affected by many factors such as electric field, charge transportation, avalanche discharge, and local motion. The change of the applied voltage will cause the change of charge transportation, and will also affect the avalanche discharge, partial discharge, etc., thus affecting the growth characteristics of the electrical tree.

From the results in Sect. 3.2.1, it can be seen that with the increase of the positive DC amplitude, the growth rate and accumulated damage of the electrical tree increase. The charge transportation under the positive polarity impulse superimposed positive DC voltage is shown in Fig. 17a. With the increase of the DC amplitude, the number of injected positive charges increase, so that the charge that can be moved after being stimulated by the injected impulse increases. The movement of the charge causes the molecular chain to break, so as the number of moving charges increases, the molecular chain breaks intensified. Due to the acceleration of the hot electrons in the free volume under the action of the electric field to hit the molecular chain, as the magnitude of the DC voltage increases, the average value of the composite voltage increases, and the energy obtained by the hot electrons accelerated in the electric field is greater, resulting in increased molecular chain breakage. At the same time, it can be seen from formula 1 that as the average composite voltage increases, the amount of avalanche discharge increases, so the growth process of the electrical tree is accelerated.

From the results in Sect. 3.2.1, it can be seen that with the increase of the negative DC amplitude, the growth rate of the electrical tree increases first and then decreases. When the +25 kV impulse superimposed −5 and −15 kV DC voltages are applied, the charge transportation process is shown in Fig. 17b. As the magnitude of the DC voltage increases from −5 to −15 kV, although the average value of the composite voltage decreases, the space charge injected in the same polarity increases, resulting in distortion of the electric field when the polarity is changed, and the more severe avalanche discharge. It can be seen from Fig. 3 that when the negative polarity DC voltage increases from −15 to −25 kV, the average value of the applied composite voltage decreases, and at the same time because of the lack of the polarity change process, there is no electric field distortion caused by space charge under +25 kV impulse superimposed −25 kV DC voltage. When the voltage is +25 kV impulse superimposed −25 kV DC, the charge transportation process is shown in Fig. 17c. It can also be seen from the charge transportation process that the energy generated by the neutralization of positive and negative charges under +25 kV impulse superimposed −25 kV DC voltage is smaller than under +25 kV impulse superimposed −15 kV DC voltage. Therefore, the growth rate of electrical tree under +25 kV impulse superimposed −25 kV DC voltage is much slower than that under +25 kV impulse superimposed −5 and −15 kV voltages.

It can be seen from the results in Sect. 3.2.3 that when the impulse voltage superimposed with a DC voltage of a certain amplitude, as the impulse amplitude increases, the growth rate and accumulated damage of the electrical tree increase. Under the negative impulse superimposed negative DC voltages, when the pulse voltage is applied, the electric field strength at the needle tip instantly increases. When the electric field at the needle tip exceeds the critical electric field, the charge carriers will change from a low energy state to a high mobility state. Charge is injected into the sample from the tip of the needle, and the amount of injected charge increases instantly, as shown in Fig. 18. In the late stage of the falling edge of the pulse voltage, the amount of injected space charge decreases due to the decrease in the field strength at the tip of the needle. During the application of the DC voltage, charge is slowly injected, and the amount of charge slowly increases. The next time the pulse voltage is applied, the total amount of injected charge continues to increase. As the amplitude of the pulse voltage increases, the amount of charge change ΔQ generated by a single pulse increases ($\Delta Q_{-35} > \Delta Q_{-25} > \Delta Q_{-15}$). The total energy generated by charge trapping and trapping increases, the number of hot electrons increases, and more polymer molecular chains are broken. Therefore, as the amplitude of the negative pulse increases, the growth rate and accumulated damage of the electrical tree will increase. Under +15 and +25 kV impulse superimposed −25 kV DC voltages, electrons are extracted on the rising edge of the pulse voltage. On the falling edge of the pulse voltage, electrons are injected again. Since $\Delta Q_{+25} > \Delta Q_{+15}$, when the pulse voltage increases from +15 to +25 kV, the growth rate of the electrical tree is

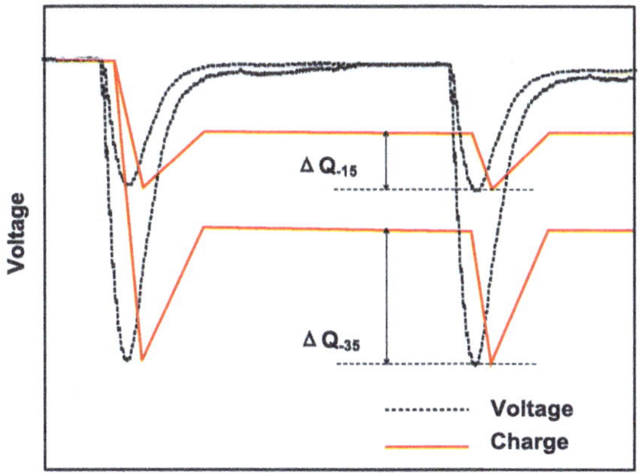

Fig. 18 Schematic of the amount of injected space charges with the change of impulse voltage [35]. Copyright Clearance Center's RightsLink® License Number: 4855741051820

accelerated. When the amplitude of the pulse voltage is +35 kV, part of the electrons is extracted on the rising edge of the pulse voltage. After the polarity changes from negative polarity to positive polarity, positive charges are injected, and these positive charges recombine with electrons to generate energy. In addition, the electric field distortion caused by the space charge will also accelerate the growth process of the electrical tree. As the pulse voltage increases from +25 to +35 kV, in addition to the energy generated by the extraction and injection of charges, the energy generated by the recombination of electrons and positive charges and the distortion of the electric field accelerate the growth of the electrical tree. As a result, the growth rate and accumulated damage of electrical tree increase.

4 Polycyclic Compounds Affecting Electrical Tree Growth Under Impulse Superimposed DC Voltage

4.1 Experimental Method

The sample used in Sect. 4 is the PP/polycyclic compounds composites sample prepared in Sect. 2.1. The experiment was conducted on the experimental platform described in Sect. 2.2. The experimental temperature was set to 30, 70, 90 °C. After the temperature reaches the set temperature during the experiment, the sample should be allowed to stand for 10 min in the thermostat to ensure that the temperature of the material is consistent with the environment before the pressure test. The experimental voltage was ±35 kV impulse superimposed −25 kV DC voltage. The frequency was 400 Hz. During the experiments, the DC voltage was applied at a rate of 1 kV/s. After 1 min, the impulse voltage was applied at a rate of 1 kV/s, and the impulse superimposed DC voltage was generated. After the electrical tree grows to breakdown, immediately turn off the power and take out the sample. In order to reduce the error, each set of experiments was repeated 20 times.

4.2 Effect of Polycyclic Compound Kind on Electrical Tree Degradation

In order to study the influence of different types of polycyclic compound on the growth characteristics of PP electrical tree, the three PP/polycyclic compound composite materials prepared in Sect. 2.1 were selected and tested under high temperature environment under impulse superimposed DC voltage. The mass fraction of the polycyclic compound added in this section is 0.1%. PP-AN0.1 is used for PP/4,4′-bis(dimethylamino)benzil composite material, and PP-BN0.1 is PP/2-hydroxy-2-phenylacetophenone composite material, PP-CN0.1 means PP/4-phenylbenzophenone composite material. Set the experiment temperature to 30, 70, 90 °C. The growth characteristics of the electrical tree under −25 kV DC superimposed ±35 kV impulse voltage were measured respectively, and the pulse voltage frequency was 400 Hz.

4.2.1 Electrical Tree Degradation Under Opposite Polarity Impulse Superimposed DC Voltage

Figure 19 is the electrical tree structure under −25 kV DC and +35 kV pulse composite voltage in PP and its polycyclic compound composite material. The pressurization time is 2 min. Among them, (a1–a3) is the electrical tree structure of the PP sample, and (b1–b3) is electrical tree structure form of the PP/polycyclic compound composite material added with 2-hydroxy-2-phenylacetophenone. It can be seen from Fig. 19 that as the temperature increases, the electrical branch channel of the pure PP and its polycyclic compound composite becomes thicker and blacker, and the number of main branches and side branches increases. At the same temperature, the length of the electrical tree of the PP-N0.1B sample is shorter than that of the PP sample, and the electrical tree channel is thinner.

Figures 20 and 21 show the growth characteristics of electrical tree with different kinds of polycyclic compounds under opposite polarity impulse superimposed DC voltage. The voltage is −25 kV DC and +35 kV pulse composite voltage. It can be seen from the figure that the 2-hydroxy-2-phenylacetophenone has a certain inhibitory effect on the length of the electrical tree and the accumulated damage of the heteropolar DC pulse composite voltage at different temperatures. The inhibition effect is the best at 30 °C. The length of the electrical tree of PP-BN0.1 is reduced by 19% and the accumulated damage by 19.6% when PP-BN0.1 was pressed for 35 min. As the temperature rises, the effect of inhibiting the growth of electrical tree gradually

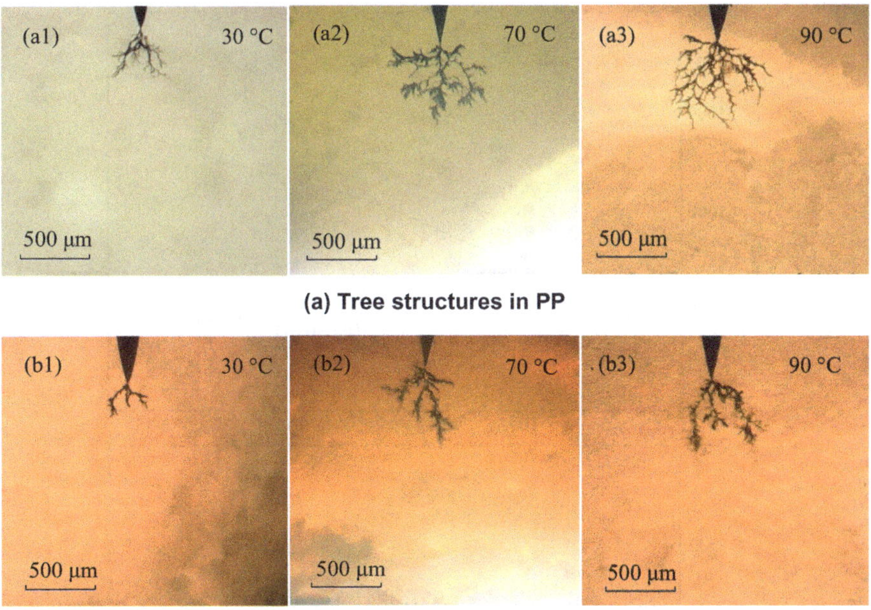

(a) Tree structures in PP

(b) Tree structures in PP-BN0.1

Fig. 19 Tree structures in PP and PP-BN0.1 under different temperatures

decreases. At a press time for 4 min at 90 °C, the electrical tree length of PP-BN0.1 is reduced by 6.6% compared with PP, and the accumulated damage is reduced. 7.7%. After the addition of 4,4′-bis(dimethylamino)benzil and 4-phenylbenzophenone, the length and accumulated damage of electrical tree increase. 4-phenylbenzophenone has the worst effect. When the pressure time at 30 °C is 35 min, the length of the electrical tree of PP-CN0.1 increases by 7.8% compared with PP, and the accumulated damage increases by 7.7%. As the temperature rises, the effect is even worse. At a press time of 90 °C for 35 min, the electrical tree length of PP-CN0.1 increases by 17.6% compared with PP, and the accumulated damage increases by 25.6%.

4.2.2 Electrical Tree Degradation Under the Same Polarity Impulse Superimposed DC Voltage

Figure 22 shows the growth characteristics of electrical tree with different kinds of polycyclic compounds. The pressurization time is 5 min. Among them, Fig. 22a1–a3 is the electrical tree structure of the PP sample, and Fig. 22b1–b3 is the electrical tree structure of the PP/polycyclic compound composite with the addition of 2-hydroxy-2-phenylacetophenone (Class B polycyclic compound). It can be seen from Fig. 22a1–a3 that with the increase of temperature, the electrical tree channel of the pure PP sample becomes thicker, the main branch and side branch of the electrical

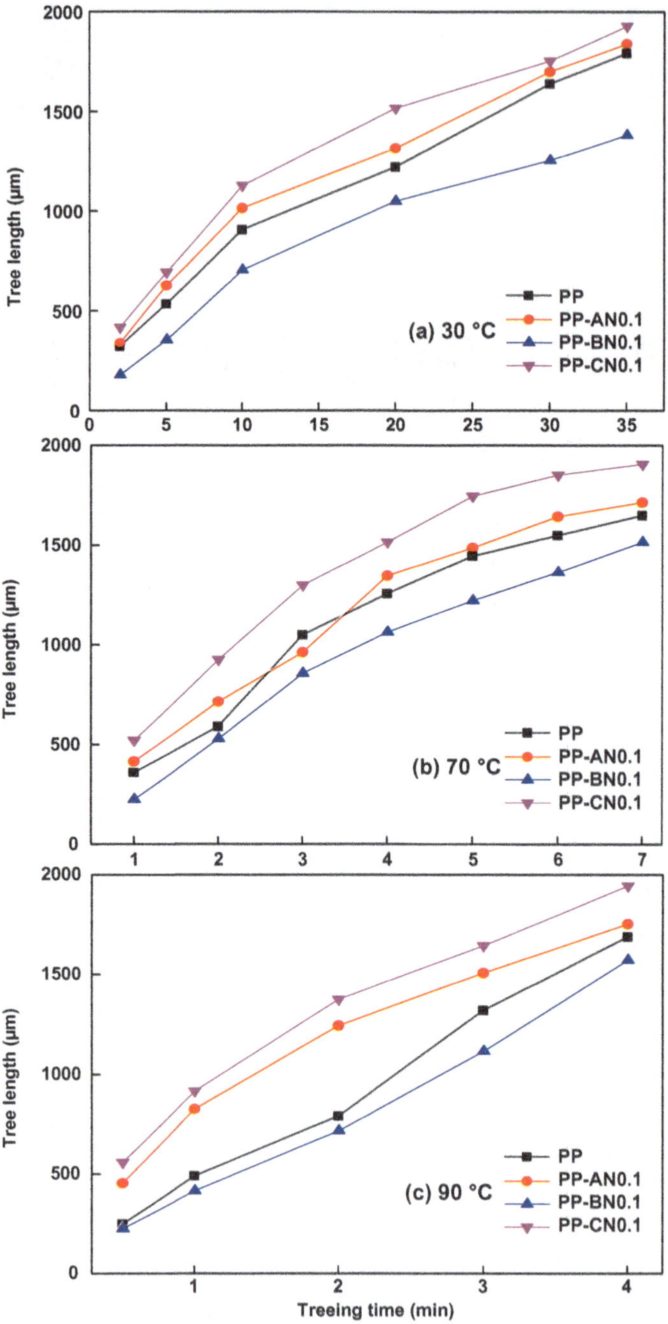

Fig. 20 The electrical tree growth characteristics (tree length) with different kinds of polycyclic compounds under different temperatures under opposite polarity impulse superimposed DC voltage

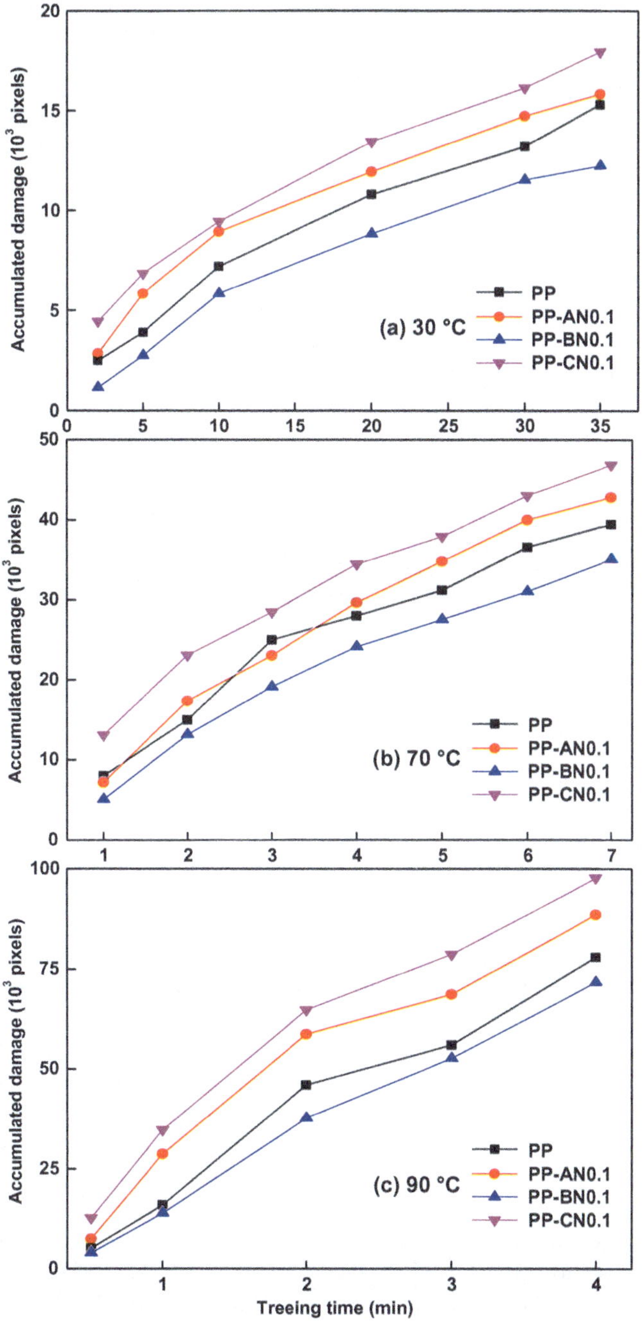

Fig. 21 The electrical tree growth characteristics (accumulated damage) with different kinds of polycyclic compounds under different temperatures under opposite polarity impulse superimposed DC voltage

Fig. 22 Tree structures in PP and PP-BN0.1 under different temperatures

branch overlap, and the shape is denser. As can be seen from Fig. 22b1–b3, as the temperature increases from 30 to 70 °C, the color of the electrical tree channel does not change significantly. As the temperature increases from 70 to 90 °C, the electrical tree channel becomes thicker. At the same temperature, the electrical tree channel of the PP-BN0.1 sample is shorter than that of the PP sample, and there is a clear black dense area.

Figures 22 and 23 show the growth characteristics of electrical tree with different kinds of polycyclic compounds under the same polarity impulse superimposed DC voltage. The voltage is −25 kV DC and −35 kV pulse composite voltage. It can be seen from the figure that the 2-hydroxy-2-phenylacetophenone has a certain inhibitory effect on the length of the electrical tree and accumulated damage under the same voltage DC pulse composite voltage at different temperatures. It has the best inhibition effect at 30 °C. As the temperature rises, its inhibition effect on the growth of electric branches gradually decreases. After adding 4,4′-bis(dimethylamino)benzoyl (class A polycyclic compound) and 4-phenylbenzophenone (class C polycyclic compound), the length and accumulated damage increase instead. The length and accumulated damage of PP-AN0.1 are not much different from the length and accumulated damage of PP. The length and cumulative damage of PP-CN0.1 are significantly larger than that of PP. When the pressure time at 30 °C was 15 min, the length of the PP-CN0.1 electrical tree increases by 13.2% compared to PP, and the accumulated damage increases by 29.1%. When the

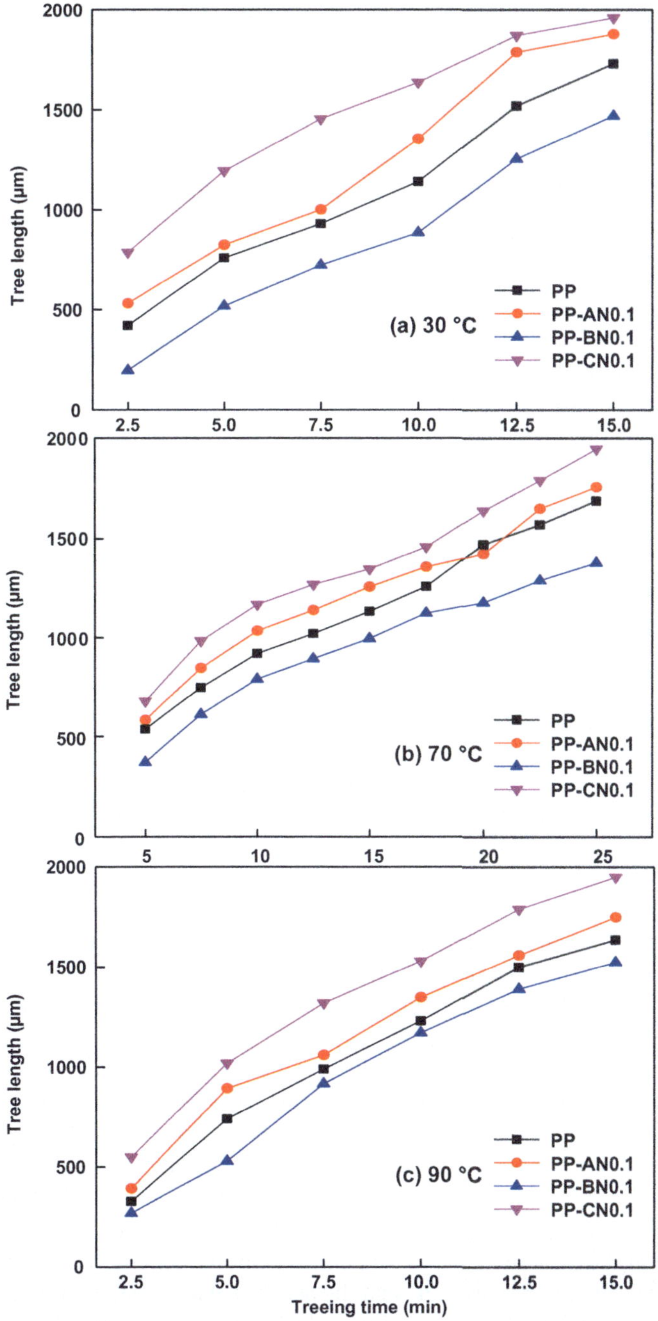

Fig. 23 The electrical tree growth characteristics (tree length) with different kinds of polycyclic compounds under different temperatures under the same polarity impulse superimposed DC voltage

pressurization time at 90 °C is 15 min, the length of the PP-CN0.1 electrical tree increases by 19.1% compared with PP, and the accumulated damage increases by 17.7%.

4.3 Effect of Polycyclic Compound Content on Electrical Tree Degradation

From the results in Sect. 4.2, it can be found that the 2-hydroxy-2-phenylacetophenone polycyclic compound can inhibit the growth of electrical tree under the same and different polarity DC-impulse composite voltages at different temperatures. In order to study the influence of the polycyclic compound content on the growth characteristics of electrical tree, in this section, PP was added with a mass fraction of 0.1, 0.2, and 0.3% of 2-hydroxy-2-phenylacetophenone and studied their electrical tree growth characteristics. Set the experiment temperature to 30, 70, 90 °C. The growth characteristics of the electrical tree under −25 kV DC superimposed ±35 kV impulse voltage were measured respectively, and the pulse voltage frequency was 400 Hz.

4.3.1 Electrical Tree Degradation Under Opposite Polarity Impulse Superimposed DC Voltage

Figure 25 shows the structure of the electrical tree with different polycyclic compounds content under −25 kV DC and +35 kV pulse composite voltage. The pressurization time is 2 min. As the temperature increases, the electrical tree channels of the PP and polycyclic compounds composite become thicker and blacker. At the same temperature, with the increase of the polycyclic compound content, the number of electrical tree channels first decreases and then increases. The length of the electrical tree decreases first and then increases.

Figures 26 and 27 show the electrical tree growth characteristics with different contents of polycyclic compounds. The voltage is −25 kV DC and +35 kV pulse composite voltage. It can be seen from the figure that the length and accumulated damage of the electrical tree with the addition of 0.1% by mass of 2-hydroxy-2-phenylacetophenone is the smallest, which can play a role in inhibiting the growth of the electrical tree. However, as the temperature increases, the suppression effect becomes worse. When the pressure time at 30 °C is 35 min, the length of the electrical tree reduces by 18.9% compared to PP, and the accumulated damage reduces by 19.6%. When the pressure time at 70 °C is 4 min, the length of the electrical tree reduces by 17.5% compared to PP, and the accumulated damage reduces by 13.2%. When the pressure time at 90 °C is 2.5 min, the length of the electrical tree reduces by 12.5%, and the accumulated damage reduces by 11.5%. The length of the electrical tree with the addition of 0.2% 2-hydroxy-2-phenylacetophenone sample is longer

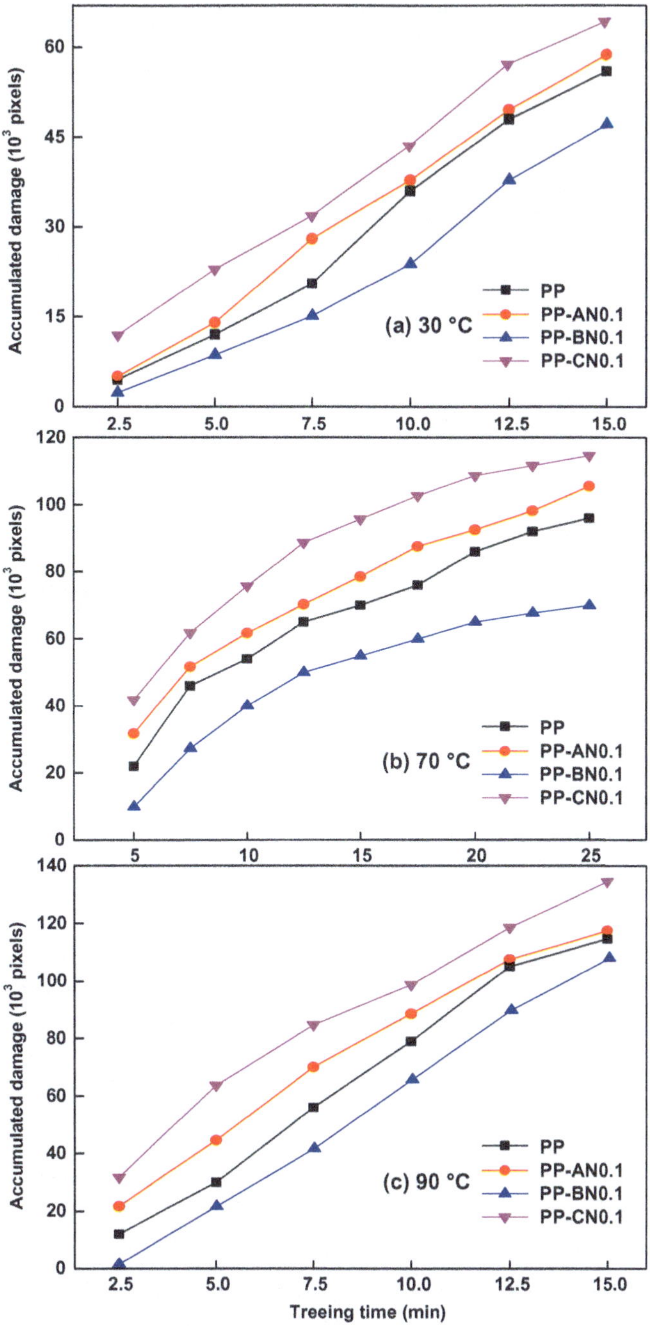

Fig. 24 The electrical tree growth characteristics (accumulated damage) with different kinds of polycyclic compounds under different temperatures under the same polarity impulse superimposed DC voltage

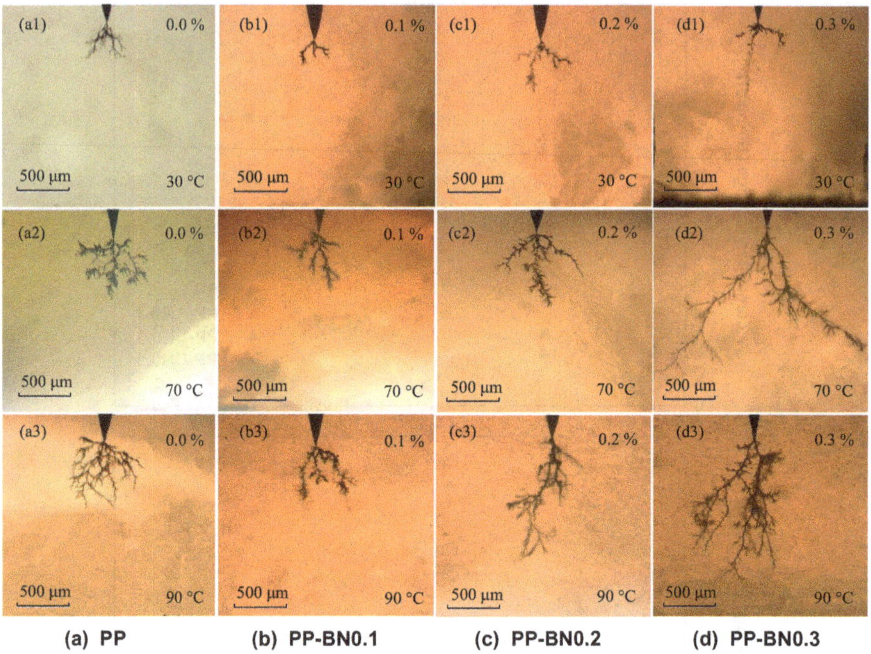

Fig. 25 Tree structures with different contents of polycyclic compounds under different temperatures

than that of pure PP, but at 30 and 70 °C, it still has a certain inhibitory effect on the accumulated damage. When the pressurization time at 30 °C is 35 min, the length of the electrical tree increases by 1.6% compared to PP, and the accumulated damage decreases by 7.5%. When the pressurization time at 70 °C was 4 min, the length of the electrical tree increases by 7.9% compared with PP, and the accumulated damage decreases by 7.1%. When the pressure time at 90 °C is 2.5 min, the length of the electrical tree increases by 66.5%, and the accumulated damage increases by 25.1%. The length of the electrical tree and the accumulated damage of the sample with 2-hydroxy-2-phenylacetophenone with a mass fraction of 0.3% are the largest, especially at high temperatures. When the pressure time at 30 °C was 35 min, the length increases by 8.9% compared to PP, and the accumulated damage increases by 11.1%. When the compression time at 70 °C is 4 min, the electrical tree length increases by 52.9% compared with PP, and the accumulated damage increases by 50.3%. When the pressure time at 90 °C was 2.5 min, the length of the electrical tree increases by 80.2% compared with PP, and the accumulated damage increases by 53.1%.

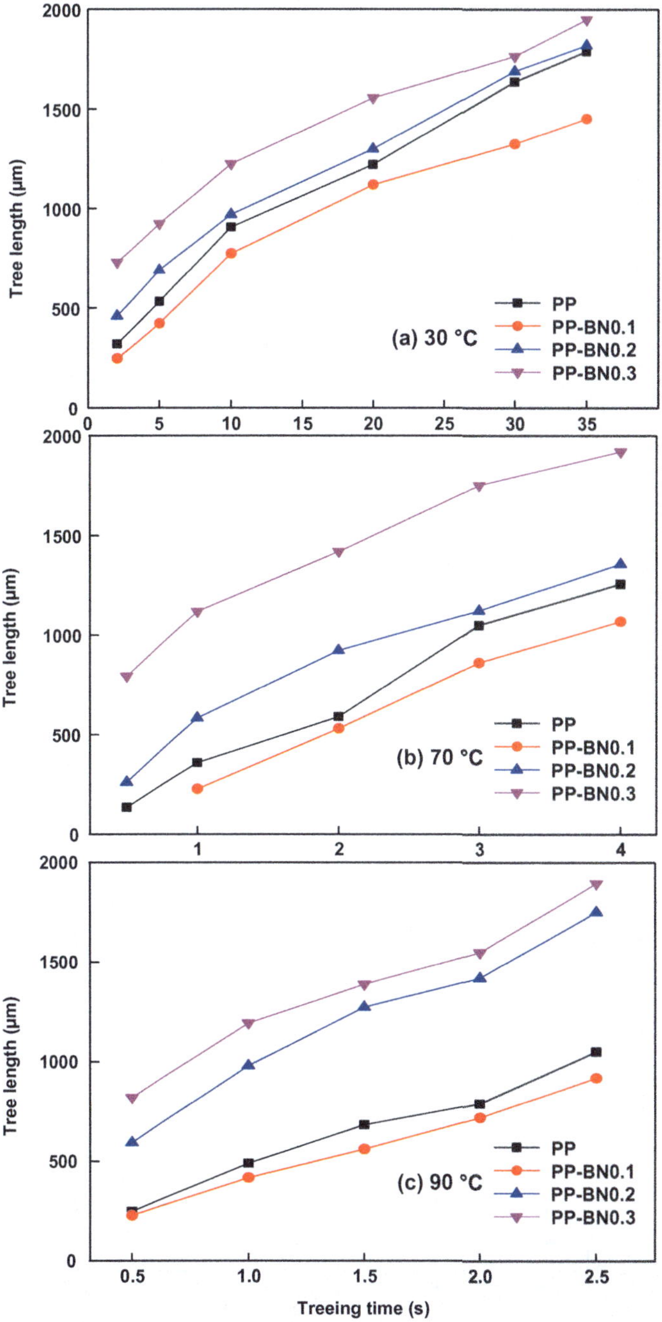

Fig. 26 The electrical tree growth characteristics (tree length) with different contents of polycyclic compounds under different temperatures under opposite polarity impulse superimposed DC voltage

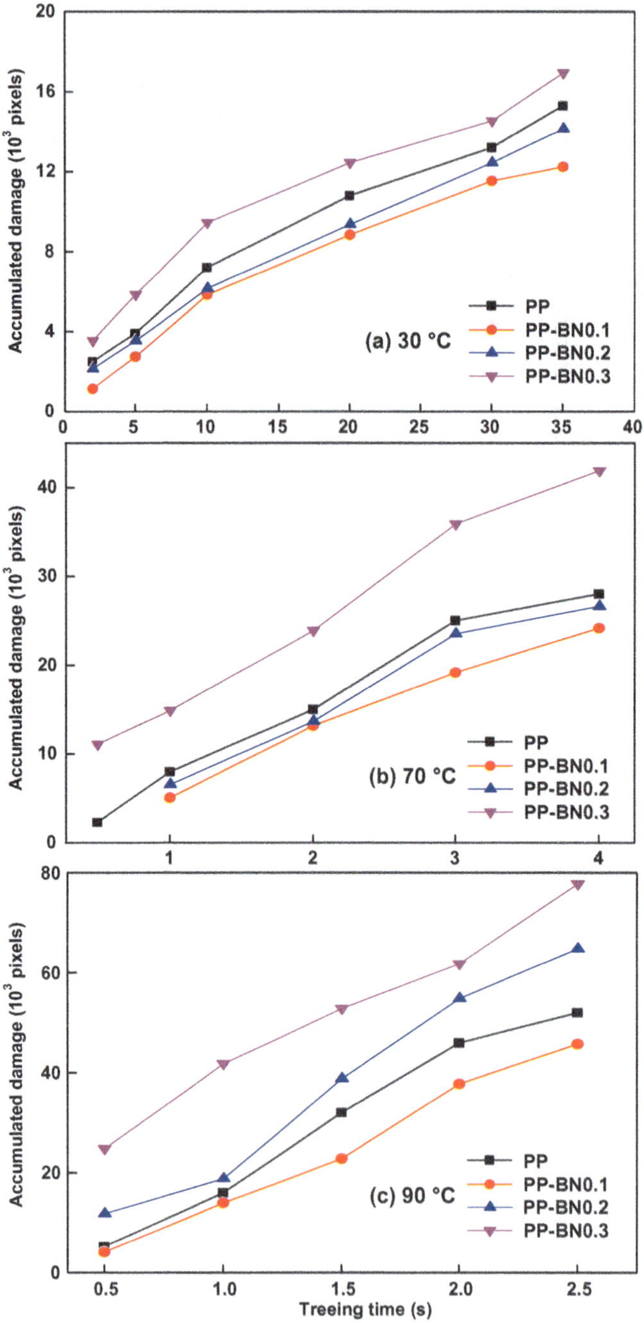

Fig. 27 The electrical tree growth characteristics (accumulated damage) with different contents of polycyclic compounds under different temperatures under opposite polarity impulse superimposed DC voltage

4.3.2 Electrical Tree Degradation Under the Same Polarity Impulse Superimposed DC Voltage

Figure 28 shows the structure of the electrical tree with different polycyclic compounds content under −25 kV DC and −35 kV pulse composite voltage. The pressurization time is 5 min. As the temperature increases, the width and number of electrical tree channels increase. At the same temperature, with the increase of the content of polycyclic compound, the number and length of electrical tree channels decrease first and then increase.

Figures 29 and 30 show the electrical tree growth characteristics with different contents of polycyclic compounds. The voltage is −25 kV DC and −35 kV pulse composite voltage. It can be seen from the figure that the addition of 0.1% by mass of 2-hydroxy-2-phenylacetophenone can play a role in inhibiting the growth of electrical tree, but as the temperature increases, the inhibition effect becomes worse. When the pressure time at 30 °C is 15 min, the length of the electrical tree reduces by 11% compared with PP, and the accumulated damage reduces by 15.5%. At a pressure of 12.5 min at 90 °C, the length of the electrical tree reduces by 10% compared with PP, and the accumulated damage reduces by 4.7%. The length of the electrical tree with the addition of 2-hydroxy-2-phenylacetophenone with a mass fraction of 0.2% is slightly longer than that of pure PP, but it still has a certain inhibitory effect on

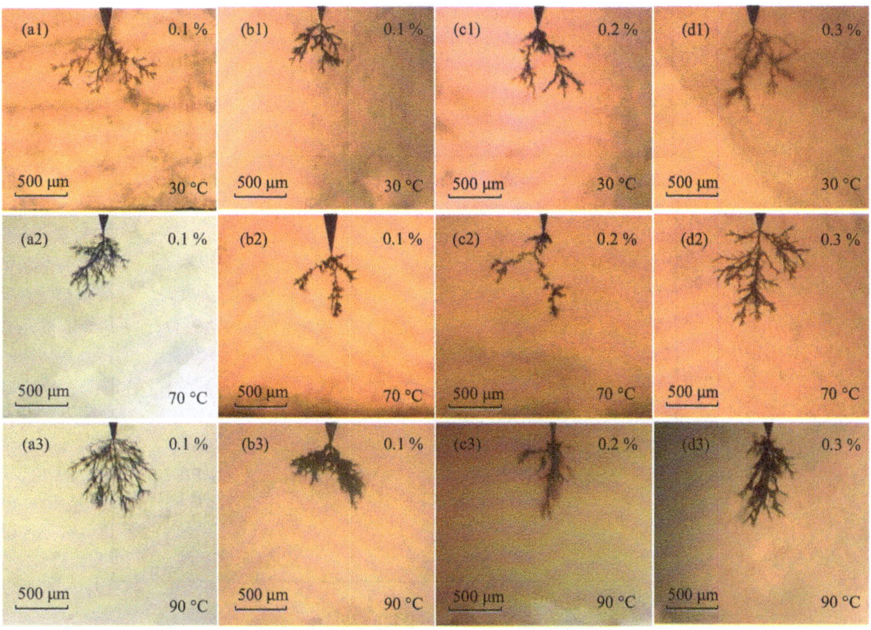

(a) PP (b) PP-BN0.1 (c) PP-BN0.2 (d) PP-BN0.3

Fig. 28 Tree structures with different contents of polycyclic compounds under different temperatures

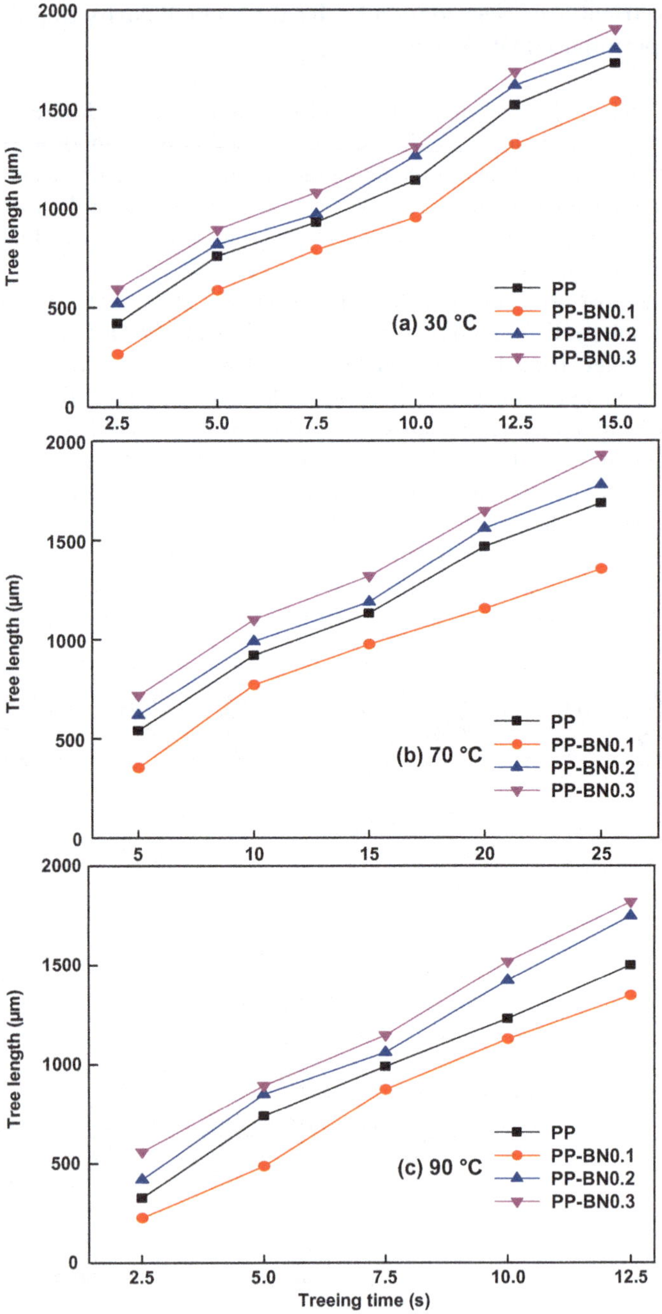

Fig. 29 The electrical tree growth characteristics (tree length) with different contents of polycyclic compounds under different temperatures under the same polarity impulse superimposed DC voltage

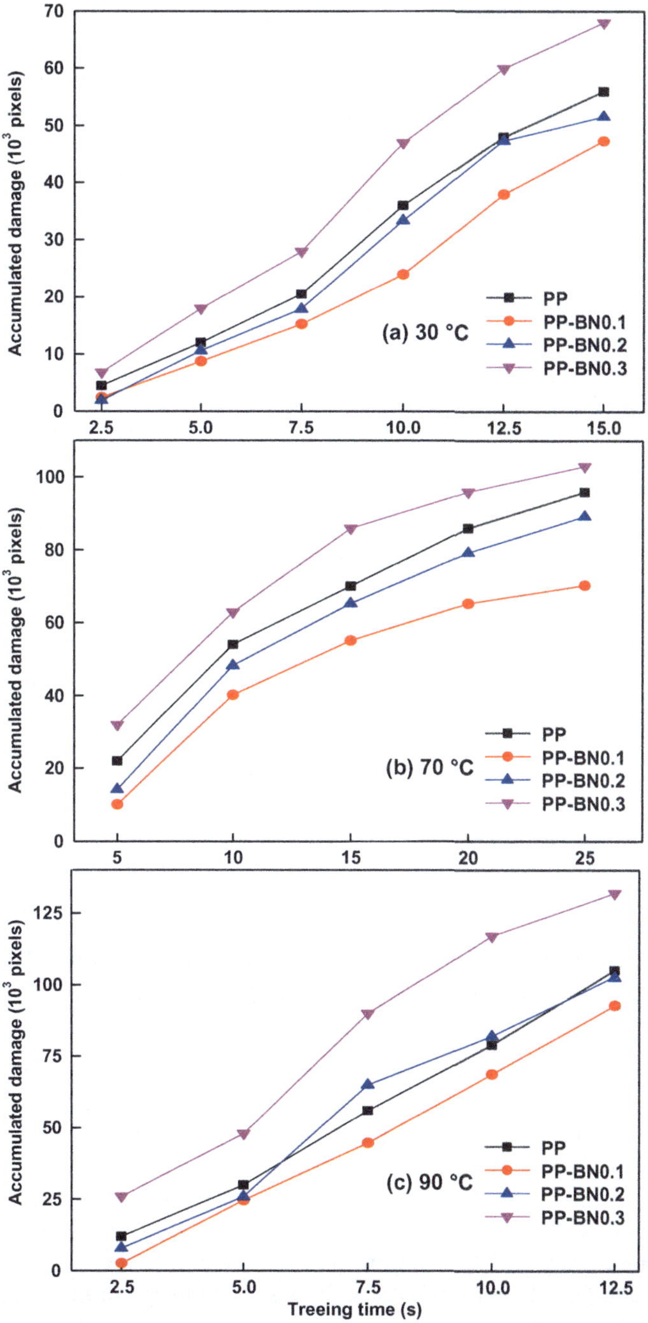

Fig. 30 The electrical tree growth characteristics (accumulated damage) with different contents of polycyclic compounds under different temperatures under the same polarity impulse superimposed DC voltage

accumulated damage. When the pressure time at 30 °C was 15 min, the length of the electrical tree increases by 4.2% compared with PP, and the accumulated damage decreases by 7.1%. When the pressure time at 70 °C was 25 min, the length of the electrical tree increase by 5.5% compared with PP, and the accumulated damage decreases by 7.3%. At a pressure of 12.5 min at 90 °C, the length of the electrical tree increases by 16.6% and the accumulated damage decreases by 1.9%. The length and accumulated damage of the electrical tree with the addition of 0.3% of the 2-hydroxy-2-phenylacetophenone sample is the largest, and it promoted the growth of the electrical tree. When the pressure time at 30 °C was 15 min, the length of the electrical tree increases by 10.1% compared with PP, and the accumulated damage increases by 23.2%. At a pressure of 70 °C for 25 min, the length of the electrical tree increases by 14.4%, and the accumulated damage increases by 7.5%. When the pressure time at 90 °C is 12.5 min, the length of the electrical tree increases by 21.3%, and the accumulated damage increases by 25.7%. Comparing the effect of the polycyclic compound content under the same polarity and heteropolar DC pulse composite voltage, it can be found that the addition of 0.1% by mass of 2-hydroxy-2-phenylacetophenone can play a role in inhibiting the growth of electrical tree. The suppression effect is better under the same polarity DC pulse composite voltage than the different polarity DC pulse composite voltage, especially at high temperature. When the pressurization time at 90 °C is 2.5 min, the length of the electrical tree decreases by 29.9% under the same polarity DC pulse composite voltage, the accumulated damage decreases by 16.7%, and the length of the electrical tree decreases by 12.5% under the DC pulse composite voltage of different polarities. The accumulated damage reduces by 11.5%. The addition of 0.2 and 0.3% mass fraction of 2-hydroxy-2-phenylacetophenone will increase the length of the electrical tree, which is more obvious under the DC voltage of the different polarity than the DC voltage of the same polarity.

4.4 The Influence Mechanism of Polycyclic Compounds on the Growth Characteristics of Electrical Tree

In Sects. 3.3 and 3.4, the charge movement process under the DC-pulse composite voltage and the relationship between the charge movement and the electrical tree growth mechanism are discussed. The growth process of the electrical tree under the DC pulse composite voltage is closely related to the thermal electron movement. When the applied voltage is −25 kV DC and −35 kV pulse composite voltage (DC pulse composite voltage of the same polarity), the change of the pulse voltage causes the charge injected by the DC voltage to generate energy, and the energy is transferred to the charge to form hot electrons accelerating the collision of molecular chains in the free volume, leading to the initiation and growth of electrical tree. When the applied voltage is −25 kV DC and +35 kV pulse composite voltage (heterogeneous DC pulse composite voltage), after the pulse voltage is applied, the positive and negative charges neutralize to generate energy and excite the hot electrons to hit the molecular chain. Since the energy required for excitation or ionization of most aromatic compounds is lower than that of PP molecular chain bonds, under the action of a strong electric field, the aromatic will first be impacted by electrons and be excited or ionized, which is not easy to damage the molecular chain of the polymer [36]. The excited state molecules formed by the excitation of aromatic compounds will release energy in a relatively harmless form through light emission or vibration, and be reduced to the initial state, while the positive radicals formed by ionization will be reduced to the initial state by interaction with electrons [37].

The addition of two polycyclic compounds, 4-phenylbenzophenone and 4,4′-bis(dimethylamino)benzoyl, accelerates the growth of the electrical tree, which may be due to that the compatibility of these two polycyclic compounds with PP molecules is not high, and some additives migrate and precipitate, resulting in a decrease in the electrical tree resistance of the PP/polycyclic compound composite material. With the addition content of 2-hydroxy-2-phenylacetophenone increasing from 0.1 to 0.3 wt.%, the growth rate of the electrical tree increases instead, due to the increase in the amount of addition, some additives will be concentrated in one place. On the contrary, it is equivalent to the introduction of impurities, which causes the performance of the electrical resistant tree to decrease. The addition of 2-hydroxy-2-phenylacetophenone with a mass fraction of 0.1% can play a role in inhibiting the growth of electrical tree, but its inhibitory effect becomes worse as the temperature increases. This is because at high temperatures, the growth mechanism of the electrical tree changes, the influence of hot electrons on the growth characteristics of the electrical tree decreases, and the partial discharge and other effects are enhanced. Therefore, the effect of the polycyclic compound that captures the hot electrons decreases with increasing temperature. After PP is added with polycyclic compound, the polar groups in the composite increase. Under DC pulse recombination voltage, polar groups are easily polarized and ionized, producing positive and negative charges [38]. Under the mixed voltage of different polarity DC pulse, the positive and negative charges injected before and after the polarity change are more likely to neutralize the negative and positive charges generated by the ionization of the polar group, generate energy, and accelerate the growth of the electrical tree. Therefore, the effect of the voltage stabilizer under the composite voltage of different polarity DC pulses is worse than that under the composite voltage of DC pulses of the same polarity. As the temperature increases, polar groups are more easily ionized, so the energy generated by the recombination of positive and negative charges increases, and the difference between the suppression effects under the same polarity and heteropolarity is more obvious [39].

5 Conclusion

In this chapter, an electrical treeing initiation and breakdown phenomenon in PP under impulse superimposed DC voltages were investigated. Then, three different polycyclic compounds were added into PP. The electrical tree growth characteristics in PP and these PP/polycyclic compounds composites under opposite and the same polarity impulse superimposed DC voltage were also studied at 30, 60 and 90 °C. Then the filling ratio of the polycyclic compound with the best effect on inhibiting electrical tree is further changed, and the influence of filling ratio on the growth rate and accumulated damage were studied. The results show that adding 0.1 wt.% 2-hydroxy-2-phenylacetophenone polycyclic compound can effectively inhibit the electrical tree degradation, which has a great potential for application in recycle HVDC cables. The conclusions could be summarized:

(1) Under the same polarity impulse superimposed DC voltage, as the DC voltage amplitude increases, the electrical tree growth rate and accumulated damage increase, and the breakdown time decreases. Under opposite polarity impulse superimposed DC voltage, as the DC voltage amplitude increases, the growth rate and accumulated damage increase first and then decrease, and the breakdown time decreases first and then increases; The breakdown time will reach a minimum value at a certain DC voltage, and the breakdown time at this time is smaller than the minimum breakdown time under the same polarity impulse superimposed DC voltage.
(2) Regardless of the DC polarity, the impulse voltage frequency has a significant effect on the structure and growth characteristics of the electrical tree. As the frequency of the impulse voltage increases, the growth rate and accumulated damage of the electrical tree increase.
(3) With the impulse amplitude increasing, the growth rate and accumulated damage increase under DC superimposed different impulse voltages, and the breakdown time decreases as the pulse voltage amplitude increases. The electrical tree is easier to breakdown under the same polarity impulse superimposed DC voltages than that under opposite polarity impulse superimposed DC voltage.
(4) The addition of 0.1 wt.% 2-hydroxy-2-phenylacetophenone polycyclic compound can inhibit the electrical tree degradation in PP under impulse superimposed DC voltage. However, the addition of 0.1 wt.% 4-phenylbenzophenone and 0.1 wt.% 4,4′-bis(dimethylamino)benzene will increase the electrical tree length and accumulated damage.
(5) As the 2-hydroxy-2-phenylacetophenone content increases from 0.0 to 0.1 wt.%, the electrical tree length and accumulated damage decrease. As the content increases from 0.1 to 0.3 wt.%, the electrical tree length and accumulated damage increase. The addition of 0.1 wt.% 2-hydroxy-2-phenylacetophenone has the best inhibitory effect on the electrical tree degradation.
(6) The effect of polycyclic compounds on the electrical tree growth characteristics is affected by the relative polarity of impulse superimposed DC voltage and temperature. Although the growth trend of electrical trees under the same

polarity and opposite polarity impulse superimposed DC voltage is similar, but the effect of polycyclic compounds is worse under opposite polarity than that under the same polarity. The effect of the polycyclic compounds at high temperatures is weaken [34].

References

1. M.G. Danikas, T. Tanaka, Nanocomposites-a review of electrical treeing and breakdown. IEEE Electr. Insul. Mag. **25**, 19–24 (2009)
2. M. Ieda, M. Nawata, Dc treeing breakdown associated with space charge formation in polyethylene. IEEE Trans. Electr. Insul. **12**, 19–25 (1977)
3. B.X. Du, L.W. Zhu, T. Han, Effect of low temperature on electrical treeing of polypropylene with repetitive pulse voltage. IEEE Trans. Dielectr. Electr. Insul. **23**, 1915–1923 (2016)
4. N.G. Hingorani, Transient overvoltage on a bipolar HVDC overhead line caused by DC line faults. IEEE Trans. Power Appar. Sys. **89**(4), 592–610 (1970)
5. B.X. Du, L.W. Zhu, Electrical tree characteristics of XLPE under repetitive pulse voltage in low temperature. IEEE Trans. Dielectr. Electr. Insul. **22**, 1801–1808 (2015)
6. Y. Liu, X.L. Cao, Electrical tree initiation in XLPE cable insulation by application of DC and impulse voltage. IEEE Trans. Dielectr. Electr. Insul. **20**, 1691–1698 (2013)
7. G. Katsuta, T. Itaya, T. Nakatsuka, H. Miyata, T. Takahashi, T. Niwa, DC and impulse treeing characteristics in insulating material for HVDC cable, in *Proceeding of the IEEE International Conference on Properties and Applications of Dielectric Materials* (1997), pp. 422–425
8. I. Iddrissu, S.M. Rowland, The impact of DC bias on electrical tree growth characteristics in epoxy resin samples, in *Proceeding of the IEEE Conference on Electrical Insulation and Dielectric Phenomena* (2015), pp. 876–879
9. H.C. Liu, Y.P. Liu, Y.D. Li, P. Zhong, H.R. Rui, Growth and partial discharge characteristics of electrical tree in XLPE under AC-DC composite voltage. IEEE Trans. Dielectr. Electr. Insul. **24**, 2282–2290 (2017)
10. G.C. Stone, R.G. van Heeswijkb, R. Bartnikas, Electrical aging and electroluminescence in epoxy under repetitive voltage surges. IEEE Trans. Electr. Insul. **27**, 233–244 (1992)
11. B.X. Du, J.G. Su, T. Han, Temperature-dependent electrical tree in silicone rubber under repetitive pulse voltage. IEEE Trans. Dielectr. Electr. Insul. **24**, 2291–2298 (2017)
12. I.L. Hosier, A.S. Vaughan, S.G. Swingler, An investigation of the potential of polypropylene and its blends for use in recyclable high voltage cable insulation systems. J. Mater. Sci. **46**, 4058–4070 (2011)
13. B.X. Du, H. Xu, J. Li, Z.L. Li, Space charge behaviors of PP/POE/ZnO nanocomposites for HVDC cables. IEEE Trans. Dielectr. Electr. Insul. **23**, 3165–3174 (2016)
14. K. Kurahashi, Y. Matsuda, A. Ueda, et al., The application of novel polypropylene to the insulation of electric power cable, in *Proceeding of the IEEE PES Transmission and Distribution Conference and Exhibition* (2002), pp. 1278–1283
15. S. Belli, G. Perego, A. Bareggi, et al., P-laser: breakthrough in power cable systems, in *Proceeding of the IEEE International Symposium on Electrical Insulation* (2010), pp. 1–5
16. M. Albertini, A. Bareggi, L. Carmi, Development and high temperature qualification of innovative 320 kV DC cable with superiorly stable insulation system, in *Proceeding of International Conference on Insulated Power Cables* (2015), p. A7.3
17. B.X. Du, L.W. Zhu, T. Han, Effect of ambient temperature on electrical treeing and breakdown phenomenon of polypropylene with repetitive pulse voltage. IEEE Trans. Dielectr. Electr. Insul. **24**, 2216–2224 (2017)
18. M. Kosaki, M. Nagao, N. Shimizu, Y. Mizuno, Solid insulation and its deterioration. Cryogenics **38**(11), 1095–1104

19. J. Holto, E. Ildstad, Electrical tree growth in extruded s-polypropylene, in *Proceedings of the 2010 10th IEEE International Conference on Solid Dielectrics, Potsdam, Germany, July 2010* (2010), pp. 1–4
20. Y. Yamano, Control of electrical tree at initiation stage in LDPE by mixed addition of Al_2O_3 nano-particle and azobenzoic compound. IEEE Trans. Dielectr. Electr. Insul. **21**(1), 209–216 (2014)
21. L.W. Zhu, B.X. Du, L.H. Na, K. Hou, Effect of polycyclic compounds fillers on electrical treeing characteristics in XLPE with DC-impulse voltage. Energies **12**(14), 1–15 (2019)
22. L.J. Heidt, Solid dielectric polyolefin compositions containing various voltage stabilizers: U.S., Patent 3, 522, 183 (1970)
23. G.H. Hunt, Cross-linked polyolefin compositions: U.S., Patent 3, 632, 680 (1972)
24. M. Jarvid, A. Johansson, J.M. Bjuggren, H. Wutzel, V. Englund, S. Gubanski, C. Muller, M.R. Andersson, Tailored side-chain architecture of benzil voltage stabilizers for enhanced dielectric strength of cross-linked polyethylene. J. Polym. Sci. Part B Polym. Phys. **52**(16), 1047–1054 (2014)
25. S. Kisin, J.D. Doelder, R.F. Eaton, Quantum mechanical criteria for choosing appropriate voltage stabilization additives for polyethylene. Polym. Degrad. Stab. **94**(2), 171–175 (2009)
26. I. Kitani, K. Arii, DC tree associated with space charge in PMMA. IEEE Trans. Electr. Insul. **22**, 303–307 (1987)
27. K. Kaneko, Effect of space charge on the breakdown strength under polarity reversal. Elec. Eng. Jap. **106**, 565–572 (1986)
28. Y. Tanaka, R. Kodera, T. Kato, H. Miyake, H. Mori, Y. Yagi, Observation of space charge accumulation behavior in cross-linked polyethylene at voltage polarity reversal, in *Proceeding of the IEEE Conference on Electrical Insulation and Dielectric Phenomena* (2015), pp. 23–26
29. L.A. Dissado, Understanding electrical trees in solids from experiment to theory. IEEE Trans. Dielectr. Electr. Insul. **9**, 483–497 (2002)
30. T. Tanaka, Charge transfer and tree initiation in polyethylene subjected to ac voltage stress. IEEE Trans. Electr. Insul. **27**, 424–431 (1992)
31. K.C. Kao, New theory of electrical discharge and breakdown in low mobility condensed insulators. J. App. Phys. **55**, 752–755 (1984)
32. B. Huang, M. Hao, J. Hao, J. Fu, Q. Wang, G. Chen, Space charge characteristics in oil and oil-impregnated pressboard and electric field distortion after polarity reversal. IEEE Trans. Dielectr. Electr. Insul. **23**, 881–891 (2016)
33. T. Tanaka, Space charge injected via interfaces and tree initiation in polymers. IEEE Trans. Dielectr. Electr. Insul. **8**, 733–743 (2001)
34. S.S. Bamji, A.T. Bulinski, R.J. Densley, Degradation mechanism at XLPE/semicon interface subjected to high electrical stress. IEEE Trans. Dielectr. Electr. Insul. **26**(2), 278–284 (1991)
35. L.W. Zhu, B.X. Du, J.G. Su, T. Han, M.G. Danikas, Electrical treeing initiation and breakdown phenomenon in polypropylene under DC and pulse combined voltages. IEEE Trans. Dielectr. Electr. Insul. **26**(1), 202–210 (2019)
36. B.X. Du, C.L. Han, Z.L. Li, J. Li, Improved DC conductivity and space charge characteristics of XLPE for HVDC cable application: effect of voltage stabilizers. IEEE Access **7**, 66576–66583 (2019)
37. B.X. Du, C.L. Han, J. Li, Z.L. Li, Effect of voltage stabilizers on the space charge behavior of XLPE for HVDC cable application. IEEE Trans. Dielectr. Electr. Insul. **26**(1), 34–42 (2014)
38. N.M. Chalashkanov, S.J. Dodd, L.A. Dissado, J.C. Fothergill, The role of bulk charge transport processes in electrical tree fformation and breakdown mechanisms in epoxy resins. IEEE Trans. Dielectr. Electr. Insul. **23**(6), 3256–3266 (2017)
39. L.W. Zhu, B.X. Du, Z.L. Li, H.N. Li, K. Hou, Polycyclic compounds affecting electrical tree growth in polypropylene under ambient temperature. IEEE Access **8**, 8886–8898 (2020)

Conduction and Charge Transport Characteristics of Silicone Rubber Composites with Nonlinear Conductivity

Zhonglei Li, Boxue Du, Zhuoran Yang, Chong Han, and Jingang Su

Abstract Silicone rubber (SiR) composites with nonlinear conductivity have been applied to modify the direct current (DC) electrical field distribution and accelerate the space charge dissipation in polymeric insulation. This chapter reviews the conduction properties of SiR incorporated with zinc oxide (ZnO) fillers varying with the electric field. The isothermal surface potential decay (ISPD) method is employed to characterize the unipolar charge transfer process under DC voltage and impulse superimposed on dc voltage. The effect of the nonlinear conductivity on the charge transport dynamic behaviours is experimental researched and a theoretical model is proposed. The results show that the enhanced nonlinear conductivity of ZnO/SiR composites results in an acceleration of the charge transport and dissipation through the percolating paths. As the surface charge dissipating, the surface-charge-induced electric field weakens, leading to the decrease of the ISPD rate. The homopolar impulse superimposed on DC voltage intensifies the surface charge accumulation, and accelerate the charge transport. Under a heteropolar impulse superimposed on a DC voltage, neutralization occurs and the heterocharges accumulate on the surface. The vertical charge transport to the sample bulk and the horizontal movement along the surface are two factors concerning the charge dissipation.

Z. Li (✉) · B. Du · Z. Yang · C. Han · J. Su
Tianjin University, No. 92, Weijin Road, Nankai District, Tianjin 300072, China
e-mail: lizhonglei@tju.edu.cn

B. Du
e-mail: duboxue@tju.edu.cn

Z. Yang
e-mail: yangzhuoran@126.com

C. Han
e-mail: 15122192795@163.com

J. Su
e-mail: sujingang@tju.edu.cn

B. Du (ed.), *Polymer Insulation Applied for HVDC Transmission*,
https://doi.org/10.1007/978-981-15-9731-2_9

1 Introduction

Insulation in the HV power equipment are often exposed to the non-uniform electric field distributions even though the hardware is well designed. Nonlinear conductive composites have been used in applications in electrical engineering for decades, in particular, for controlling the electric field distribution in DC or high voltage (HV) DC power equipment [1–3]. It is reported that a nonlinear conductive insulation layer has been arranged in HVDC cable accessories, which can effectively suppress electric field distortion at the terminal of outer semi-conductive layer of HVDC cable [4, 5]. Nonlinear conductive composite insulation is also used in solid/gaseous insulation systems to modified the disc-type insulator geometry based on the nonlinear temperature and field dependent conductivity [6, 7]. The application of insulating materials with nonlinear conductivity has also been become a research hotspot in the area of high-power insulated gate bipolar transistor (IGBT) devices in recent years [8, 9].

Nonlinear conductive insulations were generally achieved by dispersing semi-conductive or conductive fillers, for example zinc oxide (ZnO), silicon carbide (SiC) or carbon black (CB), into polymeric insulating materials [10–12]. These inorganic nonlinear fillers act as a lot of distributed microvaristors and alter the local resistance of polymeric composites under the dc field by themselves, without changing their excellent electrical properties under the lower field [13]. The nonlinear conductivity is attributed to the percolation processes through the conductive or semi-conductive fillers and the polymer matrix under a sufficient high electric field. In this case, a large number of carriers within nonlinear fillers pass through the interface between polymer and nonlinear fillers and move towards the neighboring filler with polymer matrix, resulting in a nonlinear enhancement of dc conductivity [14]. He et al. proposed a simulation model based on Voronoi network, which agreed well to the measured nonlinear conducting behaviors of ZnO-silicone rubber composites and further understood the percolation process as the stress is higher than their switching field [15]. Du and Dang et al. has reported that the space charge dynamic behaviors of nonlinear conductive insulations also showed a significantly increased carrier mobility in the initial stage of depolarization processes [16, 17]. Recently, Li et al. proposed a novel design of HVDC spacer based on the concept of adaptively controlling surface charges using nonlinear conductive materials [18–20]. It is presented that the percolation process of nonlinear conductive insulations can accelerate the charge transport under the electric field and modify the local electric field distribution, thus further enhancing the surface flashover voltage. However, the percolation process is dependent on many factors, including the electric field distribution induced by the surface charge, the field-dependent conductivity varying with the filler content. There is still a limit to our understanding of the correlation between the field-dependent conductivity and the surface charge dissipation of the nonlinear conductive composites.

For the HVDC power equipment, a transient overvoltage superimposed on the operating DC voltage is a key issue for the insulation failure in electrical equipment

[21]. It is unavoidable for an HVDC transmission system to experience a transient overvoltage caused by switching and lightning [22]. Previous studies proved that the impulse voltage leads to an obvious accumulation on the surface of the insulation, as the partial discharge or corona discharge occurs in air [23–26]. Recent research also showed that an impulse superimposed on a DC voltage could affect the surface charge accumulation and further influence the flashover characteristics of insulation [27]. The use of nonlinear conductive composites is considered to be an effective way to suppress the surface charge accumulation. However, the surface charge accumulation and dissipation behavior of nonlinear conductive composites under such a superimposed voltage have not been researched. In addition, it is significant to clarify how nonlinear conductivity affects the surface charge accumulation and dissipation under an impulse superimposed on a DC voltage.

This chapter aims to reveal the correlation between the field-dependent conductivity and the surface charge dissipation process of the nonlinear conductive composites. The materials are achieved by doping ZnO microvaristors into a silicone rubber (SiR) matrix. Different contents of ZnO provide various nonlinear conductivities for comparison. DC conductivity and isothermal surface potential decay (ISPD) tests are employed to characterize the nonlinear conductivity and charge transport dynamic behaviours of ZnO/SiR composites. A theoretical model of the ISPD process is proposed to analyze the correlation between the nonlinear conductivity and the surface potential decay of nonlinear materials under the function of the surface-charge-induced electric field. In addition, the charge transport processes under the impulse superimposed on a DC voltage is investigated by a 2-dimentional ISPD test. The effect of the polarity and amplitude of the impulse voltage on the surface charge accumulation are analyzed. More importantly, the effect of the ZnO filler content on the surface charge transport characteristics of ZnO/SiR composites are discussed.

2 Nonlinear Conductivity of SiR/ZnO Composite Insulation

2.1 Semi-Conductive Additives of ZnO

Nonlinear conductive composite insulations are generally achieved by dispersing semi-conductive or conductive fillers, for example zinc oxide (ZnO), silicon carbide (SiC) or carbon black (CB), into polymeric insulating materials [10–12]. These inorganic nonlinear fillers act as a lot of distributed microvaristors and alter the local resistance of polymeric composites under the dc field by themselves, without changing their excellent electrical properties under the lower field [13].

In this chapter, the ZnO microvaristor fillers are used as the semi-conductive additives, which are synthesized by using ZnO, Co_2O_3, MnO_2, Sb_2O_3, Cr_2O_3, Nd_2O_3 and Y_2O_3 as starting materials through the traditional solid-reaction method [27]. In

Table 1 Recipes of the ZnO microvaristors through the solid-reaction method. Copyright Clearance RightsLink® License Number: 4856251506816

Recipe	Mole percent (mol %)	Weight percent (wt %)
ZnO	95.95	63.2506
Bi_2O_3	0.7	31.2131
Sb_2O_3	1.0	1.3432
Cr_2O_3	0.5	0.3521
Co_2O_3	1.0	2.3611
MnO_2	0.5	0.6156
Nd_2O_3	0.25	0.6813
Y_2O_3	0.1	0.1829

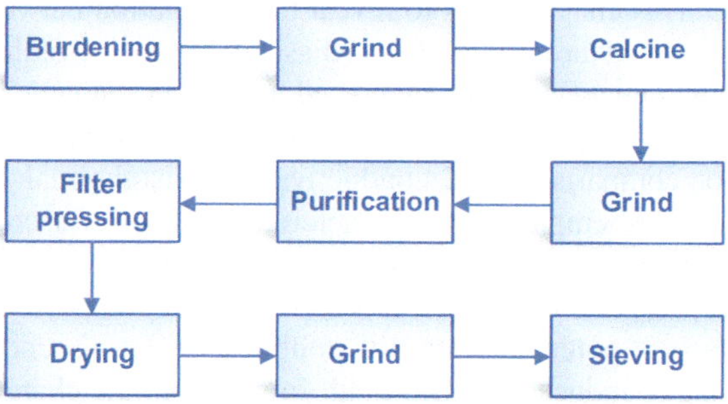

Fig. 1 Solid-reaction method process of ZnO microvaristors. Copyright Clearance RightsLink® License Number: 4856251506816

the ZnO fillers, the content of each compound is shown in Table 1, and the solid-reaction method of ZnO microvaristor is presented in Fig. 1.

2.2 *Microstructures of ZnO/SiR Composites*

The ZnO/SiR composites were composed of polydimethyl-siloxane (PDMS) of polymer, silica, micro-sized ZnO fillers and vulcanizing agent. ZnO particles, which have been washed by alcohol and dried for 12 h, are then doped in the matrix by a twin-roll mixer at room temperature (RT). Finally, the mixture is vulcanized in a stainless mold at 160 °C under 15 Mpa for 15 min. Finally, the mixture is vulcanized in a stainless mold at 160 °C under 15 Mpa for 15 min.

The scanning electron microscope (SEM) images of the cross-sections of ZnO/SiR composites with different filler contents are shown in Fig. 2, which are acquired by ZEISS GeminiSEM 500. There was no agglomeration of ZnO particles for all the composites, indicating the ZnO microvaristors are well-dispersed in the SiR matrix. It can be observed in Fig. 2a, for the 6.0 vol.% ZnO-doped SiRCs, some of the spacings between the adjacent fillers are larger than 10 μm. With the increase of the filler content from 6.0 to 12.8 vol.%, the average distance between neighbouring fillers decreases markedly. It is noted that when the filler content reaches 12.8 vol.%, the distance between the neighbouring fillers is decrease to 1–5 μm, and numbers of

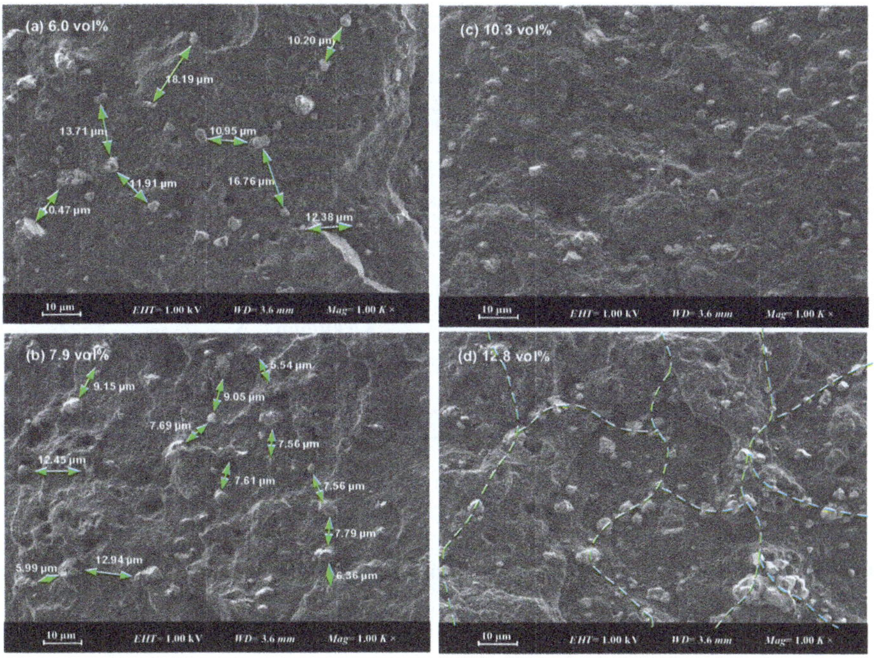

Fig. 2 Scanning electron microscope (SEM) images of the cross-sections of ZnO/SiR composites doped with different contents of ZnO microvaristors. Copyright Clearance RightsLink® License Number: 4856251506816

networks are formed by ZnO microvaristors within polymer matrix, which are marked by the dotted lines in Fig. 2d. It has been proved that the distribution of microvaristors in ZnO/SiR composites has a significant effect on the nonlinear conductivity as well as the charge transport behaviours.

2.3 Nonlinear Conductivity of ZnO/SiR Composites

The relationship between the dc current density and the electric field (J-E characteristics) of the ZnO/SiR composites shown in Fig. 3 is derived from the dc conductivities which were measured by the three-electrode arrangement. It is observed that the current density of the neat SiR rises slightly when the electric field exceeds 12 kV/mm. It is considered that the residual vulcanizing agent and its by-produces may act as impurities and are ionized under the high field, resulting in a formation of space charge. It is believed that the space-charge-limited current (SCLC) under the high field causes the raise of current density. For the 6.0 vol.% ZnO-doped SiRCs, the J-E characteristics are similar to those of the neat samples. As the nonlinear filler content is below the percolation threshold, the mean distance between neighboring

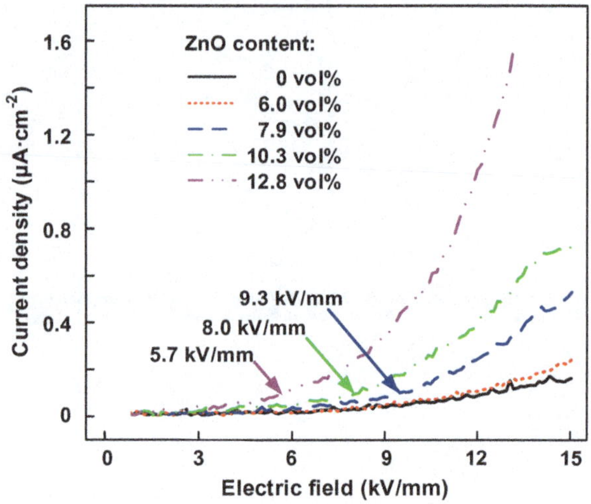

Fig. 3 J-E curves of SiRCs with different content of ZnO micro-fillers at room temperature. Copyright Clearance RightsLink® License Number: 4856251506816

fillers is so large that no percolating path forms throughout the composites, as shown in Fig. 2a.

Further increasing the volume fraction of ZnO fillers leads to a decrease of the mean distance between neighboring fillers and results in a formation of the percolating paths in the polymer material under the high electric field. It can be observed that for the SiR composites with ZnO filler content over 7.9 vol.%, there appears a nonlinear growth of the conductivities as the electric field is lower than 10 kV/mm. These field-dependent conductivities under the lower field ($\ll$10 kV/mm) are attributed to the percolation behaviours, not the SCLC effect [28].

In this study, we define the threshold field of nonlinear conductivity as the applied electric field when the current density J reached 0.1 μA/cm^2, above which J turned into the nonlinear region and increased sharply. Figure 3 presents that the threshold fields locate at 9.3, 8.0 and 5.7 kV/mm, for the SiRCs with ZnO volume fractions of 7.9, 10.3 and 12.8 vol.% respectively. It is indicated that with the increase of filler content, the threshold field of nonlinear conductivity decreases significantly and the current densities in the nonlinear regions enhance markedly.

For the nonlinear conductive ZnO/SiR composites, the SiR matrix is considered to be polymer insulation, following the hopping conduction model under a certain electric field. The ZnO microvaristors are semi-conductive fillers. Additionally, the transportation of charges between the ZnO fillers and SiR matrix is an indispensable part of dc conductivity, which is dependent on the barrier potential at the interface region between SiR matrix and ZnO fillers. As shown in Fig. 4, a pair of potential barrier containing positive space charge locates at the interfaces between ZnO fillers and SiR matrix.

When a lower field is applied, a large number of electrons in excited states perform disorderly thermal movement within ZnO microvaristors. The majority of these electrons are blocked by the interfacial barrier between ZnO and SiR phases. Even though

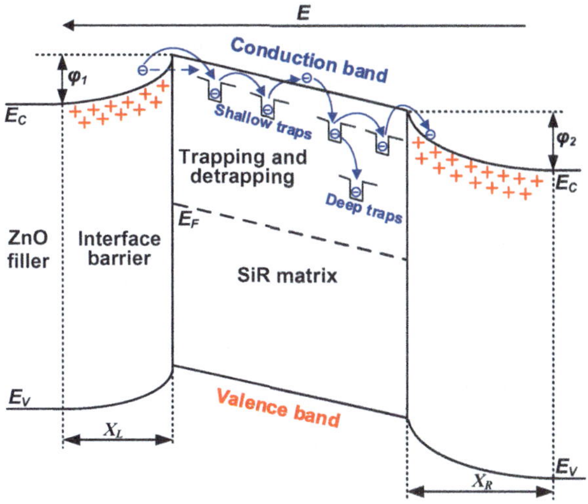

Fig. 4 Schematic model for illustrating the effect of the interface potential between ZnO and SiR matrix on the charge transport processes. Copyright Clearance RightsLink® License Number: 4856251506816

some electrons pass through the interface barrier fortunately, they may be captured by the traps in SiR matrix and unable to reach another neighboring filler. In this case, the dc conductivity of composites relies on the hopping (trapped carriers hopping from one trap site to another) and ohmic conduction (mobile electrons in the conduction band and holes in the valence band) of SiR matrix [29].

With the increase of electric field, further inclination of potential barrier on ZnO/SiR interface would happen, resulting in a further decrease of φ_1 and an increase of φ_2 [15]. In this case, a large member of electrons in extended states could pass through the barrier potential into SiR matrix, in the opposite direction to the applied electric field, as shown in Fig. 4. The electrons will be driven to the neighboring filler by the tilted potential through trapping and de-trapping. On a macro level, the dc conductivity increases nonlinearly when the stress exceeds a certain threshold field.

Additionally, with the increase of the filler content, the average distance between neighboring fillers decrease as shown in Fig. 2, resulting in an acceleration of the carrier transport and migration within SiR matrix between ZnO microvaristors. Therefore, the threshold field of nonlinear conductivity decreases with the increase of filler content.

3 Charge Transport of SiR/ZnO Composite Insulation

3.1 Isothermal Surface Potential Decay Behaviors

ISPD tests are an effective method to characterize the unipolar charge transfer process. During corona charging by a pair of needle-plate electrode, charges forming in the vicinity of needle tip are driven forward sample surface by the electrostatic field, resulting in the surface charge accumulation. Considering the geometry (thin flat) of the samples used, the measured surface potential (V_s) is proportional to the surface charge density (ρ_s), which could be expressed as

$$V_s = \frac{L}{\varepsilon_0 \varepsilon_r} \rho_s \tag{1}$$

where L is the thickness of samples, which is 300 μm in this study. ε_r is the relative permittivity of the samples, which is measured by the broadband dielectric spectrometer (BDS) of Novocontrol Concept 40. The relative permittivity is 2.55, 3.41, 3.78, 4.19, 4.32 under 50 Hz AC, for the ZnO/SiR composites with filler content of 0, 6.0, 7.9, 10.3, 12.8 vol.%, respectively.

In this chapter, the needle voltage was set as test variables in order to alter the initial surface potential immediately after corona charging (V_{s0}) of SiRCs. Figure 5 presents the relationships between the needle voltage of surface charging and the V_{s0}. With the increase of corona charging voltage, the V_{s0} increase significantly, and then tend to be stable gradually, no matter what the polarity of the needle voltage (Fig. 5). It is considered that the increased charging voltage enlarges the electric field near the needle tip, which aggravates the corona discharge and leads to more charges migrated from the needle electrode to sample surface. However, the charges deposed on the sample surface would induce an inverse field to the applied field. When the surface potentials are high enough (~3300 V as shown in Fig. 5), the charge transport towards sample surface would be suppressed, thus leading to the saturation of surface charges. It is also shown that the surface potential after the negative corona charging is higher than that after the positive corona charging with a same amplitude. This is because the ionization and migration of electrons is easier than that of positive ions [29].

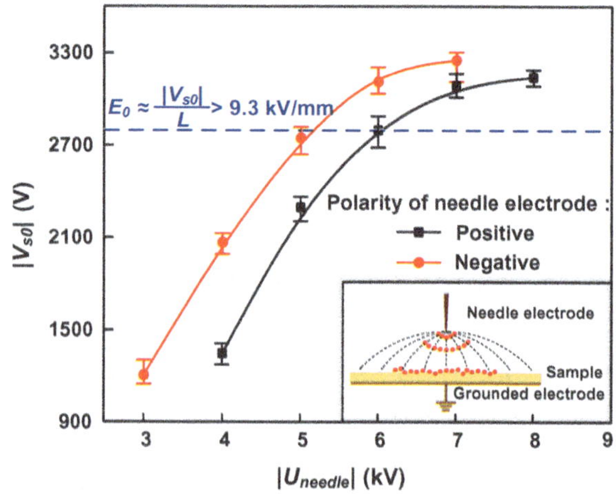

Fig. 5 Relationship between the absolute value of needle voltage and the surface potentials immediately after corona charging (V_{s0}). Copyright Clearance RightsLink® License Number: 4856251506816

After corona charging, the surface charges accumulated on SiRCs induce an electric field between the sample surface and grounded electrode. The electric field (E) in the dielectrics can be related to the potential by $E(x, t) = |V_s(t)/L|$.

In order to evaluate the effect of the nonlinear conductivities on the SPD behaviours, the surface-charge-induced field should be higher than the nonlinear threshold fields of ZnO/SiR composites. Taking the J-E characteristics of SiRCs (Fig. 3) into consideration, when the V_{s0} are higher than 2700 V, the induced fields immediately after corona charging could be higher than 9 kV/mm, in which case the conductivities of SiRCs with filler contents of 7.9, 10.3 and 12.8 vol.% turn into nonlinear conductive regions. The charging voltages, therefore, are respectively set as +8 kV and −6 kV, to ensure the nonlinear behaviours of 7.9, 10.3 and 12.8 vol.% ZnO/SiR composites.

Figure 6 presents the SPD curves of ZnO/SiR composites after corona charging with needle voltages of +8 kV and −6 kV. It is observed that, the surface potentials of these SiRCs samples decay nonlinearly. It is generally considered that three possible routes, including bulk transport from the sample surface to the grounded electrode, surface conduction along the gas–solid interface and neutralization with the ionic species in the gas, lead to the charge decay. Considering the controlled relative humidity, the gas neutralization is negative. Due to the electrode arrangement and the sample dimension, the electric field between the injected charge and the grounded electrode is much higher than that along the surface. As a consequence, bulk transport plays the leading role in the SPD process. With the increase of the ZnO filler content from 0 to 12.8 vol.%, the descending speed of surface potential increases markedly. Especially in the first stage of 600 s, the surface potentials of nonlinear conductive composites decay much faster than the undoped SiR sample.

In order to quantitatively characterize the SPD behaviours, the surface potential decay rate (R_{decay}) is defined as

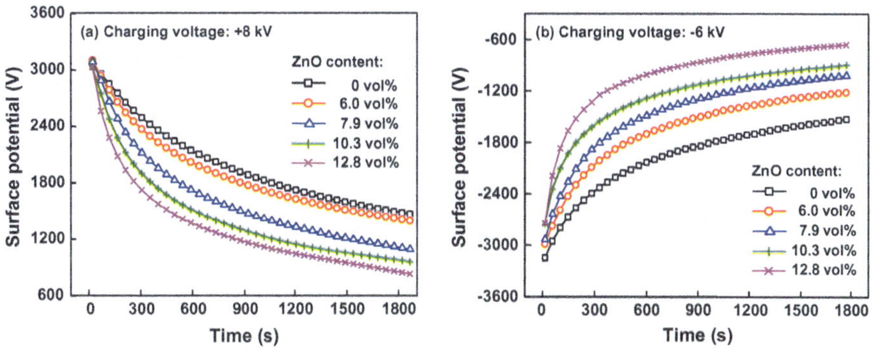

Fig. 6 SPD curves of ZnO/SiR composites, immediately after corona charging under **a** +8 kV and **b** −6 kV for 10 min, respectively. Copyright Clearance RightsLink® License Number: 4856251506816

$$R_{decay}(t) = \left|\frac{dV_S(t)}{dt}\right| = \frac{|V_S(t) - V_S(t + \delta t)|}{\delta t} \quad (2)$$

where $V_s(t)$ is the surface potential at time of t. δt is a time step, which is set as 15 s in this paper. Figures 7a and 7b present the relationships between the decay rate (R_{decay}) and the decay time of SiRCs with different filler contents. The result shows that, with the enhancement of the nonlinear conductivity, the decay rate at the same time increases no matter what the polarity of the surface charges, which could be attributed to the enhanced nonlinear conductivities of the composites. Furtherly, for the neat SiR and 6.0 vol.% ZnO/SiR composites without nonlinear conductivity, the R_{decay} almost keeps constant in the initial stage of the SPD processes, which has been reported in the literature [30]. It has been proved that, at the initial stage of the SPD process, the surface potential decay rate (R_{decay}) could be expressed as [30]

$$R_{decay}(t) = \left|\frac{dV_S(t)}{dt}\right| = \frac{\mu}{2}\left(\frac{V_{S0}}{L}\right)^2 \quad (3)$$

where μ is considered to be a constant carrier mobility in the neat SiR. Therefore, the R_{decay} keeps constant in the initial stage of SPD processes, for the SiR composites without nonlinear conductivity.

However, for the ZnO/SiR composites with nonlinear conductivity, there appears a marked decrease of R_{decay}. It can be explained as follows. At the initial stage of SPD processes, the surface-charge-induced electric field is larger than 9.3 kV/mm and therefore locates at the nonlinear conductive region of 7.9, 10.3 and 12.8 vol.% ZnO/SiR composites. In this case, the conductivity of the composites increases nonlinearly. It can be considered that the large quantities of thermal electrons and holes injected into the SiR matrix move towards the neighboring fillers through the

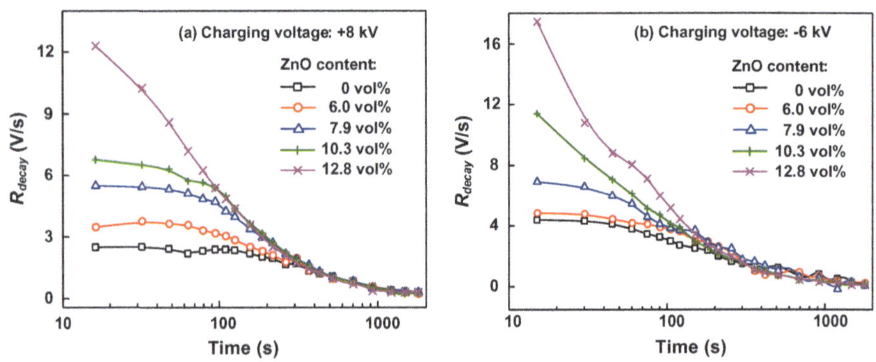

Fig. 7 Relationship between the decay rate (Rdecay) and the decay time of ZnO/SiR composites with different filler contents. Corona charging voltage: **a** +8 kV and **b** −6 kV. Copyright Clearance RightsLink® License Number: 4856251506816

trapping-detrapping processes, under the surface-charge-induced electric field. With the increase of ZnO content from 7.9 to 12.8 vol.%, the nonlinear conductivities of SiRCs are enhanced markedly as shown in Fig. 4, thus resulting in a larger R_{decay} at the same decay time of t. With the dissipation of surface charge, the surface potential drops significantly, and the field-dependent conductivities are weakened, which slows down the carrier transport processes and causes a marked decrease of the R_{decay}. When the decay time reaches 300 s, the field-dependent conductivities of 7.9, 10.3 and 12.8 vol.% ZnO/SiR composites all return to the ohm-conductive regions considering the *J-E* characteristics of SiRCs in Fig. 4, thus resulting in the overlaps between the R_{decay} curves and those of neat SiR and 6.0 vol.% ZnO/SiR composites, as shown in Fig. 7a, b.

3.2 Correlation Between Nonlinear Conductivity and Surface Charge Dissipation

The charge transport process can be estimated by a model following the approach by Sonnonstine et al. [30]. It is considered that the surface charges are stressed by the surface-charge-induced electric field and dissipate through the charge transport in the bulk, thus leading to the surface potential decay. It is assumed that the recombination of holes and electrons is negative. The current density J_T through back electrode involves the conduction current, hence [30]

$$J_T(t) = \rho_F(x, t) \cdot \mu(x, t) \cdot E(x, t) \tag{4}$$

where $\rho_F(x, t)$ is the free charge density at the location of x and the time of t in the SiRCs; $\mu(x, t)$ is the trap controlled charge carrier mobility.

For a uniform density of free charge $\rho_F(x, t)$ throughout the matrix, the conductivity $\sigma(t)$ is given as

$$\sigma(t) = \rho_F(x, t) \cdot \mu(x, t) \tag{5}$$

where $\sigma(t)$ is associated with $E(x, t)$ and temperature.

Substituting Eq. (5) into Eq. (4), then J_T can be expressed as [30]

$$J_T(t) = \sigma(t)\frac{|V_S(t)|}{L} \tag{6}$$

Additionally, the current density J_T can also be expressed by the decay rate of surface potential [31].

$$J_T(t) = \frac{\varepsilon}{L} \cdot \left|\frac{dV_S(t)}{dt}\right| = -\frac{\varepsilon}{L} \cdot \frac{d|V_S(t)|}{dt} \tag{7}$$

By combining Eqs. (6) and (7), we obtain

$$\sigma(t) = -\varepsilon \cdot \frac{1}{|V_S(t)|} \cdot \frac{d|V_S(t)|}{dt} = -\varepsilon_0\varepsilon_r \cdot \frac{d}{dt}(\ln|V_S(t)|) \tag{8}$$

Taking the nonlinear conductivities of ZnO-doped SiRCs into consideration, the current densities during SPD processes should be dependent on the surface-charge-induced electric field, then

$$\sigma(t) \propto J(E(t)) = J\left(V_s(t)/L\right) \tag{9}$$

Therefore, the relationship between the field-dependent current density and the surface potential decay process could be obtained by combining Eqs. (8) and (9).

$$J(V_s(t)/L) \ \propto \ \varepsilon_r \cdot \frac{d}{dt}(\ln|V_S(t)|) \tag{10}$$

where the $J(V_s(t)/L)$ is corresponding to the J-E curves of SiRCs under the electric field of $V_s(t)/L$.

Equation (10) suggests that, for each group of SiR composites, the $\varepsilon_r{\cdot}\mathrm{dln}|V_s(t)|/dt$ is approximatively proportional to the field-dependent conductivity at each decay time of t. It presents a quantitative relationship between the surface potential decay and the field-dependent nonlinear conductivity under the function of the surface-charge-induced electric field.

In order to further verify the Eq. (10) and make an understanding of the effects of nonlinear conductivity on the surface charge dynamic behaviours, the relationship between the field-dependent current densities and the $\varepsilon_r{\cdot}\mathrm{dln}|V_s(t)|/dt$ during the first 1000 s of decay time are calculated and shown in Fig. 8. For all the ZnO/SiR composites, there exists an approximate linear correlation between the $\varepsilon_r{\cdot}\mathrm{dln}|V_s(t)|/dt$ and

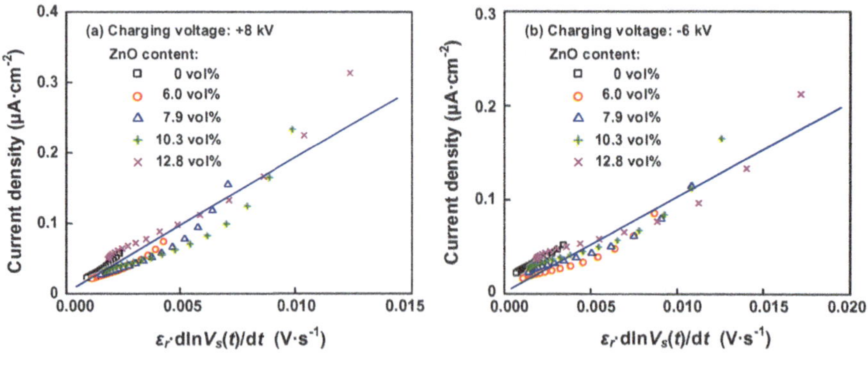

Fig. 8 Relationship between the field-dependent conductive current density and the $\varepsilon_r{\cdot}\mathrm{dln}|V_S(t)|/dt$ during the first 1000 s of decay processes. Copyright Clearance RightsLink® License Number: 4856251506816

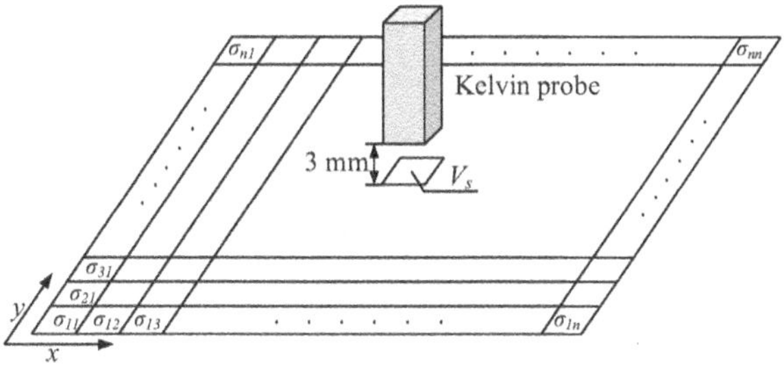

Fig. 9 Model of the 2-D surface potential measurement. Copyright Clearance RightsLink® License Number: 4856240827226

the field-dependent current densities, during the SPD process. Some differences are also found between the linear regression curves and the measured points, which may be due to the above assumptions for Eq. (10). Especially, it is required that the surface charge dissipation is only governed by the charge transport in the bulk of SiR composites, which leads to the deviations between the linear regression curves and the measured points in Fig. 9. However, from one side, these assumptions are often made to investigate the conduction and charge behaviours of polymers [30, 31]. From the other side, the relevant approximation could not influence the mechanism of the nonlinear conductivity on the surface charge dissipation. In any case, the relationship between the field-dependent current density and the surface potential decay process expressed by Eq. (10) can be used to evaluate the nonlinear conductivity dominated surface charge dissipation in ZnO/silicone rubber composites. The results shown above indicate that the ZnO/SiR composites with nonlinear conductivities have an ability of charge self-releasing. As the electric field induced by the accumulated surface charges is higher than the threshold electric field, the surface charges transport rapidly through the polymer bulk, following the quantitative relationship between the surface potential decay and the field-dependent nonlinear conductivity presented by Eq. (10). With surface charge dissipation, the surface-charge-induced field becomes lower than the threshold field, resulting in the nonlinear conductive composites turning back to the ohm-conductive conditions. Owning to the ability of charge self-releasing, the ZnO/SiR composites with nonlinear conductivity have potential applications for suppressing the surface charge accumulation.

4 Charge Transport Under Impulse Superimposed on DC Voltage

4.1 Surface Charge Distribution Under Impulse Superimposed on DC Voltage

The surface charge distribution under impulse superimposed on DC voltage of ZnO/SiR composites can be obtained by the inversion calculation of the surface potential distribution [32].

In this study, the sample surface with a dimension of 50 × 50 mm^2 was equally divided into $n \times n$ regions, as shown in Fig. 9. It was assumed that when the n is large enough, each divided square with a side length of (50/n) mm has a surface charge density of $\sigma_{(x,\ y)}$, which consists of the free charge density and the polarization charge density.

$$\sigma_{(x,y)} = \sigma_{f(x,y)} + \sigma_{p(x,y)} \tag{11}$$

where $\sigma_{f(x,\ y)}$ and $\sigma_{p(x,\ y)}$ are the density of the free surface charge and polarization charge, respectively.

For any divided region on the specimen film, its surface potential is formed by the surface charge in each divided square with an average charge density of $\sigma_{(x,\ y)}$, which can be expressed as,

$$\Phi_{(x,y)} = \frac{1}{4\pi\varepsilon_0}\sum\nolimits_{x=1}^{n}\sum\nolimits_{y=1}^{n}\sigma_{(x,y)}\int_{S(x,y)}\frac{1}{r}dS \tag{12}$$

$$E_{(x,y)} = \frac{1}{4\pi\varepsilon_0}\sum\nolimits_{x=1}^{n}\sum\nolimits_{y=1}^{n}\sigma_{(x,y)}\int_{S(x,y)}\frac{r}{r^3}dS \tag{13}$$

where ε_0 is the permittivity of vacuum, and r is the distance between the center of the region (x, y) and that of each divided square.

Numbering the divided squares from 1 to m, where m is equal to $n \times n$, then the surface potential of each divided square can be expressed by the surface charge density.

$$\begin{bmatrix}\Phi_1\\ \vdots\\ \Phi_m\end{bmatrix} = \begin{bmatrix}P_{11} & \cdots & P_{1n}\\ \vdots & \ddots & \vdots\\ P_{n1} & \cdots & P_{nn}\end{bmatrix}\begin{bmatrix}\sigma_1\\ \vdots\\ \sigma_m\end{bmatrix} = \mathbf{P}\cdot\begin{bmatrix}\sigma_1\\ \vdots\\ \sigma_m\end{bmatrix} \tag{14}$$

where P_{ij} is the contribution coefficient of σ_j to Φ_i. The matrix **P** is a correlation matrix between the matrix of the surface potential and the surface charge density. If we can calculate the inverse of matrix **P**, we can obtain the surface charge density of the sample by using the measured surface potential.

(1) When $i \neq j$, P_{ij} can be expressed as,

$$P_{ij} = \frac{1}{4\pi\varepsilon_0}\int_{S_i}\frac{1}{r_{ij}}dS = \frac{S_i}{4\pi\varepsilon_0 \cdot r_{ij}} \tag{15}$$

(2) When $i = j$, P_{ii} denotes the surface potential in the region i, which is induced by an average surface charge of 1 C/m^3. r_{ii} cannot be equal to zero; therefore P_{ii} cannot be calculated by Eq. (14). In this case, the square with a side length of (50/*n*) mm is simplified as a circle with a radius of (50/*n*) mm.

For a circle with an average charge density of σ_i, the potential of a point with a height z above the center of the circle can be expressed as

$$\Phi(z) = \frac{\sigma_i}{2\varepsilon_0}\frac{R_i^2}{z + \sqrt{z^2 + R_i^2}} \tag{16}$$

where R_i is the radius of the circle. In the limit that z infinitely approaches zero, the above equation is equal to the potential at the center of the circle.

$$\lim_{z\to 0}\Phi(z) = \frac{\sigma_i}{2\varepsilon_0}R_i \tag{17}$$

Then,

$$P_{ii} = \frac{\lim\limits_{z\to 0}\Phi(z)}{\sigma_i} = \frac{R_i}{2\varepsilon_0} \tag{18}$$

Therefore, the elements in the matrix **P** can be expressed as,

$$P_{ij} = \begin{cases} \frac{S_i}{4\pi\varepsilon_0 \cdot r_{ij}} & i \neq j \\ \frac{R_i}{2\varepsilon_0} & i = j \end{cases} \tag{19}$$

Based on the boundary condition of the electric field, the free charge density on the sample surface of the region i can be calculated by the following.

$$\sigma_{f(i)} = (\boldsymbol{D}_{i2} - \boldsymbol{D}_{i2}) \cdot \boldsymbol{n}_i = (\varepsilon_2\boldsymbol{E}_{i2} - \varepsilon_1\boldsymbol{E}_{i1}) \cdot \boldsymbol{n}_i \tag{20}$$

where $\sigma_{f(i)}$ is the average free charge density in the region *i*. $\boldsymbol{D}_{i2}$ and $\boldsymbol{D}_{i1}$ indicate the electrical displacement vector on either side of the dielectric interface. $\boldsymbol{n}_i$ is a unit normal vector pointing from the dielectrics to the gas. $\boldsymbol{\varepsilon}_2$ and $\boldsymbol{\varepsilon}_1$ are the relative permittivity of the gas and dielectrics, respectively. $\boldsymbol{E}_{i2}$ and $\boldsymbol{E}_{i1}$ indicate the electric field vector, which is a function of σ_j ($j = 1$ ~ m).

Therefore, the free charge density in the region i can be expressed as,

$$\begin{bmatrix} \sigma_{f(1)} \\ \vdots \\ \sigma_{f(m)} \end{bmatrix} = \begin{bmatrix} Q_{11} & \cdots & Q_{1n} \\ \vdots & \ddots & \vdots \\ Q_{n1} & \cdots & Q_{nn} \end{bmatrix} \begin{bmatrix} \sigma_1 \\ \vdots \\ \sigma_m \end{bmatrix} = \mathbf{Q} \cdot \begin{bmatrix} \sigma_1 \\ \vdots \\ \sigma_m \end{bmatrix} \tag{21}$$

(1) When $i \neq j$, Q_{ij} can be derived by substituting Eq. (13) into (20).

$$Q_{ij} = \frac{\varepsilon_2 - \varepsilon_1}{4\pi\varepsilon_0} \int_{S_i} \frac{\boldsymbol{r}_{ij} \cdot \boldsymbol{n}_i}{r_{ij}^3} dS = \frac{\varepsilon_2 - \varepsilon_1}{4\pi\varepsilon_0} \frac{(\boldsymbol{r}_{ij} \cdot \boldsymbol{n}_i)}{r_{ij}^3} \tag{22}$$

(2) When $i = j$, the modulus of $\boldsymbol{r}_{ii}$ is equal to zero. In this case, Q_{ii} cannot be calculated by Eq. (22). Additionally, using the method of Eq. (16), the region i is replaced by a circle with an average charge density of σ_i. The electric field of a point with a height z above the center of the circle can be expressed as

$$\boldsymbol{E}(z) = \frac{\sigma_i}{2\varepsilon_0} \left(1 - \frac{z}{\sqrt{z^2 + R_i^2}} \right) \cdot \boldsymbol{n}_i \tag{23}$$

In the limit that z infinitely approaches zero, the above equation is equal to the potential at the center of the circle.

$$\lim_{z \to 0} \boldsymbol{E}(z) = \frac{\sigma_i}{2\varepsilon_0} \cdot \boldsymbol{n}_i \tag{24}$$

Then, Eq. (20) can be expressed as

$$\sigma_{f(i)} = \left(\varepsilon_2 \frac{\sigma_i}{2\varepsilon_0} \cdot \boldsymbol{n}_i - \varepsilon_1 \frac{\sigma_i}{2\varepsilon_0} \cdot (-\boldsymbol{n}_i) \right) \cdot \boldsymbol{n}_i \tag{25}$$

Then, Q_{ii} is equal to $\frac{\varepsilon_1 + \varepsilon_2}{2\varepsilon_0}$.

Therefore, the elements in the matrix $\mathbf{Q}$ can be expressed as

$$Q_{ij} = \begin{cases} \frac{\varepsilon_2 - \varepsilon_1}{4\pi\varepsilon_0} \frac{(\mathbf{r}_{ij} \cdot \mathbf{n}_i)}{r_{ij}^3} & i \neq j \\ \frac{\varepsilon_1 + \varepsilon_2}{2\varepsilon_0} & i = j \end{cases} \tag{26}$$

Above all, the free charge density in the dielectric surface can be obtained by

$$\begin{bmatrix} \sigma_{f(1)} \\ \vdots \\ \sigma_{f(m)} \end{bmatrix} = \mathbf{Q} \cdot \mathbf{P}^{-1} \cdot \begin{bmatrix} \Phi_1 \\ \vdots \\ \Phi_m \end{bmatrix} \tag{27}$$

The schematic diagram of the 2-D surface potential measurement system in this chapter is shown in Fig. 10a. The test temperature is set to 30 °C, and the relative humidity is set at 30%. In this study, an impulse source with an alternating impulse polarity is used to obtain a 1.2/50 μs lightning impulse with a maximum amplitude of 20 kV. A coupling unit is employed to superimpose the impulse on a DC voltage. During the charge corona test, the DC voltage (V_{DC}) is set as −3 kV. After 5 min, the lightning impulse is applied, and the peak amplitude of the impulse voltage (V_I) is set as −10, −5, +10 and +15 kV. The number of impulses for each voltage is

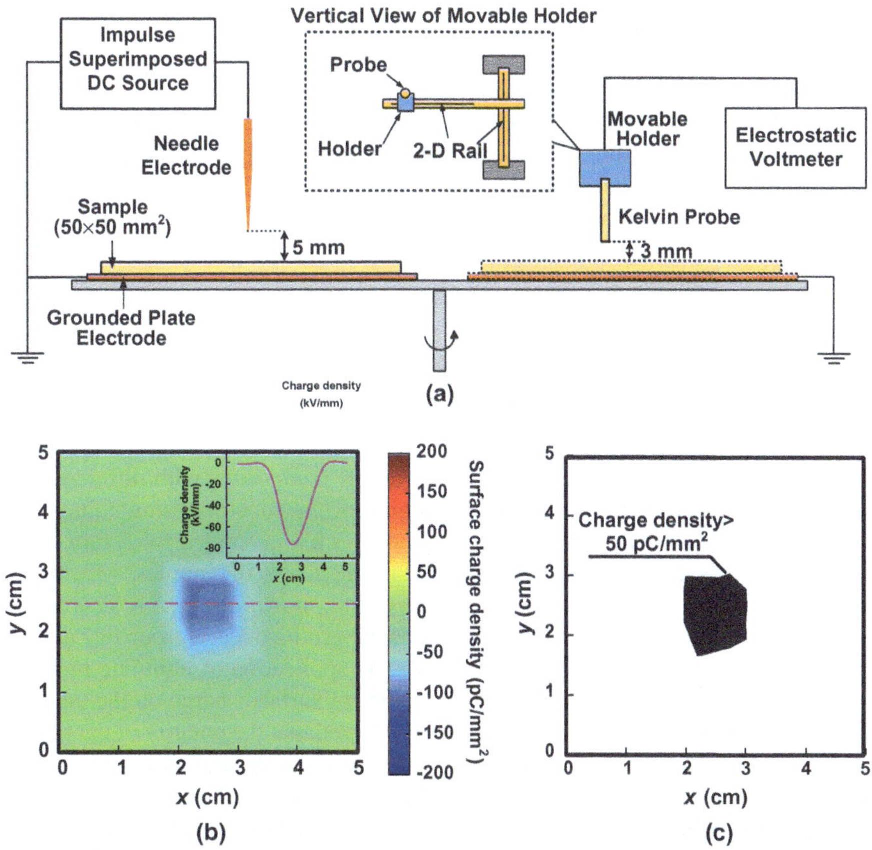

Fig. 10 a Schematic diagram of the 2-D surface potential measurement system, b typical example for a surface charge distribution on the sample, c conversion of the color picture to a black and white picture. Copyright Clearance RightsLink® License Number: 4856240827226

10. Subsequently, the samples are moved instantly below a Kelvin probe, which is connected to an electrostatic voltmeter (TREK P0865). The probe holder is a computer-controlled 2-D moveable holder driven by two motors, which could scan the sample surface within 12 s. The surface potential distributions are recorded by a computer, and the surface charge distributions are obtained by Eq. (27).

Figure 10b shows a typical distribution of surface charge density on the sample. A two-dimensional color drawing with a defined color bar presents both the polarity and amount of space charge accumulated on the sample surface with dimensions of 5×5 mm^2. The total surface charge $Q(t)$ at a depolarization time of t is defined as

$$Q(t)=\int_{s}|q(t)|dS \tag{28}$$

where $q(t)$ is the charge density at a certain location of the specimen, and S is the area of the sample surface.

To present the distribution of the surface charge density, a mountain-like profile of the surface charge density of the centerline of the sample is plotted as shown in the top right corner of Fig. 10b. In addition, the color picture in Fig. 10b can be converted to a black and white picture. Where the charge density is above 50 pC/mm^2, the area is covered by black; otherwise, it is white, as shown in Fig. 10c. The proportion of the black area to the total area is defined as P_{SC}, which can further evaluate the distribution of the surface charge.

4.2 *Surface Charge Accumulation Under Impulse Superimposed on DC Voltage*

The surface charge accumulation on the ZnO/SiR composites with different filler contents, immediately after the corona charging of the superimposing voltage, are shown in Fig. 11. When only a DC voltage of −3 kV is applied for 5 min, small quantities of negative charges are deposited in the center of samples, which form a basin-shaped profile of surface charge density as shown in Fig. 11. There is no significant difference in the surface charge density profile of each group of ZnO/SiR composites under the DC voltage of −3 kV. The application of lightning impulses can result in the accumulation of a large amount of surface charge on the undoped SiR samples, as shown in Fig. 11a. It can be observed that the polarity of the surface charges accumulated on the center of the SiR surface is dependent on the polarity of the lightning impulses. Additionally, with an increase in the peak amplitude of the impulses, both the largest surface charge density and the total charge increase markedly.

To further evaluate the accumulation behavior of the surface charge on the composites, the total surface charge and the P_{SC} of the undoped SiR sample under an impulse superimposed on a DC voltage are calculated and presented in Fig. 12. It is presented

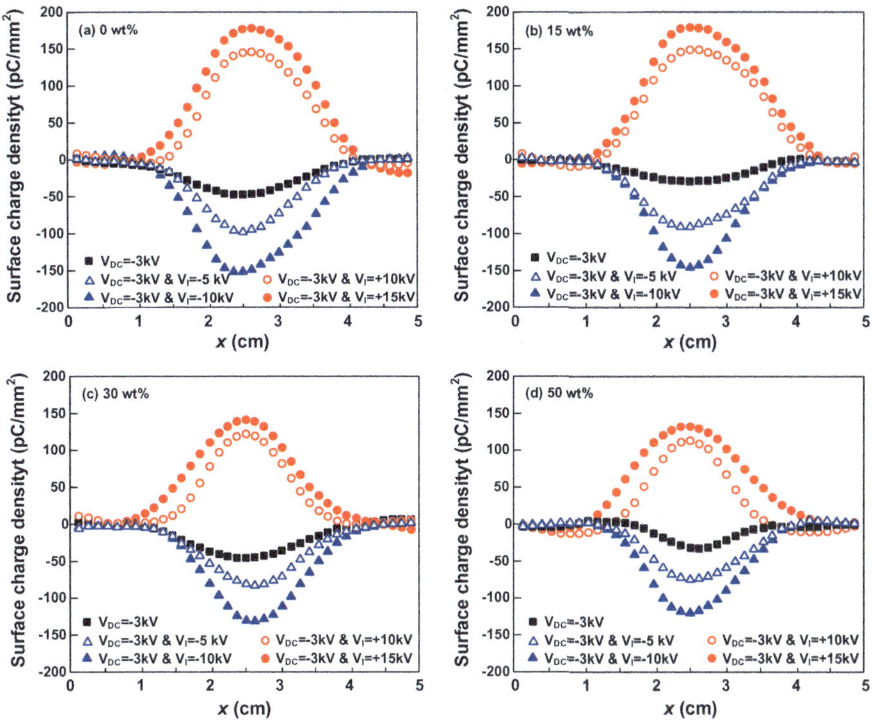

Fig. 11 Surface charge accumulation on the ZnO/SiR composites with filler contents of 0, 15, 30, and 50 wt.% immediately after corona charging under an impulse superimposed on a DC voltage. Copyright Clearance RightsLink® License Number: 4856240827226

Fig. 12 Surface charge amount and PSC of undoped SiR samples under impulse superimposed on DC voltage. Copyright Clearance RightsLink® License Number: 4856240827226

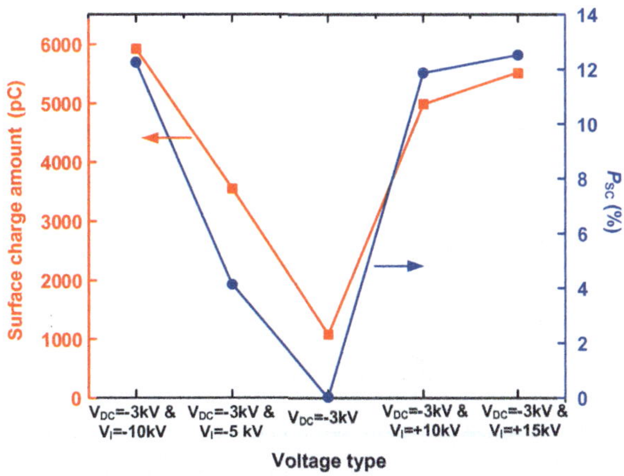

that there is an obvious consistency between the total surface charge and P_{SC} varying with the voltage type.

As the polarity of the impulses is same as that of the DC voltage, the surface charges of the undoped SiR samples under 0, −5 and −10 kV impulses are 10.73, 35.48 and 59.22 nC, respectively. The values of P_{SC} of the undoped sample under 0, −5 and −10 kV impulses are 0%, 4.15% and 12.27%, respectively. It is indicated that both the total surface charge and P_{SC} increase with an increase in the peak amplitude of the homopolar impulse. Figure 13 illustrates a schematic diagram of the effect of the impulse superimposed on a DC voltage on the surface charge accumulation. As a DC voltage of −3 kV is applied, a corona discharge occurs near the needle electrode, and negative charges migrate and deposit on the sample center, as shown in Fig. 13a. As the (-) homopolar impulses are superimposed, the corona charging of the homocharge is accelerated, leading to a more intense injection of surface charge. On one hand, the injected homocharges are captured by the carrier trap sites in the center of the sample, resulting in a larger surface charge as shown in Fig. 12. On the other hand, the injected homocharges repulse the trapped charges and free charges

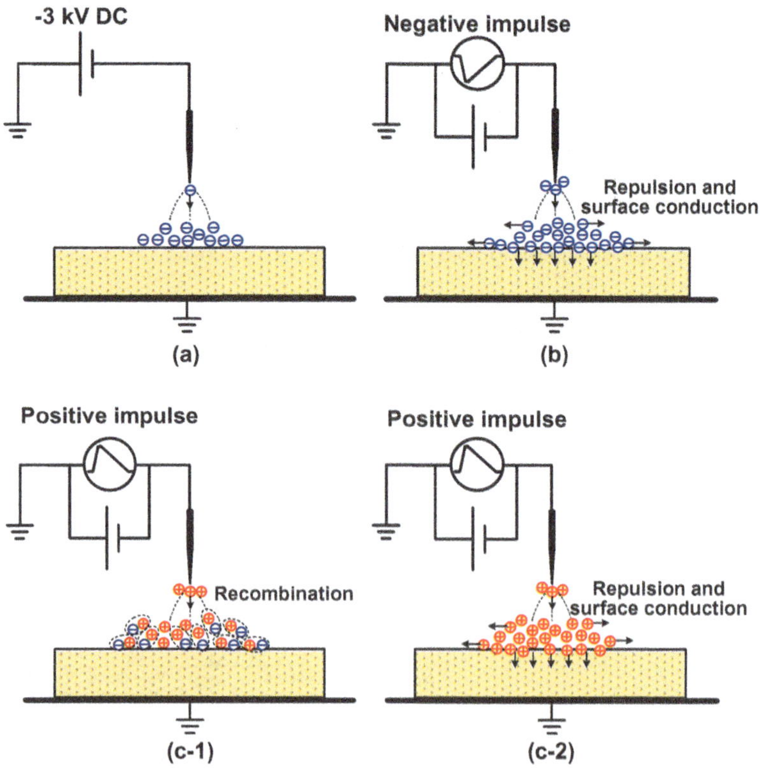

Fig. 13 Schematic diagram of surface charge accumulation under impulse superimposed on DC voltage. Copyright Clearance RightsLink® License Number: 4856240827226

and accelerate surface and bulk conduction (see Fig. 13b), which lead to an increase in P_{SC}, as shown in Fig. 12.

As impulses of positive polarity are superimposed, the polarity of the surface charge becomes positive. When the peak amplitude impulse voltage is +10 and + 15 kV, the surface charge is 49.81 and 55.16 nC, and P_{SC} is 11.87% and 12.52%, respectively. Under a positive impulse superimposed on a negative DC voltage, positive charges (positive ions) form in the vicinity of the needle electrode. The ions of positive charges migrate forward to the sample surface under the coulomb force, and then the positive charges are transferred from the ions to the hole carriers through the electron transfer process as the positive ions arrive at the sample surface. As shown in Fig. 13c-1, the recombination between positive and negative charges occurs at some trap sites of the sample surface. With a further injection of positive charges (holes), many more charges are captured by trap sites and accumulate on the sample surface, as shown in Fig. 13c-2. Meanwhile, the further injection of positive charges during the next discharge cycle creates repulsion between the original homocharges on the sample, thus accelerating the bulk conduction and the surface conduction along the sample surface following the hopping conduction model.

4.3 *Effect of ZnO Content on Surface Charge Accumulation*

As shown in Fig. 11, the ZnO content also has a significant effect on the surface charge distribution of the ZnO/SiR composites. Meanwhile, this effect also varies with the polarity of the impulse. To further quantificationally evaluate the effect of the ZnO content on the surface charge accumulation behavior, the total surface charge and the P_{SC} of the ZnO/SiR composites under an impulse superimposed on a DC voltage are calculated and presented in Fig. 14.

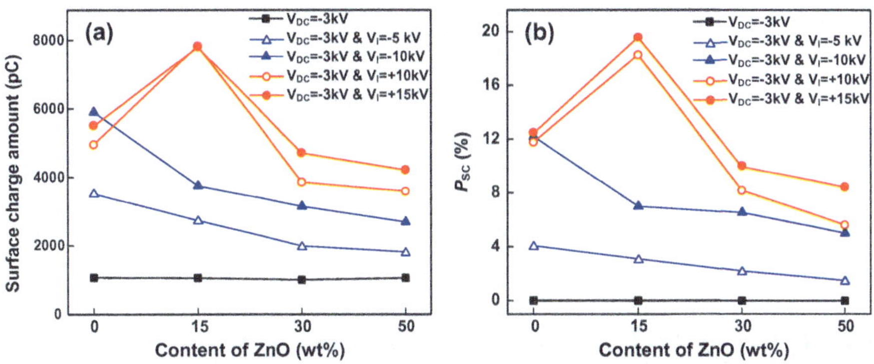

Fig. 14 Relationship between the ZnO content and the Surface charge amount as well as PSC of ZnO/SiR composites under impulse superimposed on DC voltage. Copyright Clearance RightsLink® License Number: 4856240827226

When a negative impulse superimposed on a negative DC voltage is applied, both the surface charge and P_{SC} decrease significantly with the filler content from 0 to 50 wt.%, as shown in Fig. 14a, b. The effect of nonlinear conductivity on the negative charge accumulation process is discussed in the following. As the negative impulses are applied, the electrons accumulated on the sample surface will induce an electric field, which can be decomposed into a vertical component and a horizontal component; the vertical samples leads to bulk conduction, and the horizontal component leads to surface conduction. Due to the thin thickness of the samples, the vertical component of the induced field is much higher than the horizontal component in the center of the samples, resulting in a dominant role of the vertical component. As the vertical component is high enough to exceed the threshold field of the nonlinear conductivity, a sharp increase in the conductivity will enhance the transport and dissipation of surface charges through the sample bulk to the ground. The higher the nonlinear conductivity is, the faster the charge dissipation. It is assumed that the same amount of charge reaches the surface of the sample during corona charging under the same voltage. Therefore, with an increase in the ZnO content, the total accumulated charge decreases, as shown in Fig. 7a. Meanwhile, the decrease in the surface charge density weakens the repulsive effect of the homocharges, leading to a decrease in P_{SC}.

However, as a positive impulse is superimposed on a negative DC voltage, the surface charge and PSC increase with the ZnO content from 0 to 15 wt.% and then decrease with the filler content from 30 to 50 wt.%. There appears to be an obvious "polarity effect" in the surface charge accumulation of the 15 wt.% ZnO/SiR composites, which may be ascribed to the n-type semiconductor ZnO fillers with dominant carriers of electrons. Under a negative impulse superimposed on a DC voltage, the negative surface charges (electrons) transport through the bulk, which is accelerated by the semiconductive ZnO fillers. However, when a positive impulse is superimposed on a DC voltage, ions with a positive charge accumulate on the sample surface, which then dissipate through the transport of hole carriers. According to the results shown in Fig. 14, the addition of 15 wt.% ZnO fillers can only accelerate electron transport as opposed to hole transport. Meanwhile, it is speculated that the microinterfaces between the ZnO fillers and SiR matrix introduce some trap sites or holes, which may capture more positive charges in the center of the sample surface, thus leading to an increase in the charge and PSC. When the ZnO content exceeds 30 wt.%, the markedly enhanced nonlinear conductivity can suppress the accumulation of the positive charges, as shown in Fig. 14. The "polarity effect" of ZnO is further discussed in the following section.

4.4 Effect of ZnO Content on Surface Charge Transport

It has been concluded in Sect. 3 that the high surface charge caused by repetitive impulses leads to an accelerated charge transport that suppresses surface charge accumulation, and this effect becomes more obvious in higher ZnO content samples with enhanced nonlinear conductivity.

To further analyze the effect of nonlinear conductivity on the surface charge dissipation, the surface charge density of the centerline as a function of depolarization time is calculated component between the upper and lower surfaces of the and recorded. Figures 15 and 16 present the time-dependent surface charge dissipation of the ZnO/SiR composites under a −10 kV impulse superimposed on a −3 kV DC voltage and a +15 kV impulse superimposed on a −3 kV DC voltage, respectively. The vertical axis of each drawing in Fig. 10b stands for the x direction, and the horizontal axis is time.

It can be observed that the ZnO/SiR composites with a filler content of 30 and 50 wt.% can obviously accelerate the surface charge dissipation, regardless of the polarity of the surface charge. During the depolarization process of the samples, the maximum surface charge density at the middle of the sample ($x = 2.5$ cm) decreases with time. In addition, the dark-hued region in Figs. 15 and 16, denoting the high surface charge density, is gradually narrowed, which suggests a decrease in P_{SC}. More importantly, there appears to be horizontal transport of 'charge packages' as

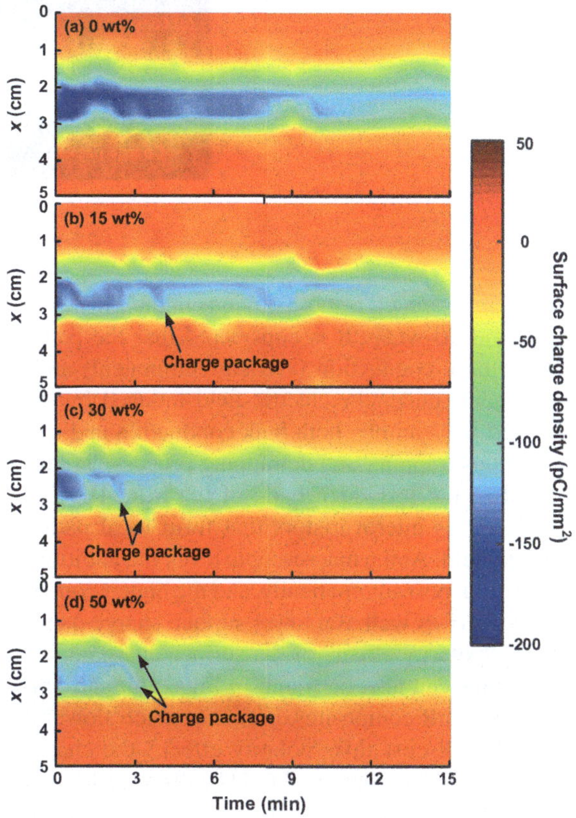

Fig. 15 Time-dependent surface charge dissipation of the ZnO/SiR composites under a −10 kV impulse superimposed on a −3 kV DC voltage. Copyright Clearance RightsLink® License Number: 4856240827226

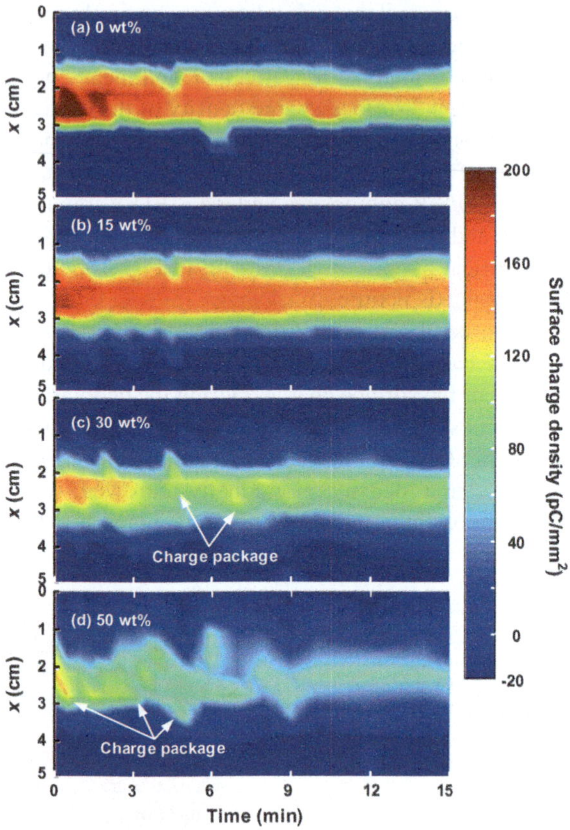

Fig. 16 Time-dependent surface charge dissipation of the ZnO/SiR composites under a +15 kV impulse superimposed on a −3 kV DC voltage. Copyright Clearance RightsLink® License Number: 4856240827226

time elapses, which is more obvious in the ZnO/SiR composites with a filler content of 30 and 50 wt.%. It is considered that the paths for surface charge dissipation include bulk conduction (vertical component) and surface conduction (horizontal component) under the electric field induced by the accumulated charges. For the undoped SiR sample, both bulk conduction and surface conduction result from the continuous trapping and detrapping process, which is associated with the trap sites located in the forbidden band of the SiR. The ZnO fillers distributed in the SiR matrix act as microvaristors that form local paths for charge transport under the induced field. At the macroscopic level, the addition of ZnO fillers can accelerate the charge transport automatically, and the higher content leads to an enhanced nonlinear conductivity as well as a lower surface charge density, as shown in Figs. 15 and 16.

The "polarity effect" of the surface charge dissipation of the 15 wt.% ZnO/SiR composites can also be found in Figs. 15b and 16b. The positive charge in the 15 wt.% ZnO/SiR composites dissipates much slower than that in the undoped sample. To further discuss this "polarity effect," the charge transport characteristics of n-type ZnO semiconductors should be discussed. The ZnO fillers consist of crystalline grains and grain boundaries between the grains. Charge transport in the ZnO fillers is mainly dominated by the grain boundaries, which have a back-to-back Schottky barrier as shown in Fig. 17a [33]. In the grain boundary, a large number of electrons are captured by the trap sites, forming a negative interface layer as well as a pair of depletion layers containing positive space charge. In addition, there are also traps that are not filled with electrons in the interface area. When the Schottky barrier of the grain boundary of ZnO is in a state of equilibrium, the depletion layers have the same width, i.e., $X_{L0} = X_{R0}$. In addition, the heights of the left and right Schottky barriers are equal. As an electric field (from right to left) is applied at the grain boundary, the height of the left barrier decreases and the height of the right barrier increases. As shown in Fig. 17b, the electrons pass through the left barrier and fill the trap sites in the interface layer. Once the electron traps are completely filled, the negative charge will accumulate in the interface layer. It becomes easier for these electrons to cross to the right grain. This is a microscopic model of the transport of electrons through the back-to-back Schottky barrier of the grain boundary. In the macroscopic view, the electrons transport in a ZnO microvaristor under the effect of the electric field.

Now, we return to the issue of charge transport in the ZnO/SiR composites. When the negative charges (electrons) accumulate at the sample surface, the electron transport in the SiR matrix follows the hopping conduction model. More importantly, the electron transport in the ZnO microvaristors occurs as discussed above under the surface-charge-induced field. Therefore, with an increase in the ZnO content, the transport of electrons is accelerated, as shown in Fig. 15. When the polarity of the impulse is positive, positive charges (positive ions) form in the vicinity of the needle electrode. The positive charges are transferred from the ions to the hole carriers through the electron transfer process as the positive ions arrive at the sample surface. It is believed that the transport of holes following the hopping conduction model plays a dominant role during the surface charge dissipation. As shown in Fig. 16b, the holes in the 15 wt.% ZnO/SiR composites dissipate slower than that in the undoped sample. The hole transport in the SiR matrix also follows the hopping

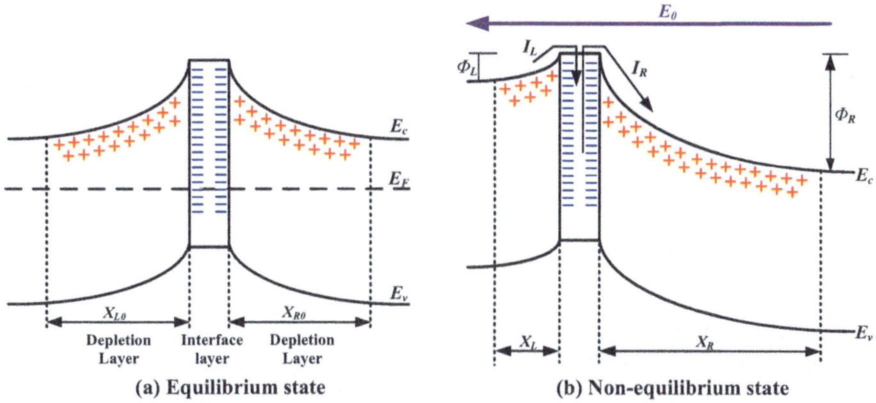

Fig. 17 Schottky barrier model of the grain boundary of ZnO microvaristors. Copyright Clearance RightsLink® License Number: 4856240827226

conduction model. However, the hole transport in the ZnO microvaristors can only be attributed to minority carriers. If the hole transport in the ZnO fillers is considered as a reverse extraction process of the electron, the electron traps located in the interface layer of the grain boundary may inhibit the charge transport. That is, a small amount of ZnO may not play a marked role in the transport of holes. As shown in Figs. 14 and 16, the positive charge in the 15 wt.% ZnO/SiR composites dissipates slower than in the undoped sample. It is speculated that the addition of ZnO forms some interface regions between the ZnO fillers and SiR matrix, which introduce trap sites for carriers and suppress the charge dissipation in the composites.

As presented above, the impulse superimposed on DC voltage will aggravate the corona charging in the vicinity of HV electrode and lead to mass accumulation of the surface charge on the SiR sample. The nonlinear conductivity introduced by the ZnO filler could accelerate the charge dissipation through the vertical charge transport to the ground and the horizontal movement on the surface, thus suppressing the surface charge accumulation. More importantly, ZnO fillers could play a more important role in the electrons dissipation than the hole dissipation.

5 Conclusion

To sum-up, this chapter reports a field-dependent conductivity dominated surface charge dissipation and understood a relationship between the field-dependent conductivity and the charge transport dynamic behaviours, under DC voltage and impulse superimposed on dc voltage. The main conclusions can be summarized as follows,

1. The ZnO/SiR composites with nonlinear conductivity show a marked resistance to the surface charge accumulation. For the SiR composites without nonlinear conductivity, the R_{decay} almost keeps constant in the initial stage of SPD processes. However, for the ZnO/SiR composites with nonlinear conductivity, there appears a marked decrease of R_{decay}, due to the decrease of the surface-charge-induced field.
2. The theoretical model and experiment results both showed that the differential coefficient of the surface potential in logarithmic coordinates were approximately proportional to the field-dependent conductivities at each instant of time during the SPD process. As the surface-charge-induced field exceeded the threshold field of the nonlinear conductive composites, the field-dependent conductivity could markedly accelerate the surface charge dissipation. As a result, the surface-charge-induced field becomes lower than the threshold field, resulting in the nonlinear conductive composites turning back to the ohm-conductive conditions.
3. The homopolar impulse intensifies the surface charge accumulation, leading to an increase in both the surface charge and area. Under a heteropolar impulse superimposed on a DC voltage, neutralization occurs first and then charges with the same polarity as the impulse accumulates on the surface.

4. With an increase in the impulse voltage, both the surface charge and area increase significantly. The larger surface charge caused by the impulse leads to a high nonlinear conductivity, which accelerates the dissipation of the surface charge, and this effect becomes more obvious in composites with a higher ZnO content.
5. The vertical charge transport to the ground and the horizontal movement on the surface are two factors concerning the surface charge dissipation. For the ZnO/SiR composites, the horizontal transport of 'charge packages' is observed.
6. There is a special case in which the positive surface charge dissipates slower in the 15 wt.% ZnO/SiR composites than in the undoped sample. It is deduced that negative charges are easier to dissipate in ZnO/SiR composites than positive charges due to the charge transport mechanism of n-type ZnO semiconductors with a back-to-back Schottky barrier of the grain boundary.

References

1. T. Christen, L. Donzel, F. Greuter, Nonlinear resistive electric field grading part 1: theory and simulation. IEEE Electr. Insul. Mag. **26**(6), 47–59 (2010)
2. L. Donzel, F. Greuter, T. Christen, Nonlinear resistive electric field grading part 2: materials and applications. IEEE Electr. Insul. Mag. **27**(2), 18–29 (2011)
3. S. Boggs, D.H. Damon, J. Hjerrild, J.T. Holboll, Effect of insulation properties on the field grading of solid dielectric dc cable. IEEE Trans. Power Deliv. **16**(4), 456–461 (2015)
4. M. Jeroense, HVDC, the next generation of transmission: highlights with focus on extruded cable systems. IEEJ Trans. Electr. Electron. Eng. **5**(4), 400–404 (2010)
5. F. Greuter, M. Siegrist, P. Kluge-Weiss, R. Kessler, L. Donzel, R. Loitzl, H.J. Gramespacher, Microvaristors: functional fillers for novel electroceramic composites. J. Electroceram. **13**(1–3), 739–744 (2004)
6. M. Tenzer, V. Hinrichsen, Investigations on microvaristors as functional fillers in insulation systems for HVDC applications, in *Proceedings of the 17th International Symposium on High Voltage Engineering* (2016), pp. 22–26
7. A. Winter, J. Kindersberger, M. Tenzer, V. Hinrichsen, Laetitia Zavattoni, Olivier Lesaint, M. Muhr, D. Imamovic, Solid/gaseous insulation systems for compact HVDC solutions, in *Cigre Science & Engineering (CIGRE)* (2014), pp. 1–12
8. L. Donzel, J. Schuderer, Nonlinear resistive electric field control for power electronic modules. IEEE Trans. Dielectr. Electr. Insul. **19**(3), 955–959 (2012)
9. B.B. Hsu, E.B. Namdas, J.D. Yuen, S. Cho, I.D. Samuel, A.J. Heeger, Split-gate organic field effect transistors: control over charge injection and transport. Adv. Mater. **22**(41), 4649–4653 (2010)
10. L. Donzel, T. Christen, R. Kessler, F. Greuter, H. Gramespacher, Silicone composites for HV applications based on microvaristors, in *IEEE International Conference on Solid Dielectrics (ICSD)* (2004)
11. D. Weida, T. Steinmetz, M. Clemens, Electro-quasistatic high voltage field simulations of large scale insulator structures including 2-D models for nonlinear field-grading material layers. IEEE Trans. Magn. **45**(3), 980–983 (2009)
12. K.P. Donnelly, B.R. Varlow, Nonlinear dc and ac conductivity in electrically insulating composites. IEEE Trans. Dielectr. Electr. Insul. **10**(4), 610–614 (2003)
13. K. Tavernier, B.R. Varlow, D.W. Auckland, M. Ugur, Improvement in electrical insulators by non-linear fillers. IEE Proc. Sci. Meas. Technol. **146**(2), 88–94 (1999)

14. C.Y. Liu, X.Q. Zheng, P. Peng, The nonlinear conductivity experiment and mechanism analysis of modified polyimide composite materials with inorganic filler. IEEE Trans. Plasma Sci. **43**(10), 3727–3733 (2015)
15. X. Yang, J. Hu, S. Chen, J. He, Understanding the percolation characteristics of nonlinear composite dielectrics. Sci. Rep. **6**, 30597 (2016)
16. Z.L. Li, B.X. Du, Z.R. Yang, J. Li, Effects of crystal morphology on space charge transportation and dissipation of SiC/silicone rubber composites. IEEE Trans. Dielectr. Electr. Insul. **24**(4), 2616–2625 (2017)
17. Z.H. Yang, P.H. Hu, S.J. Wang, J.W. Zha, Z.C. Guo, Z.M. Dang, Effect of nano-fillers distribution on the nonlinear conductivity and space charge behaviour in SiC/PDMS composites. IEEE Trans. Dielectr. Electr. Insul. **24**(3), 1735–1742 (2017)
18. C.Y. Li, C.J. Lin, J. Hu, W.D. Liu, Q. Li, B. Zhang, S. He, Y. Yang, F. Liu, J.L. He, Novel HVDC spacers by adaptively controlling surface charges–part I: charge transport and control strategy. IEEE Trans. Dielectr. Electr. Insul. **25**(4), 1238–1247 (2018)
19. C.Y. Li, C.J. Lin, Y. Yang, B. Zhang, W.D. Liu, Q. Li, J. Hu, S. He, F. Liu, X.L. Liu, J.L. He, Novel HVDC spacers by adaptively controlling surface charges–part II: experiment. IEEE Trans. Dielectr. Electr. Insul. **25**(4), 1248–1258 (2018)
20. C.J. Lin, Q. Li, C.Y. Li, B. Zhang, W.D. Liu, Y. Yang, F. Liu, J. Hu, X.L. Liu, J. Hu, J.L. He, Novel HVDC spacers by adaptively controlling surface charges–part III: industrialization prospects. IEEE Trans. Dielectr. Electr. Insul. **25**(4), 1259–1266 (2018)
21. Y. Liu, X.L. Cao, Electrical tree initiation in XLPE cable insulation by application of DC and impulse voltage. IEEE Trans. Dielectr. Electr. Insul. **20**(5), 1691–1698 (2013)
22. N. G. Hingorani, Transient overvoltage on a bipolar HVDC overhead line caused by DC line faults. IEEE Trans. Power Appar. Sys. 89(4), 592–610 (1970)
23. B. Qi, C.J. Gao, C.R. Li, L.J. Zhao, X.Q. Sun, Y.Y. Lei, The accumulation characteristic of surface charges on GIS Insulator under switching impulse voltages in SF6, in *IEEE Annual Report Conference on Electrical Insulation and Dielectric Phenomena (CEIDP)* (2015)
24. J.B. Deng, H.B. Mu, G.J. Zhang, S. Matsuoka, A. Kumada, K. Hidaka, Residual charge distribution of surface leader discharge under positive impulse voltage. IEEE Trans. Plasma Sci. **41**(4), 999–1004 (2013)
25. F. Wang, Y.C. Qiu, W. Pfeiffer, E. Kuffel, Insulator surface charge accumulation under impulse voltage. IEEE Trans. Dielectr. Electr. Insul. **11**(5), 847–854 (2004)
26. X. Jun, I.D. Chalmers, The influence of surface charge upon flash-over of particle-contaminated insulators in under impulse-voltage conditions. J. Phys. D Appl. Phys. **30**(7), 1055–1063 (1997)
27. J.T. Ma, Q.G. Zhang, H.Y. You, Z.C. Wu, T. Wen, C. Guo, G.L. Wang, C. Gao, Study on insulation characteristics of GIS under combined voltage of DC and lightning impulse. IEEE Trans. Dielectr. Electr. Insul. **24**(2), 893–900 (2017)
28. J. Castellon, H.N. Nguyen, S. Agnel, A. Toureille, M. Frechette, S. Savoie, A. Krivda, L.E. Schmidt, Electrical properties analysis of micro and nano composite epoxy resin materials. IEEE Trans. Dielectr. Electr. Insul. **18**(3), 651–658 (2011)
29. C.Y. Li, J. Hu, C.J. Lin, B.Y. Zhang, G.X. Zhang, J.L. He, Surface charge migration and dc surface flashover of surface-modified epoxy-based insulators. J. Phys. D Appl. Phys. **50**(6), 065301 (2017)
30. M.M. Perlman, T.J. Sonnonstine, J.A. St, Pierre, "Drift mobility determinations using surface-potential decay in insulators." J. Appl. Phys. **47**(11), 5016–5021 (1976)
31. S. Kumara, Y.V. Serdyuk, S.M. Gubanski, Surface charge decay on polymeric materials under different neutralization modes in air. IEEE Trans. Dielectr. Electr. Insul. **18**(5), 1779–1788 (2011)
32. H. Ootera, K. Nakanishi, Analytical method for evaluating surface charge distribution on a dielectric from capacitive probe measurement-application to a cone-type spacer in + or −500 kV DC-GIS. IEEE Trans. Power Deliv. **3**(1), 165–172 (1988)
33. K. Eda, Conduction mechanism of non-Ohmic zinc oxide ceramics. J. Appl. Phys. **49**(5), 2964–2972 (1978)

Charge Properties of SiC/SiR Composites with Nonlinear Conductivity at Different Temperatures

Zhuoran Yang, Zhonglei Li, Honghua Xu, Hao Wang, Wei Chen, and Zihe Yang

Abstract The nonlinear conductivity composites can be applied to modify the local electric field distribution and accelerate the surface and space charge dissipation in HVDC cables. This chapter focuses on the effect of the nonlinear conductivity on the charge transportation under different voltage types. The nonlinearity in the conductivity is achieved by mixing silicon carbide (SiC) particles with the SiR matrix. The conductivities of the samples are measured at temperatures of 30–90 °C. The surface-charge-accumulation property of silicone rubber (SiR)/SiC composites under impulse superimposed DC voltage and polarity-reversal voltage is studied. The space charge property is researched under DC voltage at different temperatures. The conductivity-temperature coefficient is introduced to analyze the phenomenon found in conductivity of sample at different temperatures. The space charge and surface charge property of SiR/SiC composites is discussed based in the conductivity property.

1 Introduction

THE accessories of the high voltage direct current (HVDC) cable are of significant importance to the secure operation of the HVDC transmission system [1]. These components have been proven to suffer from higher risk of failure compared with other parts in the HVDC transmission system [2, 3]. This phenomenon is basically caused by the complicated structure and the multi-layer insulation in the HVDC cable accessories. These reasons cause the distortion of the local electric field, which intensifies the charge accumulation. Modifying the permittivity distribution and the resistivity distribution of insulations are two categories of methods taken to solve

Z. Yang · H. Xu · H. Wang · W. Chen · Z. Yang
State Grid Nanjing Power Supply Company, Nanjing 210019, China

Z. Li (✉)
School of Electrical Engineering and Automation, Tianjin University, Tianjin 300072, China
e-mail: lizhonglei@tju.edu.cn

B. Du (ed.), *Polymer Insulation Applied for HVDC Transmission*,
https://doi.org/10.1007/978-981-15-9731-2_10

this problem in history [4, 5]. Now they are seen in many HVDC cable accessories in the form of stress corn and nonlinear conductivity layer respectively.

Compared with the permittivity method, the resistivity method brings a simpler structure to HVDC cable accessories [6]. Based on this technology, ABB has succeeded in manufacturing a prototype of DC cable over 300 kV by applying a field-controlling layer in the structure of joints [7]. The reported structure of HVDC cable joint is simple and the interface of different insulations are uniform. The nonlinear conductivity layer usually consists of semi-conductive filler and polymer insulation such as silicone rubber (SiR) and XLPE [8–10]. Many researches have shown the advantages of the nonlinear conductivity material in modifying the local field [11–13]. These papers pointed out some inorganic fillers in polymer matrix provide a high conductivity at high field, which automatically accelerates the dissipation of the interface and space charge.

Another factor of the accessory failure requiring for attention is the transient electric field. It is unavoidable for a HVDC transmission system to experience transient overvoltage caused by switching and lightning [14]. Some researchers showed that repetitive impulses could affect the charge accumulation and further influence the destruction of insulation such as treeing [15]. It is also unavoidable for an HVDC transmission system to experience a transient overvoltage caused by switching and lightning. Some studies have noted that lightning voltage can affect charge accumulation and further influence the initiation of failure, such as treeing. It is also important to clarify how the nonlinear conductivity insulation comes into effect under this condition.

In addition, the voltage polarity is reversed and the current direction is unchanged, when the power flow reversal is required in line commutated converter-based HVDC (LCC-HVDC) transmission. In the HVDC transmission system of $\pm$800 kV and above, LCC-HVDC transmission has significant advantages [16]. LCC-HVDC transmission has the advantages of large transmission capacity, low cable cost and low corona loss. The problems of space charge accumulation and electric field distortion under polarity-reversal voltage were studied. B. Qi et al. found that polarity-reversal voltage can enhance the electric field intensity in oil-pressboard composite insulation, leading to PD [17]. M. Fu et al. found a large amount of space charges are accumulated in cross-linked polyethylene (XLPE) power cables under polarity-reversal voltage [18]. X. Chen et al. prepared low density polyethylene (LDPE) films, showing that polarity-reversal voltage has a great influence on space charge profiles and field distortions [19]. Under the polarity-reversal voltage, the insulation material may suffer from problems such as insulation aging and breakdown [20, 21].

Temperature also has an effect on the conductivity of insulation, which is the main factor for the field distribution under DC voltage. An HVDC cable is designed to operate at 90 °C at which temperature the conductivity of the insulation material increases considerably [10]. The problem is that not all insulation materials share the same range of increasing conductivity. Some of the most widely used insulations, such as SiR and XLPE, can have an order of magnitude of difference in conductivity, which can definitely change the local field distribution [11]. The temperature also shows a gradient in the cable. The inner cable insulation experiences a higher temperature, whereas the outside insulation does not, which makes the cable exhibit a slant electric field from the center to the outside [12].

This chapter aims to study the surface and space charge behaviour considering the transient voltage together with the nonlinear conductivity of SiC/SiR composites. Different contents of SiC provide various nonlinear conductivities for comparison. The surface potential decay (SPD) and PEA methods are applied and the surface charge and space charge properties are measured at different temperatures. This chapter focuses on discussing the analyzation of experiment data, rather than the experiment process.

2 Nonlinear Conductivity

The conductivities of the SiR/SiC composites at different temperatures are shown in Fig. 1. Figure 1a presents the conductivity at 30 °C. The M0 sample conductivity increases slowly with the electric field, following Ohm's law. The number of carriers increases with the field since a high field provides high energy for carriers to hop into the conduction band. Hence, the current is proportional to the electric field. The conductivity increases with a higher rate when the field is high. This is believed to be caused by the space-charge-limited current (SCLC) [22, 23]. In the M60 and M90 samples, the conductivity exhibits a high rising speed at much lower field because it is caused by the nonlinearity in the conductivity rather than the SCLC. The semiconductive SiC particles provide more carriers in the conduction band. The number of conduction band carriers increases sharply when the field approaches the threshold. At a high field, the barrier between the SiC particle and the matrix is inclined [22]. The energy of the carriers is also enhanced by the field. Hence, a high field also makes it easier for carriers to escape from the trap and travel through the sample bulk. In addition, the filler particles can form a conductive network when the filler volume fraction is large enough according to percolation theory [23]. When the filler content increases from 60 to 90 phr, more conductive paths are formed in the bulk. With a high volume fraction, the distance between adjacent SiC particles decreases. These effects lead to a large conductivity and a high conductivity nonlinearity at 30 °C.

At 50, 70 and 90 °C, the situations for the M0 and SiC mixed samples are very different. For M0, compared with the conductivity at 30 °C, the conductivity is proportional to temperature. Most of the carriers in M0 lie in the forbidden band and can hardly be driven by the electric field. Only a small number of carriers take part in the conduction process. However, high temperature provides these carriers more energy to escape from traps and hop barriers, even though the field is not high. Hence, the conductivity increases over the entire electric field range at high temperatures, and the threshold field between the Ohm region and the SCLC region becomes unclear due to the higher conductivity at low electric fields. For M60, the conductivity is higher than that for M0 for more semi-conductive SiC particles, and the threshold field is not obvious since the conductivities appear to immediately increase rapidly at a very low field. The conductivity is similar to that of M0 when the field is approximately less

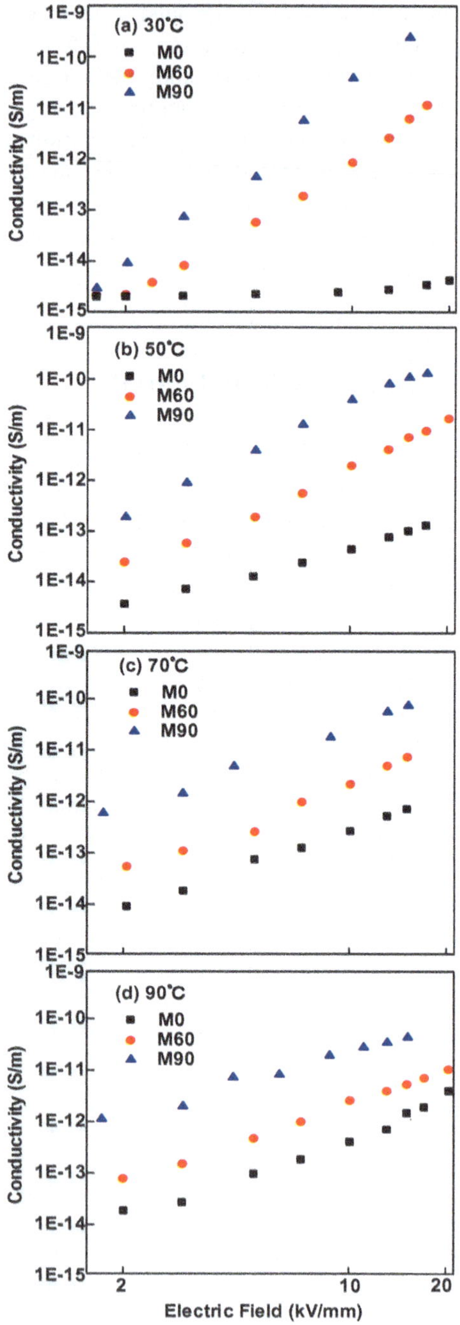

Fig. 1 Relationship between the conductivities of the SiR/SiC composites and the electric field at different temperatures (4, 856, 270, 238, 030)

than 10 kV/mm, showing that the conductivity increases with temperature. However, the increase becomes insignificant at 90 °C. At a higher electric field, the conductivity of M60 gradually decreases with temperature. The same phenomenon is also found in M90. Figure 2 presents this trend clearly with the slope of conductivity with respect to the electric field. This shows that the temperature accelerates the conductivity increase in M0, whereas it slows down the conductivity increase in M60 and M90, and the deceleration is more effective in samples with high SiC content. As a result, temperature can even play a negative role in increasing sample conductivity. A typical situation is shown in Fig. 3. The concept of the conductivity-temperature coefficient

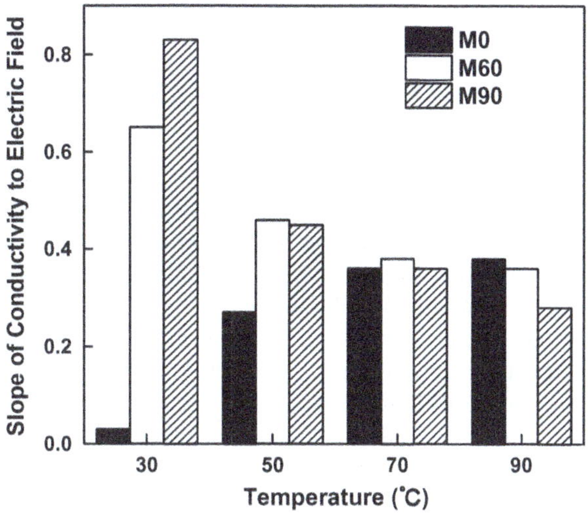

Fig. 2 Slope of the conductivity to electric field of SiR/SiC composites at different temperatures (4, 856, 270, 238, 030)

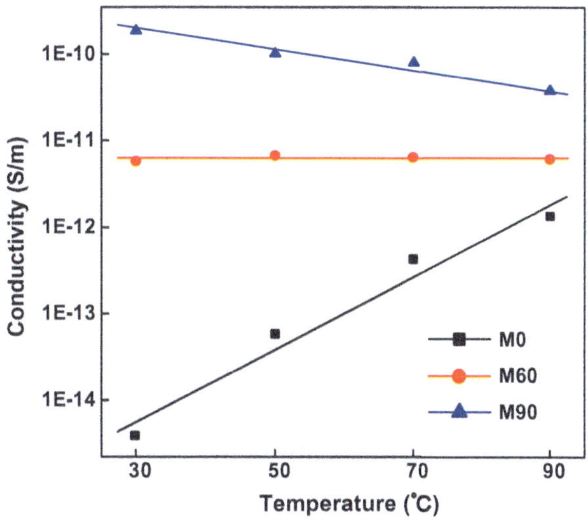

Fig. 3 Relation between the conductivity of the SiR/SiC composites and temperature at 15 kV/mm (4, 856, 270, 238, 030)

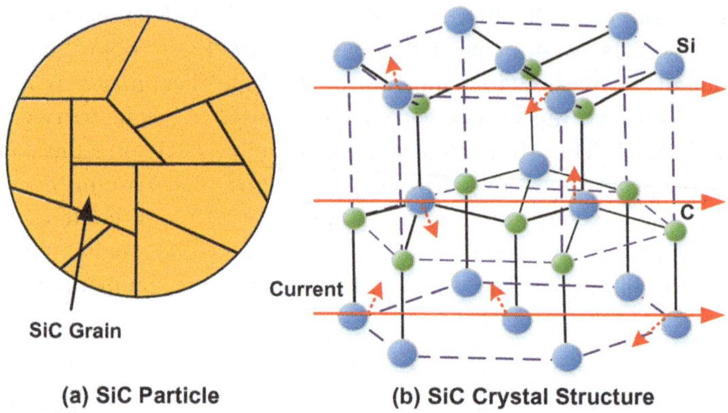

Fig. 4 Structure of the SiC particles and the SiC crystals (4, 856, 270, 238, 030)

is defined as the slope of the conductivity-temperature curve. At 15 kV/mm, the coefficient for M0 is positive, the coefficient for M60 is almost 0 and the coefficient for M90 becomes negative. This phenomenon is of significant value, since it could be the key to the solution to insulation failure resulting from the nonuniform conductivity distribution caused by the temperature gradient in the DC cable. This is mainly caused by the different conduction mechanisms of SiR and SiC. The SiR molecular structure is a noncrystalline molecular chain. Most of the carriers are restricted in the valence band in the SiR molecule, which makes SiR a great insulator. High temperature can intensify the thermal motion of the carriers, which increases the possibility of excited carriers hopping into the conduction band. These excited carriers are then driven by the electric field to form a current. This explains the positive conductivity-temperature coefficient of M0. The SiC particle is different from SiR since it consists of crystal structures as shown in Fig. 4. As a semiconductor, the energy gap between the valance band and the conduction band is relatively narrow. Most carriers are easily excited to the conduction band when the electric field is high. In this case, the excitation resulting from temperature is not as obvious. Simultaneously, the increase in temperature can also cause a lattice vibration. According to classic semiconductor theory, such a vibration is described by a lattice wave, which mainly consists of a long longitudinal acoustic wave (LLAW) and a long longitudinal optical wave (LLOW). The LLAW is used to represent the adjacent atoms that share the same vibration phase. The LLOW is used to represent the adjacent atoms that share the inverse vibration phase. The lattice vibration can scatter the electrons in the current leading to a reduction in the conductivity. The scattering probabilities are described by the following equations:

$$P_{LLAW} = c_1 T^{3/2} \tag{1}$$

$$P_{LLOW} = c_2 \frac{(h\nu)^{3/2}}{(k_0 T)^{1/2}} \cdot \frac{1}{e^{\frac{h\nu}{k_0 T}} - 1} \cdot \frac{1}{f\left(\frac{h\nu}{k_0 T}\right)} \quad (2)$$

where P_{LLAW} and P_{LLOW} are the scattering probabilities, T is the temperature, h is Planck constant, k_0 is the Boltzmann constant, f is a function of $h\nu/k_0T$ with a value between 0.6 and 1, and c_1 and c_2 are constant coefficients. Equation (1) clearly shows that P_{LLAW} increases with temperature. The relationship between temperature and P_{LLOW} mainly depends on the second term in Eq. (2); thus, P_{LLOW} also increases with temperature [24]. It can be concluded that high temperature increases the electron scattering probability, which indicates more collisions between electrons and atomic kernels. This phenomenon can also be explained by metal conduction theory. In contrast to high-purity semiconductors applied in integrated circuits and other electronic industries, the nonlinear filler for insulation is usually re-sintered to achieve an obvious nonlinearity [25, 26]. A filler particle consists of many filler grains as shown in Fig. 5a. As a result, the purity of the SiC filler is around 98%. This results in an extremely large number of carriers for not only the SiC but also the impurity. Hence the SiC particle can be seen as a degenerate semiconductor that exhibits similar resistance-temperature properties to those of metals. The atomic kernel vibration in a metal can also scatter carriers, and the resistance is described by the following empirical equation:

$$\rho(T) \approx \frac{AT}{4M\Theta_D^2} \quad (T \nu \Theta_D) \quad (3)$$

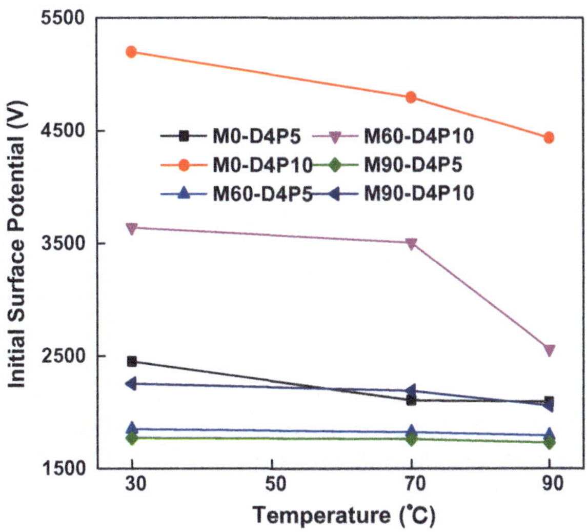

Fig. 5 ISP of the SiR/SiC composites stressed by a homopolar impulse superimposed DC voltage at different temperatures (4, 856, 270, 238, 030)

$$\rho(T) \approx 124.4 \frac{AT^5}{M\Theta_D^6} \quad (T \nu\ \Theta_D) \tag{4}$$

where ρ is the resistance, T is the temperature, A is a coefficient depending on the metal, M is the atomic mass of the metal, and Θ_D is the Debye temperature of the metal. Clearly, the resistance increases with temperature. These effects can significantly block the movement of electrons and further reduce the conductivities of M60 and M90.

In addition, the results of our relative research based on samples mixed with 0.45 μm SiC particles show different phenomena [27]. Even though the slope of conductivity versus temperature decreases, the conductivity still becomes higher at high temperatures. As shown in Fig. 5b, the carriers in the current are scattered by the lattice vibration when they go through the SiC crystal structure. If the contents of SiC are the same, the current experiences greater suppression when it flows in large SiC particles. This may indicate that a continuous SiC crystal structure is more effective in suppressing the temperature induced increase in conduction.

3 Surface Potential Property Under Homopolar Impulse Superimposed DC Voltage

This Fig. 5 shows the initial surface potential (ISP) of the samples stressed by the homopolar impulse superimposed DC voltage at different temperatures. The voltages are abbreviated in this section. For example, a −4 kV DC voltage is called D4 and an impulse superimposed DC voltage formed by the −4 kV DC voltage and the −5 kV or 10 kV impulse voltage is called D4P5 or D4-P10. With the increase in impulse voltage, all samples experience a rise in the ISP. The rise decreases with the SiC content in the sample for high conductivity, which helps to conduct surface charges to the ground. The temperature has a slight effect in reducing the ISP when the impulse voltage is −5 kV. However, the reduction becomes obvious when the impulse voltage increases to −10 kV. The field in the SiC mixed samples calculated roughly by the rate of surface potential to the sample thickness are all around or below 10 kV/mm. According to Fig. 2, the conductivity is still proportional to temperature in this electric field range. Therefore, the increase in surface potential can induce a higher nonlinear conductivity, which further results in a large reduction in ISP for M0 and M60 at 90 °C. For M90, the reduction is still not significant at 90 °C. This may result from the relatively high conductivity of M90 when the temperature is lower than 90 °C, which sufficiently suppresses the surface charge accumulation. For further study, the decay rate is defined as the potential drop per second. It is calculated by the following equation:

$$R = \frac{V_0 - V_{15}}{t} \tag{5}$$

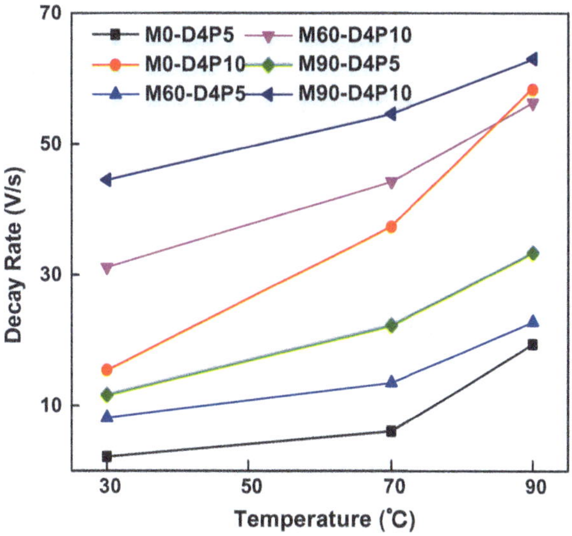

Fig. 6 Decay rate of the SiR/SiC composites stressed by a homopolar impulse superimposed DC voltage at different temperatures in the first 15 s (4, 856, 270, 238, 030)

where R is the decay rate, V_0 is the ISP, V_{15} is the surface potential at the 15th s of the decay process, and t is the time interval. Figure 6 shows the corresponding decay rate stressed by the homopolar impulse superimposed DC voltage in the first 15 s. It is clear that the increase in SiC content and the increase in impulse voltage can both accelerate the decay process. The temperature also plays a positive role in increasing the decay rate. A special phenomenon is that for M0 stressed by D4P10 at 90 °C, the decay rate is even higher than that for M60 stressed by the same voltage. This may be caused by the surface charge horizontal movement resulting from the high ISP of M0. Hence, the corresponding decay rate, which is sensitive to conductivity, is higher. It should also be noted that the SPD is actually affected by many factors and conductivity is just one of them. This phenomenon may or may not occur in experiments on similar materials, and this explanation should be seen as a hypothesis.

Another interesting comparison is shown in Table 1. It presents the increase in ISP between D4P5 and D4P10 for different samples at different temperatures. Generally, the samples mixed with SiC particles show a better surface charge suppression effect when the rise in voltage is the same. The diffusion effect related to the nonlinear conductivity can be used to explain this phenomenon. When the rise in impulse

Table 1 Increase in ISPs between the D4P5 and D4P10 stressed samples at different temperatures

Temperature/(°C)	Increase in ISP /(V)		
	M0	M60	M90
30	2750	1789	531
70	2688	1682	431
90	2341	765	283

voltage brings additional charges, the nonlinear conductivities of M60 and M90 come into effect and a considerable number of new arriving charges diffuse to adjacent regions in a short time. As a consequence, the M90 sample exhibits the smallest increase in the ISP. In addition, the temperature also exhibits a similar surface charge suppression effect, since the samples are all in the positive conductivity-temperature coefficient region.

4 Surface Potential Property Under Heteropolar Impulse Superimposed DC Voltage

Figure 7 shows the ISP of the SiR/SiC composites stressed by the heteropolar impulse superimposed DC voltage at different temperatures. Similar to the discussion in the last section, temperature can reduce the ISP due to the nonlinear conductivity regardless of the stressed voltage. The effect of SiC particles also presents the same influence on the ISP. A high-SiC-content sample always exhibits better surface charge suppression since more charges can be easily conducted to the ground. However, the heteropolar voltage does not seem to neutralize the negative charges brought by the DC voltage. The ISPs of all samples appear to increase after they are stressed by the heteropolar voltage. Moreover, for M0, the increase in the ISP at 90 °C is slightly larger than that at 30 and 70 °C, while the increases for M60 and M90 at 90 °C is much larger than those at lower temperatures. Figure 8 shows the corresponding decay rate under heteropolar impulse voltage in the first 15 s. Temperature and SiC content are still positive factors for the increase in surface charge decay for high conductivity. Simultaneously, the heteropolar impulse voltage not only intensifies the accumulation of surface charge but also slows down the decay process. It is concluded

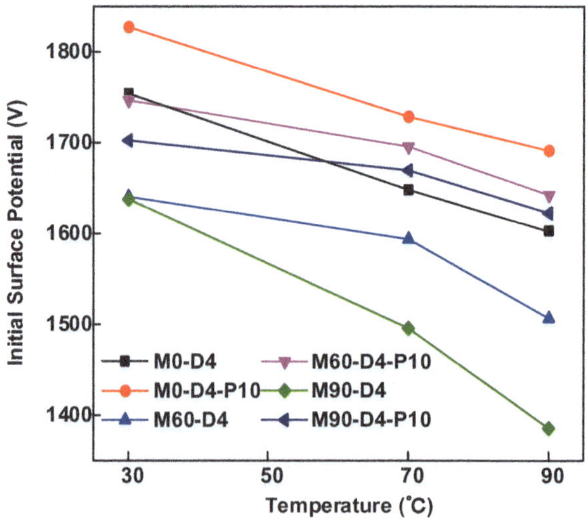

Fig. 7 ISP of the SiR/SiC composites stressed by a heteropolar impulse superimposed DC voltage at different temperatures (4, 856, 270, 238, 030)

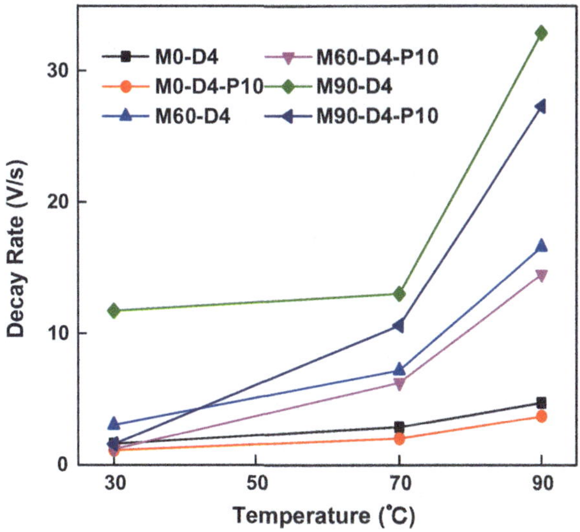

Fig. 8 Decay rate of the SiR/SiC composites stressed by a heteropolar impulse superimposed DC voltage at different temperatures in the first 15 s (4, 856, 270, 238, 030)

that all heteropolar impulse voltage stressed samples exhibit a higher ISP and lower surface charge decay rates. Obviously, the SPD property is caused by a reason other than nonlinear conductivity. Figure 9 provides a hypothetical model to explain this phenomenon. When the heteropolar impulse voltage arrives at the end of the charging process, the polarity of the voltage suddenly changes, and a number of positive charges are injected into the surface of the sample. Neutralization occurs immediately between heterocharges. However, this process cannot be completed in a short time. The positive charges are captured by traps in the region under the needle electrode and form a positive-charge region. Hence, the attraction between heterocharges continues before the neutralization is complete. Surface charges in adjacent areas are attracted to the central region, which leads to an increase in the detected ISP. This attraction movement towards the central point is also influenced by the nonlinear conductivity. High conductivity makes it easier for charges to accumulate to the central region.

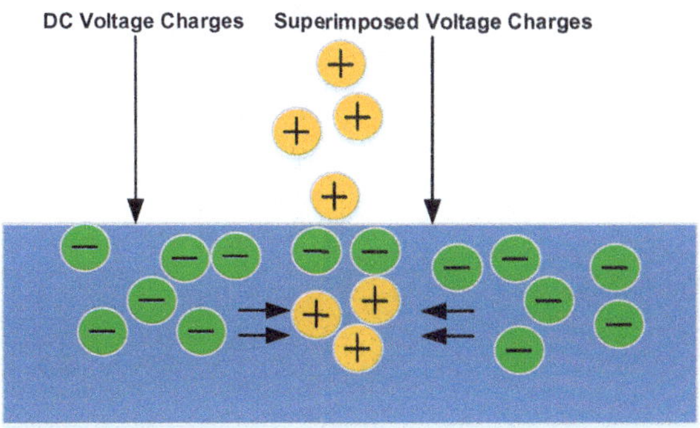

Fig. 9 Model of the surface charge movement stressed by a heteropolar impulse superimposed DC voltage (4, 856, 270, 238, 030)

Then, the number of positive charges is decreased by neutralization, and the attraction effect is strongly weakened. The surface potential begins to decay, but it is still greatly restricted by the attraction effect as long as the neutralization is not complete. This corresponds to the reduction in decay rate shown in Fig. 8. This explains the phenomenon in Fig. 7 at 90 °C. Neutralization occurs in the SPD process. When the nonlinear conductivity is not sufficiently large, the attraction can even counteract the fast nonlinear-conductivity-driven decay rate and bring the decay rate back to the level of the sample without nonlinear conductivity. The reductions in decay rates for M60 and M90 at 30 °C show the same trend as the analysis above. M0 does not exhibit nonlinear conductivity, and the decay rate for a DC voltage is also slow. Therefore, there is not a similar reduction in decay rate.

Further deduction can be made based on the model in Fig. 9. If the heteropolar impulse voltage is strong enough, more negative charges brought by a DC voltage can be neutralized immediately by a very large amount of positive charges, which may reduce the surface potential. In this case, the strong attraction effect may also last for a time, which is longer than the time taken to move the sample from the needle electrode to the probe. It is quite possible to detect the attraction effect, which is shown by the increasing surface potential in the central region. Figures 10 and 11 show data from extra experiments based on the above deduction. Figure 10 takes the M60 sample at 70 °C as an example. The origin of the horizontal axis is the central point under the electrode, and the numbers on the axis correspond to the distance between the detecting point and the central point. Obviously, when the sample is stressed by D4-P10; the surface potential increases in the central region and decreases in the surrounding areas, clearly indicating the attraction effect. Further stressing the sample by D4-P15, the surface potential in all areas decreases as deduced. Figure 11 shows the decay rate of M60 stressed by D4-P15 at 70 °C. The surface potential

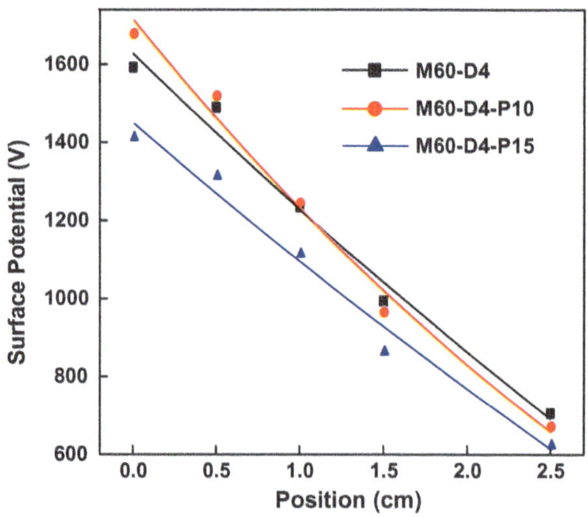

Fig. 10 Initial surface potential at different positions of M60 stressed by different voltages at 70 °C (4, 856, 270, 238, 030)

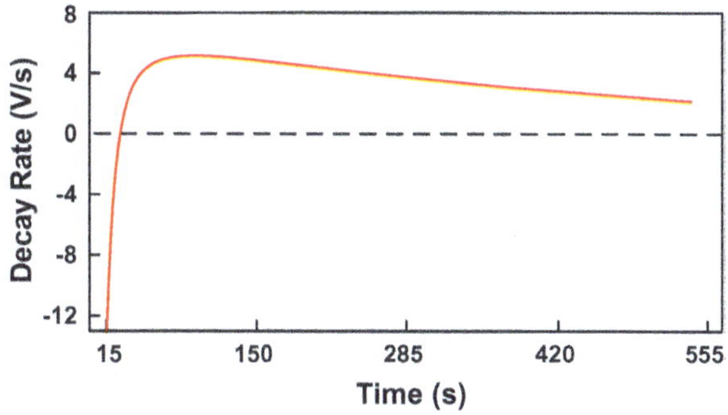

Fig. 11 Surface potential decay rate of M60 stressed by D4-P15 at 70 °C (4, 856, 270, 238, 030)

increases in the first 15 s and then begins to decay. The data indicate that the model in Fig. 9 is sufficiently convincing.

5 Surface Charge Characteristics of SiR Materials and SiC/SiR Composites Under Positive-Negative Reversal Voltage

Figure 12 shows the ISP distribution of SiR materials and SiR/SiC composites under negative DC voltage and positive-negative reversal voltage. Under negative DC voltage, the ISP at the center of M0 reaches −3350 V at 30 °C. This is due to the low conductivity of M0, which is easy to accumulate charges during corona discharge. From the center of the sample, the ISP decreases successively from 0.5

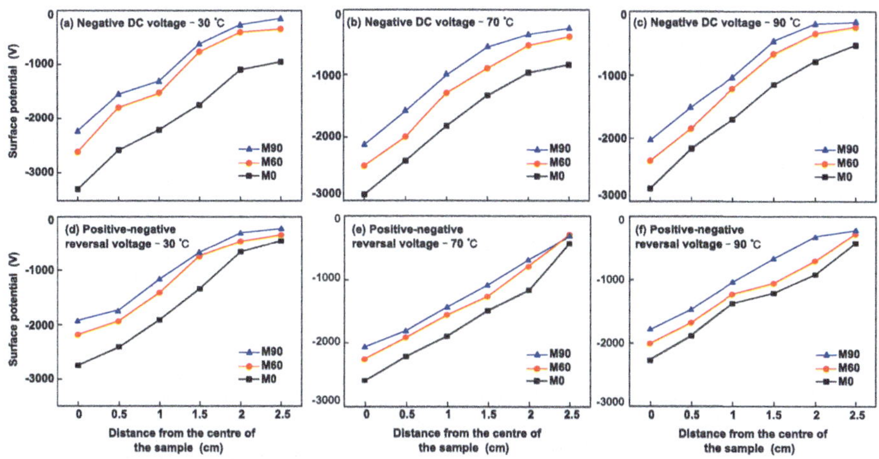

Fig. 12 ISP distribution of SiR materials and SiR/SiC composites under **a**–**c** negative DC voltage and **d**–**f** positive-negative reversal voltage

to 2.5 cm. Due to the uneven electric field distribution of the tip electrode, surface charge in the center is more than outside, resulting in the charges migrate from the center to outside. The ISP of M60 is lower than that of M0, because M60 has higher conductivity, dissipating more charges to the earth electrode during corona discharge. M90 has the highest conductivity and the lowest ISP among the three samples. At 70 °C, the conductivity of the sample increases with increasing temperature, resulting in a lower ISP than that at 30 °C. At 90 °C, the ISP of the sample is further reduced.

Under positive-negative reversal voltage, the ISP of M0 becomes negative at all positions at 30 °C, decreasing in turn from the center to outside. It can be seen that the negative charges in the radius of 2.5 cm complete the neutralization of the original positive charges accumulated on the surface of the sample, and further accumulate, so that the ISP of the central point reaches −2760 V. M60 and M90 show the same pattern, but the ISP at the center point is lower than that at M0, due to the high conductivity. The ISP of the sample at 70 °C is lower than that at 30 °C, and further decreases at 90 °C. This is due to the conductivity of the sample increases as the temperature goes up. Figure 13 compares the ISP of the central point under negative DC voltage and positive-negative reversal voltage. The ISP of M0 decreases most at 30 °C, followed by M60 and M90. The charge transport in SiR materials can be represented by simplified model shown in Fig. 14. Due to the low conductivity of M0, positive charges are accumulated after the positive corona discharge, consuming lots of negative charges in the process of surface charge neutralization. Thus, compared with that under negative DC voltage, ISP of M0 decreases the most. However, due to the high conductivity, neutralization process of M60 and M90 is weaker than M0, and the ISP of the center point decreases less. Analogously, the conductivity of the sample increases at 70 and 90 °C, which further reduces the negative charges consumed during surface charge neutralization, bridging the gap of the ISP between negative DC voltage and positive-negative reversal voltage.

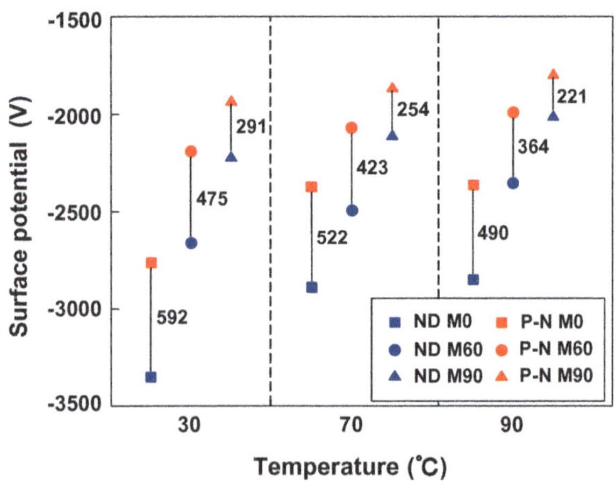

Fig. 13 ISP at the center of the sample under negative DC voltage (abbreviated in the figure as ND) and positive-negative reversal voltage (abbreviated in the figure as P-N) (the differences have been marked in the figure)

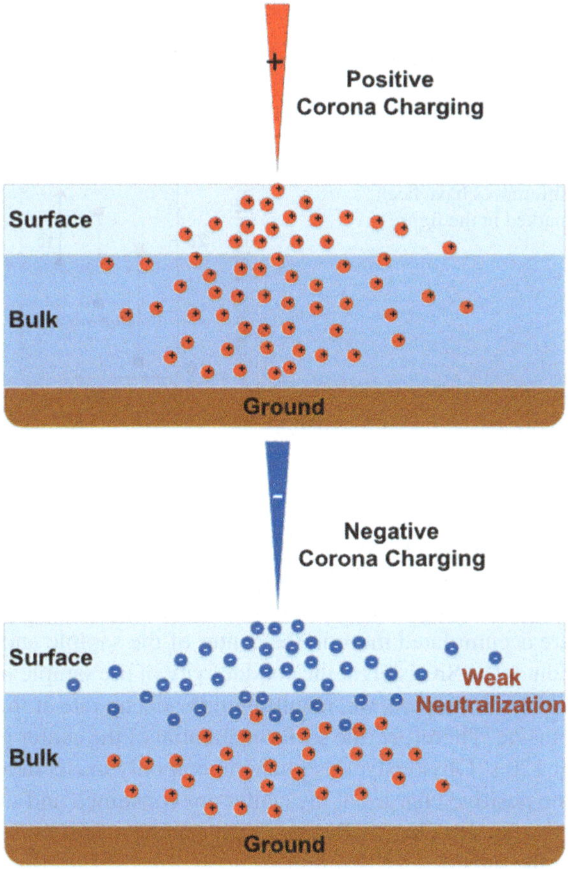

Fig. 14 Charge transport model of SiR/SiC composites

At different temperatures, the SPD of different samples has obvious similarity. Typically, as shown in Fig. 15 is the SPD curve of M90 at 70 °C. As can be seen that the surface potential of the central point decays with time, and the decay rate is continuously decreasing. The SPD at 0.5, 1 and 1.5 cm has the same pattern, with the ISP gradually decreasing. The decay rate at 2 cm reduces greatly, and the surface potential changes slightly within 300 s. The surface potential at 2.5 cm decreases little, which can be regarded as reaching a constant value. The SPD at each point is obviously correlated with the ISP, and the decay rate decreases with the reduction of ISP. Besides, no obvious horizontal charge migration is observed at the inner part of the sample. Therefore, the charge migration to the earth electrode is dominant in the whole decay process.

The effect of positive-negative reversal voltage on SPD is different from that of negative DC voltage. The surface potential at the center is higher than that at outside in amplitude at any time. However, under positive-negative reversal voltage, the surface potential at the center may be lower than that at outside. Before 30 s, the surface

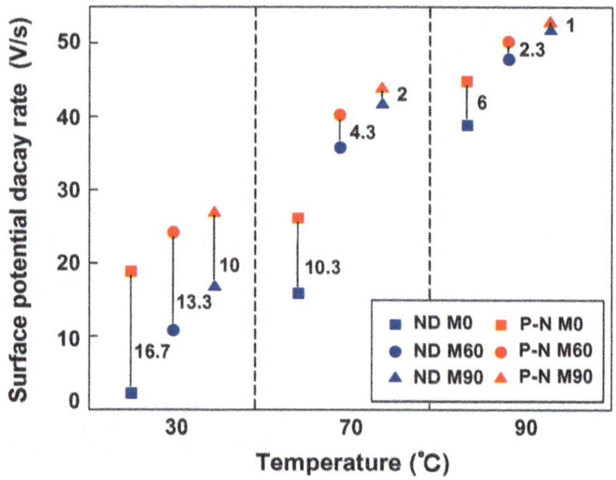

Fig. 15 Average SPD rate at the center of the sample under negative DC voltage (abbreviated in the figure as ND) and positive-negative reversal voltage (abbreviated in the figure as P-N) (the differences have been marked in the figure)

potential at the center is the highest and the surface potential decreases from inside to outside. At 120 s, the maximum surface potential appears at 0.5 cm. At 300 s, it appears at 1 cm. This phenomenon may be caused by the uneven distribution of positive charges in the sample. As can be seen in Figs. 13 and 14, positive charges are accumulated more in the center of the sample and less in the outside. When the content of SiC is high, the conductivity of the sample increases with the local electric field and temperature, resulting in the decay rate at the center higher than that at the outside. Therefore, the surface potential of the center is lower than that of the outside at 120 s. Obviously, this phenomenon only exists in a certain period of time. When the positive charges at the center are consumed and the surface potential formed by the negative charges decreases, the decay rate at the center becomes lower than that at the outside.

Finally, the surface potential difference at different positions decreases gradually, and the surface potential on the whole plane reaches a stable state. Figure 15 compares the average SPD rate of the central point within the initial 30 s under positive-negative reversal voltage and negative DC voltage. The average SPD rate of M0 increases by 16.3 V/s at 30 °C and 10.3 V/s at 70 °C, which is greater than that of M60 and M90 at the same temperature. This phenomenon may be caused by the residual positive charges in the sample. The charge transport of M0 can be explained by the model shown in Fig. 16a. After negative corona discharge, the positive charges on the surface of the sample are neutralized, with the negative charges are accumulated more in the center and less in the outside. The conductivity of M0 is relatively low, and the negative charges migrate slowly to the ground electrode. There are still large numbers of positive charges inside the sample, and the negative charge accumulation on the sample surface is serious. On the one hand, strong local electric field is formed between positive and negative charge regions, which promotes charge migration. On the other hand, the neutralization of hetero charge accelerates the dissipation of negative charges. All of these reasons make the average SPD rate at the center point of M0 increase greatly. For M60 and M90, due to the high nonlinear conductivity,

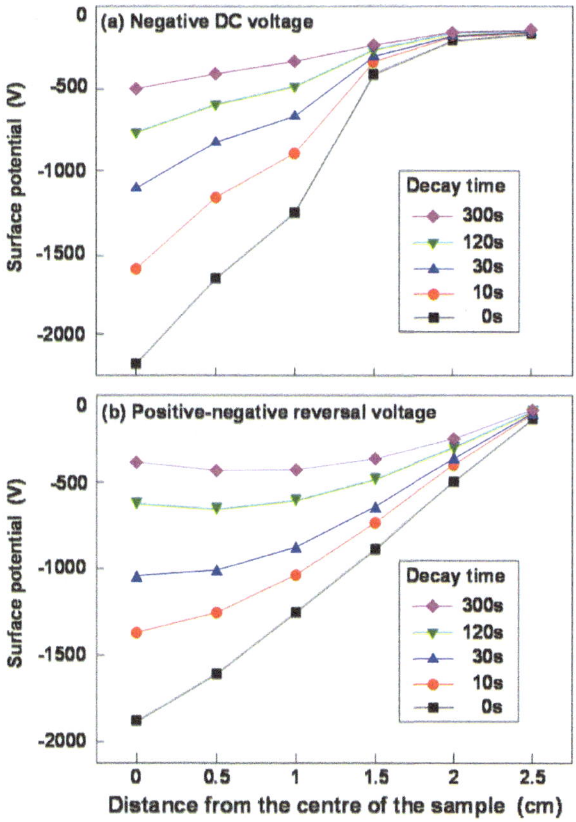

Fig. 16 SPD curve of M90 at 70 °C under **a** negative DC voltage and **b** positive-negative reversal voltage

the average SPD rate under negative DC voltage is relatively large, so the increase of average SPD rate under positive-negative reversal voltage is less than M0. It can also be seen that the average SPD rate of M60 and M90 decreases significantly at 70 °C compared with that at 30 °C. Besides, the average SPD rate of the three samples decreases to a relatively low value at 90 °C. This phenomenon can be explained by the model in Fig. 16b. The increase of temperature leads to the high conductivities of the samples. Not only the positive charges on the surface of the sample are neutralized by the negative charges, but also lots of positive charges inside the sample are consumed due to the rapid migration of negative charges. When corona discharge is over, only a small amount of positive charges in deeper layer or deep traps exist in the sample. Although these positive charges accelerate the dissipation of negative surface charges, the high conductivity caused by the electric field formed by the surface potential in the sample plays a leading role. Moreover, at high temperatures, the average SPD rate under negative DC voltage is also at a high level due to the effect of high conductivity. Therefore, when the conductivity of the samples is high, the average SPD rate at the center point increases less under positive-negative reversal voltage.

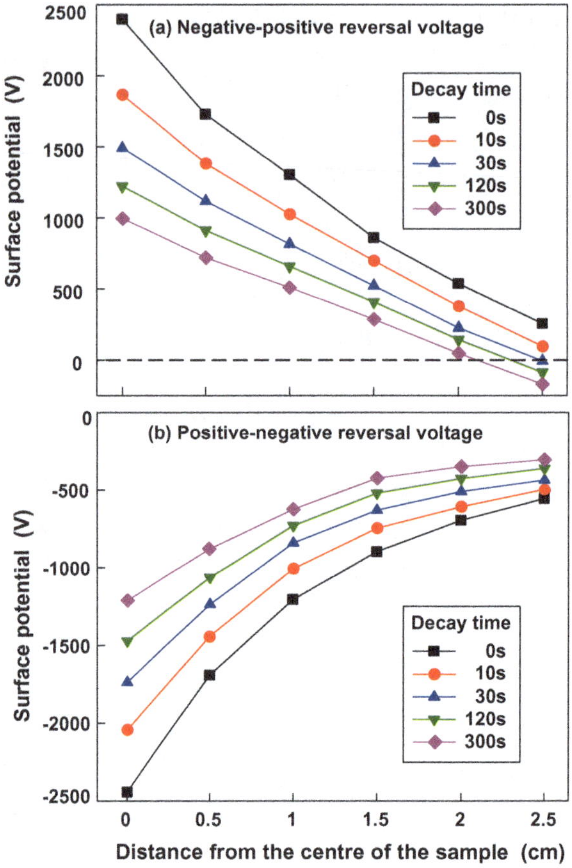

Fig. 17 SPD curve of M0 under reversal voltage at 30 °C under **a** negative-positive reversal voltage and **b** positive-negative reversal voltage

The surface charge characteristics of SiR materials and SiC/SiR composites under positive DC voltage are similar to those under negative DC voltage, but the ISP of the center point under positive DC voltage is lower. When the needle electrode has a positive voltage, the negative charges generated by air ionization are neutralized by the needle electrode, so a large number of positive charges are accumulated near the electrode, weakening the electric field near the needle. These positive charges move to the surface of the sample slowly, and eventually less positive charges reach the surface of the sample. When the needle electrode has a negative voltage, the positive charges generated by ionization gather near the needle tip, which increases the local electric field. The negative charges move to the surface of the sample quickly, resulting in more negative charges reaching the surface of the sample. Therefore, the negative DC voltage is more likely to cause the accumulation of surface charges under the same voltage level.

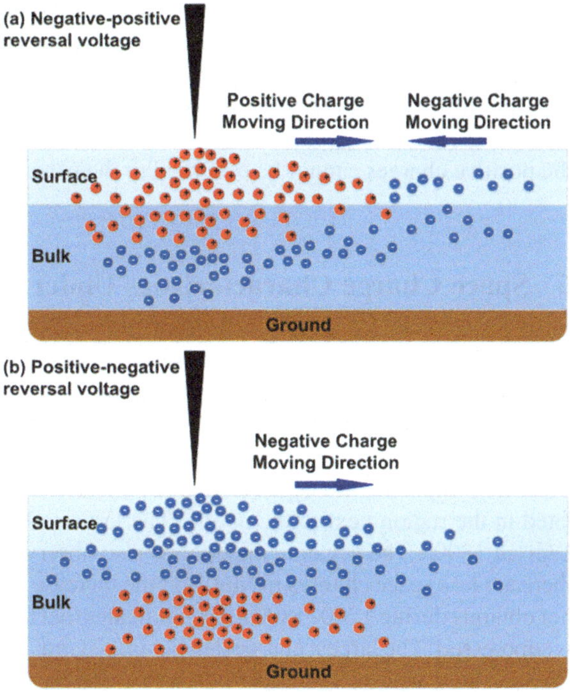

Fig. 18 Surface charge decay models of M0 under reversal voltage

Similar to the surface charge distribution under positive- negative reversal voltage, the ISP becomes positive at all positions and decreases in turn from the center to outside under negative-positive reversal voltage. Positive charges complete the neutralization of negative charges on the surface of the sample and further accumulate. Compared with M0, the surface potential of M60 and M90 is lower. The surface potential of each measuring point of the three samples at 70 °C is lower than that at 30 °C, and further decreases at 90 °C.

Unlike the samples under positive-negative reversal voltage, the negative surface charges at the lateral positions of the samples are observed when the conductivity of the sample is very low. Figure 11 is the SPD curve of M0 under polarity-reversal voltage at 30 °C. At 0 s, the surface potential is positive and decreases from inside to outside. At 10 s, the surface potential of each point decreases, which is lower than that at 0 s, but still positive. At 30 s, the surface potential at 2.5 cm decreases to 0, then it increases in reverse and becomes negative. During 30–300 s, the surface potential at 2.5 cm continues to increase in reverse. This phenomenon only occurs in the surface charge distribution of M0 at 30 °C, not occurring at higher temperature or in sample containing SiC particles.

Figure 12 can be used to explain the above phenomena. In the process of corona discharge, positive charges are accumulated more at the center and less at the outside. M0 has the lowest conductivity at 30 °C. During positive voltage charging, the positive charges of the sample extend to 2.5 cm, but there are still a certain number of negative surface charges in the outer area of the surface. In the process of surface charge dissipation, the positive and negative charges are neutralized, indicating that the negative surface charges gradually migrate to the inside. However, the negative charges diffuse rapidly on the surface of the sample. Under positive-negative reversal voltage, most of the positive charges in external region are neutralized. Therefore, the positive charges cannot be observed in the process of SPD.

6 Space Charge Characteristics Under DC Voltage

Figure 19 shows the space charge polarization of samples at 30 °C under 15 kV/mm. The space charges in M0 keep accumulating with time. It is obvious that negative charges accumulate in the region close to the anode while there is no considerable positive charges found close to the cathode. This indicates that space charges in SiC/SiR composites are negative charges injected from cathode and finally accumulated in the region next to the anode. The largest charge density in M0 reaches 3.8 C/m^3 at 1800s. The negative injection is also observed in M60. Compared with M0, there are less space charges in M60 and the increase in the number of space charge is not obvious during 1200–1800s. This indicates that the accumulation of space charge is suppressed. The largest charge density is 2.2 C/m^3 at 1800s. The local field exceeds the threshold field of nonlinear conductivity of composites when large number of space charges accumulate in M60. This leads to the increase in conductivity and space charges are driven to electrode and neutralized which in consequence reduces the amount of space charge. At 1800s, the balance between the charge injection and the suppression in charge accumulation caused by conductivity is achieved and the increase in number of space charge becomes insignificant. The least space charge accumulation is found in M90 because of the highest conductivity. The largest charge density is only 0.96 C/m^3 and there is almost no increase in number of space charge at 1800s.

Figure 20 shows the space charge polarization at 70 °C under 15 kV/mm. The temperature increases the conductivity of M0 and less space charges accumulate in sample. The injection of charges is still obvious, but the accumulation process is slowed down. The largest charge density reaches 2.5 C/m^3 at 1800s. M60 also presents a smaller largest charge density due to higher temperature. M90 presents a similar space charge distribution and the amount of space charge in the sample bulk decreases. The largest charge density in M90 is 1 C/m^3. It can be concluded that the largest charge density in M0 decreases by 34.2% and in M60 decreases by 22.7%. However, it increases by 4% in M90 at 70 °C. This can be explained by the effect of temperature on conductivity. Considering the conductivity in Fig. 1b, the conductivity of M60 presents a constant while the conductivity of M90 decreases. This indicates that temperature cannot improve the space charge accumulation property a lot since the conductivity of M60 and M90 cannot increase as much as M0.

Figure 21 shows the space charge depolarization of samples at 30 °C. It can be seen from Fig. 21a that a large amount of space charges still exist in M0 after

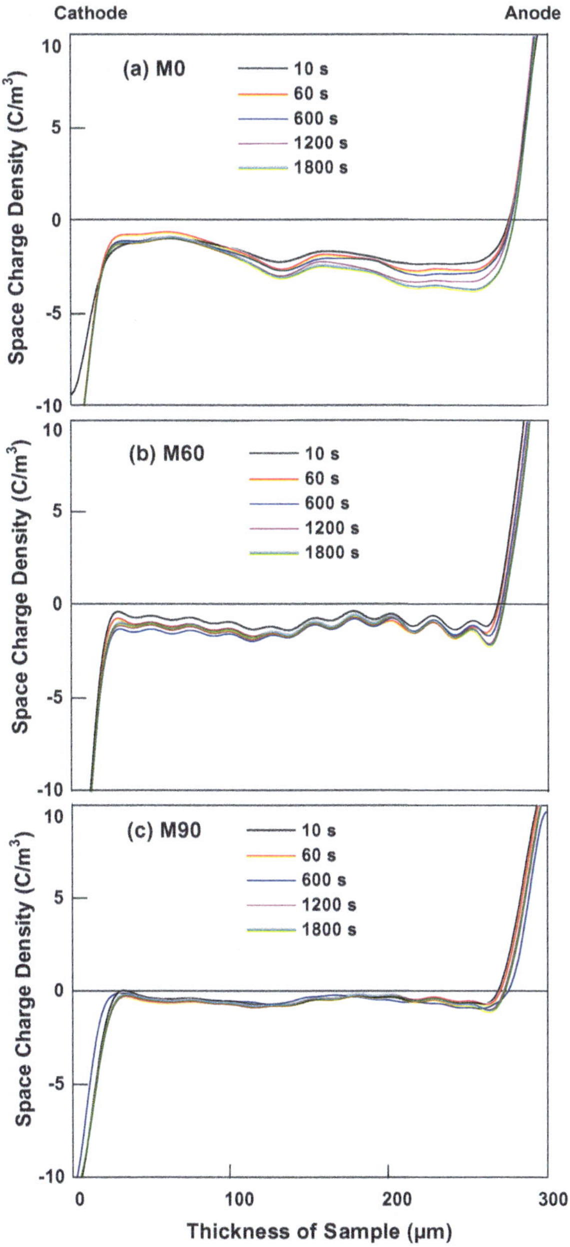

Fig. 19 The space charge polarization at 30 °C under 15 kV/mm

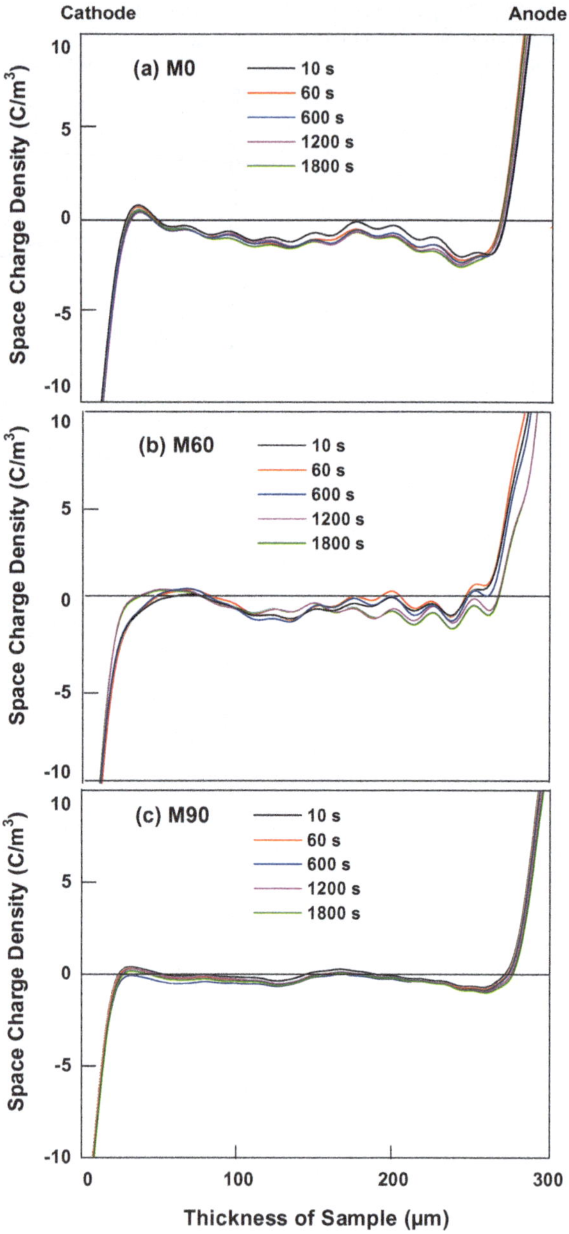

Fig. 20 The space charge polarization at 70 °C under 15 kV/mm

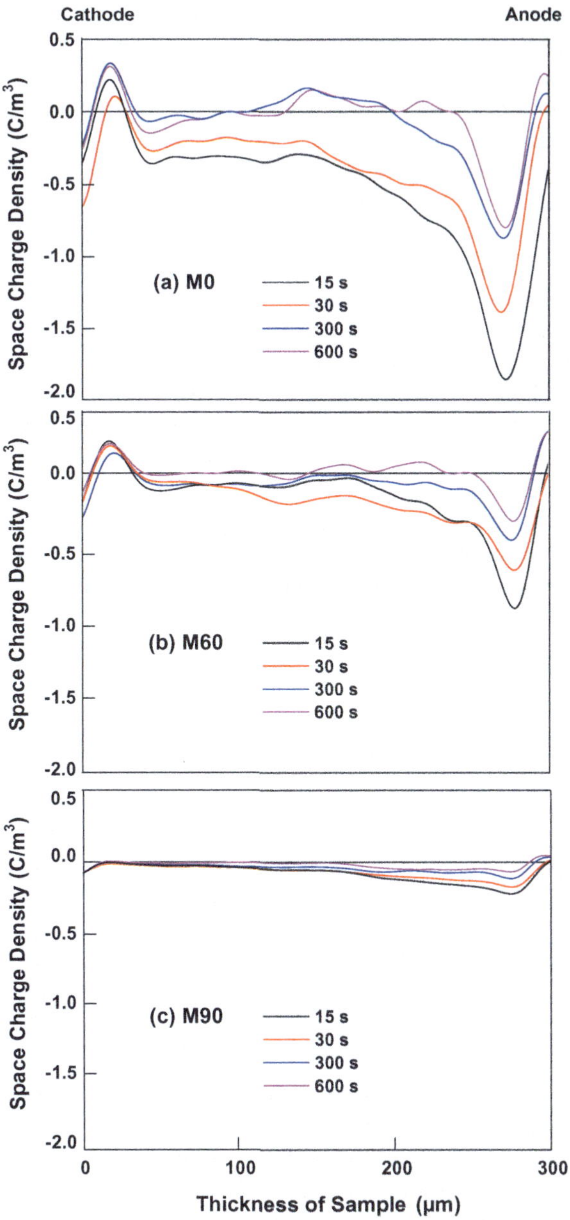

Fig. 21 The space charge depolarization at 30 °C

depolarization has lasted for 15 s. The largest charge density close to the anode is 1.9 C/m^3. The space charge distribution is similar with that in the polarization process. The largest charge density decreases by more than 50% before 300 s. After 300 s, the dissipation rate of space charge slows down and there are still large amount of negative space charges exist in the sample at 600 s. The largest charge density reduces to 0.85 C/m^3 finally. At the initial stage of dissipation, charges in shallow traps escape fast. After 300 s, The shallow trap dissipation process has finished and charges in deep traps can hardly escape. This leads to a much slower dissipation rate and a large number of remaining space charges. M60 and M90 present better dissipation property because of high conductivity. The space charge dissipates rapidly. There are less space charges remain in M60 and only a few number of space charges exist in the region close to anode in M90.

Figure 22 shows the space charge depolarization of samples at 70 °C. It is obvious that M0 presents a faster dissipation rate. The largest charge density reduces by 35.2% at 15 s. The dissipation rate is still considerable after 300 s. There are still some remaining charges at 600 s, but it is much less than that at 30 °C. M60 and M90 shows less remaining space charge compared with M0. M90 presents the most rapid dissipation. The number of negative charges decreases fast and it is hard to observe obvious space charge accumulation in the depolarization process. The space charge dissipation becomes fast in M0 at 70 °C. It is because that temperature provides more energy for carriers to escape from traps and to form stronger current in the sample. M60 presents a fast dissipation of space charge at the beginning of depolarization and the dissipation rate is still considerable at the end. The space charge in M90 dissipates within 15 s and no space charge accumulation exists at 600 s. It is believed that the fast dissipation is also caused by the trap distribution in SiR/SiC composites. The SiC particle reduces the energy level of traps. The number of shallow traps is also increased by SiC particles while the number of deep traps is reduced. These reasons make it easy for space charges to escape from traps which further forms a rapid dissipation.

7 Conclusion

1. The nonlinear conductivity is suppressed at high temperatures. The conductivity increases with temperature when the field is less than 10 kV/mm. At a higher electric field, the conductivity of M60 decreases with temperature. The same phenomenon is also found in M90. The temperature accelerates the conductivity increase in M0, whereas it slows down the increase in M60 and M90, and the deceleration is more effective in high-content samples. The suppression effect of temperature is mainly caused by the SiC particles. As a semiconductor, the energy gap between the valance band and the conduction band is relatively narrow. Most carriers are easily excited to the conduction band when the electric field is high. The excitation resulting from temperature is not obvious, but the increase in

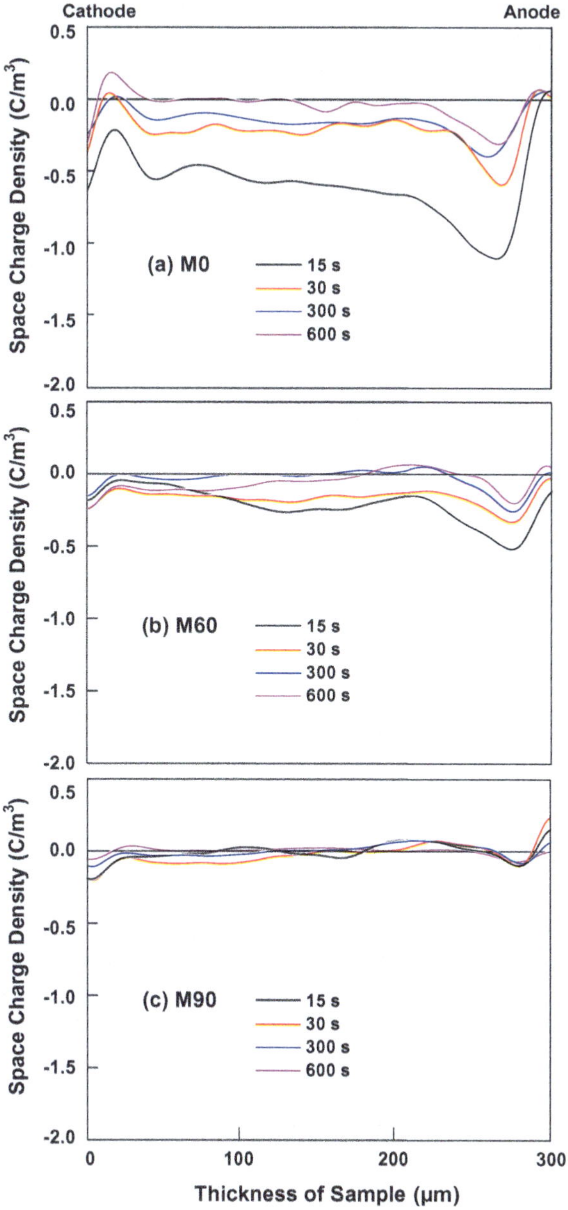

Fig. 22 The space charge depolarization at 70 °C

temperature can cause a lattice vibration, which scatters the carriers and reduces the conductivity.

2. With the increase in homopolar impulse voltage, all samples experience a rise in the ISP. It is clear that the increase in SiC content and the increase in impulse voltage can both accelerate the decay process. The heteropolar impulse voltage not only intensifies the accumulation of surface charge but also slows down the decay process. Neutralization occurs immediately between heterocharges. However, this process cannot be completed in a short time. The positive charges are captured by traps in the region, and the attraction between heterocharges is the reason for the observed phenomena.
3. Under positive-negative reversal voltage, compared with that under negative DC voltage, the ISP of the sample center point decreases, and the extent decreases with the increase of the conductivity of the samples; the SPD rate increases, and the extent decreases with the increase of the conductivity of the samples. Compared with that under positive DC voltage, the surface charge has a similar law when negative-positive reversal voltage is applied. The SPD of the samples under positive-negative reversal voltage has a case that the surface potential of the center point is lower than outside. There are more positive charges in the center than the outside, which strengthens the local electric field between the heteropolar charges, so that the SPD rate at the center is higher than outside. The SPD of the samples under negative-positive reversal voltage observes the residual negative charges remaining outside. In the process of surface charge dissipation, the positive and negative charges are neutralized. While the negative charges in SiR materials migrate more easily, and the negative charges gradually move toward the center of the sample.
4. The space charge accumulation is suppressed by SiC particles. M90 presents the least space charge accumulation because of the highest conductivity. Temperature also has a positive effect on suppressing space charge accumulation. At 70 °C, M0 shows the largest reduction in space charge density while the M90 shows the smallest. This is because that the conductivities of M0, M60 and M90 present different trends under 15 kV/mm. Both temperature and SiC content have positive effect on accelerating the dissipation of space charge. The SiC particle reduces the energy level of traps. The number of shallow traps is also increased by SiC particles while the number of deep traps is reduced. Temperature provides more energy for carriers. These reasons make it easy for space charges to escape from traps which further forms a rapid dissipation.

References

1. Y. Zhou, J.L. He, J. Hu, X.Y. Huang, P.K. Jiang, Evaluation of polypropylene/polyolefin elastomer blends for potential recyclable HVDC cable insulation applications. IEEE Trans. Dielectr. Electr. Insul. **22**(2), 673–681 (2015)

2. L. Lan, Q. Zhong, Y. Yin, X. Li, Effect of surface fluorination on space charge behavior at LDPE/EPDM interface, in *IEEE International Conference on Solid Dielectrics* (2013), pp. 444–447
3. J. Li, B.X. Du, H. Hang, Suppressing interface charge between LDPE and EPDM for HVDC cable accessory insulation. IEEE Trans. Dielectr. Electr. Insul. **24**(3), 1331–11339 (2017)
4. L. Virsberg, P.H. Ware, A new termination for underground distribution. IEEE Trans. Power Appar. Syst. **86**(9), 673–681 (1967)
5. P.N. Nelson, H.C. Hervig, High dielectric constant materials for primary voltage cable terminations. IEEE Trans. Power Appar. Syst. **103**(11), 3211–3216 (1984)
6. O. Lupo, C. Petrarca, L. Egiziano, V. Tucci, M. Vitelli, Numerical evaluation of the electric field in cable terminations equipped with non-linear grading materials, in *IEEE Conference on Electrical Insulation and Dielectric Phenomena* (1998), pp. 585–588
7. B. Jacobson, Y. Jiang-Hafner, P. Rey, G. Asplund, M. Jeroense, A. Gustafsson, M. Bergkvist, HVDC with voltage source converters and extruded cables for up to ±300 kV and 1000 MW, in *IEEE International Conference on Large High Voltage Electric Systems* (2006)
8. L. Donzel, F. Greuter, T. Christen, Nonlinear resistive electric field grading part 2: materials and applications. IEEE Trans. Electr. Insul. Mag. **27**(2), 18–29 (2011)
9. A. K. Upadhyay, C.C. Reddy, Comparative study of famous conductivity models for low density polyethylene, in *IEEE International Conference on Industrial and Information Systems* (2017), pp. 1–4
10. B.X. Du, Z.R. Yang, Z.L. Li, J. Li, Surface charge behavior of silicone rubber/SiC composites with field-dependent conductivity. IEEE Trans. Dielectr. Electr. Insul. **24**(3), 1340–1348 (2017)
11. L. Donzel, F. Greuter, T. Christen, Nonlinear resistive electric field grading part 2: materials and applications. IEEE Electr. Insul. Mag. **27**(2), 18–29 (2011)
12. S. Boggs, D.H. Damon, J. Hjerrild, J.T. Holboll, Effect of insulation properties on the field grading of solid dielectric DC cable. IEEE Trans. Power Deliv. **16**(4), 456–461 (2015)
13. Z. Wang, J.K. Nelson, H. Hillborg, S. Zhao, L. S. Schadler, Nonlinear conductivity and dielectric response of graphene oxide filled silicone rubber nanocomposites, in *IEEE Conference on Electrical Insulation and Dielectric Phenomena* (2012), pp. 40–43
14. N.G. Hingorani, Transient overvoltage on a bipolar HVDC overhead line caused by DC line faults. IEEE Trans. Power Appar. Syst. **89**(4), 592–610 (1970)
15. Y. Liu, X.L. Cao, Electrical tree initiation in XLPE cable insulation by application of DC and impulse voltage. IEEE Trans. Dielectr. Electr. Insul. **20**(5), 1691–1698 (2013)
16. M. Liu, Y. Zhang, K. Li, F.F. Shi, C. Yang, Y.T. Sun, G.H. Zhang, A model predictive control strategy for LCC-HVDC system to participate in long-term voltage stability control, in *IEEE China International Conference on Electricity Distribution (CICED)* (2018), pp. 872–877
17. B. Qi, X. Zhao, C. Li, H. Wu, Transient electric field characteristics in oil-pressboard composite insulation under voltage polarity reversal. IEEE Trans. Dielectr. Electr. Insul. **22**(4), 2148–2155 (2015)
18. M. Fu, L.A. Dissado, G. Chen, J.C. Fothergill, Space charge formation and its modified electric field under applied voltage reversal and temperature gradient in XLPE cable. IEEE Trans. Dielectr. Electr. Insul. **15**(3), 851–860 (2008)
19. X. Chen, X. Wang, K. Wu, Z.R. Peng, Y.H. Cheng, D.M. Tu, Effect of voltage reversal on space charge and transient field in LDPE films under temperature gradient. IEEE Trans. Dielectr. Electr. Insul. **19**(1), 140–149 (2012)
20. Y. Tanaka, R. Kodera, T. Kato, H. Miyake, H. Mori, Y. Yagi, Observation of space charge accumulation behavior in cross-linked polyethylene at voltage polarity reversal, in *IEEE Annual Report Conference on Electrical Insulation and Dielectric Phenomena (CEIDP)* (2015), pp. 23–26
21. S. Wang, S. Luo, Y. Tu, C. Wang, S. Qin, Effect of polarity reversal on space charge properties of CB/LDPE composite under DC field. IEEE Trans. Dielectr. Electr. Insul. **24**(3), 1349–1354 (2017)
22. J. Castellon, H.N. Nguyen, S. Agnel, A. Toureille, Electrical properties analysis of micro and nano composite epoxy resin materials. IEEE Trans. Dielectr. Electr. Insul. **18**(3), 651–658 (2011)

23. GSh. Gildenblat, A.R. Rao, S.S. Cohen, Space-charge-limited currents in materials with nonlinear velocity-field relationships. IEEE Trans. Electron Devices **34**(10), 2156–2172 (1987)
24. X. Yang, J. Hu, S. Chen, J.L. He, Understanding the percolation characteristics of nonlinear composite dielectrics. Scientific Reports **6**(1), 30597 (2016)
25. R. Strumpler, J. Glatz-Reichenbach, Conducting polymer composites. J. Electroceram. **3**(4), 329–346 (1999)
26. M. Grundmann, *The Physics of Semiconductors: An Introduction Including Nanophysics and Applications*, 2nd edn. (Springer, New York, 2010)
27. J.C. Xie, J. Hu, J.L. He, Z.C. Guo, Y. Yin, Effect of silicone rubber polymer composites on nonuniform electric fields of rod-plane gaps, in *IEEE Conference on Electrical Insulation and Dielectric Phenomena* (2013), pp. s399–402

Effect of Temperature and Mechanical Stress on Charge Transport and Ageing Properties of EPDM for Cable Accessories

Jingang Su, Boxue Du, Tao Han, Zhonglei Li, Liqiang Wei, and Peng Zhang

Abstract Effects of mechanical stress, produced by the spring clamp device and the banding force after the expansion of cable accessories, on the electrical tree growth characteristics of EPDM (Ethylene Propylene Diene Monomer) are reported in the present work. Results show that the inception time of electrical tree becomes shorter as stretching ratio increases from 0 to 30%. Tree length and accumulated damage become larger with the stretching ratio increasing. However, under the compressive stress, the treeing growth process shows a converse trend, which demonstrates that the higher compressive rate retards the electrical tree growth. Electrical tree propagation mechanism considering FFV (fractional free volume) and CED (cohesion energy density), which are respectively related to free path of electrons and molecular chains splitting during electrical tree propagation process, are also discussed and analyzed. Moreover, the main insulation materials in the cable accessories face the challenge of high temperature during the overload running process. With the temperature increasing, the initiation process of electrical tree is promoted, and appearance probability of bush tree becomes higher. The influences of mechanical stress and high temperature on electrical tree behaviors are analyzed and illustrated based on charge transport from the viewpoints of the structure–property relationship.

1 Introduction

HVDC and HVAC cables are the most important equipment for high-voltage, large-capacity and long-distance power transmission. EPDM is widely adopted in cable accessories, and it is the main insulation material in cable joints and cable terminals. The insulation materials of cable accessories have to face the challenges of mechanical stress. For example, mechanical stress can be generated by the different

J. Su (✉) · L. Wei · P. Zhang
State Grid Hebei Electric Power Research Institute, Shijiazhuang 050021, China
e-mail: sujingang@tju.edu.cn

B. Du · T. Han · Z. Li
Key Laboratory of Smart Grid of Education Ministry, School of Electrical and Information Engineering, Tianjin University, Tianjin 300072, China

B. Du (ed.), *Polymer Insulation Applied for HVDC Transmission*,
https://doi.org/10.1007/978-981-15-9731-2_11

cooling rates of inner and outer insulation materials during the cooling process of fabrication [1]. To ensure a certain interface resistance, the cable accessories are densely coated onto the main cable insulation materials under a certain amount of interference force produced by the spring clamp device [2]. During the operation of HVDC and HVAC cables, the thermo-mechanical stress can be created by the different thermal expansion of the multi-layer insulation materials. Moreover, the relaxation of molecular chains during the electrical and thermal degradation process can also affect the mechanical properties.

Electrical trees, one of the main degradation and pre-breakdown phenomenon of the insulation materials, are affected by many factors, especially the mechanical stress [3]. Mechanical stress could affect the electrical tree in insulating materials, depending on whether the mechanical stresses are of tensile or compressive nature [4]. Al-Ghamdi et al. reported the electrical tree in mechanically pre-stressed epoxy resin, it was found that electrical tree growth was decelerated by the compressive stress and a long treeing time was observed [5]. Nakanishi et al. investigated the electrical tree inception mechanisms in epoxy resins, it was found that higher inception voltage was occurred under lower mechanical stresses [6]. Crine researched the electrical properties of dielectric polymers under different electro-mechanical stress, it indicated several dielectric and electrical properties of solid dielectrics were affected by the Maxwell stress [7]. Varlow et al. found that the tensile or compressive stress had negative or positive effect on electrical tree behavior, and the electrical tree resistance was enhanced by mechanically prestressing the insulation material by the inclusion of fibers in tension [8]. Chen et al. investigated a double electrical tree structure in XLPE, it was found that the residual mechanical stress affected the electrical tree propagation and caused the insulation material around the core of XLPE cable to degrade seriously [9]. From the practical point of view, it is very important to study the treeing phenomena in EPDM considering the mechanical stress. However, the research has not been employed.

Moreover, high temperature, induced by the overload current flow of the conductor in the HVDC cable during the abnormal running process, generates a severe threat to the safe operation of cable insulation. The temperature decreases along the normal direction of the cable from the inner conductor. For the inner conductor, the rated loading temperature is designed to 90 °C. The running temperature is usually held at 50~60 °C, while it rises to 150 °C during the transient failure process in the extruded XLPE power cable system [10, 11].

When the temperature was 50 °C, the electrical tree propagated in the dense crystalline spherulite of XLPE and formed bush tree structure for 15 or 18 kV. However, when the temperature was 70 or 90 °C, the electrical tree propagated in the expanded amorphous region or the boundary of the large crystalline spherulite and formed branch tree structure for 15 or 18 kV [12]. Zhou et al. investigated the effects of temperature ranging from 20 to 180 °C on electrical tree characteristics in silicone rubber and it was found that high temperature lead to generate a denser electrical tree structure [13]. Kosaki et al. observed the initiation voltage of DC electrical trees in polyethylene, it was increased from 35 to 50 kV as the temperature was changed from 20 to −196 °C [14]. However, less researches are investigated about electrical tree of EPDM with high temperature from charge transport and physical microstructure viewpoints.

This chapter is aimed to investigate the effects of mechanical stress and temperature on treeing growth characteristics in EPDM. In this chapter, the electrical tree process is observed by the digital microscope system. Electrical tree structure and propagation characteristics are investigated under different mechanical stress and temperature. Tree length and accumulated damage analysis are employed to quantitatively analyze treeing destruction degree [15, 16]. To understand the electrical tree initiation characteristics, the tree initiation probability also researched. Moreover, the electrical tree mechanism taking into account the effects of mechanical stress and temperature is discussed based on charge transport from the viewpoints of the structure–property relationship.

2 Experiment

2.1 Sample Preparation

In this study, the employed raw materials were the EPDM (Nordel 4725, Dow Chemical) with an ethylene content of 70 wt.%, a propylene content of 25 wt.% and an ethylidenenorbornene content of 5 wt.%. The 2,5-dimethyl-2,5-di-(tert-butylperoxy)-hexane was employed as the cross-linking agent. The raw materials were mixed uniformly in a twin-roll mill at 413 K for 15 min before the cross-linking agent was added. Then the mixed raw was cured in a stainless steel mold at 443 K under a pressure of 15 MPa for 10 min. The obtained samples with thickness of 5 ± 0.2 mm were post-cured at 473 K for 4 h under ambient pressure.

2.2 Experimental Apparatus and Procedure

2.2.1 Mechanical Stretching

The experiments were carried out in the mechanical stretching apparatuse as shown in Fig. 1. For 35 and 110 kV prefabricated cable joints, the relative expansion ratio after expansion was 33.4% and 27.2%, respectively [17]. In the mechanical stretching test, the stretching ratio ξ was set to 0, 10%, 20% and 30%. The stretching ratio was defined by the formula (1):

$$\xi = \frac{\Delta l}{l} \tag{1}$$

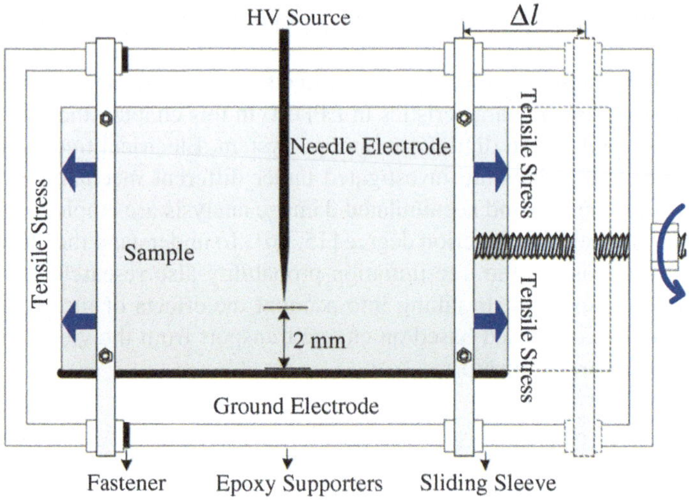

Fig. 1 Experimental setup with mechanical stretching

Where l was the length without stretching stress, Δl was the length variation under stretching stress.

2.2.2 Compressive Stress

The experiments were carried out in the compressive stress apparatuses as shown in Fig. 2. In the compressive stress test, the compressive rate ξ was set to 0, 10%, 20% and 30%. The compressive rate was defined by the formula (2):

$$\xi = \frac{\Delta d}{d} \tag{2}$$

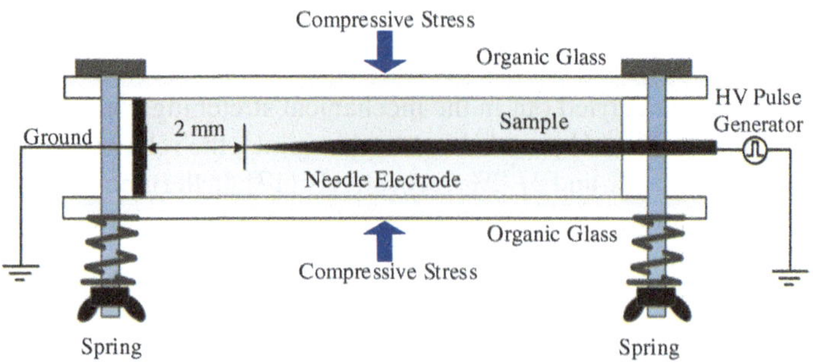

Fig. 2 Experimental setup with compressive stress

where d was the width without compressive stress, Δd was the width variation under compressive stress.

2.2.3 High Temperature

The experiment was carried out in a thermotank with the precision of 1 °C, in which isothermal conditions ranged from 10 to 300 °C could be reached. The temperature was set at 30, 45, 60, 75, 90, 105 and 120 °C, which could simulate the possible operating temperature of power cable.

In the mechanical stress and temperature test, the experimental setup and the digital observation system were the same. The tip radius of the needle was 3 μm and the needle–plane distance was 2 mm. In the experiment, the pulse frequency was 200 Hz and the calibrated pulse amplitude was ranged from 0 to 30 kV. The rise time of the waveform obtained from the high voltage pulse source was 50 μs. The details of the experimental setup and the digital observation system was reported in [16]. The temperature was controlled 20 ± 5 °C. Under each condition, 20 specimens were measured to take into account the statistical variations in the experiment.

2.3 Modelling and Simulation Methods

Molecular dynamics (MD) simulations were carried out by employing MATERIALS STUDIO software package (Accelrys Co.) [18]. Figure 3a showed an EPDM chain established in Visualizer module and the chain length was 100. Figure 3b showed the periodic cell with four EPDM chains constructed in Amorphous Cell module and the density was 0.88 g/cm^3. The implemented force field in this work was calculated by COMPASS (Condensed-phase Optimized Molecular Potentials for Atomistic Simulation Studies). The equations of motion were solved with Verlet velocity algorithm. The time step of integration was set to 1 fs. The summation methods for van der Waals and Coulomb forces were based Atom and Ewald, respectively. The cutoff, spline width and buffer width was 9.5, 1.0 and 0.5 Å, respectively. Initial velocities were randomly designated from Boltzmann distribution. Temperature was set to 293 K and controlled by Andersen method. The change of shape was obtained by applying a linear change to the atom positions [19]. CED was calculated by employing Forcite Tool, while FFV were analyzed by Atom Volume & Surfaces.

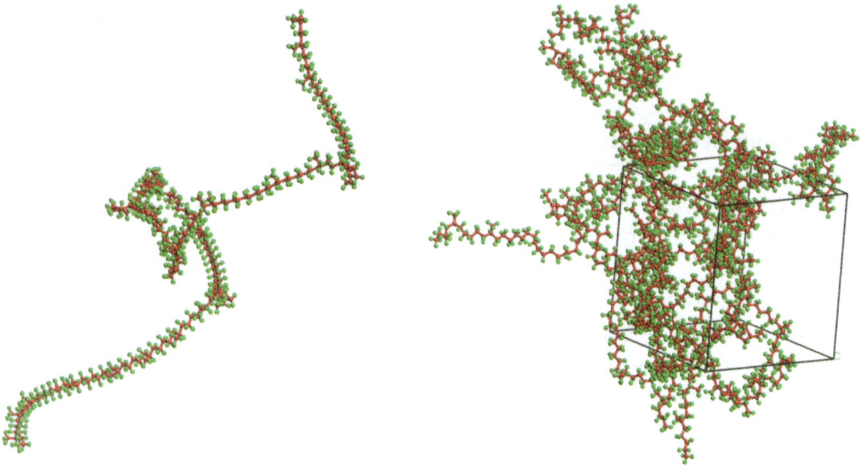

(a) EPDM polymer chain (b) Periodic cell

Fig. 3 Structure of EPDM polymer chain (**a**) and periodic cell (**b**)

3 Results and Discussion

3.1 Mechanical Stretching

3.1.1 Free Volume and Cohesive Energy

The electrons tend to be transported in a low-density region, which is mainly formed by the free volume between the molecular chains. FFV (free volume fraction) is refers to the free volume of the total polymer volume, it can be explained by the formula [20]:

$$FFV = \frac{V_f}{V_f + V_o} \tag{3}$$

where V_f is the free volume of the polymer, V_o is the occupied volume of the molecular chains in the polymer. The higher chains mobility means larger amorphous regions and FFV. FFV is calculated by molecular dynamics simulation method under different stretching ratios.

CED is mainly a measurement of the binding forces between different molecular chains during the crack growth process, which is related to the electric strength of polymer materials. The CED is defined as the energy that is needed for the material to transfer from the condensed phase to the gas phase. The CED is defined by [21]:

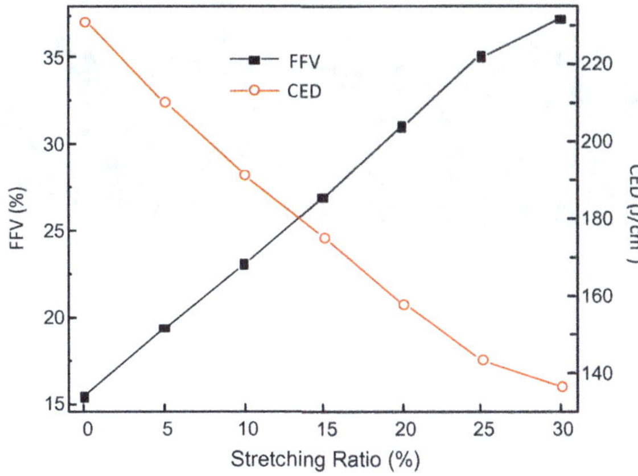

Fig. 4 FFV and CED as a function of stretching ratios

$$CED = \frac{\Delta H_{\text{vap}} - RT}{V} \quad (4)$$

where ΔH_{vap} is the heat of vaporisation, R is the gas constant, T is the temperature and V is the molar volume. CED is also equal to the energy associated with the van der Waals and electrostatic interactions between the molecules that comprise the material, which is calculated by Forcite Tool with the force field of COMPASS.

The relationship between the FFV and stretching ratio is shown in Fig. 4. FFV is increased from 15.4% to 37.23% when the stretching ratio increases from 0 to 30%. FFV becomes larger as the stretching ratio increasing, indicating the motion space of molecular chains is enlarged and the injected electron can obtain higher energy to collide with the molecular chains.

When the EPDM is yielded to mechanical stretching, the interaction energy density of van der Waals and electrostatic energy becomes smaller as the van der Waals bonds are disrupted and the inter-molecular distance becomes larger. The relationship between CED and stretching ratio can be observed in Fig. 4. CED decreases from 231.06 to 136.69 J/cm^3 as the stretching ratio increases from 0 to 30%. The smaller CED indicates that less external energy is needed for separating the binding forces between different molecular chains during the tree channel crack propagation process.

Schematic diagram of free volume distribution under different stretching ratios is shown in Fig. 5, where the blue region refers to the free volume (indicated by the arrow) and the gray region indicates the occupied volume of the molecular chains. The free volume is changed in two forms as shown in the circles of a1, b1, c1, d1 and a2, b2, c2, d2. The circles of a1 to d1 describe the free volume is enlarged as the rearrangement of molecular chains during the mechanical stretching process. While the circles of a2 to d2 reveal the molecular chains are stretched along the direction of the stretching stress, and even merged into a larger free volume as mechanical stretching ratio reaches a threshold value. In the process, the hot electron

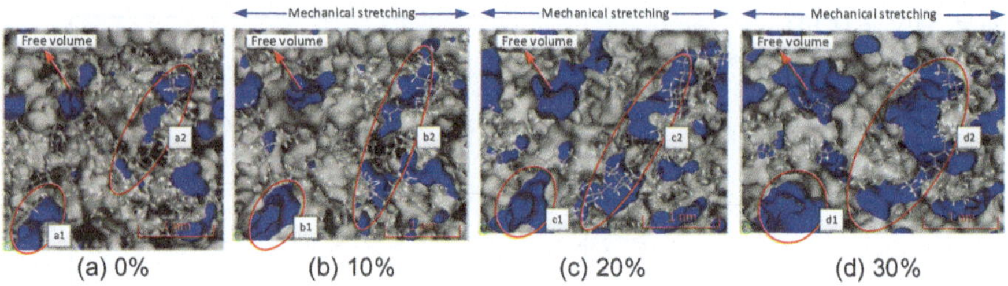

Fig. 5 Schematic diagram of free volume distribution under different stretching ratios

with relatively low energies can extend the preexisting defects and eventually increase the defect density to the level of forming electrical tree channels.

3.1.2 Electrical Tree Characteristics

With application of 20 and 25 kV, typical morphologies of electrical trees at different stretching ratios are shown in Fig. 6 (a1–a4) and (b1–b4). In the test, it is found when the stretching ratio increases from 0 to 30%, the tree channel number is increased. Under higher stretching ratio, the electrical tree shows a more densely packed structure and a larger destructive area, which indicates that the mechanical stretching accelerates the electrical tree growth.

The electrical tree propagation period consists of extension and trunk widening by erosion due to partial discharges attack in tree channels. The tree length indicates the tree extension characteristics and it is referred to the distance between the needle tip

Fig. 6 Typical morphologies of electrical trees at different stretching ratios

and the end of the longest tree branch along the electric field direction. Accumulated damage, extracting the total pixels number covering the tree area to estimate the 2D projected patterns of real electrical trees, is employed to describe treeing destructive area [15].

Tree length (a) and accumulated damage (b) as a function of different stretching ratios are shown in Fig. 7. In Fig. 7a, it notes that the mean tree length is increased from 0.5 to 0.93 mm as the stretching ratio increases from 0 to 30% at 20 kV, while it is varied from 0.82 to 1.15 mm at 25 kV. The accumulated damage also increases with increasing the stretching ratio, which indicates the mechanical stretching affects the electrical tree characteristics by accelerating the degradation of the EPDM samples. The results are related to the physical structure variation under different mechanical stretching ratios, which will be discussed in the following sections.

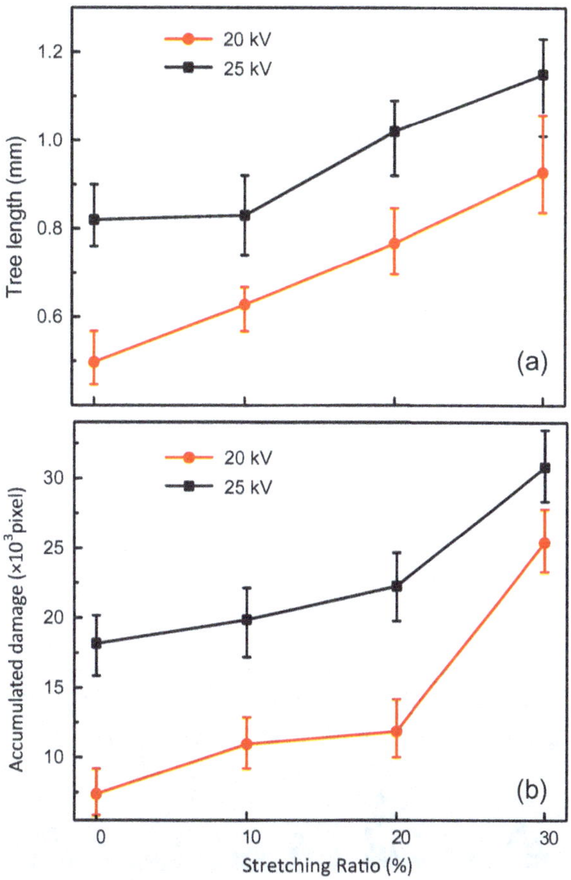

Fig. 7 Tree length (**a**) and accumulated damage (**b**) as a function of different stretching ratios

3.2 *Compressive Stress*

3.2.1 Free Volume and Cohesive Energy

When the EPDM endures the compressive stress, the molecular chains are pressed to reduce inter-molecular distance, leading to the decrease in free volume. The relationship between the fractional free volume and compressive rate is shown in Fig. 8. It notes when the compressive rate increases from 0 to 30%, the FFV is decreased from 15.4% to 4.82%. The FFV becomes smaller as the compressive rate increasing, indicating the motion space of molecular chains is restricted.

Schematic diagram of free volume distribution under different compressive rates is shown in Fig. 9. The changing process of free volume is observed under different compressive rates as shown in the red circle. The large free volume is existed in Fig. 9a, then compressed along the compressive stress in Fig. 9b and divided into

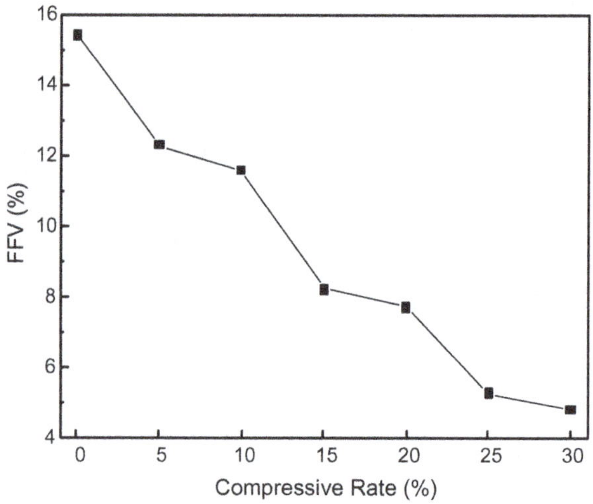

Fig. 8 Dependence of fractional free volume on compressive rate

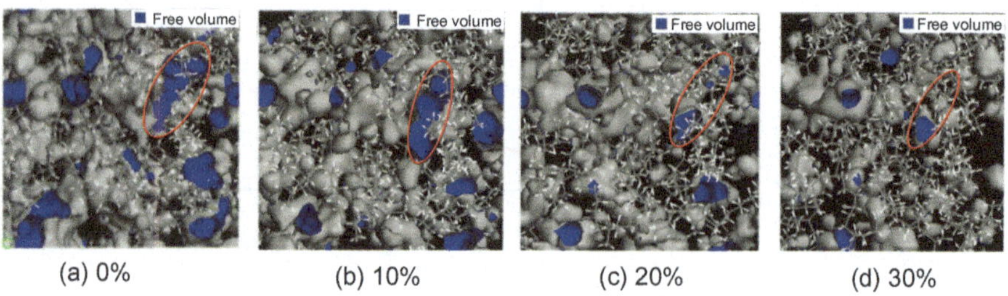

Fig. 9 Schematic diagram of free volume distribution under different compressive rates

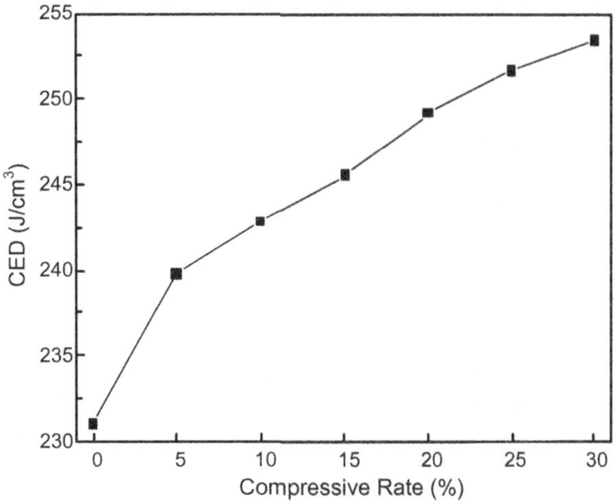

Fig. 10 Dependence of cohesive energy density on compressive rate

some small parts as shown in Fig. 9c, until the molecular chains ultimately occupy the small free volume as shown in Fig. 9d.

Dependence of cohesive energy density on compressive rates is shown in Fig. 10. The CED becomes larger as the compressive rate increases from 0 to 30%, which indicates that more external energy is needed for the crack to grow. Thus, the propagation of the tree channel crack is delayed under higher compressive rate, which restrains the degradation of the materials.

3.2.2 Electrical Tree Characteristics

The different appearance of electrical tree shapes is related to mechanical stress and electric field. With application of 20 and 25 kV, typical morphologies of electrical trees at different compressive rates are shown in Fig. 11(a1–a4) and (b1–b4). In the figure, a1 and b1 are typical bush tree structure while a4 and b4 show the typical branch tree shape. Under higher compressive rate, the electrical tree shows a more sparsely structure and a less destructive area, which obviously indicates that the compressive stress restrains the electrical tree growth.

To investigate the effects of compressive rate on electrical tree morphologies, the appearance probability of typical electrical tree morphologies at different compressive rates is shown in Fig. 12. The electrical tree shape is imaged under different compressive rates after 10 min. At 20 kV, the appearance probability of branch tree increases from 10% to 100%, while the appearance probability of bush tree decreases from 90% to 0 as the compressive rate increases from 0 to 30%. When the pulse amplitude increases up to 25 kV, the tree types also vary from bush tree to branch tree as the compressive rate increasing. The results indicate that the compressive rate affects the propagation and morphology distribution of electrical tree.

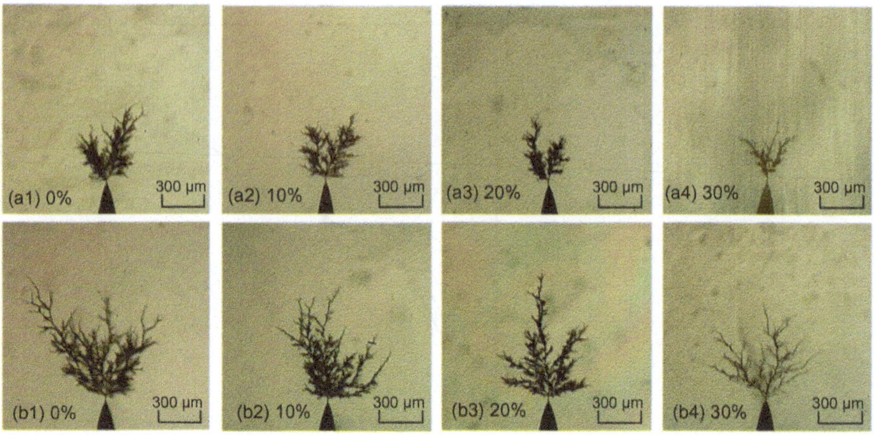

Fig. 11 Typical morphologies of electrical trees at different compressive rates after 10 min

Fig. 12 Appearance probability of typical electrical tree morphologies at different compressive rates after 10 min

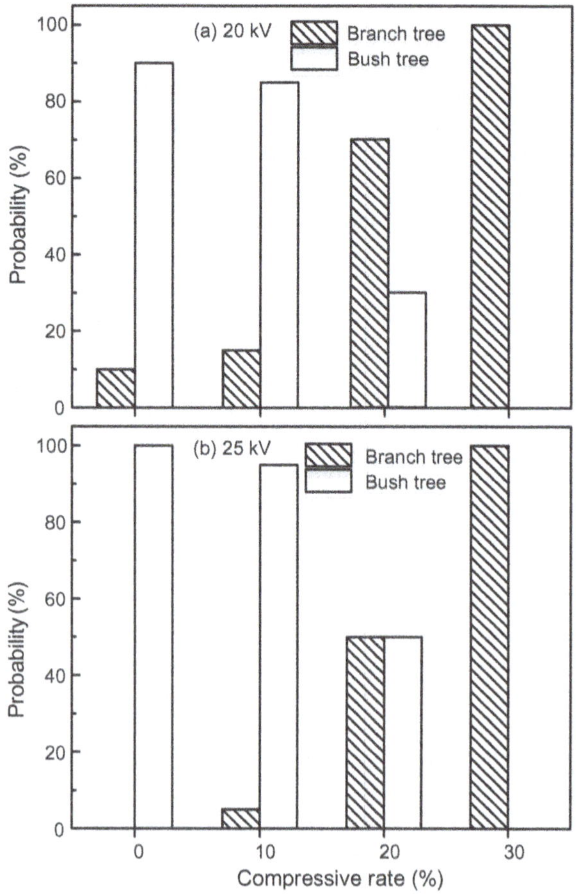

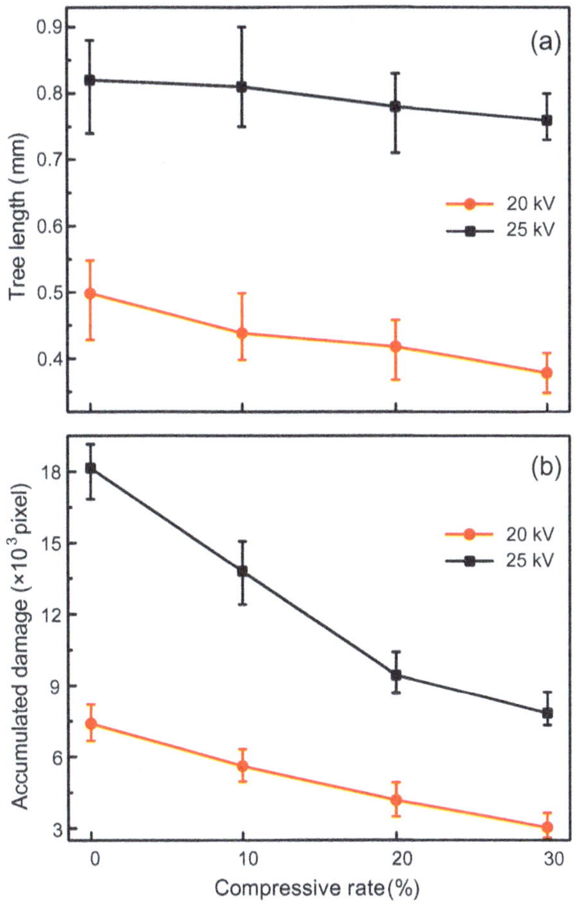

Fig. 13 Tree length (**a**) and accumulated damage (**b**) as a function of different compressive rates after 10 min

Tree length (a) and accumulated damage (b) as a function of different compressive rates after 10 min are shown in Fig. 13.The mean tree length is gradually decreased from 0.5 to 0.38 mm as the compressive rate increases from 0 to 30% at 20 kV, while it is varied from 0.82 to 0.76 mm at 25 kV. The accumulated damage also decreases with increasing the compressive rate, which indicates the compressive stress affects the electrical tree characteristics by restraining the degradation of the EPDM samples.

3.3 Electrical Tree Mechanism with Mechanical Stress

In [22], it was reported that the dielectric properties of polymer were related to the variation of free volume and the movement of molecular chains. In [23], the experimental results supported that the collision impact caused by the accelerated electrons in free volume was the dominant reason of electrical tree initiation. As the dielectric liquids could fill the free volume and cavities of XLPE to reduce the paths for the accelerated electrons acquiring sufficient energy, thus electrical tree inception voltages was increased. According to the free volume theory, the electric breakdown is associated with the longest free path. It is assumed that free volume regions are exponentially distributed in size and that the electric breakdown occurs when electrons in the largest free volume acquire enough energy to move into the neighboring free volume regions [24]. The breakdown criterion is given by the formula (5):

$$\mathrm{W} = \mathrm{e}E_b l_x \tag{5}$$

where W is the obtained energy of the electron in the free volume, E_b is the stressed electric field of the free volume and it is seriously deformed around the needle tip and the tree channel tip, e is the charge of the electron, and l_x is the longest free path of electrons.

As l_x becomes larger, the electron will obtain larger kinetic energy in the free volume. Figure 14 shows the free path of electrons in the free volume without mechanical stretching (a) and with mechanical stretching (b), (c). In Fig. 14a, the molecular chain occupies the free volume, dividing the free volume into two small parts and restraining the electron to pass through the free volume. In Fig. 14b, the curly molecular chain is stretched along the direction of mechanical stretching, inducing the small free volume merges into a large one. When the mechanical stretching stress continues to increase as shown in Fig. 14c, the curly molecular chains are straightened along the direction of the mechanical stretching, leaving an enlarged free volume. With the mechanical stretching ratio inceasing, the free paths become larger for injected electrons to obtain higher energy to break the molecular chain, accelerating the electrical tree process.

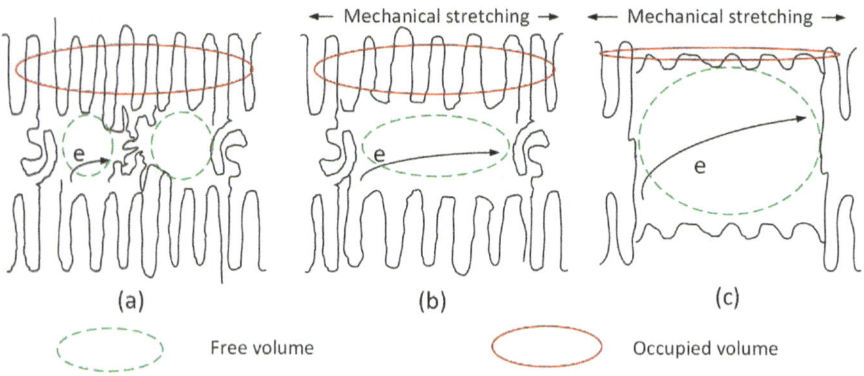

Fig. 14 Free path of electrons in the free volume without mechanical stretching (**a**) and with mechanical stretching (**b**), (**c**)

In the process, the electron with sufficient energies can extend the preexisting defects and eventually increase the defect density to the level of forming tree channels. To break the C–C bond, the energy is reported to be about ~3.8 eV. According to Mason's formula [25], the electric field is about 1690 kV/mm at the needle tip, l_x is estimated to be 2.25 nm for disruption of the C–C bond calculated by Eq. (5). It is reasonably proposed that the electrical tree inception is most likely caused by the enlarged l_x, which is increased from ~1 to ~3 nm as the stretching ratio increases from 0 to 30%.

The channel crack growth of electrical tree is related to chemical bonds breaking and molecular chains splitting, which are caused by electromechanical, thermal and chemical damage of the polymers. The material weakening behavior is considered as the result of bond density reduction induced by the local deformation, which is caused by mechanical stretching. When the EPDM is yielded to mechanical stretching, the bond density is reduced as the local deformation increases, which induces a smaller maximum cohesive strength [26]. Therefore, the traction-displacement curve is shifted from $f(x)$ to $f(x1)$ and the cohesive strength is varied from σ_C to σ_{C1} under mechanical stretching in Fig. 15. The crack toughness with mechanical stretching is smaller than that without mechanical stretching, which indicates less energy is needed for the crack to grow. The cohesive strength of EPDM is also reduced with mechanical stretching, thus the same partial gas pressure, released during partial discharge process, can generate more serious accumulated damage in the samples.

However, under the compressive stress, the free volume shows a converse trend. Under higher compressive stress, the molecular chain occupies the free volume and

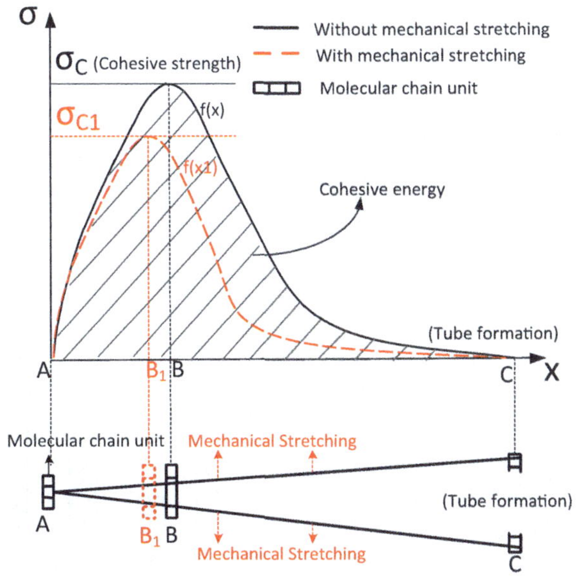

Fig. 15 A typical traction-displacement curve of the crack in cohesive zone

reduces the free path of the electron, which weakens the collision influence of electron during the partial discharge process. While CED becomes larger as the EPDM endures higher compressive stress, thus the energy threshold for the growth of tree channel crack is improved. The two factors restrain the electrical tree propagation in EDPM under higher compressive stress.

3.4 High Temperature

3.4.1 Electrical Tree Characteristics

EPDM is in high elastic state, and it is an amorphous polymer at room temperature. The ethylene and propylene monomers are randomly arranged on the main chain. The physical structure and electrical properties are different from the semi-crystalline polymers of polyethylene and polypropylene, which are with the regular arrangement of molecular chains. As the temperature rises, the main molecular chains of EPDM are unable to move freely, but the thermal motion of the molecular chain has been able to overcome the internal rotation barrier and change the physical structure by adjusting the C–C single bond in the main chain. Therefore, the charges are enable to easily break the molecular chains, accelerating the degradation process of electrical tree.

Figure 16 shows the relationship between the electrical tree inception probability and temperature. The tree inception probability is defined the proportion of samples with electrical tree structures in 20 samples. It can be seen that the inception probability of electrical tree is changed from 50% to 90% as the temperature is increased from 30 °C to 120 °C. The inception probability of electrical

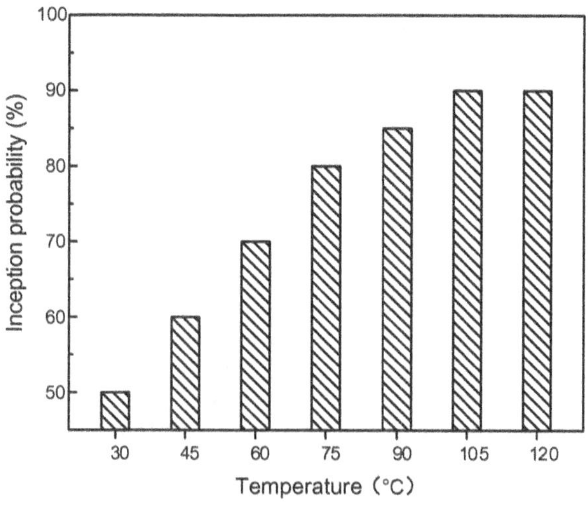

Fig. 16 Relationship between the inception probability and temperature

Fig. 17 Typical tree morphologies at different temperatures

tree increases monotonously with temperature and gradually reaches a steady state. The high temperature can promote the initiation of electrical tree and accelerate the degradation process of insulation materials.

The electrical tree structures are depended on the temperature. Figure 17 shows typical tree morphologies at different temperatures. When the temperature is 30 °C, the electrical tree channel is sparsely distributed, and the degradation area is small. When the temperature is 90 °C, the electrical tree channel is the longest and densely distributed, and the degradation area becomes larger. Compared with 90 °C, the electrical tree channel is a little shorter with 120 °C.

With the same temperature condition, the electric tree morphologies are different after the same time, and they can be divided into branch tree and bush tree structures. The relation between the electrical tree morphologies and temperatures is shown in Fig. 18. When the temperature is 30 °C, the probability of branch tree is 85%, and the probability of bush tree is 15%. When the temperature is 60 °C, the probability of branch tree and bush tree is 45% and 55%, respectively. When the temperature is 120 °C, the probability of bush tree is increased to 100%. With the temperature increasing, the molecular chains are relaxed with the mechanical property reducing, which is beneficial to generate new electrical tree channels around the needle tip. Therefore, bush tree is easily formed with higher temperature.

The relation between the tree length and temperatures is shown in Fig. 19. The tree length becomes longer with the temperature increasing. When the temperature is 90 °C, the tree length reaches the maximum value. When the temperature increases to 120 °C, the tree length becomes smaller. The reason is that the probability of bush tree increases at the temperature of 120 °C, which is with a shorter tree length. For HVDC and HVAC cable accessories, the high temperature promotes electrical tree propagation, which reduces the reliability insulation materials.

3.4.2 Electrical Tree Mechanism with High Temperature

The EPDM are amorphous polymers at room temperature, they are in the elastic state with a low elastic modulus [27]. With the increase of temperature, the enhanced movement causes local molecular chain relaxation, thus the electrical and mechanical strength are decreased. Meanwhile, the free volume will be expanded due to the local molecular chain relaxation, and the mean free path of hot electron increases. Thus,

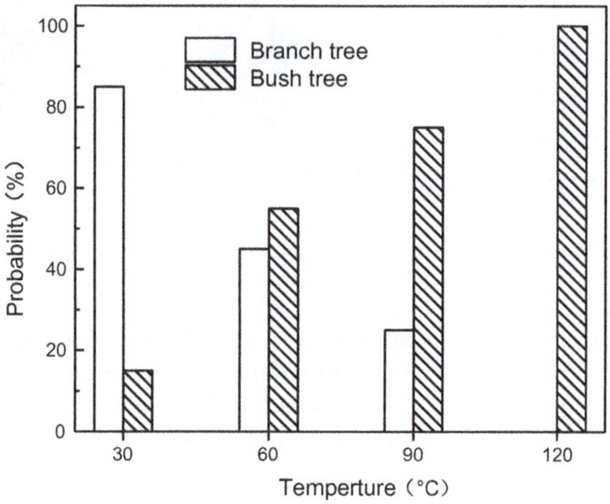

Fig. 18 Relation between the electrical tree morphologies and temperatures

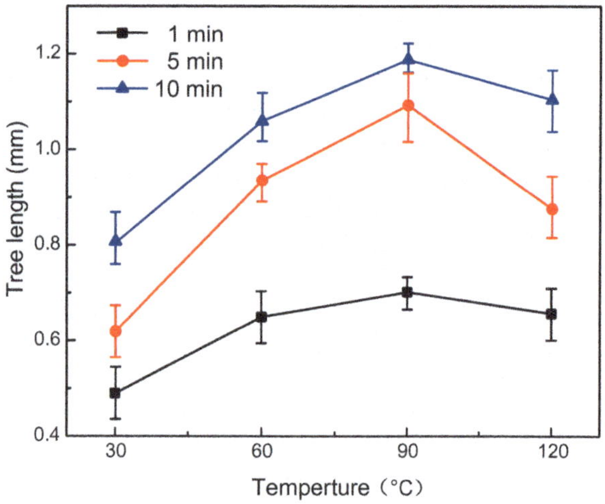

Fig. 19 Relation between the tree length and temperatures

the hot electrons can obtain more energy to collide with molecular chains, resulting in chain scission and promoting electrical tree growth.

Figure 20 shows the trap density and trap level with different temperatures by the modified isothermal discharge current method [28]. As the temperature increases, the trap density and trap level become larger. At 30 °C, the trap centre is located at 0.88 eV, and the trap density is 2.7×10^{20} eV^{-1} m^{-3}. When the temperature is 120 °C, the trap centre is located at 1.16 eV, and the trap density is 6.3×10^{20} eV^{-1} m^{-3}. With high temperature, the molecular chain movement is enhanced, and the carrier activation energy is increased. The charges captured by deep trap are easily escaped with the thermal-electrical coupling field. Thus, the trap level and trap density become larger with the increasing temperature. As the charges in deep traps are easily

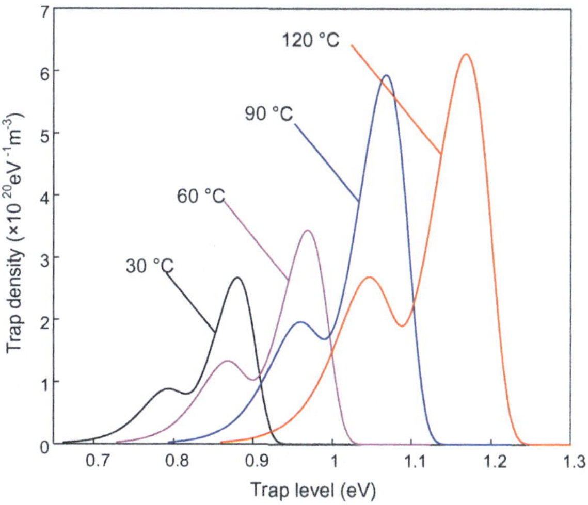

Fig. 20 Trap distribution with different temperatures

detrapped with the high temperature, the carrier density is also increased, which is easy to break the molecular chains of the insulation materials and generate electrical tree channels.

Moreover, the adjustment of molecular chains will change the potential energy distribution and charge transport behaviors as shown in Fig. 21 [29], which affects the electrical degradation of insulation materials. High temperature also affects the space charge dynamics, the ionic dissociation, charge mobility and electrical conductivity are enhanced [30]. Because of the increased charge mobility and increased carrier density at an elevated temperature, more hot electrons are generated. Thus, the partial

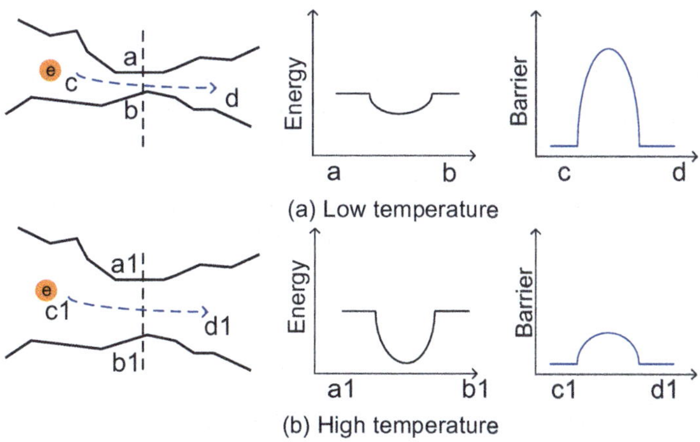

Fig. 21 The relation between potential energy, potential barrier and charge transport in intermolecular chains

discharges increase in the insulation materials, promoting electrical tree initiation and propagation.

4 Conclusion

Effects of temperature and mechanical stress on electrical tree is investigated in EPDM. Experimental results reveal that tree propagation characteristics are depended on the temperature environment and mechanical stress. The following conclusions are obtained in the chapter.

(1) The electrical tree propagation characteristics are associated with the yielded mechanical stretching of the test samples. Tree length and accumulated damage are increased when the stretching ratio is elevated from 0 to 30%, implying a larger electrical tree destruction under higher stretching ratio. However, tree length and accumulated damage are reduced when the compressive rate is elevated from 0 to 30%, indicating a smaller electrical tree destruction under higher compressive ratio
(2) When EPDM is yielded to the mechanical stretching, two small free volume units can merge into a large one. The enlarged free volume not only extends the free path of the hot electron to enhance the collision with the molecular chain. CED becomes smaller as the mechanical stretching increases, which indicates a reduction of the binding forces between different molecular chains. Thus, main molecular chains are either directly fractured or easily broken by the high electrical field, accelerating electrical tree inception and propagation. However, the free volume and CED show a converse trend with the compressive stress, which restrain electrical tree inception and propagation.
(3) The high temperature promotes the initiation process of electrical tree. When the temperature changes from 30 °C to 120 °C, the inception probability of electric tree increases from 50% to 90%. The temperature is closely related to the morphology of electrical tree, and the probability of bush tree increases at higher temperature. The movement ability of molecular chain is improved with high temperature, which leaves a large free volume for the charges to obtain sufficient energy. The charges are easily detrapped in deep traps, then the charge transport ability is enhanced, which can accelerate electrical tree deterioration process.

References

1. B.X. Du, H. Xu, J. Li, Effects of mechanical stretching on space charge behaviors of PP/POE blend for HVDC cables. IEEE Trans. Dielectr. Electr. Insul. **24**(3), 1438–1445 (2017)
2. X. Wang, C.C. Wang, K. Wu et al., An improved optimal design scheme for high voltage cable accessories. IEEE Trans. Dielectr. Electr. Insul. **21**(1), 5–15 (2014)

3. M.H. Ahmad, N. Bashir, H. Ahmad, A.A. Abd Jamil, A.A. Suleiman, An overview of electrical tree growth in solid insulating material with emphasis of influencing factors, mathematical models and tree suppression, TELKOMNIKA Indonesia. J. Electr. Eng. **12**(8), 5827–5846 (2014)
4. B.R. Varlow, Mechanical prestressing improves electrical strength. IEEE Electr. Insul. Mag. **18**(1), 12–25 (2002)
5. S.A. Al-Ghamdi, B.R. Varlow, Treeing in mechanically pre-stressed electrical insulation. IEEE Trans. Dielectr. Electr. Insul. **11**(4), 130–135 (2004)
6. K. Nakanishi, S. Hirabayashi, Y. Inuishi, Phenomena and mechanisms of tree inception in Epoxy Resins. IEEE Trans. Electr. Insul. EI-14(6), 306–314 (1979)
7. J.P. Crine, Influence of electro-mechanical stress on electrical properties of dielectric polymers. IEEE Trans. Dielectr. Electr. Insul. **12**(4), 791–800 (2005)
8. B.R. Varlow, G.J. Malkin, Electrical treeing in mechanically prestressed insulation. IEEE Trans. Dielectr. Electr. Insul. **7**(6), 721–724 (2000)
9. X.Q. Zheng, G. Chen, Propagation mechanism of electrical tree in XLPE cable insulation by investigating a double electrical tree structure. IEEE Trans. Dielectr. Electr. Insul. **15**(3), 800–807 (2008)
10. N. Shimizu, Y. Shibata, K. Ito, K. Imai, M. Nawata, Electrical tree at high temperature in XLPE and effect of Oxygen. IEEE Conf. Electr. Insul. Dielectr. Phenomena (CEIDP) 329–332, (2000)
11. R. Bono, C. Gemnie, F. Guastavino, The effects of temperature on the tree growth phenomena and relevant PD. IEEE Conf. Electr. Insul. Dielectr. Phenomena (CEIDP), 69–72 (1995)
12. X.R. Chen, Y. Xu, X.L. Cao et al., Electrical treeing behavior at high temperature in XLPE cable insulation samples. IEEE Trans. Dielectr. Electr. Insul. **22**(5), 2841–2851 (2015)
13. Y.X. Zhou, R. Liu, F. Hou et al., Morphology of electrical trees in silicon rubber. J. Electrostat. **71**(3), 440–448 (2013)
14. M. Kosaki, M. Nagao, N. Shimizu et al., Solid insulation and its deterioration. Cryogenics **38**(11), 1095–1104 (1998)
15. G. Chen, C.H. Tham, Electrical treeing characteristics in XLPE power cable insulation in frequency range between 20 and 500 Hz. IEEE Trans. Dielectr. Electr. Insul. **16**(1), 179–188 (2009)
16. B.X. Du, L.W. Zhu, Electrical tree characteristics of XLPE under repetitive pulse voltage in low temperature. IEEE Trans. Dielectr. Electr. Insul. **22**(4), 1801–1808 (2015)
17. X. Wang, C.C. Wang, K. Wu, D.M. Tu, S. Liu, J.K. Peng, An improved optimal design scheme for high voltage cable accessories. IEEE Trans. Dielectr. Electr. Insul. **21**(1), 5–15 (2014)
18. K.L. Yin, D.H. Zou, J. Zhong, D.J. Xu, A New Method for calculation of elastic properties of anisotropic material by constant pressure molecular dynamics. Comput. Mater. Sci. **38**(3), 538–542 (2007)
19. J. Diani, K. Gall, Molecular dynamics simulations of the shape-memory behaviour of polyisoprene. Smart Mater. Struct. **16**(5), 1575–1583 (2007)
20. J. Artbauer, Electric strength of polymers. J. Phys. D: Appl. Phys. **29**, 446–456 (1996)
21. L.A. Dissado, J.C. Fothergill, *Electrical Degradation and Breakdown in Polymers* (London, UK, Peter Peregrinus for IEEE, 1992)
22. S.T. Li, W.W. Wang, S.H. Yu, H.G. Sun, Influence of hydrostatic pressure on dielectric properties of Polyethylene/Aluminum Oxide Nanocomposites. IEEE Trans. Dielectr. Electr. Insul. **21**(21), 519–528 (2014)
23. N. Shimizu, H. Tanaka, Effect of liquid impregnation on electrical tree initiation in XLPE. IEEE Trans. Dielectr. Electr. Insul. **8**(2), 239–243 (2001)
24. M. Roy, J.K. Nelson, R.K. MacCrone, L.S. Schadler, C.W. Reed, R. Keefe, W. Zenger, Polymer nanocomposite dielectrics—the role of the interface. IEEE Trans. Dielectr. Electr. Insul. **12**(4), 629–343 (2005)
25. J.H. Mason, Breakdown of solid dielectrics in divergent fields. Proc. IEE-Part C: Monogr. **102**(2), 254–263 (1955)
26. Z.N. Zhang, X.R. Ge, Micromechanical consideration of tensile crack behavior based on virtual internal bond in contrast to cohesive stress. Theoret. Appl. Fract. Mech. **43**(3), 342–359 (2005)

27. B.X. Du, T. Han, J.G. Su, Effect of low temperature on tree characteristics in silicone rubber with different power frequency. IEEE Trans. Dielectr. Electr. Insul. **21**(4), 1680–1686 (2014)
28. J.G. Simmons, M.C. Tam, Theory of isothermal currents and the direct determination of trap parameters in semiconductors and insulators containing arbitrary trap distributions. Phys. Rev. B **7**(8), 3706–3713 (1973)
29. J.P. Jones, J.P. Llewellyn, T.J. Lewis, The contribution of field-induced morphological change to the electrical aging and breakdown of polyethylene. IEEE Trans. Dielectr. Electr. Insul. **12**(5), 951–966 (2005)
30. Y.L. Chong, G. Chen, Y.F.F. Ho, Temperature effect on space charge dynamics in XLPE insulation. IEEE Trans. Dielectr. Electr. Insul. **14**(1), 65–76 (2007)

Theoretical Model and Suppressing Method of Interface Charge Accumulation in HVDC Cable Accessory: A Review

Jin Li, Chenlei Han, Boxue Du, and Tatsuo Takada

Abstract HVDC cable has been widely used with the development of flexible HVDC transmission. Cable accessories are the weakest part of HVDC cables, while the most serious problem of DC cable joint is the uneven distribution of electric field caused by space charge accumulation, which may lead to the ageing and failure of joint insulation, especially at the interface between different materials. In this paper, the progress of the interface charge research of the HVDC cable accessories is discussed, the existing theoretical model and the suppression method of the interface charge are analysed, and the research trend of the interface charge of the accessories in the future is prospected. The results show that space charges accumulate at the interface between different material are closely related to temperature, applied electric field and working conditions such as polarity reversal, etc.; the Maxwell-Wagners-Sillars theory and bipolar charge transport model are discussed in detail, and the method of using quantum chemical theory to calculate and analyse interface charges is introduced; the methods of interface charge suppression based on geometry design, surface modification, field grading materials and material selection are analysed and compared. The summary and conclusion of these research results can provide reference for the future research on the interface charge behaviours and suppression methods of HVDC cable accessories

1 Introduction

The innovation of next-generation power electronics technology promotes the development of large-scale energy transmission by high voltage direct current (HVDC), which is built to interconnect networks composed of renewable energy sources with the main power grid or regions of consumption [1–3]. Particularly, as the major

J. Li · C. Han · B. Du (✉)
School of Electrical and Information Engineering, Tianjin University, Tianjin 300072, China
e-mail: duboxue@tju.edu.cn

T. Takada
Tokyo City University, Tokyo 158-8557, Japan

B. Du (ed.), *Polymer Insulation Applied for HVDC Transmission*,
https://doi.org/10.1007/978-981-15-9731-2_12

equipment in the HVDC transmission network, the HVDC cable has been one of the best options for long-distance large-capacity transmission and urban power supply, which expanded the application range of polymeric insulated cables considering the great advantages of their lower losses and unique practical feasibility [4]. The insulation properties of cable accessories are vital for the safe operation of HVDC cable transmission networks, as well as for the responsibility of HVDC transmission systems [5–7].

It is generally known that the electric field distribution under AC conditions is basically capacitive and depends on the permittivity reciprocal of the material, which is insensitive to the electric field or ambient temperature [8]. However, it becomes resistive under DC conditions, and the conductivity is drastically dependent on the temperature and exhibits nonlinear characteristics under a high electrical field [9, 10]. Figure 1 exhibits the schematic cross section of an HVDC cable joint. Under the condition of a DC electric field, the space charge at the interface between dielectrics is likely to accumulate because of the various characteristics of different dielectric materials, and this accumulation may accelerate the degradation of HVDC cable accessory systems under combined stresses [11, 12]. The complex geometry and doping of fillers coupled with nonlinear effects render the interface charge performance difficult to evaluate in operation [13]. To alleviate this problem, it is necessary to establish an accurate theoretical model and evaluate the interface charge behaviour of HVDC cables by using specific measurement techniques and suppression methods, which is of great significance for insulation design of HVDC cable accessories [14]. In this paper, interface charge behaviours together with Maxwell-Wagners-Sillars (MWS) theory, bipolar charge transport model, quantum chemical model and suppression methods will be reviewed to better understand and regulate the interface charge behaviours in HVDC cable accessories.

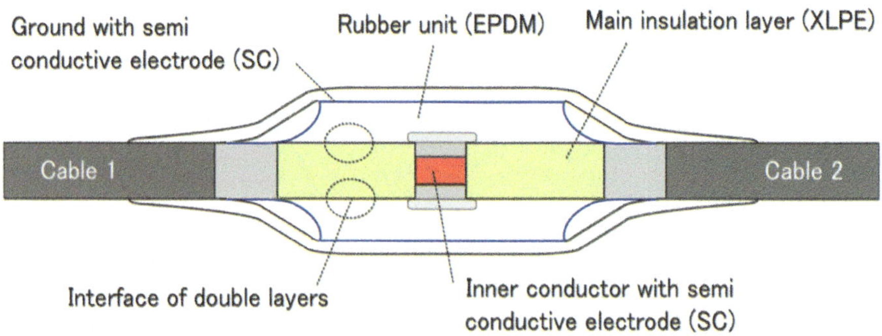

Fig. 1 Cross-sectional view of an HVDC cable joint. [6] Copyright Clearance Center's RightsLink® License Number: 4857331488009

2 Interface Charge Accumulation

2.1 Space Charge Characteristics in Double-Layer Dielectrics

The double-layer dielectric structure composed of polyethylene and rubber is a classical design structure of accessories for HVDC cables [6]. A typical interface charge distribution in EPDM/LDPE double-layer dielectrics measured by the PEA method under an electric field ranging from 10 to 60 kV/mm is shown in Fig. 2. The negative charge accumulates at the EPDM/LDPE interface under a positive applied electric

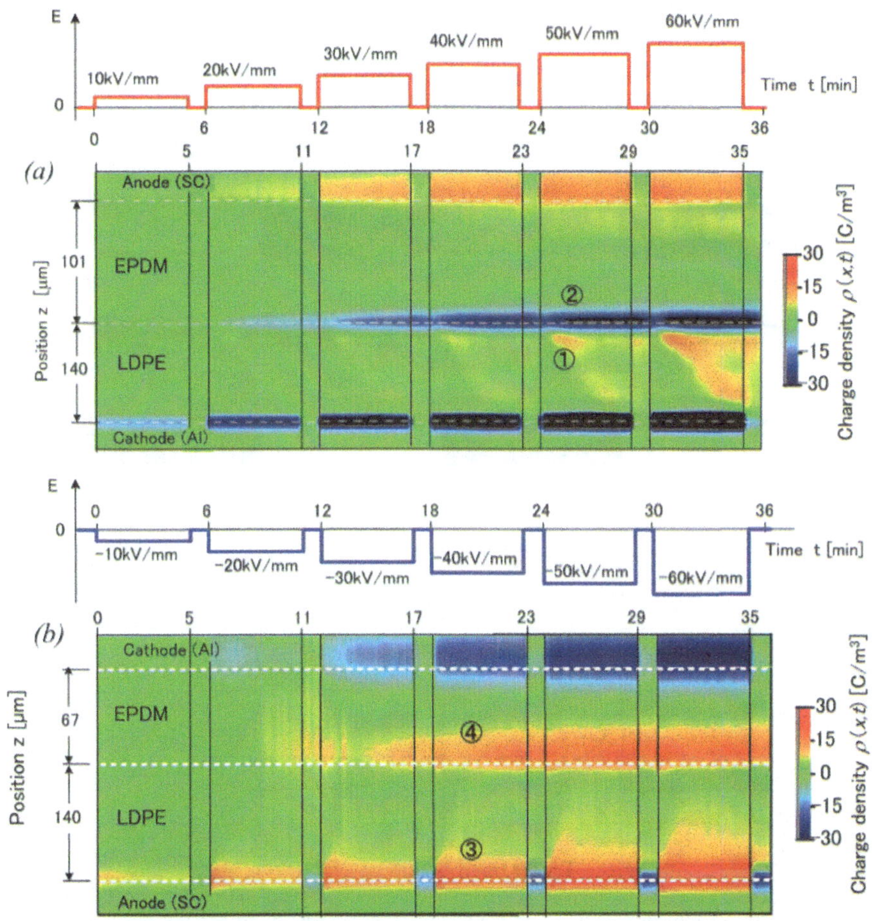

Fig. 2 Space charge distribution in the EPDM/LDPE double-layer structure under various applied electric fields. [6] Copyright Clearance Center's RightsLink® License Number: 4857331488009

field (Fig. 2a) above 30 kV/mm (shown in blue, ②). The positive charge in the LDPE component drifts to the cathode under the external electrical stress (shown in yellow, ①), while the negative charge remains at the interface. In addition, under the negative polarity electric field (Fig. 2b) of 20 kV/mm and above, the positive charge is distributed at the anode and interface (shown in yellow, ③ and ④).

2.2 *Interface Charge Characteristics Under Voltage Polarity Reversal*

Space charge tends to accumulate at the interface between materials as shown above. Accordingly, polarity reversal, which occurs occasionally in HVDC cable systems, is extremely important as the electric field in the space charge region will become higher instantaneously, which may worsen the insulation and even lead to its breakdown [8]. Space charge behaviours in EPDM/LDPE samples at 40 kV/mm before and after voltage reversal are shown in Fig. 3a, while Fig. 3b shows the temporal variation of

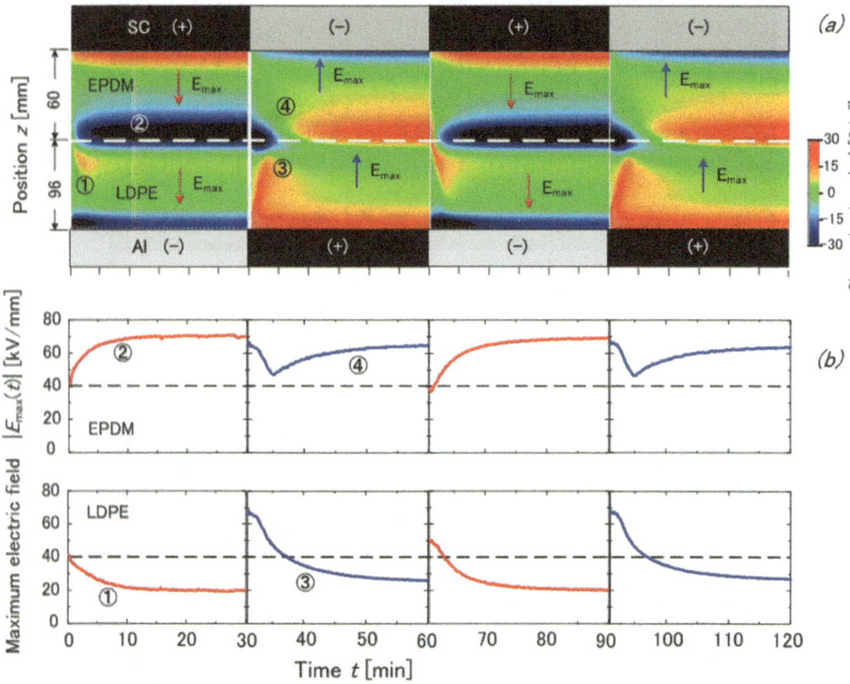

Fig. 3 **a** Space charge distribution and **b** maximum electric fields in EPDM and LDPE layers under voltage polarity reversals. [6] Copyright Clearance Center's RightsLink® License Number: 4857331488009

the maximum electric field intensity of EPDM and LDPE after 30 min of poling for each series voltage reversal [6].

Obviously, regardless of the polarity of the applied voltage, the maximum electric field on the EPDM side occurs over time.

As shown, space charge behaviours in double-layer media exhibit four important characteristics. These are as follows:

(1) Positive charges are generated at the interface and move faster in the medium than negative charges at the cathode (Fig. 3a, ①).
(2) The negative charge generated at the interface is trapped on the side of EPDM and has difficulty migrating (Fig. 3a, ②).
(3) The positive charge injected from the anode is packet-like and drifts rapidly to the interface (Fig. 3a, ③).
(4) The injected positive charge accumulates at the interface and is trapped on the side of the EPDM again (Fig. 3a, ④).

3 Theoretical Model of Interface Charge

3.1 *Maxwell-Wagners-Sillars Theory*

Maxwell-Wagners-Sillars (MWS) theory, as the most common mechanism for explaining interface charge accumulation along with the stress time, takes interfacial polarization as the core [15]. Concretely, the HVDC cable accessory is simplified into a dual-layer model consisting of two dielectrics, as shown in Fig. 4. +V is the external applied DC voltage, and Σs is the space charge density at the interface.

According to the Maxwell equations:

$$\sigma_1 E_1 = \sigma_2 E_2 \tag{1}$$

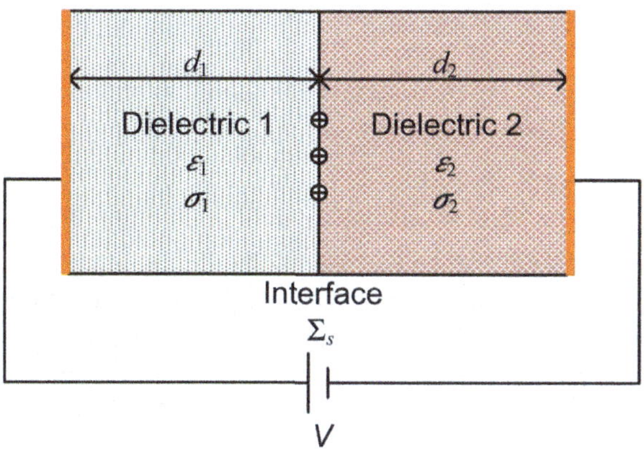

Fig. 4 Interfacial polarization in the dual-layer dielectrics according to the Maxwell-Wagner-Sillars model

$$d_1 E_1 + d_2 E_2 = -U \tag{2}$$

$$\varepsilon_1 E_1 - \varepsilon_2 E_2 = \Sigma_s \tag{3}$$

where d_1 and d_2 are the thickness of dielectric 1 and 2, ε_1 and ε_2 are the permittivity and σ_1 and σ_2 are the conductivity. Then, the interface charge density Σs of the dual-layer dielectrics with differential ε and σ can be transformed into Eqs. (9) and (4):

$$\Sigma_s = \frac{\varepsilon_1 \cdot \sigma_2 - \varepsilon_2 \cdot \sigma_1}{\sigma_1 \cdot d_2 + \sigma_2 \cdot d_1} \cdot U \cdot (1 - e^{-t/t_c}) \tag{4}$$

$$t_c = \frac{(d_1 \cdot \varepsilon_2 + d_2 \cdot \varepsilon_1)}{(d_1 \cdot \sigma_2 + d_2 \cdot \sigma_1)} \tag{5}$$

where t_c is the characteristic time depending on the physical size and dielectric properties of the two materials [16]. It can be speculated from Eq. (9) that more space charges will be produced if the mismatch in the conductivity and permittivity between the two dielectric materials is higher, and the interface charge build-up procedure is decided by the time constant tc. In other words, the polarity and quantity of the interface charge are functions of the electrical field and ambient temperature, which can be calculated from the measured permittivity and conductivity values using the MWS theory.

According to the measured conductivity value and MWS theory, Vu et al. predicted the functional relationship between the quantity of interfacial charges in double-layer XLPE/rubber media and the applied voltage and temperature in a DC electric field [15]. Figure 5 illustrates the relationship between the polarity of the calculated

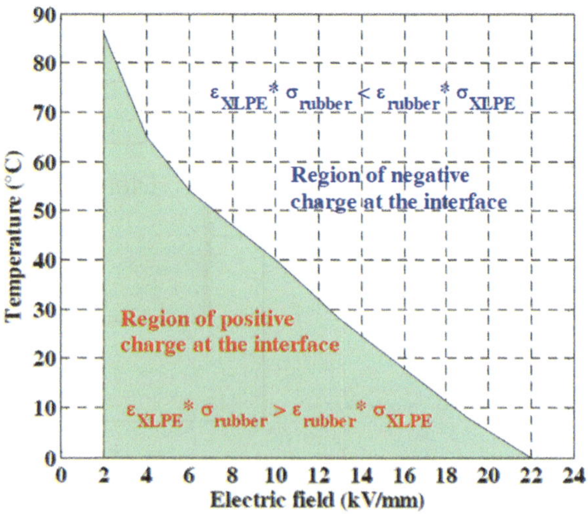

Fig. 5 Predicted landscape for the sign of the interfacial charge for the dual-layer system of a cable accessory [15]

interface charge and temperature and the average applied electric field under steady-state conditions.

For electric fields ranging from 2 to 22 kV/mm and temperatures from 0 to 86 °C, there are two distinct positive and negative charge regions at the interface between the two dielectrics, separated by the dependence demarcation line of the temperature and electric field.

3.2 *Bipolar Charge Transport Model*

In solid dielectrics, carriers are governed by a set of self-consistent equations as follows [17, 18]:

- Current continuity equation:

$$\partial q_f^-/\partial t + \nabla \cdot \mathbf{j}^- = s_f^- \tag{6}$$

$$\partial q_f^+/\partial t + \nabla \cdot \mathbf{j}^+ = s_f^+ \tag{7}$$

$$\partial q_t^-/\partial t = s_t^- \tag{8}$$

$$\partial q_t^+/\partial t = s_t^+ \tag{9}$$

where qf− and qf+ are the charge densities of the mobile negative and positive carriers, C/m^3; qt− and qt+ are the charge densities of the trapped negative and positive carriers, C/m^3; and j− and j+ are the conduction current densities of the mobile negative and positive carriers, A/m^2. The charge trapping-detrapping and recombination processes are considered in the following source terms, where sf− and sf+ are the source terms of the mobile negative and positive carriers, C/(m^3·s); and st- and st+ are the source terms of the trapped negative and positive carriers, C/(m^3·s).

$$s_f^- = -P_{tr} \cdot q_f^-(1 - q_t^-/Q_t^-) + P_{de} \cdot q_t^- - R \cdot q_f^-(q_f^+ + q_t^+) \tag{10}$$

$$s_f^+ = -P_{tr} \cdot q_f^+(1 - q_t^+/Q_t^+) + P_{de} \cdot q_t^+ - R \cdot q_f^+(q_f^- + q_t^-) \tag{11}$$

$$s_t^- = P_{tr} \cdot q_f^-(1 - q_t^-/Q_t^-) - P_{de} \cdot q_t^- - R \cdot q_t^- \cdot q_f^+ \tag{12}$$

$$s_t^+ = P_{tr} q_f^+(1 - q_t^+/Q_t^+) - P_{de} q_t^+ - R q_t^+ q_f^- \tag{13}$$

where Qt− and Qt+ are the trap densities, C/m^3; Ptr and Pde are the charge trapping and detrapping coefficients, 1/s; and R is the recombination coefficient, $m^3/(C{\cdot}s)$

- Charge transport equation:

$$\mathbf{j}^- = q_f^- \cdot \mu_- \cdot \mathbf{E} - D_- \cdot \nabla q_f^- \tag{14}$$

$$\mathbf{j}^+ = q_f^+ \cdot \mu_+ \cdot \mathbf{E} - D_+ \cdot \nabla q_f^+ \tag{15}$$

where μ− and μ+ are the mobility of negative and positive carriers, $m^2/(V{\cdot}s)$; D− and D+ are the diffusion coefficients of negative and positive carriers, m^2/s; and E is the electric field strength, V/m.

- Poisson's equation:

$$-\nabla^2 V = (-q_f^- - q_t^- + q_f^+ + q_t^+)/(\varepsilon_0 \varepsilon_r) \tag{16}$$

$$\mathbf{E} = -\nabla V \tag{17}$$

where V is the electric potential, V; ε_0 is the permittivity of vacuum, F/m; and ε_r is the relative permittivity of a dielectric.

- Boundary conditions:

At the boundaries between the dielectric and the metal electrode, the normal current densities can be described as:

$$-\mathbf{n} \cdot \mathbf{j}^- = -\alpha \cdot J_s^- - (1 - \alpha) \cdot \mathbf{n} \cdot \mathbf{E} \cdot \mu_- \cdot q_f^- \tag{18}$$

$$-\mathbf{n} \cdot \mathbf{j}^+ = (1 - \alpha) \cdot J_s^+ - \alpha \cdot \mathbf{n} \cdot \mathbf{E} \cdot \mu_+ \cdot q_f^+ \tag{19}$$

where α = 1 at the cathode and α = 0 at the anode; and Js− and Js+ are the Schottky injection current densities, which can be described as follows:

$$J_s = AT^2 \cdot \exp[(-\phi + \sqrt{E \cdot e^3/4\pi\varepsilon_0\varepsilon_r})/k_B T] \tag{20}$$

where A is the Richardson constant, $C/(m^2{\cdot}s{\cdot}K^2)$; ϕ is the contact potential barrier between the dielectric and the metal electrode, eV; kB is the Boltzmann constant; and T is the thermal temperature, K.

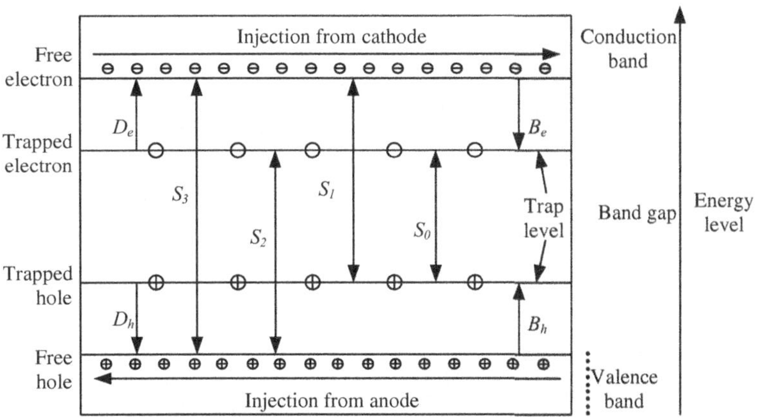

Fig. 6 Mechanism of bipolar charge transportation in a dielectric

The simulation results show that the interface charge density in the HVDC cable accessory is increased by the deep surface trap energy level, high interface barrier and large carrier mobility difference [19]. The accumulation of interface charge causes distortions in the electrical field distribution (Fig. 6).

3.3 Interface Charge Model Based on Quantum Chemical Calculations

The Gaussian 09 quantum chemical program can be used to calculate the energy bands of LDPE and EPDM, while B3LYP was chosen as the exchange correlation function and 6–31 G* was chosen as the basis set [20, 21]. To assist in the analysis of the complex space charge behaviour observed at the interface between the two dielectrics, the electronic energy levels of the double-layer EPDM/LDPE system are calculated [6]. The calculated electronic energy levels in Fig. 7 are obtained in the case of electrode contact. The results show that when EPDM and LDPE are in contact, the dielectric system will polarize and be exposed to voltage reversal. In the absence of applied voltage, considering that the Fermi level of LDPE (EF = −3.0 eV) is lower than that of EPDM (EF = −2.6 eV), positive charges will accumulate on the side of LDPE (Fig. 7b ③), and negative charges will accumulate on the side of EPDM (Fig. 7b ④). In Fig. 7a, the electronic energy levels of the double dielectrics are shown when the positive voltage is applied to the SC electrode. Positive charges can drift more easily through shallow traps in LDPE. However, negative charges in the double-layer dielectrics can be trapped in deep traps on the surface of EPDM (Fig. 7a ②, electron trap depth $\phi te = 0.9-1.1$ eV), so it is not easy to drift through EDPM.

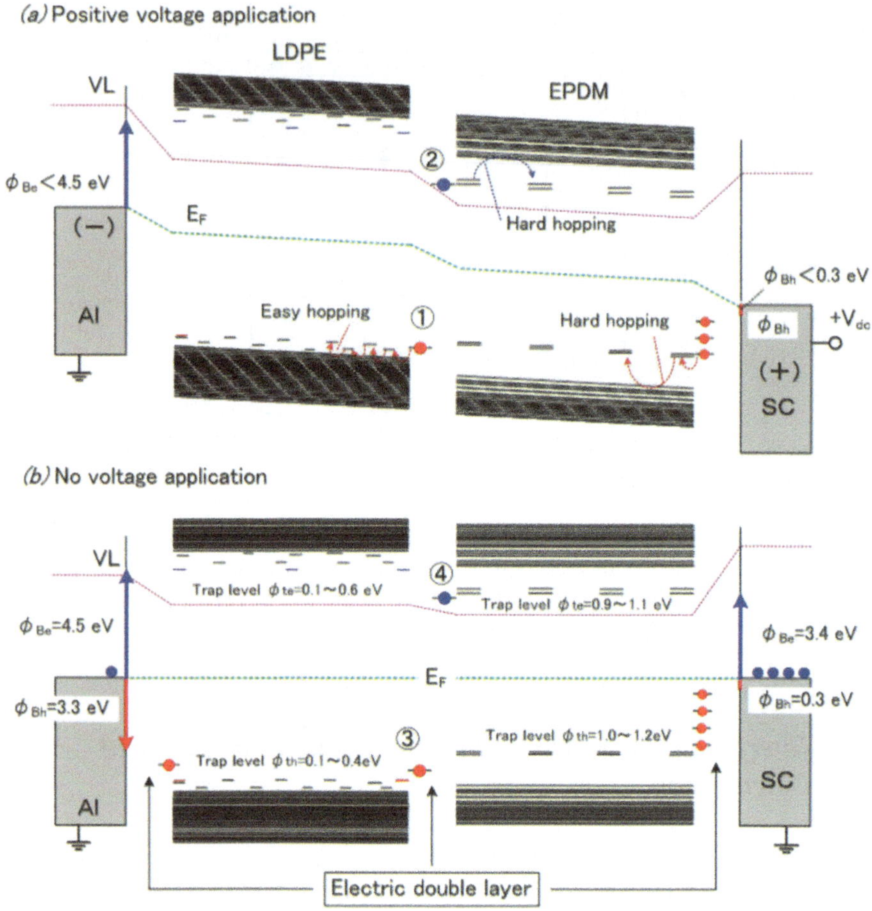

Fig. 7 Energy levels within the double-layer LDPE/EPDM system with different voltage application. [6] Copyright Clearance Center's RightsLink® License Number: 4857331488009

The polarity of the injected interface charge is consistent with that of the voltage applied on the LDPE side. Compared with the negative charge, the polarity of the injected charge is dominated by the positive charge. The difference in the distribution of holes and electron traps leads to polar-dependent charge motion, which is determined by means of QCC. Deep traps in EPDM materials are the main reason for the accumulation of interfacial charges on the LDPE side. With the reversal of the applied voltage, the residual interfacial charge at the interface is neutralized by the heteropolar charge injected from the LDPE side. After that, additional injected charge will be trapped at the interface.

4 Suppression Methods

Interface charge behaviours are rather complex, especially under various stresses, such as electrical field, ambient temperature, mechanical deformation and different contact states. Researchers have tried many methods to suppress the accumulation of interfacial charge at the interface to improve the reliability of HVDC cable accessories and systems based on the above theory.

4.1 Geometry Design

The electric field has a great influence on the interface charge characteristics. Therefore, a more optimized field distribution can effectively regulate the space charge injection or accumulation and can be effectively regulated. One of the methods is to use a deflector that is made of conductive rubber, but this form of "geometric" FG takes up more space and material, and it is obviously "static" even when the electric field distributions vary with time (e.g., load cycles). Therefore, it must be designed for the worst case. In this way, geometric FG can lead to over-design of the whole device, resulting in a significant increase in cost, especially for equipment used in high-pressure environments. Actually, the application of "geometry" FG in EPDM joints with voltage grades up to 320 kV has achieved satisfactory results, which is also the operating limit of current HVDC extruded cable joints (see Sect. 1) [22].

4.2 Surface Modification

As mentioned in Sects. 3.2 and 3.3, an interface barrier will be induced because of the mismatch of the two materials, which includes their intrinsic surface state. Thus, a proper surface treatment is needed to improve the contact of two dielectrics [23]. Li et al tried to modulate the surface molecular structure of EPDM and LDPE [24]. They first built a gradient fluoride layer on EPDM treated with a gas mixture of N2 and F2, as shown in Fig. 8.

The obvious change in the cross-sectional morphology of the sample occurred after fluorination. It can be observed that fluorine (5.66 weight percent) is introduced into the surface of the sample after fluorination, which indicates that a fluoride layer with a thickness of ~2 μm formed on the surface of the sample. For further analysis, the interface trap depth was calculated, and the results show that the modified surface state decreases the surface deep trap blocking interface charge in the HVDC cable accessory.

Element	Wt%	At%
CK	78.76	96.16
OK	02.57	02.36
AuL	18.67	01.48
Matrix	Correction	ZAF

Element	Wt%	At%
CK	66.91	90.11
OK	03.04	03.07
FK	05.66	04.82
AuL	24.40	02.00
Matrix	Correction	ZAF

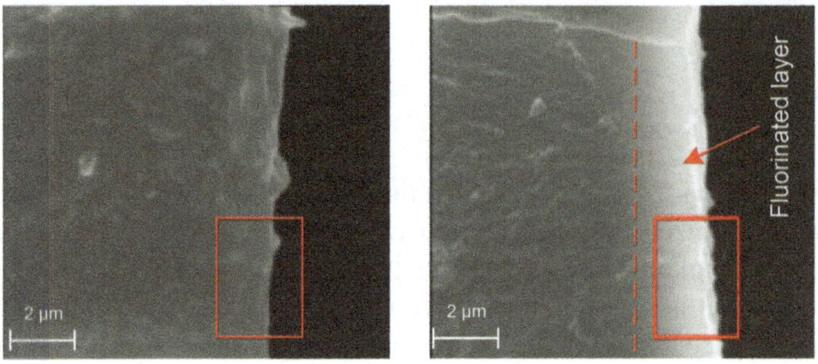

a b

Fig. 8 Comparison of the SEM and EDS of the original and fluorinated EPDM fracture surfaces. [24] Copyright Clearance Center's RightsLink® License Number: 4857330490610. **a** Original sample. **b** Surface fluorinated sample

4.3 Field Grading Materials

Generally, when the applied DC voltage and temperature increase, the DC characteristics of different insulation materials (joints and cables) will cause unstable field distribution in the joint area of prefabricated joints.

Nonlinear resistive field gradient materials (FGMs) are therefore proposed to solve the above problems [25, 26]. FGMs can be obtained by adding special fillers with nonlinear conductivity to the polymer matrix under composite stress [27]. When the electric field exceeds the (critical) threshold, the conductivity of these materials can be reversibly changed from high insulation to high conductivity [25]. Therefore, for HVDC extruded cables, the latest generation of prefabricated joints may add a fourth layer of FGM (Fig. 9) [27] between the internal and external deflectors. The geometry of joint insulation is designed according to the requirements of the impact test. The distribution of the DC electric field is controlled by the non-linear FGM between the two layers of insulation, so that the joint is not sensitive to the change in the conductivity of the two layers caused by differences in production and temperature.

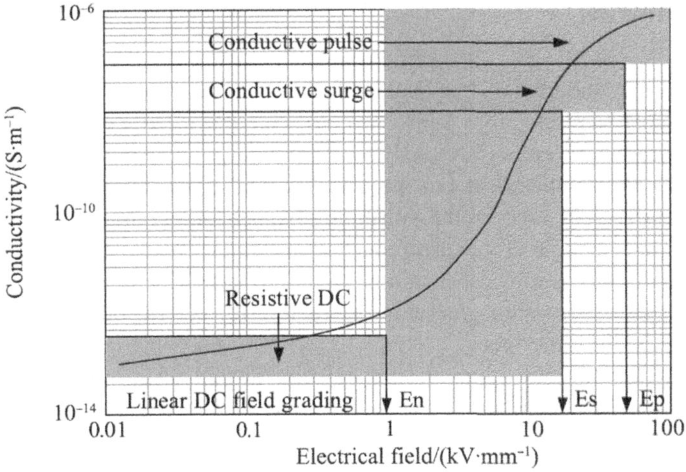

Fig. 9 Typical field grading material with nonlinear conductivity

4.4 *Material Selection*

The properties of a joint insulation material should match those of the main cable insulation material, mainly in the temperature dependence of the insulation resistance. Otherwise, the electric field at the insulating interface may be distorted, which will accelerate the ageing of the insulation and lead to a reduction in the breakdown strength or even premature failure of the material, especially after polarity reversal. Zhong presented a novel concept for DC cable accessory design; i.e., molded crosslinked polyethylene material has properties similar to those of cable insulation. Cable insulation and the enhanced insulation of the accessory match well at their interface, which eliminates the interface to an extent and suppresses the accumulation of space charge fundamentally. Following this concept, interface-free accessories for ±200 kV DC power cables were developed and passed the test recommended by CIGRE TB 496, which verified the proposed design concept [28].

The key to the design of a prefabricated joint body lies in electric field control of the cable/joint interface and the entire medium. In the design, not only the DC voltage and different load conditions but also the switching pulse and possible lightning pulse superimposed with the DC voltage, as well as voltage interruption and voltage polarity reversal, should be considered. The tangential field of the insulation materials of the joint body and the cable under different loads should be the focus. In fact, the electric field increases seriously at the semiconductor edge at the end of the electrical grounding layer of the cable, resulting in the uneven distribution of the electric field at the junction of the high voltage cable [27]. To obtain a reasonable tangential field distribution along the interface, the temperature and electric field dependence of the conductivity of the two insulating materials must be matched [29–31].

5 Conclusions

The macroscopic parameters can be used to simulate the charge behaviour at large-scale interfaces such as cable joints and terminals to a certain extent. It must be noted that only when we know exactly how these parameters depend on the applied electric field and temperature can we use this method appropriately. In addition, once the electric field exceeds the threshold of space charge injection, the injection and extraction of charge at the electrode-polymer interface should be considered to improve the model. Among other things, the trap distribution needs to be modelled. However, to date, there is no general model to describe and explain the experimental results for dielectric interfaces.

Potential barriers and trapping sites existing in dielectric interfaces should be further investigated. In particular, the characteristics of interfacial contacts should be clarified, which are Ohmic, blocking and/or injecting in nature. The kids of carriers should be clarified.

Acknowledgements This work is supported by the National Key Research and Development Program of China under the Grant 2016YFB0900701; National Natural Science Foundation of China (51807136 and 51537008); Natural Science Foundation of Tianjin City (18JCQNJC07300); National Postdoctoral Program for Innovative Talents (BX201700168); China Postdoctoral Science Foundation (2017M621070); Project SKLD19KM08 supported by China State Key Lab. of Power System.

References

1. Y.H. Liu, J. Arrillaga, N.R. Watson, Addition of four-quadrant power controllability to multi-level VSC HVDC transmission. IET Gen. Trans. Distribu. **1**(6), 872–878 (2007). https://doi.org/10.1049/iet-gtd:20070097
2. W. Long, S. Nilsson, HVDC transmission: Yesterday and today. IEEE Power Ener. Mag. **5**(2), 22–31 (2007). https://doi.org/10.1109/MPAE.2007.329175
3. T. An, G.F. Tang, W.N. Wang, Research and application on multi-terminal and DC grids based on VSC-HVDC technology in China. High Volt. **2**(1), 1–10 (2017). https://doi.org/10.1049/hve.2017.0010
4. J.L. He, G. Chen, Insulation materials for HVDC polymeric cables. IEEE Trans. Dielectr. Electr. Insul. **24**(6), 1307–1307 (2017). https://doi.org/10.1109/TDEI.2017.006721
5. P. Han, J.W. Zha, S.J. Wang et al., Theoretical analysis and application of polymer-matrix field grading materials in HVDC cable terminals. High Volt. **2**(1), 39–46 (2017). https://doi.org/10.1049/hve.2016.0067
6. T. Takada, T. Tohmine, Y. Tanaka, J. Li, Space charge accumulation in double-layer dielectric systems—measurement methods and quantum chemical calculations. IEEE Electr. Insul. Mag. **35**(5), 36–46 (2019)
7. Y. Sakai, H. Niinobe, T. Hirasawa, *Development of pre-molded accessories for HVDC extruded cable system* (Versailles, France, June, Inter. Conf. Insul. Power Cables, 2011), p. A23
8. B.X. Du, J. Li, Effects of ambient temperature on surface charge and flashover of heat-shrinkable polymer under polarity reversal voltage. IEEE Trans. Dielectr. Electr. Insul. **23**(2), 1190–1197 (2016). https://doi.org/10.1109/TDEI.2015.005182

9. B.X. Du, Z.R. Yang, Z.L. Li et al., Surface charge behavior of silicone rubber/SiC composites with field-dependent conductivity. IEEE Trans. Dielectr. Electr. Insul. **24**(3), 1340–1348 (2017). https://doi.org/10.1109/TDEI.2017.006137
10. D.W. Auckland, W. Su, B.R. Varlow, Nonlinear fillers in electrical insulation. IEE Proc.-Sci. Measur. Technol. **144**(3), 127–133 (1997). https://doi.org/10.1049/ip-smt:19971085
11. D. Fabiani, G.C. Montanari, C. Laurent et al., Polymeric HVDC cable design and space charge accumulation. Part 1: Insulation/semicon interface. IEEE Electr. Insul. Mag. **23**(6), 11–19 (2007). https://doi.org/10.1109/MEI.2007.4389975
12. C.Y. Li, C.J. Lin, G. Chen et al., Field-dependent charging phenomenon of HVDC spacers based on dominant charge behaviors. Appl. Phys. Lett. **20**, 202904 (2019)
13. S. Delpino, D. Fabiani, G. C. Montanari, et al., Polymeric HVDC cable design and space charge accumulation. Part 2: insulation interfaces. IEEE Electr. Insul. Mag. **24**(1), 14–24 (2008). https://doi.org/10.1109/mei.2008.4455499
14. G. Chen, M. Hao, Z. Xu et al., Review of high voltage direct current cables. CSEE J. Power Ener. Sys. **1**(2), 9–21 (2015). https://doi.org/10.17775/CSEEJPES.2015.00015
15. T.T.N. Vu, G. Teyssedre, S.L. Roy, et al., Maxwell–wagner effect in multi-layered dielectrics: interfacial charge measurement and modelling. Tech. **5**(2), 27 (2017). https://doi.org/10.3390/technologies5020027
16. P.H.F. Morshuis, R. Bodega, D. Fabiani et al., *Dielectric interfaces in DC constructions: space charge and polarization phenomena* (Winchester, United kingdom, July, IEEE Int. Conf. Solid Dielectr., 2007), pp. 450–453
17. Lan, L.: 'Effect of temperature on space charge distribution in polymer insulation', Shanghai Jiaotong University, 2015
18. L. Lan, Q. Zhong, Y. Yin et al., *Effect of surface fluorination on space charge behavior at LDPE/EPDM interface* (Bologna, Italy, June, IEEE Inter. Conf. Solid Dielectr., 2013), pp. 444–447
19. J. Li, H.C. Liang, B.X. Du et al., Numerical simulation of interface charge behaviors between LDPE/EPDM for HVDC cable accessory insulation based on the bipolar charge transport model. High Volt. Eng. **44**(5), 1443–1449 (2018)
20. J. Li, H.C. Liang, M. Xiao, B.X. Du, T. Takada, Mechanism of deep trap sites in epoxy/graphene nanocomposite using quantum chemical calculation. IEEE Trans. Dielectr. Electr. Insul. **26**(5), 2028–2031 (2019)
21. J.G. Su, B.X. Du, T. Han et al., Multistep and multiscale electron trapping for high-efficiency modulation of electrical degradation in polymer dielectrics. J. Phy. Chem. C **123**(12), 7045–7053, (2019)
22. G. Mazzanti, J. Castellon, G. Chen et al., The insulation of HVDC extruded cable system joints. Part 1: review of materials, design and testing procedures. IEEE Trans. Dielectr. Electr. Insul. **26**(3), 964–972 (2019)
23. R. Bodega, P.H.F. Morshuis, J.J. Smit, Space charge measurements on multi-dielectrics by means of the pulsed electroacoustic method. IEEE Trans. Dielectr. Electr. Insul. **13**(2), 272–281 (2006). https://doi.org/10.1109/TDEI.2006.1624272
24. J. Li, B.X. Du, H. Xu, Suppressing interface charge between LDPE and EPDM for HVDC cable accessory insulation. IEEE Trans. Dielectr. Electr. Insul. **24**(3), 1331–1339 (2017). https://doi.org/10.1109/TDEI.2017.006073
25. T. Christen, L. Donzel, F. Greuter, Nonlinear resistive electric field grading part 1: theory and simulation. IEEE Electr. Insul. Mag. **26**(6), 47–59 (2010). https://doi.org/10.1109/MEI.2010.5599979
26. L. Donzel, F. Greuter, T. Christen, Nonlinear resistive electric field grading part 2: materials and applications. IEEE Electr. Insul. Mag. **27**(2), 18–29 (2011). https://doi.org/10.1109/MEI.2011.5739419
27. A. Gustafsson, M. Jeroense, P. Sunnegardh, M. Saltzer, H. Ghorbani, H. Rapp, New developments within the area of extruded HVDC cables, in *Proceedings of the 11th IET International Conference on AC and DC Power Transmission* (2015), pp. 1–5

28. H. Zhong, P. Wang, J. Wang et al., Design of dc cable accessories for suppressing interface space charge. High Volt. Eng. **41**(4), 1140–1146 (2015). https://doi.org/10.13336/j.1003-6520.hve.2015.04.011
29. B.X. Du, J. Li, Interface charge behaviors between LDPE and EPDM filled with carbon black nanoparticles. IEEE Trans. Dielectr. Electr. Insul. **23**(6), 3696–3703 (2017). https://doi.org/10.1109/TDEI.2016.005719
30. J. Li, B. Du, Y. Liu et al., *Interface charge distribution between LDPE and carbon black filled EPDM* (Montpellier, France, July, IEEE Int. Conf. Dielectr., 2016), pp. 406–409
31. J. Li, B.X. Du, X.X. Kong, Z.L. Li, Nonlinear conductivity and interface charge behaviors between LDPE and EPDM/SiC composite for HVDC cable accessory. IEEE Trans. Dielectr. Electr. Insul. **24**(3), 1566–1573 (2017)

Polymer Insulation for HVDC GIS/GIL

Epoxy Resin Insulating Composites for Vacuum Cast Electrical Insulators of GIS

Yushun Zhao, Kerong Yang, Song Zhang, Bin Du, Xuepei Wang, and Yuanhan He

Abstract Vacuum casting basin insulator and supprorting insulator are the key insulation components of GIS system. Their insulation performance, mechanical performance and aging resistance directly affect the safe operation of power grid. As the castable of insulation components, the performances and pouring process of epoxy resin composite are the key factors affecting the performance of insulation components. The problems of residual stress and various defects are easy to occur in the process of casting and molding of insulation components. The aging phenomenon of insulation components will occur due to equipment heating during the operation. In this chapter, the residual stress distribution and curing defects of epoxy resin casting insulation components during casting molding are studied. In addition, the thermal aging mechanism of epoxy composites is studied, and the comprehensive performance of epoxy composites is improved by chain-extended modification.

1 Introduction

Gas-insulated high-voltage switchgears (GIS) are commonly used in HVDC substations [1, 2], because of its space saving, outstanding reliability, operational safety and environmental compatibility. Vacuum cast basin type insulators and supporting insulators are key insulation components for GIS systems. Dielectric strength, mechanical strength, and aging resistance are key performances of these insulation components [3, 4]. Highly filled castable epoxy resin insulating composites are employed to manufacture these insulation components. The intrinsic properties of the epoxy resin composites, such as dielectric strength, mechanical strength, and aging properties, determine the performances of these insulation components directly.

For the dielectric strength of the composites, filling of inorganic particles can change the dielectric strength of the epoxy composites [5, 6]. Chemical bonds of epoxy resin and curing agent determine high voltage withstand properties of the epoxy resin composites [7]. Curing defects, such as shrinkage marks, can generate

Y. Zhao (✉) · K. Yang · S. Zhang · B. Du · X. Wang · Y. He
Hefei University of Technology, 193 Tunxi Road, Hefei, Anhui, China
e-mail: yushunzhao@126.com

B. Du (ed.), *Polymer Insulation Applied for HVDC Transmission*,
https://doi.org/10.1007/978-981-15-9731-2_13

a distortion electrical filed to the epoxy composites [8], and subsequently deteriorate insulation performances of the insulation components [9]. For the mechanical strength of the composites, average molecular weight of the BPA type epoxy resin determines the epoxy group content in the uncured epoxy resin composites [10], and subsequently influences the heat release rate of the epoxy resin composites during its curing process. The residual stress depends to a large extent on the heat release properties of the epoxy resin composites [11, 12]. A large or non-uniform residual stress in the epoxy resin composites will reduce the mechanical strength of the insulation components [13]. For the aging properties of the composites, due to the serious heating and poor heat dissipation of the GIS equipment, the epoxy resin composites will be damaged by heat accumulation [14, 15], which will affect the safe operation of the insulation components [16]. Therefore, it is of great significance to study the effect of high temperature on the aging behaviors of epoxy resin composites, so as to better evaluate the status of insulation performance of the insulation components in GIS equipment.

In this chapter, the variation of residual stress in the epoxy resin composites during cure are measured using resistance strain gage, surface insulation performances of epoxy resin composites with different shrinkage marks are studied. Furthermore, thermal aging mechanism of epoxy-tetrahydrophthalic anhydride thermoset is studied based on molecular simulation. Finally, molecular of liquid epoxy resin are extended with Me-THPA to reduce the heat release rate of the castable epoxy resin composites. The heat release properties, residual stress, and DC-insulation properties of the chain-extended epoxy resin composites are investigated.

2 Residual Stress of Epoxy Composites Under Different Cooling Methods

2.1 Experimental

2.1.1 Preparing Samples for the Residual Stress Measurement

There are great differences in molecular structure between solid epoxy resin and liquid epoxy resin. The stress of the two kinds of epoxy resin generated in their curing process is also different. In this work, the stress test samples of epoxy composites of solid epoxy resin and liquid epoxy resin were prepared respectively for residual stress measurements.

- **Raw Materials**

Liquid epoxy resin E-51, chemically pure, viscosity (25 °C) 6000~10000 mPa•s. Methyl hexahydrophthalic anhydride (MHHPA), chemically pure, viscosity (25 °C) 40~50 mPa•s. Accelerator DMP-30, chemically pure, 2,4,6-tris (dimethylamino

Table 1 Difference in preparation techniques of two kinds of resin samples [17]. Copyright Clearance Center's Rightslink® License Number: 4861230296090

Preparation conditions	Liquid epoxy	Solid epoxy
promoter	DMP-30	/
Mixing temperature (°C)	50	120
Curing temperature (°C)	100	170
Cure time (h)	12	24

methyl) phenol. Solid epoxy resin type GE170262, epoxy value 0.325 eq/100 g, viscosity (120 °C) 407 mPa•s. Anhydride curing agent model GH17066, viscosity (25 °C) 93 mPa•s. α-Al_2O_3 micron particles, spherical in shape, median particle size 15 μm.

- **Preparation Method of the Sample from Solid Epoxy Resin**

(1) Epoxy resin and curing agent were preheated for 2–3 h at 120 °C, and the reaction molds were preheated for 30 min.
(2) 100 parts of micro Al_2O_3 filler were added to the composite. The amount of curing agent is calculated by the equivalent of anhydride group and epoxy group. Then, the resin, filler and curing agent were poured into the reactor and stirred at 400 rpm for 10 min at 120 °C, and the vacuum degree was lower than 100 Pa.
(3) Pour the castable into the reaction molds and conduct vacuum treatment for 10 min at 120 °C and 100 Pa.
(4) The cured epoxy composite samples were prepared by curing at 120 °C for 4 h and then curing at 170 °C for 20 h.

- **Preparation Method of the Sample From Liquid Epoxy Resin**

The preparation method of liquid epoxy resin sample was similar to that of solid epoxy resin, and the difference of some process parameters was shown in Table 1.

2.1.2 Methods of Residual Stresses Measurements

- **Principle of Residual Stress Measurement**

The residual stress of the epoxy resin/anhydride cured product is measured by a resistive stress piece. For the resistance type stress piece, the resistance is calculated according to Eq. (1).

$$R = \rho \frac{L}{S} \tag{1}$$

Generally, the sensitive grid of the stress plate uses copper-chromium al-loy, the resistance change rate is constant, and stress should be proportional to the stress plate, as shown in Eq. (2).

$$\frac{\Delta R}{R_0} = K\varepsilon \tag{2}$$

Therefore, the stress can be calculated flowing Eq. (3).

$$\sigma = E\varepsilon = E\frac{\Delta R}{KR_0} \tag{3}$$

In the above equations, σ is the residual stress, E is the elastic modulus, ΔR is the resistance deviation of the stress piece, R_0 is the initial resistance of the stress piece, K is the sensitivity coefficient of the stress piece, and ε is the strain of the stress piece.

- **Internal Residual Stress Measurement**

Firstly, the resistance stress piece is vertically embedded into the uncured liquid epoxy resin. After the curing reaction is completed, the residual stress in the center of the sample is measured. The principle of residual stress measurement is shown in Fig. 1.

- **Surface Residual Stress Measurement**

The principle of residual stress test on the sample surface is almost the same as that of internal stress test. The difference is that the stress sheet is pasted on the sample surface when measuring the residual stress on the sample surface, and the principle is shown in Fig. 2.

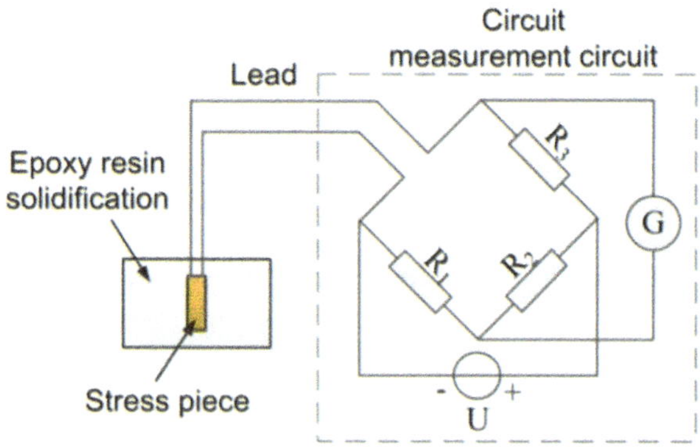

Fig. 1 Schematic of sample internal residual stress measurement [17]. Copyright Clearance Center's Rightslink® License Number: 4861230296090

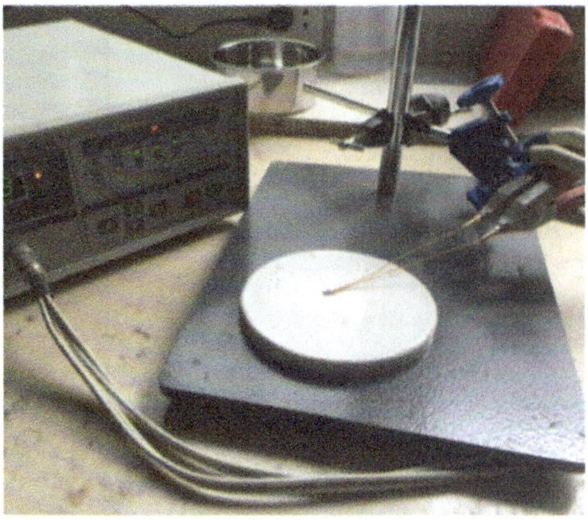

Fig. 2 Surface residual stress measured by stress sheet [17]. Copyright Clearance Center's Rightslink® License Number: 4861230296090

2.2 *Results and Analysis*

2.2.1 Internal Residual Stress of the Liquid Epoxy Resin Composite

The liquid epoxy composite was heated in an oven at 100 °C for 30 min. Then, take out the sample and cool it in air to room temperature. The temperature change of the sample from heating and curing to cooling to room temperature is shown in Fig. 3.

As shown in Fig. 3, when the sample had been heated for 30 min, the temperature of the epoxy composite firstly reached the highest value of 134 °C. This indicates that the curing reaction of epoxy composite belongs to exothermic reaction.

When the temperature was higher than 80 °C, the measurement error of the stress plate was too large to be accepted. Therefore, the residual stress of the sample was measured only when the temperature was lower than 80 °C. The variation of the internal residual stress during this process is shown in Fig. 4.

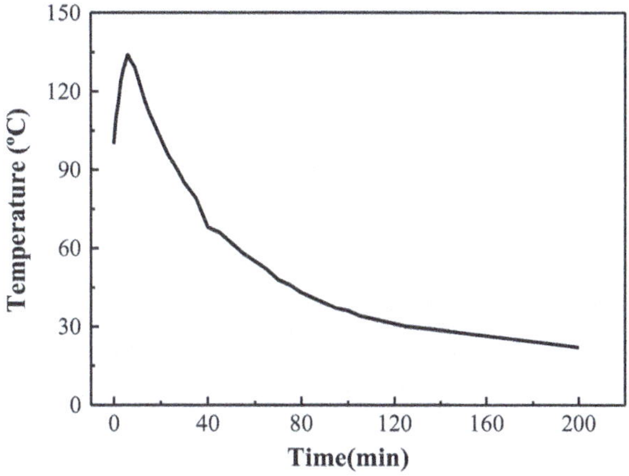

Fig. 3 Curing temperature of epoxy resin after the curing reaction [17]. Copyright Clearance Center's Rightslink® License Number: 4861230296090

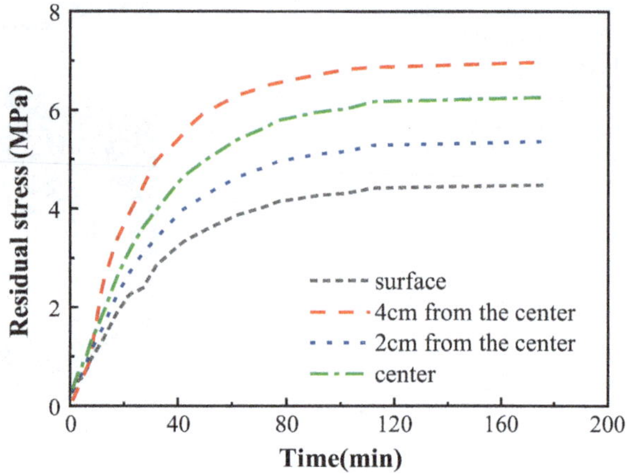

Fig. 4 Distribution of residual stress in the sample during cooling [17]. Copyright Clearance Center's Rightslink® License Number: 4861230296090

As shown in Fig. 4, during the whole cooling process, the change trend of residual stress in different positions of the sample is the same, but the residual stress in the center of the sample is the largest. With the increase of cooling time, the residual stress tends to a stable value. According to the change trend of stress in different positions, it can be inferred that the residual stress in the uniform curing product decreases with the increase of the distance from the center, and reaches the minimum value on the surface. Therefore, the residual stress of the epoxy composite can be evaluated by measuring the stress change of the sample surface.

2.2.2 Effect of Epoxy Value on Residual Stress of Epoxy Composites

The alumina fillers of 100 parts were dispersed into solid epoxy resin curing system and liquid epoxy resin curing system, respectively. The relationship between epoxy concentration and residual stress was studied. The results are shown in Table 2.

In the Table 2, the epoxy concentration of solid epoxy resin is lower than that of liquid epoxy resin, and the residual stress is smaller. The principle is that the crosslinking density of the cured product will be reduced when the concentration of epoxide in the epoxy resin is reduced. During the curing process, the volume shrinkage caused by the reaction will also decrease, so the residual stress will decrease.

Table 2 Residual stress of two epoxy resins with different epoxy values [17]. Copyright Clearance Center's Rightslink® License Number: 4861230296090

Type of epoxy resin	Liquid epoxy	Solid epoxy
The epoxy value (eq/100 g)	0.51	0.325
Residual stress (MPa)	2.4	1.6

2.2.3 Residual Stresses of Epoxy Composites Cooled with Different Methods

The epoxy resin composite needed to be heated in the curing process. After curing, the temperature of the composite was higher than room temperature, and the composites need to be cooled using certain method. In order to study the influence of different cooling methods on the stress of the composites, the sample after curing was firstly taken out of the oven and then cooled in air and water respectively. The measured residual stress results in the sample are shown in Fig. 5. Because the resistance strain gauge was greatly affected by the temperature, the temperature to start timing was 80 °C.

As shown in Fig. 5, the residual stress in the two cooling methods increases with the decrease of temperature. The temperature drop of the sample in air is slower than that in water, but the residual stress is smaller when the sample is cooled to room temperature in air. This may be because the thermal conductivity of water (0.6 W/(m·K)) is higher than that of air (0.026 W/(m·K)), while that of epoxy resin is generally about 0.2 W/(m·K). When the sample is in direct contact with water, the heat on the surface of the sample will be released quickly, and the surface temperature of the sample will drop rapidly. In this case, it will lead to uneven shrinkage of the sample, which will lead to large residual stress. However, the thermal conductivity of air is small, the overall heat dissipation of the sample in the air is relatively stable, and the residual stress generated is small as a whole.

After the temperature of the cured epoxy composite (sample) is reduced to room temperature, the residual stresses can be adjusted again by heating. The sample is first cooled to room temperature in water, and the relationship between the stress and temperature of the sample is shown in the purple curve in Fig. 6. Then the temperature of the sample is heated to 80 °C, and finally the sample is cooled to room temperature in air. At this time, the relationship between the stress and temperature of the sample is shown in the yellow curve in Fig. 6. The residual stress of the sample cooled in

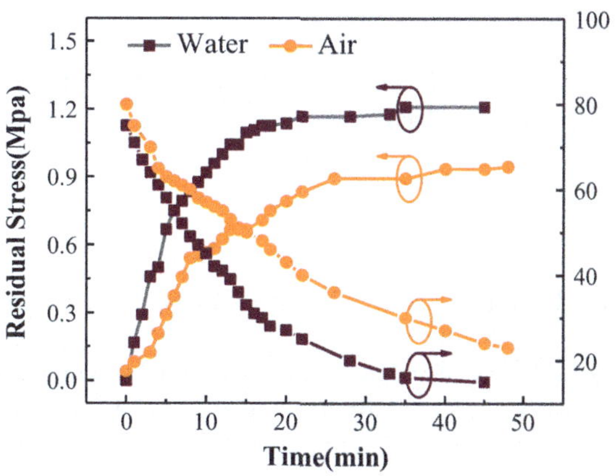

Fig. 5 Residual stress of sample in water and air [17]. Copyright Clearance Center's Rightslink® License Number: 4861230296090

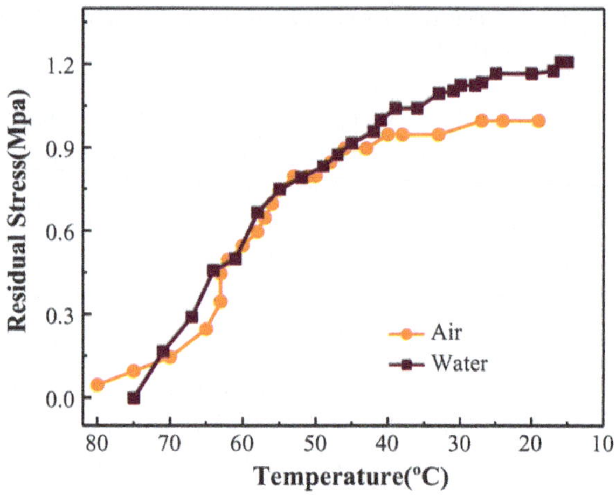

Fig. 6 Adjustment of residual stress

air is reduced compared to that of cooled in water. It shows that the residual stress is reversible, and the residual stress can be adjusted by heating.

2.3 Conclusion

(1) After curing, the residual stress in the sample increases with the decrease of temperature. The further the distance from the measurement point of residual stress to the center of sample, the smaller the residual stress.
(2) The cooling environment has a great influence on the residual stress of the sample. The higher the thermal conductivity of the medium around the sample, the greater the residual stress. The residual stress in the sample can be adjusted by reheating.

3 Insulating Properties of Epoxy Composites with Curing Shrinkage Marks

3.1 Experimental

3.1.1 Preparing Samples with Curing Shrinkage Marks

The solid epoxy resin has the characteristics of small curing shrinkage. For the convenience of studying the effect of curing shrinkage marks on the insulating properties of epoxy resin composite, the samples with cured shrinkage marks were prepared by solid epoxy resin.

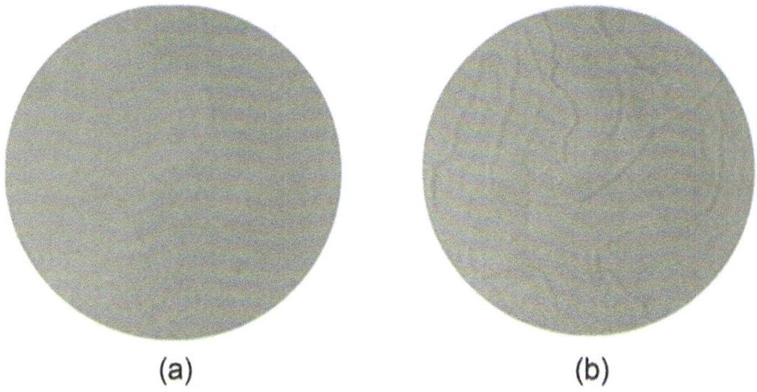

Fig. 7 Experimental samples: **a** without defects; **b** curing shrinkage marks

- **Raw Materials**

Araldite CT 5531 CI resin, epoxy value 0.24 eq/100 g, viscosity (120 °C) 450 mPa•s, purchased from Huntsman Company, USA. Aradur HT 903 CI curing agent, density (130 °C) 1.22 g/cm3, purchased from Huntsman Company, USA.

- **Preparation Method of the Sample With Curing Shrinkage Marks**

The preparation process of the sample with curing shrinkage marks is the same as that of the stress sample except that no filler is added. Therefore, it will not be described in detail here. The prepared samples are shown in Fig. 7.

3.1.2 Test Principle of Surface Flashover Voltage and Tolerance Time of Samples

In order to simulate the actual working condition, the epoxy composite (sample) with curing shrinkage marks were tested in SF_6 conditions with gas presser of 0.1 MPa. As shown in Fig. 8, the angle between the electrodes and shrinkage marks were

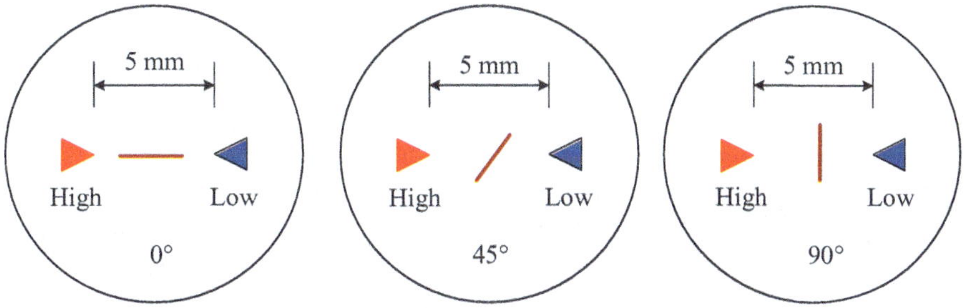

Fig. 8 The angle between the electrodes and shrinkage marks

adjusted to 0°, 45 °, and 90 °, respectively. Two tapered electrodes were used in this work, and the distance between the two electrodes was 5 mm.

3.2 *Results and Analysis*

3.2.1 Surface Flashover Voltage

Eight times of surface flashover voltage tests were carried out on each of the defective samples and the samples without defects. The surface flashover voltage of each sample was analyzed by Weibull distribution, and their surface flashover voltage values with probability of 63.2% were compared. The statistical results of Weibull distribution of surface flashover voltage are shown in Fig. 9, and the corresponding characteristic parameters are shown in Table 3.

As shown in Fig. 9 and Table 3, when the angles between the electrode and the curing shrinkage mark are 0° and 45° respectively, the surface flashover voltages of defective samples are not much different from that of the sample without defects, or even slightly higher. When the angle is 90°, the surface flashover voltage of the sample decreased slightly. However, the maximum voltage of the defective sample is 1.52% higher than that of the sample without defects, and the minimum voltage

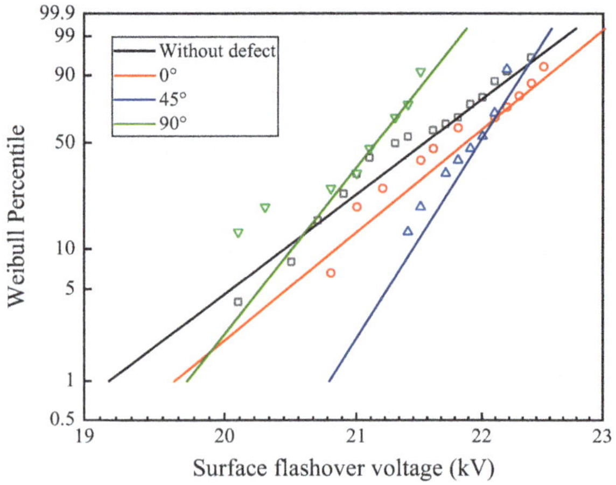

Fig. 9 Weibull distribution of surface flashover voltage of epoxy composite samples

Table 3 Weibull characteristic parameters of flashover voltage of epoxy composite samples

Defect types	α(kV)	β
Without defects	21.75	36.46
0°	22.06	39.48
45°	22.08	76.43
90°	21.28	60.80

of defective sample is 2.16% lower than that of sample without defects. The effect of curing shrinkage marks on the surface flashover voltage isn't obvious.

3.2.2 Tolerance Time at 95% Surface Flashover Voltage

The initial value of the voltage is set at 95% of the average surface flashover voltage of the sample. Each sample is also tested for 8 times. Record the tolerance time from the beginning of the test to the appearance of surface flashover, and use Weibull distribution for statistical analysis tolerance time. The results are shown in Fig. 10 and Table 4.

In the Fig. 10 and Table 4, the tolerance time at 95% surface flashover voltage of the sample with curing shrinkage marks is significantly shorter than that of the samples without defects. When the angle is 0°, curing shrinkage mark has little effect on the tolerance time. When the angle is 90°, the effect of the shrinkage mark on the tolerance time of the sample is most obvious. At this time, the tolerance time of the sample with curing shrinkage mark is only 33.9% of that of the sample without defects. Although SF_6 gas insulating properties is good, to some extent, it can make up for the effect of curing shrinkage mark, but the volume of SF_6 filling between electrodes decreases with the increase of included angle. As a result, the tolerance

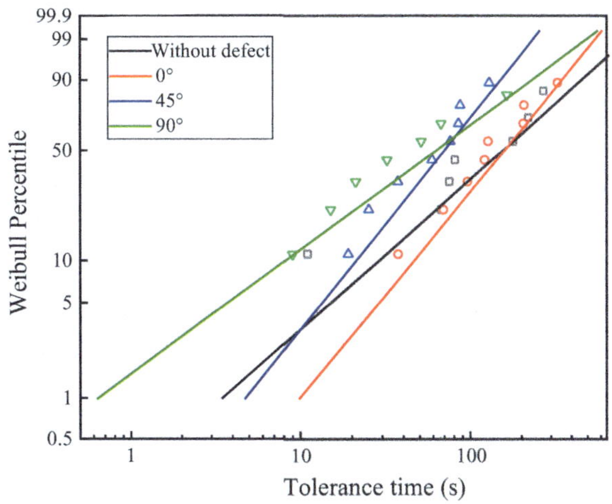

Fig. 10 Weibull distribution of tolerance time at 95% surface flashover voltage of epoxy composite samples

Table 4 Weibull characteristic parameters of tolerance time at 95% flashover voltage of epoxy composite samples

Defect types	α(s)	β
Without defects	216.59	1.11
0°	201.85	1.52
45°	88.35	1.57
90°	73.32	0.92

time at 95% surface flashover voltage of the sample with shrinkage marks decreases with the increase of the angle.

3.2.3 Surface Resistivity and Volume Resistivity

The volume resistivity and surface resistivity of the samples before and after the surface flashover are tested. For the purpose of visual comparison, their values are converted into logarithmic form, and the results are shown in Fig. 11.

As shown in Fig. 11, after the sample surface flashover, the volume resistivity of both samples decreased by one order of magnitude, and the surface resistivity decreased by two orders of magnitude. After the surface flashover, the volume resistivity of the sample decreases, which may be due to the ablation of the sample by the arc during the surface flashover. However, before and after the sample surface

Fig. 11 Volume resistivity and surface resistivity of samples: **a** volume resistivity; **b** surface resistivity

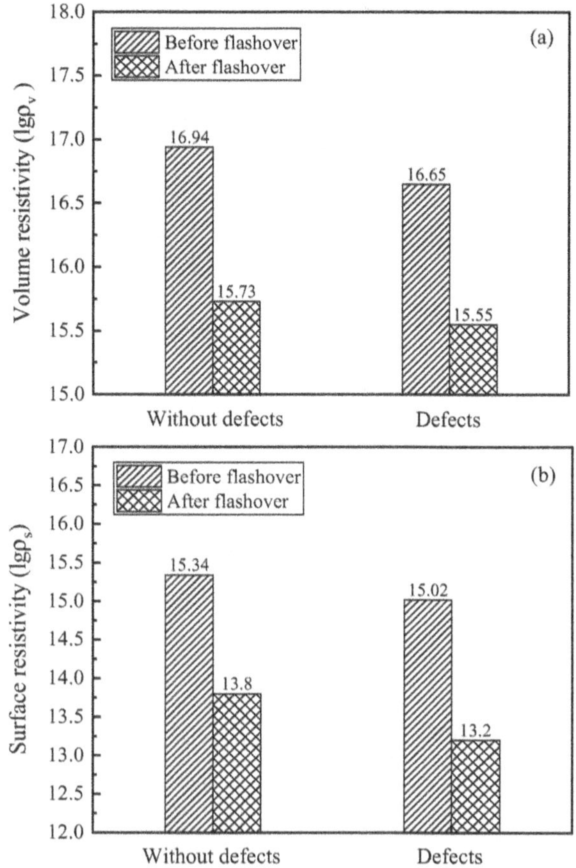

flashover, the volume resistivity and surface resistivity of the sample without curing shrinkage mark are both higher than those of the sample with curing shrinkage mark.

3.3 Conclusion

(1) The curing shrinkage mark has little effect on the surface flashover voltage of the sample surface, but has a great effect on the tolerance time of the sample. When the curing shrinkage mark is 90° to the electrode, the tolerance time of the sample is the shortest.
(2) Surface flashover will reduce the volume resistivity and surface resistivity of the sample. And the volume resistivity and surface resistivity of the sample with curing shrinkage marks are smaller than those of the sample without curing defects, indicating that the curing shrinkage marks will lead to the decline of the insulating properties of the epoxy composite.

4 Thermal Aging Mechanism of DGEBA Epoxy Resin for Vacuum Cast Electrical Insulators

4.1 Modeling and Calculation

The composite model of bisphenol A epoxy resin (DGEBA) and tetrahydrophthalic anhydride THPA was established by molecular dynamics simulation technology [18, 19], as shown in Fig. 12.

After the dynamic balance of the composite model, the epoxy resin and curing agent are crosslinked to form a three-dimensional network structure. The crosslinked model is shown in Fig. 13.

In order to accelerate the thermal aging process of epoxy composite, the thermal aging temperature of epoxy resin was set to 1600 K [21, 22]. Using NVT ensemble, the simulation time was 100 ps, during the simulation calculation.

4.2 Results and Analysis

4.2.1 The Number of Macromolecular Chains During the Thermal Aging Process of Epoxy Composite

Taking C21 and C29 molecular chains as examples, the variation of molecular chain number with thermal aging time was analyzed. The results are shown in Fig. 14.

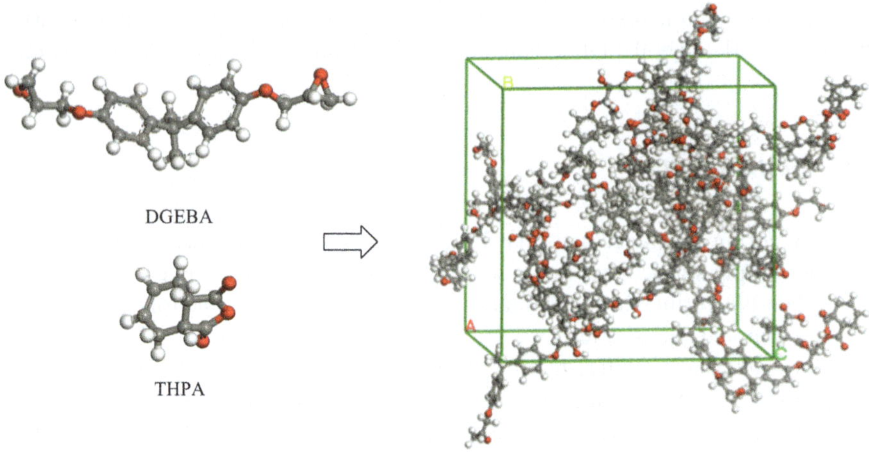

Fig. 12 Composite model of tetrahydrophthalic anhydride and bisphenol A epoxy resin [20]. Copyright Clearance Center's Rightslink® License Number: 4861240798739

Fig. 13 Cross linking composite model [20]. Copyright Clearance Center's Rightslink® License Number: 4861240798739

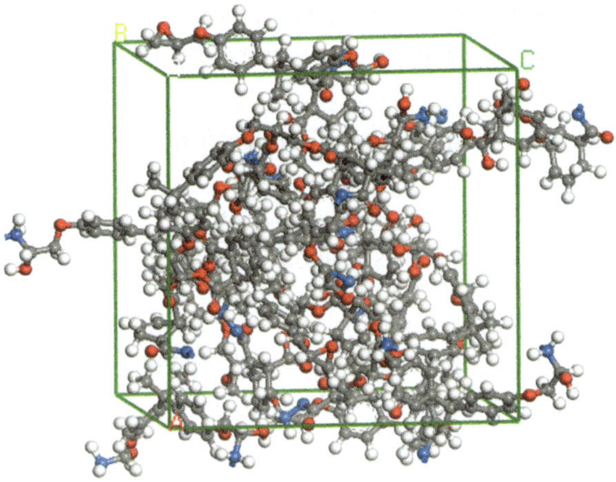

As shown in Fig. 14, with the increase of thermal aging time, the number of two kinds macromolecular chains decreases. In a short period of accelerated aging, the number of molecular chains has a significant reduction trend. When the macromolecular chain breaks, it will form shorter macromolecular chain, which may lead to the degradation of mechanical and insulating properties of epoxy composite.

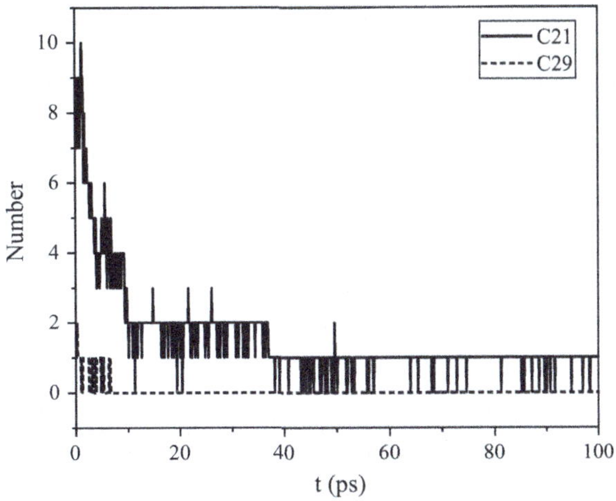

Fig. 14 The relationships between the number of molecular chains and the thermal aging time [20]. Copyright Clearance Center's Rightslink® License Number: 4861240798739

4.2.2 The Number of Small Molecule Chains During the Thermal Aging Process of Epoxy Composite

The types of small molecular substances produced in the process of thermal aging of epoxy resin are counted, and the results are shown in Fig. 15. Water, free hydroxyl and free hydrogen will be produced in the molecular chain of epoxy resin during thermal aging. Free hydroxyl and free hydrogen will recombine to form water, and the existence of water will lead to the degradation of the insulating properties of epoxy composite.

Figure 15 shows that the carbon dioxide content is increasing in the range of thermal aging time less than 38 ps, which is related to the decreasing number

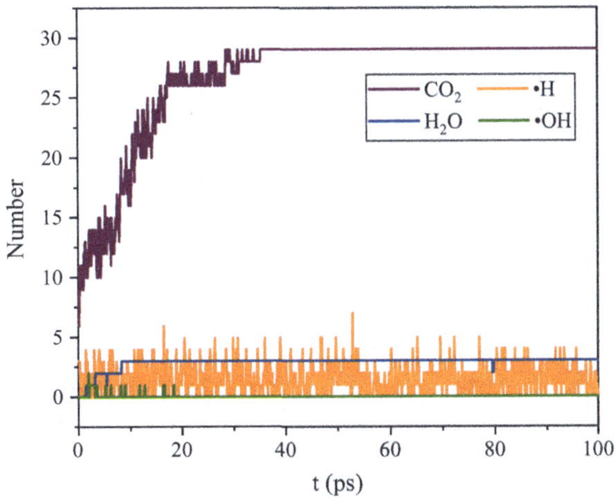

Fig. 15 The relationships between thermal aging time and the number of CO_2, H_2O, free hydroxyl groups and free hydrogens [20]. Copyright Clearance Center's Rightslink® License Number: 4861240798739

of macromolecular chains. Macromolecular chains break to form small molecular chains, and small molecular chains break to form carbon dioxide. However, when the number of macromolecular chains no longer changes, the amount of carbon dioxide tends to be basically stable. In the initial stage of aging, water content is increasing, hydroxyl content is fluctuating, and then tends to be stable. This process may be caused by the combination of free hydroxyl, free hydrogen and unsaturated molecular structure.

4.2.3 Bond Length Distribution During Thermal Aging of Epoxy Composite

In Fig. 16, A and B represent the bond length distribution of epoxy resin composite under the condition of non-aging and aging respectively. The bond length distribution strength of O–H, C–H, C–O and C–C bonds in the aging epoxy composite is significantly weaker than that in the non-aging epoxy composite. In the aging state, the bond length distribution strength of C = O chemical bond of epoxy composite is stronger than that of non-aging epoxy composite, which indicates that there are chemical bond fracture and new chemical bond formation in the aging process of epoxy composite.

4.3 Conclusion

(1) During the thermal aging process of epoxy composite, the chemical bond will break, and small molecules such as water, free hydroxyl and free hydrogen will be formed gradually. The water formed by recombination of free hydroxyl and

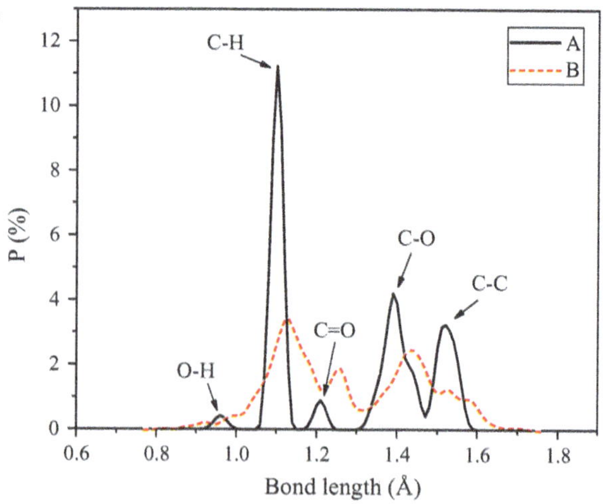

Fig. 16 Bond length distribution curves of epoxy composite in aging and non-aging models [20]. Copyright Clearance Center's Rightslink® License Number: 4861240798739

free hydrogen and the water directly produced by aging of epoxy resin will reduce the insulating properties of epoxy composite.

(2) The bond length distribution characteristics of epoxy composite in aging state and non-aging state are different, which proves that there are chemical bond fracture and formation phenomenon in the aging process of epoxy composite.

5 Improving the Properties of Epoxy Composite Insulation Materials by Chain Extension

5.1 Experimental

5.1.1 Chemical Reaction Principle of Chain Extension

The chemical reaction underlying the chain extension of liquid epoxy resin with Me-THPA is shown in Fig. 17. The reaction consists of three steps. The first is the formation of ester and carboxyl groups by an esterification reaction between a hydroxyl group (–OH) in the epoxy resin and an anhydride group, as shown in Fig. 17a. The second step is the formation of ester and hydroxyl groups by reaction between a carboxyl group and an epoxy group, as shown in Fig. 17b. In the third step, a newly formed hydroxyl group reacts with another anhydride group and another epoxy group at the same time, as shown in Fig. 17c and 17d. The reaction continues, forming a cross-linking system composed mainly of ester bonds and ether bonds.

5.1.2 Chemical Reaction Principle of Chain-Extended Epoxy Resin Curing

The chemical reactions underlying the Me-THPA, DMP-30 and Al_2O_3 fillers added to the chain-extended epoxy resin for curing are shown in Fig. 18. The reaction process consists of three steps. The first is the formation of carboxylate anion through the reaction between a tertiary amine group in DMP-30 and an anhydride group, as shown in Fig. 18a. The second step is the formation of oxygen anions through the reaction between a carboxylate anion and an epoxy group, as shown in Fig. 18b. The third step is the reaction between an oxygen anion and another anhydride group to form carboxylate anion again, as shown in Fig. 18c. The chain-unextended epoxy resins would form a cross-linking system composed mainly of ester bonds.

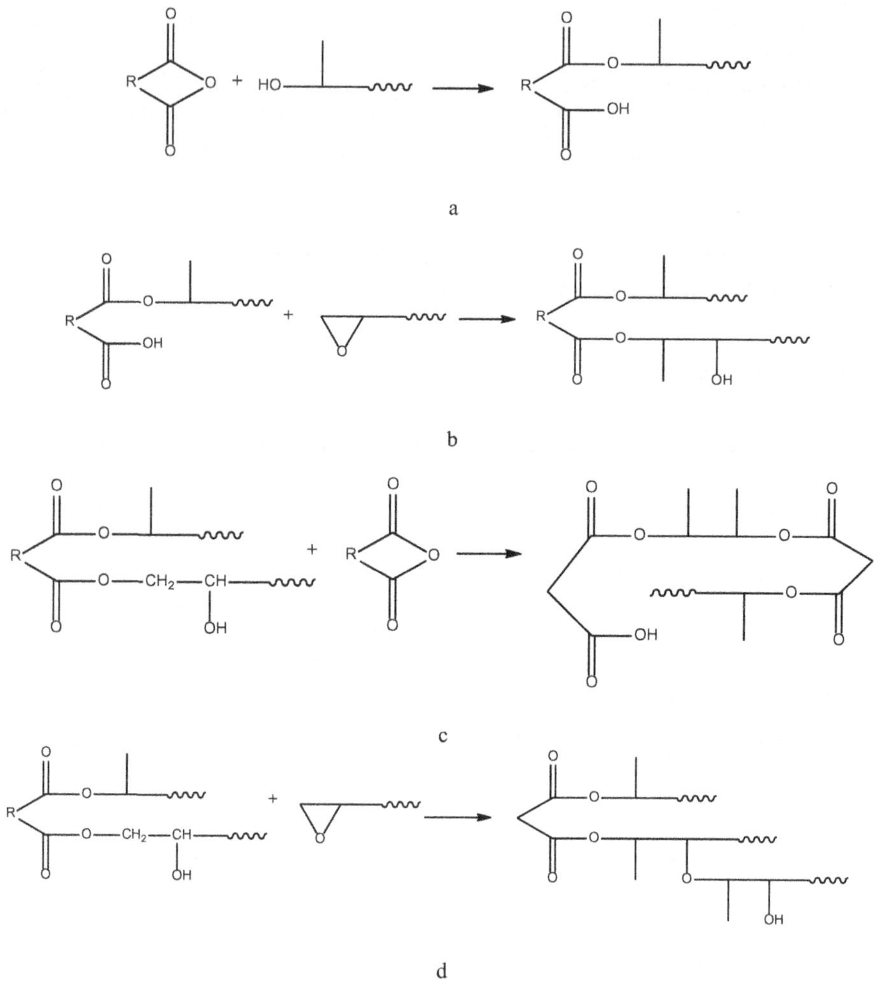

Fig. 17 Chemical reaction underlying extension of the chain of liquid epoxy resin with Me-THPA

5.1.3 Raw Materials

Liquid epoxy resin (EP): Chemically Pure, purchased from South Asia Resin Co., Ltd. Its relative epoxy resin equivalent was 182–192 g/equiv; Methyl tetrahydrophthalic anhydride (Me-THPA): Chemically Pure, purchased from Puyang Huicheng Electronic Material Co., Ltd.; Micro α-Al_2O_3 fillers: average diameter 15–20 μm, provided by Zhengzhou Light Metals Research Institute of CHINALCO.; Curing accelerator: 2,4,6-tris(dimethylaminomethyl)phenol (DMP-30), purchased from Shang Hai SSS Reagent Co., Ltd.

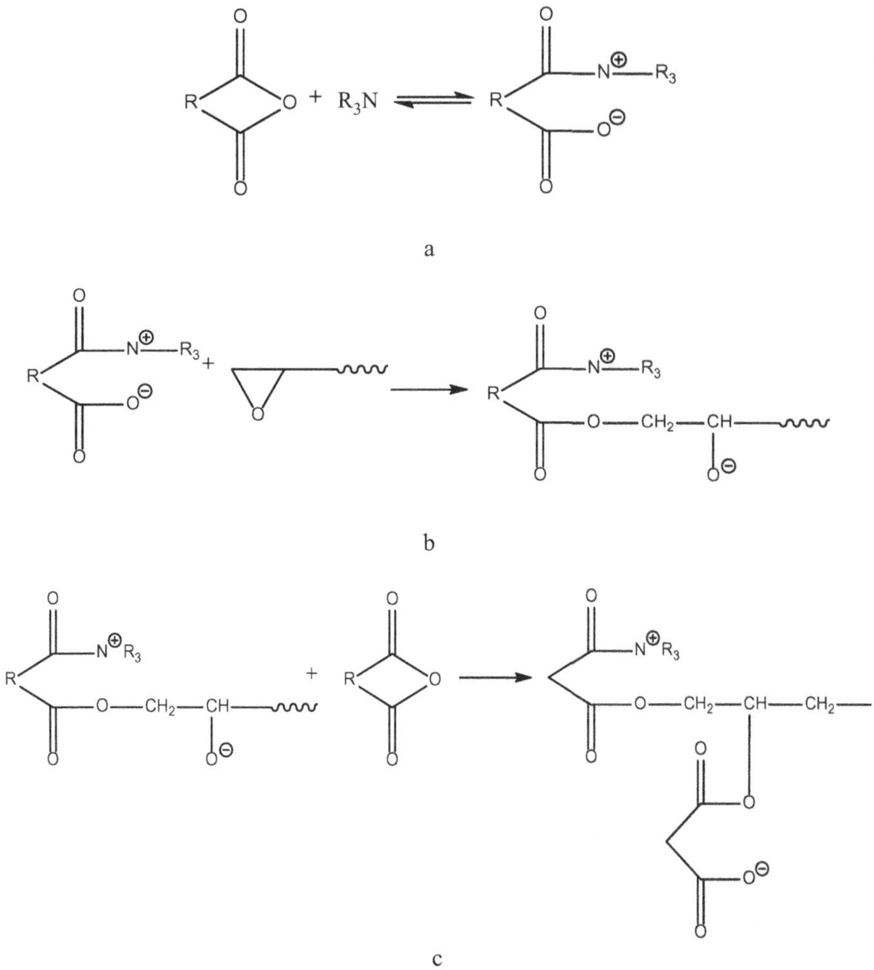

Fig. 18 Chemical reaction underlying of the chain-extended epoxy resin curing

5.1.4 Preparation of the Chain-Extended Epoxy Resin

Five different chain-extended epoxy resins were prepared by changing the molar ratio of liquid epoxy resin to Me-THPA. The molar ratios used are listed in Table 5. The chain extension reaction was carried out in a vacuum reactor at a constant temperature of 170 °C for 8 h. During the reaction, continuous mechanical stirring was applied.

Using the weight ratios shown in Table 6, the chain-extended epoxy resin, Me-THPA, and α-Al_2O_3 fillers were successively added to a flask. The amount of Me-THPA was determined by the quantity of liquid epoxy resin in the chain-extended epoxy resin. The mixture was dispersed uniformly using a high-speed mechanical stirrer in a vacuum at 70 °C.

Table 5 Molar ratios of liquid epoxy resin to Me-THPA

No.	EP/Me-THPA (mol)	EP (g)	Me-THPA (g)
1	0.1	392	33.2
2	0.2	392	66.4
3	0.3	392	99.6
4	0.4	392	132.8
5	0.5	392	166.0

Table 6 Formulas of chain-extended epoxy resin/Al_2O_3 composites

No.	Chain-extended epoxy resin (g)	Me-THPA (g)	α-Al_2O_3 (g)	DMP-30 (g)
1	100	84.7	300	1
2	100	71.4	276.6	0.92
3	100	58.1	256.5	0.86
4	100	46.5	239.1	0.80
5	100	38.2	224.1	0.75
6	100	29.9	210.9	0.7

5.1.5 Preparation of the Chain-Extended Epoxy Resin/Al_2O_3 Composite Material

After the chain-extended epoxy resin and Al_2O_3 fillers had dispersed sufficiently, the curing accelerator (DMP-30) was added to the mixture. The mixture was continued dispersing for 5 min to prepare the castable epoxy resin composites (formula(1) in Table 6 is to prepare the chain-unextended epoxy resin/Al_2O_3 composite material). The six castable epoxy resin composites were added to moulds designed in accordance with GB/T 1408.1-2006. The cured samples were disks of diameter 100 mm and thickness 1 mm. The preparations of the six epoxy resin/Al_2O_3 composites are shown in Figs. 19 and 20. Before testing, the cured samples were placed in an ultrasonic cleaner, then, washed sequentially with ethyl acetate, ethanol and deionized water, before finally being dried in a vacuum oven.

5.1.6 Testing and Characterization

The weight-average molar weight (M_w) and molecular weight distribution of the chain-extended epoxy resin composites were measured by Gel Permeation Chromatography (GPC) (Waters 1515, Waters, USA). The functional groups of the composites were studied by Fourier Transform Infrared Spectroscopy (FTIR) (Nicolet 67, Thermo Nicolet, USA). The rate of heat release during the curing reaction of the composites was measured using a Differential Scanning Calorimeter

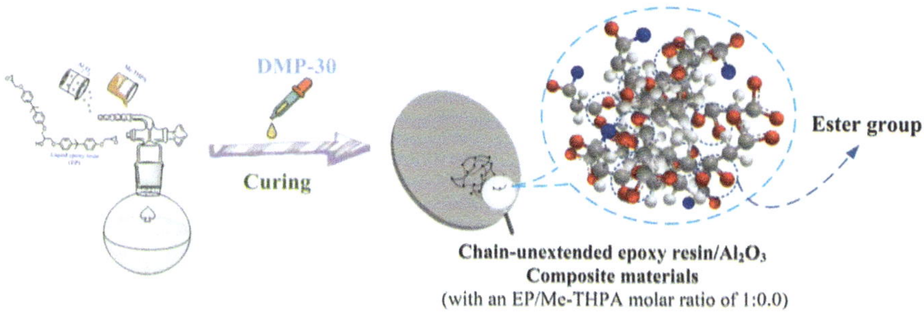

Fig. 19 Preparation of the chain-unextended epoxy resin/Al_2O_3 composite material

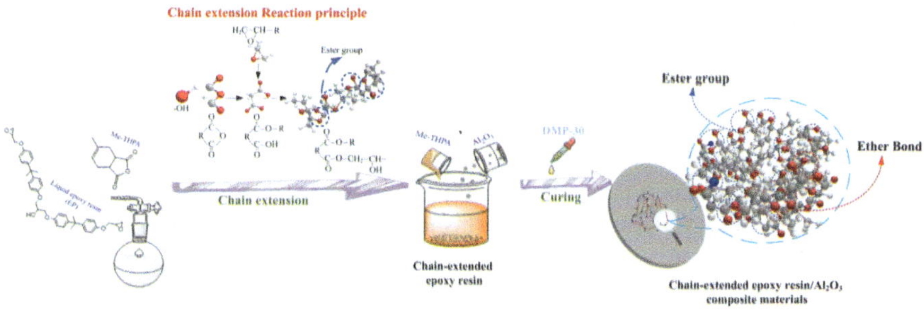

Fig. 20 Preparation of the chain-extended epoxy resin/Al_2O_3 composites

(DSC) (Q2000, TA Instruments, USA). The surface resistivity and volume resistivity were measured using a insulation resistance tester (ZC-36, Shanghai No. 6 m Factory Co., Ltd., China).

The process of generating curing stress during the curing process of epoxy resin composites is as follows. When the curing temperature of an epoxy resin composite material rises to the glass transition temperature (T_g), the material gradually changes from a liquid to a gel, and the curing stress begins to appear. Over time, the material changes from a gel to a solid, the curing stress continually established throughout the curing process [23]. Moreover, the main process, during which the curing stress established, is the cooling process of the epoxy resin composite material from the curing temperature to room temperature [24]. In the present study, the curing stress was measured during cooling by circuit shown as Fig. 21 [25].

The testing method of curing stress is introduced as follows: 500 g of each kind of castable epoxy resin composite material was prepared according Table 6. The composite material was cast into and cured in a cylindrical stainless steel mould with bottom diameter 110 mm. After demolding, one stress sensor was adhered on the sample surface, resistance of the sensor changed with the variation of curing stress during the cooling process of the sample, and a resistance tester was employed to measure the resistance of this stress sensor. The measurement results showed a similar linear growth during the cooling of each sample. After a certain time had

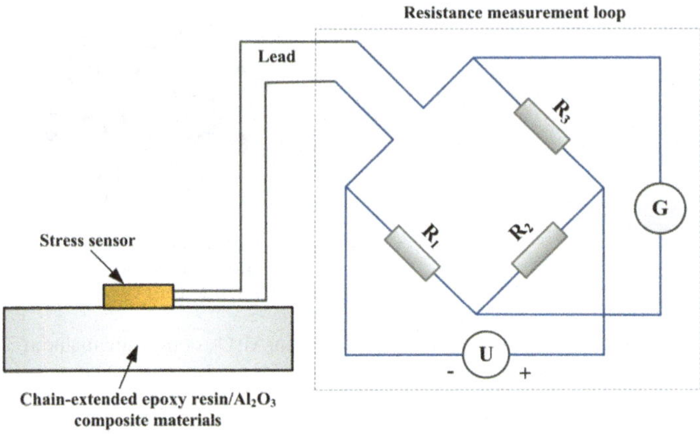

Fig. 21 Circuit for measuring curing stress of chain-extended epoxy resin/Al_2O_3 composites

passed, the results reached a constant value, and this value was taken as the curing stress. The curing stress is calculated by Eq. (3) in Sect. 1.2.

The test circuit for measuring surface flashover characteristics is shown in Fig. 22. The negative output DC voltage of the test transformer ranged from 0 to −70 kV. A voltage divider (K = 1000/1) and test cavity were connected in parallel with the test transformer through a high-voltage rectifier stack and a protection resistor (1 MΩ), respectively. To simulate the discharge of a basin-type insulator inside a GIS system, the surface flashover characteristics were tested in a sulphur hexafluoride (SF_6) atmosphere [26]. Generally, the pressure of SF_6 in a GIS system is less than 0.45 MPa, and the pressure in laboratory equipment (0.1–0.4 MPa) is less than that in field equipment [27]. In this study, the tests were performed at an SF_6 pressure of 0.2 MPa. After the test cavity had been completely evacuated, SF_6 gas of 99.99% purity was added up to the given pressure. The boost speed of the test transformer was 0.5 kV/s. A rod–rod electrode system was used, and the surface flashover characteristics of the chain-extended epoxy resin composites were tested by changing the distance between the two electrodes.

5.2 *Results and Analysis*

5.2.1 Molecular Weight Distribution

The weight-average molar weight (M_w) and molecular weight distribution of chain-extended epoxy resins were tested. The test samples were chain-extended epoxy resins with EP/Me-THPA molar ratios of 1/0.1, 1/0.3, and 1/0.5. Tetrahydrofuran was used as a solvent. The results in Fig. 23 show that, with the increase of EP/Me-THPA molar ratio, the peak of the curve moves to the left, indicating that the molecular

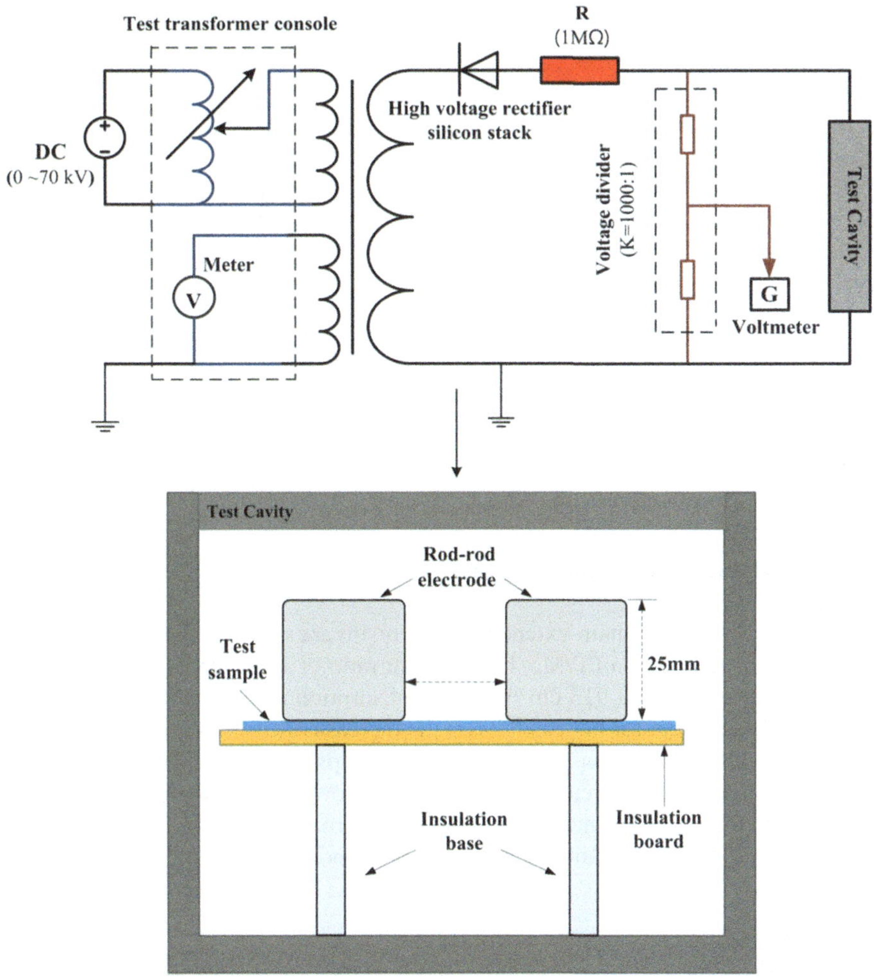

Fig. 22 Circuit for determining surface flashover characteristics of chain-extended epoxy resin/Al_2O_3 composites

chain of the modified resin was lengthened owing to the production of material with higher molecular weight. The chain-extended epoxy resin with the largest M_w of 717 was produced with a molar ratio of 1:0.5. These results show that the molecular weight distribution of the liquid epoxy resin was increased by chain extension with Me-THPA. Thus, the aim of chain extension of the low-molecular-weight epoxy resin was achieved.

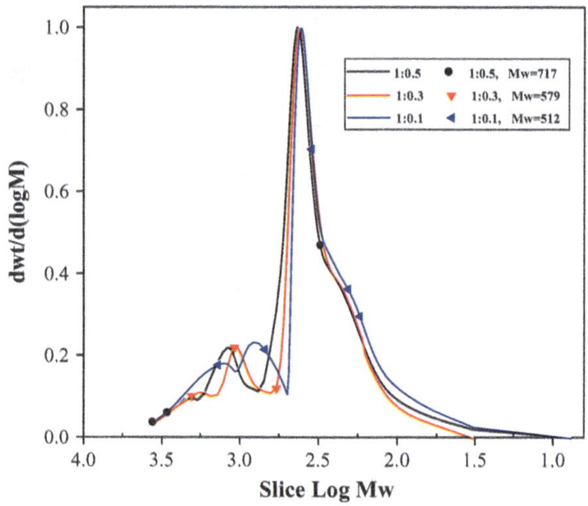

Fig. 23 M_W and molecular weight distribution of chain-extended epoxy resins

5.2.2 Variation of Functional Groups

The FTIR spectra of the chain-extended epoxy resins are shown in Fig. 24. The chain-unextended resin (with an EP/Me-THPA molar ratio of 1:0.0) produced the highest epoxy absorption peak at 913 cm^{-1}, but the absorption peak of carbonyl groups at 1750 cm^{-1} was absent for this resin. With the increase of EP/Me-THPA molar ratio, the infrared absorption intensity of the epoxy groups gradually decreased and that of the carbonyl groups increased. These results show that epoxy rings in the resin were opened through the reaction with anhydride groups, resulting in the formation of carbonyl groups and obtaining chain-extended epoxy resins.

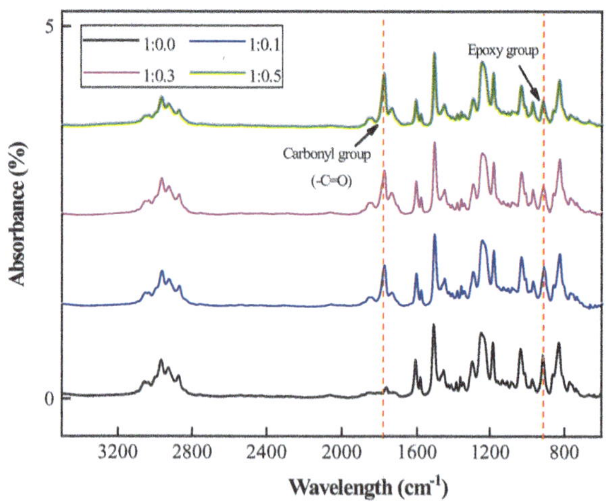

Fig. 24 FTIR spectra of chain-extended epoxy resins

5.2.3 Curing Characteristics

- **Curing Exothermic Properties**

The curing exothermic rates of chain-unextended and chain-extended epoxy resins were tested at a constant temperature of 170 °C. Curves of heat flow versus time are shown in Fig. 25. The heat flow rates of the three kinds of chain-extended resin were significantly lower than that of the chain-unextended resin. Moreover, with the increase of EP/Me-THPA molar ratio, the heat flow rate decreased continuously. The 1:0.5 chain-extended resin exhibited the lowest heat flow rate, about 0.143 mW/mg.

The ring-opening polymerization of epoxy groups is an exothermic reaction, and lower epoxy-group contents are result in lower curing exothermic rates. For the chain-extended epoxy resins, the epoxy groups were partially consumed during the chain extension reaction process, so the content of these groups in the chain-extended resins was reduced. These stated results and analysis showed that chain extension of epoxy resins resulted in lower curing exothermic rates.

- **Curing Stress Characteristics**

Results for the curing stress in the chain-extended epoxy resin/Al_2O_3 composites are shown in Fig. 26. It can be seen that the curing stress decreased with the increase of EP/Me-THPA molar ratio.

Taking the curing stress of the epoxy resin composites without chain extension (EP/Me-THPA molar ratio of 1:0.0) as a reference, an EP/Me-THPA molar ratio of 1:0.1 reduced the curing stress by 13.99%; at 1:0.3, it was reduced by 21.24%; and at 1:0.5, the curing stress was 0.75 MPa, a decrease of 64% compared with the resin without chain extension. The reduction in curing stress can be attributed to the low curing exothermic rate of the curing system. The curing exothermic test results showed that the curing exothermic rate was lowest for the chain-extended epoxy

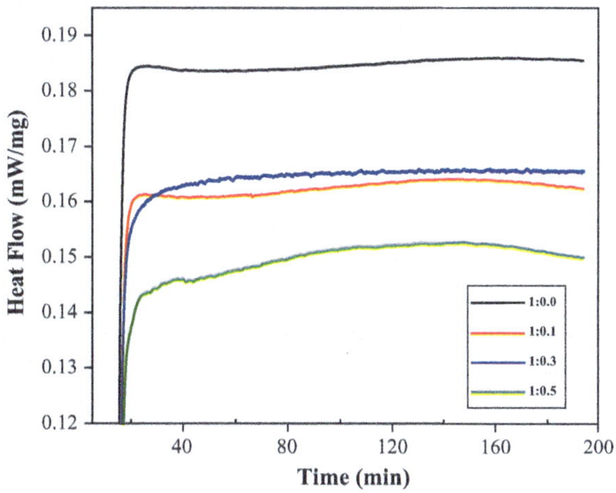

Fig. 25 Curing heat flow versus time for different chain-extended epoxy resins at a constant temperature of 170 °C

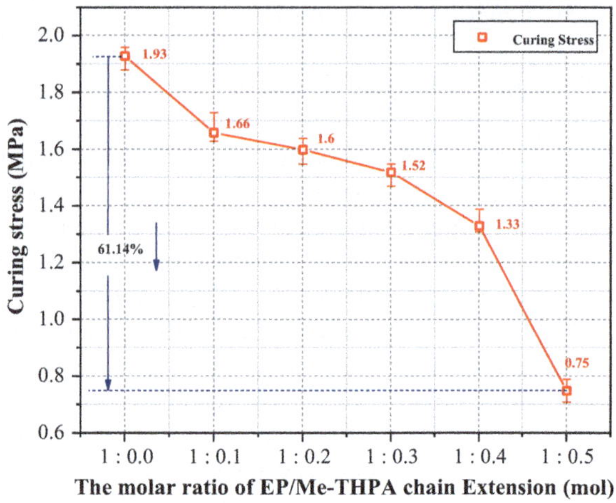

Fig. 26 Curves of curing stress of chain-extended epoxy resin/Al_2O_3 composites

resin with a molar ratio of 1:0.5, as a consequence of which the curing reaction of the chain-extended epoxy resin was milder. The explosive phenomenon caused by rapid reactions in the curing system could thus be prevented, thereby helping to reduce the formation of curing defects in cured epoxy resin composites [28].

5.2.4 Surface Flashover Characteristics Under Negative DC Voltage

- **Surface Resistivity and Volume Resistivity**

Surface resistivity (ρ_s) and volume resistivity (ρ_v) are important characteristics of insulation materials, and are typically material-dependent. The curves of ρ_s and ρ_v for epoxy resin composites with different EP/Me-THPA molar ratio are shown in Fig. 27.

It can be seen that the curves of ρ_s and ρ_v decline with respect to the increase of molar ratio, and so both ρ_s and ρ_v are lower in the chain-extended epoxy resin composites than in the chain-unextended material (EP/Me-THPA molar ratio of 1:0.0). With the increase of EP/Me-THPA molar ratio, ρ_s and ρ_v decreased continuously. At an EP/Me-THPA molar ratio of 1:0.0 (chain-unextended material), $\rho_s = 16.43648 \times 10^{14}\ \Omega$ and $\rho_v = 16.7277 \times 10^{15}\ \Omega\cdot\text{cm}$. Taking the EP/Me-THPA molar ratio of 1:0.0 as a reference, a molar ratio of 1:0.5 causes ρ_s to decrease by 8.74% and ρ_v to decrease by 6.14%. A decrease in resistivity can accelerate electric charge dissipation and reduce charge accumulation on the insulation materials surface. To some extent, this is beneficial for improving the surface flashover voltages of epoxy resin composites [29].

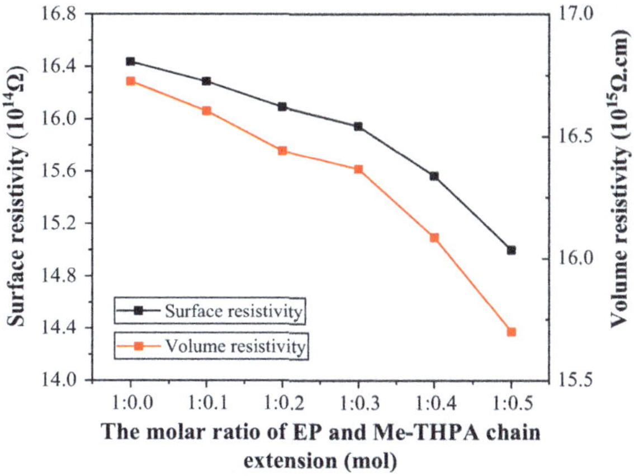

Fig. 27 Surface resistivity (ρs) and volume resistivity (ρv) for different chain-extended epoxy resins

- **Surface Flashover Voltage**

The Weibull distribution is widely used in reliability analysis. In particular, the Weibull distribution of two parameters has been used to analyse the insulation properties of insulating materials [30], and such an approach was applied to the test results in the present study. Surface flashover tests on the chain-extended epoxy resin/Al_2O_3 composites were carried out by applying a negative DC voltage with the distance between the rod electrodes set to 5 mm [31]. Figure 28 shows the results, after Weibull transformation, for the surface flashover voltage at different values of the EP/Me-THPA molar ratio.

Figure 28 shows that the flashover voltage of the chain-extended epoxy resin composites is in good agreement with the Weibull distribution. It can be seen that the curves move to the right as the EP/Me-THPA molar ratio increases, indicating that

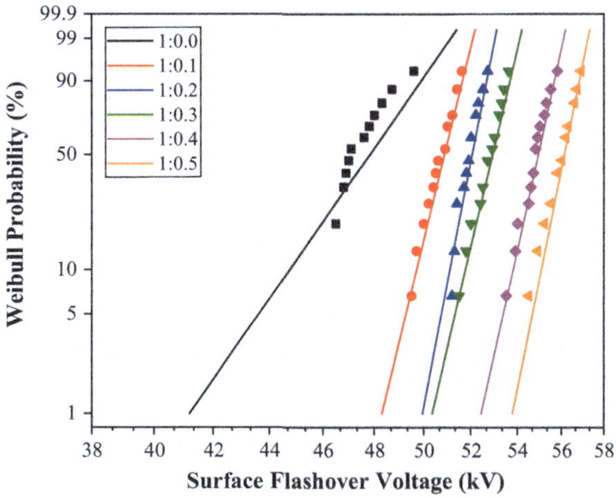

Fig. 28 Weibull fitting curves of surface flashover voltage of chain-extended epoxy resin/Al_2O_3 composites under application of a negative DC voltage

Table 7 Weibull fitting curve parameters of chain-extended epoxy resin/Al2O3 composites under application of a negative DC voltage

EP/Me-THPA molar ratio (mol)	Scale parameter, α (kV)	Shape parameter, β
1:0.0	48.45	28.3194
1:0.1	51.10	81.4228
1:0.2	52.23	103.2753
1:0.3	53.15	85.59
1:0.4	55.16	90.7208
1:0.5	56.34	98.4294

chain extension can improve the surface flashover voltage of epoxy resin composites: as the EP/Me-THPA molar ratio increases, so does the flashover voltage. The Weibull fitting curve of the 1:0.5 chain-extended epoxy resin/Al_2O_3 composite material is at the far right, and has the highest flashover voltages.

Table 7 presents the scale parameters (α) and shape parameters (β) of the Weibull fitting curves for the chain-extended epoxy resin/Al_2O_3 composites. These parameters take their lowest values when the EP/Me-THPA molar ratio is 1:0.0 (chain-unextended), reflecting the lowest flashover voltages and the greatest dispersion. In the 1:0.2 case, α and β had increased by 7.8% and 265%, respectively. β represents the data dispersion. The β values of all chain-extended epoxy resin/Al_2O_3 composites were greater than that in the 1:0.0 case, indicating that chain extension reduces the surface flashover voltage dispersion of the epoxy resin composite material.

- **Tolerance Time at 95% Surface Flashover Voltage**

Based on the test results for the surface flashover voltage, the tolerance time at 95% surface flashover voltage of the chain-extended epoxy resin/Al_2O_3 composites were tested by applying a negative DC voltage. The applied negative DC voltage was set to 95% surface flashover voltage. The distance between the rod electrodes was set to 4 mm. The tolerance time test results are shown in Fig. 29.

With the increase of EP/Me-THPA molar ratio, the tolerance time gradually became longer. The tolerance time of the chain-unextended epoxy resin/Al_2O_3 composite material (EP/Me-THPA molar ratio of 1:0.0) was 21 min, whereas that of the chain-extended material with a molar ratio of 1:0.1 was 30 min. At the molar ratio of 1:0.5, the tolerance time was 58 min, which is 1.76 times that of the chain-unextended material. These results show that chain extension of epoxy resins significantly prolongs the tolerance time of the composite material under a negative DC voltage, thereby improving the resistance of its insulation aging performance.

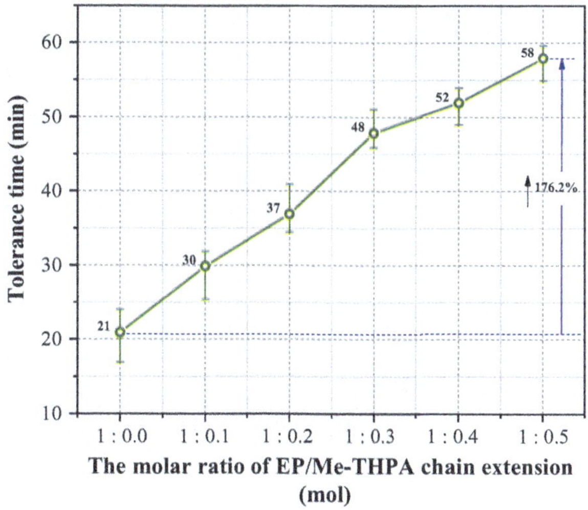

Fig. 29 Tolerance time of chain-extended epoxy resin/Al_2O_3 composites under application of a negative DC voltage

5.3 *Discussion*

The surface flashover characteristics of chain-extended epoxy resin/Al_2O_3 composites show that chain extension clearly improves tolerance time, but has little effect on the surface flashover voltage. To further investigate the effects of chain extension on surface flashover characteristics, the surface morphology and surface molecular functional groups of the chain-extended epoxy resin/Al_2O_3 composites were studied.

In the presence of defects in epoxy resin composites, such as bubbles and shrinkage defects, it is easy for charge to accumulate in the defect area under the action of an electric field. This accumulation of charge distorts the electric field in defect areas and adjacent regions, which is likely to lead to partial discharge from an insulating material [32]. Such a partial discharge would not develop rapidly to form penetrating channels, so the electric field would not have a great impact on the insulating material over a short time period. The electric field strength of bubbles in the insulating materials is much greater than the dielectric field strength, and the electric field strength of the air is lower than that of insulating materials. Thus, the electric field lines will concentrate near the boundary of the bubble, enhancing the likelihood of flashover. Shrinkage cavities cause a depression inside the insulating materials. Previous simulation results [33] have shown that the electric field strength will gradually increase throughout the depression, reaching a peak at the end of the depression, so the region of the shrinkage cavity will produce a discharge. With the prolonged application of a voltage, bubbles and shrinkage cavities have a significant impact on the voltage withstand capability of an epoxy resin composite material. Therefore, the voltage withstand capability of epoxy resin composites could be improved by reducing the number of bubbles and shrinkage cavities.

Scanning electron microscopy (SEM) was used to observe the surface morphology of the composites obtained by chain extension at different molar ratios before surface

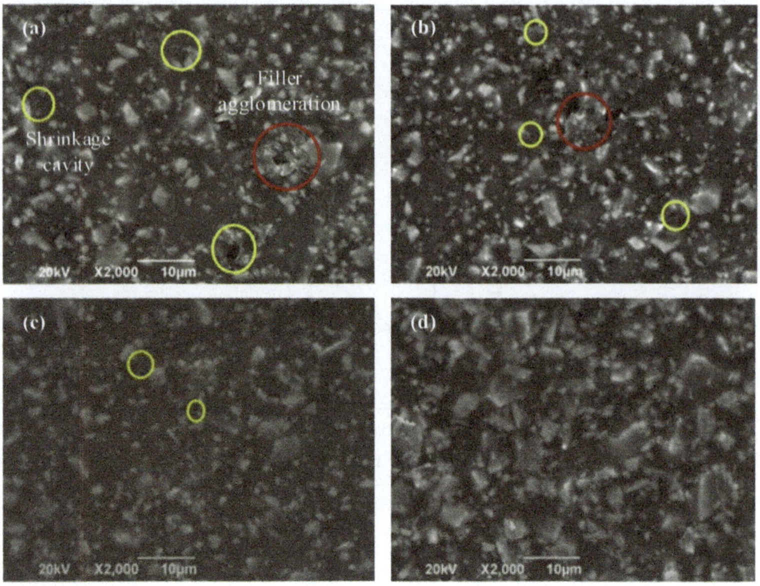

Fig. 30 Surface topography of chain-extended epoxy resin composites before surface flashover for EP/Me-THPA molar ratios of **a** 1:0.0, **b** 1:0.1, **c** 1:0.3, **d** 1:0.5

flashover. The results are shown in Fig. 30, with bubbles and shrinkage cavities marked by yellow circles. As can be seen in Fig. 30a, b, there were three bubbles or shrinkage cavities in the composite material prepared from chain-unextended epoxy resin (EP/Me-THPA molar ratio of 1:0.0) and that prepared from 1:0.1 molar ratio chain-extended resin, whereas only two bubbles or shrinkage cavities were found in the material formed from the 1:0.3 chain-extended resin (Fig. 30c). Moreover, there were almost no bubbles or shrinkage cavities on the surface of the material based on the 1:0.5 chain-extended resin (Fig. 30d). These results indicate that the number of defects in epoxy resin/Al_2O_3 composites can be significantly reduced by the increasing EP/Me-THPA molar ratio.

During the curing of a chain-extended epoxy resin/Al_2O_3 composite material, the curing exothermic rates tends to decrease, suppressing the phenomenon of rapid curing in local areas. The curing stress of the composite material becomes lower, leading to a decrease in the number of voids produced by liquid shrinkage and solidification shrinkage in the supplementary solidified layer. Thus, the number of internal defects in the composite material also decreases. As the areas where charge may accumulate are reduced in number, there is less distortion of the electric field. Thus, the voltage withstand capability of the composite material is improved and the tolerance time is prolonged.

Filler agglomeration is another internal defect that deteriorates the insulation properties of insulating materials [34]. When a composite material is formed from filler and epoxy resin, there are obvious gas–solid interface defects. These prevent uniform dispersion of the filler, leading to its agglomeration. Air gaps and impurities often exist in the areas of filler agglomeration. Partial discharge from air gaps and impurities will occur under the action of an electric field, degrading the insulation properties of epoxy resin composites [35]. Therefore, reducing filler agglomeration will help to improve the performance of insulating materials.

As can be seen from Fig. 30, there were obvious filler agglomerations on the surface of the composites based on chain-unextended and 1:0.1 chain-extended epoxy resins (red circles). However, there were no filler agglomerations in the composites based on the 1:0.3 and 1:0.5 chain-extended resins, in which the fillers were dispersed uniformly. In these latter materials, the compatibility at the interface between the epoxy resin modified by chain extension and the filler was improved. The filler dispersed uniformly into the epoxy resin, reducing the occurrence of agglomeration. Therefore, the regions of charge accumulation were reduced, and the voltage withstand capability of the composites was improved.

FTIR spectra of chain-extended epoxy resin/Al_2O_3 composites before and after surface flashover are shown in Fig. 31. From Fig. 31a, b, it can be seen that no new functional groups were produced during surface flashover. Carbon–oxygen double bonds (–C = O, wavenumber range 1750–1680 cm^{-1}), benzene rings (1600–1500 cm^{-1}), and carbon–oxygen single bonds (–C–O, 1275–1020 cm^{-1}) were present in both the chain-unextended and chain-extended epoxy resin/Al_2O_3 composites.

The number of benzene rings on the molecular chains of the six epoxy resin composites did not change significantly as a result of surface flashover. This is because the benzene rings are chemically stable, and bond breakage does not easily occur under the action of an electric field. Therefore, the respective absorption peaks of benzene rings before and after surface flashover could be used as references. The absorption peaks of each functional group were compared with these, and the relative contents of different functional groups in each system before and after surface flashover were thereby obtained. The results are presented in Tables 8 and 9.

From Tables 8 and 9, it can be seen that the –C=C, –OH, –C–C and –C=O contents in the molecular chains were relatively unchanged by surface flashover. Therefore, it can be assumed that these functional groups are not significantly affected by the electric field. However, the content of –C–O bonds in the molecular structure of the epoxy resin was reduced by chain extension, and, after surface flashover, the content of –C–O in the composites had reduced, with the reduction increasing with the EP/Me-THPA molar ratio.

The bond energy of the –C–O single bond is 326 kJ/mol. This is less than the bond energies of –C=C, –OH, –C–C and –C=O bonds, and thus requires the least energy to break the bond. Therefore, under the action of an electric field, the –C–O bonds break more easily, which reduces the electrical insulating performance of epoxy resin/Al_2O_3 composites. Note that only the –C–O bonds change significantly in number as a result of flashover, and it can be concluded that the –C–O functional groups had a major influence on the surface flashover characteristics of the composites. From the above test results, it can be seen that chain extension reduces the content of –C–O bonds in the molecular structure of epoxy resin composites,

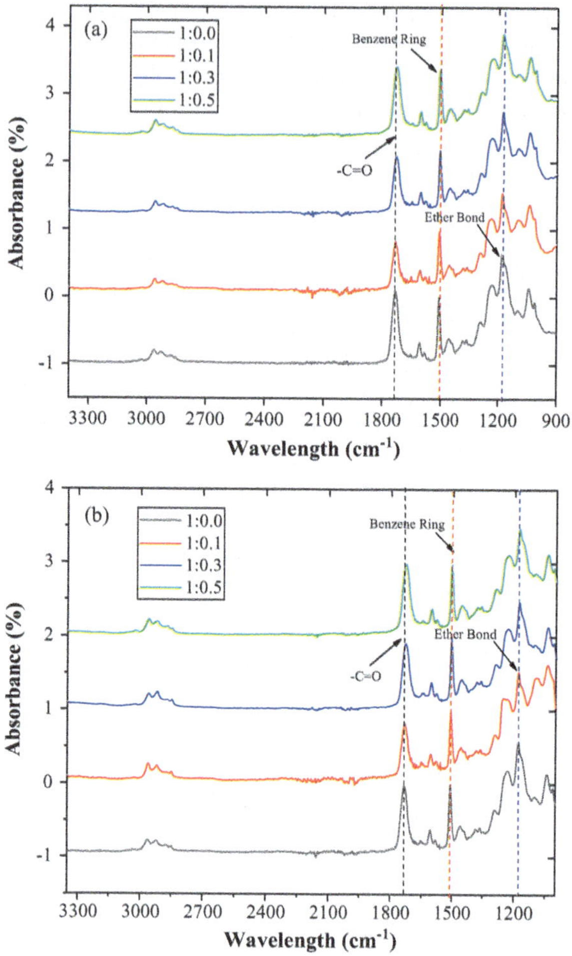

Fig. 31 FTIR spectra of chain-unextended and chain-extended epoxy resin/Al_2O_3 composites **a** before and **b** after surface flashover

Table 8 Infrared absorption peaks of functional groups relative to that of benzene rings before surface flashover

EP/Me-THPA molar ratio (mol)	– C=C	– OH	– C–C	– C=O	– C–O
1:0.0	0.3321	0.093	1.5528	1.1022	1.6528
1:0.1	0.4237	0.1132	1.4631	0.8494	1.5964
1:0.3	0.3733	0.1041	1.4709	0.9097	1.5709
1:0.5	0.3682	0.0928	1.5216	1.0201	1.5216

Table 9 Infrared absorption peaks of functional groups relative to that of benzene rings after surface flashover

EP/Me-THPA molar ratio (mol)	– C=C	– OH	– C–C	– C=O	– C–O
1:0.0	0.3808	0.0922	1.573	0.9761	1.5730
1:0.1	0.4233	0.1103	1.5419	0.839	1.5419
1:0.3	0.3926	0.114	1.4866	0.9235	1.4866
1:0.5	0.3926	0.079	1.4800	1.0035	1.3872

as a result of which the molecular structure became more stable. This enhances the resistance to degradation and improves the surface flashover characteristics. The 1:0.5 chain-extended epoxy resin/Al_2O_3 composite material had the lowest content of –C–O bonds, and therefore the best surface flashover characteristics.

Basin-type insulators are made by chain-extended epoxy resin/Al_2O_3 composites, which are indoor insulation components. The chain-extended epoxy resin is prepared from liquid bisphenol-a epoxy resin. Both benzene rings and ester groups exist in the molecules of bisphenol-a epoxy resin, and therefore exist in chain-extended epoxy resin/Al_2O_3 composites. The benzene rings are easy to breakdown under outdoor UV radiation. The ester groups are easily hydrolysed, resulting in poor water ageing under outdoor conditions. For outdoor insulation materials, excellent hydrophobicity, UV-radiation resistant and weather resistance abilities are required. The outdoor insulation materials are usually prepared by epoxy resins and curing agents which contain no unsaturated functional groups in molecular chains, such as hydrogenated bisphenol-a epoxy resins, cycloaliphatic epoxy resin and methylhexahydrophthalic anhydride (Me-HHPA). In our opinion, the chain-extended epoxy resins prepared in this work should not be used as outdoor insulation materials, but the method of chain extension could potentially be employed for the hydrogenated bisphenol-a epoxy resins. There have been many studies on the mechanical and thermal properties of chain-extended epoxy resins, but few reports on its electrical performance. It is hoped that this study will provide a reference for preparing indoor and outdoor insulation materials.

5.4 Conclusion

During the process of producing basin-type insulators, there is rapid heat release from the epoxy resin, leading to a concentration of curing stress and eventually to defects in the form of bubbles and contraction cavities. In this study, low-molecular-weight epoxy resins were modified by chain extension using Me-THPA with the aim of reducing this problem. The conclusions are as follows:

(1) The curing stress is closely related to the curing exothermic rates. Through chain extension, the curing exothermic rates can be decreased, thereby reducing the curing stress in chain-extended resin/Al_2O_3 composites. In addition, both the

curing exothermic rates and curing stress decreased gradually with increasing EP/Me-THPA molar ratio.

(2) Under the application of a negative DC voltage, the surface flashover voltage of the chain-extended epoxy resin/Al_2O_3 composites did not increase significantly. However, as the EP/Me-THPA molar ratio increased, the voltage withstand capability of the composites improved. The composite material based on the chain-extended epoxy resin with a 1:0.5 molar ratio achieved the longest tolerance time.

(3) There were many bubbles and shrinkage cavities on the surfaces of the chain-unextended epoxy resin composites, and filler agglomeration was also observed. For the epoxy resin composites modified by chain extension, the number of internal defects (bubbles and shrinkage cavities) was significantly reduced, which greatly improved the voltage withstand capability of the composites.

(4) The content of –C–O bonds in the molecular structure of the epoxy resin was reduced by the chain-extension modification. The –C–O bond energy is low, so this bond breaks easily when a voltage is applied. Thus, the –C–O bonds were the main functional group affecting the surface flashover properties of the epoxy resin composites. Reducing the content of –C–O bonds in the resin had a beneficial effect on the surface flashover properties of the composites.

References

1. S. Okabe, S. Yuasa, S. Kaneko et al., Simulation of propagation characteristics of higher order mode electromagnetic waves in GIS[J]. IEEE Trans. Dielectr. Electr. Insul. **13**(4), 855–861 (2006)
2. D. Chernoskutov, Development and research of switching performance of high voltage Gas insulated disconnecting switch[C], in *International Siberian Conference on Control and Communications* (IEEE, 2015)
3. G. Ueta, S. Okabe, T. Utsumi et al., Electric conductivity characteristics of FRP and epoxy insulators for GIS under DC voltage[J]. IEEE Trans. Dielectr. Electr. Insul. **22**(4), 2320–2328 (2015)
4. P. Li, J. Li, Y. Qin et al., Wormholes effect in flashover process of alumina filled epoxy resin surfaces in SF_6[J]. IEEE Trans. Dielectr. Electr. Insul. **22**(4), 2142–2147 (2015)
5. J.Y. Shin, I.B. Jeong, H.M. Choi et al., Electrical insulating properties of nano-composites according to additive[C], in *IEEE International Conference on Properties and Applications of Dielectric Materials* (IEEE, 2009)
6. M. Awais, R. Sundararajan, I.A. Sajjad et al., Investigation on optimal filler loadings for dielectric strength enhancement of Epoxy/TiO_2@SiO_2 Nanocomposite[J]. Mater. Res. Express (2019)
7. Y. Zhao, Y. He, K. Yang et al., Improving the Surface Insulating Performance of Epoxy Resin/Al_2O_3 Composite Materials by Extending Chain of Liquid Epoxy Resin with Me-THPA[J] (2019)
8. S. Zhang, Y. Zhao, X. Wang et al., Study on the effect of curing shrinkage marks on electrical properties of epoxy composites[M], in *Proceedings of the 21st International Symposium on High Voltage Engineering*, vol 2 (2019)
9. G. Ueta, Insulation performance of three types of micro-defects in inner epoxy insulators[J]. IEEE Trans. Dielectr. Electr. Insul. **19**(3), 947–954 (2012)

10. K. Ravindranath, K.S. Gandhi, Molecular weight distributions in epoxy resins[J]. J. Appl. Polym. Sci. **24**(4), 1115–1123 (1979)
11. Atsushi Sudo, Yukio Isobe, Takeshi Endo, Epoxy curing process with small shrinkage based on Binary Nucleophilic Reagent System consisting of Amine and Carboxylic Acid[J]. J. Appl. Polym. Sci. **112**(2), 836–842 (2009)
12. X.Q. Li, G. Liu, Z.L. Zhang et al., Curing shrinkage and process parameter optimization of epoxy resin based on FBG strain monitoring[J]. Acta Mater. Compos. Sinica **33**(10), 2151–2157 (2016)
13. S.S. Kim, H. Murayama, K. Kageyama et al., Study on the curing process for Carbon/Epoxy composites to reduce thermal residual stress[J]. Compos. A **43**(8), 1197–1202 (2012)
14. P. Preetha, M.J. Thomas, R. Ranjan, Electrothermal ageing of epoxy nanocomposites[J]. IEEE Trans. Dielectr. Electr. Insul. **19**(6), 2081–2089 (2012)
15. Y. Yang, G. Xian, H. Li et al., Thermal aging of an anhydride-cured epoxy resin[J]. Polym. Degrad. Stab. **118**, 111–119 (2015)
16. W.X. Sima, Z. Yin Z, P. Sun et al., Effects of arc ablation on surface electrical performance of epoxy resin insulation[J]. J. Phys. D Appl. Phys. **51**(26) 2018
17. Y.S. Zhao, T.Q. Zheng, K.R. Yang, et al., Study on residual stress of epoxy resin under different cooling methods[C], in *2018 IEEE International Conference on High Voltage Engineering and Application (ICHVE)*
18. S. Zhang, C. Tang, J. Hao et al., Thermal Stability and Dielectric Properties of Nano-SiO_2-doped cellulose[J]. Appl. Phys. Lett. **111**(1), 012902 (2017)
19. C. Tang, S. Zhang, X.B. Wang et al., Enhanced mechanical properties and thermal stability of cellulose insulation paper achieved by doping with melamine-grafted Nano-SiO_2[J]. Cellulose **25**(6), 1–15 (2018)
20. S. Zhang, Y.S. Zhao, K.R. Yang et al., Study on aging mechanism of Bisphenol a epoxy resin based on molecular simulation[C], in *2018 IEEE 2nd International Electrical and Energy Conference (CIEEC)*
21. J. Ding, L. Zhang, Y. Zhang et al., A reactive molecular dynamics study of N-heptane Pyrolysis at high temperature[J]. J. Phys. Chem. A **117**(16), 3266–3278 (2013)
22. K.L. Yin, G.Y. Wu, C.L. Chen, Investigation of thermal cracking process of n-decane by molecular dynamics simulation[J]. Acta Petrolei Sinica **17**(2), 77–82 (2001)
23. Y. Deshui, Formation and Slowing of Internal Stress in Epoxy Casting Components[J]. High Volt. Appar. **4**, 48–54 (1986)
24. M. Shimbo, M. Ochi, Y. Shigeta, Shrinkage and internal stress during curing of epoxide resins[J]. J. Soc. Mater. Sci. Jpn. **26**(7), 2265–2277 (1981)
25. W. Hongbin, L. Shanjun, Y. Tongying, Epoxy resin internal stress and its determination[J]. Thermosetting Resin **4**, 30–35 (1990)
26. H. Qifeng, L. Shengtao, S. Jian et al., Influences of trap levels on surface flashover of XLPE in Vacuum[J], in *2009 IEEE 9th International Conference on the Properties and Applications of Dielectric Materials*, Harbin, China, pp. 679–682, Sept. 2009
27. T. Ju, H. Yao, Y. Qiang et al., Decomposition characteristics of SF6 under partial discharge at different gas pressures[J]. High Volt. Eng. **40**(8), 2257–2263 (2014)
28. Z. Lizhen, T. Jiansheng, Formation and reduction of internal stress in epoxy resin composite products[J]. Guangdong Sci. Technol. **5**, 45–47 (2004)
29. M. Yiyang, Z. Cheng, K. Fei et al., Surface treatment of epoxy resin by sub-atmospheric-pressure dielectric barrier discharge: the effect on surface charge dissipation and aging characteristics[J]. Trans. China Electrotech. Soc. **33**(22), 5168–5177 (2018)
30. Z. You, J. Jun, L. Yingting et al., Statistical analysis of AC Breakdown Voltage for transformer oil[J]. Insul. Mater. **48**(3), 73–77 (2015)
31. M. Linkun, X. Dengming, Experimental study of the insulation performance of c-C4F8 in a sphere-plate electrode gap and its surface flashover characteristics around an insulator[J]. High Volt. Appar. **52**(12), 66–70 (2016)
32. W. Yongqiang, H. Fangfang, X. Jun et al., Simulation study on electric field change rule of GIS disc insulator with air gap defects[J]. Insul. Mater. **49**(1), 40–45 (2016)

33. W. Haoran, G. Zihao, Z. Siyu et al., Influence of UHV AC GIS spacer defects on electrical field distribution[J]. High Volt. Eng. **44**(3), 982–992 (2018)
34. Y. Xiangxiao, N. Pinghong, Research and application of epoxy modified heat-resistant polyester solventless imprenating varnish[J]. Insul. Mater. (6), 3–6 + 10 (2003)
35. K. Linlong, C. Pengfei, Z. Hui et al., Effect of surface modification for Micro-Al_2O_3 on insulating properties of epoxy resin of electrical grade[J]. Insul. Mater. **51**(6), 35–41 (2018)

HVDC Spacers by Adaptively Controlling Surface Charges

Chuanyang Li, Chuanjie Lin, and Tohid Shahsavarian

Abstract A novel design of HVDC spacer was introduced based on the concept of adaptively controlling surface charges by using nonlinear materials. Unlike the commonly used traditional approaches, in this method, a novel idea for controlling the location of accumulated charges and properly decaying of these charges has been proposed initiatively, which fundamentally solves the problem of surface charge accumulation. The mechanical and electrical test results show that the charge adaptively controlling spacer has high operating capability under the DC voltage and has great industrial application potential.

1 Introduction

The spacers operating in DC application face the problem of surface charge, which is a serious limiting factor since they accumulate on the spacer surface, distort local electric field, and potential cause of surface flashover [1]. Many studies have been conducted in last few decades regarding the surface charge origin and suitable solutions for decaying these charges [2, 3]. In a recently published paper, a field dependence theory was introduced that determines the origin of surface charges depending on the electric field strength [4]. It is illustrated that the majority of surface charges are from volume conduction current at low electric field, while charges of hetero polarity may be generated if there are local defects on the surface of the grounded enclosure [4]. Meanwhile, a lot of efforts have been made by researchers to find the most suitable solution for the decay of surface charges [5–7]. Although the origin of these charges can be speculated, it is still quite difficult to hinder them or even modify them without affecting the other insulation properties, i.e. the leakage current due to a surface modification by creating a more conductive surface layer [7, 8].

C. Li (✉) · C. Lin
Tsinghua University, Beijing, China
e-mail: lichuanyangsuper@163.com

T. Shahsavarian
University of Connecticut, Storrs, USA

B. Du (ed.), *Polymer Insulation Applied for HVDC Transmission*,
https://doi.org/10.1007/978-981-15-9731-2_14

In this chapter, we introduce a novel designed HVDC spacer model, named charge adaptive control spacer (CACS) that is based on the concept of adaptively controlling surface charges. In case of a CACS, the charges from the volume or from the gas are controlled by the electric field distribution and accumulate on the surface of the charge adaptive region (CAR). The doping of nonlinear material in CAR ensures the limitation of surface charge amount on the surface of the CAR. As long as the existing charge amount exceeds a certain value, the CAR adaptively dissipates charges without being concerned about the location of accumulated charges and their amount. Meanwhile, this spacer can be prepared by one-shot molding without subsequent surface treatment procedures. In addition, the leakage current can be well controlled under a suitable electric field [9].

2 Design Concept

2.1 Shape Optimization

Cone-type spacers are designed to separate gas compartments, and thereby are manufactured generally into a cone shape to increase the mechanical stress caused by the imbalance of the gas pressure during the operation and/or the gas handling process. This conical structure inevitably leads to a normal electric field component on the spacer, which causes charge accumulation at the gas–solid interface.

Unlike the regular cone-shaped spacer (RS), a CACS is formed like a bowl, which is composed of two parts, namely the insulation region (IR) and the charge adaptive control region (CACR), as shown in Fig. 1. By doing so, CACSs can meet the requirements of the longitudinal mechanical force and more importantly, the normal component of the electric field can be concentrated on the side wall of the spacer near the ground enclosure, i.e. CACR [9].

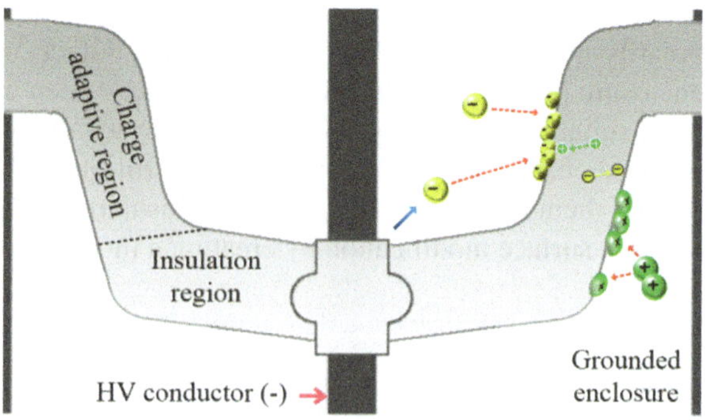

Fig. 1 The sketch of the spacer structure and the space charge migration [9]. Copyright Clearance Center's RightsLink© License Number: 4857300749179

2.2 *Material Modification*

The IR and the CACR consist of epoxy resins blended with Al_2O_3 and SiC in different ratios. Unlike the traditional manufacturing method of epoxy based spacers, the manufacturing method and the curing process of the CACS have been separated into two stages.

In this manner, when the local surface charge in the CACR exceeds a certain value and the electric field increases beyond a certain level, the local dielectric conductivity increases up to one to two orders of magnitude, which dissipates the surface charges and limits the local field intensification to a safety limit. Thus, when a polarity reversal or over-voltage occur, the conductivity of the surface charge region can be increased accordingly, which optimizes the local electric field distribution and releases the local surface charges. In addition, the spacer compact of the IR and CACR exert better control in limiting the leakage current, compared to the spacer doped with high mass fraction of nonlinear materials.

3 Nonlinear Conductivity

The materials used for the spacers were epoxy resins blended with 99.9% pure α-Al_2O_3 insulating powders; the particle size distribution was analyzed using a Malvern laser particle size analyzer. Figure 2 shows the particle size fraction and cumulative particle size fraction as a function of the particle size. The size of the Al_2O_3 particles is ranged from 0.5 to 61 μm with a volume mean diameter of around 21 μm. For the SiC particles, the size is ranged from 2.4 to 19.5 μm with the volume mean diameter of around 11 μm [9].

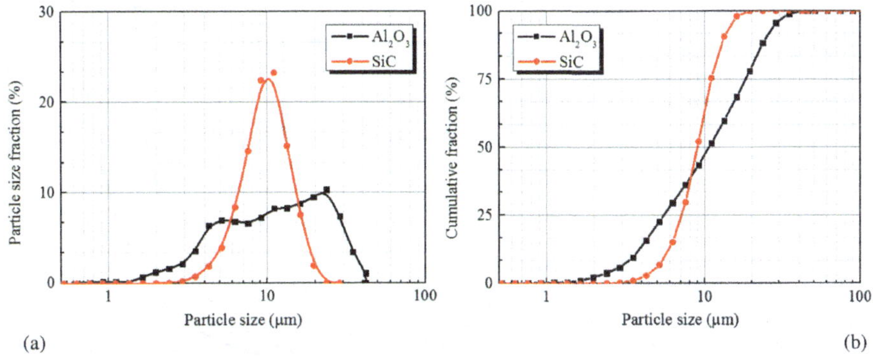

Fig. 2 Particle size distribution of Al_2O_3 and SiC (800 mesh particle size): **a** Particle size fraction versus particle diameter. **b** Cumulative particle size fraction versus particle diameter. Copyright Clearance Center's RightsLink© License Number: 4857300749179

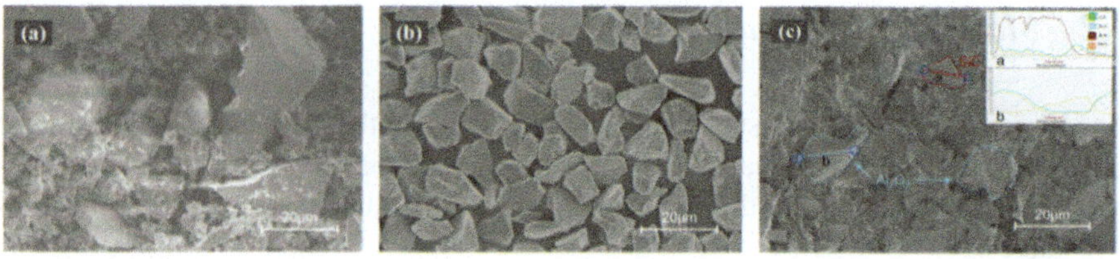

Fig. 3 SEM images of experimental samples: **a** Al_2O_3 particles, **b** SiC particles, and **c** surface of epoxy based Al_2O_3 and SiC composite (SiC weight ratio = 20%) [9]. Copyright Clearance Center's RightsLink© License Number: 4857300749179

Figure 3 shows the SEM images of the experimental particles before and after they were blended into the epoxy resins. In Fig. 3a, it can be observed that the shape of the Al_2O_3 particles varied widely and include some particles with sharp angular edges. The shape of the SiC particles was much more uniform, as shown in Fig. 3b. Figure 3c shows the epoxy resin composite with an Al_2O_3 to SiC ratio of 8/2 by weight. It was found that the particles were uniformly distributed within the matrix.

SiC is a type of nonlinear material whose bulk resistivity decreases as the electric field strength increases. It has many advantages over other fillers, such as a high thermal conductivity, low thermal expansion coefficient, high chemical stability, and high hardness, which makes it suitable for a wide range of applications in electrical engineering. However, it is not well specified in terms of electrical properties. It is known that even being produced by the same supplier, the SiC manufactured in different batches may possess uncontrolled electrical properties, which remains as a potential problem. Figure 4 shows the J(E) characteristics of 800-mesh-sized SiC particles of different groups under different temperature. Each group was manufactured by the same supplier but produced in different batches. It can be found that the SiC particles have nonlinear properties under electric fields ranging from 3 to

Fig. 4 J(E) characteristics of 800-mesh-sized SiC particles of different groups under different temperature [9]. Copyright Clearance Center's RightsLink© License Number: 4857300749179

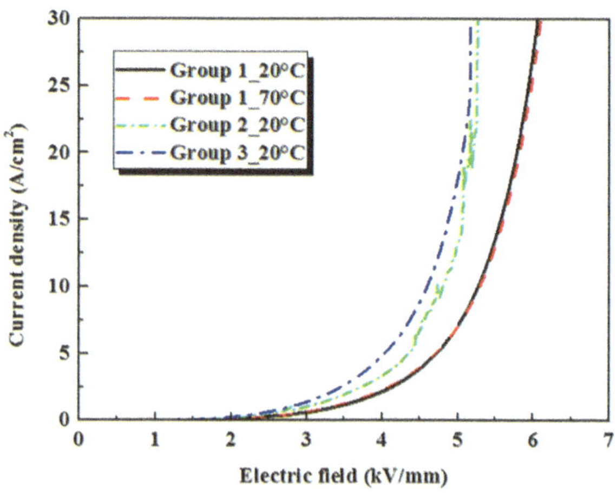

5 kV/mm. However, there are differences in the nonlinear characteristics of different batches of samples. Our results show that the J(E) characteristic curve in the nonlinear range of group 3 has shifted to the left with an offset of 1 kV/mm compared with group 1. In addition, it is interesting to note that the change in temperature can slightly affect the nonlinear characteristic of SiC particles.

Figure 5 shows the bulk conductivity changes in insulators doped with SiC and Al_2O_3 fillers. It can be seen that the local conductivity of the modified material remains below 10^{-14} S/m when the electric field strength is in the range of 1–4 kV/mm. When the electric field strength exceeds 5 kV/mm, the conductivity of the modified dielectric increases beyond 1×10^{-14} S/m, which reaches one order of magnitude higher than that of the pure epoxy resins. In addition, it can be seen that the increasing bulk conductivity is positively correlated with the doping concentration and the particle size. It can be concluded that the SiC doped composite shows better nonlinear properties when the grain particle size is smaller. The volume fraction of SiC fillers is less than 23%, which is below the percolation threshold of SiC/epoxy composites [9]. Additionally, Al_2O_3 fillers may block the conduction path formed by SiC fillers. Under this circumstance, smaller SiC fillers are easier

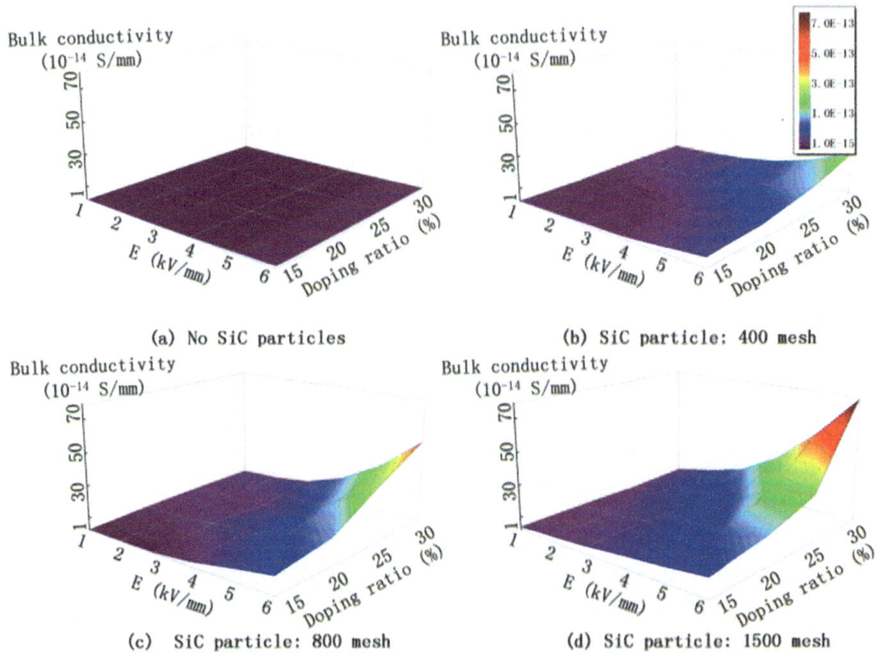

Fig. 5 The non-linear bulk conductivity of epoxy -based materials using SiC fillers under different electric field stresses for an epoxy resin blended with: **a** Al_2O_3, **b** Al_2O_3 and SiC (400 mesh: with most of the particle size ranging from 36 to 38 μm), **c** Al_2O_3 and SiC (800 mesh: with most of the particle size ranging from 15 to 18 μm), and **d** Al_2O_3 and SiC (1500 mesh: with most of the particle size ranging from 6 to 10 μm) [9]. Copyright Clearance Center's RightsLink© License Number: 4857300749179

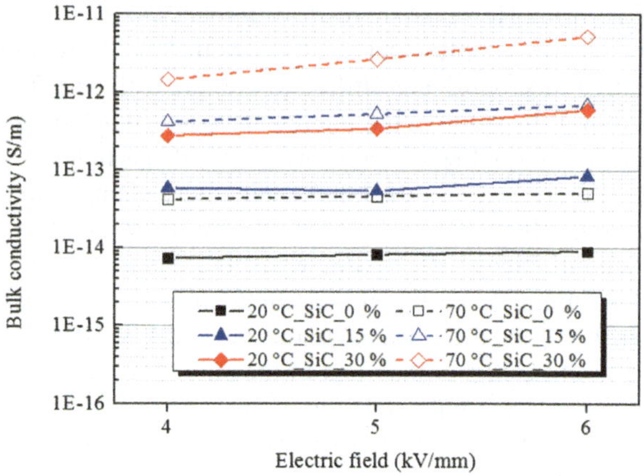

Fig. 6 The temperature dependency of bulk conductivity of experimental samples with different doping amount of SiC fillers under different electric field stresses [9]. Copyright Clearance Center's RightsLink© License Number: 4857300749179

to create local contact interfaces which could contribute to nonlinear conductivity property. Therefore, under relatively high electric field, composites with a smaller SiC fillers possess more active contact interfaces and exhibit considerable nonlinear conduction behavior.

However, it was found that the smaller grain can increase the viscosity of the epoxy-based material in the blending process, which makes the casting process harder. As a consequence, we determined that SiC particles with 800-mesh-sized particles are optimal to use in the manufacture the proposed novel spacers.

Due to the fact that the temperature gradient from inner to outer conductor is typically 50 K and has severe effect on the local conductivity distribution, the temperature dependency of the SiC filled composite should be discussed. The bulk conductivity of the epoxy with and without SiC particles of 800-mesh-sized particles were tested and the results are shown in Fig. 6. It is found that for the experimental samples, the bulk conductivity increases up to one order of magnitude when the temperature is increased from 20 °C to 70 °C. However, the increase in temperature does not have significant impact on the nonlinear properties of experimental samples.

4 Sample Preparation

The fillers used in the IR and the CACR were prepared separately, and the mass ratio of the fillers in the two regions is given in Table 1. The mold was cleaned very carefully using the release agent. After that, the fillers, epoxy resins, and the mold were placed in an oven at 130 °C for over 4 h. The fillers were mixed separately with the epoxy and stirred at 130 °C for more than 20 min. Then, the mixed epoxy-based composite was placed inside a vacuum oven for over 1 h. The epoxy-based composite was blended with a hardener and stirred for 10 min before it was poured into the mold. Then, the mold was placed back into the oven for the initial curing stage of

Table 1 Mechanical parameters of the materials [11]. Copyright Clearance Center's RightsLink© License Number: 4857310188716

Parameter	Epoxy	Aluminum
Young's modulus	6.2 GPa	72 GPa
Poisson's ratio	0.313	0.33
Density	2.69 g/cm^3	2.3 g/cm^3
Bending strength	69–100 MPa	–
Tensile strength	120–136 MPa	–

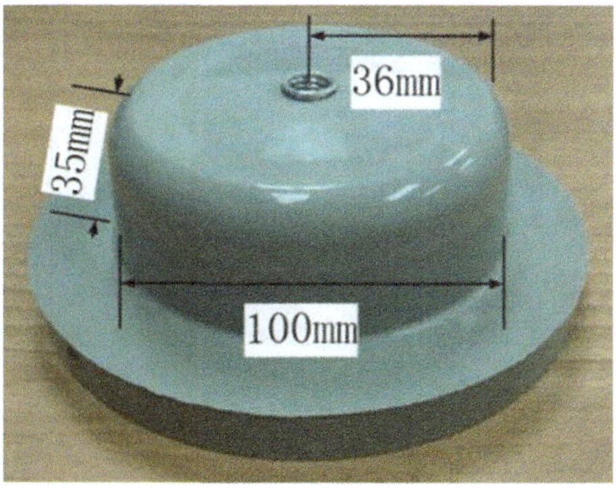

Fig. 7 Profile of the charge adaptively control spacer [10]. Copyright Clearance Center's RightsLink© License Number: 4857301321085

the insulation region. After 4 h, the epoxy-based composite was blended with the hardener and stirred for 10 min prior to pouring into the mold. The final curing step lasted 24 h. After the spacer was removed from the mold, the surface local defects were polished using 1000 # sandpaper. The profile of the CACS can be found in Fig. 7.

5 Mechanical Property

5.1 Mechanical Stress Simulation

Table 1 lists the mechanical properties of the used epoxy resin and aluminum. The bending strength and the tensile strength used in the paper are 69 and 120 MPa respectively. The force conditions of the RS and the CACS are shown in Fig. 8. The main forces exerted on the cone-type spacers are those from the SF6 gas pressure and the fixed force of the electrodes.

The designed pressure value, the routine and destruction-level test values of the pressure are 0.75, 1.5, and 2.25 MPa respectively. The spacers will work normally if their maximum tolerance stress exceeds 2.25 MPa. According to the four strength

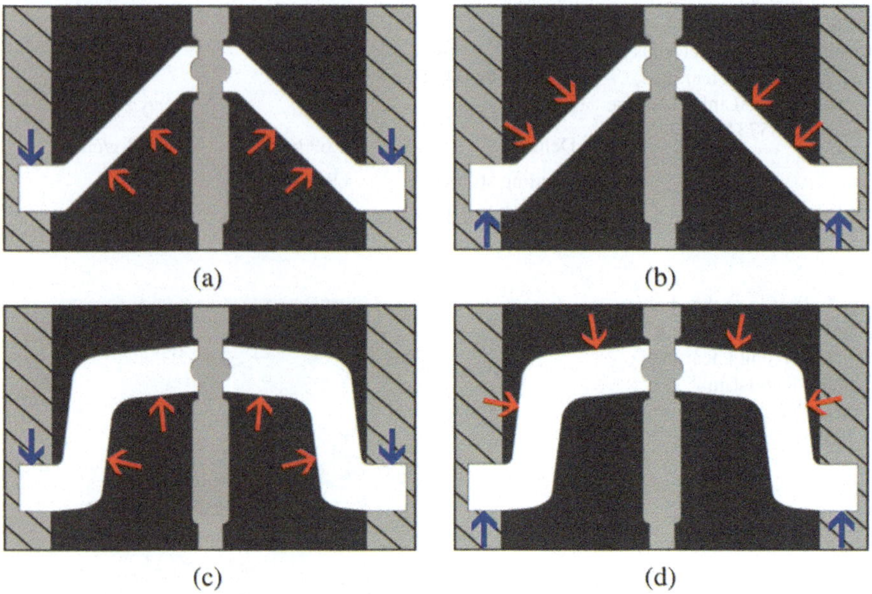

Fig. 8 Mechanical stress conditions of the spacers: **a** and **b** show the cases where the stresses are exerted on the convex and concave sides of the RS, while **c** and **d** show the cases where the stresses are exerted on the convex and concave sides of the CACS, respectively. The red arrows indicate the direction of the force exerted by the SF6 gas pressure. The blue arrows indicate the direction of the counterforce exerted at the contact point with the external enclosure

Table 2 Maximum tolerable stresses of the spacers: (Units: MPa)

Position	RS	CACS
	FPS	FPS
Convex	10.16	15.33
Concave	7.68	7.56

theories in the fundamental of mechanics of materials, for a physically brittle structure, the load-carrying capacity is dominated by the first principal stress (FPS) and the tensile strength of the material.

The simulation results are shown in Table 2, in which all the values exceed the 2.25 MPa. The maximum tolerable stresses on the convex side are higher, and the geometric improvements lead to higher tolerable stresses on the convex. For the RS, the maximum tolerable stresses on the convex and the concave sides are 10.16 and 7.68 MPa respectively. For the CACS, the maximum tolerable stresses on the convex and the concave sides are 15.33 and 7.56 MPa respectively.

Table 3 The mass ratio of the fillers in the insulation region and in the charge adaptive control region

Sample	Mass ratio (Al_2O_3:SiC)	
	Insulation region	Charge decay region
B1 _0% _0%	Al_2O_3:SiC = 330:0	Al_2O_3:SiC = 330:0
B2 _0% _30%	Al_2O_3:SiC = 330:0	Al_2O_3:SiC = 231:99
B3_10%_25%	Al_2O_3:SiC = 297:33	Al_2O_3:SiC = 247:83
B4_15%_35%	Al_2O_3:SiC = 280:50	Al_2O_3:SiC = 214:116
B5_20%_30%	Al_2O_3:SiC = 264:66	Al_2O_3:SiC = 231:99
B6_25%_35%	Al_2O_3:SiC = 247:83	Al_2O_3:SiC = 214:116

5.2 *Water Pressure Test*

The mechanical properties are tested by the water pressure test to verify the pressure bearing capacity of the CACS. Figure 9 shows the experimental setup. First, the water holder is filled with water. Then, the experimental sample is fixed to the water holder and compressed using a top cap. Six bolts are then used to clamp the spacer to the water holder. A hydraulic pump is then used to raise the pressure in the water holder until a crack appears in the spacer [11].

Figure 10 shows the water pressure test results for RS and CACS. The results suggest that the ultimate failure pressure for the CACS is slightly lower than the AC model spacer when the concave surface is under water pressure. However, when the convex surface is under this pressure, the ultimate failure pressure of the CACS is much higher, at above 12 MPa, compared to the AC model spacer (less than 11 MPa). As Fig. 10b, c show, the cracks in all RS and CACS samp sturdy enough.

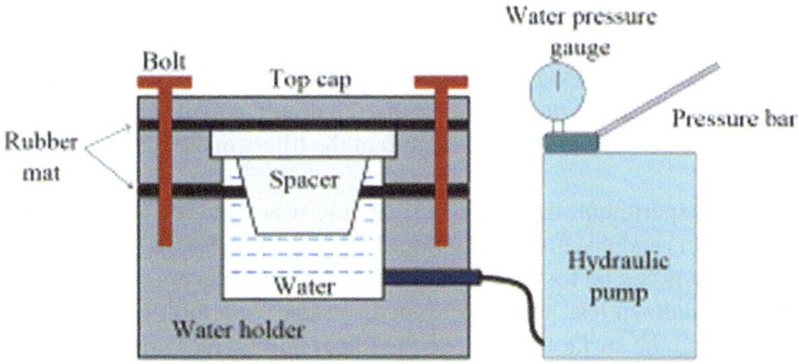

Fig. 9 Experimental setup for water pressure test [11]. Copyright Clearance Center's RightsLink© License Number: 4857310188716

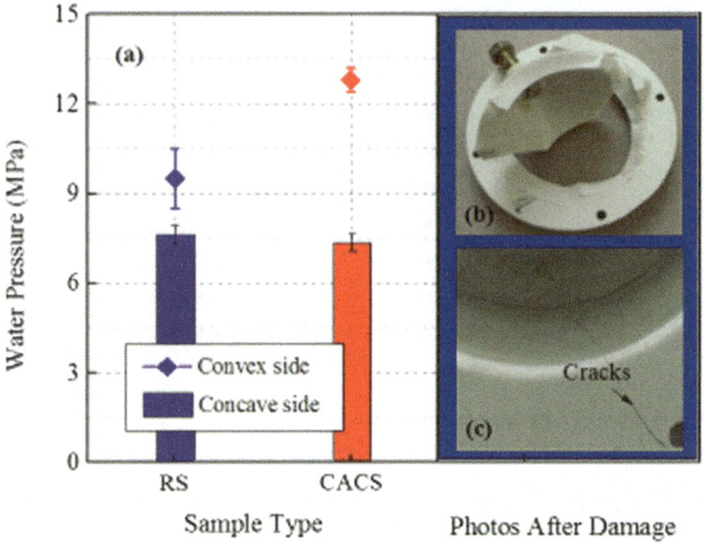

Fig. 10 Water pressure test results: **a** water pressure results for the RS, and the CACS with different doping concentrations in the charge adaptive region; **b** photograph shows the spacer damage mode when the concave side is under pressure; **c** photograph shows the spacer damage mode when the convex side is under pressure [11]. Copyright Clearance Center's RightsLink© License Number: 4857310188716

6 Electrical Property

6.1 DC Surface Flashover Results

The manufacturing methods and the curing process of the experimental samples were the same as for real-sized industrial spacers. The manufacturing process details can be referred to our previous published articles [9–12]. For the experimental bowl -shaped spacers, the fillers used in the insulation region and the charge decay region were prepared separately, and the mass ratio of the fillers in the two regions are given in Table 3.

During the experiment, the spacer test sample was fixed into a model enclosure (shown in Fig. 11) and this model was placed inside a 220 kV GIL [10]. The details of experimental tests can be found in Table 4. For the pure spacers, we used a new one after a DC breakdown test, while the doped spacers were multiply used. After DC test, we modified the surface to remove the traces and surface charges from surface flashover and reused the sample.

Figure 12 shows the DC surface flashover test results of the experimental samples. It can be found that the doping of nonlinear material can affect the surface flashover voltage, even so, it is not always the case that the doping of nonlinear material can increase the surface flashover voltage. For sample B2, i.e. samples without doping

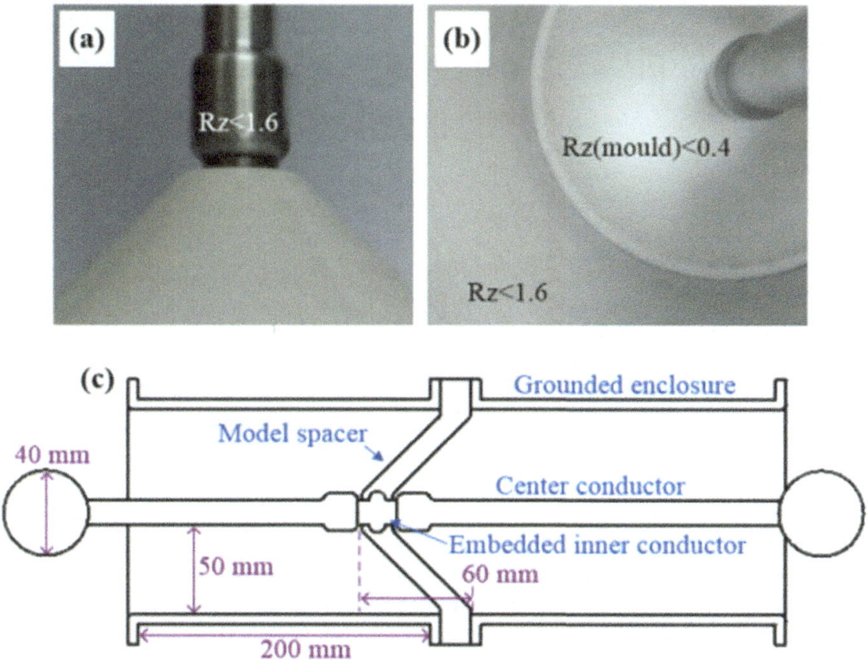

Fig. 11 Photographs showing a magnified view of the experimental sample model (panels (**a**) and (**b**)). Panel (**c**) shows a diagram of the installed spacer in the system [10] Copyright Clearance Center's RightsLink© License Number: 4857301321085

Table 4 Experimental samples profile and applied voltage procedure.[10] Copyright Clearance Center's RightsLink© License Number: 4857301321085

Experimental samples		Voltage procedure
Spacer type	Side view	
Novel shaped spacer (NSS)	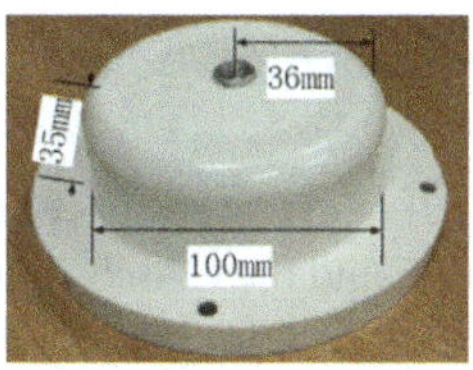 	**DC test:** To apply a DC voltage up to −240 kV with a rate of −3 kV per second for 30 min, and then, increase the DC voltage with a rate of −1 kV per second up to −250 kV. Repeat this process until the surface flashover occurs
Charge adaptive spacer (CACS)	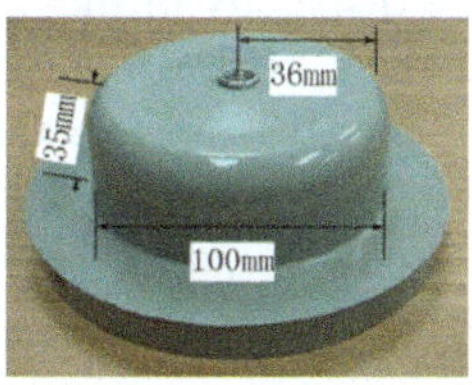 	**AC test:** To increase the AC voltage with a rate of −1 kV per second to measure the surface flashover voltage

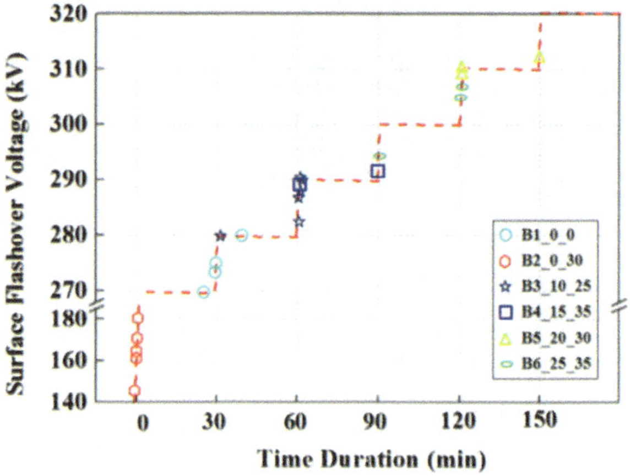

Fig. 12 DC surface flashover test results of experimental samples [12]. Copyright Clearance Center's RightsLink© License Number: 4857310380945

of nonlinear material in the dielectric region, the surface flashover voltage shows the lowest values which are less than 180 kV. This value is lower compared with pure samples B1 whose surface flashover voltage values range from 270 to 280. When the insulation region and the charge adaptive control region are both doped with nonlinear material, the surface flashover can be increased, however, only when the doping ratio of the insulation region increases to above 15%, surface flashover voltage can be higher than 300 kV. The DC surface flashover voltage of tested spacers can be stabilized at 310 ± 5 kV (more than about 10% compared with that of the traditional cone type spacer [10]) when the mass fraction of nonlinear materials doped in the insulating region and the charge adaptive region are 20 and 30% respectively. It is experimentally verified that the samples group with the optimal doping ratio is B5_20_30.

6.2 AC Surface Flashover Results

Figure 13 shows the AC surface flashover test results of the experimental samples. The surface flashover voltage of the NSS and the CACS (between the value of 200 kV and 225 kV) is increased to 185 kV compared to the RS whose surface breakdown value is around 160 kV. In addition, the dispersity of the surface flashover voltage value for the RS and NSS is larger than the CACS, while CACS shows good convergence in its surface flashover voltage range. It is to note that during the experiment, the lowest value of the NSS marked by the green box is less than 160 kV, which is due to the defects near the embedded conductor caused by uneven heating at the pouring stage.

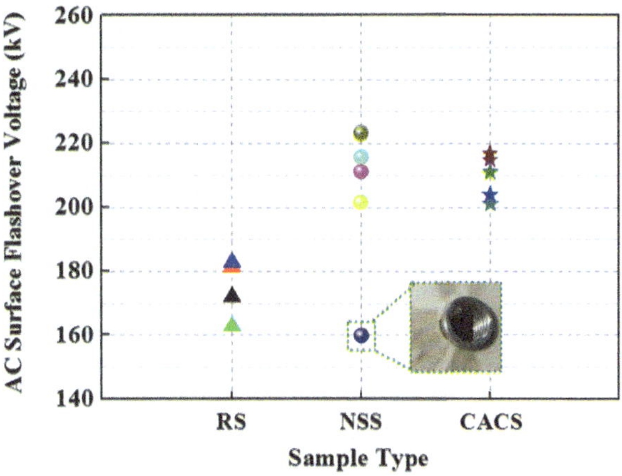

Fig. 13 AC surface flashover test results of experimental samples [11]. Copyright Clearance Center's RightsLink© License Number: 4857310044176

6.3 *Discussion*

The doping of nonlinear material dramatically affects the DC electrical property of the bowl-shaped spacer. It can be found in Fig. 14 that the electric field of the B2_0_30 and the B3_10_25 show similar curves with a mutation at the interface of the spacer and with a higher value near the HV conductor. This is due to the reason that the nonlinear property of epoxy doped with 10% mass ratio of SiC is not obvious. Meanwhile, the B1_0_30 shows the worst performance in the DC surface flashover test. This is due to the high electric field strength in the triple junction near the high voltage electrode. However, for sample with a doping ratio of more than 15% in the insulation region, the doping of nonlinear particles can introduce nonlinear property at high electric field near the high voltage conductor. As a consequence,

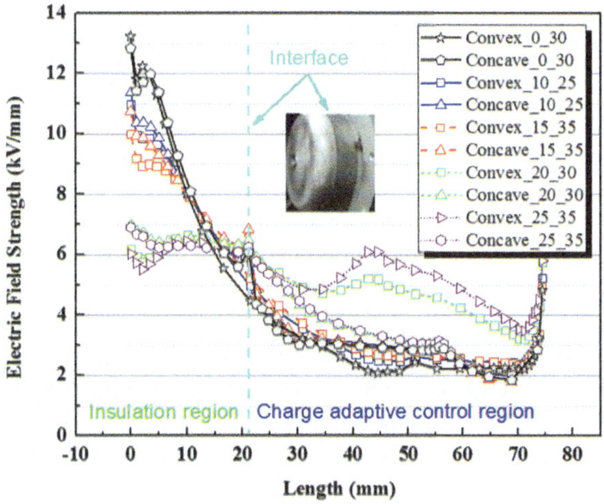

Fig. 14 The electric field strength of the bowl-shaped spacer under application of 250 kV [12]. Copyright Clearance Center's RightsLink© License Number: 4857310380945

the DC surface flashover values of the B5_20_30 and B6_25_35 are higher than the B4_15_35 [12].

The DC surface flashover voltage of the pure sample is higher than that of B2_0_30. This is due to the reason that the voltage potential drops mostly at the charge adaptive control region for B2_0_30, resulting in a high electric field near the triple junction at the high voltage conductor, as shown in Fig. 14. Meanwhile, the electric field mutation in the interface could also initiate a discharge activity which introduces more charges accumulated on the spacer surface. When the doping ratio of the nonlinear material increases up to 20% for the insulation region and the charge adaptive control region, the electric field near the high voltage conductor and the interface region are controlled by the conductivity. Meanwhile, the conductivity of the charge adaptive control region is increased when charges accumulate on the surface of the charge adaptive control region. This increase can accelerate the decay rate of surface charges, and decrease the surface charge amount in this region. As a result, the B5_20_30 and B6_25_35 show higher DC surface flashover performance. However, due to a large increase of leakage current, B6_25_35 could not be a proper model for manufacturing enlarged spacers.

7 Future Work

7.1 Controlling the Electrical Performance of Composites based on SiC Doped Epoxy

It was mentioned in part I that some problems might exist in the doping of SiC particles. For example, introducing the SiC filler may bring uncertain electrical performance to dielectrics, which can hardly be solved. In order to further demonstrate the impact of the electrical properties instability of SiC composites on the nonlinear property of the CACS, J(E) curve offset was shifted using the 800-mesh-sized SiC particle doped composite. Taking the electric field axis of the J(E) curve as a reference, the J(E) curve was shifted 20% to the left or right, and the corresponding electric field strength curves along the surface of the CACS_15_30 and the CACS_20_30 were simulated and analyzed. The results are shown in Fig. 15.

The electrical properties of nonlinear particles can affect the nonlinear properties of composites. However, we know from Fig. 15 that for the CACS, the field distribution is not affected that much after the shift in J (E) curve. When the J(E) curve is shifted by 20% to the left, the maximum distortion point of the electric field along the surface of CACS_15_30 occurs on the convex side. The distorted electric field value is 7.1 kV/mm, which is less than 6% distortion amount compared with the original electric field value which is 6.7 kV/mm. For the CACS_15_30, the maximum electric field distortion point appears on the concave side when the J(E) curve is shifted by 20% to the left. At this point, although the maximum electric field value can reach to

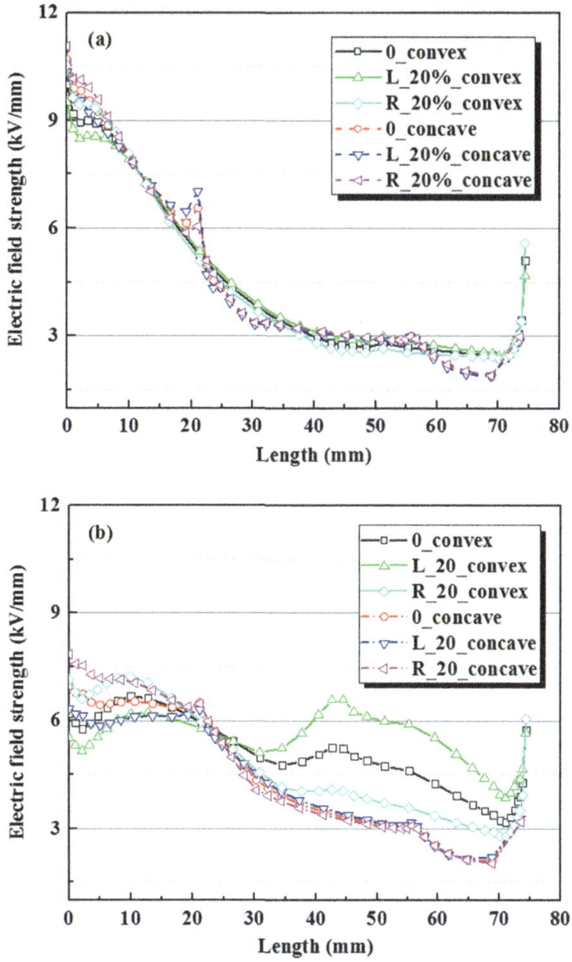

Fig. 15 The electric field strength of **a** CACS_15_30 and **b** CACS_20_30 under a DC bias of 250 kV: (X_Y_Convex: X represents the the offset direction of J(E) characteristics of SiC doped composites, L stands for Left and R stands for Right; Y represents the offset percentage of J(E) characteristics of SiC doped composites) [10]. Copyright Clearance Center's RightsLink© License Number: 4857301321085

the nearly 25% distortion rate compared with that of its original electric field value, the maximum field strength value is only 6.9 kV/mm.

These results indicate that although the SiC particles have undefined electrical properties, they still do not have much impact on the electrical performance of the CACS. Similar deduction has also been drawn by researchers for nonlinear materials doped to the HVDC bushings [13]. In addition, the distortion that occurs due to the shift in J(E) curve could be properly controlled by appropriately improving the level of non-linear material preparation process.

7.2 Electrical Performance Equivalence Considering Size Effect and Tests Under Temperature Gradient

This chapter provides performance test results and reliability analysis of charge adaptive control spacer models. However, the migration feasibility of downsizing spacer models to real sized spacers as well as nonlinear effect of fillers need to be considered carefully. Prior to the industrialization of this model, more tests must be carried out for the enlarged spacers, including routine tests considering PD test [14]. Moreover, for DC systems, the temperature is one of the significant test parameters influencing the dielectric strength and has to be assessed [15, 16]. Some important properties—long-term thermal and electrical aging, as well as the scalability of "fast" polarization currents to real sized components under temperature gradient—have to be further investigated.

8 Conclusion

Surface modification methods to enhance surface flashover property of HVDC spacers were discussed. A novel design concept of a HVDC spacer was introduced, by adaptive control of surface charges.

The mechanical and electrical test results show that the charge adaptively controlling spacer has high operating capability under DC voltage and has great industrial application potential. Admittedly, the content of this study primarily shows a novel design concept from a new perspective for the control of surface charge accumulation on the spacers using the nonlinear materials. To meet the standards requirements for industrial HVDC spacer application, many other corresponding effects would have to be taken into account.

References

1. C.M. Cooke, Charging of insulator surfaces by ionization and transport in gases. IEEE Trans. Electr. Insul. **EI-17**(2), 172–178 (1982)
2. C.Y. Li et al., Understanding surface charge accumulation and surface flashover on spacers in compressed gas insulation. IEEE Trans. Dielec. Electr. Insul. **25**, 1152–1166 (2018)
3. J.K. Berger, C. Ederle, Surface charge decay on insulators in air and sulfurhexafluorid—part II: measurements. IEEE Trans. Dielec. Electr. Insul. **15**(4) 949–957 (2008)
4. C.Y. Li et al., Field-dependent charging phenomenon of HVDC spacers based on dominant charge behaviors. Appl. Phys. Lett. **114**(20), 202–904 (2019)
5. T. Shao, W.J. Yang, C. Zhang, Z. Niu, P. Yan, E. Schamiloglu, Enhanced Surface Flashover Strength in Vacuum of Polymethylmethacrylate by Surface Modification using Atmospheric-Pressure Dielectric Barrier Discharge. Appl. Phys. Lett. **105**, 071607 (2014)
6. B.X. Du, J. Li, W. Du, Surface charge accumulation and decay on directfluorinated polyimide/Al_2O_3 nanocomposites. IEEE Trans. Dielectr. Electr. Insul. **20**(5), 1764–1771 (2013)

7. C.Y. Li, J. Hu, C.J. Lin, J.L. He, The control mechanism of surface traps on surface charge behavior in Alumina-filled Epoxy Composites. J. Phys. D: Appl. Phys. **49**, 445304 (2016)
8. Y. Liu, Z. An, J. Cang, Y. Zhang, F. Zheng, Significant suppression of surface charge accumulation on Epoxy resin by direct fluorination. IEEE Trans. Electr. Insul **19**, 1143–1150 (2012)
9. C.Y. Li et al., Novel HVDC spacers by adaptively controlling surface charges-Part I: charge transport and control strategy. IEEE Trans. Dielectr. Electr. Insul. **25**(4), 1238–1247 (2018)
10. C. Li et al., Novel HVDC spacers by adaptively controlling surface charges—Part II: experiment. IEEE Trans. Dielectr. Electr. Insul. **25**(4), 1248–1258 (2018). https://doi.org/10.1109/TDEI.2018.007056
11. C. Lin et al., Novel HVDC spacers by adaptively controlling surface charges—Part III: industrialization prospects. IEEE Trans. Dielectr. Electr. Insul. **25**(4), 1259–1266 (2018). https://doi.org/10.1109/TDEI.2018.007555
12. C. Li et al., Novel HVDC spacers in GIS/GIL by adaptively controlling surface charges—insulation compounding scheme, in *2019 2nd International Conference on High Voltage Engineering and Power Systems (ICHVEPS), Denpasar, Bali, Indonesia*, pp. 268–271 (2019). https://doi.org/10.1109/ICHVEPS47643.2019.9011050
13. X. Yang, X.L. Zhao, J. Hu, J.L. He, Grading electric field in high voltage insulation using composite materials. IEEE Electr. Insul. Mag. **34**, 16–26 (2018)
14. H. Koch, et al. High Power Underground Transmission for HVDC, CIGRE-21, Paris, France, B3—104 (2016)
15. C.Y. Li, J. Hu, C.J. Lin, J.L. He, The neglected culprit of DC surface flashover-electron migration under temperature gradients. Sci. Rep. **7**, 1–11 (2017)
16. A. Winter, J. Kindersberger, M. Tenzer, V. Hinrichsen, L. Zavattoni, O. Lesaint, M. Muhr, D. Imamovic, Solid/gaseous insulation systems for compact HVDC solutions, in *Cigre Session 45*, Paris, France, pp. D1_103 (2014)

Surface Charge Accumulation on Insulators in HVDC Gas-Insulated Systems: Measurement, Characteristics, and Suppression

Boya Zhang and Guixin Zhang

Abstract Charge accumulation on the insulator surface is one of the critical factors for the development of DC gas-insulated equipment since it will lead to the overstress of solid insulation due to local field distortion and enhancement. In this chapter, the measurement, characteristics, and suppression methods of surface charge were investigated. For the measurement, Kelvin probe method was used to detect the surface potential distribution of a cone-shaped insulator in a downsized gas-insulated line (GIL) and the inverse calculation algorithm of surface charge was introduced. Based on this method, the charge accumulation characteristics of an epoxy insulator under DC voltage in air and SF_6 was studied. The results show that there are two patterns in the charge distribution: dominant uniform charging and random charge speckles. The mechanism of these two patterns were discussed and verified by experiments. Finally, two material modification methods to suppress surface charge accumulation were proposed. One is fullerene-doping, for the purpose of reducing the bulk conductivity of insulators to decrease the bulk current in the solid and thus suppress the uniform charging. Another is nano-laminar coating, to promote charge dissipation rate and thus reduce charge speckles.

1 Introduction

High voltage direct current (HVDC) offers efficient non-synchronized bulk electric power transmissions with economic benefits of reduced power loss and enhanced stability. It has become the optimal solution for the increasing demand of transmission capacity and the growing geographical distance between energy generation and consumption. Meanwhile, the growth of urbanization and new energy integration has

B. Zhang (✉)
State Key Laboratory of Electrical Insulation and Power Equipment, Xi'an Jiaotong University, No. 28 Xianning West Road, Xi'an 710049, Shaanxi, China
e-mail: zhangby@xjtu.edu.cn

G. Zhang
Department of Electrical Engineering, Tsinghua University, Beijing 100084, China
e-mail: guixin@tsinghua.edu.cn

B. Du (ed.), *Polymer Insulation Applied for HVDC Transmission*,
https://doi.org/10.1007/978-981-15-9731-2_15

spurred the installation of gas-insulated transmission lines and substations (GIL/GIS) owing to their compactness and environmentally-friendliness.

However, it is widely recognized that insulators in GIS/GIL, usually made of Al_2O_3-filled epoxy, suffer from the problem of surface charge accumulation under DC stress, because the charges migrate in one direction and can be easily trapped on the gas–solid interface, especially in a closed and dry environment. These residual charges will, if of significant magnitude, strengthen and distort the local electric field, which may eventually initiate a flashover at the gas–solid interface particularly at the time of abrupt polarity reversal. It has become a major concern that affects the reliability of HVDC gas-insulated systems, falling significantly short of the ever increasing demands for a robust power system.

Since the late 1970s, researchers in the United States, Japan, and Europe have successively carried out research and development of DC GIS/GIL, with particular attention to the surface charging on the solid insulating spacers. Matsubishi Electric Corp. initiated the development of ±125 kV HVDC gas insulated converter station jointly with Kansai Electric Power Co. in 1980 [1]. Following the successful development, since 1982 Mutsubishi Electric had performed a three-year R&D project on the ±500 kV HVDC GIS and carried out series of design and testing work, including measurement of the surface charge distribution on a real cone-type spacer for the first time [2]. In June 2000, the world's first ±500 kV DC GIS was successfully adopted in Anan converter station in Japan (in fact, soon after its application, it was then operated down to ±250 kV), which was developed by Mitsubishi Electric Corp. and Toshiba Corp.. In the same years, ABB corp. performed a feasibility study of SF_6 GIS for HVDC at 500 and 600 kV based on existing AC materials (insulating parts). Finally, their long term performance required special spacers modified for DC technology. However, criteria for design and optimization of HVDC GIS are not well understood up to now. Thus, existing devices were usually over-sized to ensure a safe operation.

Among all these R&D projects, the subjects related to surface charge accumulation were always put into first priority. However, due to the invisibility of charges and the difficulty in online monitoring surface charge distribution, assessing charge transport phenomena and their influence on electrical filed distortion and the resulting flashover have always been a tricky problem. Although numerous studies have been done by previous researchers regarding surface charge accumulation, there still exists some issues that need to be solved.

First, how to obtain the accurate results of surface charge distribution remains an important subject. At present, the Kelvin probe is commonly used by most researchers to measure surface charge distribution. However, the output of the probe is surface potential instead of surface charge. If analysis was performed based on the surface potential distribution, a lot of detailed information would be ignored, which is not conducive to the study of surface charging phenomenon. To solve this problem, multipoint potential measurements are often required to establish inverse calculation for the determination of an unknown surface charge distribution. So far, extensive studies have been conducted on this subject; nevertheless, the methods are either too complicated and time consuming, or only applicable for specific geometric configuration, or with poor accuracy. In order to obtain the two-dimensional distribution of the surface charge density with high resolution, a precise measurement platform should be built and a more accurate inverse algorithm should be developed.

Second, a more comprehensive understanding of the surface charge patterns is needed and the charge transport mechanism behind these patterns should be uncovered. Up till now, the evaluation of the physical origin of the surface charges and the mechanism for trapping them is still being discussed, because it is not easy to measure either the magnitude, the location or the distribution of the charges with high precision, nor to correlate the results with different kinds of mechanisms which can be assumed to be responsible for charge accumulation. Further, abundant literature has shown that the surface charge distribution is rather inhomogeneous and not very reproducible, which is owing to high-field effects, such as partial discharge, micro-corona or field emission. This is very different and random, depending on the design and cleanliness of the system. It is generally accepted that surface charging is a complicated process that involves electric conduction in the solid dielectric (i.e. both surface and bulk conduction) and charge drift and trapping in the gas. Depending on the different charge transport mechanisms and their driven forces, the surface charge patterns must be studied separately.

Third, effective and practical methods should be proposed to suppress charge accumulation. Starting from optimizing the electric field distribution, researchers have proposed methods and established principles to optimize the insulator structure in order to suppress the surface charge accumulation. This method is easy to implement in practical, but its effect is somewhat limited. Material modification is another strategy and has been considered from two aspects. One is to suppress charge injection from the conductor into the bulk of the solid insulator. The other one is to increase the lateral charge dissipation rate so as to inhibit charge accumulation and homogenize the charge distribution. Although the effect has been proved significant and increased surface flashover voltage has been achieved with some of the proposed methods, none of them have been put into practical yet due to different reasons. Therefore, for now, exploring new materials and new technologies to realize active control and suppression of the surface charge accumulation is still the key to develop HVDC gas insulated apparatus.

In the past decade, we have made tremendous efforts to investigate the surface charging phenomena in HVDC gas-insulated system from the above three aspects. In this chapter, the surface charge measurement techniques, surface charge accumulation patterns and mechanisms, as well as some materials modification methods to suppress the charge accumulation will be introduced in detail.

2 Surface Charge Measurement

2.1 Basic Principle of Measurement

The Kelvin probe together with an electrostatic voltmeter is the most widely-used method for surface potential measurement. The schematic diagram of the system is shown in Fig. 1. Assuming the sensor and the tested surface can be modelled as a parallel-plate capacitor C. In this configuration, an electric current *I* is being induced in the vibrating sensor. This current can be expressed as:

$$I = \Delta U \cdot \frac{\mathrm{d}C}{\mathrm{d}t} = \Delta U \cdot \frac{\mathrm{d}}{\mathrm{d}t}\left(\frac{\varepsilon S}{D_0 + D_1 \cdot \sin(\omega t)}\right) = -\Delta U \cdot \varepsilon S \cdot \frac{D_1 \omega \cos(\omega t)}{[D_0 + D_1 \cdot \sin(\omega t)]^2} \quad (1)$$

Here ΔU is the potential difference between the vibrating sensor and the tested surface; *S* is the surface area of the sensing electrode; ε is the permittivity of the material between the electrode and the tested surface; *D0* is a constant representing the separation between the sensor and the tested surface when the sensor is not vibrating; *D1* and ω is the amplitude and circular frequency of vibrations.

The current signal *I* is amplified and demodulated by a phase-sensitive demodulator to produce a voltage proportional to the amplitude of *I*. The output of the integrator drives a high voltage amplifier circuit to replicate the voltage on the tested surface. The amplified voltage is applied to the probe thus nullifying the electric field between the tested surface and the sensor. The current *I* vanishes if the probe reaches the same potential as the tested surface potential, which serves as the voltmeter's output. This voltage-following technique makes the measurement independent of the distance D_0, at least within a certain range of D_0.

As mentioned in the Introduction, to obtain surface charge distribution requires multipoint probe outputs to establish inverse calculation. The probe output, therefore,

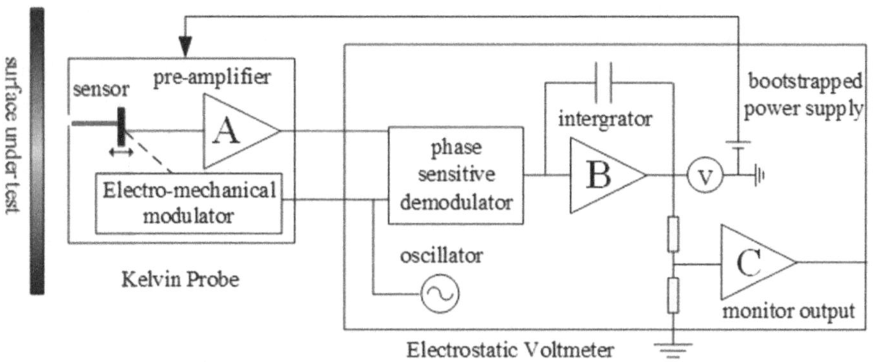

Fig. 1 Schematic diagram of the Kelvin probe and electrostatic voltmeter

corresponds to a linear superimposition of the effects of all the deposited charges on the measured object. Suppose the entire measuring surface is subdivided into N elements and each element$j$ acquires a charge density σ_j. Then, the probe output φ_i obtained above the center of element i can be given by:

$$\varphi_i = \sum_{j=1}^{N} h_{ij} \cdot \sigma_j \tag{2}$$

The above equation can be also expressed in matrix form, i.e.:

$$\boldsymbol{\varphi} = \boldsymbol{H} \cdot \boldsymbol{\sigma} = \begin{pmatrix} h_{11} & \ldots & h_{1N} \\ \ldots & \text{O} & \ldots \\ h_{N1} & \ldots & h_{NN} \end{pmatrix} \begin{pmatrix} \sigma_1 \\ \ldots \\ \sigma_N \end{pmatrix} \tag{3}$$

Here, $\boldsymbol{H}$ is the transfer function matrix, in which the coefficient hij is the probe output at the point i caused by a unit charge at the element j. Once the components in matrix $\boldsymbol{H}$ are obtained, the electric charge density $\boldsymbol{\sigma}$ can be calculated by solving the matrix-related formula.

Considering the spatial property of different measured objects, the measuring system can be divided into two categories, shift-variant system and shift-invariant systems. In a shift-invariant system, the relational function between a probe output and a charge on the measured object is given as a function of the relative position of the probe and the charge [3]. For such a configuration, the relational matrix equation Eq. (3) can be re-expressed as a convolution equation so that it can be processed in spatial frequency domain by two-dimensional Fourier transformation. While for a shift-variant system, the relational function differs for the measuring position, thus the data from a spatial frequency domain cannot be used. In the following sections, the inversion algorithm for the shift-variant system will be interpreted. For shift-invariant system, the detailed algorithm can be found in Ref. [4].

2.2 *Inversion Algorithm for Shift-Variant Systems*

2.2.1 Measuring System

In this section, a downsized cone-type insulator is used, which can be treated as a "shift-variant" system, where the transfer function matrix between a probe output and a charge on the measured object is not given as a function of the relative position of the probe and the charged. The insulator is made of epoxy resin with 100 mm outer diameter, 12 mm inner diameter and 17.5 mm height. It is mounted between a grounded hollow cylindrical enclosure and a central conductor, which forms a downsized model of a gas-insulated transmission line (Fig. 2, upper-left). The insulator is stressed under continuous dc voltage. For multipoint measurement, a platform is

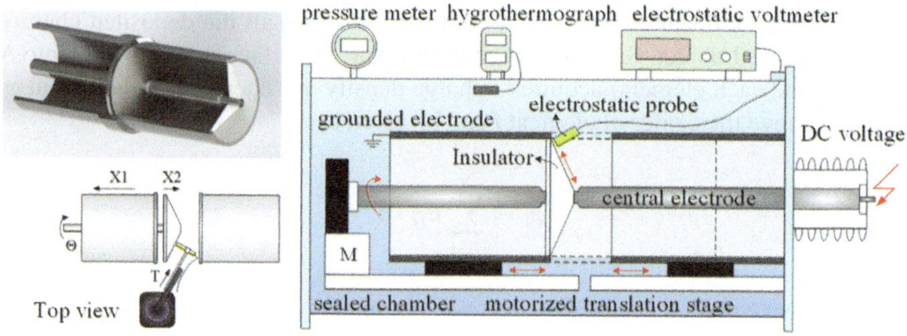

Fig. 2 Surface charge measurement platform and the tested insulator

designed with a motion control system as shown in Fig. 2. After the removal of dc voltage, the enclosure opens automatically and the Kelvin probe is driven to scan along the slope surface of the insulator within an exact separation of 2 mm.

2.2.2 Transfer Function Matrix

After measuring the potential distribution on the insulator surface, all the components in the transfer function matrix is needed in order to solve the Eq. (3). From Eq. (3), we can see that if just one element in $\boldsymbol{\sigma}$ equals to 1 and others are set to zero, i.e. just $\sigma_i = 1$, the obtained $\boldsymbol{\varphi}$ can be expressed as:

$$\boldsymbol{\varphi}_{N\times 1} = \boldsymbol{H}^{i}_{N\times 1} \tag{4}$$

where the superscript i represents the *i*-th column of matrix $\boldsymbol{H}$. Therefore, theoretically, the N^2 components of matrix $\boldsymbol{H}$ can be obtained by numerical field computation for N times. However, if high spatial resolution is required, the elements N on the tested surface can be such a huge number that the computation is very time-consuming and even unachievable.

Nevertheless, for a rotational symmetric system, the matrix $\boldsymbol{H}$ is proven to be a block circulant matrix, which can be used to reduce the computation times significantly. Take the cone-type insulator as an example. Suppose the measured surface is divided into N elements, $N = r \times d$, where r is the number of cycles in radial direction and d is the number of elements in each cycle. Then, $\boldsymbol{H}$ can be expressed as:

$$\boldsymbol{H}_{N\times N} = \begin{bmatrix} \boldsymbol{H}_{11,d\times d} & \boldsymbol{H}_{12,d\times d} & \cdots & \boldsymbol{H}_{1r,d\times d} \\ \boldsymbol{H}_{21,d\times d} & \boldsymbol{H}_{22,d\times d} & \cdots & \boldsymbol{H}_{2r,d\times d} \\ \cdots & \cdots & \text{O} & \cdots \\ \boldsymbol{H}_{r1,d\times d} & \boldsymbol{H}_{r2,d\times d} & \cdots & \boldsymbol{H}_{rr,d\times d} \end{bmatrix} \tag{5}$$

Fig. 3 Electrostatic field computation to obtain the components in matrix H

where $\boldsymbol{H}_{11}, \boldsymbol{H}_{12},\ldots\boldsymbol{H}_{rr}$ are the block matrixes of the matrix $\boldsymbol{H}$ and each of the block matrix is a $d \times d$ dimensional circulant matrix. A circulant matrix is a special matrix where each row vector is rotated one element to the right relative to the preceding row vector. Therefore, based on formula (5) and the method described by formula (4), the numerical field computation can be reduced to only r times.

For accuracy, the electrostatic field computation is implemented by a 1:1 scale model built in a finite element software, which consists of the GIL unit, surrounding gas and the chamber, as shown in Fig. 3a. In this measuring system, the translation of the probe and the rotation of the insulator are precisely controlled by interpolation motion with 44 × 360 points, i.e. $N = 15{,}840$, $r = 44$, $d = 360$, as shown in Fig. 3b. For sampling, the pitch in radial direction is set to 1 mm, and the rotation pitch is set to 1°. When conducting the computation, take turns to set the surface density of each element in radial direction equal to 1 C/m^2, and get the surface potential values of all the elements after each computation, as shown in Fig. 3c. After 44 times, all the components in matrix $\boldsymbol{H}$ can be obtained.

2.2.3 The Ill-Posed Character of Transfer Function Matrix

Obviously, once the matrix $\boldsymbol{H}$ is obtained, the surface charge distribution can be calculated from measured surface potential, i.e.:

$$\sigma_{N\times 1} = \boldsymbol{H}^{-1}_{N\times N}\varphi_{N\times 1} \tag{6}$$

However, the inverse calculation becomes instable when the dimension of matrix $\boldsymbol{H}$ is so large that it probably becomes ill-posed. That means even though the measurement error $\delta\varphi$ and the calculation error $\delta\boldsymbol{H}$ are very small, the solution can diverge easily. This situation can be expressed as:

$$\sigma + \delta\sigma = (\boldsymbol{H} + \delta\boldsymbol{H})^{-1}(\varphi + \delta\varphi) \tag{7}$$

The bound of perturbation $\delta\sigma$ can be estimated by the following formula:

$$\frac{\|\delta\sigma\|}{\|\sigma\|} \leq \frac{\|H\|\left\|H^{-1}\right\|}{1-\|\delta H\|\left\|H^{-1}\right\|} \cdot \left(\frac{\|\delta H\|}{\|H\|} + \frac{\|\delta\varphi\|}{\|\varphi\|}\right) \tag{8}$$

Here the symbol $\| \|$ denotes the Euclidean norm for vectors, and the spectral norm for matrices. The ill-posed characteristics of the transfer function can be diagnosed according to the conditional number of the matrix: $\mathrm{Cond}(\boldsymbol{H}) = \|\boldsymbol{H}\|\left\|\boldsymbol{H}^{-1}\right\|$. The conditional number of matrix $\boldsymbol{H}$ is calculated, and the value is 648. The experience in statistical application shows that when $100 \leq \mathrm{Cond} \leq 1000$, the matrix is considered to be ill-posed. It can also be seen from Eq. (8) that when the value of $\mathrm{Cond}(\boldsymbol{H})$ is large, even if the disturbance is small, the deviation of the result will be very large. Even worse, the solution may become absolutely meaningless.

2.2.4 Wiener Filter and Tikhonov's Regularization

In order to reduce the influence of disturbances in the measurement process on the calculation result, we draw on the Wiener Filter in the digital image restoration technology to solve Eq. (3). Wiener filter is an optimal filter based on minimum mean square error (MMSE) and is widely used to extract image information contaminated by stationary noise. Based on the Tikhonov regularization principle, the filter minimizes the mean square error between the output and the expected output [5]. A regularization term for Eq. (3) can be included in this minimization:

$$\min: \left\|\boldsymbol{H}\hat{\boldsymbol{\sigma}} - \boldsymbol{\varphi}\right\|^2 + \alpha\left\|\hat{\boldsymbol{\sigma}}\right\|^2 \tag{9}$$

The estimated solution of surface charge density obtained by this criterion is:

$$\hat{\boldsymbol{\sigma}} = (\boldsymbol{H}^{\mathrm{T}}\boldsymbol{H} + \alpha\mathbf{I})^{-1}\boldsymbol{H}^{\mathrm{T}}\boldsymbol{\varphi} = \boldsymbol{G}\boldsymbol{\varphi} \tag{10}$$

where α is the optimal Tikhonov parameter and is also called the ridge parameter, which can be determined by the L-curve method and other approaches.

The L-curve method was proposed by Hansen [6] and its principle is briefly described as follows. In Eq. (10), $\boldsymbol{H}\hat{\boldsymbol{\sigma}} - \boldsymbol{\varphi}$ and $\hat{\boldsymbol{\sigma}}$ are all functions of the ridge parameter α. Choose different α values, draw a scatterplot with $\left\|\boldsymbol{H}\hat{\sigma} - \varphi\right\|$ as the abscissa and $\left\|\hat{\sigma}\right\|$ as the ordinate, and get a curve like L after fitting. The ridge parameter α can be get from the corresponding point where the curvature on the curve is the largest, as shown in Fig. 4. Using this method, we can obtained the ridge parameter as 1.58×10^{17}. And the resulting condition number of matrix $\boldsymbol{H}^{\mathrm{T}}\boldsymbol{H} + \alpha\boldsymbol{I}$ is only 2.978. Compared with the condition number of the original transfer function matrix, the stability of the inverse process is greatly improved. In image processing, the ridge parameter can be regarded as the ratio of noise and signal variance, i.e. $\alpha = {\sigma_n}^2/{\sigma_s}^2$. The value of α can be determined by measuring the power density spectrum of noise and signal in advance. However, in this study, it is difficult to obtain σ_n and

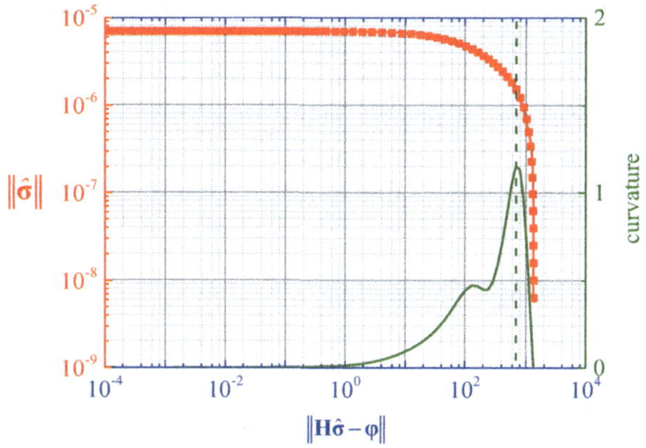

Fig. 4 Determination of the ridge parameter by L-curve

σ_s through experiments. Reference [3] gives a recommended range of α, i.e. 0.03 ~ 1% of the largest eigenvalue λ_{max} of the matrix $\boldsymbol{H}^{\mathrm{T}}\boldsymbol{H}$.

In fact, from the perspective of digital image processing, the measurement of surface potential and the inverse calculation of the measured data is similar to the image deterioration and restoration process, which can be expressed by the model in Fig. 5. The estimated surface charge $\hat{\boldsymbol{\sigma}}$ is generated by the true value $\boldsymbol{\sigma}$ via the measurement (deterioration) process $\boldsymbol{H}$ and the restoration process G :

$$\hat{\boldsymbol{\sigma}} = \boldsymbol{G}\boldsymbol{H}\boldsymbol{\sigma} = (\boldsymbol{H}^{\mathrm{T}}\boldsymbol{H} + \alpha\mathbf{I})^{-1}\boldsymbol{H}^{\mathrm{T}}\boldsymbol{H}\boldsymbol{\sigma} \tag{11}$$

It can be proved that the eigenvectors $\boldsymbol{v}_1$ ~ $\boldsymbol{v}_N$ of the matrix $\boldsymbol{H}^{\mathrm{T}}\boldsymbol{H}$ are the same with the eigenvectors of matrix $\boldsymbol{GH}$. Therefore, $\boldsymbol{\sigma}$ and $\hat{\boldsymbol{\sigma}}$ can be decomposed as:

$$\sigma = \sum_{i=1}^{N} k_i v_i, \hat{\sigma} = \sum_{i=1}^{N} k_i' v_i \tag{12}$$

where k_i and k_i' are coefficients with the relationship:

$$k_i^{'} = \left[\lambda_i / (\lambda_i + \alpha)\right] k_i \; (i = 1, 2, \ldots, N) \tag{13}$$

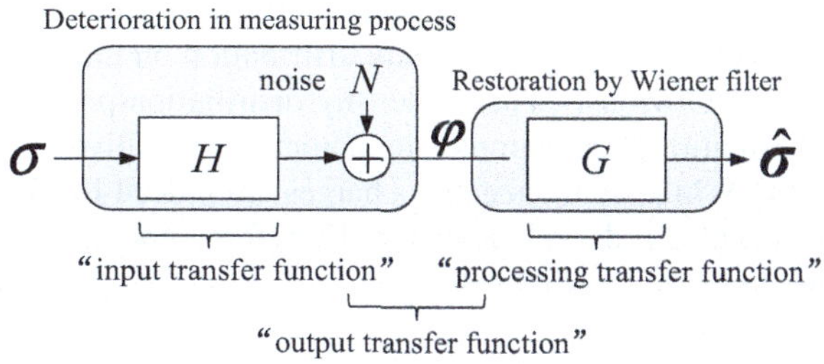

Fig. 5 Schematic diagram of the image deterioration/restoration model [4]. Copyright © 2017, IEEE

λ_i is the eigenvalue corresponding to the eigenvectors $\boldsymbol{v}_i$. When λ_i is much greater than α, k_i'/k_i is approximately 1; when λ_i is much less than α, k_i'/k_i reaches 0. In other words, the component $k_i\boldsymbol{v}_i$, whose eigenvalue $\boldsymbol{v}_i$ is small, will be filtered out in the inversion calculation. In the spatial frequency domain, these smaller eigenvalues correspond to the high-frequency part of the charge in the spatial distribution, so the Wiener filter designed in this way is a low-pass filter.

2.2.5 Spatial Resolution

In the study of surface charge, higher spatial resolution will provide more detailed information. The Point Spread Function (PSF) is usually used to evaluate the spatial resolution for an optical imaging system. It is the light field distribution of the output image when the input object is a point light source. Obviously, the more concentrated the light field distribution of the output image, the better the imaging effect. Similarly, in Eq. (11), the estimated value $\hat{\boldsymbol{\sigma}}$ and the true value $\boldsymbol{\sigma}$ of the charge distribution can be regarded as the "image" and "object" in the imaging system, respectively. And the "imaging" relationship is determined by the output transfer function $\boldsymbol{GH}$. Therefore, the PSF of the system can be obtained by setting a unit surface charge $\boldsymbol{\sigma} = 1\text{C/m}^2$ on the insulator surface and getting the estimated surface charge distribution $\hat{\boldsymbol{\sigma}}$ through Eq. (11).

Figure 6a shows the calculated results of $\hat{\boldsymbol{\sigma}}$ with different values of α. Then, Fourier transform is applied to Fig. 6a to obtain the charge distribution in the spatial frequency domain, as given in Fig. 6b, where the amplitude of the curves are normalized. The Fourier transformation of PSF indicates the spatial frequency transmission characteristics of the measuring system [22]. The spatial resolution of the measuring system can be defined as the reciprocal of the cut-off frequency where the intensity of the normalized curve decreases to 0.5. The vertical dotted lines in Fig. 6b indicate the cut-off frequency, which shows that the spatial resolution becomes worsen with the increase of α. When α ranges from $0.003\%\lambda_{\max}$ to $0.1\%\lambda_{\max}$, the spatial resolution is from 1.8 to 2.9 mm.

2.2.6 Comparison of Surface Potential and Surface Charge Distribution

By using the above measurement platform and inversion algorithm, the surface potential and surface charge density distribution on the insulator is obtained, as shown in Fig. 7. Obviously, charge density distribution provides more details than potential distribution. For example, the potential is positive on the whole surface of the insulator. While, some negative charges are indeed locally distributed in the vicinity of center electrode and the flange. Therefore, more accurate behaviors of surface charge are likely to be obtained.

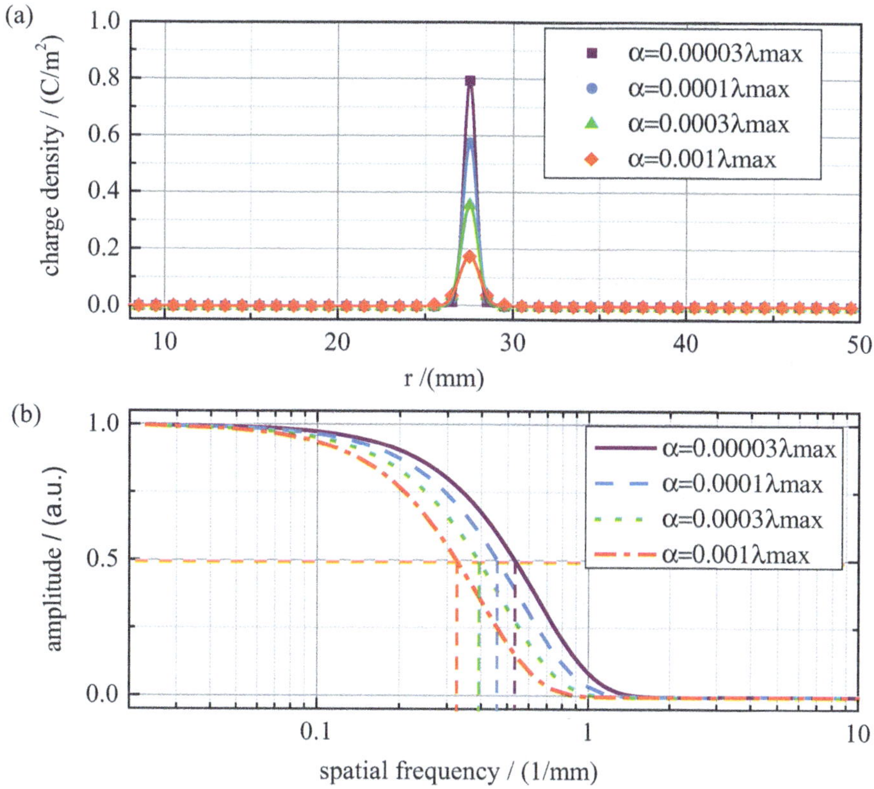

Fig. 6 **a** Estimated surface charge along the radial direction by inverse calculation for a unit surface charge on r = 28 mm. **b** Fourier transformation of PSF [4]. Copyright © 2017, IEEE

3 Characteristics of Surface Charge Accumulation

3.1 Surface Charge Accumulation in Air and SF_6

The surface charge distribution of the cone-type insulator in 0.1 MPa air when it is stressed under +30 kV and −30 kV DC voltage are shown in Fig. 8. The measurements in each group are performed six times for different durations, ranging from 15 to 1000 min. The surface charge patterns show a dominant uniform distribution which is centrosymmetric and also a lot of charge speckles which are randomly distributed from group to group. The polarity of the uniform charging is the same as that of the applied voltage. Charge speckles can be dots and streaks, most of which is of the opposite polarity to the applied voltage.

The same experiment is also conducted in 0.1 MPa SF_6. The results are given in Fig. 9. Similar to the results in air, the polarity of the overall surface charge is the same as the applied voltage and the charge density maximum is around the central

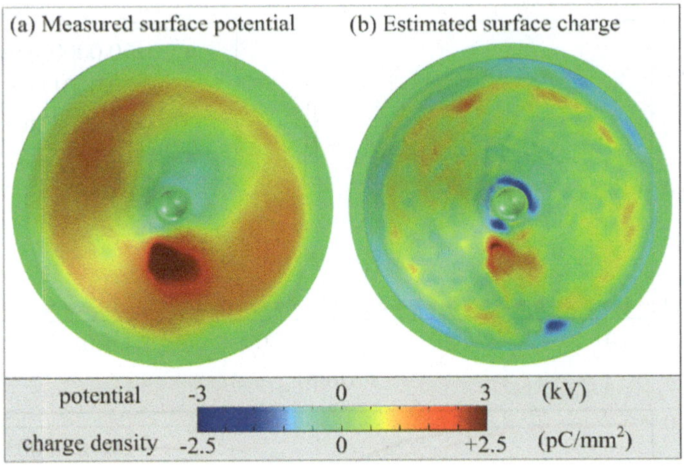

Fig. 7 Surface potential and surface charge distribution on a cone-type insulator after application of −20 kV dc voltage for 6 h in air [4]. Copyright © 2017, IEEE

conductor, which is referred to as dominant uniform charging. The charge speckles are also observed in this situation. They are mainly distributed near the junction points between insulator and the flange. It seems that both the number and the density of the charge speckles are less in SF_6 than that in air (noting that the scale of the color bars are different in Figs. 8 and 9). In fact, if we compare the results with those under ±10 and ±20 kV (not given in this chapter), charge speckles with extreme density appear with the increase of the applied voltage. So we infer that the formation of charge speckles should be related with the micro discharges in the gas domain, which highly depends on the electric filed strength and the gas insulation property.

3.2 *Surface Charge Accumulation Patterns*

Form the above experiment results, we can see that there are two patterns, *uniform charging* and *charge speckles*, coexisting in the surface charge accumulation. The former one shows a centrosymmetric distribution with smooth variation. While the charge speckles are highly localized pockets of charge and distributed randomly in the form of dots and streaks. This phenomenon can be also apparently observed in the original data of the measured potential, as shown in Fig. 10.

Many studies have revealed that surface charge distribution is nonhomogeneous and not reproducible, which is also confirmed by our measurement, thus, it is very hard to obtain the regularity of charge accumulation patterns. However, if the uniform charging and the charge speckles are investigated separately, some interesting phenomena can be discovered.

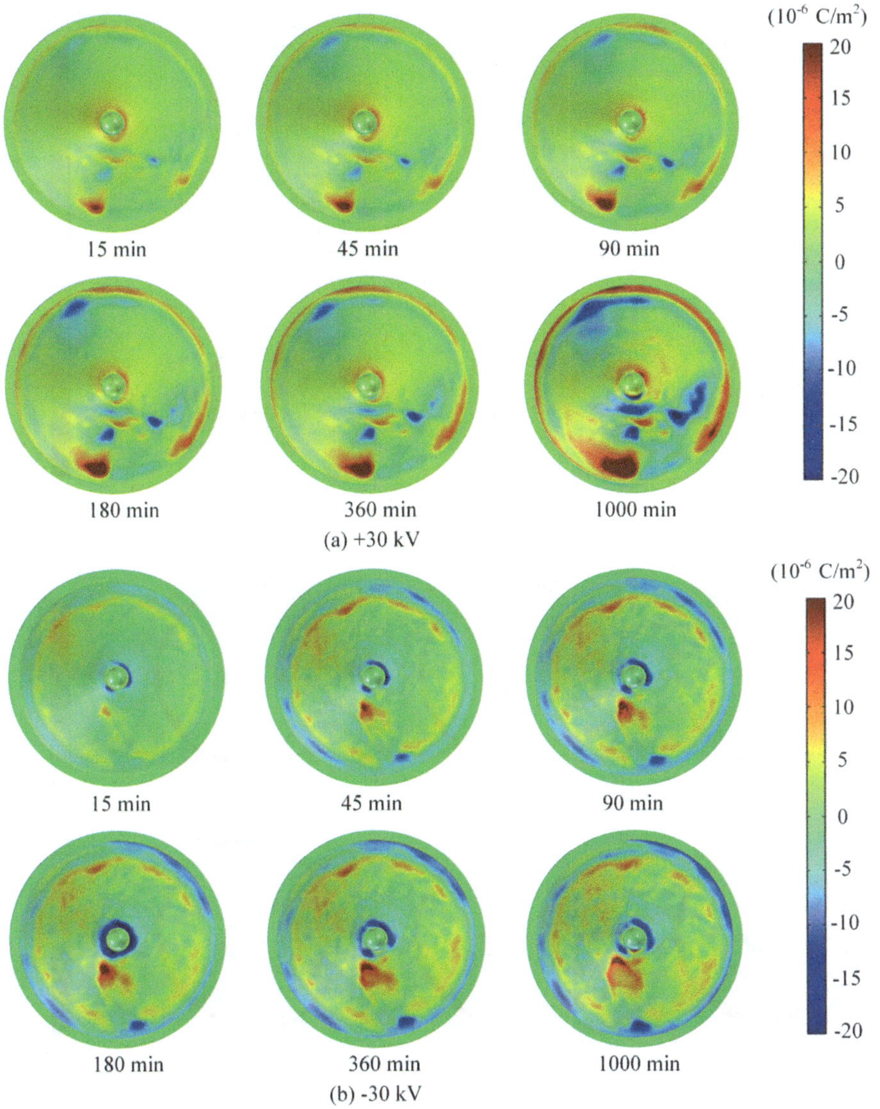

Fig. 8 Surface charge distribution in 0.1 MPa air

The model shown in Fig. 11 can be used to explain the mechanism of the uniform charging. The temporal change of surface charge density ρ_S is related to three parts, namely the bulk current flow from solid insulator (J_I), form the gas (J_G), and along the gas–solid interface (J_S), which can be described by the following continuity equation (the direction of unit normal vector $\mathbf{n}$ on the interface is defined from the bulk to the gas):

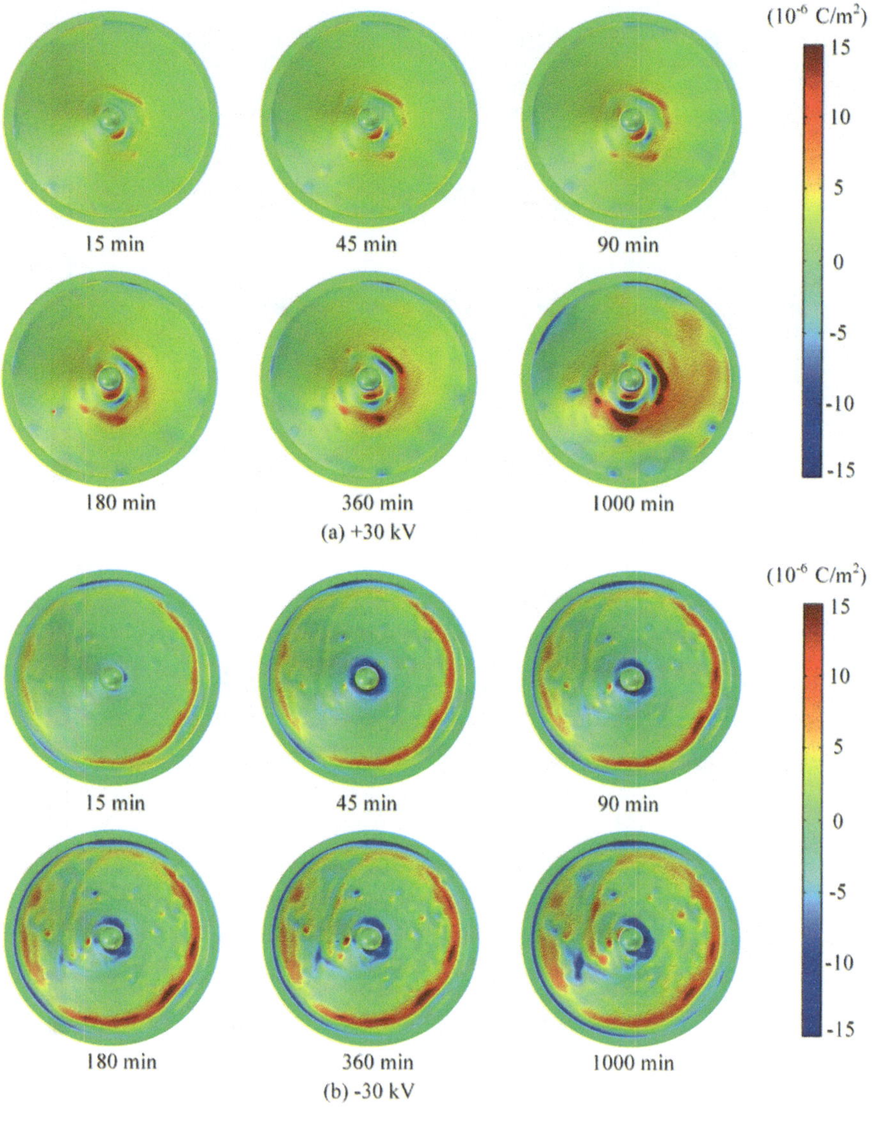

Fig. 9 Surface charge distribution in 0.1 MPa SF_6

$$\partial \rho_S / \partial t = \mathbf{n} \cdot J_I - \mathbf{n} \cdot J_G - J_S \tag{14}$$

In engineering applications, the surface current J_S is usually much less than J_{In} and J_{Gn} due to the limited surface conductivity of the epoxy insulator, so its contribution to the surface charge can be neglected. Therefore, the formation of surface charge ρ_S is determined by the relative magnitude of the current flow on both sides of the

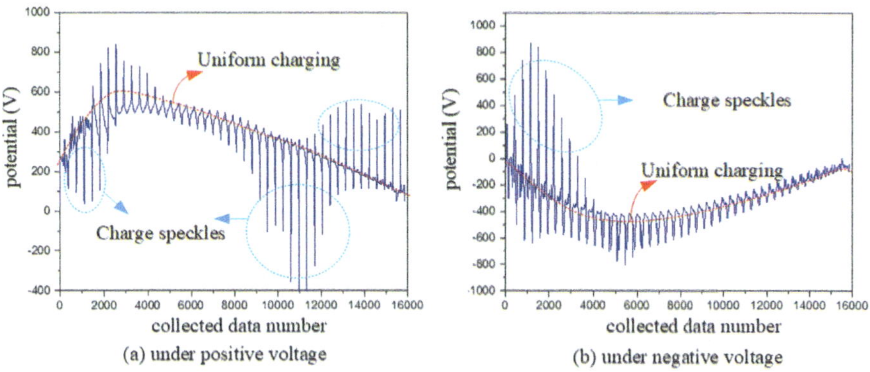

Fig. 10 A typical measurement results of the surface potential when the Kelvin probe scans the insulator surface, which clearly shows two different patterns

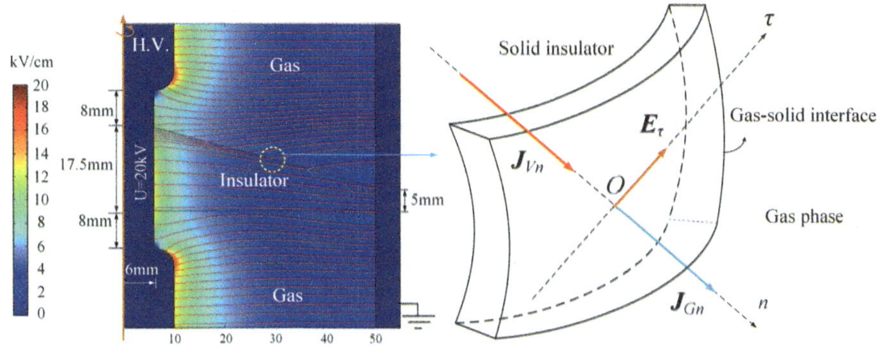

Fig. 11 Electric field distribution of the GIL unit (left) and the surface charging model at the gas–solid interface (right)

gas–solid interface. Considering the field distribution at the interface and the polarity of the uniform charging in these experiments, bulk conduction of the solid insulator is more dominant than the gas conduction, i.e. $J_{In} > J_{Gn}$, because the polarity of the overall charge distribution is the same as the applied voltage. In fact, in recent studies, more and more experimental and simulation evidence have revealed that bulk conduction of the solid insulating material determines the insulator surface charging under DC voltage [7, 8].

However, it should be noted that the above conclusion is valid for most of the engineering applications in which the electric field is moderate and no irregular ionization happens in the gas phase. *Li* et al. has proposed a field-dependent surface charging model to explain different surface charge features obtained by former researchers [9]. When the applied electric filed strength is high enough and exceeds a certain value, due to strong ionization in the gas phase, the relative magnitude of the current flow on the gas–solid interface would change to $J_{In} < J_{Gn}$. Thus, the charge distribution pattern would be a surface covered by hetero-charges, which will be shown in Sect. 3.3.3.

The second obvious pattern of the charge accumulation distribution is the highly localized pockets of charge. This pattern is referred to as "speckles" due to the randomly scattered patterns which appear on otherwise uniform charging pattern. The charge speckles are usually found in two forms: dots and streaks. Dots can be two types: single unipolar charge and bipolar charge pair. Based on extensive experiment results, we find some regularities for these charge speckles. First, most single unipolar charges are of the opposite polarity to the applied voltage. Second, if there is a charge dot with the same polarity as the applied voltage, there probably be a hetero-charge next to it to form a bipolar charge pair. Third, bipolar charge pairs always have an orientation that the hetero-charge is faced to the central electrode and the homo-charge is away from it. Figure 12 shows some typical examples in the experiments for the three types of charge speckles.

We are certain that at least some of these charge dots can be caused by the micro discharges from tiny metal particles or shavings that are inadvertently introduced into the GIL unit and adhered to the spacer surface. In such cases, electric field is enhanced at the tip of the particle so that it may trigger partial discharges (PDs). A schematic diagram is shown in Fig. 13 to help interpret the possible mechanism of the charge dots. When a positive voltage is applied (Fig. 13a), positive ions generated by PD will drift into the gas phase under the normal component of the electric field (case 1). Therefore, negative charges will remain on the surface. On the other hand, positive and negative charges produced by PD will also move along the insulator surface in opposite directions following the tangential component of the electric force (case 2). In this way, negative charges will appear at the end of the metal

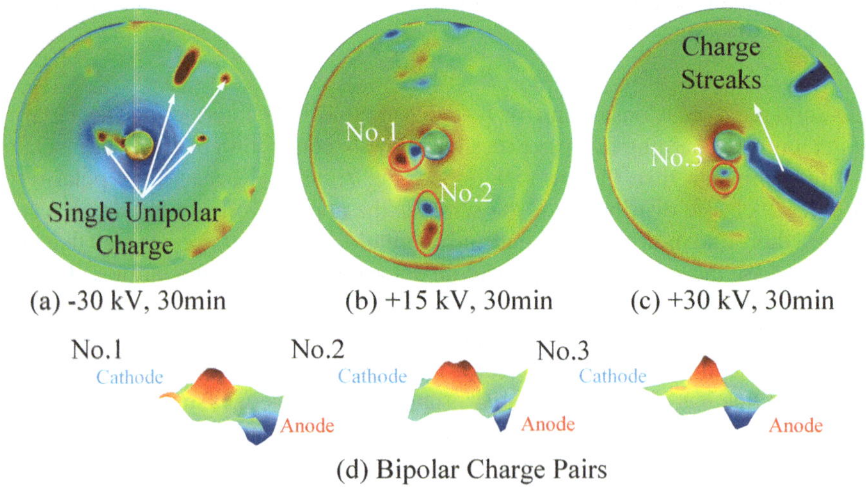

Fig. 12 Typical examples of charge speckles

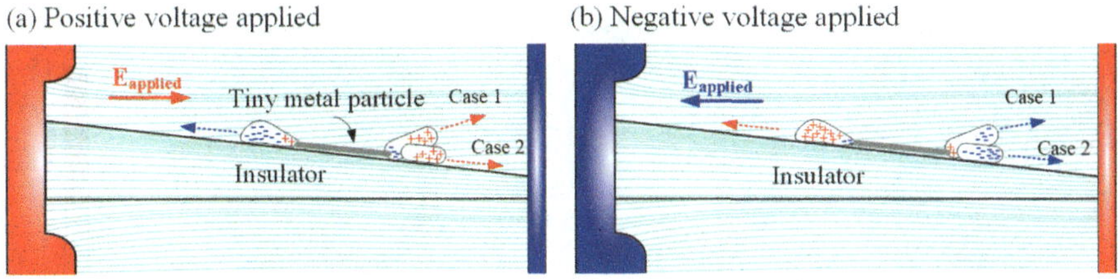

Fig. 13 A Possible mechanism illustrating the formation of charge dots

particles near the center electrode, while positive charges will appear away from the center electrode, forming a bipolar charge pair. When negative voltage is applied (Fig. 13b), similar situation can be deduced. The above explanation has also been verified experimentally in our previous studies [10]. By artificially placing metal particles on the insulator surface before the experiment, we found bipolar charge pairs where the particles were placed as expected.

Admittedly, some other possible reasons can also result in the bipolar charge pairs, such as micro cracks and bulges, subsurface (cast-in-solid) metal or resistive particles, inhomogeneous conductivity of the solid or surface region, etc. They may arise from impurities of the filler, casting mold, or release agent. For example, in Ref. [11], scanning electron microscopy images of the surface microstructure of the insulator show that, in addition to the Al_2O_3 filler, a metal particle can be clearly observed. Chemical analysis shows that it is a 95:5 Al–Mg alloy with a size of about 50 μm. Such metal impurities in the surface layer are likely to cause local polarization, which will also lead to charge speckles.

The charge streaks are usually found in the vicinity of outer flange and their polarity is usually opposite to the applied voltage. They usually originate from the outer flange and drift to the central electrode. These charges probably arise from the micro discharges owing to the defects (cracks or bubbles) in the so-called "triple junction" points. Special attention should be paid to this kind of charge speckles, because some of the streak may traverse the entire insulator surface, from the ground enclosure to the central conductor (Fig. 14). It is conceivable that once this situation develops, flashover across the insulator surface will easily occur.

It should be noted that stronger electric field strength which approaches the dielectric limit of the gas for a given pressure tends to trigger more charge speckles with higher charge density. As a result, these randomly distributed charge speckles spread and become contiguous. Eventually, they may obliterate and mask the uniform charging pattern. In Ref. [12], *Suda* et al. have conducted a similar experiment and some representative results are presented, as shown in Fig. 15. We can see that under −30 kV, the insulator surface was completely covered by homo-charges, indicating a uniform charging pattern, which is dominated by bulk current inject from the solid insulator. With the increase of voltage, charge speckles with opposite polarity appeared and the uniform charging pattern could still be observed under −60 kV. When the voltage is increased to −90 kV, the insulator surface was mostly covered

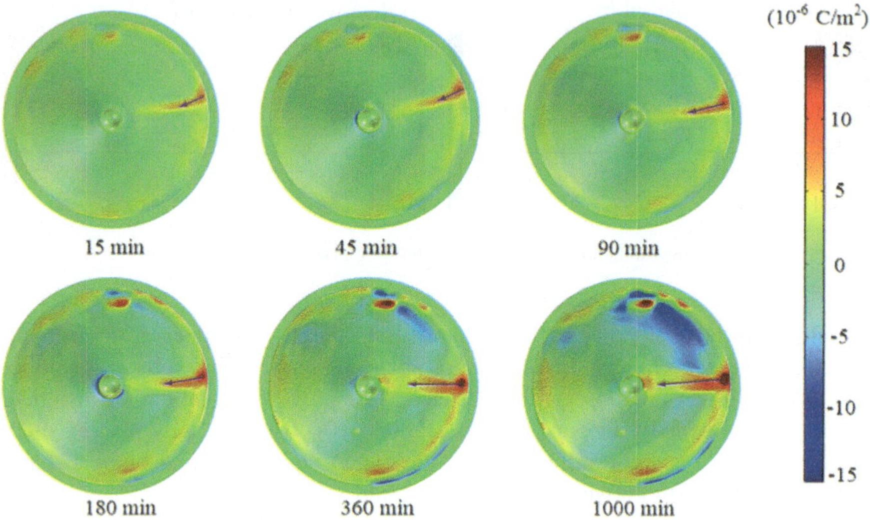

Fig. 14 Charge streaks traverse the entire insulator surface, from the outer flange to the central conductor. (−20 kV, 0.1 MPa SF_6)

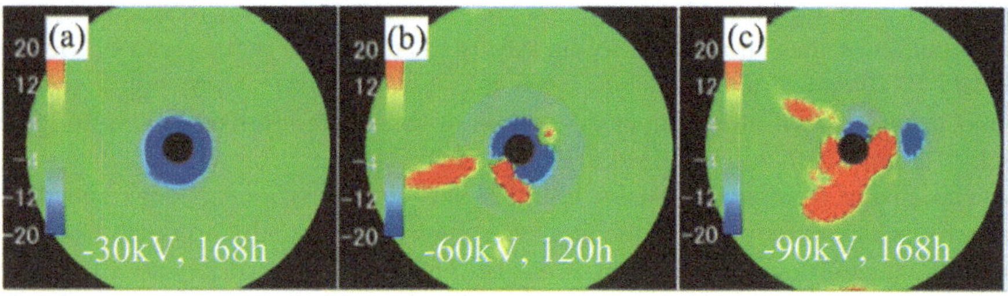

Fig. 15 Surface charge distribution on the sloping surface of the cone-type insulator in 0.5 MPa SF_6 after different voltage application [12]

by charge speckles and the uniform charging pattern was masked, which indicates sever gas ionization happens in the gas-phase.

3.3 Verification of the Uniform Surface Charging Pattern

Based on the analysis in Sect. 3.2, the dominant uniform charging pattern mainly depends on the relative magnitude of the current density from both sides of the gas–solid interface. For the test insulator in our experiment, the bulk current in the solid-phase is larger than that in the gas-phase (i.e. $J_{In} > J_{Gn}$), so homo-charge accumulated on the insulator surface. In order to verify this mechanism, several tests have been performed and the results are discussed in the following sections.

3.3.1 Influence of the Bulk Conductivity of the Insulator

The bulk conduction current is highly determined by the bulk conductivity of the material. Therefore, changing the bulk conductivity of the insulator material will change the bulk injection current and thus change the relative relationship between J_{In} and J_{Gn}.

Here, four different insulators are selected for the test which are made of silicone rubber, polytetrafluoroethylene (PTFE), polymethyl methacrylate (PMMA), and epoxy resin, respectively, as shown in Fig. 16. Among these materials, the silicone rubber has the highest bulk conductivity, in the order of magnitude of 10^{-14} S/m. The bulk conductivity of PMMA and epoxy resin is approximately 10^{-17} ~ 10^{-16} S/m. The bulk conductivity of PTDE is the lowest, which is in the order of magnitude of 10^{-18} ~ 10^{-17} S/m. The four insulators were stressed separately under DC voltage for 180 min in 0.1 MPa air and their surface potential was measured and surface charge density distribution was obtained. The results are given in Fig. 17. It should be noted that the applied voltage for PTFE, PMMA and epoxy resin insulator is ±20 kV; while for silicone rubber, the applied voltage is ±5 kV, otherwise the surface potential will exceed the measuring range of the Kelvin probe.

It can be seen from the results that the measured overall surface potential is of the same polarity as the applied voltage, which is the dominant uniform charging pattern. Among these four insulators, the charge accumulated on the surface of silicone rubber insulator is the largest, and the charge density is much higher than that of the other three materials. Surface charging level on the PTFE insulator is the lowest (neglecting the charge speckles). This is because the bulk conductivity of silicone rubber is the largest, the current injection from the solid-phase is much large than the current in the gas-phase; While for PTFE, whose bulk conductivity is the smallest, the difference between the conduction current in the solid-phase and the gas-phase is relatively small, so the charging level at the gas–solid interface is the lowest.

From the above analysis, we can conclude that in a moderate electric field and defect-free condition, the uniformly distributed homo-charges is in domination and

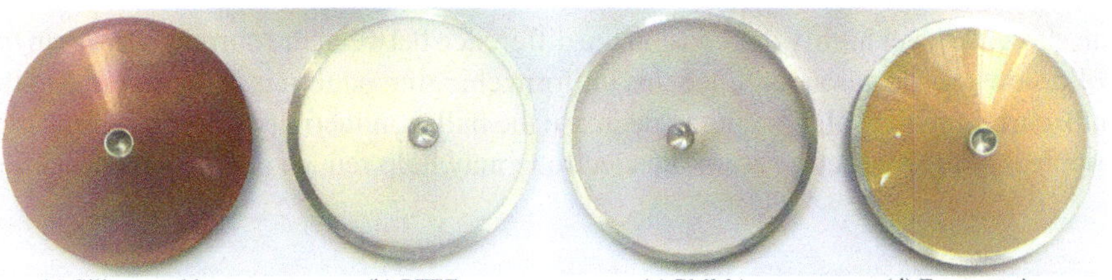

(a) Silicone rubber (b) PTFE (c) PMMA (d) Epoxy resin

Fig. 16 Test insulators made of four different materials

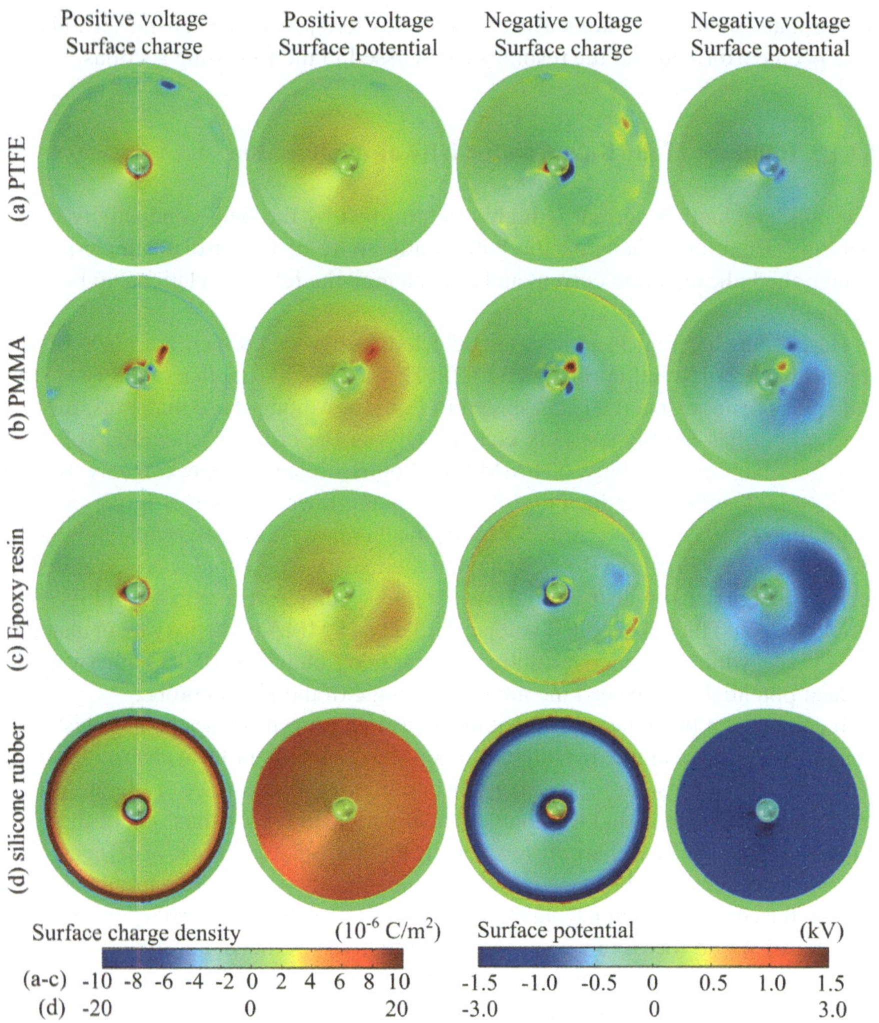

Fig. 17 Surface potential and surface charge distribution of insulators made of four different materials under DC voltage for 180 min in 0.1 MPa air

the charging level highly depends on the difference between the current density in the solid side and the gas side. Since this uniform charging pattern is primarily due to the bulk current in the solid, trying to decrease the bulk conductivity of the material or to suppress charge injection from the electrode may help reduce charge accumulation.

3.3.2 Influence of the Thermal Gradient of the Insulator

In a real HVDC gas insulated system, operation is connected with ohmic heating and, therefore, non-uniform temperature distributes inside the insulator. Thus, due to the thermal gradient, there is also a gradient in the electric conductivity, which may lead to different surface charging than compared to the case without thermal gradient within the insulator. Therefore, in this section, we tried to investigate the effect of thermal gradient on the surface charge accumulation.

At normal operating condition, typical temperatures of central and outer conductor of a horizontal GIS-compartment are 343 K and 313 K, respectively. To achieve this, the central conductor is made hollow with heating oil inside. The external-circulation oil is driven by a heating system to control the temperature of the conductor to 343 K. Under such condition, the measured temperature of outer enclosure is around 303 K due to natural radiative and convective cooling. Both simulated and observed temperature of the insulator are shown in Fig. 18.

The volume and surface conductivity of the material specimen under different temperature from 25 °C to 90 °C was measured using three electrodes method (1 kV/mm) according to IEC 60093. The currents are collected with a high resolution electrometer (Keithley 6517B). Figure 19 shows the measured volume current and the temperature-dependent conductivity. We can see that the material's conductivity is varying by order of magnitudes in the relevant temperature range. The temperature dependence of conductivity κ_V can be fitted by the Arrhenius equation:

$$\kappa_V = \mathrm{A} \cdot \exp(-\mathrm{B}/T) \tag{15}$$

Where the constant are chosen to A = 0.013 S/m and B = 9225 K corresponding to the measured volume conductivities.

In Eq. (14), the current flow from the insulator is $J_\mathrm{I} = \partial_t \mathbf{D} + \kappa_v \cdot \mathrm{E}$ and the current along the surface is $J_s = div(\kappa_s \cdot \mathbf{E}_\tau)$. This time, the volume and surface conductivity

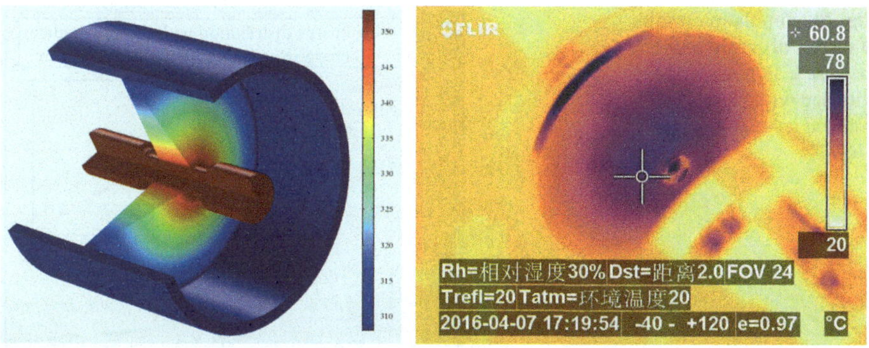

Fig. 18 Thermal gradient of the test insulator by simulation using COMSOL (left) and observation using an infra-red thermal imager (right)

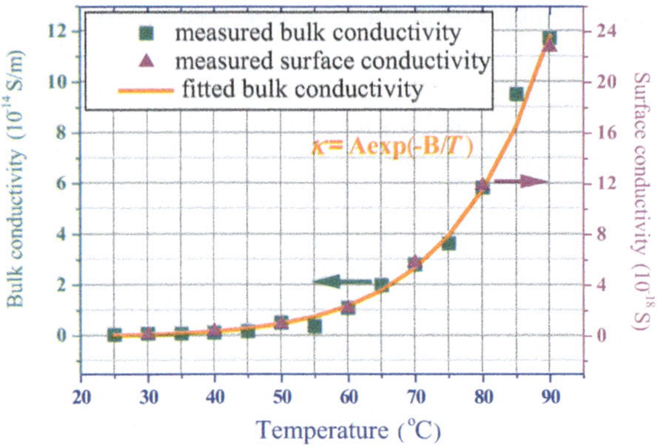

Fig. 19 Measured and fitted conductivity under different temperatures

are temperature-dependent rather than constant. The details of a simulation model for surface charge accumulation dynamics is described in Ref. [10]. After performing the simulation by considering the thermal-gradient in the insulator and the temperature-dependent conductivity, the surface charge density distribution can be obtained, as shown in Fig. 20. We can see that when the volume and surface conductivities are constant at room temperature, the accumulated charges are found to be of the same polarity as the applied voltage, which indicates that bulk conduction dominates the surface charging under the experiment condition. Furthermore, the profile (blue line) actually also coincides with the distribution of normal component of the electric field along the gas–solid interface, so the effect of surface conduction is limited since it is driven by the tangential component of the electric field. As for the charge distribution with thermal gradient, interestingly, the surface charge profile (green line) follows an exponential distribution, which is in accord with the thermal-dependent bulk conductivity distribution. Charge injection due to bulk conduction is enhanced and dominate the surface charging on the insulator. The maximal surface charge density

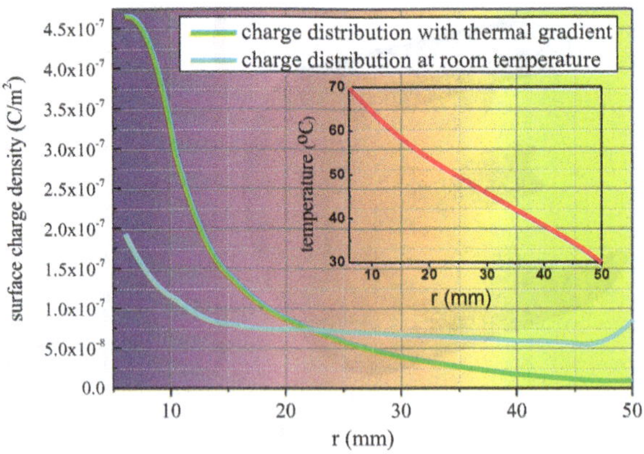

Fig. 20 Simulation results of charge distribution along the spacer sloping surface with and without thermal gradient

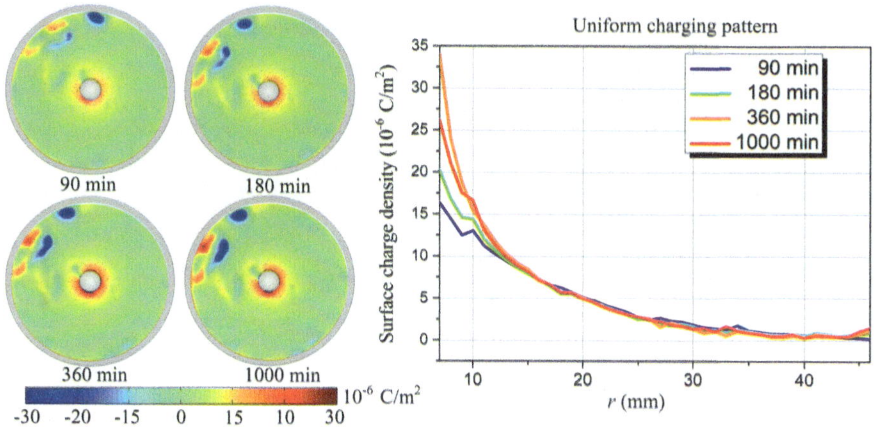

Fig. 21 Measured surface charge distribution under +20 kV with thermal gradient (left) and the averaged charge density along the radial direction (right)

is near the central electrode and can be more than twice that of the surface charge density without thermal gradient.

Figure 21 shows the measured surface charge distributions under a positive dc voltage of 20 kV with thermal gradient from 70 °C in the center to 30 °C in the edge. Accumulated charges are positive under positive voltage and the charge accumulation is much more sever near the central electrode, since charge injection from the electrode into the insulator bulk is further enhanced when temperature is higher. In the right, we can see that the charge density is higher than that at room temperature and follows an exponential distribution, which coincides with the simulation results. It should also be noted that the randomly distributed charge speckles near the central electrode disappear or reduce in its density. These charge speckles are inferred to come from micro discharges in the gas phase near the triple-junctions. The smoothing of the charge speckles in the center area should be due to the high surface conductivity in the high temperature region.

3.3.3 Influence of the Ionization Rate in the Gas Phase

We have realized that the uniform charging pattern mainly is determined by the difference between the current injection in the solid-phase and the current from the gas-phase. And the former one dominates the obtained homo-charge distribution. Therefore, if the current in the gas-phase is artificially increased, will the opposite result occur? To this end, the ion fan is placed inside the chamber, so that it can continuously provide high concentration of positive and negative ions in the gas when the insulator is under DC stress. After switching off the DC voltage and the ion fan, the surface charge on the insulator is measured.

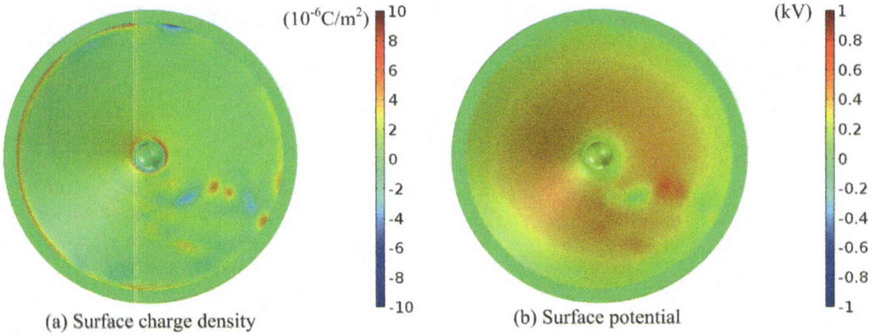

Fig. 22 Measured surface charge distribution under +20 kV with thermal gradient (left) and the averaged charge density along the radial direction (right)

Figure 22 shows the surface charge distribution and surface potential distribution of the insulator under −10 kV for 30 min. It can be seen that the accumulated charge is basically positive, which is opposite in polarity to the applied voltage. This is because the ion fan provides a large number of positive and negative ions in the gas, and the positive ions are drift along the electric field line to the surface of the insulator and the negative ions are away from the insulator surface. In this case, the current in the gas-phase becomes large than the bulk current in the solid-phase. As a result, the uniformly distributed surface charge becomes hetero-charge. This test again demonstrates that the uniform charging pattern depends on the relative magnitude of the current from both sides of the gas–solid interface. And this relationship is not constant but is influenced by many factors.

4 Surface Charge Suppression Methods

For an untreated epoxy insulator surface (which means the surface conduction is very limited), both the solid bulk conduction and the gas conduction determine the surface charge accumulation. According to previous discussions, it is found that by reducing the bulk conductivity of the insulator material and thus the bulk current, the difference between the current in the solid-phase and the gas-phase can be reduced. Consequently, the uniform charging level can be reduced. In addition, an appropriate increase in the charge dissipation rate on the insulator surface may inhibit the formation of locally distributed charge speckles. From these two aspects, many attempts have been made in the past decades. In this section, we proposed two representative materials modification methods in terms of bulk modification and surface modification.

4.1 Suppress Bulk Current to Decrease the Uniform Charging Level

Polymer-inorganic nanocomposites is one of the most common modification methods for dielectric materials. Many studies showed that doping a small amount of functional nanoparticles can effectively reduce the bulk conductivity of polymers [13–15]. However, due to the high specific surface area and the surface energy of inorganic nanoparticles, doping 1 ~ 10 wt% nanoparticles will greatly increase the viscosity of the polymer matrix, bringing much inconvenience to the preparation of polymer composites. Buckminster-fullerene (C60) and its soluble derivatives have shown great potential in a range of application areas, such as organic electronic, polymer technology, pharmaceutics and biomedicine. Due to its three-dimensional conjugated structure, small recombination energy, and the strong electron affinity (~2.8 eV), fullerenes possess a unique set of electronic properties. Studies have shown that fullerene C60 can significantly affect the charge recombination when involved in the electron transfer reaction [16] and have confirmed the electron transfer reaction of fullerene C60 occurs in the polymeric matrix [17]. Researchers also used the fullerene-based structure to prepare ultra-low *k* dielectrics [18]. Recently, researchers found that fullerenes are promising voltage stabilizers that can scavenge high-energy electrons due to their high electron affinity and may be used in polyethylene power cables [19].

Inspired by these studies, we believe that doping trace amount of nano-C60 in epoxy resin can help capture high-energy electrons and inhibit the electron migration, and thus achieve a reduced bulk conductivity. The details of the preparation processes of C60-doped epoxy resin insulator can be found in Ref. [20]. The bulk conductivity of epoxy resin insulators with different C60 filler loadings were measured under 10 kV/mm dc field at room temperature. The results are shown in Fig. 23 and the data presented is the average value of three measurements. One can see that the

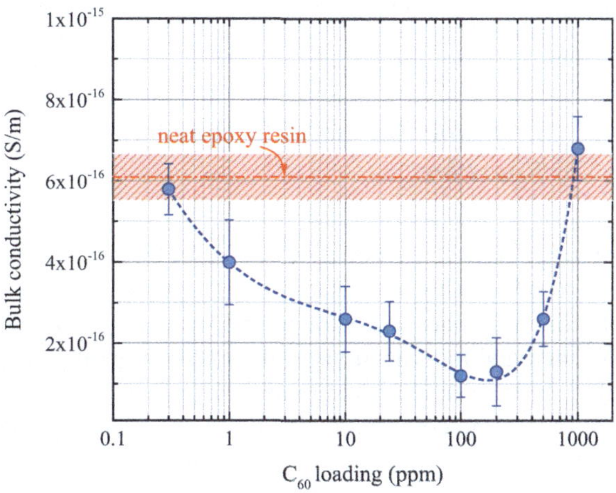

Fig. 23 Measured bulk conductivity of neat epoxy resin and C60/epoxy composites with different loadings [20]. Copyright © 2018, IEEE

conductivity of the neat epoxy is 6.1×10^{-16} S/m. With the addition of tiny amount of C60, the conductivity decreases significantly with the loading and reaches its minimum around 100 ~ 200 ppm. The conductivity of 100 ppm C60/epoxy composite is only 1.0×10^{-16} S/m, which is no more than 20% that of pure epoxy. However, further increase the C60 loading will increase the conductivity. When the C60 loading exceeds 1000 ppm, the conductivity of the composite become higher than that of the unfilled epoxy.

The decreased conductivity in C60/epoxy composite with lower filler loadings results from the hindrances in charge carrier transport through different chains and the polymer-filler interfaces. On the one hand, there is strong interface dynamics near the nanoparticle surface, i.e. the so-called interaction zone, at low filler loadings, including stable bonding and polymer chain entanglements. On the other hand, the strong electron affinity of the fullerene cage can effectively scavenge high-energy electrons in the composites under high electrical field [19]. Due to these physico-chemical changes in the local structure, the mobility and energy of charge carriers will be reduced and thus the conductivity will decrease. However, further increase the filler loading will decrease inter-particle distance and thus the interaction zones may overlap with each other, which leads to the percolation of charge carriers, causing an increase in the electrical conduction. Besides, C60 is a semi-conductive material which has much higher conductivity than the epoxy, so increasing the concentration of C60 to a certain amount will increase the conductivity of the base polymer.

The surface charge accumulation for the different insulators was then measured, as shown in Fig. 24. One can see that the charge patterns show a centrosymmetric uniform distribution together with charge speckles that are randomly distributed. Regardless of the charge speckles, if only the dominant uniform charging patterns are compared between the neat epoxy and the C60/epoxy composites, one can see that the charging level of the C60/epoxy insulators are apparently lower than that of the neat epoxy insulator, especially near the central electrode, where the electric field strength is the highest and thus charge injection is the most. The 200 ppm C60/epoxy insulator has the lowest charging level due to its lowest bulk conductivity, which effectively restrained the charge injection.

4.2 Promote Surface Charge Dissipation to Reduce Charge Speckles

Many attempts have been made to tailor the surface properties of the insulation material to achieve higher surface conductivity and fast charge dissipation rate, including fluorination, plasma treatment and surface coating. Most of these approaches focus on chemical modifications of the insulator, which are neither eco-friendly nor scalable. Moreover, chemical modifications mostly involve molecular mechanisms that are difficult to control and are highly material-dependent. A question is then raised: is it possible to simply re-engineer the physical structure of the insulator surface to

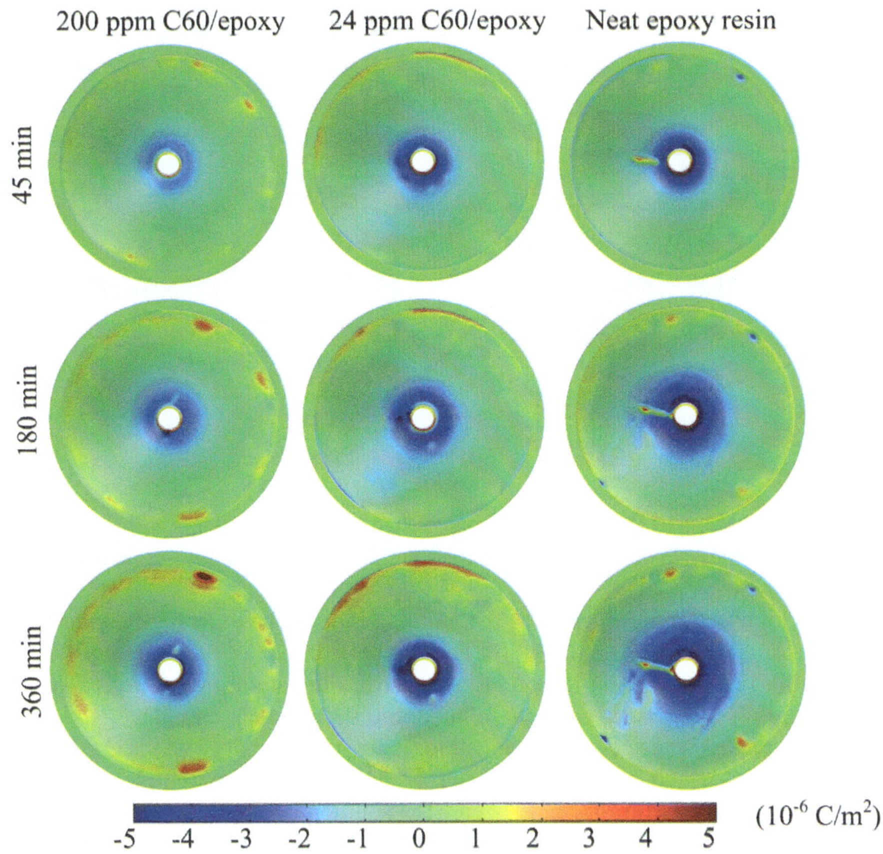

Fig. 24 Surface charge distribution under −25 kV for different insulators [20]. Copyright © 2018, IEEE

offer a more effective strategy for suppressing surface charge accumulation? If the answer is yes, what would be the optimal structure?

In fact, through its long evolutionary process, nature itself has ingeniously realized perfect biological structures with superior properties. One of the most outstanding biological models is the amazing hierarchical "bricks-and mortar" architecture of nacre, in which aragonite platelets are alternatingly packed in a matrix of biopolymers, as shown in Fig. 25. It has been demonstrated that nacre-mimetic structures possess a unification of exceptional mechanical and barrier properties, which can significantly dissipate both mechanical stress and thermal stress [21, 22] because of their highly orientated inorganic/organic interfaces. Inspired by this, we boldly predict that mimicking the hybrid microstructure of nacre on an insulator surface may also dissipate "electrical stress", namely accumulation of electrical charges. Therefore, we have proposed a nacre-mimetic coating on epoxy insulators. Without

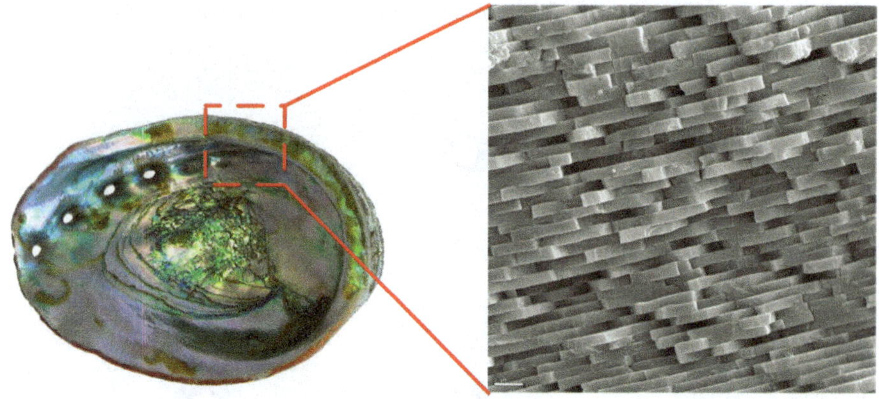

Fig. 25 Nacre and its micro-structure

changing the chemical structure of insulators, we are able to use this add-on laminar structure to achieve faster charge dissipation and less charge accumulation under DC stress.

Here, we use montmorillonite as the building blocks due to its high aspect ratio and relatively high ionic conductivity. The crystal structure of the montmorillonite is shown in Fig. 26. It is an important member of the natural inorganic 2:1 smectite clay family and can be easily exfoliated into ~1 nm thick monolayer sheets in water with the assistance of ultra-sonication. Polyvinyl alcohol (PVA) is used as the polymer binder, which is a colorless, water-soluble synthetic resin employed principally as adhesives and film-forming agent. The solution intercalation method is used to prepare the PVA/MMT coating and the nano-laminar structure is achieved by a simple flow-induced co-assembly technique. The details about the materials preparation is described in Ref. [23] and the preparation strategy is illustrated in Fig. 27. In Fig. 27e, the high resolution transmission electron microscopy (TEM) image clearly shows strikingly well-spaced alternating dark (MMT) and gray (PVA) layers within the coating. The inter-layer distance is characterized as ~2.88 nm.

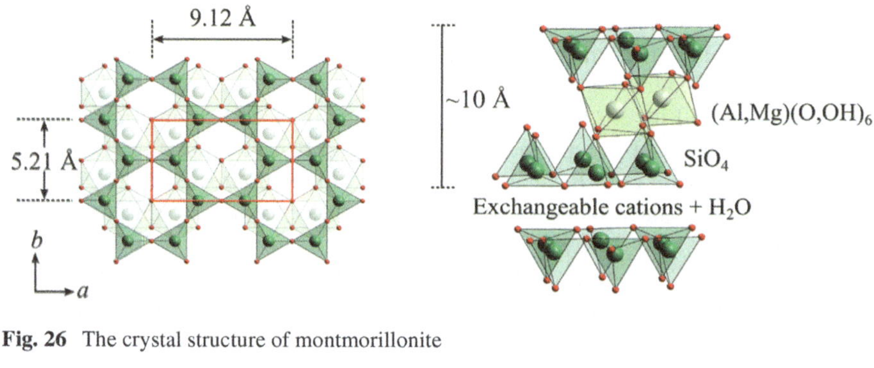

Fig. 26 The crystal structure of montmorillonite

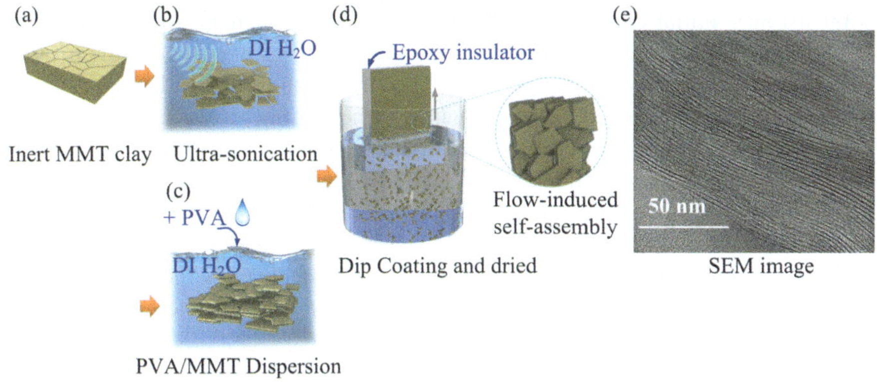

Fig. 27 The preparation of the PVA/MMT coating and its micro-structure

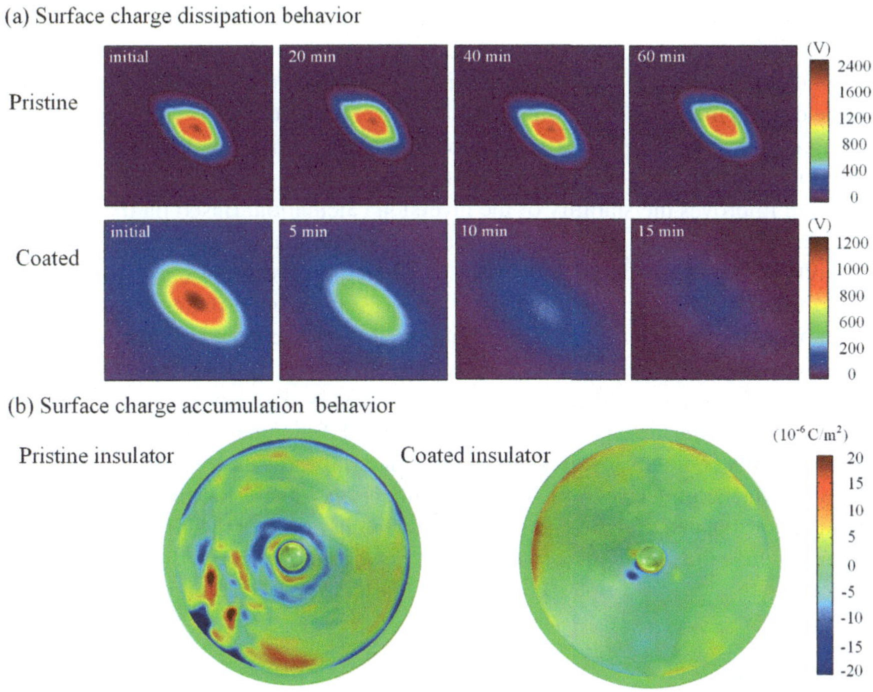

Fig. 28 Surface charge dissipation and accumulation behaviors of the pristine insulator and the coated insulator. Reproduced from Ref. [23] with permission from the Royal Society of Chemistry

The charge dissipation behavior of the neat epoxy insulator and the one with PVA/MMT coating is analyzed by using a corona discharge source to pre-charge the insulator surface. The results are shown in Fig. 28a. The spatial–temporal evolution of the surface potential distribution reveals that the coated epoxy insulator leads to much lower initial charge magnitude and much faster charge dissipation rate. This is owing to the unique lateral orientation of the MMT nanosheets, the trapped charges are more likely to be released in all directions along the abundant PVA/MMT interfaces. We then evaluated the performance of the coating on disc insulators under dc stress. Figure 28b shows the charge density distributions of the pristine and coated insulators after a 30 min-application of −20 kV dc voltage. The results indicate significantly lower surface charge density on the coated insulator than on the pristine one. Besides, the pristine insulator shows rather severe locally distributed charge speckles. While for the coated insulator, due to the effective charge dissipation on the PVA/MMT coating layer, the charge density distribution is quite uniform.

5 Summary

In this chapter, a systematic and comprehensive study of the surface charge characteristics under DC voltage is presented and effective methods to suppress the charge accumulation at the gas–solid interface is proposed, which is of great significance for the development of DC gas-insulated power equipment.

Based on the digital image processing technique, the surface charge inversion algorithm for the "shift-variant" systems is introduced, which greatly reduces the system noise and improves the stability of the inversion calculation. Based on the measurement results of the surface charge distribution of insulators in air and SF_6 under DC voltage, two patterns of the charge accumulation are proposed, i.e. dominant uniform charging and charge speckles. The uniform charging pattern is determined by the bulk current injection in the solid material and the charge speckles are related with the micro-discharges under high electric field strength caused by the impurities and defects. Material modification methods are proposed to suppress surface charge accumulation on the insulator. For bulk modification, fullerene doping can effectively reduce the bulk conductivity of the insulating material and therefore reduce the dominant uniform charging level. For surface modification, a nacre-mimetic nano-laminar coating is introduced and the coating can effectively promote the charge dissipation rate and reduce the locally distributed charge speckles.

References

1. T. Nitta, K. Nakanishi, Y. Arahata and M. Shimada, Development of SF6 Gas Insulated for 5500 kV HVDC Switchgears, in *Conference Record of the 1985 International Conference on Properties and Applications of Dielectric Materials*, vol. 1 (1985), pp. 340
2. H. Ootera, K. Nakanishi, Analytical method for evaluating surface charge distribution on a dielectric from capacitive probe measurement-application to a cone-type spacer in ±500 kV DC GIS. IEEE Trans. Power Deliv. **3**(1), 165–172 (1988)
3. A. Kumada, S. Okabe, Charge distribution measurement on a truncated cone spacer under DC voltage. IEEE Trans. Dielectr. Electr. Insul. **11**(6), 929–938 (2004)

4. B. Zhang, W. Gao, Z. Qi, Q. Wang, G. Zhang, Inversion algorithm to calculate charge density on solid dielectric surface based on surface potential measurement. IEEE Trans. Instrum. Meas. **66**(12), 3316–3326 (2017)
5. W.K. Pratt, *Digital Image Processing* (Wiley, New York, 1978), pp. 378–388
6. P.C. Hansen, D.P. O'Leary, The use of the L-curve in the regularization of discrete ILL-posed problems. SIAM J. Sci. Comput. **14**(6), 1487–1503 (1993)
7. A. Winter, J. Kindersberger, M. Tenzer, V. Hinrichsen, L. Zavattoni, O. Lesaint, M. Muhr, D. Imamovic, *Solid/Gaseous insulation systems for compact HVDC solutions, in Cigré Session 45* (Paris, France, 2014), pp. D1_103
8. R. Gremaud, C. B. Doiron, M. Baur, P. Simka, V. Teppati, M. Hering, J. Speck, S. Grossmann, B. Källstrand, K. Johansson, U. Riechert, U. Straumann, *Solid-gas insulation in HVDC gas-insulated system: measurement, modeling and experimental validation for reliable operation, in Cigré Session 46* (Paris, France, 2016), pp. D1_101
9. C. Li, C. Lin, G. Chen, Y. Tu, Y. Zhou, Q. Li, B. Zhang, J. He, Field-dependent charging phenomenon of HVDC spacers based on dominant charge behaviors. Appl. Phys. Lett. **114**(20), 202904 (2019)
10. B. Zhang, Z. Qi, G. Zhang, Charge accumulation patterns on spacer surface in HVDC gas-insulated system: Dominant uniform charging and random charge speckles. IEEE Trans. Dielectr. Electr. Insul. **24**(2), 1229–1238 (2017)
11. H. Iwabuchi, S. Matsuoka, A. Kumada, K. Hidaka, Y. Hoshina, T. Yasuoka, M. Takei, Influence of Tiny metal particles on charge accumulation phenomena of GIS model spacer in high-pressure SF6 Gas. IEEE Trans. Dielectr. Electr. Insul. **20**(5), 1895–1901 (2013)
12. R. Suda, S. Matsuoka, et al., Charge accumulation phenomena of silica-filled epoxy resin spacer under dc field, in *Proceedings of the 19th International Symposium on High Voltage Engineering* (2015)
13. T. Tanaka, Dielectrics Nanocomposites with Insulating Properties. IEEE Trans. Dielectr. Electr. Insul. **12**(5), 914–928 (2005)
14. J. Katayama, Y. Ohki, N. Fuse, M. Kozako, T. Tanaka, Effects of nanofiller materials on the dielectric properties of epoxy nanocomposites. IEEE Trans. Dielectr. Electr. Insul. **20**(1), 157–165 (2013)
15. B. Zhang, W. Gao, P. Chu, Z. Zhang, G. Zhang, Trap-modulated carrier transport tailors the dielectric properties of alumina/epoxy nanocomposites. J. Mater. Sci. Mater. Electron. **29**(3), 1964–1974 (2018)
16. D.M.G. And, K.D. Asmus, Electron transfer from C76 (C2v') and C78 (D2) to radical cations of various arenes: evidence for the Marcus inverted region. J. Am. Chem. Soc. **119**(24), 5744–5745 (1997)
17. E. Nakamura, H. Isobe, Functionalized fullerenes in water. The first 10 years of their chemistry, biology, and nanoscience. ChemInform **25**(8), 807–815 (2004)
18. H. Hermann, K. Zagorodniy, A. Touzik, M. Taut, G. Seifert, Computer simulation of fullerene-based ultra-low k dielectrics. Microelectro. Eng. **82**(3), 387–392 (2005)
19. M. Jarvid, A. Johansson, R. Kroon, J.M. Bjuggren, H. Wutzel, V. Englund, S. Gubanski, M.R. Andersson, C. Muller, A new application area for fullerenes: voltage stabilizers for power cable insulation. Adv. Mater. **27**(5), 897–902 (2015)
20. B. Zhang, W. Gao, Y. Hou, G. Zhang, Surface charge accumulation and suppression on fullerene-filled epoxy-resin insulator under DC voltage. IEEE Trans. Dielectr. Electr. Insul. **25**(5), 2011–2019 (2018)
21. W. Xie, S. Tadepalli, S.H. Park, A. Kazemi-Moridani, Q. Jiang, S. Singamaneni, J. Lee, Extreme mechanical behavior of nacre-mimetic graphene-oxide and silk nanocomposites. Nano Lett. **18**, 987–993 (2018)
22. P. Das, C.K. Yeo, J. Ma, K.D. Phan, P. Chen, M.B. Chan-Park, H. Duan, Nacre mimetic with embedded silver nanowire for resistive heating. ACS Appl. Nano Mater. **1**, 940–952 (2018)
23. B. Zhang, Q. Wang, Y. Zhang, W. Gao, Y. Hou, G. Zhang, A self-assembled, nacre-mimetic, nano-laminar structure as a superior charge dissipation coating on insulators for HVDC gas-insulated systems. Nanoscale **11**, 18046 (2019)

Nano-Modification for Enhancing the DC Surface Insulation Strength of Epoxy Resin

Qing Xie, Haoou Ruan, Jun Xie, Qijun Duan, Zhenyu Zhan, Kai Yin, and Fangcheng Lü

Abstract The flashover under a DC electric field can be a serious problem for epoxy-made insulators in gas insulated transmission line (GIL). The threat of that can be attributed to factors including the special charge transfer process at the gas—solid interface, the unreasonable structure of the electric field, and the unsatisfactory performance of the epoxy insulating materials. In this chapter, from the perspective of material modification, the method of nano-doping and filler functional modification are adopted to reinforce the DC surface insulation strength of insulating epoxy materials, including fluoridation modification, morphology regulation, rare earth element doping, etc. The feasibility of nano-modification on surface insulation of epoxy resin was studied, and its modified mechanism was further discussed.

Q. Xie (✉) · H. Ruan · Z. Zhan · K. Yin · F. Lü
School of Electrical and Electronic Engineering, North China Electric Power University, Beijing 102206, China
e-mail: xieqing@ncepu.edu.cn

H. Ruan
e-mail: whiteghost@foxmail.com

Z. Zhan
e-mail: zhanzhy@ncepu.edu.cn

K. Yin
e-mail: kyi@et.aau.dk

F. Lü
e-mail: lfc0818@sohu.com

J. Xie · Q. Duan
Hebei Provincial Key Laboratory of Power Transmission Equipment Security, North China Electric Power University, Baoding 071003, China
e-mail: junxie@ncepu.edu.cn

Q. Duan
e-mail: duan_ncepu@163.com

B. Du (ed.), *Polymer Insulation Applied for HVDC Transmission*,
https://doi.org/10.1007/978-981-15-9731-2_16

1 Introduction

As the most commonly used composite matrix for insulating materials, epoxy resin (ER) has been used in insulation scenarios across a wide range of voltages. However, surface flashover under a DC electric field can be a significant problem for ER [1]. Whether operating in air, in SF_6 or in vacuum, the continuous DC field can cause surface flashover in epoxy insulators [2, 3] (Fig. 1).

Methods have been proposed to overcome this adverse factor, including optimising the surface structure to increase the creep distance [4, 5], and applying chemical treatment on the insulating surface [6, 7], to ensure its stability under flashover arc, etc. Surface flashover can be regarded as the response of an insulator to the external electric field, whose influencing factors can be finally attributed to the physico-chemical parameters of the material itself [8]. In this aspect, ER can be readily combined with other materials to form a composite structure, thus offering the potential to adjust its performance parameters. Studies have shown that, by doping with fillers including SiO_2, Al_2O_3, BN, AlN, SiC, etc., specific properties of ER can be reinforced, including insulating properties [9, 10], thermal conductivity [11], mechanical properties [12], and corrosion resistance [13]. Seeking to different application backgrounds, we need to choose different fillers on an ad-hoc basis.

As electrical insulation, epoxy insulators working under moderate voltages are doped with micro-SiO_2 to improve the mechanical and insulation strength of the epoxy. As for high-voltage insulators working in SF_6, micro-Al_2O_3 fillers [14, 15] are chosen to improve their thermal conductivity, and to ensure that such insulators will not be corroded by the decomposition of SF_6. According to current technical experience, the focus of epoxy fillers has always been on the micron-scale. As the study moving on, nano-fillers have a unique size effect, with better intrinsic properties and a more complex interface structure to bestow improved modification probability. Nano-fillers can improve the performance of the matrix even at a low dopant content (as used for energy storage and other frontier science scenarios); however, studies of the insulating application of nano-fillers are rare, a fact that can be attributed to the gap in research involving nanotechnology in this cognate area and the high threshold imposed by the manufacturing industry for new product entry. For the electrical engineering scenario herein, the application of nano-fillers warrants long-term research from both the economic and scientific perspectives. In North China

Fig. 1 A flashover accident involving GIL disc spacers

Electric Power University, Prof. Xie and Prof. Lü have investigated the effects brought by nano-filler doping on DC insulating performance of epoxy composites, with some of the results shared in this chapter.

In order to accurately characterize the DC surface flashover performance of epoxy composites, a test platform was built to test the DC flashover voltage and dynamic charge characteristics under different field structure, gas atmosphere and voltage waveform. In this paper, o facilitate the comparison of different materials, it mainly presents the flashover and charge test results under open atmospheric environment with pair needle electrodes, as shown in Fig. 2.

The spacing of the electrodes is 7 mm. The radius of curvature of the electrode tip is 0.2 mm. All the electrical experiments were carried out in a sealed stainless cylindrical chamber as Fig. 2 shows and 10 breakdown tests for each sample after unified tailoring. A DC high voltage source with a range of 0–50 kV was applied on the stainless steel pin-pin electrode to explore the surface insulation performance of the composites in an extremely inhomogeneous electric field, and local maximum field strength exceeds 10^7 V/m in Fig. 3a. The voltage rising is controlled by digital

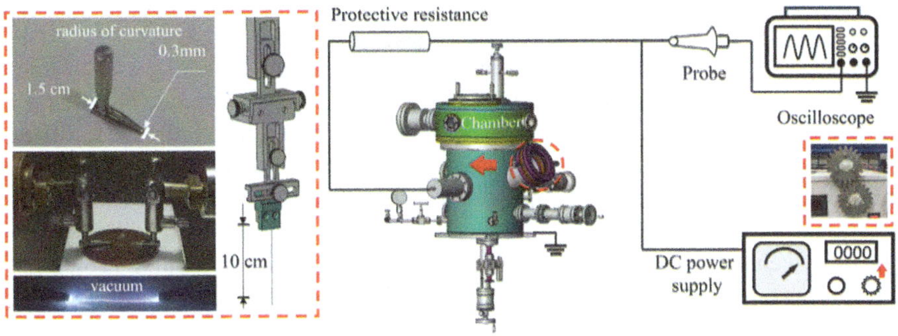

Fig. 2 The schematic of flashover and potential decay testing platform, including parameters of electrodes and dimensions of flashover

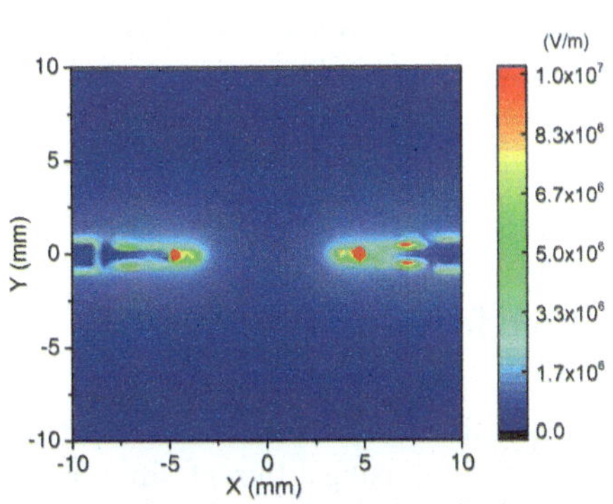

(a) Typical electric filed before flashover

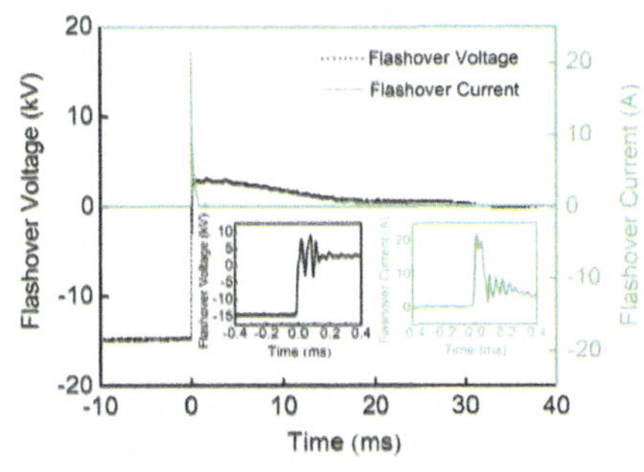

(b) typical voltage wave form for flashover

Fig. 3 The typical electric filed distribution and voltage waveform of flashover test mechanical gears. The rate is 25 kV/min uniformly, and the voltage waveform before and after flashover on the sample is shown in Fig. 3b.

Surface charge and trap distribution are measured by ISPD (Isothermal surface potential decaying) method. By testing the attenuation curve of the material's surface potential after corona charging, the carrier trap distribution of the material under the double-exponential trap model was calculated according to Eqs. (1) and (2). The method of isothermal surface potential decay (ISPD) has pointed out that the trap depth E_T can be regarded as a function of dissipation time t, which can be shown as follow

$$E_T = k_B T \ln(vt) \tag{1}$$

Where, k_B is the Boltzmann constant, T is the kelvin temperature of the medium, and v is the escape frequency of charge (for DC electric field, it is set as a constant of $4.17 \times 10^{13}\ \mathrm{s}^{-1}$). Meanwhile, the isothermal relaxation current *I(t)* can be calculated by the decay of surface potential, which can be shown as follow

$$I(t) = (\varepsilon_0 \varepsilon_r / eL) \cdot tdU/dt \tag{2}$$

Where, ε_0 is the permittivity of vacuum, ε_r is the relative dielectric constant of medium, e is the elementary charge, L is the thickness of samples, and dU/dt is the decay rate of surface potential. Regardless of the constant coefficients, the trap density is considered to be proportional to the variable of *tdU/dt*. Accordingly, the trap depth of the medium can be presented by the curve of *tdU/dt*- E_T. Considering that we have used double exponential function to fit the *U-t* data, there will always be one or two characteristic peaks which can be observed in the trap curve [16]. Peaks with lower energy level were regarded as shallow trap peak, and that with higher energy were deep trap peaks. According to the trap effect, shallow traps can facilitate the process of charge dissipation, while deep ones would hinder that. However, the relationship between trap and flashover is still inconclusive. Many of the studies are contradictory. This is exactly what we are interested in and worth studying. A large number of preparation and testing experiments involved in this chapter have shown us the macroscopic law of the flashover voltage with material parameters, and also helped us put forward a more reasonable and comprehensive explanation from the perspective of traditional flashover theory.

2 Nano Doping for Enhancing the Surface Insulation of ER

When we consider adopting nano-fillers to improve the DC surface insulation performance of ER, the simplest modification method is to dope using different, single types of fillers. Although filler doping is a form of bulk modification, flashover is mainly affected by the surface properties of the material, and the relationship between them is compact and complex. There are many factors to be studied.

The main preparation process used for epoxy nano-composites is similar to that used for conventional resin matrix composites. It is worth noting that a strong dispersion step is needed to minimise the agglomeration of nano-fillers in the slurry, to prepare homogeneous composites. Acetone is often needed to reduce the viscosity of the resin slurry, and the shear cavitation and high-power stirring equipment is used for its dispersal. The most representative ER used in insulating materials is DGEBA (bisphenol-A-diglycidyl-ether), and the matched curing agent is MeTHPA (mMethyl-tetrahydrophthalic-anhydride) with promoter DMP-30 (tris(dimethyl-aminomethyl) phenol). In insulator manufacturing, the most commonly used moulds are made of stainless steel, which must be treated via high temperature bluing to ensure the density and compactness of the inner surface. Release agent for most insulators is the organic silicon series, whose main film-forming component is silicon oil. The curing process is conducted at 90 °C for 2 h and 110 °C for 1 h; post-curing treatment is often set at 140 °C for 10 h. Without post-curing treatment, the electrical and mechanical properties of epoxy composites will be greatly impaired. The aforementioned process can be deemed representative of the synthesis approach for epoxy insulating materials.

2.1 Nano-SiO_2

SiO_2 can be used as a traditional filler for ER. Micro-SiO_2 fillers are widely used to improve the mechanical and thermal conductivity of resin materials in insulation devicesof moderate voltages, such as cabinet bushings and contact boxes. Nano-SiO_2 has potential for uses in these devices [17, 18].

Sphere-type nano-SiO_2 with an average size of 20, 50, 80, and 100 nm was selected for doping modification, with its doping mass fraction set as 1 wt%-7wt%. It is worth noting that when the SiO_2 mass fraction reaches 7wt%, the viscosity of the slurry will exceed the recommended industrial value about 2500 mPa·s, which is not conducive to the pouring process and limits the doping content.

Figure 4 shows the SE micrographs of the bottom surface of the composites doped with 20-nm SiO_2 with 1 and 3 wt% doping contents respectively. When the doping content is 1 wt%, the SiO_2 buried in the epoxy generally presents a uniform distribution. As the 3 wt%, the SiO_2 particles appear as local agglomerations.

The test results of flashover voltages are shown in Fig. 5. In most cases, the DC flashover voltage of the doped composite is higher than that of pure epoxy, which evinces the efficiency of nano-SiO_2 doping. We also noted that, for most samples, the flashover voltage showed a rising-falling trend with a change at 3 wt%. The flashover voltage also showed a rising-falling trend with increasing particle size, and its inflection point appeared at 50–80 nm.

In summary, by merely doping with nano-SiO_2, the promotion of DC flashover can be conditionally limited. Both the mass fraction, particle size of fillers can affect the DC flashover voltage of nano-SiO_2/epoxy composites.

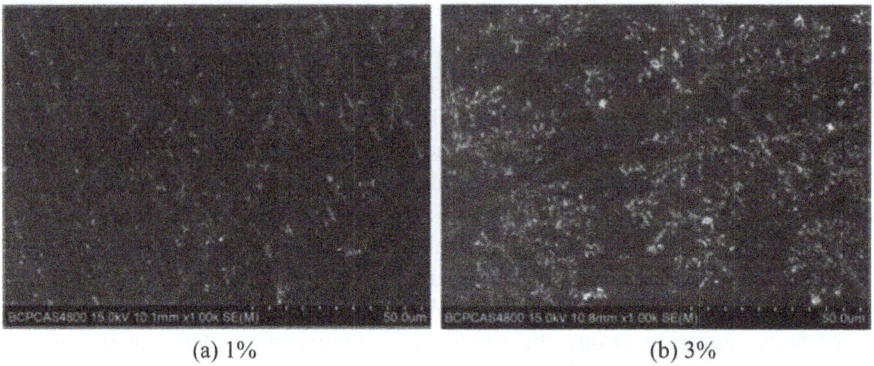

Fig. 4 SE micrographs of SiO_2/epoxy composites with doping contents 1 and 3%

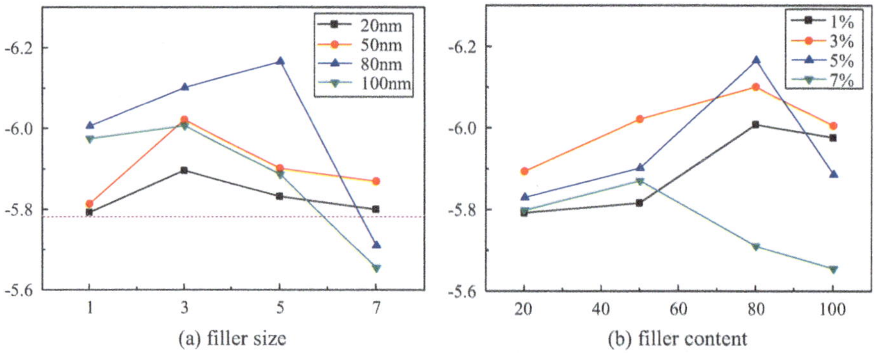

Fig. 5 The flashover voltage of nano-SiO_2/epoxy composites with different filler parameter

2.2 *Nano-Al_2O_3*

Alumina is the most commonly used filler in medium- and high-voltage insulation devices, such as the disc spacers in gas insulated transmission line (GIL) and gas insulated switch-gear (GIS) [19]. Micro-Al_2O_3 with a mass fraction of 60–70 wt% is used as the main filler to ensure the mechanical, thermal, and insulation performance of the epoxy insulator. For the aforementioned devices, it is necessary to discuss the application value of the doping of nano-sized Al_2O_3 from the perspective of size replacement and micro-nano co-doping.

In this section, ER was doped with spherical nano-Al_2O_3 particles with sizes of 20–100 nm, and the composite materials with mass fractions of 1–7 wt% were studied. As shown in Fig. 6, based on the microscopic characterisation of cured products, it can be found that the nano-Al_2O_3 fillers appeared as a sediment at the bottom of samples and some of the filler was exposed outside the matrix.

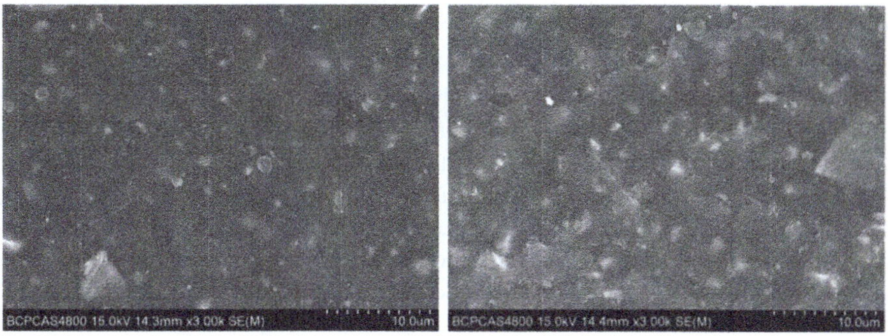

(a) 1wt% (b) 3wt%

Fig. 6 SEM images of Al_2O_3/epoxy composites with doping contents

Figure 7 shows DC surface flashover test results: the DC flashover voltage of the composite material first increased, then decreased with increasing mass fraction of fillers, and its inflection point appeared at 3 wt%. For composites with the same mass fraction, the DC flashover voltage also showed a trend of rising-and-falling, and inflection points appeared at 50 and 80 nm.

In summary, the flashover promotion of ER by nano-Al_2O_3 is conditionally limited. In the particle size range of 20–100 nm and concentration range of 1–7 wt%, choosing the optimal particle size and mass fraction can improve the DC flashover voltage of the composite material the most.

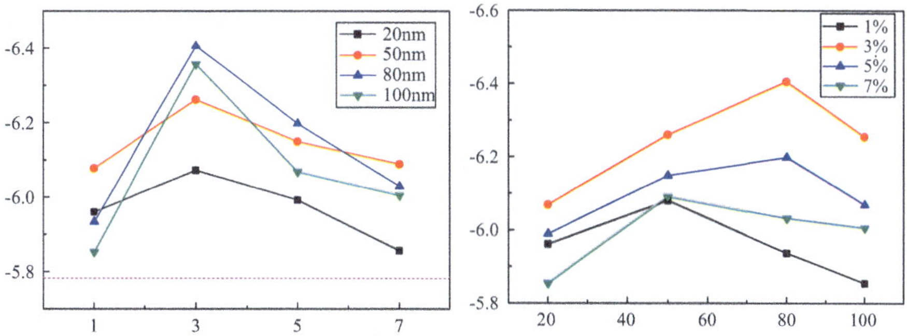

(a) filler size (b) filler content

Fig. 7 The flashover voltage of Al_2O_3/ER composites with different filler parameter

2.3 Nano-TiO_2

Nano-TiO_2, a semiconductor filler widely used in the manufacture of refractory materials, photocatalysts, batteries, and other special materials, is also regarded as an ideal filler for polymer composite materials. The semiconductor properties are very useful for the homogenization of distorted electric field, but for nano-TiO_2, the doping amount is usually low, far below the seepage threshold. In this case, whether nano-TiO_2 can still achieve flashover enhancement is worth studying. The microscopic morphology of TiO_2 filler powder with a 20-nm particle size and its composites is shown in Fig. 8. Nano-TiO_2 has poor compatibility with ER and a treating process of filler-resin interface is require.

The DC surface flashover testing is shown in Fig. 9. When the mass fraction of TiO_2 is about 1 wt%, the flashover voltage fluctuates with the doping content and has not showed an obvious regulation. When the doping content exceeds 2 wt%, the flashover voltage increases monotonically with the TiO_2 mass fraction. By comparing the samples of different particle sizes, it can be found that, the flashover voltage corresponding to a 100-nm particle size is most less than that when using 20-nm and 500-nm particles. It may be that 100 nm is an inappropriate particle

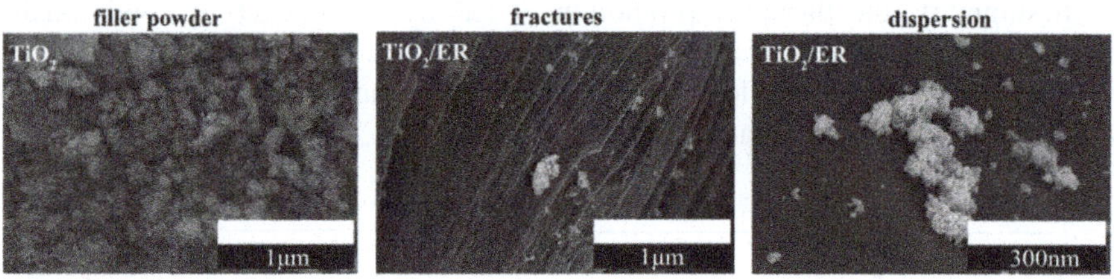

Fig. 8 SEM micrographs of TiO_2/ER composites

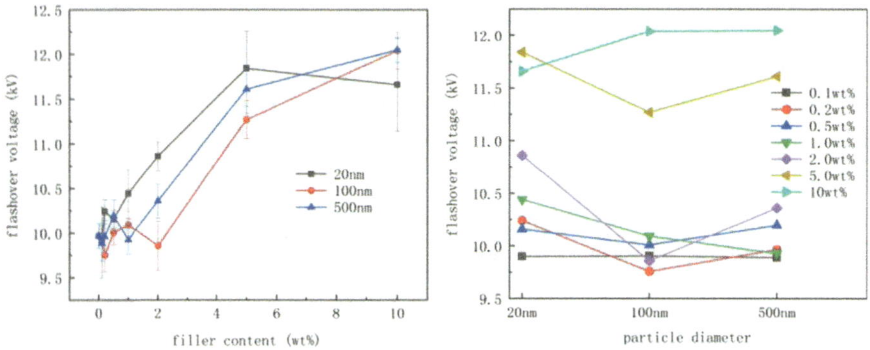

(a) filler size (b) filler content

Fig. 9 The flashover voltage of TiO_2/ER composites with different filler parameter size, which suffers from both difficult nanometer technology and poor theoretical enhancement [20].

Needle type electrode used in flashover test generates an uneven electric field. The addition of TiO_2 will modify the electric field distribution to a certain extent and reduce the electric field intensity in the areas of the strong electric field, at the expense of improving the electric field intensity in areas at lower intensity. Of course, with such a low mass fraction added, the component of TiO_2 will fail to form a percolating structure, and the ER hinders charge transport between TiO_2 particles, ensuring the basic insulation strength of such composite materials.

2.4 BN Nanosheets

Boron nitride (BN), with its low dielectric constant and ideal insulating strength, has been an important filler for epoxy and other polymer matrices [21]. By ball-milling and stripping, BN nanosheets (BNNS) with two-dimensional honeycomb hexagonal structure can be obtained.

As shown in Fig. 10, after stripping treatment, the BNNS presents a distorted and dispersed form, with a thickness of about 10 nm, and were dispersed well in the ER matrix. The DC flashover test results show that a small amount of doping BNNS can improve the flashover voltage of the ER. When the doping content is 0.5 wt%, the flashover voltage of BNNS/ER composite is optimised to about 20% higher than that of pure ER. However, with the increase of the dopant, the improvement effect of the flashover voltage of epoxy composite is decreased, and then tends to be stable. We think that the contribution of BNNS to the improvement of flashover of epoxy composite is mainly due to its good electrical insulation performance. Moreover, when the mass fraction of fillers is excessive high, the damage brought by the filler to the original charge transfer channel of the matrix will increase, and it is not conducive to improvement of the flashover voltage (Fig. 11).

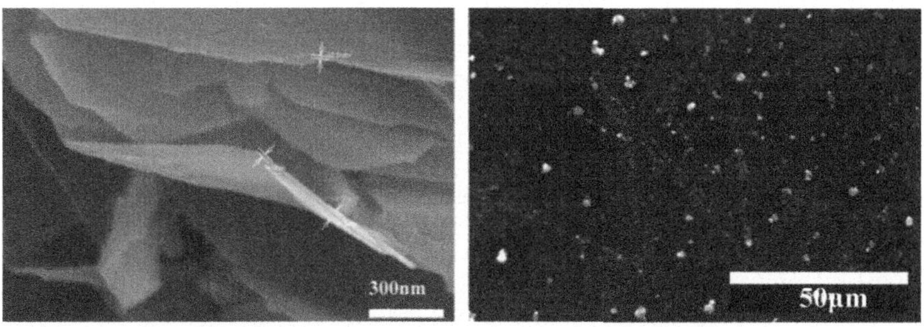

(a) BNNS (b) BNNS/ER

Fig. 10 SEM micrographs of BNNS and BNNS/ER

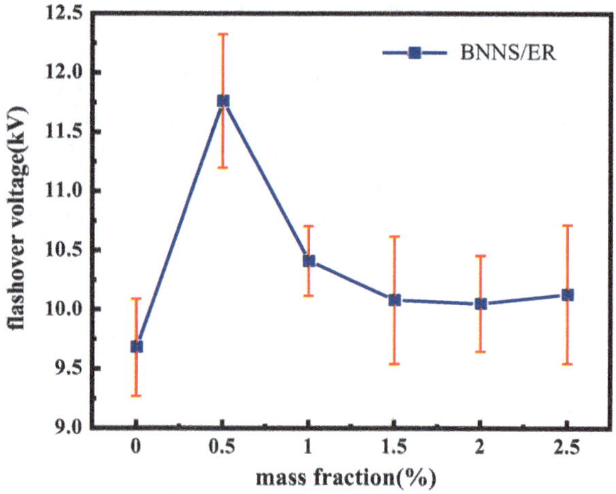

Fig. 11 Flashover voltage of BNNS/ER

In summary, the DC flashover voltage of the composite can be effectively increased under appropriate BNNS filling concentration.

2.5 *Other Nano-Fillers*

In addition to the above fillers, nano-fillers such as SiC, AlN, BTO, etc., are also used as the dopant for epoxy resins. SiC is a wide-gap semiconductor material with nonlinear conductive characterization. On the premise of ensuring the uniformity of the composite materials, the doping content of SiC must be controlled at a low level. The surface flashover voltage test results of epoxy composites containing 1, 3, and 5 wt% nano-SiC fillers are shown in Fig. 12.

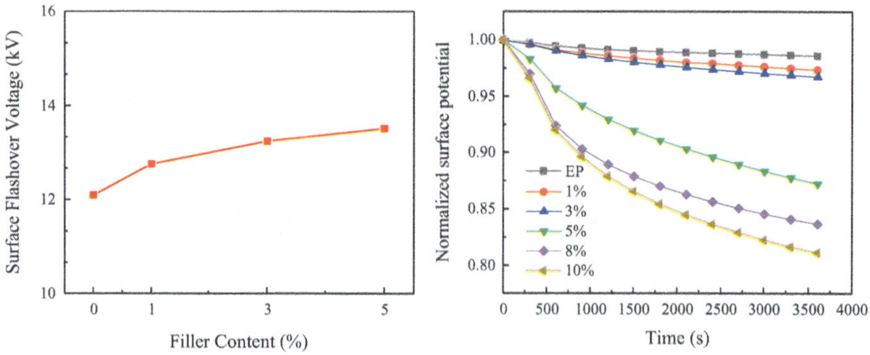

(a) flashover voltage (b) normalised surface potential

Fig. 12 Surface insulation strength of nano-SiC/ER composites

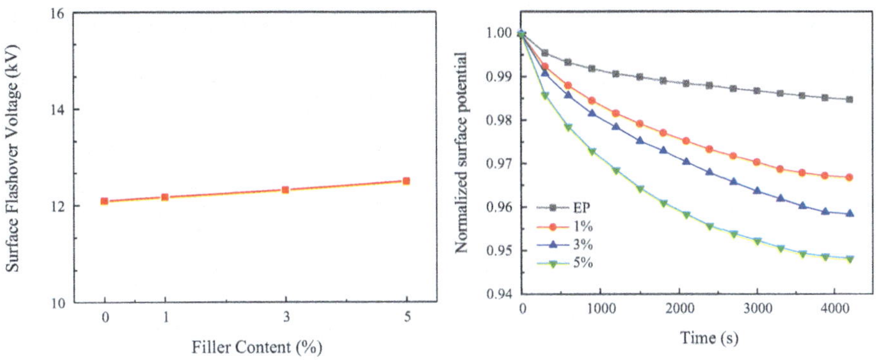

(a) flashover voltage
(b) normalised surface potential

Fig. 13 Surface insulation strength of AlN/ER composites

AlN has better thermal conductivity and insulation strength than Al_2O_3. Despite its high cost of preparation, it is not prevented from becoming a candidate filler for insulating materials. The surface flashover test was performed on ER samples doped with nano-AlN fillers at 1, 3, and 5 wt%, respectively. As shown in Fig. 13, nano-AlN filler doping can slightly increase the flashover voltage of epoxy composites, but the effect is not significant. The rate of dissipation of normalised surface potential increases with the content of AlN fillers. This shows that AlN doping is harmless to the flashover performance of epoxy insulator, and has little effect on the electrostatic dissipation performance. Considering the improvement of thermal conductivity of AlN itself, the filler has some value of application in epoxy insulator [22, 23].

Graphene nanosheets are two-dimensional carbon nano-material with high electrical and thermal conductivity. From a general point of view, epoxy doped conductor is not a normal way to improve flashover. However, a very small amount of conducting powder can improve the dielectric and charge evacuation performance of the material while preserving the overall insulation. Through ball-milling, graphene nanosheets with an average scale-length of about 500–800 nm, and a thickness of about 100 nm can be conveniently prepared in laboratory. Figure 14 shows the typical SEM micrographs of the morphology. Graphene nanosheets are doped into ER in mass ratios of 0.2, 0.4, 0.6, 0.8, and 1%.

The experimental results of DC flashover of epoxy composites doped with graphene nano-fillers are shown in Fig. 15a. The flashover voltages of the modified epoxy composites are higher than that of the pure material. When the mass fraction of epoxy composite is 0.2%, it generates the optimal improvement. The surface trap characteristics of GN-doped epoxy composites were studied by ISPD test method. As shown in Fig. 15b, GN doping makes the deep-trap concentration of the ER composite increase slightly, while the shallow-trap concentration increases by a significant amount. It is found that shallow traps play a positive role in improving the flashover characteristics of ER, but the flashover characteristics are also affected

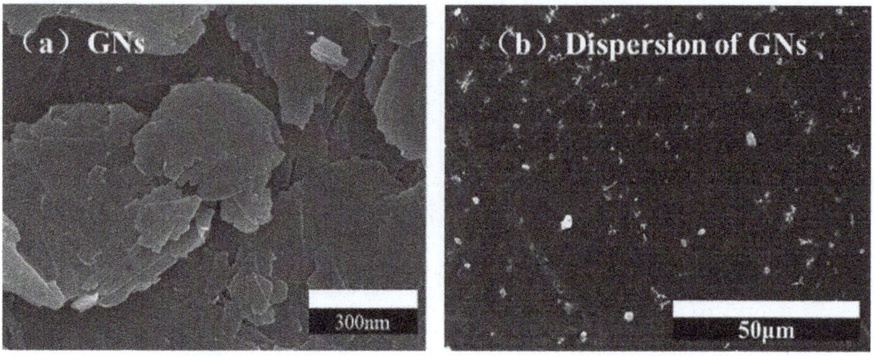

(a) GN (b) GN/ER

Fig. 14 SEM micrographs of GN and GN/ER composites

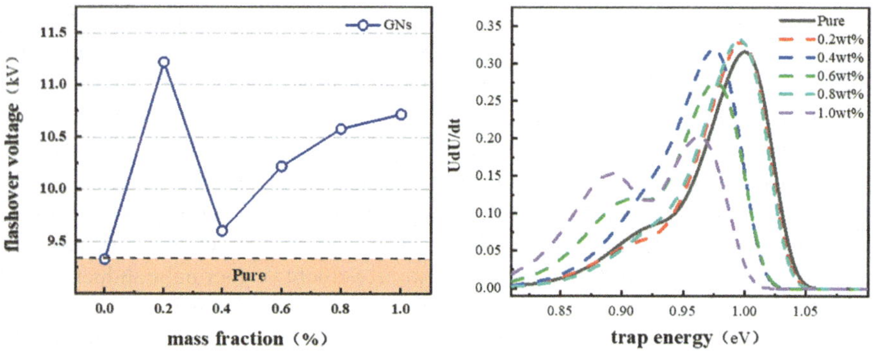

(a) Flashover voltage (b) trap distribution

Fig. 15 Flashover and trap distribution of GN/ER composites

by many factors, such as: filler type, dispersion, material conductivity, and dielectric properties [24, 25].

Bismuth ferrite ($BiFeO_3$, BFO) is a multi-ferroic ceramic. As the band-gap and dielectric properties of BFO can be controlled by thermodynamics and doping, respectively, it has become a proper filler to manipulate the dielectric performance of resin composites [26]. Moreover, the semiconductor properties of BFO brought the potential to limit the charge distribution that is detrimental to insulation. We prepared 5% mol yttrium-doped $BiFeO_3$/ER with different filler contents. A reduction in the dielectric constant and leakage current of particles can be seen by Y-doping, which is useful for increasing the flashover voltage. We synthesised BY5FO particles of approximately 600 nm diameter. As shown in Fig. 16, the TEM images illustrate the crystalline domain spacing of 3.96 Å and 2.84 Å, corresponding to the (012) and (104) planes of a rhombohedral phase crystal. The epoxy in the presence of BY5FO caused a slight granular protuberance, as seen on the composite surface. The change

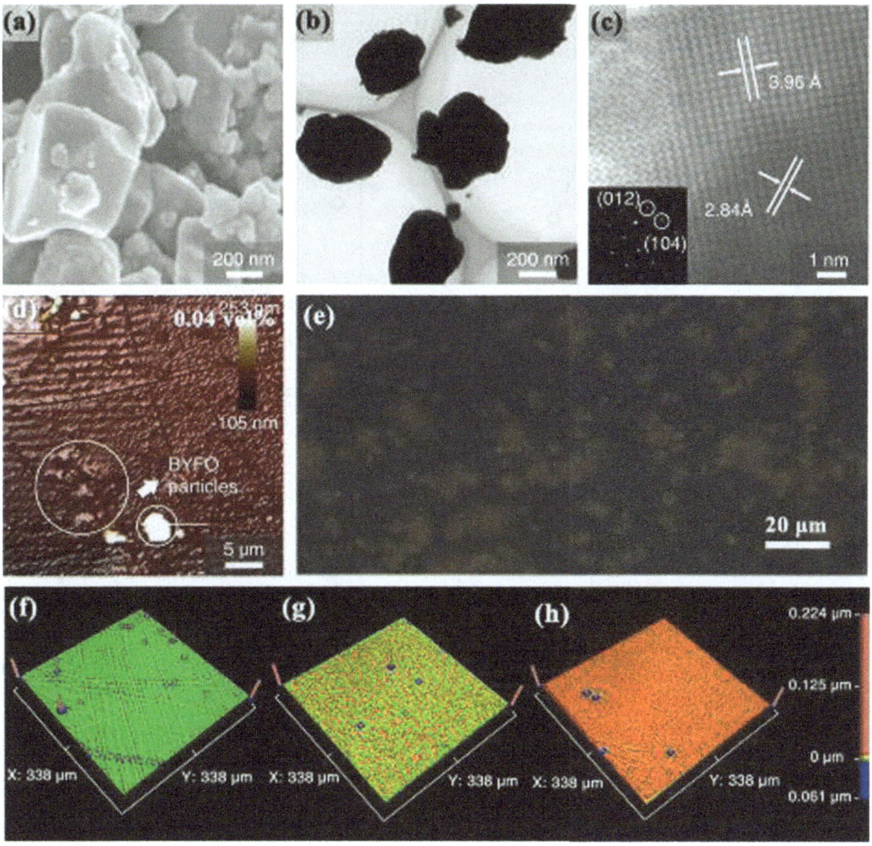

Fig. 16 **a** SEM, **b** Low-resolution and **c** high-resolution TEM images and the SAED pattern (inset) of dispersed BY5FO particles. **d** AFM image and **e** optical image of the surface of BY5FO/ER composite. 3-d surface topography of the BY5FO/ER composite with **f** 0 vol%, **g** 0.24 vol%, and **h** 5.6 vol% BY5FO content

of the surface morphology of BY5FO/ER was related to the filler content: for the composite materials of 0.24 vol% fillers, the BY5FO were scattered on the epoxy base. In contrast, the protrusion of the filler dominated the surface of the 5.6 vol% BY5FO/ER composite [27, 28].

The surface flashover voltages (in negative polarity) and the standard deviation of the BY5FO/ER composites are presented in Fig. 17 and show a regular sinusoidal trend with increasing filler content.

In a highly non-uniform electric field, the DC vacuum flashover of the BY5FO/ER with the 13.6 vol% filler content is 111% higher than that of pristine epoxy, without any modification of the surface structure or electrode configuration. To sum up, the modification effect of nano doping of Al_2O_3, SiO_2, TiO_2, BN, SiC, AlN, GN and BFO(BYFO) are studied. For most kinds of nano-filler doping, the flashover voltage of the composite materials would show a non-monotonic change trend with

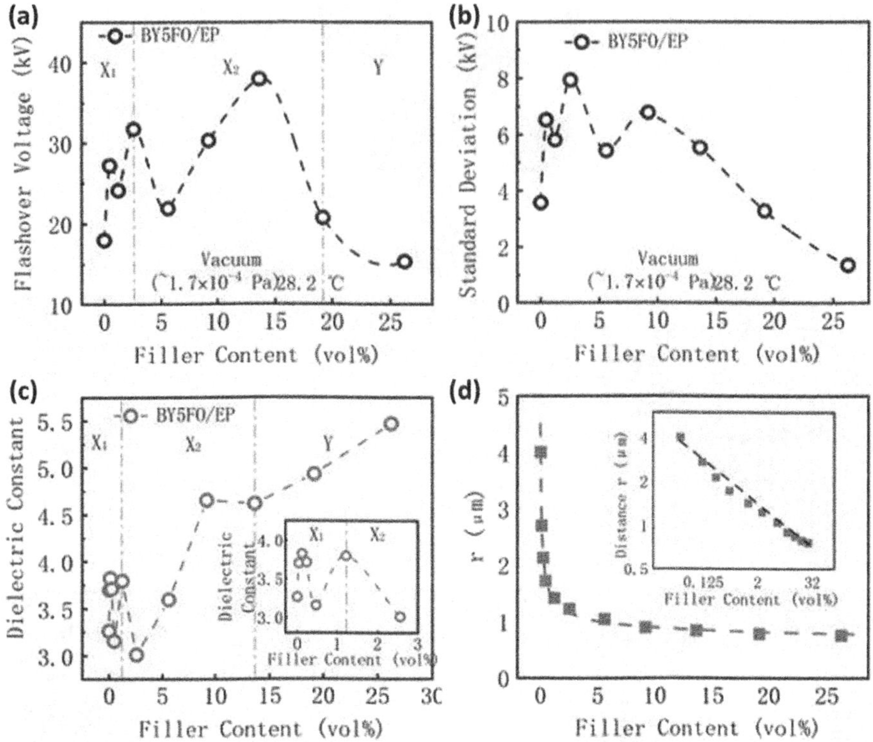

Fig. 17 **a** Flashover voltages, **b** standard deviation and **c** dielectric constant (inset shows enlarged plot at the low volume content of filler) of BY5FO/ER composites in a vacuum. **d** Spacing between filler particles *r* as a function of the filler content. *r*-filler content curve in log-log coordinates (inset)

the increase of the mass fraction of fillers, and there was more than one inflection point. This is a common and interesting phenomenon for nano doping of polymer composites, which need to be further analyzed and explained. Figure 17d shows that the spacing of fillers shrinks rapidly at first and then more slowly as the filler content increases [29].

3 Nano Functionalization for Enhancing the Surface Insulation of ER

3.1 Coupling Agent Treatment of Nano Fillers

Coupling agents (CAs) can be used to improve the interfacial properties of ER matrices and fillers, and can be achieved by simply presoaking the filler in industry.

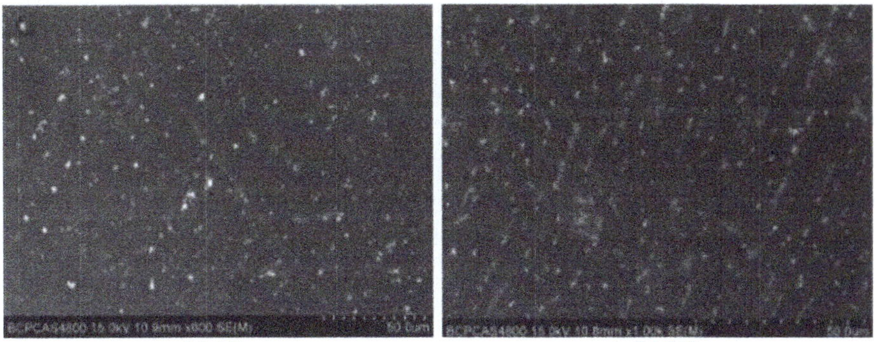

(a) 1% (b) 3%

Fig. 18 SEM micrographs of nano-SiO_2/ER (1 and 3 wt%)

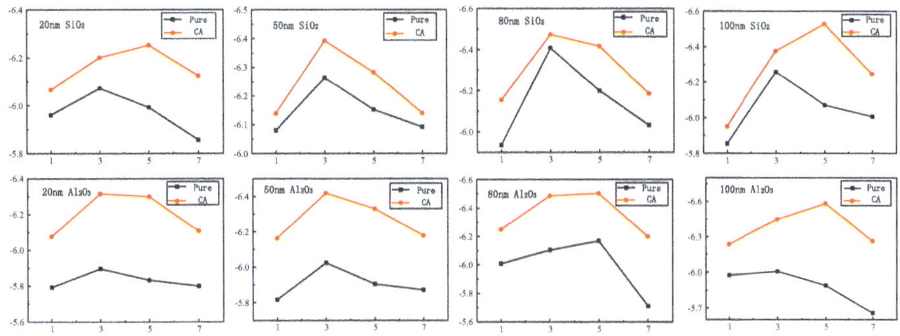

Fig. 19 Flashover voltages for SiO_2/ER and Al_2O_3/ER

KH550 is a common used CAs for epoxy composites [30]. In this section, nano-Al_2O_3 and nano-SiO_2 were modified by KH550 and prepared into ER composites. Through the comparison of Figs. 18 and 2, it can be concluded that the agglomeration of fillers can be reduced through KH550 treating.

As shown in Figs. 19 and 20, the flashover voltage of the composite was increased after CAs treatment, and the dielectric constant of these composite materials can be reduced. The chemical bond generated between the modified nanoparticles and the resin matrix increased the interaction between them, which hindered the steering of the polar polymer chain segment or side group in the interface region, resulting in a resistance to polarisation, and a reduction of the dielectric constant of the composite material. We also noticed in earlier studies that the dielectric constant of ER without post-curing and resin with imperfect preparation process would be high, which was also related to the molecular chain constraint characteristics [31].

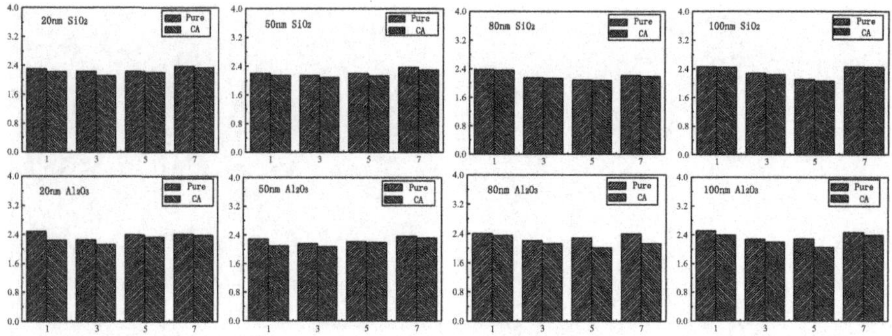

Fig. 20 Dielectric constants for SiO_2/ER and Al_2O_3/ER

3.2 *Morphology Control of Nano Fillers*

Filler morphology is also an important parameter in the selection of industrial formula, which has an effect on the electrical insulation performance inside the material. Doping of BTO nanofillers can improve the surface insulation performance of ER composites. BTO is generally composed of inhomogeneous spherical nanoparticles. One-dimensional BTO nanofibres can be prepared by electrospinning [32]. Compared with nanoparticles, nanofibre materials impart a stronger nano-size effect, which is mainly manifest in its higher specific surface area and surface energy. A filler network forms readily in the matrix, theoretically conferring a better nano-modification effect. We synthesised BTO nanoparticles and nanofibres, filled the ER with 1, 2, 4, and 8% BTO nano-fillers with two different morphologies, and further characterised the nano-fillers and their dispersion in the ER matrix by SEM. As shown in Fig. 21, the diameter of BTO nanoparticles is about 500–700 nm, and the average diameter of BTO nanofibres is about 400 nm. Due to the presence of more active sites on the surface of fibrous nano-fillers, the bonding between nano-fillers and ER matrix is easier, and the agglomeration of granular nano-fillers is larger. Based on the surface flashover test, the influences of the filler concentration of two types of BTO nano-fillers on the flashover characteristics of epoxy composites were analysed [33, 34].

The experimental results are shown in Fig. 22: the flashover voltages of BTO nanofibre systems are higher than that of BTO nanoparticle systems. When increasing the concentration of BTO nanoparticles, the flashover voltage of the two composite materials presents a similar N-type trend. The flashover voltage of the composites doped with 1% BTO nanoparticles decreases slightly. When the mass fraction is 2%, the flashover voltage reaches its maximum. This is similar to the aforementioned experimental results, that is, at a low dopant concentration, the surface flashover characteristics of ER composites improve most [35].

The flashover voltages of the BTO nanofibre system are similar to that of BTO nanoparticle systems at low dopant concentrations. But the flashover voltage of BTO nanofibre systems increases significantly with increasing filler concentration. This

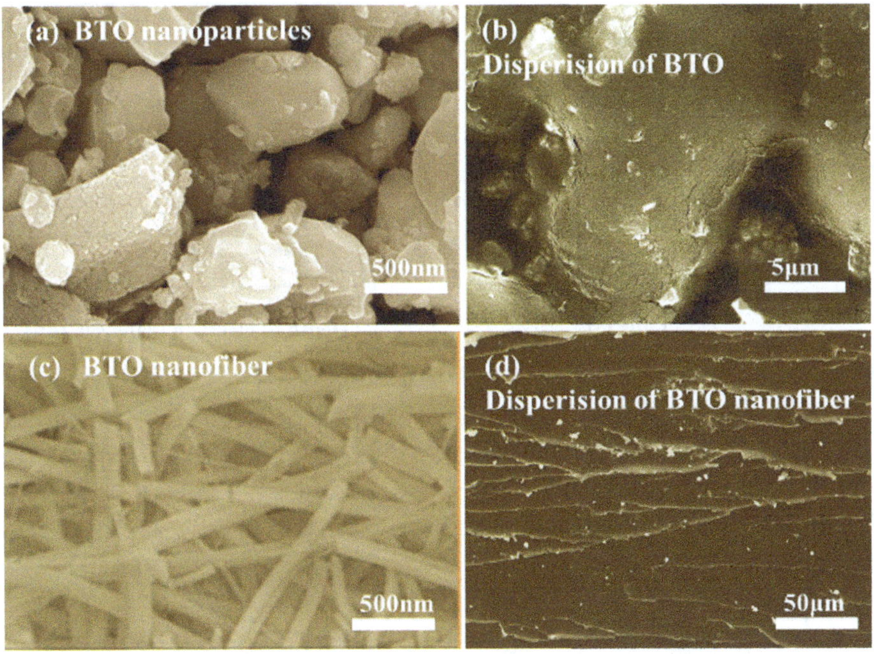

Fig. 21 **a** BTO nanoparticles, **b** BTO/ER, **c** BTO fibres (XBTO), **d** XBTO/ER

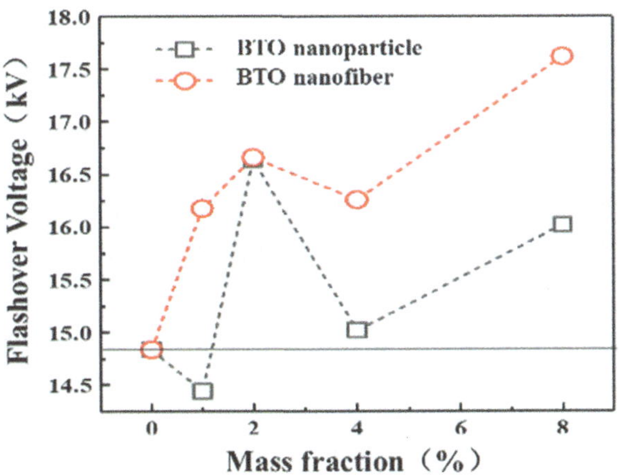

Fig. 22 The flashover voltage of BTO/ER composites

shows that, with the increase of the filler concentration, the change of the interface area between the nanoparticles and the matrix affects the material properties, resulting in the corresponding change of the flashover characteristics. By studying the surface charge dissipations of epoxy composites, the trap distribution characteristics of the material were calculated (Fig. 23).

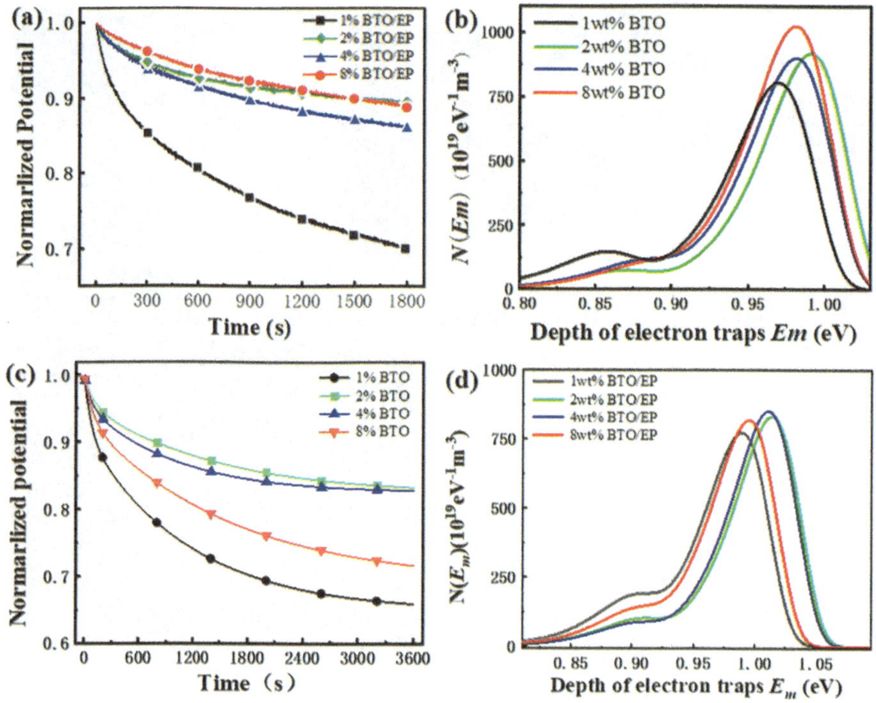

Fig. 23 Normalized surface potential of **a–b** BTO and **c–d** BTO fibres

It can be found that doping with BTO nano-fillers can reduce the rate of dissipation of surface charge on the composite to a certain extent. This indicates that the addition of packing destroys the charge transfer channel of the original ER matrix. However, with the increase of the filler concentration, the rate of charge dissipation of fibrous nano-BTO tends to increase, while that of granular nano-BTO is lower. This means that BTO nanofibres more readily show the effect of filler modification. Both granular and fibrous BTO doping will generate many deep traps to the surface of the material. When the filler concentration reaches 8%, the deep-trap concentration of BTO nanofibre system begins to decrease, and the shallow-trap concentration begins to increase. But the equivalent plot for the BTO nanoparticle system still shows no inflection point at this concentration [36].

BFO, a semiconductor material, is widely used in digital storage devices, magnetoelectric induction equipment, and other devices [37]. The semiconductor properties of BFO have the potential to limit charge accumulation. In addition, adjustment of the BFO preparation process was found to be able to control the morphology and particle size of the filler. These can affect its microscopic distribution parameters in the polymer matrix and many aspects of the electrical performance thereof [38].

BFO with different particle sizes were prepared by the sol-gel route, and bismuth ferrite nanofibres (fBFO) were synthesised by electrospinning method. The fillers

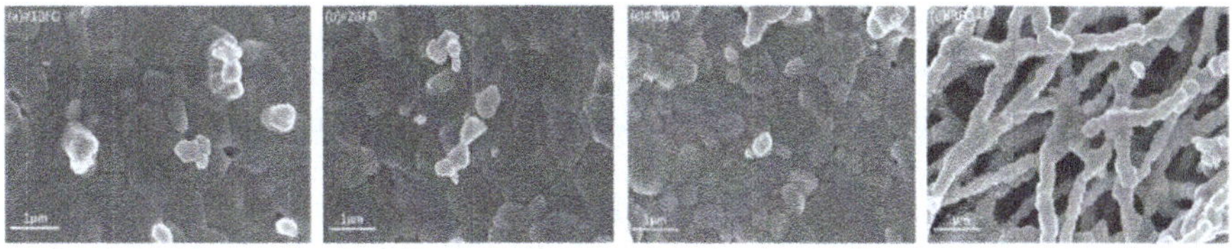

Fig. 24 SEM micrographs **a** #1BFO **b** #2BFO **c** #3BFO **d** BFO fibres

were then mixed with ER to prepare epoxy composites with a variety of filler concentrations.

SEM micrographs of BFO with different particle sizes and fBFO nanofibres are presented in Fig. 24a–c. The average particle size of BFO particles is around 600 nm, the particle size distribution is non-uniform. And the surface of the particles has a visible ridge-like morphology, which may be the result of the preferential growth of the crystal plane. The SEM micrographs and laser particle size test analysis show that the average particle sizes of #1, #2, and #3 BFO are 871 nm, 593 nm, and 406 nm, respectively. After the filaments are annealed at high temperature, the surface of the BFO nanofibre is formed by connecting many small particles of uniform particle size and regular morphology. The fibre diameter is around 300 nm, and the length is, on average, 4 μm [39].

The DC surface flashover voltage variation curve of the ER composites is presented in Fig. 25. The flashover voltage of ER composites doped with #1BFO particles or fBFO fibres is higher than that of pure ER. The flashover voltage of BFO/ER composites exhibits M-shaped variation with increasing mass fraction. The flashover voltage of fBFO/ER composites shows a monotonically increasing trend. The analysis shows that, compared with BFO particles, nanofibre surface particles lack coordination of adjacent atoms, and they more readily combine with the ER matrix to form a more stable cross-linked structure. As the dopant concentration

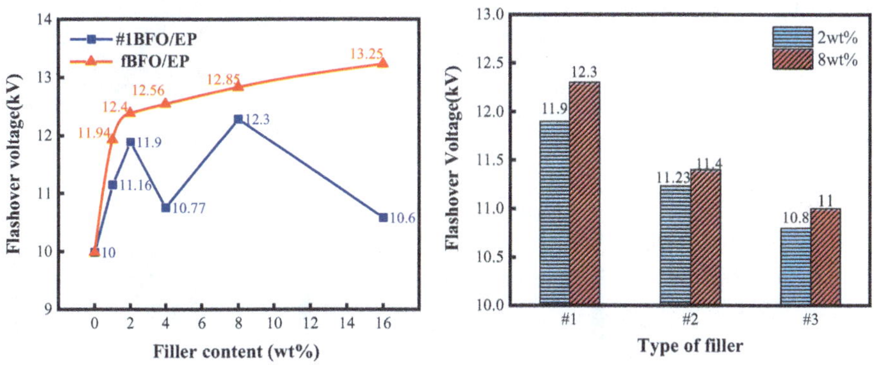

Fig. 25 Surface flashover voltage test: flashover voltages of BFO/ER composites with different particle sizes

increases, fBFO nanofibres (with their high aspect ratios) make contact with each other in the polymer matrix, creating more charge transport paths within the ER composites, which reduces surface electric field distortion and increases the flashover voltage.

To study the effect of the filler particle size on the surface insulation of composites, BFO with different particle sizes were selected as fillers to prepare BFO/ER epoxy composites with mass fractions of 2 and 8 wt%, and their flashover voltages were measured (Fig. 25). The results show that the flashover voltage decreases with the particle size at a given dopant concentration.

In summary, reasonable selection of filler morphology can greatly improve the compatibility of filler and matrix, thus affecting the charge transport performance of the ER.

3.3 *Rare Earth Element Doping of Nano-Fillers*

By controlling the chemical component of the filler, the band gap characteristics can be regulated. The conductivity of BFO can be regulated within a wide range by A-site ion doping with different valence states was used to prepare pristine BFO, and Ca^{2+} and Sm^{3+}-doped BFO as a filler blended with ER. The surface of the BFO sub-particles was characterised by SEM (Fig. 26a–c). These show the BFO sub-particles as highly crystalline with sizes of between 500 nm and 1 μm. As shown in Fig. 27, the direct band-gap values for BFO ceramic samples were determined from the square of the Kubelka–Munk function at different energies and extrapolation of the linear part of F2(R) to zero value. It is found that the band gap values for BC5FO, P-BFO, BS5FO are 2.1819, 2.2011, and 2.2170 eV, respectively. [40]

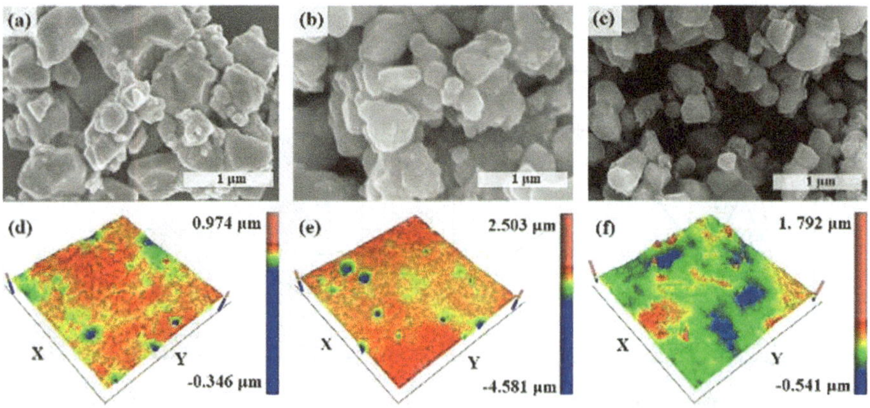

Fig. 26 SEM micrographs of **a** P-BFO, **b** BC5FO, and **c** BS5FO; Surface morphology of **d** P-BFO, **e** BC5FO, and **f** BS5FO

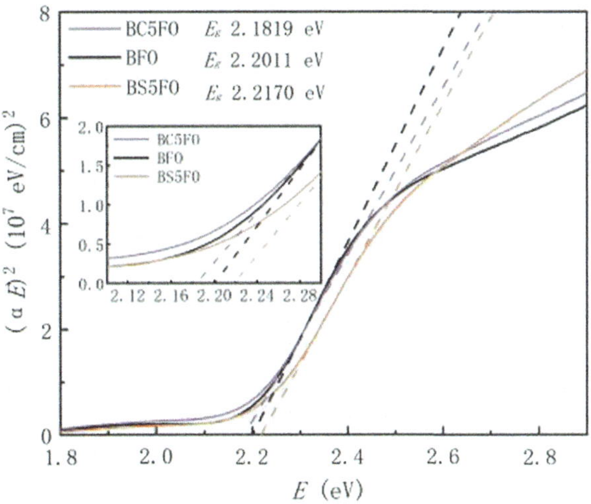

Fig. 27 Plots of $(\alpha E)^2$ against photon energy E

The flashover voltages of BFO/ER composites in the air and a vacuum are shown in Fig. 28. All samples exhibit sinusoidal variations regardless of the test environment under extremely uneven E-field conditions. In ambient air, the flashover voltages of P-BFO/ER, BC5FO/ER, and BS5FO/ER composites are enhanced by 22.4%, 40.7%, and 11.1%, separately. In addition, the maximum flashover voltages generally emerge within a high filler-loading range. In contrast, the enhancements under vacuum flashover conditions are 61.9%, 13.2%, and 1.1%. Vacuum flashover

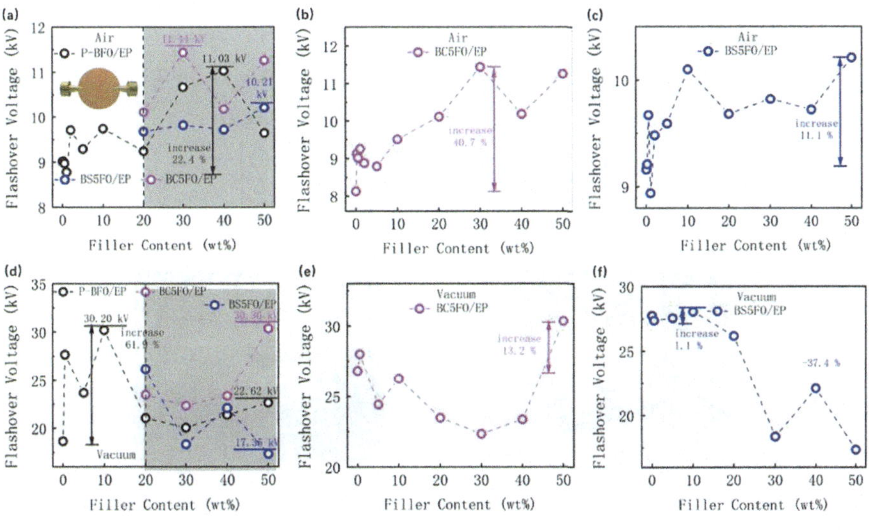

Fig. 28 The flashover voltages of BFO/ER in **a**–**c** air and **d**–**f** vacuum

is more sensitive to filler conductivity, and $BiFeO_3$ with higher conductivity (when used as the filler) is more conducive to surface insulation, whether in the air or in a vacuum. Finally, the optimal $BiFeO_3$ bandgap of 2.1819 eV is determined in terms of enhancing the surface insulation of such composites [41, 42].

3.4 Plasma Fluoridation Treatment of Fillers

3.4.1 Filler Fluoridation Platform

To improve the electrostatic dissipation characteristics of epoxy insulators, Prof. Xie and Prof. Lü in North China Electric Power University proposed a treatment method for fluoridation of fillers. By means of plasma grafting treatment, the filler surface can obtained fluorine-rich groups before blending into composites matrices. Fluoridation of fillers can be used as a pre-treatment procedure in insulator manufacturing to improve the static dissipation performance together with its DC flashover resistance. The plasma treatment method uses CF_4 as the fluorine source gas, to generate fluorine-containing active groups through ionisation of CF groups, thus to modify the surface of the filler powder. This method can be applied to fillers with a certain tolerance to F element, such as Al_2O_3, AlN and other insulating filler powder, such as: TiO_2, SiC, and other semiconductor powder. This method is not recommended for reactive powders such as SiO_2 that will react with fluorine. As shown in Fig. 29a, the plasma fluoridation treatment platform mainly consists of an electrical circuit

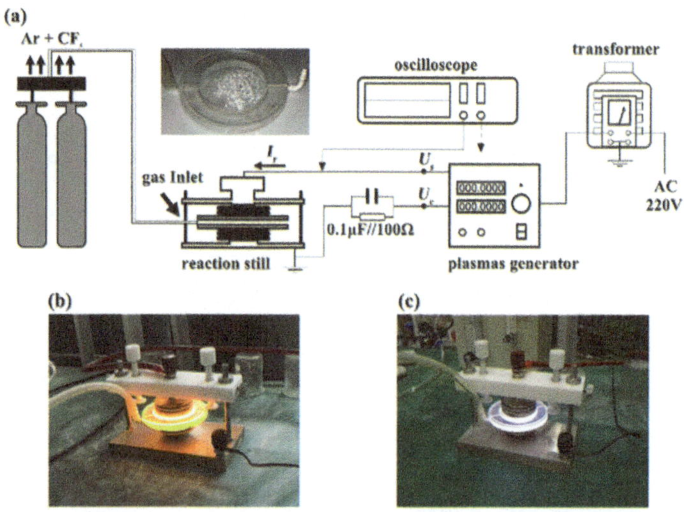

Fig. 29 **a** The main devices of plasma fluoridation. The discharge phenomenon of **b** Ar/CF_4 plasma and **c** Ar plasma

that generates the dielectric barrier discharge and a gas circuit that provides fluorine source gas. During the discharge, the mixed gas of CF_4 and Ar gas enters the reactor through the gas inlet, and a stable dielectric barrier discharge is formed between the high-voltage end and the ground electrode by applying an external high-frequency electric field. Fine-tuning the frequency causes the dielectric barrier discharge reactor to produce a stable filamentous arc accompanied by an orange halo, as shown in Fig. 29b and c, the orange colour of the light is due to the properties of the plasma of CF_4, and the pure Ar plasma emits a blue-purple light. The treated fluorinated nano-filler will be blended with ER and prepared as an ER composite material [43, 44].

3.4.2 Fluoridation of Nano-Al_2O_3

As shown in Fig. 30, the plasma treatment with and without CF_4 both can relieve the particle agglomeration of nano-Al_2O_3 in ER matrix. In the inset (c), we see many fractions on the surface of fluorinated nano-Al_2O_3, with diameters much less than the original value of tens of nanometers. For 20-nm Al_2O_3, plasma fluoridation can introduce fluorine at a mass ratio of 28.43%.

Figure 31 has presented the FTIR of the Al_2O_3 filler powder and composite samples. For powders, the appeared absorption peak of CF_2 can be found at 507 cm^{-1} and the absorption peak of CF_3 at 1170 cm^{-1}. The absorption peaks of surface hydroxyl corresponding to 3465 cm^{-1} and 1650 cm^{-1} are significantly weakened by plasma fluoridation (with Ar/CF_4). The fluorinated composite materials only show the new absorption peak of the CF_2 groups at 507 cm^{-1}. Therefore, it can be speculated that plasma fluoridation has replaced the hydroxyl on the surface of the Al_2O_3 with fluorine-containing groups, including CF_2 and CF_3. During curing, the CF_3 groups have reacted with the epoxy matrix, thus turning into CF_2.

Flashover testing was conducted to assess the final modifying effect. It can be found that, for composites filled with untreated Al_2O_3, the flashover voltage shows two rise-fall cycles as the filler content increases, and the maximum occurs at a filler

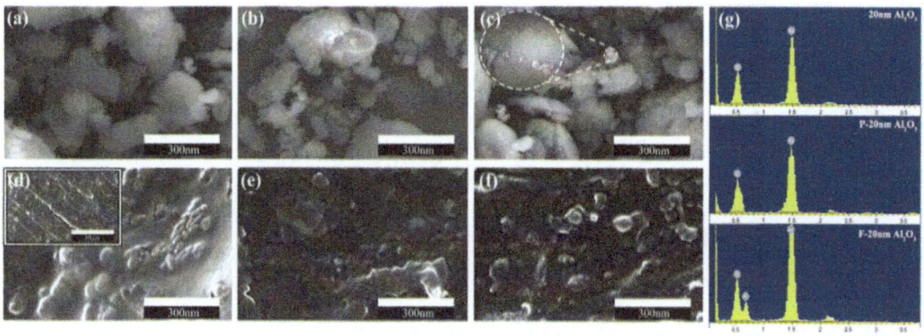

Fig. 30 SEM micrographs and EDS results of F-Al_2O_3/ER

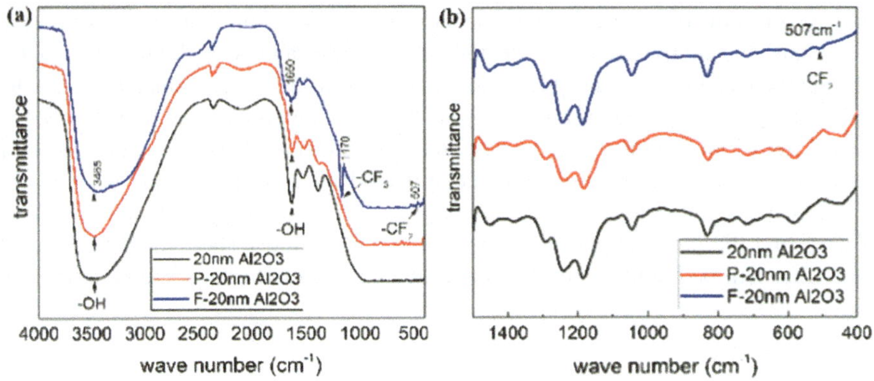

Fig. 31 FTIR results of **a** filler powder and **b** composite samples

content of 5 wt%, further doping will lead to a decrease in flashover voltage. After filler fluoridation, it presents an monotonic increasing trend in the range from 0 to 10 wt%, and appears to be unsaturated at a filler content of 10 wt%. That is to say, when doped with fluorinated Al_2O_3, flashover promotion no longer mainly depends on the complex filler-ER interaction, but is positively correlated with the upper limit of doping content. The flashover voltage will be further improved if the tech of nano doping are raised.

As shown in Fig. 32, the charge dissipation of the ER is rather poor. The Al_2O_3 particles, when used without fluoridation, have a little effect in promoting charge dissipation. After being placed in air for 15 min, the composites filled with 20-nm Al_2O_3 can reduce the surface potential to 90% of the initial value. The fluorinated Al_2O_3 can significantly promote charge dissipation in such epoxy composites. When

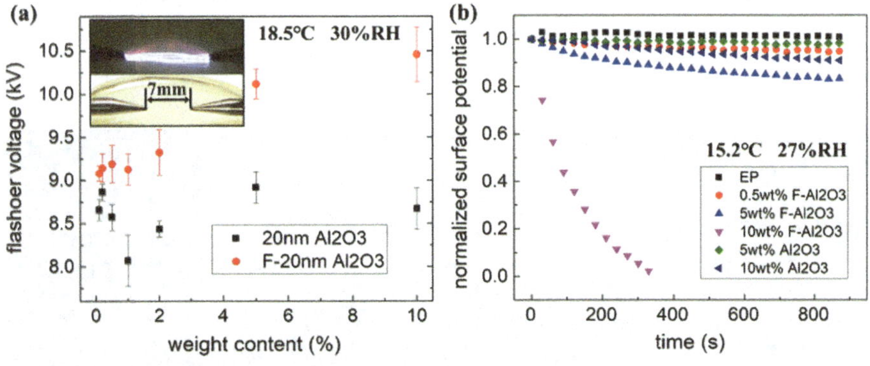

(a) flashover voltage (b) normalized surface potential

Fig. 32 Surface insulation properties of Al_2O_3/ER composites

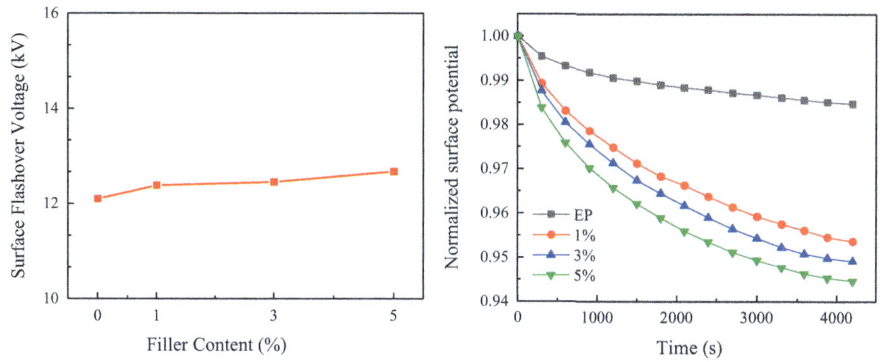

Fig. 33 Flashover and normalised surface potential of F-AlN/ER

the added amount is 10 wt%, the composites doped with 20-nm Al_2O_3 can reduce the surface potential to 0 V in 355 s.

3.4.3 Fluoridation of Other Nano-Filler

Plasma fluoridation treatment was performed on AlN with a particle size of 50 nm, and prepared into composites with different filler contents (1, 3, and 5%). In Fig. 33, when the filler content is 5%, the flashover voltage of the sample is 12.68 kV, which is 5% higher than that of pure ER. From the right-side pic results, the fluoridation of nano-AlN can slightly accelerate the potential decay rate [45, 46].

Nano-SiC of 40 nm diameter was fluorinated and then used to prepare ER composites with different mass fractions of 1, 3, and 5%. Figure 34 shows the flashover voltage and the normalised surface potential of composite samples. When 5% F-SiC is added, the flashover voltage of the sample reaches 1.3 times of that of the pure ER.

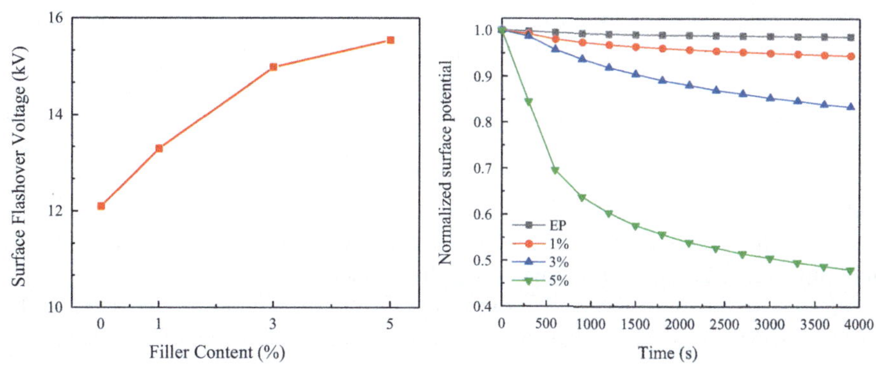

Fig. 34 Flashover and normalised surface potential of F-SiC/ER

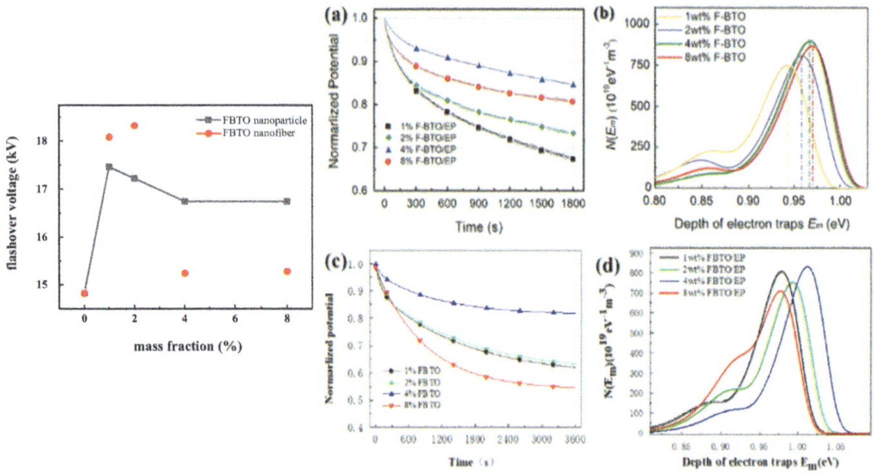

Fig. 35 Flashover voltage and charge–trap properties of F-BTO and F-FBTO

When the mass fraction of the fluorinated filler is 5%, the surface potential decreases to 48% of the initial value in 1 h [47].

Barium titanate ($BaTiO_3$, BTO) is a strong dielectric compound material, which has a high dielectric constant and low dielectric loss. It is widely used as a filler to enhance the dielectric constant, power storage, and mechanical and thermodynamic properties of polymer nanocomposites. As shown in Fig. 35, the optimal modification effect of FBTO nanoparticle system was achieved at the 1% mass fraction. And the flashover voltage firstly decreased, then stabilised, when increasing the filler content. The flashover voltage of the FBTO nanofibre system is the highest at the 2% mass fraction, which is significantly higher than that of the FBTO nanoparticle system. The trend in flashover voltages after the inflection point is the same as that of the FBTO nanoparticle system, and the decrease in flashover voltages is much greater than that in the FBTO nanoparticle system. The fluoridation treatment can thus improve the flashover characteristics of epoxy composites along the surface, but the modifying effect will be limited when the doping concentration is high. After fluoride treatment, the shallow-trap concentration on the surface of the composite increases, while the deep-trap concentration decreases. Fluorinated grafting can narrow the band-gap of nano-BTO and introduce many capture traps with lower energy levels. The increased shallow-trap concentration promotes carrier migration on the surface of the material and enhances its ability to dissipate charge. This reduces charge accumulation on the surface of the epoxy composites, alleviates electric field distortion in the material, and improves the flashover voltage along the surface [48].

Plasma treatment of GN will result in the formation of many active sites on the surface. SEM results show that there are fine granular adherents on the surface of FGN after fluoridation treatment, and FGN disperse more evenly in an ER matrix

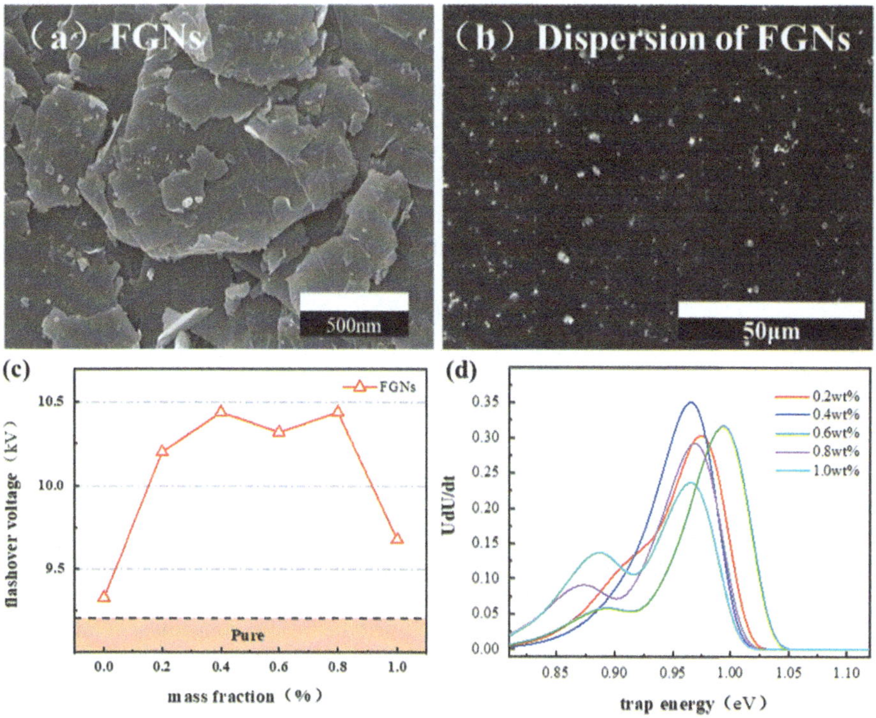

Fig. 36 SEM micrographs of **a** FGN and **b** FGN/ER; The **c** flashover voltage and **d** trap distribution of FGN/ER

than GN, indicating that the presence of fluorine-containing groups can improve filler-matrix compatibility [49].

The DC flashover test results of FGN/ER system show that fluoridation modification can improve the surface flashover voltage of epoxy composites, but the extent of the improvement is limited. It is found that fluorinated fillers can improve the shallow-trap concentration and reduce the deep-trap concentration of epoxy composites. On the one hand, it is easier to combine FGN and ER matrices in the form of bond cooperation after fluoridation. On the other hand, fluoridation modification introduces many low-energy capture traps, improves the rate of dissipation of charge on the surface of the composite, alleviates electric field distortion, and improves its flashover characteristics (Fig. 36).

4 Mechanism of Filler Doping and Its Functionalization

Through the comparison and analysis of the above nano doping researches, it can be found that the flashover voltage showed special non-monotonic trends with the filler content increasing, as shown in Figs. 3, 5, 7, 9, 10, 13, 15, 17, 20, 23, 26, 30, 33, 34. We found that the flashover voltage when plotted against the filler load, exhibit regular sinusoidal variations. Take the research of BY5FO/ER as the example, the specific variation range can be defined as 'X' type curve and divided into two phases: in X_1, the flashover voltages increase and then decrease, before finally increasing. This stage of the curve is N-shaped. In contrast, X_2 exhibits an inverted N-shape. The 'X' curve can be approximately decomposed into a variable-frequency sinusoid (Eq. 3 $f_1(x)$) and a double exponential function Eq. (4) $f_2(x)$.

$$f_1(x) = a_1 \sin(a_2 \sqrt[a_4]{(x + a_3)}) \tag{3}$$

$$f_2(x) = b_1 e^{b_2 x} + b_3 e^{b_4 x} \tag{4}$$

where a_i, and b_i (i is an integer) are constants, and x is the filler content. a_1 is the amplitude of $f_1(x)$. a_2, and a_4 are used to adjust the frequency, and a_3 is a translation parameter for the curve. $f_1(x)$ describes the variation corresponding to X_1-X_2 as a result of the interaction with filler particles. The trend in $f_2(x)$ describes whether, or not, the introduction of the filler is beneficial. The frequency of $f_1(x)$ decreases as the filler content increases, which can be attributed to the non-linear relationship between the spacing of the adjacent particles and the filler content. The steadily declining stage is defined as 'Y', integrating this with stage X_2 reveals the percolation phenomenon of the flashover. The same situation is seen for the dielectric constant at filler contents exceeding the first turning point. When b_3 is equal to 0, $f_2(x)$ becomes a single exponential function. If $b_1 = k_1$ and $b_2 = \ln k_1$-$\ln k_2$, $f_2(x)$ can be rewritten as Eq. (3), namely, Lichtenecker's equation of effective dielectric constant. Where k_1 and k_2 are the dielectric constant of fillers and matrices, respectively. That means that $f_2(x)$ is driven by a diphasic dielectric model. Compared with the monotonous trend of the Lichtenecker model, bi-exponential $f_2(x)$ can be used as an extended model to illustrate other properties which show different trends before, and after, percolation.

$$\ln f_2(x) = x \ln k_1 + (100\% - x) \ln k_2 \tag{5}$$

The explanation method of this kind of fluctuation curve were mainly through the perspective of multi-core model in previous researches. Professor Q.M. Zhang [50] studied the influence of the interaction of filler on dielectric polarization, and simulated the polarization fluctuation caused by the interaction of multiple layers by phase field method. Further, polarization was closely connected to flashover characteristic in composite materials. We think the changed topological structure of fillers will cause electrical parameter fluctuations, here mainly refers to the dielectric polarization properties, and the electrical parameters in participating in the process of flashover, with some other important factors in the process of actual electrical interaction, such as carrier traps, etc., ultimately cause the fluctuation of flashover strength.

Next, we will try to disassemble flashover fluctuation from two aspects of dielectric polarization fluctuation and trap fluctuation.

According to ETPR theory [51], the occurring of flashover can be determined by the storage and release of trap polarization energy. To provide further insight into molecular and nano-scale mechanisms responsible for the observed sine-shaped dielectric variations, we computed the local polarization of the composites based on a well-established multi-layer core model In this model, the interfacial region between the filler particles and the polymer matrix consists of a bonded layer, a bound layer, and a loose layer, each with different electrical properties. The loose layer K_2 is the outermost layer and contains more active dipoles than the bound layer K_1. The properties of the bonded layer K_M are similar to the intrinsic properties of the filler. Besides, we characterized the distribution of filler in the matrix, as shown in Figs. 37(a) and 4(b). When filler appears continuous phase, the particle attribute of filler will weaken sharply, while the matrix shows the particle effect. Equation (5) is introduced to demonstrate the polarization around the interfacial region. When the filler concentration is relatively high, the polarization of the discontinuous epoxy

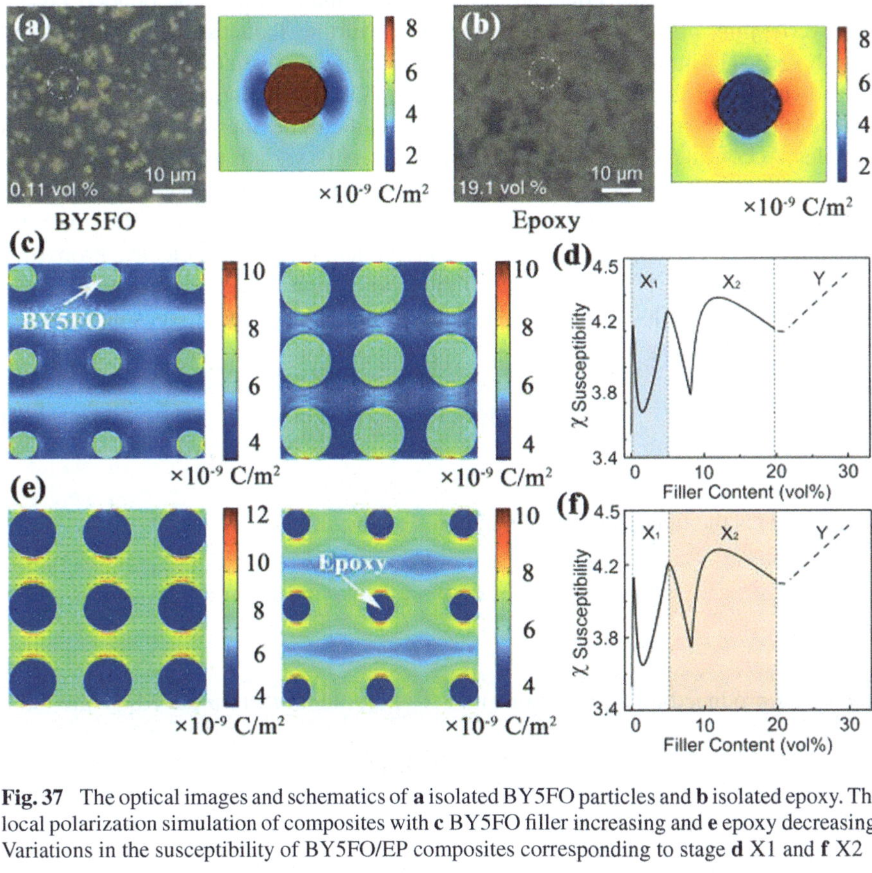

Fig. 37 The optical images and schematics of **a** isolated BY5FO particles and **b** isolated epoxy. The local polarization simulation of composites with **c** BY5FO filler increasing and **e** epoxy decreasing. Variations in the susceptibility of BY5FO/EP composites corresponding to stage **d** X1 and **f** X2

is taken into consideration, and the distribution of polarization can be obtained by Eq. (6).

$$K_{X_1}(r) = K_M + (K_2 g - K_1)e^{-g}, g = (r/r_0)^2 \tag{6}$$

$$K_{X_2}(r) = K'_{X_1} - (K_2 g - K_1)e^{-g}, g = (r/r_0)^2 \tag{7}$$

where r is the distance from the surface of the nanofiller, and r_0 is the characteristic width of the interfacial region. K_{X1}´is the final value of BY5FO/EP in stage X1. For the epoxy, the matrix can be considered as a homogeneous medium. When the filler is introduced at a low concentration, the composites first exhibit stage X1. The interplay between the matrix and the outer layer contributes to the polarization.

According to ETPR theory, the initial matrix defects change from cross-linked defects (of magnitude Å) to inclusions (of micrometer magnitude), which favors the surface insulation. As the filler concentration increases, the effect of the loose layer becomes saturated, and the bound layer comes into play. The polarization worsens with the increase in filler concentration until the interaction of the bonded layers. The overlapping of the innermost bound layers further reduces the dipole rotation barriers [52]. This process corresponds to Fig. 37c and the variations in susceptibility are shown in Fig. 37d. If the filler content continues to increase, the filler and matrix reverse roles; the epoxy then acts as the filler and can be approximately treated as particles. According to the symmetry of the X curve, the variation tendency reappears, corresponding to stage X2 (Fig. 37e and f). The corresponding transition can be regarded as changing from the macromolecular matrix to the inorganic filler. As filler content increases continuously, the percolation phenomenon of dielectric properties appears at the stage of Y. As a result of the polarization induced by the field of the trapped charges, the local field E(i) produced on the site i by charges and dipoles based on ETPR theory is

$$\mathbf{E}(i) = \mathbf{E}_q(i) + \sum_{i \neq j} \mathbf{E}_j(i) \tag{8}$$

where Eq(i) is the field produced by charges on the site i and

$$\mathbf{E}_j(i) = \frac{1}{4\pi\varepsilon_0}[3(\mathbf{P}_i \cdot \mathbf{r}_{ij}) - r_{ij}^2 \mathbf{P}_j]\frac{1}{r_{ij}^5} \tag{9}$$

is the field produced by dipoles j on site i. The dipole moment induced on site i is

$$\mathbf{P}_i = \varepsilon_0 \alpha_i \mathbf{E}(i) \tag{10}$$

where α_i is the polarizability of the spaces placed on the site itself. The total polarization energy stored in the medium per charge is proportional to the square of the local electric field:

$$W_p = \sum_i \frac{1}{2}\mathbf{P}_i \cdot \mathbf{E}(i) = \sum_i \frac{1}{2}\varepsilon_0\alpha_i E(i)^2 \tag{11}$$

For a polar medium, Wp can be written as

$$W_p \approx \frac{2}{3}E_R\frac{a_0}{a}\chi K(\alpha_{or}) \approx 3\chi K(\alpha_{or}) \tag{12}$$

where ER = 13.56 eV, and a0/a is the Bohr radius and the distance of the first neighbor from the charge, respectively. χ is the static susceptibility, and $K(\alpha_{or})$ is a function of α_{or}, the orientational polarizability. This means that the electrostatic energy contained in composites with certain filler contents is in proportion to the polarization of composites.

The polarisation energy release is closely related to trapping stability [53]. The charge stability can be affected mainly by the surface-trap parameters of these composites. We used the isothermal surface potential decay method to calculate the surface trap depth of BY5FO/ER composites from the potential decay curve (Fig. 38a). The normalised potential, which is the ratio of dynamic surface potential to initial maximum potential, can represent the charge decay capacity of composites (Fig. 38b). There is a significant increase in the amplitude of characteristic peaks

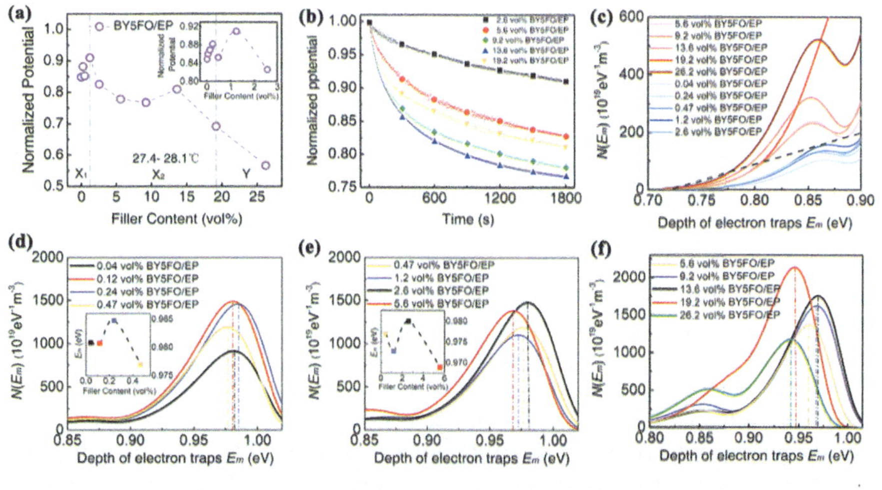

Fig. 38 **a** The potential decay over 1800s of BY5FO/ER composites with a high filler concentration. **b** The normalised potential decay of BY5FO/ER composites as a function of the filler content. **c** Shallow electron trap distributions on the surface of BY5FO/ER composites. **d–f** Deep electron trap distributions on the surface of BY5FO/ER composites

corresponding to shallow traps (Fig. 38c), indicative of an increase in the conductivity of BY5FO/ER composites. The deep-trap distributions of BY5FO/ER with filler loading of 0 ~ 0.47 vol%, 0.47 ~ 5.6 vol%, and 5.6 ~ 26.2 vol% are shown in Fig. 38d–f, respectively. Moreover, deep traps also exhibit regular variations. For composites with a high filler concentration, the corresponding value of E_t increases from 0.961 eV for 5.6 vol% BY5FO/ER to 0.969 eV for 13.6 vol% specimens, and the number of deep traps then decreases when the filler concentration exceeds 13.6 vol%. According to the localised state structure and the potential barrier model, there are different potential barriers between multiple layers of particles in the matrix due to the type of fillers used and the different Fermi levels. Under a high-strength electric field, potential barriers tend to become inclined due to the Poole–Frenkel effect. The electrons are released from shallow traps to the conduction band, and the conductivity of such composites can be expressed thus:

$$\sigma \propto N(E_m)^{1/2} e\mu \exp(-\frac{E_m}{2kT})\exp\left(\frac{e^{3/2}E^{1/2}}{(4\pi\varepsilon_0\varepsilon_r)^{1/2}kT}\right) \quad (13)$$

where $N(E_m)$ is the density of electric traps. k is Boltzmann's constant, and E is the applied field strength. Moreover, electrons in deep traps are unlikely to escape or tunnel between deep-trap states. The mobility of carriers in deep traps is given by

$$\mu = \frac{\mu_0}{1+\exp\left[\left(E_m - \beta_F\sqrt{E}\right)/kT\right]} \quad (14)$$

where β_F is a coefficient related to the Poole–Frenkel effect. The increased shallow-trap density $N(E_m)$ increases the conductivity of BY5FO/ER composites with high filler loading, which is consistent with the $f_2(x)$ of normalised potential decay in Fig. 38b; however, the increased number of deep traps causes a reduction in the amount of charge released from BY5FO/ER with ~0.24, 2.6, and 13.6 vol% filler contents, resulting in enhanced charge-trapping stability. The variations in the number of deep traps correspond to $f_1(x)$ of normalised potential decay.

Thus, the match of the polarisation and trapping stability of BY5FO/ER with increased filler loading contributes to similar variations of flashover voltages. This can be verified by voltage conditioning and the consensus is that the final stage of flashover is caused by gas desorption and ionisation. Consecutive breakdowns can desorb gases on the surface of materials and provide a free-state plasma. Declining levels of residual gases lead to a gradual increase in flashover voltages (Fig. 39a). According to the aforementioned dielectric constant and deep-trap variations, high trapping energies stored in BY5FO/ER composites with certain filler contents such as 13.6 vol% are not easily destabilised to initiate voltage breakdown. Once the breakdown occurs, the energy released by a one-off breakdown of the composite is sufficiently large to desorb most of the gases near the electrode. The flashover voltage rapidly reaches a stable state after several such breakdown events thus, the

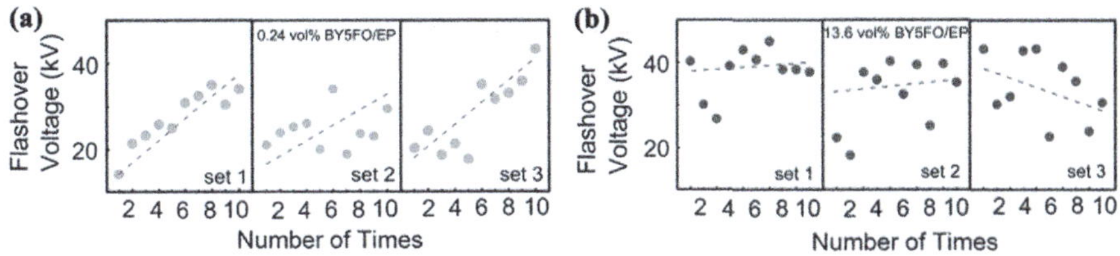

Fig. 39 Relationship between flashover voltages and the number of breakdown events for **a** 0.24 vol% and **b** 13.6 vol% BY5FO/ER composites

dependence on the number of breakdowns is weak (Fig. 39b), which also explains the correlation of the variations in flashover error and filler content.

Whether in the air or the vacuum, the amplitude of flashover within the interval of X_1 varies little with the filler content, revealing that the ER is the essence of the insulating medium, although the surface-trap parameters are changed by the addition of various fillers. Traps in the epoxy form 3D cross-link points of Å order of magnitude (Fig. 40a); meanwhile, relatively high surface resistance engenders a charge accumulation response to the applied voltage (Fig. 40b), which reduces the surface insulation performance. When filler loading exceeds 20 wt%, the function of the filler begins to appear. All samples' maximum flashover voltages appear in the range of X_2. In this case, the main component of the matrix is BFO particles, which are isolated by a thin epoxy interlayer (Fig. 40b). Numerous BFO particles alleviate the E-field distortion due to their semiconducting characteristics. We also measured the surface breakdown strength of solid BFO, another extreme case to compare against pristine ER, as shown in Fig. 40c. The flashover voltage thereof is about 15 kV in a vacuum. The absorbed gases on the dielectric surface are readily thermally activated to engage in development of flashover as a result of the high rate of gas leakage from solid BFO (Fig. 40f): this then harms the vacuum insulation. In

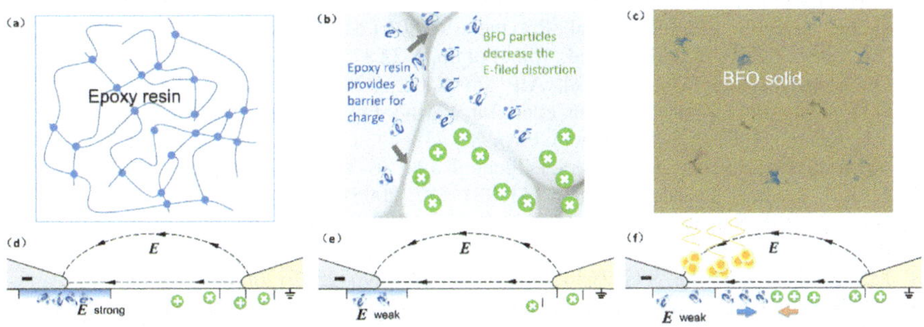

Fig. 40 Schematic illustration of the interaction between nanoparticles and polymeric matrices: **a** the cross-linked structure of pristine ER, **b** BFO particles surrounded by epoxy barrier, and **c** composites with an extremely high BFO content. The schematic representation of the surface E-field distribution and flashover process for **d** pristine epoxy, **e** BFO/ER with high filler loading, and **f** BFO materials with a small amount of ER

conclusion, the combination of the trap effect and a uniform electric field is essential when trying to increase the flashover voltage of such composites [54, 55].

References

1. J.Y. Xue, J.H. Chen, J.H. Dong, et al., J. Phys. D Appl. Phys. **52**, 405502 (2019)
2. C.Y. Li, C.J. Lin, G. Chen, et al., Appl. Phys. Lett. **114**, 202904 (2019)
3. B. Zhang, Q. Wang, Y. Zhang, et al., Nanoscale **11**, 18046–18051 (2019)
4. B.X. Du, Z.X. Liu, M. Xiao, IEEE Trans. Dielectr. Electr. Insul. **22**, 3481–3487 (2015)
5. L. Que, Z. An, Y. Ma, et al., IEEE Trans. Dielectr. Electr. Insul. **24**, 1153–1161 (2017)
6. A. Mohamad, G. Chen, Q. Chen, IEEE Trans. Dielectr. Electr. Insul. **24**, 3786–3793 (2017)
7. G.Y. Sun, B.P. Song, B.H. Guo, et al., J. Phys. D Appl. Phys. **51**, 295201 (2018)
8. Q. Xie, H.F. Lin, S. Zhang, et al., Plasma Sci. Technol. **20**, 025504 (2018)
9. J.R. Harris, M. Kendig, B. Poole, et al., Appl. Phys. Lett. **93**, 241502 (2008)
10. R.D. Zhou, G.Y. Sun, B.P. Song, et al., J. Phys. D Appl. Phys. **52**, 375304 (2019)
11. Y.P. Tu, F.W. Zhou, H. Jiang, et al., IEEE Trans. Dielectr. Electr. Insul. **25**, 1308–1317 (2018)
12. B. Wetzel, F. Haupert, K. Friedrich, Eng. Fract. Mech. **73**, 2375–98 (2004)
13. X.M. Pang, R.Q. Wang, Q. Wei, et al., Mater. Res. Express. **5**, 2053–1591 (2017)
14. C. Zhang, Y. Zhou, T. Shao, et al., Appl. Surf. Sci. **311**, 468–77 (2014)
15. H. Althues, J. Henel, S. Kaskel, Chem. Soc. Rev. **36**, 1454–65 (2009)
16. R. Agarwal, Y. Sharma, S. Hong, et al., J. Phys. D Appl. Phys. **51**, 275303 (2018)
17. Y. Yao, N. Zhao, J.J. Feng, et al., Ceram. Int. **39**, 4735–4738 (2013)
18. X.L. Song, N. Jiang, Y.K. Li, et al., Mater. Chem. Phys. **110**, 128–135 (2008)
19. B.X. Du, H.C. Liang, J. Li, et al., IEEE Trans. Dielectr. Electr. Insul. **26**, 1956 (2019)
20. H.O. Ruan, Q. Xie, F.C. Lu, et al., J. Phys. D: Appl. Phys. **53**, 145204 (2020)
21. Z. Zhang, X. M. Duan, B. F. Qiu, et al., J. Alloys Compd. **852**, 156765 (2021)
22. G.S. Li, C.H. Wang, H.Y. Zhang, et al., Mater. Lett. **280**, 128598 (2020)
23. W.R. Liu, J.S. Tian, D.P. Xiang, et al., J. Clean Prod. **274**, 122799 (2020)
24. J.A. Guerra, J.R. Angulo, S. Gomez, et al., J. Phys. D Appl. Phys. **49**, 195102 (2016)
25. C.Y. Li, J. Hu, C.J. Lin, et al., Sci Rep. **7**, 3271, (2017)
26. T. Tanaka, M. Kozako, N. Fuse, et al., IEEE Trans. Dielectr. Electr. Insul. **12**, 669–681 (2005)
27. S.N.A. Rusly, I. Ismail, K.A. Matori, et al., Ceram. Int. **46**, 737–746 (2020)
28. W.B. Zhao, G.J. Zhang, Y. Yang, et al., IEEE Trans. Dielectr. Electr. Insul. **14**, 170–178 (2007)
29. Q. Xie, K.X. Fu, S.D. Liang, et al., Polymers. **10**,801 (2018)
30. Y.F. Chen, Z.Z. Bian, Y. Fan, et al., Iran. Polym. J. **22**,377–383 (2013)
31. Q. Xie, S.D. Liang, B.O. Liu, et al., AIP Adv. **8**, 075332 (2018)
32. C.H. Sheng, H.D. Zhang, S. Chen, et al., Int. J. Mod. Phys. B, **29**,1550066 (2015)
33. Y.C. Kang, L.I. Chan, Mater. Chem. Phys. **85**, 438–443 (2004)
34. Y. Hu, G. Du, N. Chen, Compos. Sci. Technol. **124**, 36–43 (2016)
35. M. Kok, J. Mater. Process. Technol. **161**, 381–387 (2005)
36. B.J. Ash, R.W. Siegel, L.S. Schadler, J. Polym. Sci. Part B Polym. Phys. **42**, 4371–4383 (2003)
37. P. Rittigstein, J.M. Torkelson, J. Polym. Sci. Part B Polym. **44**, 2935–2943 (2010)
38. L.X. Zhao, L.Y. Zheng, S.G. Zhao, Mater. Lett. **60**, 2590–2593 (2006)
39. O. Bin Sohail, M.J. Khan, P.A. Sreekumar, et al., Polym. Eng. Sci., **54**, 1941–1946 (2014)
40. Z. Li, K. Okamoto, Y. Ohki, et al., IEEE Trans. Dielectr. Electr. Insul. **17**, 653–661 (2010)
41. H. Tao, J. Lv, R. Zhang, et al., Mater. Design. **120**, 83–89 (2017)
42. Z.X. Cheng, X.L. Wang, S.X. Dou, et al., J. Appl. Phys. **104**, 116109 (2008)
43. K. Yin, J.Y. Yan, H.C. Zhang, J. Phys. D Appl. Phys., **53**, 395502 (2020)
44. H.H. Zhou, Q. Xie, H.O. Ruan, et al., Polym. Composite. **41**, 2881–2892 (2020)
45. F.C. Lu, H.O. Ruan, J.X. Song, et al., J. Phys. D Appl. Phys. **52**, 155201 (2019)
46. C. Dai, X.R. Chen, T. Jiang, et al., Polym. Test. **86**, 106502 (2020)

47. C.B. Liu, Z.Y. Zhan, F.C. Lu, et al., Colloid. Surface A **585**, 124029 (2020)
48. H. Wu, Q. Xie, Q.J. Duan, et al., J. Mater. Sci. **4** 1499–1510 (2020)
49. J. Li, H.C. Liang, M. Xiao, et al., IEEE Trans. Dielectr. Electr. Insul. **26**, 1577–1580 (2019)
50. Y. Thakur, T. Zhang, C. Lacob, et al., Nanoscale **9**, 10992–10997 (2017)
51. G. Blaise, C. Legressus, J. Appl. Phys. **69**, 6334–6339 (1991)
52. R.A. Anderson, J.P. Brainard, J. Appl. Phys. **51**, 1414–1421 (1980)
53. M. Akahane, K. Kanda, K. Yahagi, J. Appl. Phys. **44**, 2927–2927 (1973)
54. K. Yin, Q. Xie, H.O. Ruan, et al., Compos. Sci. Technol. **197**, 108199 (2020)
55. T. Shao, W.J. Yang, C. Yang, et al., Appl. Phys. Lett. **105**, 0761607 (2014)

Electric Field Regulation Along Gas–Solid Interface in HVDC GIL with Nonlinear Conductivity Material

Hucheng Liang, Boxue Du, Cheng Zhang, and Jin Li

Abstract The development of HVDC gas insulated transmission lines (GIL) is the key direction of power transmission for large-scale renewable energy integrations and in specific situations. Under complex conditions, unpredictable surface charge accumulations on spacer surfaces easily cause electric field distortions and even flashovers, threatening the safe operation of the power transmission system. Therefore, regulating the electric field distribution is one of the critical problems for DC-GIL. Nonlinear conductivity materials have the function of regulating the dc electric field distribution, and such insulating materials can be prepared by doping SiC or ZnO particles into polymers. In this chapter, epoxy was doped with SiC particles to prepared epoxy/SiC composites with nonlinear conductivities. Then, the surface nonlinear conductivity spacer was proposed to relax the electric field distortions on spacer surfaces by coating spacers with epoxy/SiC composites. The regulation effects of surface charge and electric field distributions under complex conditions were investigated by simulation. Finally, novels spacers were fabricated and their insulation improvements were verified by flashover tests.

1 Introduction

GAS insulated lines (GILs) have been widely used in electric power systems because of, among other factors, their high reliability, large capacity, compact size, and environmentally friendliness [1–4]. GIL spacers are made of EP/Al2O3 composites, which support the conductor, provide electrical insulation and isolate the chambers.

H. Liang (✉) · B. Du · C. Zhang · J. Li
Tianjin University, No.92 Weijin Road, Nankai, Tianjin, China
e-mail: hcliang@tju.edu.cn

B. Du
e-mail: duboxue@tju.edu.cn

J. Li
e-mail: lijin@tju.edu.cn

B. Du (ed.), *Polymer Insulation Applied for HVDC Transmission*,
https://doi.org/10.1007/978-981-15-9731-2_17

However, a recent survey showed that more than half of the GIS/GIL faults occurring from 2009 to 2013 were caused by spacer failures [5]. Surface flashover triggered unpredictably inside GILs is one of the common spacer failures [6, 7]. To improve the insulation performance, there are usually two approaches: one is to enhance the ability of dielectrics to resist discharge when discharge occurs [8, 9], the other is to reduce the occurrence of discharge.

The nonuniform E-field distribution along the spacer surface is typically considered an important cause of flashover faults. To reduce the E-field distortion, several complex techniques have been used in practical GILs, such as optimizing the spacer shape, installing shielding electrodes and introducing embedded electrodes. Considering the limited effects and higher manufacturing cost of these complex techniques, simple and effective methods to reduce the occurrence of flashover faults in GILs are needed. In recent years, a considerable amount of work on functionally graded material (FGM) has been reported [10–17]. And such spacers have been proven to exhibit the field grading function and have a considerably higher breakdown voltage (BDV) than conventional spacers under lightning impulse voltage [10]. The disadvantages of FGM are that the ε/ρ-distributions of FGM spacers must be carefully designed and their fabrication techniques are not sufficiently simple for FGM spacers to be brought into application.

Recently, the nonlinear conductivity materials have become popular because of their E-field regulation function in DC system [18–20]. The volume conductivities of such materials increase substantially when the E-field strength reaches a certain value, which in turn reduces the E-field strength. In this chapter, the electrical conductivity properties of epoxy/SiC composites were studied first. Then the concept of the surface nonlinear conductivity spacer was proposed to regulate the electric field distributions along the spacer surfaces. By simulation, the surface charge and the electric field distributions of spacers were investigated under complex conditions, i.e. temperature gradient, electrode roughness, metal particle and polarity reversal voltage. Finally, novel spacers were fabricated by dip coating, centrifugation and sputtering, flashover tests were conducted to evaluate the electrical performances of novel spacers.

2 Electrical Conductivity of Epoxy/SiC Composites

2.1 Sample Preparation

In order to obtain the epoxy/SiC composite samples, the bisphenol A epoxy resin was selected as the matrix, the hardener (low molecular weight polyamide resin, HY-165) was selected as curing agent, two types of micro-sized SiC (α-crystal and β-crystal, ~12 μm diameter, density of 3.2 g/cm3, shown in Fig. 1) were selected as fillers. The epoxy resin matrix used in this experiment is a high-activity, low-viscosity liquid epoxy resin, which has good electrical insulation and infiltration. The curing agent HY-165 can be mixed with epoxy resin and various additives to obtain products for

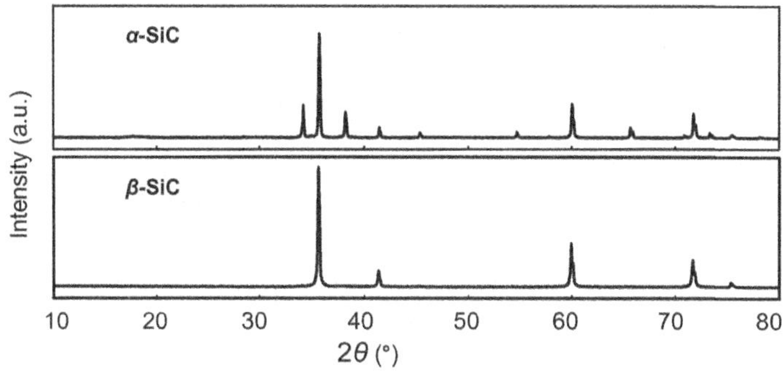

Fig. 1 X-ray diffraction results of SiC particles

various applied conditions. It can cure the epoxy resin at room temperature or heating conditions without by-products and the cured product performance is better than that of general amine or anhydride curing agent. Table 1 lists the epoxy resin technical index and Table 2 lists curing agent technical index.

Figure 2 shows the flow chart of sample preparation of epoxy/SiC composites. The micro-sized SiC was dispersed into the epoxy matrix after being dried at 110 °C for 12 h, and the mixture was stirred for 40 min, and then ultrasonicated for 40 min to make SiC particles disperse uniformly in epoxy matrix. Next, the hardener was added into the mixture and stirred for 20 min. And then vacuum evacuation was performed to avoid bubbles in the mixture. Finally, the mixture was poured into a pre-heated tablet press mold and cured for 15 min. The pressure and temperature of the tablet press were set to 10 MPa and 120 °C, respectively. After the mold was naturally cooled to the room temperature, the epoxy/SiC composite samples were

Table 1 Epoxy resin technical index

Item	Index
Appearance	Colorless clear liquid
Colour (number)	≤ 2
Density (g/mL, 25 °C)	1.02–1.11
Viscosity (MPa/25 °C)	100–500
Epoxide equivalent (g/eq)	178–195
Shear strength	≥ 10

Table 2 Curing agent technical index

Item	Index
Appearance	Light yellow viscous liquid
Amine value (mgKOH/g)	400
Viscosity (mPa.s/25 °C)	200–1000
Density (kg/m^3)	0.99

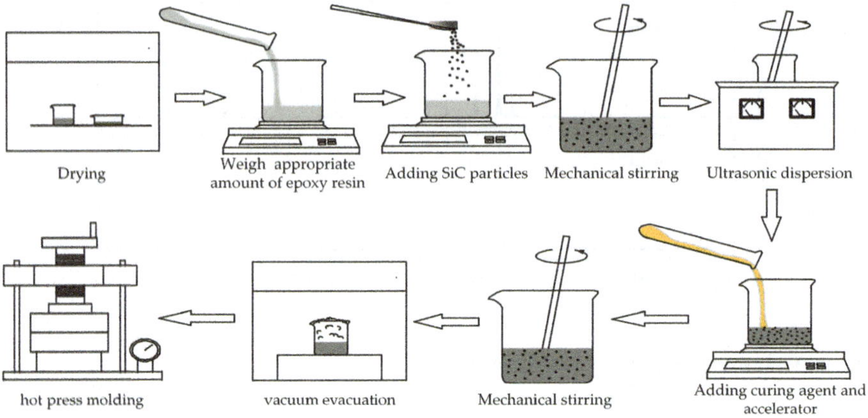

Fig. 2 Flow chart of sample preparation of epoxy resin/SiC composites

Table 3 The unit conversion of SiC content in epoxy resin/SiC composites

SiC content (phr)	60	90	120	150
wt%	37.50	47.37	54.55	60.00
vol%	18.22	25.05	30.82	35.78

obtained. The shape of sample was square with a thickness of 0.5 mm and a length of 90 mm.

In this experiment, the epoxy/α-SiC composite samples with α-SiC content of 60, 90, 120, and 150 phr were prepared; the epoxy/β-SiC composite samples were prepared with β-SiC content of 60, 90 phr, 120 phr. The contents are described in terms of weight percent, i.e. 60 phr means that the epoxy matrix is 100 g, and the SiC fillers are 60 g. The conversion of the corresponding mass fraction and volume fraction is shown in Table 3.

2.2 Conductivity Measurement

2.2.1 Experimental Arrangement

The DC bulk conductivity of the epoxy/SiC composites was measured by the three-electrode system. The high voltage electrode was connected to a dc source, which had an output voltage of 0 ~+ 30 kV. The main electrode was connected to an electrometer Keithley 6517B, which has a measuring range from 1 pA to 20 mA. A protective resistance was connected to the circuit to prevent the device form being damaged during measurement. The ring-shape electrode was grounded for avoiding the effect of the leakage current along the sample surface. The whole system was set in a constant temperature and humidity chamber to improve the measurement accuracy and prevent the influence of temperature fluctuation on the measurement data. The current I was detected after applying dc voltage for 35 min.

The sample conductivity σ can be calculated by the following equation

$$\sigma = \frac{I}{U}g\frac{4L}{\pi(d+g)^2} \tag{1}$$

where U is the dc voltage stressed on the samples, L is the thickness of the samples, d is the diameter of the main electrode, 50 mm, and g is the gap between the main electrode and ring-shape electrode, 2 mm.

The non-linear characteristic between I and U can be expressed as

$$I = KU^{\alpha} \tag{2}$$

where K is the coefficient related to material properties, α is the nonlinear coefficient of the material properties.

The relationship between current density J and electric field strength E is

$$J = K_1 E^{\alpha - 1} \tag{3}$$

where K_1 is the constant related to material properties. The relationship between the conductivity γ and the electric field E is

$$\gamma = AE^k \tag{4}$$

where $k = \alpha - 1$, is the nonlinear conductive coefficient.

By logarithmic transformation of Eq. (3), a linear equation can be derived

$$\lg \gamma = 1gA + k1gE \tag{5}$$

In this paper, the non-linear conductivity of composites can be characterized by the *J-E* relationship between current density J and electric field E, and non-linear conductive coefficient k can be calculated by the following equation

$$\mathrm{k} = \frac{\lg(J_1/J_2)}{\lg(E_1/E_2)} \tag{6}$$

where the current density J_1, J_2 is corresponding to the electric field E_1, E_2, respectively, and usually the value of J1/J2 is greater than 10.

2.2.2 Measurement Results

(a) Effects of Crystal Morphology

The dc volt-ampere characteristics at room temperature (20 °C) were measured, and the nonlinear coefficient k and threshold field E_b were calculated according to the measurement results.

Figure 3 shows the dc conductivity volt-ampere characteristic curves of epoxy/SiC composite samples with different crystal morphology. As shown in Fig. 3, the samples with different crystal morphology have the non-linear conductivity characteristics obviously. The current density of the samples increases across an order of magnitude when the electric field exceeds the threshold field. The current density of the epoxy/β-SiC composite sample increases more rapidly compared with the epoxy/α-SiC composite sample, and the corresponding electric field strength of epoxy/α-SiC composite sample is smaller than that of epoxy/β-SiC composite sample after the nonlinear characteristics appeared.

The nonlinear coefficient k and threshold field E_b of samples with different contents are shown in Tables 4 and 5. As the increase of filler content, the nonlinear coefficient of epoxy/SiC composite samples shows an upward trend and the

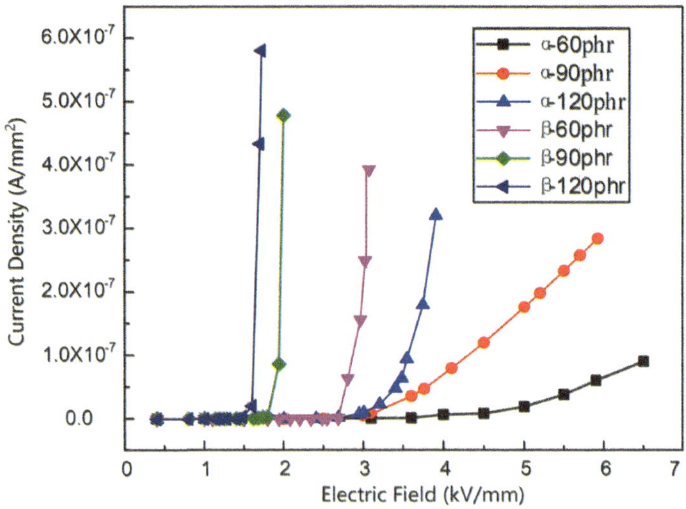

Fig. 3 J-E curves of epoxy/SiC composite samples with different SiC crystal morphology

Table 4 The nonlinear coefficient k and threshold field E_b of the samples with different α-SiC contents

α-SiC	60	90	120
k	3.12	5.56	11.55
E_b (kV/mm)	4.32	3.36	3.15

Table 5 The nonlinear coefficient k and threshold field E_b of the samples with different β-SiC contents

β-SiC	60	90	120
K	30.22	44.76	45.28
E_b (kV/mm)	2.83	1.96	1.64

threshold field shows a downward trend. The nonlinear coefficient of the epoxy/β-SiC composite is significantly greater than that of the α-SiC composite with the same filler content. Obviously, when the SiC content is 60 phr, the nonlinear coefficient of epoxy/β-SiC composite reaches 30.22, which is more than 9 times that of α-SiC composite with the same content.

(b) Effects of Temperature

In order to study the effect of temperature on the conductivity of epoxy/SiC composites, the dc volt-ampere characteristics of composite samples with α-SiC content of 60 and 150 phr were measured at 20, 40, 60 and 80 °C, respectively, and calculated the nonlinear coefficient k and threshold field E_b.

Figure 4 presents the dc conductivity volt-ampere characteristic curves of the composite samples at different temperature, where Fig. 4a is the experimental result of samples with α-SiC content of 60 phr, and Fig. 4b is 150 phr. Tables 6 and 7 are the nonlinear coefficient k and threshold field E_b of two samples at different temperatures. When the SiC content is low (60 phr), the current density of the composite sample

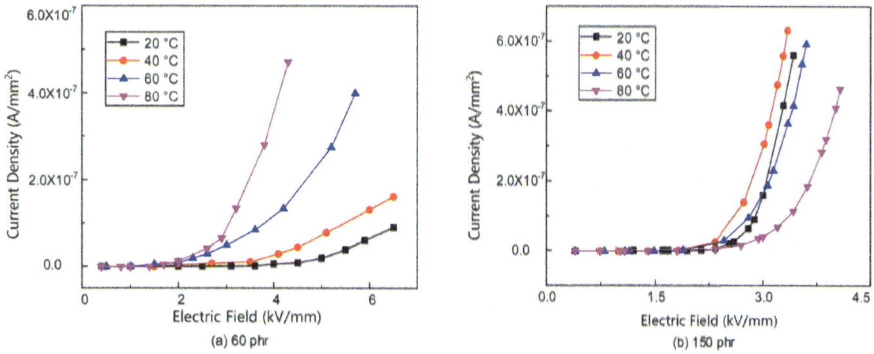

Fig. 4 J-E curve of epoxy/SiC composite samples at different temperatures with fillers content of **a** 60 phr **b** 150 phr

Table 6 The nonlinear coefficient k and threshold field E_b of the sample at different temperatures with α-SiC content of 60 phr

Temperature (°C)	20	40	60	80
k	3.12	3.66	3.56	4.67
E_b (kV/mm)	4.32	3.50	3.28	2.66

Table 7 The nonlinear coefficient k and threshold field E_b of the sample at different temperatures with α-SiC content of 150 phr

Temperature (°C)	20	40	60	80
k	11.73	8.84	7.91	7.95
E_b (kV/mm)	2.67	2.55	2.83	3.02

increases with the temperature raising under the same electric field strength. Meanwhile, the nonlinear threshold field decreases and the nonlinear coefficient increases. When the SiC content reaches 150 phr, the difference between current density is unconspicuous at 20, 40, 60 °C. Furthermore, the nonlinear conductivity characteristics of epoxy/SiC composites decreases with increasing temperature, different from the former composites with low filler content.

(c) Discussion

The crystal morphology can affect the non-linear conductivity characteristics of epoxy/SiC composites. The β-SiC have a one-dimensional crystal morphology, which can form a large number of charge transportation channels as the electric field exceeds the threshold field. The β-SiC (rod-shaped) is easier to form percolation path than that of α-SiC (granular-shaped) under a certain electric field, and increases the non-linear conductivity of composites. Besides, the layer array structure causes the differences in conductivity between different crystal morphology. The layer array property in α-SiC leads to irregular array sequence and gaps among internal voids, and the layer array of tetrahedron parallel bonding in β-SiC structure causes regular array sequence and gaps among internal voids. This β-SiC structural regulation determines that as the voltage increases, more voids get into work, thereby enhancing the non-linear conductivity.

The temperature can also affect the non-linear conductivity characteristics of epoxy/SiC composites. As the fillers content is small, there is no contact between the SiC particles in the epoxy/SiC composite to form a percolation path, and a few carriers participate in the hopping conduction process, which makes the conductivity increase slowly with the electric field. With the temperature increases, the carrier concentration increases as well as the thermal excitation process of the carriers, leading to the rapid increase of non-linear conductivity characteristic. When the fillers content reaches a certain high level, a perfect permeation path has formed in the composite bulk. As a semiconductor, because the forbidden band gap between the valance band and the conduction band is relatively narrow, the carriers are easily excited to the conduction gap, so that the conductivity of SiC has low temperature dependence. Besides, the increasing temperature can cause a lattice vibration, which results in carriers scattering and eventually affects the conductivity.

3 Application of Epoxy/SiC Composites in Novel Spacers

3.1 Concept of Surface Nonlinear Conductivity Spacer

Figure 5 shows the equivalent-circuit models for the conventional spacer (Fig. 5a), spatially nonlinear spacer (Fig. 5b) and EP/SiC coated IER spacer (Fig. 5c); these

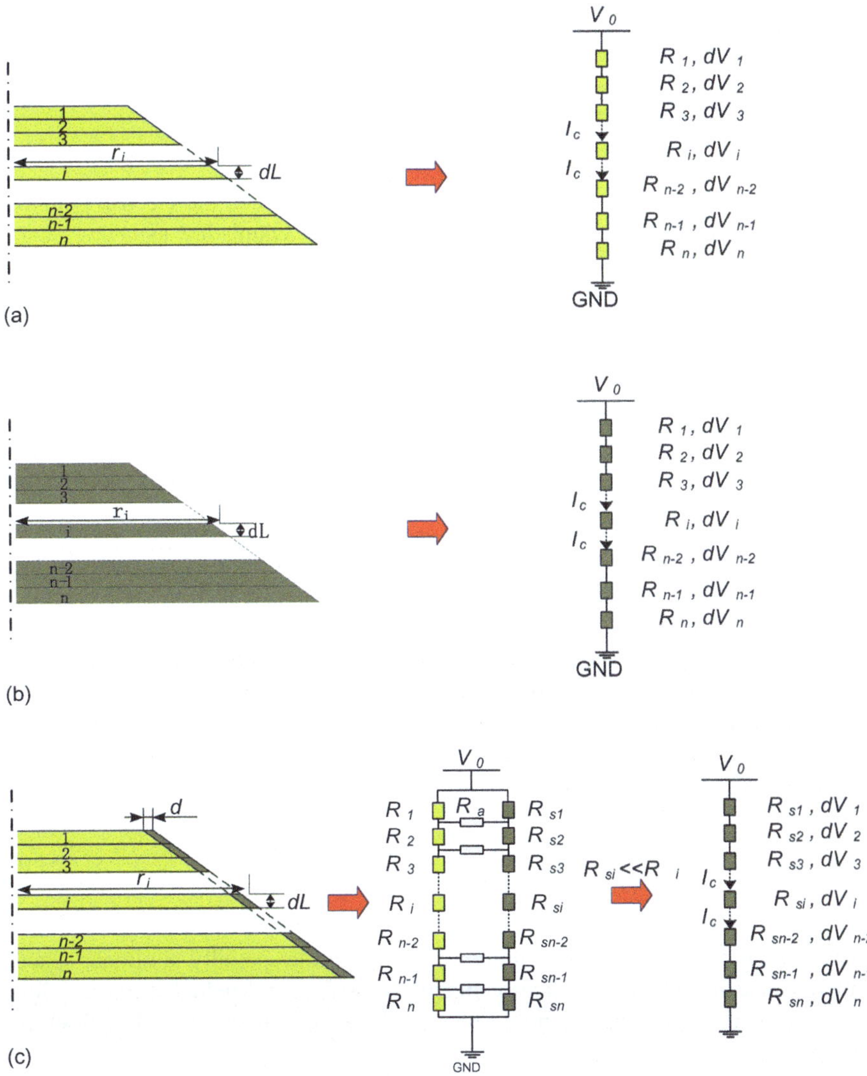

Fig. 5 Equivalent-circuit models for **a** the conventional spacer, **b** the spatially nonlinear spacer and **c** the IER spacer

models provide some insight into the steady-state E-field distributions [21]. The conventional spacer is made of EP/Al_2O_3 composites, whose volume resistivity can be regarded as a constant. The spatially nonlinear spacer is made of EP/SiC composites. The EP/SiC coated IER spacer consists of two parts: the body, which is made of EP/Al_2O_3 composites, and the skin layer, which is made of EP/SiC composites.

As shown in Fig. 5a, a cone-type spacer can be divided into n ($n \to \infty$) elements. Each element i can be regarded as a cylinder with a radius of r_i and a height of dL. Denoting the resistance value of each element i as R_i, the conventional spacer can be approximately regarded as a resistor series ΣR_i. According to the law of resistance, the resistance value of element i can be calculated as:

$$R_i = \rho \frac{dL}{\pi r_i^2} \tag{7}$$

where ρ is the resistivity of the spacer, Ω·m. In steady state, the electric potential drop dV_i on element i can be obtained according to the Ohm's law as:

$$dV_i = I_c R_i \tag{8}$$

where I_c is the steady-state conductive current, A. Based on the above analysis, the E-field strength on element i can be described as:

$$E_i^a = \frac{dV_i}{dL} = \frac{I_c}{\pi} \frac{\rho}{r_i^2} \tag{9}$$

According to Eq. (3), the E-field distribution along the cone-type spacer surface is extremely uneven. The maximum E-field strength appears near the HV electrode and declines from the HV electrode to the GND electrode.

For the spatially nonlinear spacer, as shown in Fig. 5b, the volume resistivity of spacer is E-dependent. When the E-field strength reaches a certain value, the rapidly decreased volume resistivity reduces the E-field distortion. Similar to the conventional spacer shown in Fig. 1a, the E-field strength on element i of the spatially nonlinear spacer can be described as:

$$E_i^b = \frac{I_c}{\pi} \frac{\rho(E)}{r_i^2} \tag{10}$$

where $\rho(E)$ is the E-dependent volume resistivity of the spatially nonlinear spacer, Ω·m. Because of the E-field regulation function of $\rho(E)$, the E-field distribution along the spatially nonlinear spacer should be slightly more uniform than that along the conventional spacer.

For the EP/SiC coated IER spacer, as shown in Fig. 5c, both the spacer and skin layer can be divided into n ($n \to \infty$) elements. Each element i of the skin layer can be regarded as a column ring with a radius of r_i, a thickness of d and a height of dL. According to the law of resistance, the resistance value of element i of the skin layer can be approximately calculated as:

$$R_{si} = \rho_s(E)\frac{dL}{2\pi r_i d} \tag{11}$$

where $\rho_s(E)$ is the E-dependent volume resistivity of the skin layer, Ω·m. In the equivalent-circuit model for the EP/SiC coated spacer, the body resistor series ΣR_i and the skin layer resistor series ΣR_{si} are combined in parallel through a contact resistance R_c. When $R_{si} < < R_i$, the equivalent circuit can be simplified to the skin layer resistor series ΣR_{si} only. The E-field strength on element i of the EP/SiC coated spacer can be described as:

$$E_i^c = \frac{I_c}{2\pi d}\frac{\rho_s(E)}{r_i} \tag{12}$$

The E-field distribution along the EP/SiC coated spacer surface should be more uniform than that along the spatially nonlinear spacer surface because E_i^c is proportional to $\rho_s(E)/r_i$ and E_i^b is proportional to $\rho_s(E)/r_i^2$.

3.2 *Parameter Optimization of Surface Nonlinear Conductivity Spacer*

3.2.1 Simulation Model

(a) Geometry model

A geometry model simplified from the vertically laid 500 kV DC-GIL was used for simulation, and some of the critical dimensions were marked in Fig. 6 [22]. To investigate the influences of micro discharges on the surface charge and the electric field distributions, a metal particle was set on the convex surface of the spacer, and a rough region was set on the inner wall of the pipe. The length of the metal particle was 10 mm, and the height of the rough region was 20 mm. To reduce the computation burden, this simulation was conducted in a rotationally symmetric system. Therefore, the particle was technically not a stick but a ring, and the rough region was tubular in shape.

(b) Spacer parameter

Three types of spacers were used in simulation: (1) an EP/Al_2O_3 spacer made of the EP/Al_2O_3 composite, (2) a bulk nonlinear conductivity (BNC) spacer made of an EP/SiC composite, and (3) a surface nonlinear conductivity (SNC) spacer. The bulk

Fig. 6 Geometry model of the simplified 500 kV DC-GIL

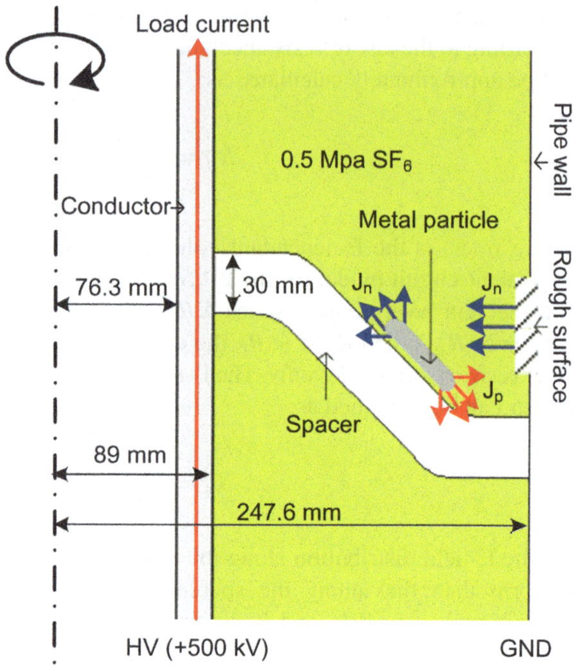

of the SNC spacer was made of the EP/Al_2O_3 composite and the surface layer was made of the EP/SiC composite.

The volume conductivity of the EP/Al_2O_3 composite can be described as [22]:

$$\kappa_{v,1} = A \cdot e^{-B/T} \tag{13}$$

where T is the temperature of the spacer; A and B are constants. In this paper, $A = 12.3$ S/m, $B = 1.05 \times 10^4$ K.

The volume conductivity of the EP/SiC composite has a nonlinear relationship with the electric field, which can be described as:

$$\kappa_{v,2} = a \cdot e^{b \cdot E} \tag{14}$$

where E is the electric field; a and b are temperature dependent factors. a can be regareded as the Ohmic conductivity because it represents the conductivity under a very low electric field; b can be regarded as the nonlinear coefficient because it represents the dependence of the conductivity upon the electric field. Figure 7 presents the volume conductivity of an EP/SiC composite at 20, 40 and 60 °C. As the temperature increases, the Ohmic conductivity increases, but the nonlinear coefficient decreases. Unless otherwise stated, the bulk of the BNC spacer and the surface layer of the SNC spacer are both made of this EP/SiC composite in the following.

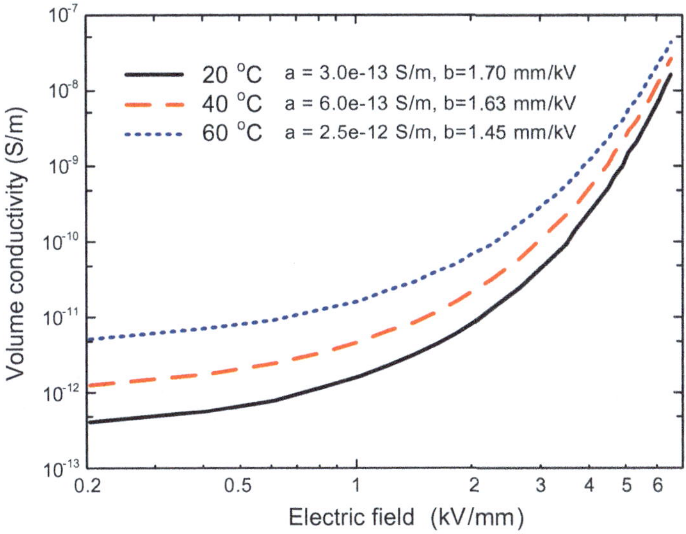

Fig. 7 Volume conductivity of the EP/SiC composite at different temperatures

In this paper, all the spacer surfaces are assumed to be dry and clean, the surface conductivities of the EP/Al_2O_3 and the BNC spacers are assumed to be very low.

$$\kappa_{s,1} = 1 \times 10^{-20}\ \mathrm{S} \tag{15}$$

For the SNC spacer, the thickness of the EP/SiC surface layer is usually less than 100 μm, which is much smaller than the size of the spacer. The thickness of the surface layer can be ignored in the geometry structure by setting the surface conductivity of the SNC spacer as the product of the volume conductivity and the thickness of the EP/SiC layer [23]

$$\kappa_{s,2} = \kappa_{v,2} \cdot d \tag{16}$$

where d is the thickness of the surface layer, which is set to 50 μm in this paper.

(c) Electric field calculation

The surface charge and electric field distributions inside GIL were calculated considering the following physical processes: (1) electric conduction along the surface of the spacer, (2) electric conduction in the gas domain, and (3) electric conduction in the spacer domain. The fluid model was used to describe the generation, recombination and motion of charge carriers in 0.5 MPa SF6 gas [23] A positive voltage of + 500 kV was applied on the conductor and the pipe wall is grounded.

(d) Micro discharge

When a metal particle or a rough pipe surface is assigned inside the DC-GIL, gas discharges will occur in very small regions. The micro discharge fluxes from the boundary to the gas domain were described as following [24]:

$$\Gamma_p = k \cdot \beta_{eff} E \cdot (\beta_{eff} E - p(E/p)_{crit}) \cdot \alpha \quad (17)$$

$$\Gamma_n = k \cdot \beta_{eff} E \cdot (\beta_{eff} E - p(E/p)_{crit}) \cdot (1 - \alpha) \quad (18)$$

where Γ_p and Γ_{n} are the fluxes of positive and negative carriers; β_{eff} is the microscopic field enhancement factor caused by micro protrusions, which is set to 30 and 4 on the rough pipe surface and the metal particle surface, respectively; E is the macroscopic electric field; p is the pressure; $(E/p)_{\mathrm{crit}} = 85$ kV/(mm·MPa) is the critical field of SF_6 gas; k is an adjustment coefficient, which is assumed to be 1.56×10^{-3} $V^{-2} \cdot s^{-1}$ in this paper; α is set to one and zero on the anode and the cathode, respectively.

(e) Temperature calculation

The temperature distribution was calculated considering the thermal conduction, thermal radiation and thermal convection processes inside the DC-GIL. The load current on the conductor is set to 0 and 3000 A, and the environmental temperature is set to 30 °C.

3.2.2 Electric Field Regulation Under Complex Conditions

(a) Temperature gradient

For an operating DC-GIL, when the load current flows through the conductor, Joule heat will be produced, leading to a graded temperature distribution inside the pipe. Figure 8 presents the gas flow velocity and the temperature distributions inside the pipe when the load current is 3000 A. The temperature on the conductor is 43.3 °C, and the temperature on the pipe wall is 35.6 °C. Because the heated gas has a trend of going up in the vertical direction, the gas convection in the lower gas domain helps to bring heat from the conductor to the spacer surface, causing the temperature on the concave surface to be higher than that on the convex surface.

Figure 9 presents the tangential electric field distributions along surfaces of the EP/Al_2O_3 spacer when the load current is 0 and 3000 A. $r = 89$ mm is the location of the conductor, and $r = 247.6$ mm is the location of the pipe wall. When the load current is zero, the temperature inside the pipe is uniform. The maximum electric field is located near the conductor on both convex and concave surfaces of the spacer. When the load current is 3000 A, the temperature gradient causes the volume conductivity of the spacer to be graded. The volume conductivity of the spacer is higher near

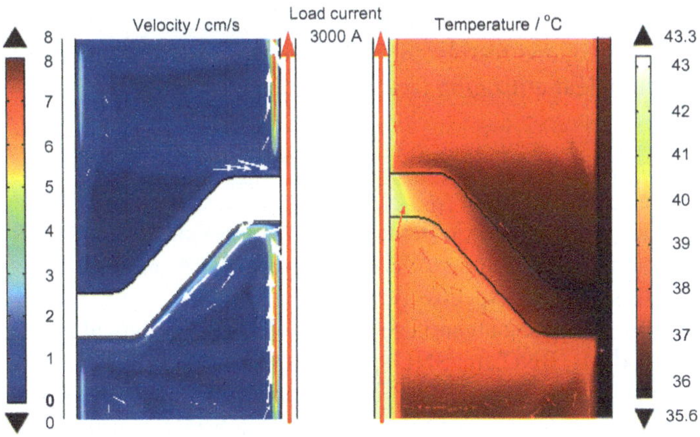

Fig. 8 Gas flow velocity and temperature distributions at 3000 A

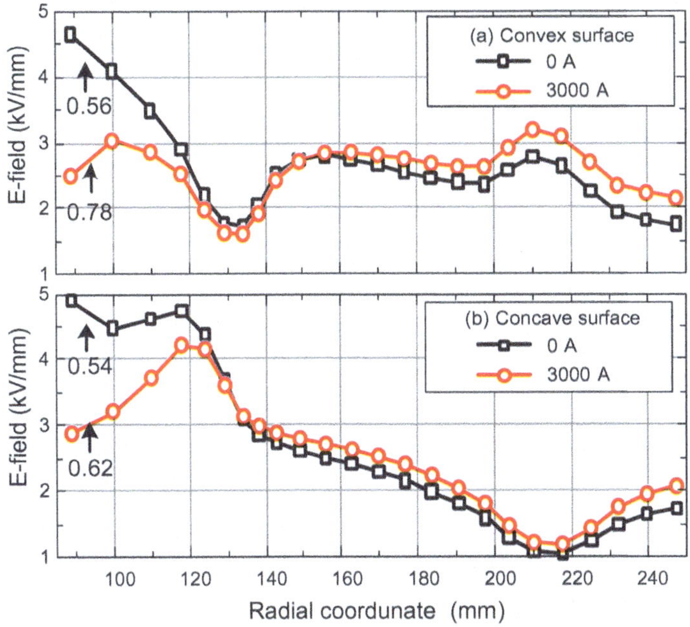

Fig. 9 Tangential electric field distributions along convex and concave surfaces of the EP/Al_2O_3 spacer at 0 and 3000 A

the conductor but lower near the pipe wall. Therefore, the electric field near the conductor is reduced, and the electric field near the pipe wall is increased. In this paper, the electric field utilization factor is defined as:

$$f = E_{av}/E_{\max} \tag{19}$$

where E_{av} is the average electric field; E_{max} is the maximum lectric field. At 3000 A, the electric field utilization factor is increased by ~39% and ~15% on the convex and the concave surfaces, respectively.

Figure 10 presents the tangential electric field distributions along the concave surfaces of EP/Al_2O_3, BNC and SNC spacers at 0 and 3000 A. Both BNC and SNC spacers can regulate the electric field distributions whether there is a temperature gradient or not. However, the electric field utilization factor of the SNC spacer is higher than that of the BNC spacer.

Table 8 lists the loss powers of different spacers. For the EP/Al_2O_3 and the BNC spacers, the majority of the conduction loss is in the spacer bulk because their surface conductivities are very low. The SNC spacer has a much higher surface conductivity; therefore, its surface loss power takes the dominant position. When there is a temperature increment at 3000 A, the total loss powers of all spacers are increased due to the increased conductivities. The total loss powers of BNC and SNC spacers are higher than that of the EP/Al_2O_3 spacer because the EP/SiC composite has higher volume

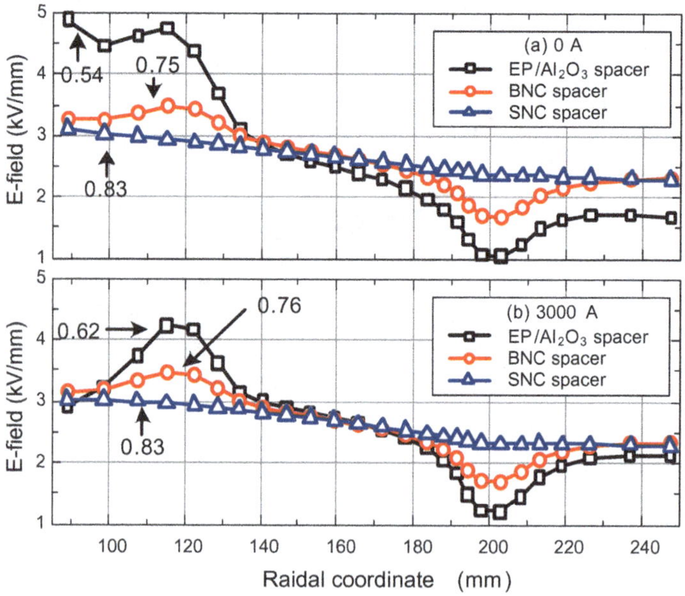

Fig. 10 Tangential electric field distributions along the concave surfaces of EP/Al_2O_3, BNC and SNC spacers at 0 and 3000 A

Table 8 Loss power of different spacers

Type	0 A		3000 A	
EP/Al_2O_3 spacer	S:2.7 × 10^{-8}	T: 5 × 10^{-4}	S:2.8 × 10^{-8}	T: 1.2 × 10^{-3}
	B:5 × 10^{-4}		B:1.2 × 10^{-3}	
SNC spacer	S:4.2 × 10^{-3}	T: 4.8 × 10^{-3}	S:5.7 × 10^{-3}	T: 7.1 × 10^{-3}
	B:6.2 × 10^{-4}		B:1.4 × 10^{-3}	
BNC spacer	S:2.8 × 10^{-8}	T: 1.5	S:2.8 × 10^{-8}	T: 2
	B:1.5		B:2	

S: surface loss power, W; B: bulk loss power, W; T: total loss power, W

conductivity than the EP/Al_2O_3 composite. The electric field regulation effects of BNC and SNC spacers are at the cost of increasing the conduction loss. Compared with the BNC spacer, the SNC spacer has a better electric field regulation effect and a much lower loss power, which is only 6–10 times higher than that of the EP/Al_2O_3 spacer.

(b) Rough electrode surface

There are strict standards on the surface roughness of the conductor ($R_a < 2\ \mu m$), but very few attentions have been paid to the surface microstructure of the pipe wall. Surface protrusions on the inner pipe wall due to scratches, welds, particulate trap, will cause micro discharges and unexpected surface charge accumulations. Figure 11 presents the surface charge distributions on EP/Al_2O_3 and SNC spacers when there

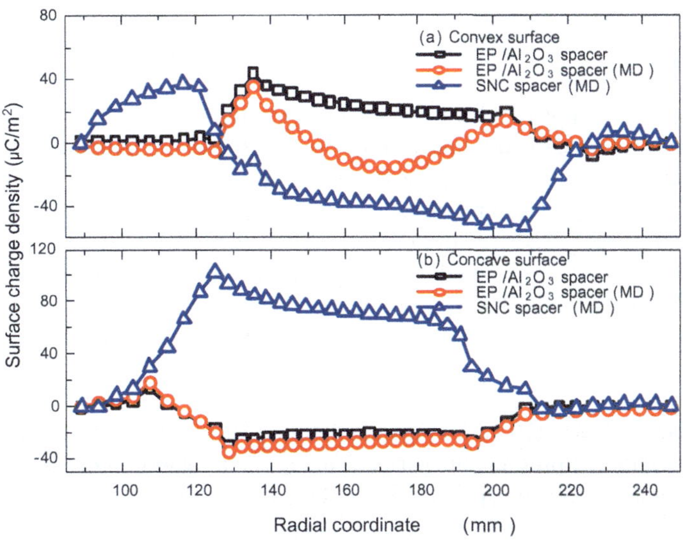

Fig. 11 Surface charge distributions on EP/Al_2O_3 and SNC spacers with micro discharges on the pipe wall

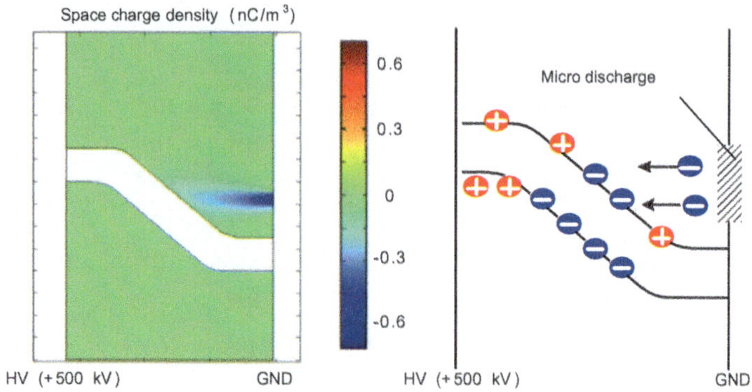

Fig. 12 Surface charge accumulation process with micro discharges on the pipe wall

are micro discharges (MD) on the pipe wall. When there is no micro discharge, the majority of surface charges come from the spacer side, causing the positive surface charge accumulation on the convex slope and the negative surface charge accumulation on the concave slope. When micro discharges occur on the pipe wall, negative ions will transport to the convex slope under electric field force, as shown in Fig. 12. Positive charges on the convex slope are reduced, and negative surface charges can be observed in a small region. Because micro discharges occur in the upper gas domain, surface charge distributions on the concave surfaces of EP/Al_2O_3 and SNC spacers are barely influenced.

Figure 13 presents the tangential electric field distributions along the surfaces of EP/A12O3 and SNC spacers with micro discharges on the pipe wall. Negative charges on the convex surface increase the electric flux from the conductor, causing electric field concentrations near the conductor. The electric field utilization factor is reduced by ~ 21% and ~ 13% on the convex and the concave surfaces, respectively. When the SNC spacer is applied, electric field concentrations are significantly reduced on both surfaces because the nonlinear-conductivity layer can intelligently regulate the surface charge distributions. As shown in Fig. 6, positive charges on the convex surface near the conductor are greatly increased. Moreover, the majority of surface charges on the concave surface of the SNC spacer are positive. The electric flux from the conductor is reduced as well as the tangential electric field near the conductor. Compared with the EP/A12O3 spacer, the electric field utilization factors of the SNC spacer are improved by ~ 9% and ~ 52% on convex and concave surfaces, respectively.

(c) Metal particle

Metal particles inside GIL are unavoidable during transportation, installation and operation. Figure 14 presents the surface charge accumulation process with a metal particle levitating over the spacer surface. Positive and negative ions are produced at the particle tips due to micro discharges, then transport under electric field force and accumulate on the spacer surface. Besides, there are strong normal components

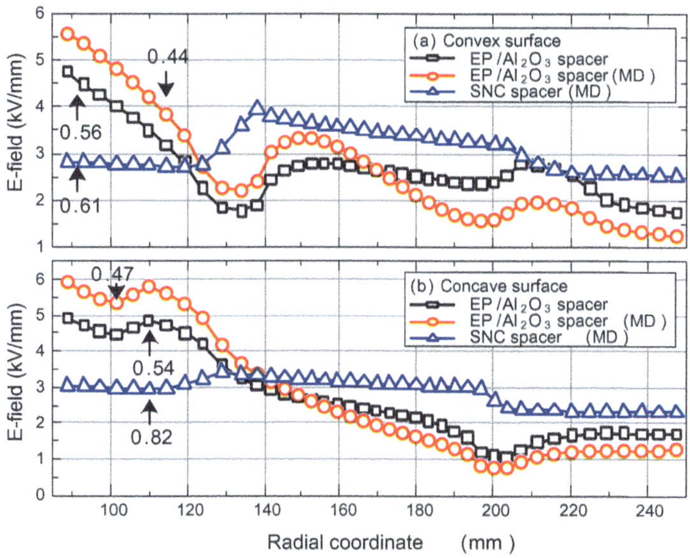

Fig. 13 Tangential electric field distributions along the surfaces of EP/Al_2O_3 and SNC spacers with micro discharges on the pipe wall

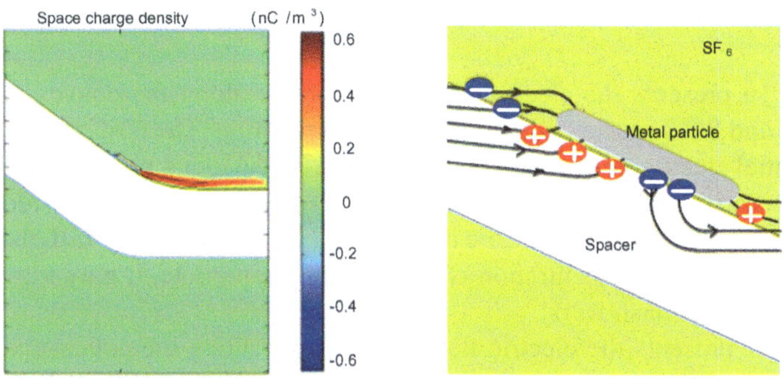

Fig. 14 Surface charge accumulation process with a metal particle levitating over the spacer surface

of electric field in the middle of the metal particle, which causes surface charge accumulations from the spacer side.

Figure 15 presents the surface charge distributions along the surfaces of EP/Al_2O_3 and SNC spacers with metal particle levitation. On the convex surface of the Al_2O_3 spacer, surface charges caused by the metal particle can be divided into two parts: the dipole charge at the tips caused by micro discharges and the dipole charge underneath the particle caused by the normal electric field. The SNC spacer can eliminate the dipole charge caused by micro discharges but not the normal electric field.

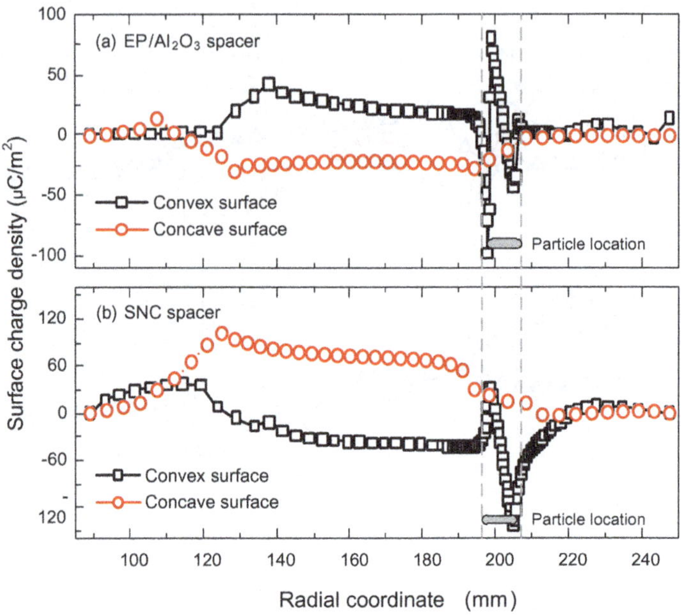

Fig. 15 Surface charge distributions along the surfaces of EP/Al_2O_3 and the SNC spacers with metal particle levitation

Figure 16 presents the electric field distributions along the convex surfaces of EP/Al_2O_3 and SNC spacers with metal particle levitation. The SNC spacer can relax the tangential electric field concentration caused by the metal particle. However, the normal electric field concentration becomes a bit more severe in the gas side. Even so, the SNC spacer can still improve the insulation performance of DC-GIL, because the normal electric field concentration is underneath the particle, it may trigger partial discharges but not flashovers.

Figure 17 presents the electric field distributions along the concave surfaces of EP/Al_2O_3 and SNC spacers with metal particle levitation. Positive charges on the concave surface of the SNC spacer reduce the electric flux from the conductor; both tangential and normal electric field distributions become more even than those of the EP/Al_2O_3 spacer.

(d) Polarity reversal voltage

Figure 18 presents the tangential electric field distributions along the convex and the concave surfaces of the EP/A12O3 spacer under a polarity reversal voltage. The electric field distributions before polarity reversal are in the DC steady state. The voltage is reversed from +500 kV to −500 kV in two minutes. After polarity reversal, the electric field near the conductor is strengthened on both convex and concave surfaces.

Figure 19 presents the charge distributions under the polarity reversal voltage. Before polarity reversal, surface charges near the conductor are positive on both

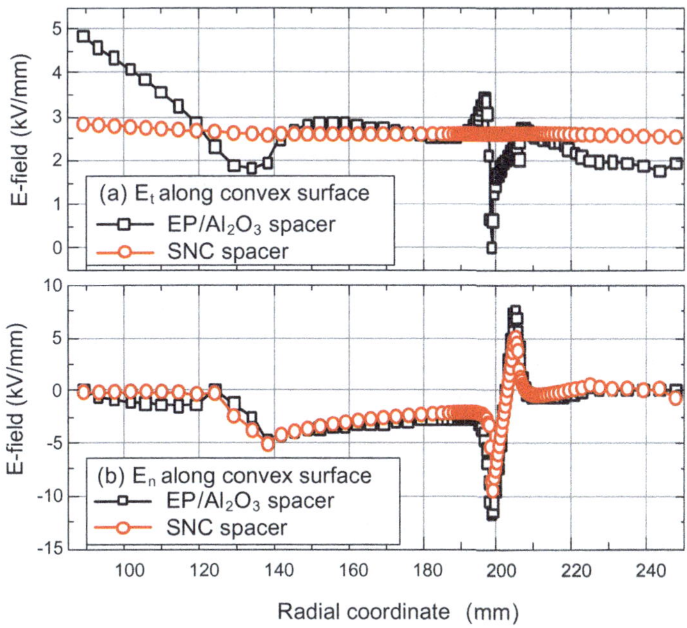

Fig. 16 Electric field distributions along the convex surfaces of EP/Al_2O_3 and SNC spacers with metal particle levitation

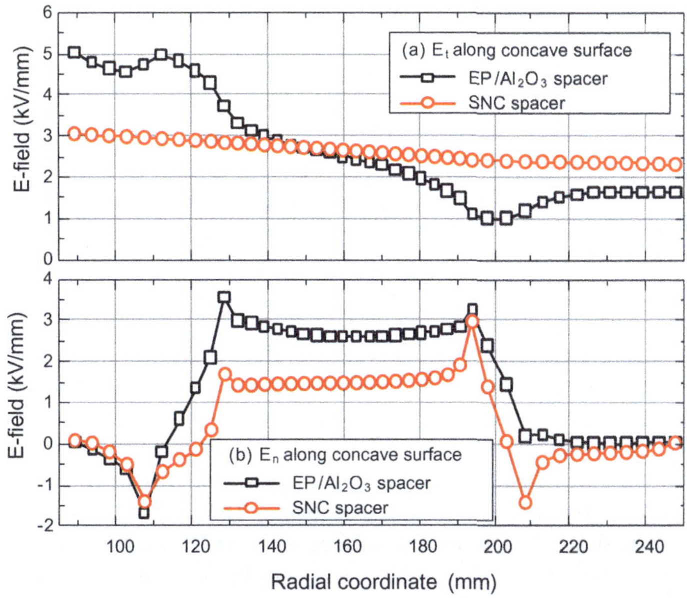

Fig. 17 Electric field distributions along the concave surfaces of EP/Al_2O_3 and SNC spacers with metal particle leitation

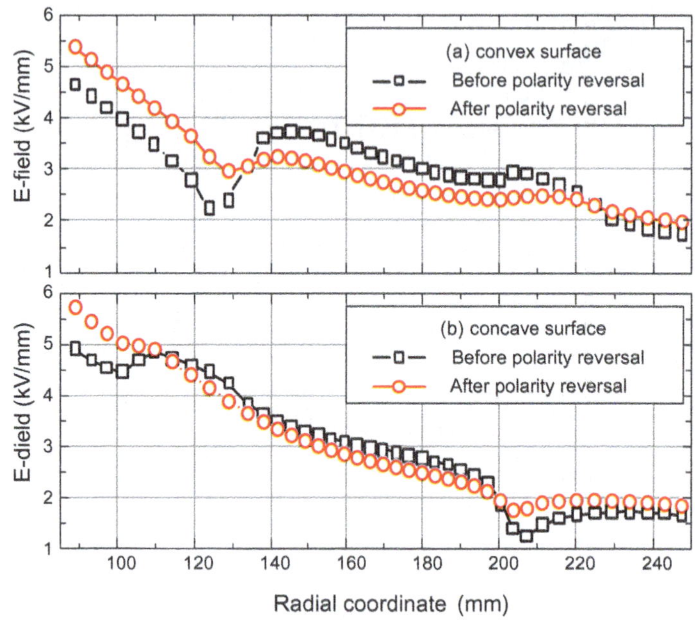

Fig. 18 Electric field distributions along the convex and the concave surfaces of the EP/Al_2O_3 spacers under the polarity reversal voltage

Fig. 19 Charge distributions under the polarity reversal voltage

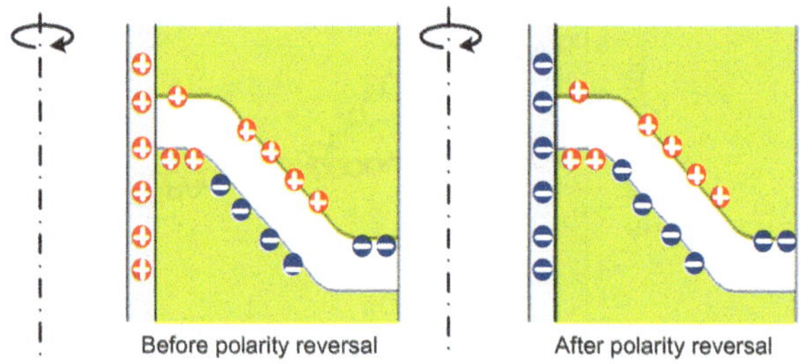

convex and concave surfaces, which reduce the electric flux from the conductor and the electric field near the conductor. Although negative charges on the concave slope of the spacer may cause an electric flux increment, it is not dominant. After polarity reversal, the charge polarity on the conductor turns from positive to negative, but the polarity of surface charges near the conductor remains positive because it takes a long time for surface charges to be redistributed. Therefore, the electric field near the conductor is strengthened on both surfaces of the EP/Al_2O_3 spacer.

From the above analyses, it can be concluded that the electric field distribution after polarity reversal is closely related to the response time of redistributing the surface/space charges from a state to another state. Figure 20 presents the transient electric field variations at the triple junction of the conductor, the concave spacer

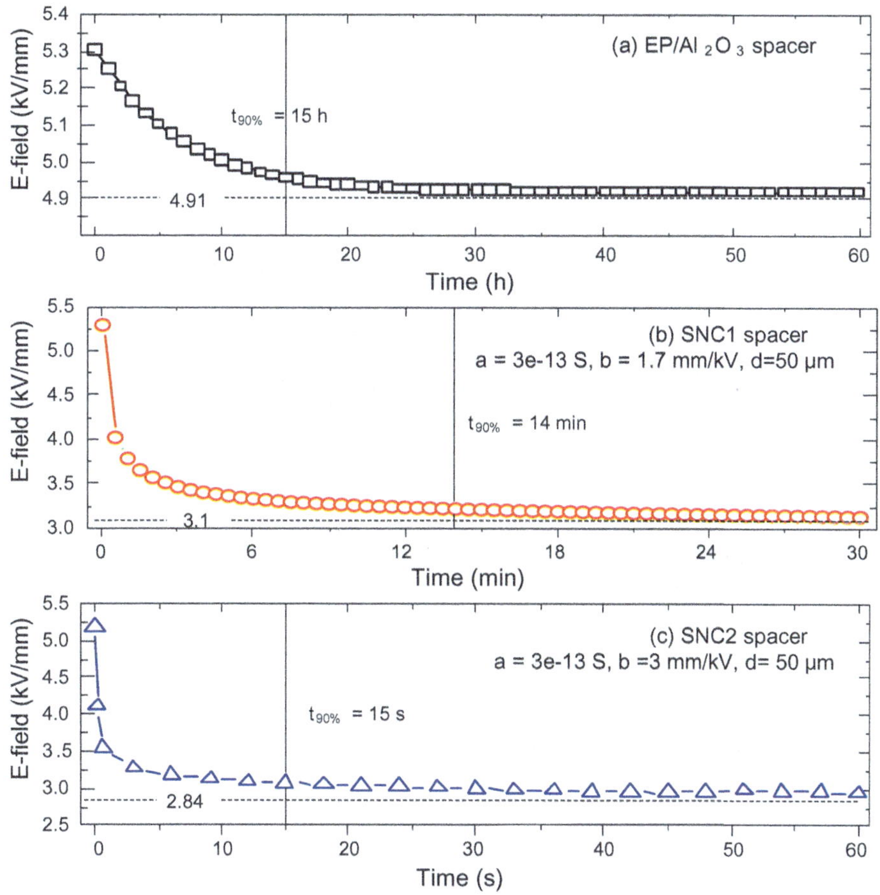

Fig. 20 Transient electric field variations at the triple junction after applying a DC voltage of + 500 kV

surface and the SF_6 gas after applying a DC voltage of +500 kV. The electric field distribution in the transient process changes from the capacitive state to the resistive state. During this process, surface charges accumulate on the spacer surfaces, reducing the electric flux from the conductor. The electric field strength at the triple junction point declines with time and finally reaches the steady state.

The time constant $t_{90\%}$ of a spacer is defined as:

$$|[E(t_{90\%}) - E(t_{\infty})]/[E(t_0) - E(t_{\infty})]| = 0.1 \tag{20}$$

where $E(t_0)$ is the initial electric field strength; $E(t_{\infty})$ is the steady-state electric field strength; $E(t_{90\%})$ is the electric field strength in the critical saturation state. The time constants of EP/Al_2O_3, SNC1 and SNC2 spacers are 15 h, 14 min and 15 s. It means

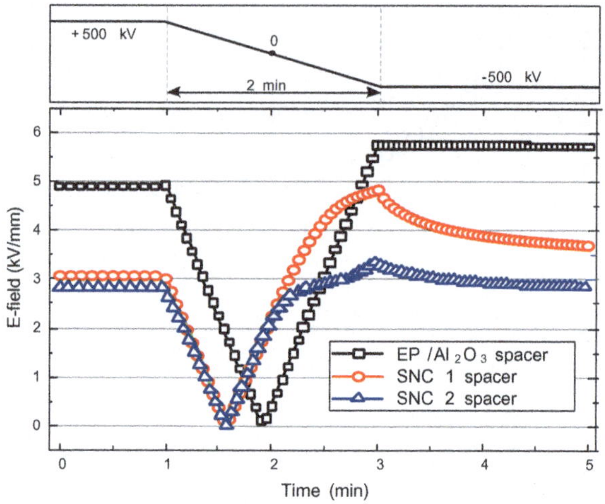

Fig. 21 Transient electric field variations at the triple junction under the polarity reversal voltage

that a higher surface conductivity leads to a short transient process of the surface charge redistribution.

Figure 21 presents the transient electric field variations at the triple junction of the conductor, the concave spacer surface and the SF_6 gas under the polarity reversal voltage. After polarity reversal, the electric field strengths at the triple junction points of different spacers are higher than their steady-state values before the polarity reversal. Therefore, the over-shoot electric field strength must be taken into consideration while designing a DC-GIL spacer. The lower the time constant of a spacer is, the lower the over-shoot electric field amplitude. Although the SNC1 and the SNC2 spacers have comparable field regulation effects in steady state, the time constant of the SNC2 spacer is much lower. For the SNC2 spacer, the over-shoot electric field is small and the electric field distribution reaches a new steady state immediately after reversing the voltage polarity. The SNC2 spacer would have a better insulation performance than the SNC1 spacer when there is a polarity reversal voltage.

To investigate the influence of the time constant on the over-shoot electric field, the overshoot rate is defined as:

$$r_{os} = \left|(|E_{bef}| - |E_{aft}|)/|E_{bef}|\right| \tag{21}$$

where E_{bef} and E_{aft} are the electric fields before and after polarity reversal, respectively.

Figure 22 presents the relationship between the overshoot rate and the time constant of the SNC spacer. Increasing the inversion time and reducing the time constant of the SNC spacer are both helpful to restrain the electric field overshoot after polarity reversal. If an acceptable overshoot rate is defined as 15% under polarity reversal voltages, the time constants of SNC spacers should be lower than 0.1, 1, 10

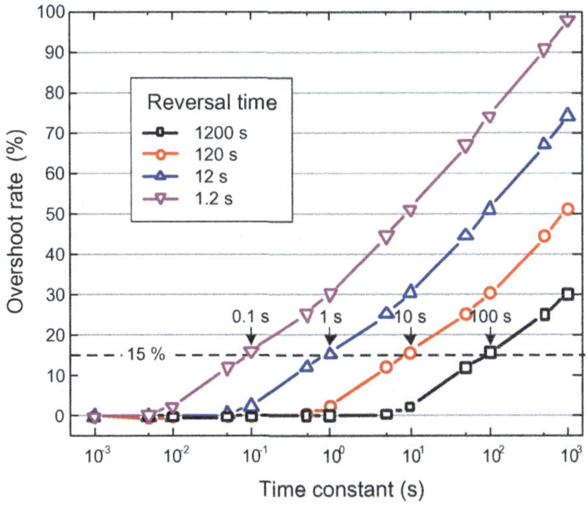

Fig. 22 Relationship between the overshoot rate and the time constant of the SNC spacer

and 100 s when the inversion times are 1.2, 12, 120 and 1200 s. There is a relationship between the inversion time and the time constant:

$$\tau_{pr}/\tau_c \geq 12,\ if\ r_{os} \leq 15\% \tag{22}$$

where τ_{pr} is the inversion time of the polarity reversal voltage; τ_c is the time constant of the SNC spacer.

3.2.3 Parameter Optimization

Simulation results in Sect. 3 have proved that the SNC spacer can regulate the electric field distributions under complex conditions in DC-GIL. In this section, parameter analyses will be conducted to determine the optimal parameters of a SNC spacer. As mentioned in Sect. 2, the product of the volume conductivity and the thickness of the nonlinear-conductivity surface layer can be regarded as the surface conductivity of the SNC spacer. Therefore, the value of a •d is regarded as the Ohmic surface conductivity in the following.

Figure 23 presents the variation of the electric field utilization factor with parameters. The electric field utilization factor of the SNC spacer increases with the Ohmic surface conductivity and the nonlinear coefficient. When the utilization factor exceeds 0.9, it is almost saturated. The corresponding parameter region is defined as the saturation region.

Figure 24 presents the variation of the surface loss index with parameters. The surface loss power of the SNC spacer keeps growing with increasing parameters, even in the saturation region of the electric field utilization factor. Therefore, the

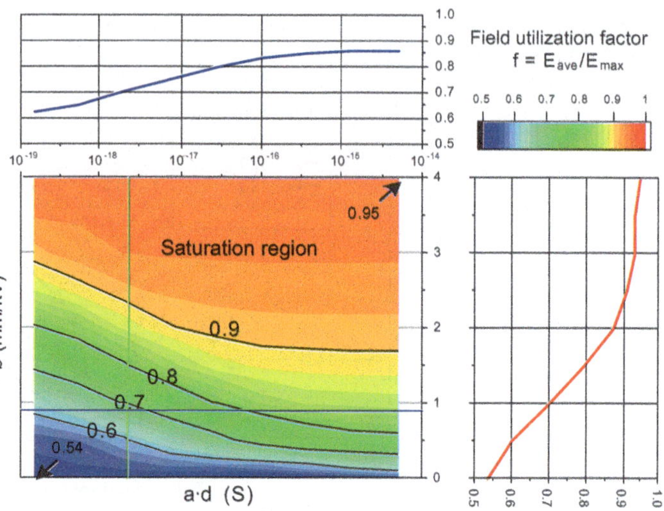

Fig. 23 Variation of the electric field utilization factor with parameters

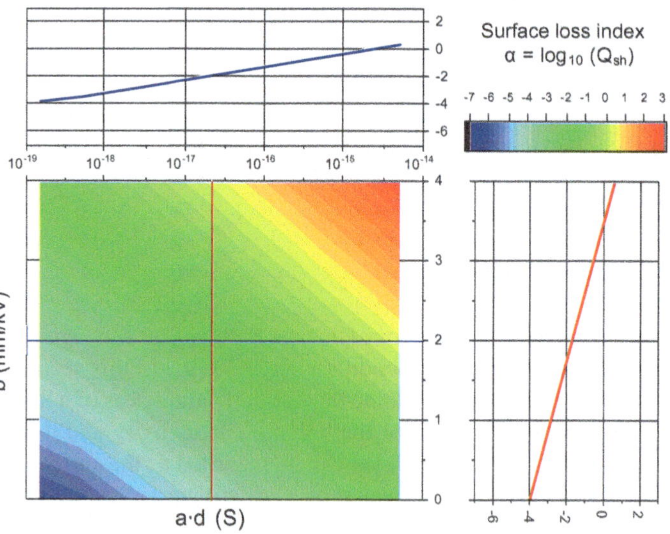

Fig. 24 Variation of the surface loss index with parameters

parameters of the SNC spacer should not be in the saturation region, because the growing surface loss causes no further reduction of electric field.

Figure 25 presents the time constants of SNC spacers with different parameters. When the nonlinear coefficient is higher than 1 mm/kV, there is an inverse proportional relationship between the time constant and the Ohmic surface conductivity of

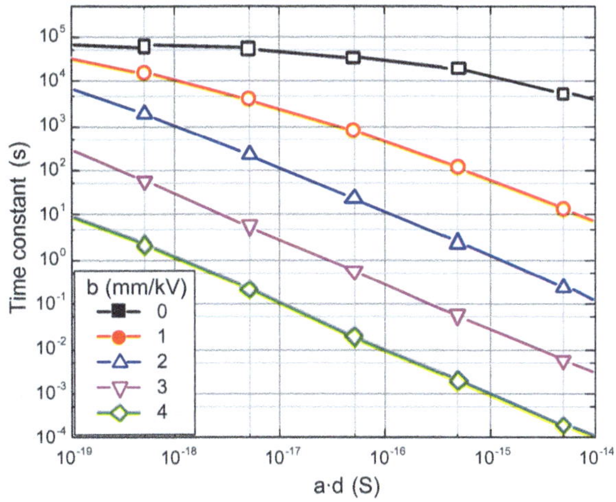

Fig. 25 Time constants of SNC spacers with different parameters

the SNC spacer. The time constant of the SNC spacer can be reduced by increasing the Ohmic surface conductivity and the nonlinear coefficient.

To design a SNC spacer, some indexes should be determined first according to the actual operation conditions. (1) The first index is the electric field utilization factor, which should be as high as possible but not in the saturation region. (2) The second index is the surface loss power, which is constrained below 0.1 W in this paper. (3) The last index is the time constant, which should be determined according to the inversion time of the polarity reversal voltage. Assuming the common inversion time for a DC-GIL is 2 min and an acceptable overshoot rate is less than 15%, the time constant of the SNC spacer should be lower than 10 s. After all these indexes are fixed, the objective parameter region can be obtained by drawing a parameter map, as shown in Fig. 26.

4 Fabrication and Electrical Evaluation of Novel Spacers

4.1 Dip Coating

The cone-type spacer was coated with the EP/SiC composites to fabricate the EP/SiC coated spacers [21]. First, different contents (15, 20, and 25 vol%) of micro-sized β-SiC were dispersed in the EP matrix to obtain the 15, 20 and 25% EP/SiC coatings, respectively. After fully mixing, the EP/SiC coatings were placed in a vacuum for 1 h to remove the bubbles. The cone-type spacers were made of EP/Al_2O_3 composites, whose production process was identical to the industrial manufacturing process. Before soaking the spacers, the basic surface cleaning and surface roughness pretreatment processes were performed to enhance the interfacial adhesion between the

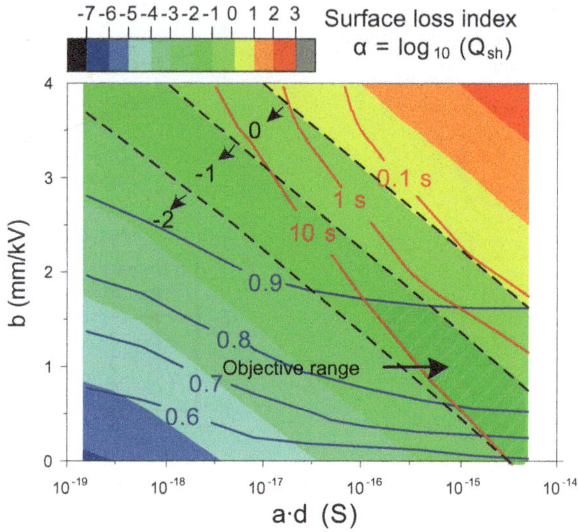

Fig. 26 Parameter map of the SNC spacer for the 500 kV DC-GIL

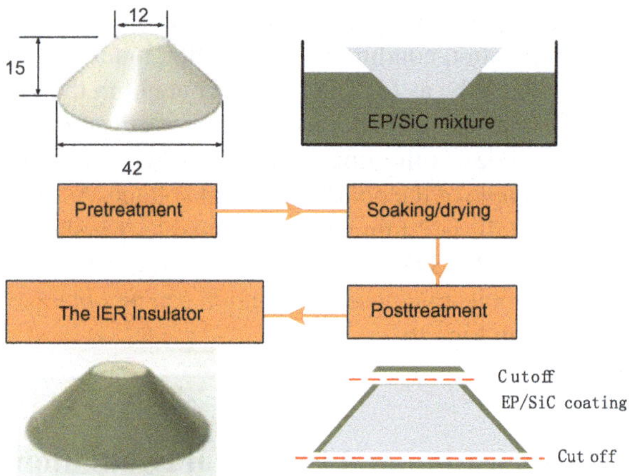

Fig. 27 Schematic diagram of the dip coating method [12]. Copyright Clearance Center'RightsLink® License Number: 4858111175808

spacers and coatings. After soaking, drying and posttreatment, the EP/SiC coated spacers were obtained, as shown in Fig. 27.

4.2 Centrifugation

Figure 28 shows a diagram of a novel spacer fabrication technique [25]. A diagram of the centrifugal machine is presented in Fig. 28a, which mainly consists of a motor speed control module, a temperature control module and a rotation platform. A sealed mold filled with liquid-state epoxy/SiC mixture (epoxy matrix (Al-6057)

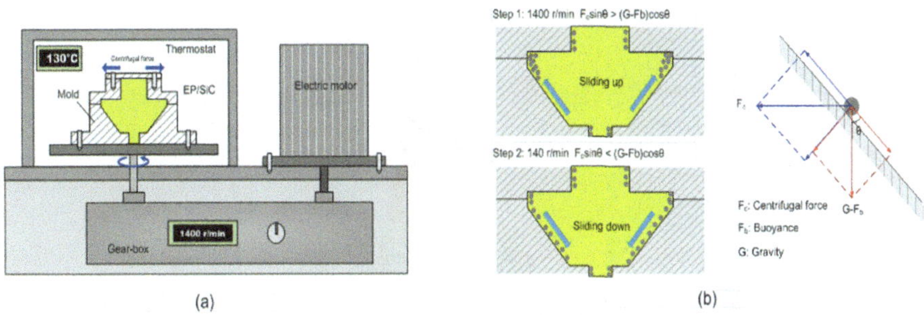

Fig. 28 Diagram of the IER spacer fabrication technique; **a** centrifugal machine, **b** two-step centrifugation method [25]. Copyright Clearance Center'RightsLink® License Number: 4858130345337

to curing agent (Al-3010B) to accelerating agent to SiC particle (α-15 μm) ratio = 100:85:0.2:x) is fixed on the rotation platform so that SiC particles can be pushed to the surface of the mold under centrifugal force. Before centrifugation, the epoxy/SiC mixture must be evenly blended, and bubbles must be removed. Figure 28b presents the two-step centrifugation method of fabricating the novel spacer, and the speeds and times of centrifugation at different stages were determined after several experiments. First, the speed of the rotation platform is set to 1400 r/min for 2 min at 130 °C. At this stage, particles tend to slide up along the mold surface because the tangential component of centrifugal force is much higher than that of gravity minus buoyancy. After 2 min, most of the particles concentrate at the top edge of mold. Next, the speed of the rotation platform is set to 140 r/min for 30 min at 130 °C. At this stage, the tangential component of gravity minus buoyancy is higher than that of the centrifugal force. Particles slide down along the mold surface during curing process, forming a uniform layer of highly concentrated SiC particles. Then, the mold is put in an oven for 2 h to finish the second curing process.

4.3 Gradient Sputtering

By magnetic sputtering, the surface conductivity of the deposited ZnO film has a nonlinear relationship with both the sputtering time and E-field strength [26]. Based on the scattered measurement data, the nonlinear surface fitting method was used to obtain a two-dimensional surface, as shown in Fig. 29.

By depositing a ZnO film onto a spacer surface, the E-field distortion can be reduced due to the nonlinear surface conductivity of the ZnO film. By grading the sputtering time in radial direction of the spacer, a further E-field relaxation is expected. And the iteration method was used to optimize the thickness distribution of the ZnO film. Figure 30 shows a diagram of fabricating the optimized spacer. To control the sputtering time distribution along the spacer surface, a baffle with a designed gap was placed above the rotating spacer during the magnetron sputtering

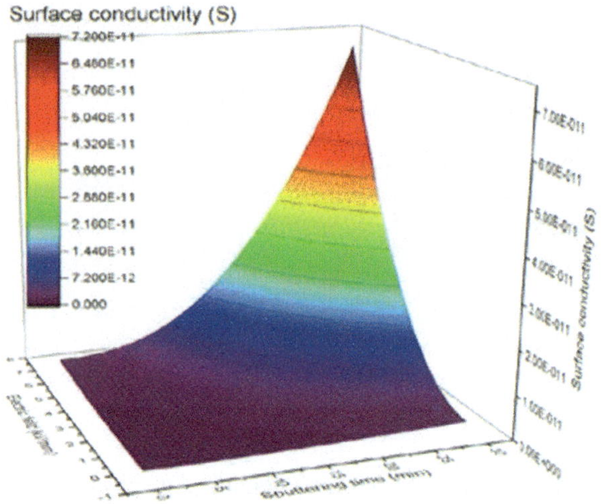

Fig. 29 Fitted two-dimensional surface of surface conductivity of ZnO film [26]. Copyright Clearance Center'RightsLink® License Number: 4858121482497

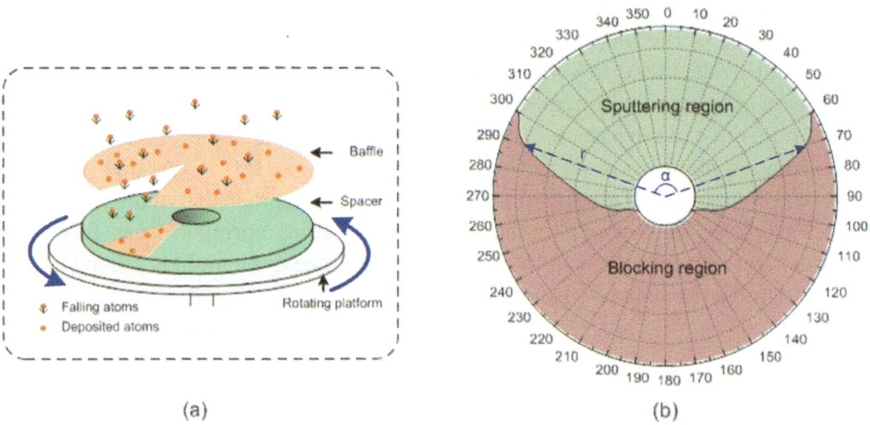

Fig. 30 Diagram of fabricating the optimized novel spacer: **a** baffle placement in sputtering, **b** sketch map of the designed baffle [26]. Copyright Clearance Center'RightsLink® License Number: 4858121482497

process, as shown in Fig. 30a. Because of the gradually-changed gap width, the probability of target atoms getting through the baffle gap and depositing onto the spacer surface is graded in radial direction. At each radial position, the relationship between the effective sputtering time and total sputtering time can be described as,

$$\alpha = 2\pi \frac{T_e}{T_t} \tag{23}$$

where Te is the effective sputtering time; Tt is the total sputtering time; α is the gap radian at each radial position. According to the optimal sputtering time distribution, i.e., the effective sputtering time distribution, in Sect. 3 and a determined total sputtering time, the gap radian distribution can be obtained. In this paper, the determined total sputtering time is 45 min, a sketch map of the designed baffle is presented in Fig. 30b. The red region represents the blocking region of the baffle, the green region represents the sputtering region, i.e., the baffle gap.

4.4 *Flashover Test*

In this section, flashover tests under DC and polarity reversal voltages were both conducted to evaluate the insulation performances of different spacers. Table 9 lists some detailed information of the voltage waveforms and operation steps for the flashover test. Each test was repeated for ten times for data accuracy.

(1) For the novel spacer fabricated by dip coating, the DC flashover voltage under negative voltage was improved by up to 40%.
(2) For the novel spacer fabricated by centrifugation, the DC flashover voltages were improved by up to 14.8 and 28% under positive and negative voltages, and the flashover voltage under polarity reversal voltage was improved by up to 40%.
(3) For the novel spacer fabricated by sputtering, the DC flashover voltages were improved by up to 15.9 and 17.4% under positive and negative voltages, and the flashover voltage under polarity reversal voltage was improved by up to 15%.

Table 9 Voltage waveforms and operation steps in flashover test

Test type	Waveforms	Operation steps
DC	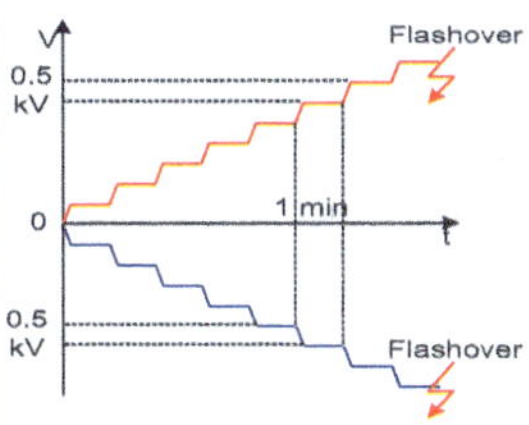	Raise the voltage amplitude by 0.5 kV/min until a flashover occurs
Polarity reversal	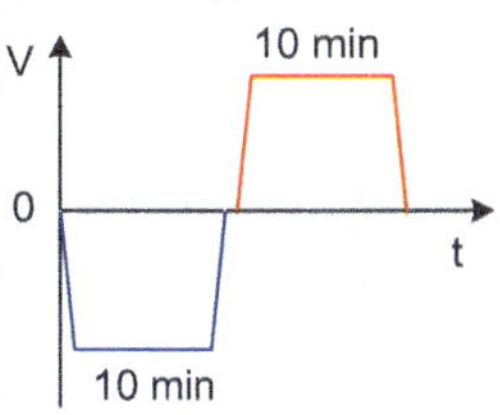	1. Apply a negative voltage for 10 min, then turn off the power source and apply a positive voltage of the same amplitude for another 10 min. The interval time between reversing the voltage polarity is 2 min 2. If no flashover occurs, raise the voltage amplitude by 1 kV and repeat the above step until a flashover occurs

5 Conclusion

In this chapter, the nonlinear conductivity properties of epoxy/SiC composites were investigated, and the surface nonlinear conductivity spacer was proposed to regulate the electric field distribution along spacer surfaces. The main conclusions are as following:

(1) By doping epoxy with α-SiC and β-SiC particles, epoxy/SiC composites gain nonlinear conductivities, which barely changes in the Ohmic region but grows rapidly in the nonlinear region. When the filler contents are the same, the nonlinear characteristics of epoxy doped with β-SiC is higher than that of epoxy doped with α-SiC.
(2) In a GIL, epoxy/SiC composites can be applied on spacer surfaces, i.e. the surface nonlinear conductivity spacer, to relax the surface electric field distortions. Simulation results show that such novel spacers can regulate the surface charges caused by rough electrodes and metal particles, and uniform the electric field distributions under temperature gradient and polarity reversal voltage.
(3) Novel spacers were fabricated by dip coating, centrifugation and gradient sputtering. Flashover tests were conducted under DC and polarity reversal voltages, indicating that such novel spacers have better insulation performances than the conventional spacer.

References

1. A. Winter, J. Kindersberger, Stationary resistive field distribution along epoxy resin insulators in air under DC voltage. IEEE Trans. Dielectr. Electr. Insul. **19**, 1732–1739 (2012)
2. U. Straumann, M. Schuller, C.M. Franck, Theoretical investigation of HVDC disc spacer charging in SF_6 gas insulated systems. IEEE Trans. Dielectr. Electr. Insul. **19**, 2196–2205 (2012)
3. B.X. Du, H.C. Liang, J. Li, Z.Y. Ran, Electric field distribution along SF_6/N_2 filled DC-GIS/GIL Epoxy Spacer. IEEE Trans. Dielectr. Electr. Insul. **25**, 1202–1210 (2018)
4. A.D. Lorenzi, L. Grando, A. Pesce, R. Specogna, Modeling of epoxy resin spacers for the 1 MV DC gas insulated line of ITER neutral beam injector system. IEEE Trans. Dielectr. Electr. Insul. **16**, 77–87 (2009)
5. X. Zhang, G.Y. Li, R.H. Huang, Z. Xiang, Y.M. Zhao, S.J. Yao, X.H. Wang, Statistical analysis of defects and maintenace advice for GIS in different operating years above 110 kV. High Volt. Appar. **52**, 184–188 (2016). (in Chinese)
6. T. Nitta, K. Nakanishi, Charge accumulation on insulating spacers for HVDC GIS. IEEE Trans. Electr. Insul. **26**, 418–427 (1991)
7. C. Li, C. Lin, B. Zhang, Q. Li, W. Liu, J. Hu, J. He, Understanding surface charge accumulation and surface flashover on spacers in compressed gas insulation. IEEE Trans. Dielectr. Electr. Insul. **25**, 1152–1166 (2018)
8. M. Kozako, R. Kido, T. Imai, T. Ozaki, T. Shimizu, T. Tanaka, Surface roughness change of epoxy/TiO_2 nanocomposites due to partial discharges. IEEE. Int. Sym. Electr. Insul. Mater. (ISEIM), 661–664 (2005).

9. C. Charalambous, M. Danikas, Y. Yin, N. Vordos, J. W. Nolan, A. Mitropoulos, Study of the behavior of water droplets under the influence of a uniform electric field on conventional polyethylene and on crosslinked polyethylene (XLPE) with MgO nanoparticles samples. Eng. Technol. Appl. Sci. Res. **7**, 1323–1328 (2017)
10. K. Kato, M. Kurimoto, H. Shumiya, A. Hiroshi, S. Shuichi, O. Hitoshi., Application of functionally graded material for solid insulator in gaseous insulation system. IEEE Trans. Dielectr. Electr. Insul. **13**, 362–372 (2006)
11. N. Hayakawa, Y. Miyaji, H. Kojima, K. Kato, Electric field reduction by functionally graded materials (FGM) with permittivity and conductivity distribution, in *Annual Report Conference on Electrical Insulation and Dielectric Phenomena (CEIDP)* (2012), pp. 627–630
12. N. Hayakawa, J. Ishiguro, H. Kojima, K. Kato, H. Okubo, Fabrication and simulation of permittivity graded materials for electric field grading of gas insulated power apparatus. IEEE Trans. Dielectr. Electr. Insul. **23**, 547–554 (2016)
13. J. Ishiguro, M. Kurimoto, H. Kojima, K. Kato, H. Okubo, N. Hayakawa, Electric field control in coaxial disk-type solid insulator by functionally graded materials (FGM), in *Annual Report Conference on Electrical Insulation and Dielectric Phenomena (CEIDP)* (2014), pp. 663–666
14. N. Hayakawa, J. Shimomura, T. Nakano, M. Hanai, Fabrication technique of permittivity graded materials (FGM) for disk-type solid insulator, in *Annual Report Conference on Electrical Insulation and Dielectric Phenomena (CEIDP)* (2012), pp. 32–35
15. H. Shumiya, K. Kato, H. Okubo, Feasibility study on FGM (Functionally Graded Materials) application for gas insulated equipment, in *Annual Report Conference on Electrical Insulation and Dielectric Phenomena (CEIDP)* (2004), pp. 360–363
16. Z. Liu, W.D. Li, Y.B. Wang, G.Q. Su, G.J. Zhang, Y. Gao, D.C. Li, Topology optimization and 3D-printing fabrication feasibility of high voltage FGM insulator, in *IEEE International Conference on High Voltage Engineering and Application* (2016), pp. 1–4
17. Z. Liu, W.D. Li, L.Y. Zhang, X.R. Li, J.B. Deng, G.J. Zhang, Simulation design and 3D-printing fabrication of conductivity graded insulator, in *IEEE International Conference on Electrical Materials and Power Equipment* (2017), pp. 171–174
18. T. Christen, L. Donzel, F. Greuter, Nonlinear resistive electric field grading, Part 1: Theory and simulation. IEEE Electr. Insul. Mag. **26**, 47–59 (2010)
19. J. Li, B.X. Du, X.X. Kong, Z.L. Li, Nonlinear conductivity and interface charge behaviors between LDPE and EPDM/SiC composite for HVDC cable accessory. IEEE Trans. Dielectr. Electr. Insul. **24**, 1566–1573 (2017)
20. B.X. Du, H.C. Liang, J. Li, C. Zhang, Temperature-dependent surface potential decay and flashover characteristics of epoxy/SiC composites. IEEE Trans. Dielectr. Electr. Insul. **25**, 631–638 (2018)
21. B.X. Du, H.C. Liang, J. Li, Interfacial E-Field self-regulating insulator considered for DC GIL application. IEEE Trans. Dielectr. Electr. Insul. **26**, 801–809 (2019)
22. H.Y. Zhou, G.M. Ma, C.R. Li, C. Shi, S.C. Qin, Impact of temperature on surface charges accumulation on insulators in SF6-filled DC-GIL. IEEE Trans. Dielectr. Electr. Insul. **24**, 601–610 (2017)
23. A. Winter, J. Kindersberger, Transient field distribution in gas-solid insulation systems under DC voltages. IEEE Trans. Dielectr. Electr. Insul. **21**, 116–128 (2014)
24. M. Schueller, R. Gremaud, C.B. Doiron, C.M. Franck, Micro discharges in HVDC gas insulated systems. IEEE Trans. Dielectr. Electr. Insul. **22**, 2879–2888 (2015)
25. B.X. Du, H.C. Liang, J. Li, Z.H. Wang, Z.Y. Ran, H.L. Liang, Fabrication and electrical evaluation of interfacial E-field regulating insulator for DC-GIL application. IEEE Trans. Dielectr. Electr. Insul. **26**, 1956–1964 (2019)
26. B.X. Du, H.C. Liang, J. Li, Novel spacer coated with functionally graded ZnO film for HVDC gas insulated line. IEEE Trans. Dielectr. Electr. Insul. **26**, 2061–2068 (2019)

Surface Molecular Structure Modified Epoxy Resin Materials for HVDC GIL Spacer

Zhaoyu Ran, Boxue Du, Wenbo Zhu, and Jin Li

Abstract In this chapter, insulators with the surface functional graded material (σ-SFGM) was proposed, prepared by the surface molecular structure regulation. Based on the simulation results, it can be seen that the electric field at the triple junction was highly decreased. In addition, the electrical resistance properties under the direct current (DC) condition were improved. For the insulator with a gradient of the surface conductivity, the flashover voltage was improved obviously by 36.3%, compared to that of the traditional ones. Then, a novel procedure was conducted to fabricate continuous surface FGM (SFGM), regulating fluorination degrees from the upper to the bottom of the insulator surface by a spatial temperature gradient. The parameter of the continuous SFGM is optimized considering the field non-uniformity and the leakage current of the insulator. Flashover characteristics of the insulator were analyzed under DC voltage, DC voltage with preload and polarity reversal voltage. The promotion effect of the continuous SFGM was proved and the mechanism were based on the trap distribution and carrier transport process.

1 Introduction

In recent years, gaseous insulated pipelines (GILs) have been applied in many power transmission projects with the continuous increasing need of electrical power all over the world. The development of GILs plays a significant role especially for the special occasions, because of their various superiors of large capacity, flexible laying mode, long distance and environmental friendliness compared with the traditional transmission modes [1–4].

In the system of GIL, insulators casted by epoxy resin composite are essential for the support and insulation. The surface of the insulator, as the gas/solid interface, is the weak point which attracts the attention of researchers. Electric field distortion along the insulator surface, especially at the triple junction of the gas, insulator and high voltage electrode, may increase the rate of flashover faults

Z. Ran (✉) · B. Du · W. Zhu · J. Li
Tianjin University, Tianjin, China
e-mail: ranzhaoyu@tju.edu.cn

B. Du (ed.), *Polymer Insulation Applied for HVDC Transmission*,
https://doi.org/10.1007/978-981-15-9731-2_18

and even cause the paralysis of the entire system. Direct current (DC) GILs bear a long-term unipolar voltage, the electric field distribution is affected by the charge accumulation and the insulator structure, which will be more serious compared with alternating current (AC) condition[5–9]. Researches shows the flashover voltage of the basin insulator reduced by up to 22.8% under the effect of the electric field unevenness, which can be attributed to the hetero-charge accumulation [10]**. What's more, the insulators are confronted with many problems, such as the phenomenon of polarity reversal. Thus, the existing AC GIL equipment cannot be directly installed and applied in the DC environments** [11]**.**

Aiming to increase the flashover properties, reduce the device size and save the costs of the DC GIL, the electric field along the insulator surface needs to be regulated and optimized. The effective control of the electric field is of great significance to the development of GIL transmission in the DC system. Traditional methods, such as electrode adjustment, have gradually lost the attention of scholars due to their limited effects and complicated procedures [12, 13]. Instead, functionally graded material (FGM) began to be applied to insulator body modification. FGM is a heterogeneous composite material that controls the spatial distribution of microscopic variables such as its composition or structure, thereby making the material function change continuously (or quasi-continuously) in space. FGM has some history in electric field regulation and control, and has shown great development potential and research prospects in gas/solid composite insulation systems [14–21].

Most studies have focused on the preparation process and electric field distribution simulation of different insulators with FGM. The lamination process was first used to prepare FGM columnar insulators to make them have a gradient distribution of conductivity or dielectric constant [22]. Researchers have centrifuged an insulation structure with a gradient of dielectric constant. In the SF_6 gas environment of 0.1–0.4 MPa, the lightning impulse flashover properties significantly increased [16]. The concept of flexible mixed casting is put forward to prepare the material with a dielectric gradient, by which the simulation results show that the electric field on the surface of the insulator can be relaxed and regulated. The improvement of the initial discharge voltage is up to about 35% [23]. The researchers used the Fused Deposition Modeling (FDM) process to perform 3D-printed FGM insulators. The simulation results showed that the insulator with a gradient of conductivity, the electric field distortion will be greatly reduced at the three junctions; in addition, the team also proposed to control the mixing ratio of two resin systems with different dielectric constants through a mixing extrusion device. Then the photo-curing three-dimensional molding technology was used to accumulate from point to layer finally an FGM structure was formed. The simulation results showed that the insulator structure with conductivity gradient distribution could greatly reduce the electric field concentration at the triple junction [24, 25].

Based on the above, although FGM is designed and manufactured through various manufacturing processes such as centrifugation, multilayer coating and many other methods hoping to improve electrical resistance properties of the insulator. There exist some questions that cannot be solved: 1. The above FGM insulators require a brand-new production line in actual industrial production and face the problems of complex process and limited preparation materials; 2. High dielectric constant filler in the insulator will cause higher leakage current and not Evenly distributed fillers may cause degradation of mechanical properties; 3. The range of increasing the dielectric constant through particle doping is limited, which will limit the effect of FGM insulator electric field regulation.

Unfortunately, the above methods are mainly obtained through most of the data in the form of simulation. The voltage applied in the flashover test for verification is AC voltage, impulse voltage or lightning impulse. At the same time, little attention has been paid to the flashover characteristics under DC condition. Intuitively speaking, the actual operation condition is much more complicated. The research on DC Gil is far from enough. Therefore, it is necessary to simplify the manufacturing of conductivity FGM insulator. A new conductivity FGM is urgently needed in the DC GIL system.

It is worth noting that the conductance partial pressure under a DC electric field, at the same time, the electric field distribution has a great correlation with the material surface. It is a practicable method to optimize the electric field by adjusting the physical and chemical properties and electrical parameters of the material surface, because compared with the bulk doping method, it can significantly reduce the conduction loss and keep the existing insulator shape, manufacturing process, and casting production line all unchanged. The surface treatment method has relatively mature research basis and simple program, which provides a new idea for the research on σ-FGMs [26–28].

The results show that Fluorine atoms has special characteristics, such as the small atomic radius, high electronegative, high bond energy, which result in strong ability to bind electrons, thermal oxidation stability and special surface properties. The surface fluorination of insulating materials can effectively adjust the surface conductivity and improve the interface performance without huge cost and complex treatment process. Based on our previous studies, the introduction of fluorine into the surface of materials and the regulation of electrical properties by surface molecular structure control technology can be used to solve the problem of electric field distortion and help to increase the voltage of electrical resistance of the insulator [29–33]. Therefore, the concept of insulator with surface FGM (SFGM) with a conductivity gradient through surface fluorination technology is proposed.

In this chapter, a mathematical model was established to study the influences of conductivity SFGM design on DC insulator. In addition, fluorinated samples were prepared to investigate the microstructure and measure the surface conductivity of the epoxy. The electric field distributions of uniform and gradual conductivity along the surface of the insulator are simulated and compared with the theoretical calculations. The effectiveness of conductivity SFGM is studied by flashover test under DC condition. Furthermore, the continuous conductivity SFGM is applied to the cone-type insulator by the surface molecular regulation based on the temperature-control technology. The surface conductivities of the treated epoxy plate samples were measured. The nonuniformity factor and leakage current of different SFGM were compared and the parameters of SFGM were optimized. The role of the SFGM in the regulation of electric field is analyzed by simulation. Flashover test was conducted under different

DC applied voltages. Moreover, the polarity reversal phenomenon was considered. In order to further understand the mechanism of flashover process, the trap distribution, carrier mobility and charge transfer were focused on, based on which, the flashover model was constructed for insulator with continuous SFGM. It can be concluded that SFGM is of great value for improving flashover characteristics and can be referred to in the application of DC GIL insulator.

2 Insulator with Graded Conductivity SFGM

2.1 *The Concept of Novel Insulator with SFGM*

Figure 1 shows the original insulator equivalent circuit model, which consists of epoxy resin and alumina and is divided into n parts. When n tends to infinity ($n \to \infty$), each part is approximately equal to a small cylinder. The resistance of cylinder i can be calculated as below based on the law of resistance.

$$R_i = \rho \frac{h}{\pi r_i^2} \tag{1}$$

where ρ is the resistivity of insulator in Ω·m, h is the height of the small cylinder in m, and r_i is the upper radius of the small cylinder ($n \to \infty$) in m. Under applied DC voltage, the numerous small cylinders can be considered to be connected in series. The voltage and electric field of each cylinder can be obtained by calculation.

$$U_i = I R_i \tag{2}$$

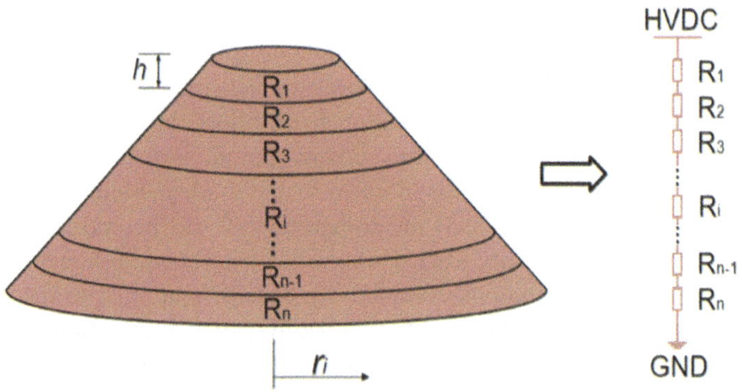

Fig. 1 Equivalent circuit model of the original insulator. Copyright Clearance Center's RightsLink® License Number: 4854530389301

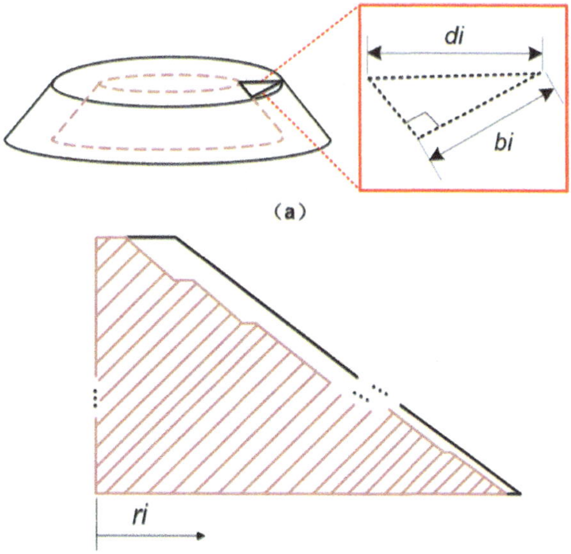

Fig. 2 **a** Example of fluorinated layer; **b** fracture diagram of graded fluorination. Copyright Clearance Center's RightsLink® License Number: 4854530389301.

$$E_i = \frac{U_i}{h} = \frac{I}{\pi} \frac{\rho}{r_i^2} \tag{3}$$

where U_i is the voltage that the cylinder i withstands, V. E_i is the electric field that the cylinder i withstands, V/m. I is the steady current in the circuit, A. Based on the above calculation process, the electric field near the high voltage electrode is significantly high due to the small value of radius.

After fluorination treatment, a certain thickness of fluoride layer is formed on the side of the small cone, which is far smaller than the radius of the upper surface. As Fig. 2a shows, the thickness of fluoride layer is defined as b_i, which is perpendicular to the tangent direction of insulator surface. It can be seen clearly that $b_i < d_i$. However, when such cone is small enough, the thickness of fluoride layer can be close to the horizontal distance on the upper surface of the cone. Figure 2b shows the half section of the whole insulator. The small cones with layers of different thicknesses constitute the insulator with conductivity SFGM. Insulator has a thin fluorinated surface layers changing from the upper surface to the lower surface, which can be realized later by fluorination technology.

The equivalent circuit model of insulator with SFGM by gradual fluoride layer is shown in Fig. 3. The division mode of insulator is the same as the above. The insulator consists of numerous small cone, each of which consists of a fluoride layer and an original interior.

$$R_{fi} = \rho_{fi} \frac{h}{2\pi r_i' d_i} \tag{4}$$

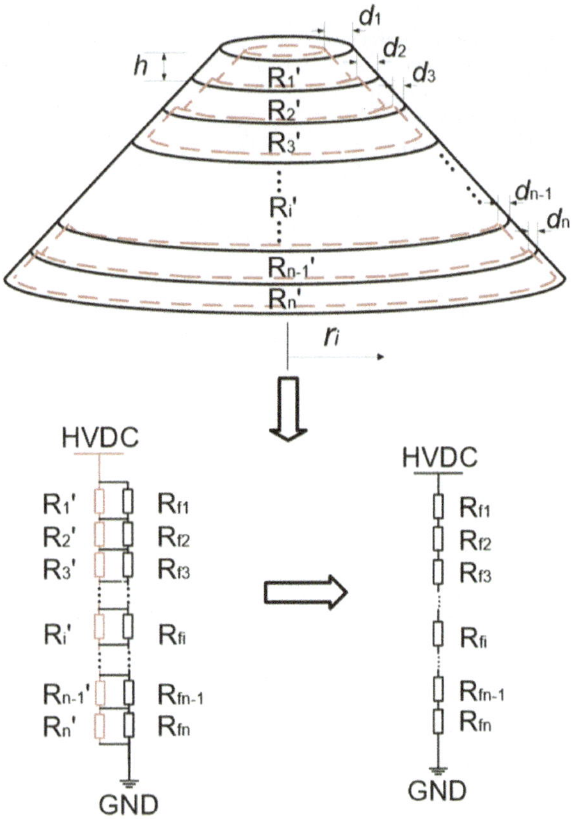

Fig. 3 Equivalent-circuit models for the insulator with SFGM. Copyright Clearance Center's RightsLink® License Number: 4854530389301

$$R_i' = \rho \frac{h}{\pi r_i'^2} \tag{5}$$

where R_{fi} is the resistance of the fluoride layer of cylinder i, Ω. R_i' is the resistance of theinterior of cylinder i, Ω. ρ_{fi} is the resistivity of the fluoride layer, whose value depends on conditions of the fluorination process, Ω·m. r_i' is the upper radius of the original interior of cylinder i, m.

The equivalent resistance of each cylinder is the parallel connection of R_{fi} and internal R_i', then such countless equivalent resistors are connected in series to form the final equivalent circuit. For the reason that R_{fi} is much smaller than R_i', the parallel units can be simplified to R_{fi} during the calculation. Then the voltage and electric field can be expressed as:

R_{fi} and R_i' are connected in parallel to form a cylinder, and then countless cylinders are connected in series to form the equivalent circuit of the insulator with SFGM. The parallel element can be simplified to R_{fi} in the calculation process, because the value of R_i' is far bigger than the value of R_{fi}. The voltage and electric field are calculated as follows:

$$U_i = IR_{fi} \tag{6}$$

$$E_i^{'} = \frac{I}{2\pi d_i} \frac{\rho_{fi}}{r_i^{'}} \tag{7}$$

It can be concluded that the electric field distribution depends on the resistance parameter and the size value. There is an inversely proportional relationship between the electric field and radius r_i. By comparing the results of original insulator and the insulator with SFGM, the reduction effect of the SFGM, especially at the triple junction with a small radius, can be revealed. And the potential of electric field homogenization by SFGM through fluorination is verified.

2.2 *Fabrication of Insulator with Conductivity SFGM*

2.2.1 Fabrication Process

In this chapter, the conical insulator is cast by Shandong taikai high voltage switch Co., Ltd. according to the weight ratio of Al_2O_3, epoxy resin and curing agent 300:100:33. In addition, 500 μm thick plate samples with the same composition mixture were prepared in the laboratory for the measurement of dielectric parameters and the comparative observation of micro morphology. The specific steps of fluorination are as below.

(1) Before fluorination, wipe the surface of epoxy resin sample and conical insulator with ethanol, and dry in a vacuum oven at 80 °C for 1 h.
(2) After more than five times of evacuation and purification with nitrogen, the purpose of cleaning the sealed stainless steel reactor had been achieved. The F_2/N_2 mixture with 20 vol% F_2 was input through the gas pipeline, and the pressure was 0.05 MPa.
(3) Fluorination treatment was carried out at a constant temperature of 25 °C until the set time was reached. After fluorination treatment, residual fluorine-containing gas was absorbed by tail gas treatment device [34].
(4) In order to make different regions of insulator surface have different fluorination treatment time, gradient fluorination treatment was carried out in different regions. Finally the insulator with conductivity SFGM was obtained.

After 15, 30, 45 and 60 min treatment, the insulators are named respectively as F-15, F-30 and F-45, while the insulators with SFGM of three and four gradients are F-30-15-0 and F-45-30-15-0. Figure 4 shows the treatment of the insulator.

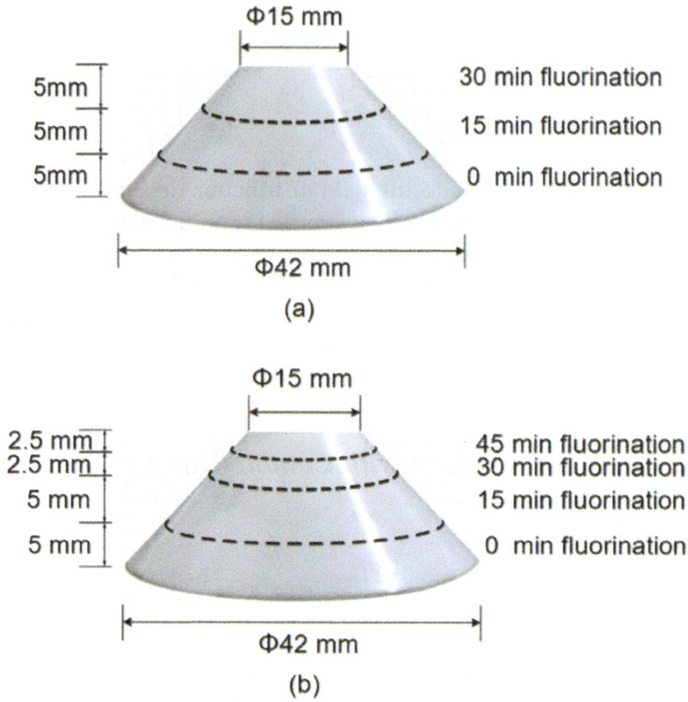

Fig. 4 Insulators with SFGM. **a** F-30-15-0, and **b** F-45-30-15-0. Copyright Clearance Center's RightsLink® License Number: 4854530389301

2.2.2 Micro Morphology of Epoxy

SEM (scanning electron microscope, Hitachi s4800) was used to observe and characterize the microstructure and morphology of fluorinated epoxy composite. Figure 5 shows the cross section of the samples treated with different fluorination. Fluorination causes the change of the physical and chemical composition of epoxy/alumina composite, and the surface layer is obviously different from the internal structure, which directly shows the function of fluorination. The corresponding relationship between the thickness of the fluoride layer and the fluorination treatment time is shown in Fig. 6. The longer the fluorination time is, the deeper the reaction degree is in the insulator surface layer, so the thickness of the surface layer gradually becomes thicker, from 0 to hundreds of nanometers.

2.2.3 Surface Conductivity Test

In this chapter, the surface conductivity of the epoxy plate samples with different fluorination time were measured by the three-electrode system at 25 °C. The ambient humidity is below 30% to prevent the effects of moisture. The surface conductivity

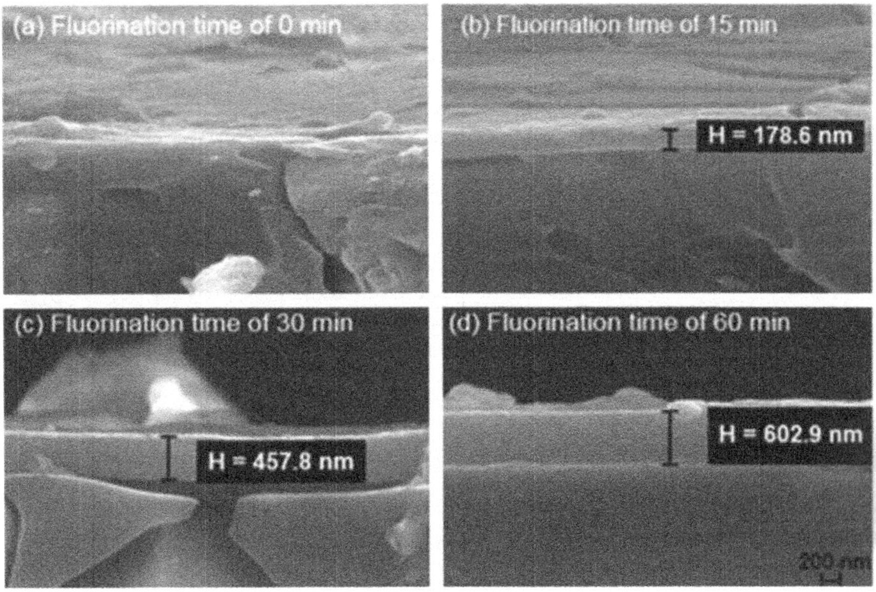

Fig. 5 Cross-section of original samples and fluorinated samples with different fluorination time. Copyright Clearance Center's RightsLink® License Number: 4854530389301

Fig. 6 Relation between the thickness of fluorination layer and the fluorination time. Copyright Clearance Center's RightsLink® License Number: 4854530389301

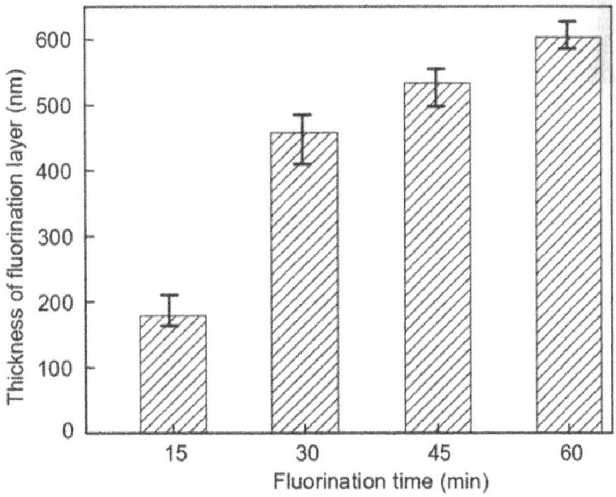

can be calculated from the following formula,

$$\sigma_s = \frac{ln\frac{D_2}{D_1}}{2\pi R_s} \tag{8}$$

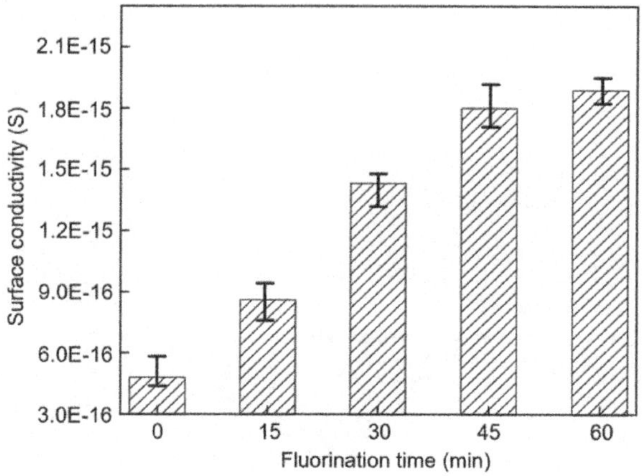

Fig. 7 Relation between the surface conductivity and the fluorination time. Copyright Clearance Center's RightsLink® License Number: 4854530389301

where D_1 is the diameter of the measuring electrode, D_2 is the inner diameter of the protective electrode, cm. R_s is the surface resistance in Ω.

The variation of surface conductivity of epoxy with fluorination time is shown in Fig. 7. After fluorination, the surface conductivity of the epoxy composite sample is significantly higher than that of the pure sample, nearly several times. With the increase of fluorination time, it is worth noting that the surface conductivity of the sample is saturated after 45 min. Therefore, the surface conductivity of 60 min fluorinated sample is similar to that of 45 min fluorinated sample. Considering economy and conductance loss comprehensively, 60 min fluorinated sample is no longer the research object in the follow-up study. The processing range from 0 to 45 min is used to construct the conductivity gradient of subsequent SFGM.

2.3 *Electrical Field Distribution Based on the Simulation*

COMSOL Multiphysics software was used to model insulator and simulate electric field distribution. In the simulation, the finite element method is used to study the influence of SFGM on the electric field distribution in GIL under DC environment. Figure 8 shows the simulation model in the rotational symmetry model. The physical dimensions of the insulator and electrode are expressed in millimeter scale coordinates. The applied voltage on HV electrode is negative high voltage of 15 kV, and the voltage of grounding electrode is 0 kV. A spherical region with an external radius of 1 m of the insulator is set as an air region with no electric field line through the boundary. The following is the governing equation for steady-state electric field analysis. The inputs of the fluoride layer and surface conductivity are based on the tested data discussed before.

$$\nabla \cdot J = 0 \tag{9}$$

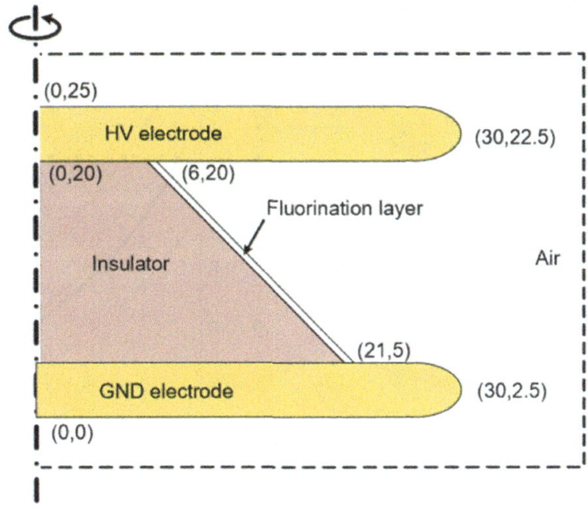

Fig. 8 Simulation model for the E-field calculation. Copyright Clearance Center's RightsLink® License Number: 4854530389301

$$J = E/\rho \tag{10}$$

$$E = -\nabla V \tag{11}$$

where J is the current density, A/m^2; E is the E-field, V/m; ρ is the volume resistivity of the involved dielectrics, Ω·m; V is the electric potential, V.

The surface electric field distribution of uniform fluorinated insulator with different fluorination time is shown in Fig. 9. For the original insulator, the electric field distortion is very serious at the triple junction. The electric field intensity

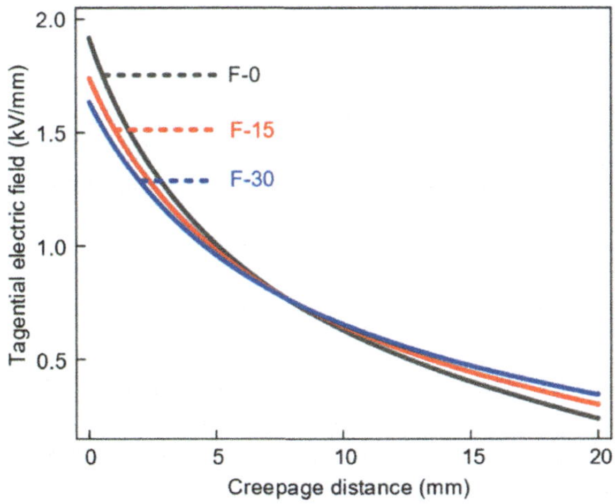

Fig. 9 Electric field distribution along the original and the homogeneous fluorinated insulators. Copyright Clearance Center's RightsLink® License Number: 4854530389301

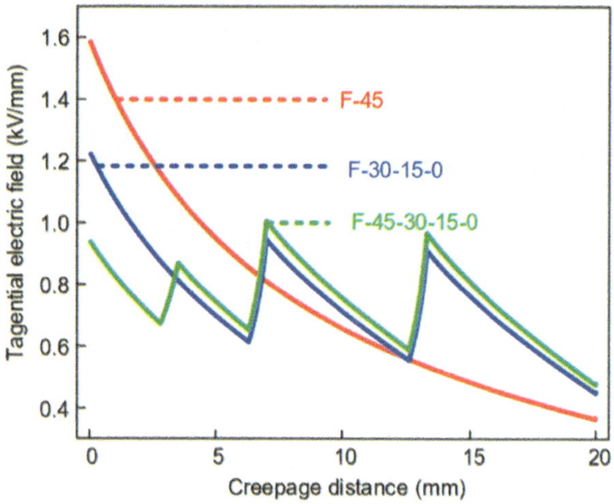

Fig. 10 Electric field distribution along the homogeneous and the insulators with SFGM. Copyright Clearance Center's RightsLink® License Number: 4854530389301

near the earth electrode is relatively small. Compared with the fluorinated insulator, it can be seen that the electric field along the insulator surface is obviously reduced due to the increase of surface conductivity.

In order to analyze the function of SFGM, Fig. 10 compares the electric field distribution of uniform fluorinated insulator and insulator containing SFGM. The overall electric field distribution of F-45 insulator is close to F-30. It can be seen that different degrees of uniform fluorination have similar mitigation effect on electric field. However, at the triple junction of F-30-15-0 and F-45-30-15-0 insulator, the electric field distortion is obviously further restrained. This can be attributed to the close relationship between the potential drop of different parts of insulator and its DC conductivity.

However, three or four layer fluorinated insulators have the same problem. At the boundary transition of gradient fluorination treatment, the electric field intensity will change sharply, which is due to the existence of interface due to the sudden change of thickness of fluorination layer. Compared with F-30-15-0, F-45-30-15-15 has a similar mutation, but the whole trend of electric field distribution becomes smoother. It is feasible to realize electric field homogenization by further increasing the number of gradients. Therefore, it can be reasonably inferred that surface design has a broad prospect in the regulation of insulator electric field.

2.4 *Improved Flashover Voltage*

The DC flashover test circuit of insulator is shown in Fig. 11. In this paper, the flat electrode is used in a sealed container for DC pressure operation, with a constant temperature of 25 °C and a constant humidity of 30%.

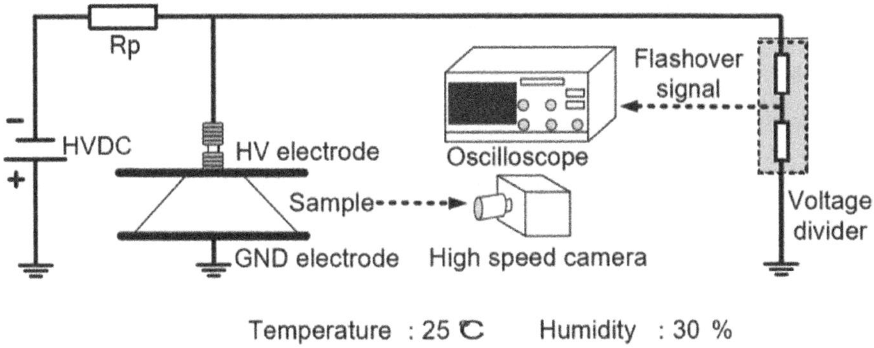

Fig. 11 Schematic diagram of the experiment setup for DC flashover test. Rp-protection resister 1 MΩ. Copyright Clearance Center's RightsLink® License Number: 4854530389301

The method of increasing the applied DC voltage and the voltage waveform when flashover occurs are shown in Fig. 12. Figure 12a shows that a fixed value voltage −15 kV lower than the actual flashover voltage is preloaded first, and the application time is selected as 20 min, in order to cause charge accumulation under DC voltage and obtain relatively stable electric field distribution. Then, gradually increase the DC voltage value at the rate of 1 kV/min until the insulator flashover occurs along the surface. Read the flashover instantaneous value by observing the waveform surge of the resistance voltage divider with an oscilloscope [35]. Observing the voltage waveform when flashover occurs, we can see that the flashover is accompanied by the sharp change of voltage. Repeat the test 15 times for each insulator according to the same operation to eliminate the influence of environmental factors. Before each test, the insulator shall be wiped with ethanol and dried for more than 3 h to avoid the effect of residual surface charge.

Figure 13 is a flashover path along the insulator surface captured by a high-speed camera. It is observed that the path starts from the HV electrode at the triple junction then gradually grows until reaches the ground electrode, which further confirms that the electric field distortion is the most serious at the triple junction and may trigger the flashover faults.

The flashover experiment results show the similar rule under the positive and negative DC voltage, so this paper omits the case under the positive voltage and only discusses the experimental data under the negative voltage. Weibull diagram is used to analyze flashover voltage data, showing flashover voltage and probability distribution of different insulators [4]. Figure 14 shows the effect of fluorination time on flashover voltage of insulator. As shown in the figure, the horizontal dotted line is used to represent the 63.2% confidence probability, where the corresponding flashover voltage is defined as the scale parameter to compare the performance of different insulators. There is a positive correlation between flashover voltage and fluorination treatment time. The flashover voltage of fluorinated insulator is higher than that of original insulator. At the same time, with the increase of time, the data of flashover

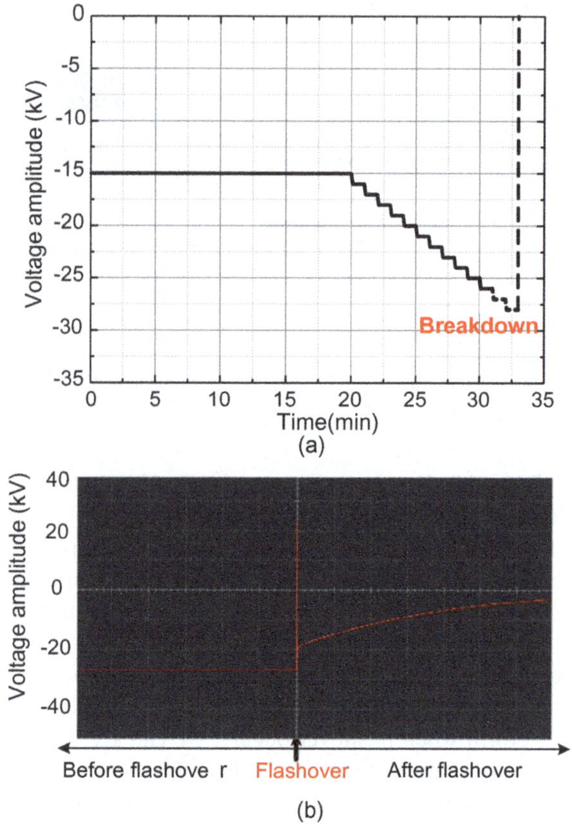

Fig. 12 Voltage during experiment **a** applied test voltage, **b** voltage waveform during flashover. Copyright Clearance Center's RightsLink® License Number: 4854530389301

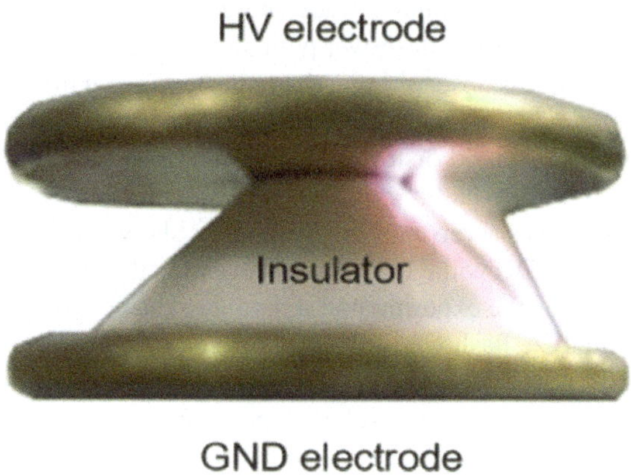

Fig. 13 Flashover path along the surface of the insulator. Copyright Clearance Center's RightsLink® License Number: 4854530389301

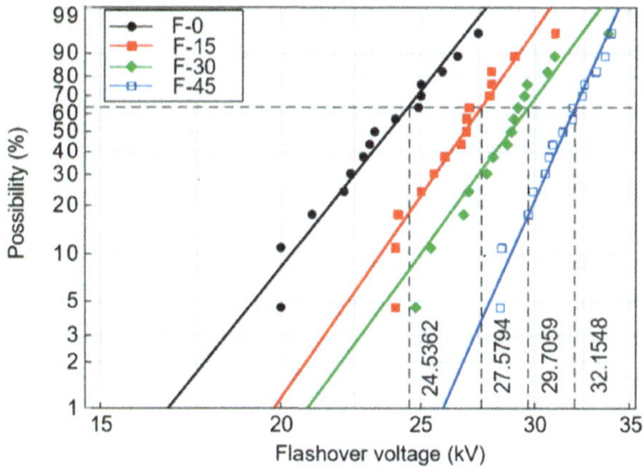

Fig. 14 Flashover voltages of the original and the homogeneous fluorinated insulators. Copyright Clearance Center's RightsLink® License Number: 4854530389301

voltage is more centralized, which shows that the stability of insulator voltage withstand performance has been improved to some extent. Figure 15 shows the flashover voltage distribution of insulators with different SFGM designs. Compared with the original insulator F-0, the flashover voltage of F-30-15-0 is increased by 26.0%, and the flashover voltage of F-45-30-15-0 is increased by 36.3%, which shows that the performance of SFGM insulator is superior, and the gradient fluorinated insulator is better than the uniform fluorinated insulator.

In addition, F-45-30-15-0 has higher surface conductivity and more uniform electric field distribution, so its flashover voltage is higher than F-30-15-0, indicating that the optimal design is multi-level gradient distribution. Such experimental results can be correlated with the previous theoretical analysis based on the circuit model.

Through theoretical analysis, simulation and experimental verification, conductivity SFGM can be helpful for electric field regulation. The electric field distortion at the triple junction is an important factor leading to the flashover of insulator surface.

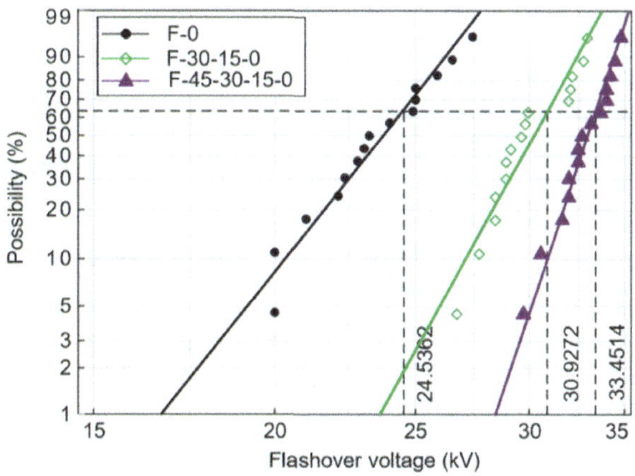

Fig. 15 Flashover voltages of the homogeneous and the insulators with SFGM. Copyright Clearance Center's RightsLink® License Number: 4854530389301

In the initial stage of surface flashover, micro discharge may occur in the triple junction, or in the vicinity of defects on insulator surface and suspended metal particles. Due to the discharge, the electric field is distorted, which leads to electron emission and ionization. Both the original point and the newly generated charge may be trapped in the trap of the insulator, which increases part of the electric field. After that, further ionization leads to electron avalanche and finally the formation of the conductive channel, which is the flashover path.

For the insulator treated by surface fluorination, the surface conductivity increases obviously and the surface state becomes better, which promotes the charge collapse and dissipation, which can restrain the initial charge causing flashover. The electric field homogenization based on the optimization of surface conductivity distribution can suppress charge injection and gas ionization. In the flashover process, the charge accumulation is less and the electric field distribution is uniform, the development of micro discharge will be hindered, and the flashover characteristics along the insulator surface will be effectively improved.

3 Insulator with Continuous Conductivity SFGM

3.1 SFGM Fabrication by Temperature-Controlled Fluorination

3.1.1 Surface Fluorination Method Based on Temperature Control

The concept diagram of continuous SFGM manufactured by temperature controlled fluorination technology is shown in Fig. 16, which is used for surface modification of cone type insulator. Different from the above, the temperature control system is used to construct the gradient distribution of temperature from high to low on the surface of insulator. The fluorination reaction is stronger in the part with high

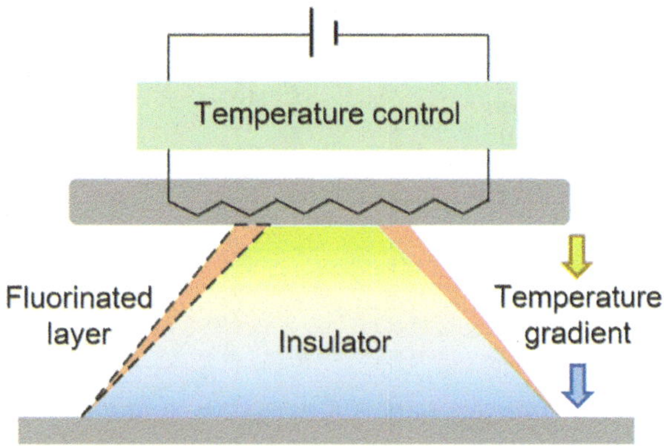

Fig. 16 Diagram of SFGM fabrication by fluorination process. Copyright Clearance Center's RightsLink® License Number: 4941331285590

local temperature. Due to the temperature difference between the upper part and the lower part of insulator surface, different degrees of fluorination reaction can be formed in different parts, and the difference transition is continuous. The steps of the preparation process can be summarized as follows.

(1) From a safety point of view, the temperature distribution in the air is now determined. Move the probe to measure the surface temperature of insulator in the air. When the insulator forms a relatively stable temperature gradient under the action of temperature control device, the air is completely discharged to start the fluorination operation.
(2) Use nitrogen to remove existing air and moisture. Then open the gas valve and introduce the fluorine-containing gas required for fluorination operation, which has been described in detail above.
(3) After a certain time of treatment, a continuous fluorinated surface layer is formed on the insulator surface. Finally, the residual gas is treated.

3.1.2 Basic Parameters Test of Epoxy

The size of the conical insulator studied in this chapter is 15 mm high, the upper radius is 5 mm, and the lower radius is 20 mm. The size of the insulator in the simulation model is consistent with the size of the sample. At the same time, the epoxy composite sheet samples with the same formula as the insulator were prepared. In order to manufacture insulator with continuous conductivity SFGM, the parameters of SFGM and its corresponding surface treatment conditions, such as temperature and time, need to be determined first. After fluorination treatment at different temperature and time, the microstructure of epoxy composite was observed by SEM. Figure 17 shows the structure of the sample surface layer.

According to previous studies, there is a close relationship among surface conductivity, charge accumulation and flashover [35, 36]. The influence of fluorination conditions is reflected by the measurement of the surface conductivity of epoxy composite samples. The operation and requirements of the experiment are consistent with the above.

The relationship between the thickness of fluorinated layer and the surface conductivity of epoxy samples with different fluorination treatments is shown in Fig. 18. As can be seen from any curve in Fig. 18, the distribution of four points on each curve represents the surface conductivity of the sample at four different treatment temperatures, and the thickness of the fluorinated layer increases with the increase of temperature. With the increase of treatment time or temperature, the surface conductivity of epoxy composite samples will be improved. Due to the deepening of fluorination reaction, the fluorinated layer on the surface will also become thicker. After fitting the values of the test points, the relationship between the surface conductivity and the thickness of the fluorinated layer was obtained. This is to construct the gradient distribution of surface conductivity based on the thickness gradient of fluorinated layer. However, for samples with 10 or 30 min of fluorination time, the curves of the cross section are not completely coincident. The same thickness of fluorinated

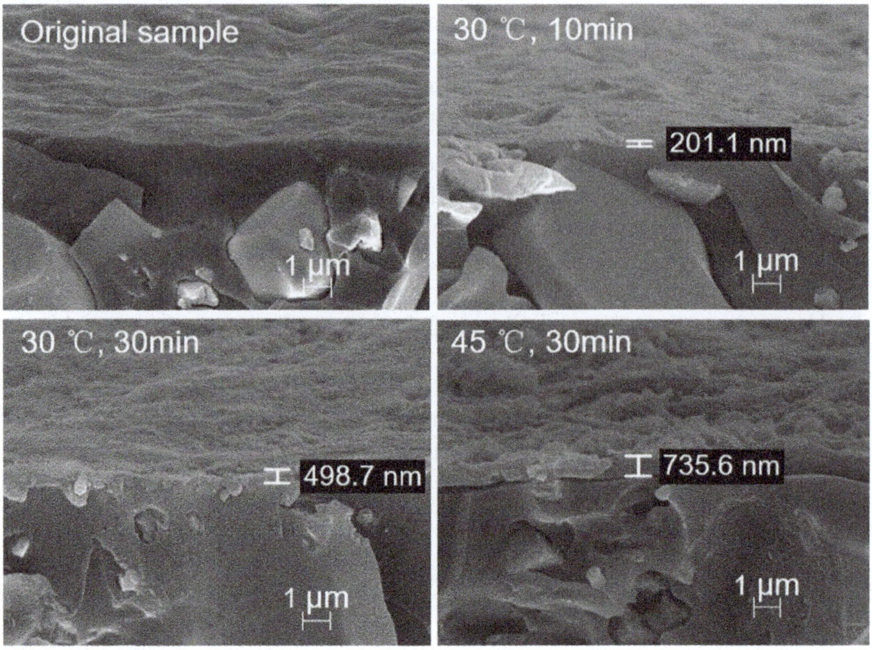

Fig. 17 Micromorphology of original and fluorinated epoxy samples. Copyright Clearance Center's RightsLink® License Number: 4941331285590

Fig. 18 Relationship between fluorinated layer thickness and surface conductivity of the epoxy samples. Fluorination time: 10 or 30 min. Fluorination temperature: 25–60 °C. Copyright Clearance Center's RightsLink® License Number: 4941331285590

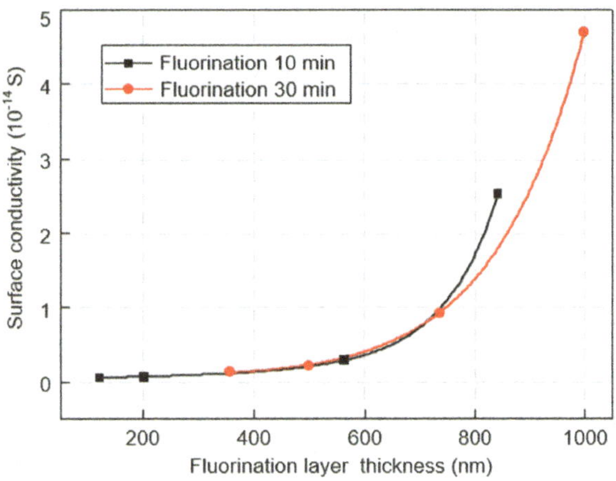

layer may correspond to different combinations of treatment time and temperature. The surface conductivity of different samples with the same thickness of fluorinated layer is not necessarily the same.

3.2 *Optimization of the Insulator with Continuous SFGM*

3.2.1 Insulator Model in Simulation

Based on the basic parameters, the finite element method in COMSOL is used in this chapter. The purpose of the simulation is to study the effect of dielectric parameters on the electric field distribution of insulator and to select the best parameters. The simulation model of insulator is shown in Fig. 19, and the details of boundary conditions and control equations of the model refer to the above. The specific input parameters are shown as Table 1. In this simulation, the fluorinated surface layer is defined as an electrical shielding layer with no thickness and only volume conductivity. Surface conductivity is the product of bulk conductivity and surface layer thickness.

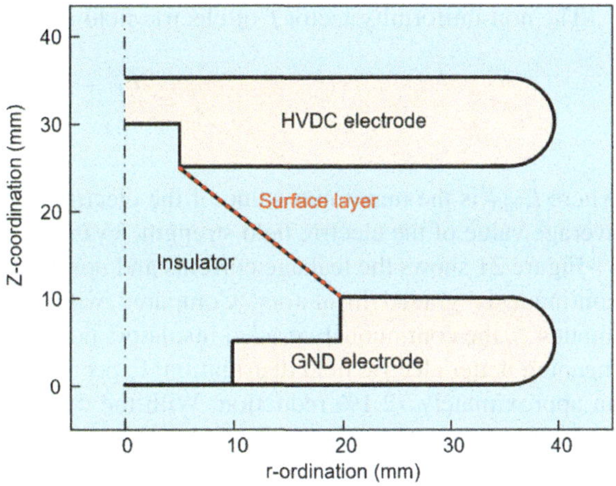

Fig. 19 Geometric structure of the insulator model. Copyright Clearance Center's RightsLink® License Number: 4941331285590

Table 1 Parameters used in the simulation

Parameter	Material	Value
Volume conductivity/S·m^{-1}	Spacer	1×10^{-15}
	Air	1×10^{-20}
	Fluorinated layer	σ_b
Relative permittivity	Spacer	6
	Air	1
	Fluorinated layer	7

3.2.2 Optimal Parameters of Continuous SFGM

According to the existing conditions and feasibility of the experiment, the temperature of the insulator is chosen between 30 and 60 °C. The corresponding range of surface conductivity is 7.12×10^{-16}–4.70×10^{-14} S. This paper chooses 5 types of insulators with continuously gradient surface conductivity aiming to obtain the optimal parameter for the insulator. The surface temperature of the insulator was measured under different heating conditions to obtain the corresponding surface conductivity and fluorinated layer thickness. Insulator 1 or 4 has a fluorination treatment time of 10 min while insulator 2, 3, or 5 has a fluorination treatment time of 30 min. By fitting the temperatures of five test points equidistant on the surface of the insulator, the relationship between the temperature and z-coordination is obtained. Combining with the conductivity-fluorinated layer thickness curves in Fig. 18, the surface conductivity distributions of insulators with different treatments are obtained, shown in Fig. 20.

The non-uniformity factor f of electric field is defined as,

$$f = \frac{E_{max}}{E_{av}} \tag{12}$$

where E_{max} is the maximum value of the electric field strength, kV/mm; E_{av} is the average value of the electric field strength, kV/mm.

Figure 21 shows the leakage currents and non-uniformity factors of five different continuously graded insulators. Compared with the original insulator with f of about 4.3, the continuously graded insulators possess lower non-uniformity factors, meaning better electric field distribution. Especially for insulator 4 or 5, f possesses an approximately 72.1% reduction. With the decrease of surface conductivity, the leakage current is decreased. The leakage currents of insulator 4 is 5.06×10^{-11} A,

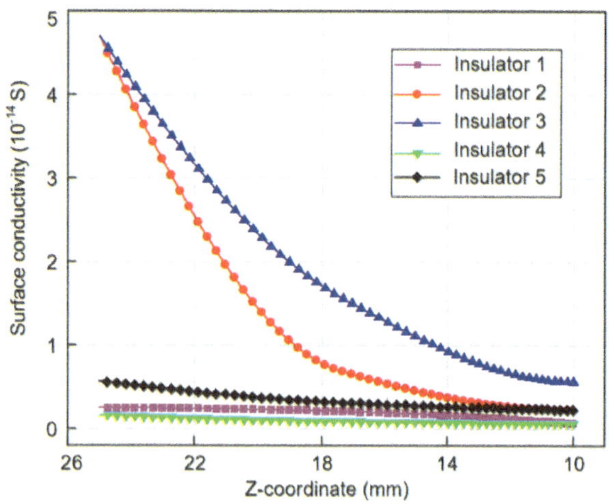

Fig. 20 Surface conductivity distributions of insulators with different treatments. Copyright Clearance Center's RightsLink® License Number: 4941331285590

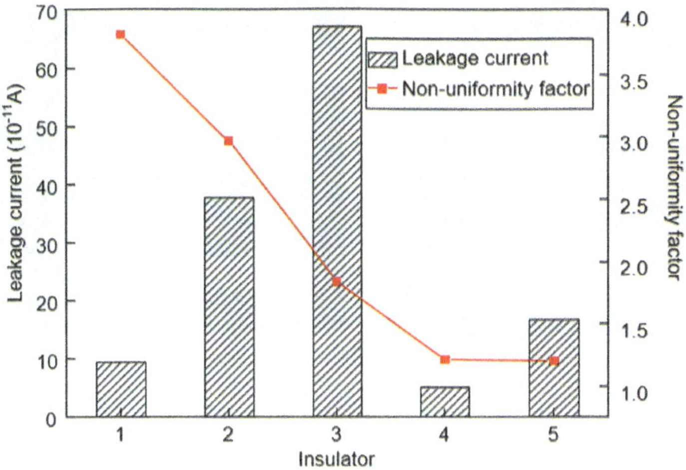

Fig. 21 Leakage currents and non-uniformity factors of insulators with different treatments. Copyright Clearance Center's RightsLink® License Number: 4941331285590

which is the lowest value among all the designs. According to the calculation of the electric field non-uniformity coefficient and the leakage current, insulator 4 possesses the most superior properties. The optimal parameters of SFGM are finally determined from the perspective of economy and ease of operation: 7.12×10^{-16}–1.66×10^{-15} S from the bottom to the upper of the insulator.

3.3 *Electric Field Calculation*

3.3.1 Steady State Field Distribution

To make further comparison in the ability for electric field regulation, the insulators with different fluorinated layers are shown in Table 2. Insulator E has the same treatment with insulator 4, which represents the optimal insulator with continuously SFGM. The results of stable state electrical simulation are shown in Fig. 22. As Fig. 22a shows, the electric field strength reaches its highest value near the tipple junction of the original insulator A, where is thus believed to be the start of the flashover. As for the insulator E with optimal SFGM, it can be known from Fig. 22b

Table 2 Insulators with different fluorinated layers

Insulator	A	B	C	D	E
Fluorinated layer	None	Uniform	Uniform	Discretely graded	Continuously graded

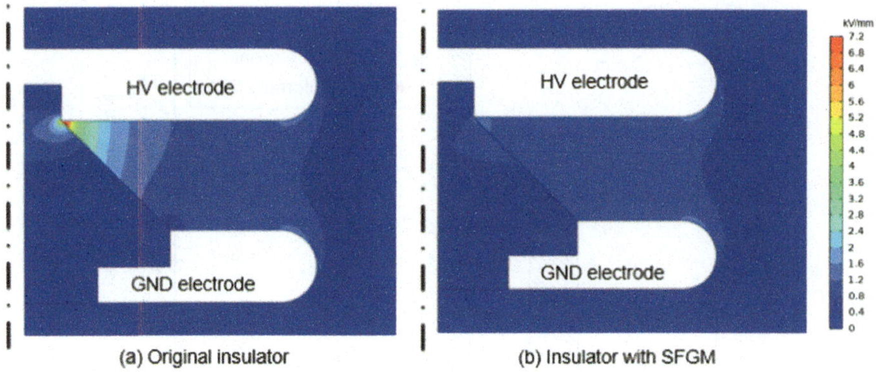

Fig. 22 Electric field distribution under DC stable state. Copyright Clearance Center's RightsLink® License Number: 4941331285590

that the serious electric field distortion is highly eased by the effect of the continuously gradient surface conductivity distribution.

To compare the effects of different types of fluorinated layer, the relationships with surface conductivity and z-coordination of different insulators are shown in Fig. 23a. Figure 23b shows the tangential electric field strength along the insulator B, C, D and E. The electric field gradually weakens from the high voltage electrode to the ground electrode. It is worth noting that the unevenness of insulator B and C is almost coincident, which shows the similar mitigation effect of the uniform fluorinated layer on the electric field distribution. The reason can be attributed to the huge gap between conductivity of the surface layer and internal of the insulator. As for insulator D, there are three distortions of electric field corresponding to the mutations of surface conductivity. The interface between fluorinated layers with different surface conductivity limits the effectiveness of the layer. Such problem of local electric field distortion at the demarcation of the insulator with discretely graded layer is solved in insulator E, which possesses continuous SFGM. The electric field around the high voltage electrode is definitely reduced and the distribution is more pronounced uniform, indicating the superior of the optimal design.

3.3.2 Transient Electric Field Distribution

The Transient electric field distribution is calculated under three types of DC electric field. The tangential electric field of insulator E with SFGM compared with the original insulator A is shown in Fig. 24 . The hollow curves in Fig. 24 shows the electric field under the DC voltage from 0 to 15 kV in 2.5 min under the constant increasing voltage. While the solid curves represent the situation with preload: firstly, the HV electrode is applied with the preload voltage of 10 kV for 20 min and then the voltage is raised by 1 kV in 6 s and is kept for 54 s, which is repeated until 15 kV is reached. Whether there is preload or not, the electric field near the HV electrode of insulator E

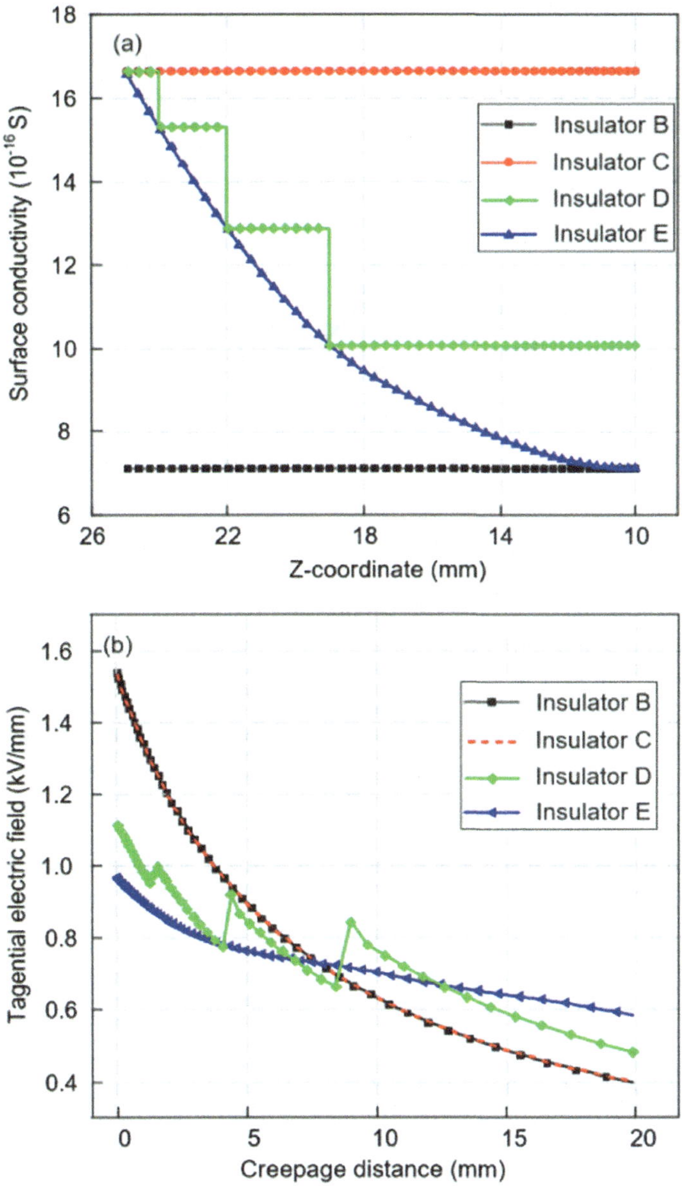

Fig. 23 **a** Relationship with surface conductivity and z-coordination of different insulators; **b** Tangential electric field strength along the creepage distance of the insulators. Copyright Clearance Center's RightsLink® License Number: 4941331285590

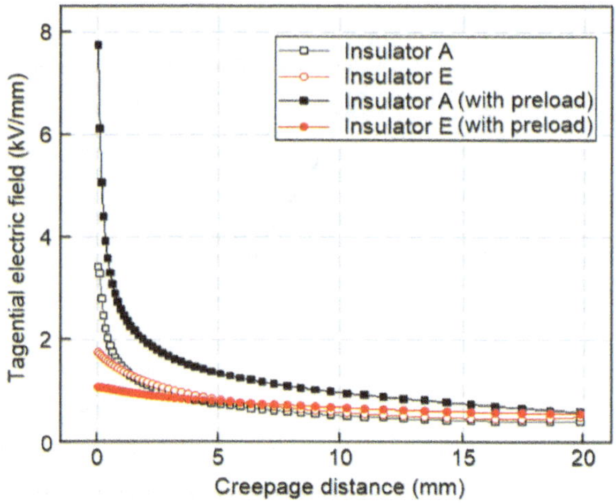

Fig. 24 Transient tangential electric field distribution along different insulators under DC voltage with or without preload. Copyright Clearance Center's RightsLink® License Number: 4941331285590

is greatly reduced. The electric distortion of the original insulator A becomes much more severe due to the accumulation of the surface insulator caused by the long-term application of the same polarity DC voltage to some extent. Nevertheless, it can be seen the law is different for insulator E. The highest electric field value under preload is slightly smaller than that without preload, which is owing to the enough time for SFGM to play its role in field regulation.

Figures 25 shows the variation of maximum electric field with time for different insulators under polarity reversal phenomenon. The voltage is applied in 5 stages: (1) from 0 to 15 kV in 1 min, (2) keep 15 kV for 18 min, (3) from 15 to −15 kV in 2 min, (4) keep −15 kV for 18 min, (5) from −15 to 0 kV in 1 min. It can be seen in Fig. 10 that for the insulator A, the change tendency of the maximum electric field strength with time is basically consistent with the waveform of the applied voltage.

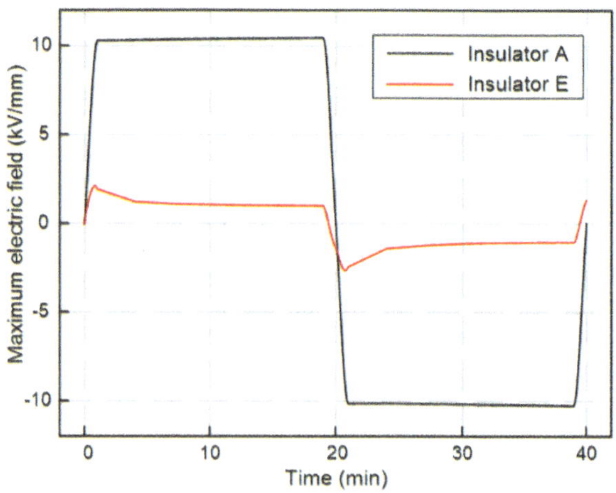

Fig. 25 Maximum electric field of different insulators under polarity reversal. Copyright Clearance Center's RightsLink® License Number: 4941331285590

The maximum electric field strength of insulator E is significantly smaller than that of insulator A by approximately 79.69%. It is worth noting that the zero crossing of the voltage is at 20 min and the moment when the polarity of the maximum field strength of the insulator E changes is at about 19.5 min. From 20 to 20.1 min, insulator A possesses the lower value of electric field strength. Until 20.2 min, the positive role of field regulation played by SFGM is once again revealed. Based on the above analysis, it takes approximately within 18 s for SFGM to overcome the dilemma of charge accumulation. The above three different electric field distributions correspond to different applied voltages of the insulators, which provides a valuable reference to the improvement of flashover voltage.

3.3.3 Improved DC Flashover Test

The flashover experimental system was built by the laboratory. The flashover characteristics were tested in a sealed box where the temperature was kept as 28 °C and the humidity remained constant at 35%. The resting time between two tests was more than 3 h to substantially eliminate the effect of residual surface charges. Based on our research, there were same trends in results under both negative and positive voltage, thus the case under negative voltage is discussed in this paper. To make a thorough inquiry for the effects of the SFGM, three voltage forms were chosen to be applied on the insulator, shown in Table 3.

As is shown in Table 3a, the applied DC voltage was increased directly until flashover happened when the voltage waveform changed abruptly. Each of the insulators was tested for 10 times and the finally the average value was obtained to represent the flashover property under transient DC voltage. Figure 26 shows the results of different insulators. For any insulator with fluorinated layer, the flashover voltage is obviously improved compared with that of the insulator A. Furthermore, the insulator E with SFGM possesses the highest flashover voltage, indicating the positive role played by the continuously graded fluorinated layer. The experimental data are dispersive, which may be attribute to the effect of the transient DC electric field.

To obtain the stable electric field distribution, the voltage was applied as Table 3b. Firstly, –10 kV was kept for 20 min to reach a comparatively steady state. Then the voltage was raised by –1 kV/min and when the flashover occurred, the voltage value was recorded. The results were analyzed and shown in Fig. 27. The average flashover voltage of insulator E is 17.71 kV, which is higher than that of insulator A by approximately 33.16%. In addition, the whole trend of the flashover voltage is slightly reduced in contrast to that under directly increased voltage, related to the more serious electric field distribution brought by charge accumulation.

In the real power transmission, the polarity reversal is inevitable which may trigger the surface flashover and greatly threatens the security and economy of the system. In this paper, such condition was taken into account. The voltage form is shown in Table 3c. Negative and positive voltages of the same value were applied alternately

Table 3 Different methods of increasing voltage

(a) Direct raising voltage	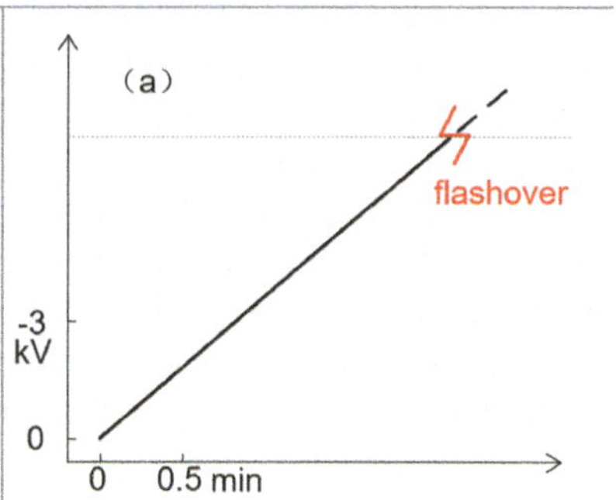
(b) Preload before raising voltage	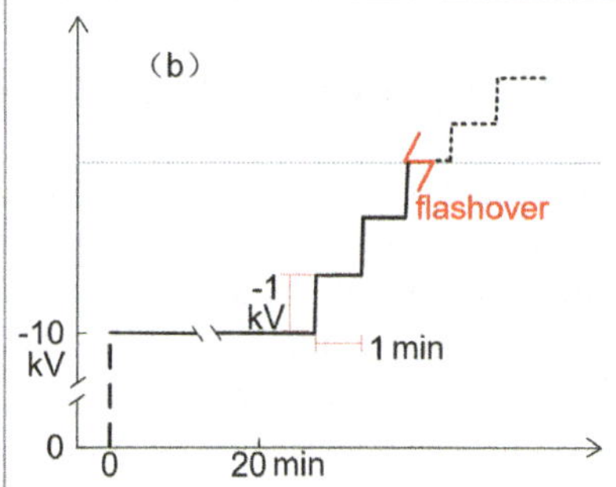
(c) Polarity reversal voltage	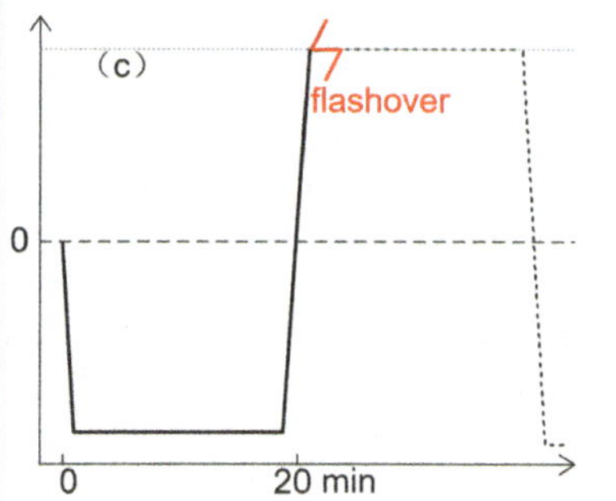

and if the insulator could bear the voltage, change the polarity before increasing the voltage by –1 kV. As shown in Fig. 28, Flashover mechanisms of the original insulator and the insulator with SFGM, the flashover voltages of insulator B and C are close. As for insulator E, the flashover property is slightly improved, meaning the consequent of the SFGM is not that obvious under such severe condition. The reverse voltage process should be paid on more attention in the subsequent studies.

3.3.4 Improved DC Flashover Test

By corona discharge at 8 kV for 3 min, the charges were injected into the epoxy samples with the same formulation and treatment as the corresponding insulator, and

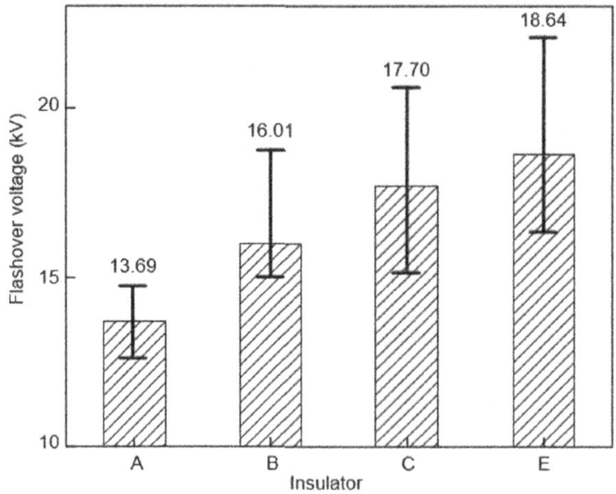

Fig. 26 Flashover voltage of insulator under directly raising voltage. Copyright Clearance Center's RightsLink® License Number: 4941331285590

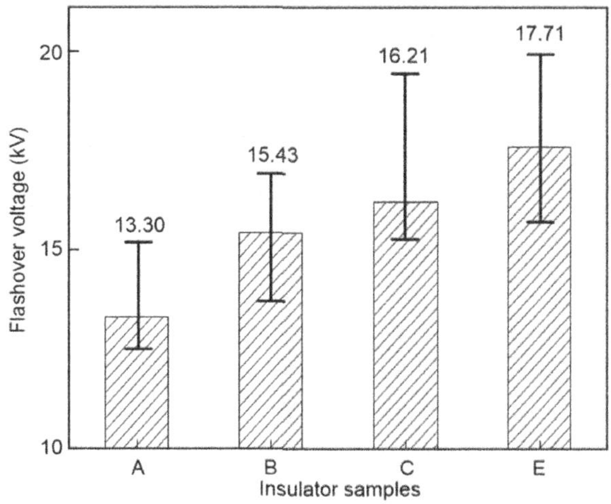

Fig. 27 Flashover voltage of insulator under raising voltage with preload. Copyright Clearance Center's RightsLink® License Number: 4941331285590

then the surface potential decay (SPD) was measured [25]. Table 4 shows the trap distribution and carrier mobility of different epoxy samples, which were calculated from the results of SPD. Original epoxy sample mainly contains deep traps and as for the fluorinated sample, there are obvious double peaks in the trap distribution. After the fluorination treatment, the carrier mobility is enhanced. The charges trapped by the shallow trap are more likely to escape and then transfer to the ground electrode more quickly.

Based on the above discussion, the differences between the flashover mechanism of the insulator A and E are shown in Fig. 29. For the original insulator, the initial charges induced by high field emission near the triple junction may collide with solid surface to excite secondary electrons. The charges move under the effect of the electric field, constantly leading to generate more electrons from the insulator and

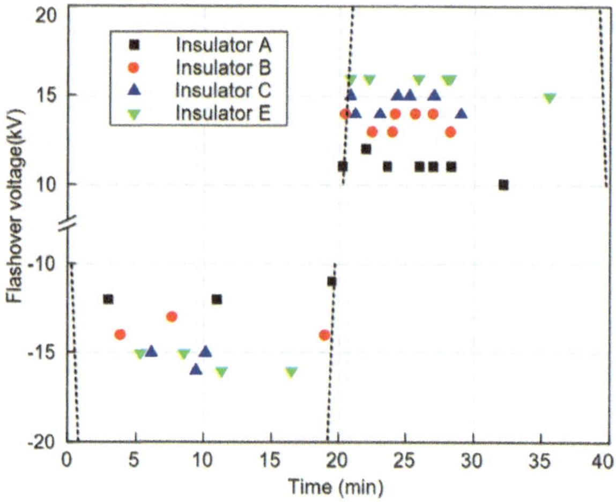

Fig. 28 Flashover voltage of insulator under polarity reversal voltage. Copyright Clearance Center's RightsLink® License Number: 4941331285590

Table 4 Trap distribution of different epoxy samples corresponding to different insulators

	Trap energy (eV)	Trap density ($\times 10^{21}$ eV^{-1} m^{-3})	Carrier Mobility ($10^{-13} m^2/Vs$)
Insulator A	0.89	4.3	5.7
Insulator B	0.81	0.6	8.9
	0.89	3.4	
Insulator C	0.80	1.0	11.8
	0.88	2.5	
Insulator E	0.80–0.81	0.6–1.0	8.9–11.8
	0.88–0.89	2.5–3.4	

ionize more positive charges and electrons from the air near the insulator. Thus, large amount of charges is caught in the deep traps in the epoxy insulator, which cannot move easily, further aggravating the unevenness of the electric field distribution. Electron collapse will happen and finally the path of surface flashover is formed.

After the fluorination treatment, the surface graded fluorinated layer is formed. Due to the high conductivity of the surface layer near the HV electrode, the electric field in the triple junction is highly homogenized and the seed charges that may trigger flashover are suppressed, which prevents the start of flashover to some extent. In addition, most of the charges pushed by the electric field are trapped in the shallow traps of the surface layer and will not enter the interior of the insulator, which significantly promotes the escape and transport of the charges and reduces the probability of impact ionization, which helps to hinder the formation of flashover paths (Fig. 29).

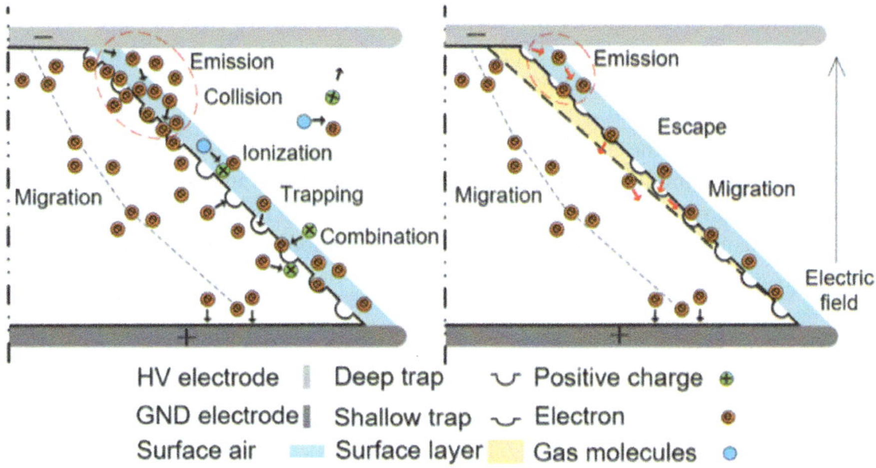

Fig. 29 Flashover mechanisms of the original insulator and the insulator with SFGM. Copyright Clearance Center's RightsLink® License Number: 4941331285590

4 Conclusion

In this chapter, an insulator surface modification method applied to DC Gil system is proposed. Through equivalent circuit theory analysis, electric field distribution simulation and experiment, it is proved that the insulator with SFGM has a broad application prospect. A temperature-controlled method is proposed to fabricate insulator with continuous SFGM and the effectiveness of the continuous SFGM is evaluated by flashover tests under different forms of DC voltages. Considering the complex manufacturing process and conduction loss of the existing modified methods, the new insulator with conductivity SFGM proposed in this chapter has great applicability and industrial advantages. The main conclusions are as follows:

(1) The equivalent circuit model of insulator shows that the electric field distortion at the triple junction can be greatly reduced by staged fluorination, which theoretically analyzes the effectiveness of SFGM in electric field regulation.
(2) After fluorination, a certain thickness of fluoride layer was formed on the surface of epoxy composite. With the increase of the treatment time or temperature, the conductivity increased obviously.
(3) Based on the conductivity measurement and the simulation calculation, the optimal design of SFGM is chosen as 7.12×10^{-16}–1.66×10^{-15} S from the bottom to the upper of the cone-type insulator.
(4) Both the graded and the continuous SFGM present an ability of electric field regulation under different DC voltage forms. The maximum electric field strength of insulator with continuous SFGM is reduced by approximately 79.69% under the polarity reversal voltage.

(5) The flashover voltage of the insulator is increased by over 33% under DC voltage, whether there is a preload or not. Besides, when polarity reversal happens, the continuous SFGM also plays a positive role in limiting the surface flashover.
(6) The trap level of the epoxy sample is lowered and the carrier mobility is increased after surface modification. For the gradient fluorinated insulator, there forms a gradient in surface conductivity and trap distribution from the top to the bottom of the insulator. Therefore, the charge accumulation and electric field distortion are restricted, which contributes to the hinder of the flashover process.

References

1. T. Magier, M. Tenzer, H. Koch, Direct current gas-insulated transmission lines. IEEE Trans. Power Del. **33**, 440–446 (2018)
2. H.Y. Zhou, G.M. Ma, C.R. Li, C. Shi, S.C. Qin, Impact of temperature on surface charges accumulation on insulators in SF6-filled DC-GIL. IEEE Trans. Dielectr. Electr. Insul. **24**, 601–610 (2017)
3. S. Kumara, B. Ma, Y.V. Serdyuk, S.M. Gubanski, Theoretical investigation of HVDC disc spacer charging in SF6 gas insulated systems. IEEE Trans. Dielectr. Electr. Insul. **19**, 2196–2205 (2012)
4. H.C. Liang et al., Mechanical stress distribution and risk assessment of 110 KV GIS insulator considering Al_2O_3 settlement. High Volt. **4**, 65–71 (2019)
5. B.X. Du, Q. Du, J. Li, H.C. Liang, Carrier mobility and trap distribution dependent flashover characteristics of epoxy resin. IET Gener. Trans. Distrib. **12**, 466–471 (2018)
6. J. Li et al., Surface functional graded spacer for compact HVDC gaseous insulated system. IEEE Trans. Electr. Insul. **26**, 664–667 (2019)
7. H.C. Liang, B.X. Du, J. Li, Mechanical stress distribution and risk assessment of 110 KV GIS insulator considering Al_2O_3 settlement. High Volt. 4, 1–7 (2019)
8. B.X. Du, H.C. Liang, J. Li, Z.Y. Ran, Electrical field distribution along SF6/N2 filled DC-GIS/GIL epoxy spacer. IEEE Trans. Dielectr. Electr. Insul. **14**, 1202–1210 (2018)
9. J. Li, H.C. Liang, B.X. Du, Progress in electrical field distribution along gas-solid interface in compact gas insulated system and its regulation methods. High Volt. Eng. **45**, 2619–2628 (2015)
10. B. Qi, C.J. Gao, L.J. Zhao, Influence of surface charge on flashover voltage of gas insulated substation basin insulator under AC and DC voltage. High Volt. Eng. **43**, 915–922 (2017). (in Chinese)
11. E. Volpov, HVDC gas insulated apparatus: electric field specificity and insulation design concept. IEEE Electr. Insul. Mag. **18**, 7–36 (2002)
12. M. Abdel-Salam, E.K. Stanek, Optimizing field stress on highvoltage insulators. IEEE Trans. Electr. Insul. **EI-22**, 47–56 (1987)
13. N.L. Allen, P.N. Mikropoulos, Profile effect on surface flashover in a uniform field. *IEEE 11th Symposium on High Voltage Engineering and Application (ISHVEA)*, (1999), pp. 216–219
14. K. Kato, M. Kurimoto, H. Shumiya, H. Adachi, S. Sakuma, H. Okubo, Application of functionally graded material for solid insulator in gaseous insulation system. IEEE Trans. Dielectr. Electr. Insul. **13**, 362–372 (2006)
15. N. Hayakawa, Y. Miyaji, H. Kojima, K. Kato, Electric field reduction by functionally graded materials (FGM) with permittivity and conductivity distribution. *Annual Report Conference on Electrical Insulation and Dielectric Phenomena. (CEIDP)*, (2012), pp. 32–35

16. N. Hayakawa et al., Fabrication and simulation of permittivity graded materials for electric field grading of gas insulated power apparatus. IEEE Trans. Dielectr. Electr. Insul. **23**, 547–554 (2016)
17. Z. Liu, W.D. Li, Y.B. Wang, G.Q. Su, G.J. Zhang, Y. Cao, D.C. Li, Topology optimization and 3D-printing fabrication feasibility of high voltage FGM insulator. *IEEE International Conference on High Voltage Engineering and Application (ICHVEA)*, (2016), pp. 1–4
18. J. Ishiguro, M. Kurimoto, H. Kojima, K. Kato, H. Okubo, N. Hayakawa, Electric field control in coaxial disk-type solid insulator by functionally graded materials (FGM), in *Proceedings of the IEEE International Conference on Electrical Insulation and Dielectric Phenomena* (2014), pp. 663–666
19. H. Shumiya, K. Kato, H. Okubo, Feasibility study on FGM (Functionally Graded Materials) application for gas insulated equipment. *Annual Report Conference on Electrical Insulation and Dielectric Phenomena (CEIDP)*, (2014), pp. 360–363
20. H. Shumiya, K. Kato, H. Okubo, Application feasibility of permittivity graded FGM (Functionally Graded Materials) for gas insulated equipment. Trans. Inst. Electr. Eng. Jpn. B **125-B**, 695–700 (2007)
21. M. Koizumi, FGM activities in Japan. Compos. B Eng. **28**, 1–4 (1997)
22. R.H. Brealey, P.W. Juneau, J.E. Zlupko, Fabrication design and electrical evaluation of bulk-graded filled polymer post insulators. *IEEE International Conference on Eletrical Insulation (ICEI)*, (1982), pp. 225–228
23. N. Hayakawa et al., Simulation on discharge inception voltage improvement of GIS spacer with permittivity graded materials (ε-FGM) using flexible mixture casting method. IEEE Trans. Dielectr. Electr. Insul. **25**, 1318–1323 (2018)
24. Z. Liu, W.D. Li, L.Y. Zhang, X.R. Li, J.B. Deng, G.J. Zhang, Simulation design and 3D-printing fabrication of conductivity graded insulator. *IEEE International Conference on Electrical Materials and Power Equipment (ICEMPE)*, (2017), pp. 171–174
25. W.D. Li, X.Y. You, H.B. Mu et al., Numerical optimization and 3D-printing fabrication concept of high voltage FGM insulator[C]. *IEEE PES Asia-Pacific Power and Energy Engineering Conference (APPEEC)* (Brisbane, IEEE, 2015), pp. 1–4
26. T. Shao, F. Liu, B. Hai, Y.F. Ma, R.X. Wang, C.Y. Ren, Surface modification of epoxy using an atmospheric pressure dielectric barrier discharge to accelerate surface charge dissipation. IEEE Trans. Dielectr. Electr. Insul. **24**, 1557–1565 (2017)
27. C. Zhang, H.F. Lin, S. Zhang, Q. Xie, Plasma surface treatment to improve surface charge accumulation and dissipation of epoxy resin exposed to DC and nanosecond-pulse voltages. J. Phys. D: Appl. Phys. **50**(40), 405203 (2017)
28. J. Li, B.X. Du, J.G. Su, H.C. Liang, Surface layer fluorination-modulated space charge behaviors in HVDC cable accessory. Polymers **10**, 500 (2018)
29. J. Li, B.X. Du, J.G. Su, H.C. Liang, Y. Liu, Surface layer fluorination-modulated space charge behaviors in HVDC cable accessory. Polymers **10**, 500 (2018)
30. J. Li, B.X. Du, H. Xu, Suppressing interface charge between LDPE and EPDM for HVDC cable accessory insulation. IEEE Trans. Dielectr. Electr. Insul. **24**, 1331–1339 (2017)
31. B.X. Du, J. Li, Surface charge coupling behavior of fluorinated polyimide film under DC and pulse voltage. IEEE Trans. Dielectr. Electr. Insul. **24**, 567–573 (2017)
32. B.X. Du, Z.L. Li, Hydrophobicity, surface charge and DC flashover characteristics of direct-fluorinated RTV silicone rubber. IEEE Trans. Dielectr. Electr. Insul. **22**, 934–940 (2015)
33. J. Li, Z.Y. Ran, H.C. Liang, B.X. Du, Review of fluorination modification of epoxy resin insulators for gas insulated transmission pipeline. *IEEE International Conference on Electrical Materials and Power Equipment (ICEMPE)*, pp. 106–109
34. A. Li, J. Li, B.X. Du, Z.L. Li, M.M. Zhang, M.L. Fu, S. Hou, Surface charge characteristics of direct-fluorinated epoxy/TiO_2 nanocompsite. IEEE *International Conference* Dielectrics *(ICD)*, (2016), pp. 330–333
35. B.X. Du, J. Li, Q. Du, M.L. Fu, Surface charge and flashover voltage of EVA/CB nanocomposite under mechanical stresses. IEEE Trans. Dielectr. Electr. Insul. **23**, 3734–3741 (2016)

36. C.Y. Li, C.J. Lin, B. Zhang, Q. Li, Understanding surface charge accumulation and surface flashover on spacers in compressed gas insulation. IEEE Trans. Dielectr. Electr. Insul. **25**, 1152–1166 (2018)

Plasma Surface Treatment of Al_2O_3-Filled Epoxy Resin for Accelerating Surface Charge Dissipation

Cheng Zhang, Fei Kong, and Tao Shao

Abstract Al_2O_3-filled epoxy resin (Al_2O_3-filled ER) is widely used as insulating materials in HVDC transmission. Improving the surface properties are important to safe operation of high voltage system. In this chapter, atmospheric-pressure plasma deposition was used for the modification of the Al_2O_3-ER surface. The surface charge dissipation was compared before and after the treatment. For comparison, plasma etching and direct fluorination were also conducted. The experimental results showed that the maximum surface potentials after plasma deposition, direct fluorination and plasma etching decreased to 20.49%, 43.75%, and 49.18%, respectively, compared to that of the untreated sample. The decay rates of the surface charges on the plasma deposited and fluorinated samples exceeded 98%, while they were 54.43% and 0.75% on the plasma etched and untreated samples. The average flashover voltage increased the most (10%) after plasma deposition, followed by the direct fluorinated and plasma etched samples. Surface morphology revealed that the surface roughness of the plasma deposition was smaller than the plasma etching. A flat surface improved the uniformity of electric field, thereby reducing the surface charge accumulation. The surface conductivity after plasma deposition increased three orders of magnitude higher than that of the untreated sample, which effectively improved the surface insulating property of the Al_2O_3-ER sample. This method may open new avenues for the development of next-generation organic–inorganic insulation materials with customized charge dissipation properties.

C. Zhang · F. Kong · T. Shao (✉)
Beijing International S&T Cooperation Base for Plasma Science and Energy Conversion, Institute of Electrical Engineering, Chinese Academy of Sciences, P. O. Box 2703, Beijing 100190, China
e-mail: st@mail.iee.ac.cn

C. Zhang
e-mail: zhangcheng@mail.iee.ac.cn

F. Kong
e-mail: kongfei@mail.iee.ac.cn

B. Du (ed.), *Polymer Insulation Applied for HVDC Transmission*,
https://doi.org/10.1007/978-981-15-9731-2_19

1 Introduction

Al_2O_3-filled epoxy resins (Al_2O_3-ER) are commonly used as dielectric insulator in the power system [1]. However, interface charge accumulation can induce insulation breakdown, which significantly restricts the development and industrialization of DC transmission equipment [2, 3]. For example, accumulated charges may be generated on the Al_2O_3-ER surface between two electrodes after the continuous application of DC voltage for a long time, resulting in the distortion of electric field distribution and initiating a flashover or breakdown of the insulation system [4, 5].

Scientists have been made considerable efforts to suppress the generation of accumulated surface charges on Al_2O_3-ER insulators. Optimizing the insulator geometry is a common solution [6]. However, due to the limitation of the insulator dimension, it is difficult to reduce the charge-induced flashover by increasing the surface creepage distance. Recently, scientists have proposed surface modification methods to reduce surface charge accumulation or increase surface charge dissipation, including direct fluorination, electron beam irradiation, functionally graded material and nanoparticle additives [7–10]. Du et al. had conducted fluorination and irradiation to accelerate surface charge dissipation of the Al_2O_3-ER insulators. He also developed a modified model for epoxy spacer to uniform the electric field distribution caused by joule heating of the conductor [11, 12]. However, electron beam irradiation, surface functionally graded material, nanoscale particles additive need accurate control of parameters and depend on strict operation conditions. An et al. modified the Al_2O_3-ER insulators by using direct fluorination in F_2/N_2 mixture, and the results showed that the surface conductivity of the sample increased to 4 orders of magnitude after the modification, and thus the electric field near the conductor was uniformed [7, 13]. Note that although fluorination technology has been developed for almost forty years, the research on direct fluorination of ER is still under investigation due to its complex process and environmental pollution.

Plasma processing is a high-technology discipline by ionized gas [14, 15]. It activates a chain of chemical reactions at a substrate surface, which deliver activation energy to the surface without heat. Therefore, chemically reactive plasma discharges are widely used to modify the surface properties of materials [16, 17]. They provide a promising approach because of no hazardous solvents, no environmental pollution and simple operation. Since 2007, Institute of Electrical Engineering, Chinese Academy of Sciences (IEE CAS) has investigated the gas discharge plasmas and plasma applications in surface modification [18, 19]. Non-thermal plasmas are generated by different discharge types, including dielectric barrier discharge (DBD), atmospheric-pressure plasma jet (APPJ), gliding discharge and diffuse discharge [14, 20, 21]. Surface modification of polymers and PMMA has been investigated by the aforementioned non-thermal plasmas. For example, hydrophilic and hydrophobic surface on the polymer were achieved by introducing oxygen-containing and fluorine-containing species respectively on the surface after plasma modification [22]. Different species change the surface energy on the polymers, and thus affect the surface properties of the polymer. Shao et al. modified the PMMA by APPJ in CF_4/Ar mixture, and the results showed the surface flashover voltage of treated samples increased more than 30% in vacuum [23]. The improvement of the surface insulation of PMMA contributed to an increase in the surface roughness and a decrease in secondary electron emission yield. Since 2015, plasma modification has extended to Al_2O_3-ER surface treatment for suppressing surface charge acceleration. This chapter will review the plasma surface treatment of Al_2O_3-ER for accelerating surface charge dissipation in IEE CAS [24–29]. For comparison, results of plasma etching, plasma deposition and direct fluorination will be compared by surface morphology observation, chemical component analysis and electrical parameter measurement.

2 Experimental Setups

2.1 Sample Preparation

Al_2O_3-ER samples were provided by manufactured by Shandong Taikai Electrical Insulation Co. Ltd. They were epoxy resins blended with microscale ranged alumina particles. The ratio of epoxy resin, Al_2O_3 and curing agent was 100:330:38. The sample, whose thickness was 2 mm, was a parallelepiped of 50 mm in length and width. Before the treatment, all Al_2O_3-ER samples were cleaned twice in an ultrasonic bath (KQ2200DB, Kunshan Hechuang, China) with distilled water and alcohol at 40 °C for 30 min. Then, these samples were dried in a vacuum drying cabinet for 24 h at a temperature of 70 °C and a gas pressure of 3000 Pa. After the treatment, samples were tested after 2 days.

2.2 Sample Treatment

In the experiments, Al_2O_3-ER samples were modified by non-thermal plasma and direct fluorination. The schematic picture of the plasma treatment and fluorination were shown in Fig. 1 [28]. Non-thermal plasmas were produced in a DBD reactor. It consisted of two pieces of circular aluminum plates of 40 mm diameter, which severed as a high-voltage electrode and a ground electrode, respectively. The upper electrode was covered by a piece of K9 glass and the lower electrode was covered by the Al_2O_3-ER sample to be treated. Both glass and Al_2O_3-ER sample had a thickness of 2 mm and a length of 50 mm. The discharge gap was fixed at 2 mm. In the treatment, different working gas in the treatment resulted in different effects on the Al_2O_3-ER surface. When the working gas was pure argon (Ar), the treatment was plasma etching. When the working gas was Ar/tetra ethoxy silane (TEOS) mixture, the treatment was plasma deposition. The experimental conditions for plasma etching were as follows: DBD reactor was sustained by a microsecond-pulse generator. The

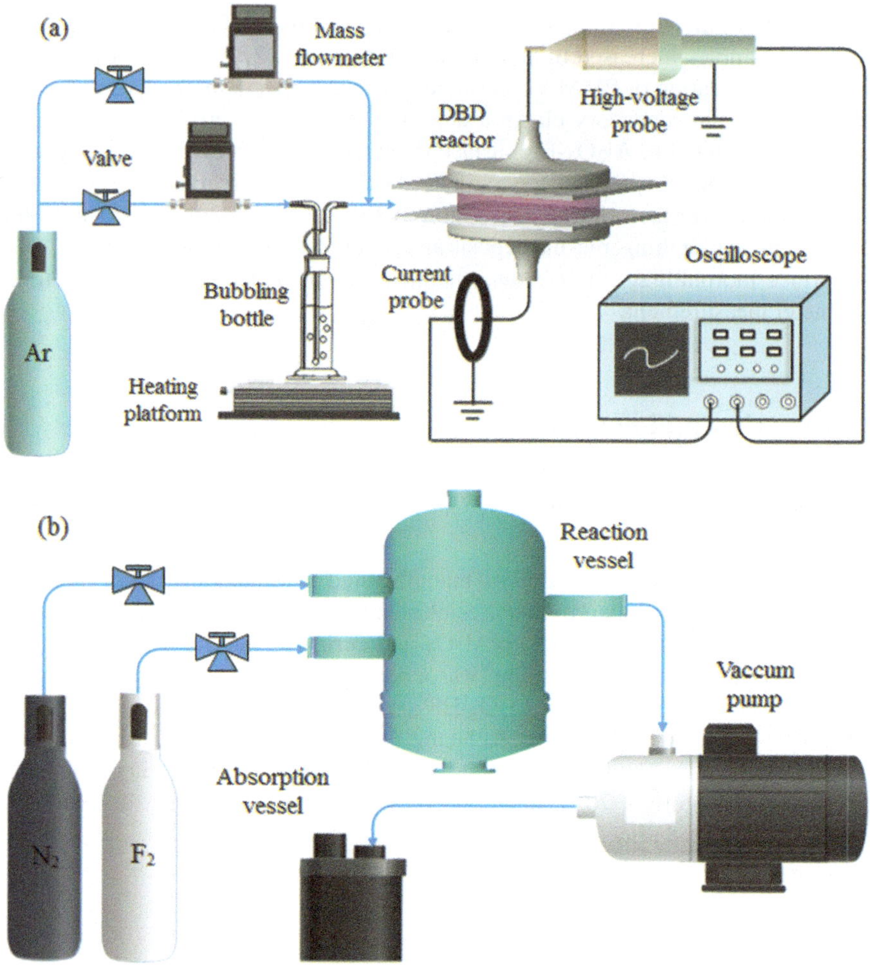

Fig. 1 Schematic picture of (a) plasma treatment and (b) fluorination [28]. Copyright Clearance Center's RightsLink® License Number: 4864701191116

output pulses had a rise time of 500 ns and a full width at half maximum of 8 μs. The voltage amplitude was 9 kV, and the repetition rate was 1.5 kHz. A lateral blowing device generated an Ar flow at a flow rate of 2 standard liters per minute (slm). The treatment time was set to 5 min. Figure 2 shows the electrical characteristics and discharge image of the treatment. It could be seen the current amplitude was 0.8 A. The instantaneous power reached 9.5 kW, and the energy per pulse was approximately 1.5 mJ. Note that the discharge was uniform.

The experimental conditions for plasma deposition were as follows: The same microsecond-pulse generator was used. The voltage amplitude was 11 kV, and the repetition rate was 1.5 kHz. The TEOS was placed in a bubbling bottle and heated

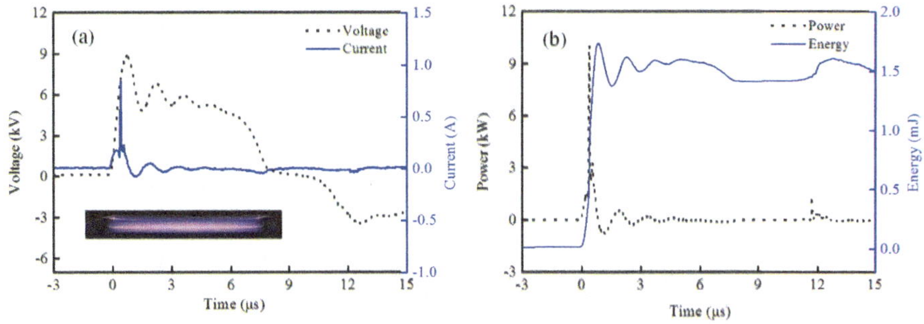

Fig. 2 Electrical characteristics and discharge image in the plasma deposition [26]. Copyright Clearance Center's RightsLink® License Number: 4864710579632

at a fixed temperature of 60 °C. Ar was pumped into the TEOS with a flow rate of 0.5 slm. The Ar/TEOS mixture then was mixed with 2 slm pure Ar and entered into the discharge gap. The treatment time was fixed to 10 min.

For comparison, Al_2O_3-ER samples were treated by direct fluorination, and the setup was shown in Fig. 1b. The experimental conditions were as follows: Al_2O_3-ER samples were placed in a stainless-steel reaction vessel. After evacuation and purification with N_2, the surface fluorination was carried out using a F_2/N_2 mixture (10% F_2 by volume) at 50 °C, 0.1 MPa, and the treatment time was fixed to 30 min. The exhaust gas was collected by an absorption vessel.

2.3 Surface Potential and Dielectric Parameters

Figure 3 showed the schematic picture and photograph of the surface potential measurement system. The aim of the system was to obtain the characteristics of surface charges on the Al_2O_3-ER samples. It consisted of a charging system and a Kelvin type probe (TREK Model 6000B-6, USA), which were placed in a sealed PMMA chamber. The accumulation and dissipation of surface charges were evaluated by measuring the change of the surface potential on Al_2O_3-ER after charging. A corona discharge with a needle-to-plane electrode configuration was used to charge the Al_2O_3-ER sample. The needle electrode was connected to the high voltage DC generator (71030 N, Genvolt). The needle electrode was made of stainless steel, with a diameter of 1 mm and a curvature radius of 80 μm. The plane electrode was grounded, which was made of a copper foil with a dimension of 5 cm × 5 cm. The sample was placed on the grounded electrode when it was charging. The gap distance was 5 mm. The typical charging voltage was 5 kV. Then, the charged sample was moved to below a Kelvin probe by a stepper motor. The moving time was 12 s. Surface potentials were measured by the probe coupled with an electrostatic voltmeter (TREK Model-347, USA). It had a measurement accuracy of ±3 V and a

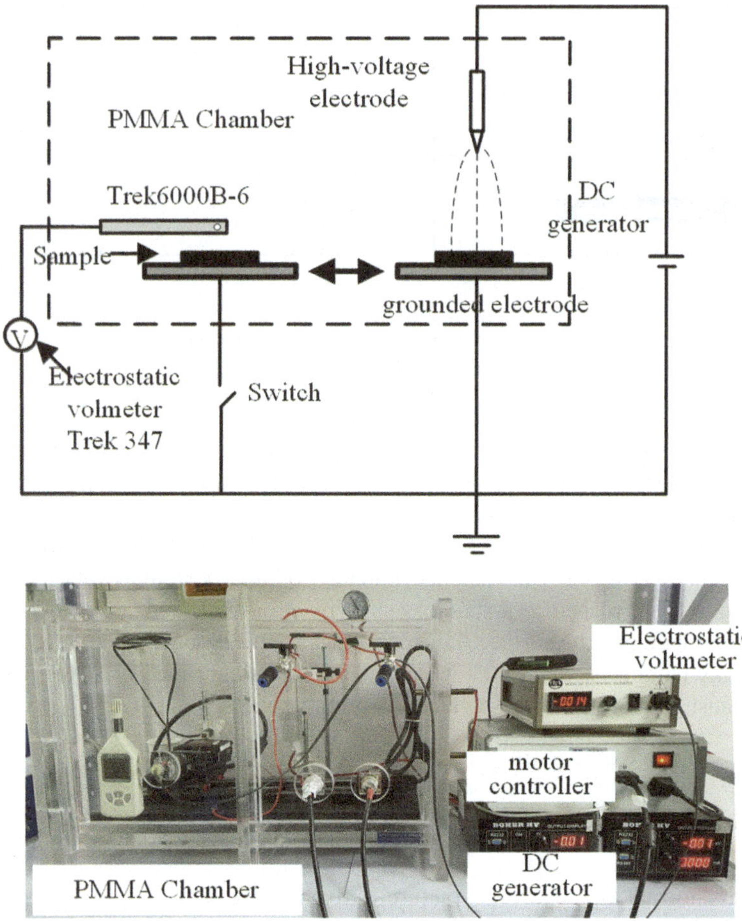

Fig. 3 Schematic picture and photograph of surface potential measurement system

spatial resolution of 3 mm. Finally, the data were captured by an acquisition card (USB5935) and stored.

In order to obtain the distribution of surface charges, multipoint measurement was necessary. In our experiment, the Al_2O_3-ER sample was moved in a zigzag pattern through the probe area by using two stepper motors along the X-axis and Y-axis. Thus, there were 441 points on the Al_2O_3-ER sample with a dimension of 20 × 20 mm. The total scanning time of the above measurement process was 210 s. The decay rates were calculated as follows [30]:

$$D = \frac{V_{(t_0)} - V_{(t)}}{V_{(t_0)}} \times 100\% \quad (1)$$

where $V_{(t0)}$ was the surface potential at the time t_0, and $V(t)$ was the surface potential at time t. The temperature and the relative humidity could be controlled in the PMMA chamber, and they were fixed at 20 °C and 40%, respectively.

Surface and bulk conductivities of the samples were measured by a Keithley 6517B electrometer (Keithley, Ohio, USA) based on the three-electrode method. The internal high-voltage DC power supply in the electrometer applied a certain voltage on the Al_2O_3-ER sample, the corresponding current was measured by an internal micro-ammeter. Thus, the surface resistivity and bulk resistance could be calculated according to the ohm formula. The surface and bulk conductivity could be obtained.

2.4 Flashover Voltage Test

In order to measure the flashover voltage before and after the treatment, a finger-like electrode configuration was used, which was similar to the one used in Ref. [24]. Two hemispherical copper electrodes worked as high-voltage and ground electrodes, respectively. The diameter of the electrodes was 10 mm. Al_2O_3-ER sample was sandwiched between two electrodes, and the gap distance was fixed at 5 mm. A DC generator (TD2202P0-2000) was used for testing the flashover voltage. The range of the output voltage was 0—50 kV. During the test, the step rate of the applied voltage was set to 5 kV/min in the Labview program. In order to facilitate comparison, the measured flashover voltage values under various conditions were arranged in order from small to large, and the average flashover voltage was be obtained.

2.5 Surface Morphology and Chemical Composition

The surface morphology of the Al_2O_3-ER sample was measured by a field emission scanning electron microscope (FE-SEM, Zeiss Sigma, Germany) and atomic force measurement (AFM, Bruker Dimension Icon). The area of the AFM image was 25 × 25 μm. The mean roughness (R_a) and the root mean square roughness (R_q) were calculated by Nanoscope Analysis software. The smoothing function of the software was used to restore the surface morphology before calculation. The uniform height range was used to compare the morphology of different samples, and the height range slightly affected the calculation results. The height range was from –300 to 300 nm in this experiment.

The chemical composition of the Al_2O_3-ER sample was measured by Fourier Transform Infrared Spectroscopy (FTIR) and X-ray photoelectron spectroscopy (XPS). FTIR (ThermoNicolet iS10) worked in the attenuated total reflection mode. The spectra ranged from 4000 to 650 cm^{-1}, and the spectra resolution was set to 2 cm^{-1}. The average value of 16 scanning data were taken as the measurement results. XPS (Thermo Fisher ESCALAB 250Xi) used non-monochromatic Mg Kα-radiation operating at 300 W. The pressure in the analyzing chamber was maintained at 3×10^{-9} mbar, and the binding energies were referred to the C1s line at 284.7 eV from adventitious carbon. Curve of the C1s peak was fitted using XPS peak 4.1 software.

3 Accumulation and Dissipation Characteristics

3.1 Surface Charge Accumulation Before and After the Treatments

Because of the limitation of the processing technology, small defects may exist in the process of casting and processing the Al_2O_3-ER. These defects will lead to the flashover along the Al_2O_3-ER surface due to the local electric field distortion and the surface charge accumulation under the continue DC stress. Figure 4 shows three-dimensional distribution of surface potential after the Al_2O_3-ER sample is exposed to DC voltage. The polarity is negative and the amplitude is 5 kV. It can be seen that the surface potential distribution appears like an umbrella. It has a peak with a potential value of −3280 V near the center and then gradually decreases in the radial direction.

Figure 5 shows three-dimensional distribution of surface potential after the treatments. In the case of plasma etching, surface potential appears to be a flat plane, which means the slight changes of the potential during the scanning time. The maximum potential is 1613 V, which is 49.18% of the value before treatment. However, differences between the maximum and minimum surface potentials are only 254 V. In the

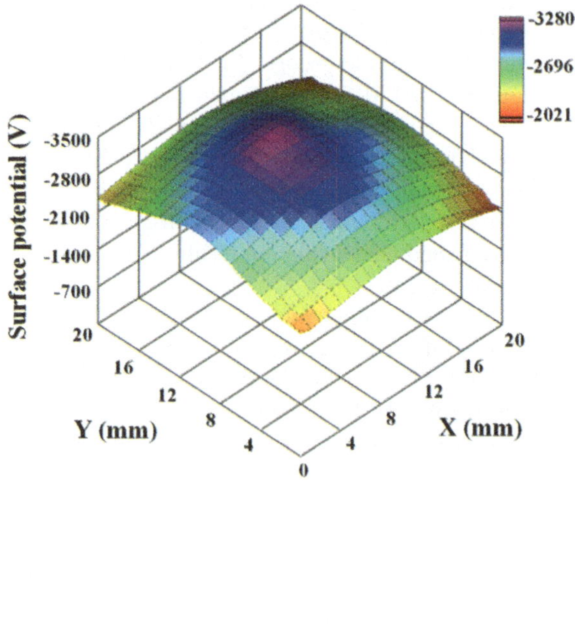

Fig. 4 Three-dimensional distribution of surface potential after charging by DC voltage [28]. Copyright Clearance Center's RightsLink® License Number: 4864701191116

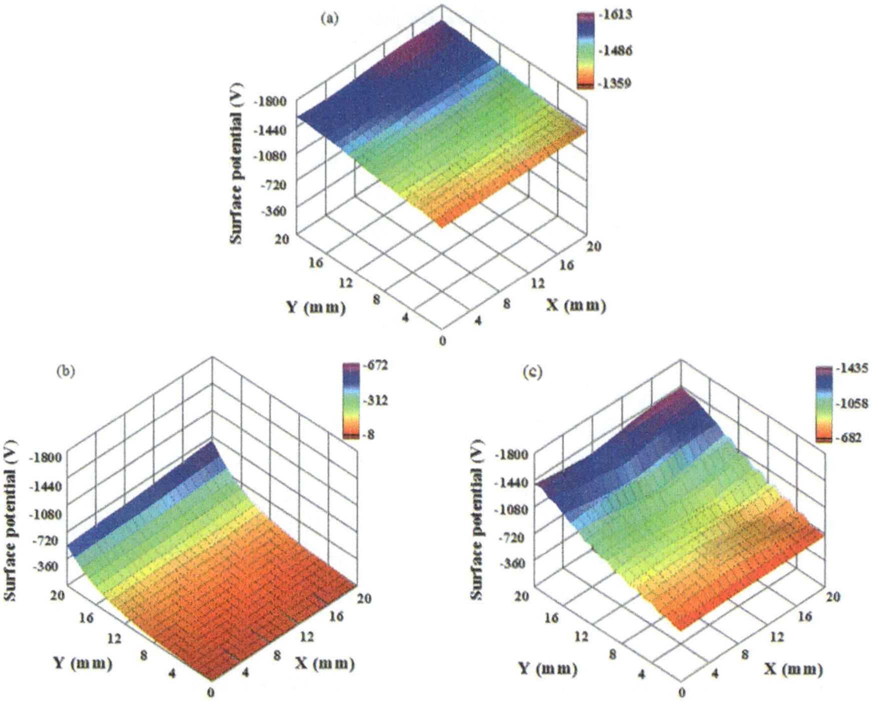

Fig. 5 Three-dimensional distribution of surface potential after the treatments [28]. (**a**. plasma etching **b**. plasma deposition and **c**. direct fluorination) Copyright Clearance Center's RightsLink® License Number: 4864701191116

case of plasma deposition, surface potential appears to be a bend curved surface. The maximum potential is 672 V, which is 20.49% of the value before treatment, and the minimum potential is nearly zero. In the case of direct fluorination, surface potential appears to be an uneven curved surface. The maximum potential is 1435 V, which is 43.75% of the value before treatment, and the minimum potential is 682 V. It should be mentioned that the surface charge distributions under different treatment conditions were measured three times for each sample, and the results showed no obvious difference. From the above data, although all three methods significantly decrease the initial surface potential after the corona charge, plasma deposition is the best solution to suppress the surface charge accumulation.

3.2 Surface Charge Dissipation Before and After the Treatments

When the corona discharge distiguishes, surface charges accmulated on the Al_2O_3-ER surface begins to dissipate. In atmospheric air, there are three dissipation paths: (1) electric conduction through the bulk; (2) charge neutralization by gas ions in the air; (3) electric conduction along the surface. The characteristic time for the charge decay τ_v (s) through the volume resistivity ρ_v (Ω/m) is given by [31]:

$$\tau = \varepsilon_0 \varepsilon_r \rho_v \tag{2}$$

where ρ_v(Ω/m) is the bulk resistance of the insulator. For ER, $\varepsilon_r = 4.95$, the measured bulk conductivity of the ER is 8.84×10^{-17} S/m. The corresponding ρ_v is 1.13×10^{16} Ω/m. Thus, the calculated decay time through the ER is 4.95×10^5 s. Dissipating through the bulk is not the dominant factor. Moreover, surface charge dissipation cosumes more than 8.64×10^5 s under the sole condition of charge neutralization by gas ions [32, 33]. Therefore, electric conduction along the surface is the most likely path for the accumulated surface charges.

Figure 6 shows the suface charge density and the decay rate of the Al_2O_3-ER samples before and after the treatments. It can be seen that surface charge density on the untreated sample is relatively stable. The dacay rate is only 0.75% after 1800s, indicating the charge dissipation is slow. After plasma etching, during 900-s decay, surface charge density decreases from 32.92 to 17.92 pC/mm^2. The decay rate is 54.43%. After plasma deposition and direct fluorination, surface charge densities decrease rapidly in the first 300 s. The decay rates are approximately 84% for both treatments, and finally they exceed 98% in 900 s. It should be pointed out that well stabilized surface charge density for the untreated sample means surface charges are difficult to dissipate along the Al_2O_3-ER surface before the treatment. Both plasma

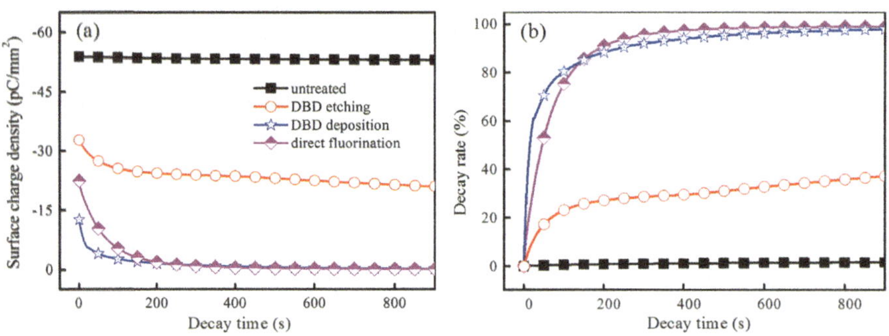

Fig. 6 Surface charge density and decay rate of Al_2O_3-ER samples before and after the treatments [28]. (**a**. surface charge density **b**. decay rate) Copyright Clearance Center's RightsLink® License Number: 4864701191116

depostion and direct fluorination present excellent peformance for accerlerating charge dissipation of Al_2O_3-ER samples.

3.3 Surface Charge Dissipation After the Storage

In order to study the stability of the treatments, Al_2O_3-ER samples are stored in ambient air for several days. Figure 7 shows surface charge density and decay rate of Al_2O_3-ER samples after the storage. Note that the surface charge density of the plasma etched sample goes back after storage. When the storage time reaches 5 days, the surface charge density is close to the untreated sample. The corresponding decay rate is only 5.72%. The surface charge densities of plasma deposited and fluorinated samples slightly change after 5 days of storage. The corresponding decay rate exceeds 95%. It indicated that the ageing time slightly affected the dissipation characteristics of the Al_2O_3-ER samples after plasma deposition and direct fluorination.

The different dissipation rates of plasma etching and plasma deposition are due to the different reaction gases. In the case of plasma etching, Ar is the working gas. The Ar excited state is populated by a one-step process [34]:

$$Ar + e \rightarrow Ar^* \tag{3}$$

The subsequent radiative decays generate Ar species:

$$Ar^* \rightarrow Ar + hv \tag{4}$$

In the plasma etching, owning to the Penning effect, metastable state Ar can dissociate oxygen molecule and excite atomic oxygen to the excited state [35]:

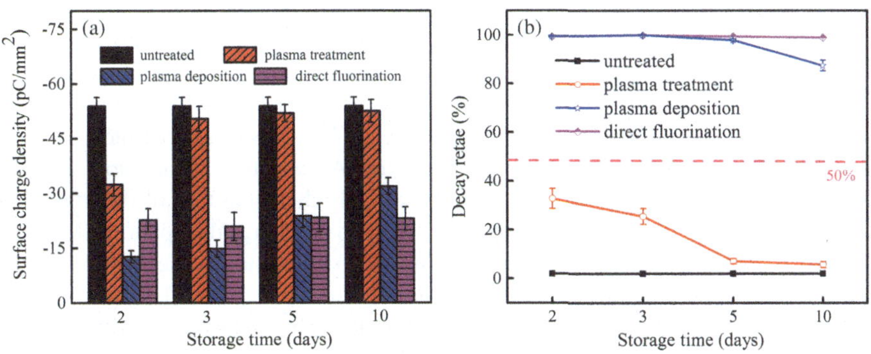

Fig. 7 Surface charge density and decay rate of Al_2O_3-ER samples after the storage [28]. (**a**. surface charge density **b**. decay rate) Copyright Clearance Center's RightsLink® License Number: 4864701191116

$$Ar * + O_2 \rightarrow O + O^* + Ar \tag{5}$$

$$Ar * + O \rightarrow O^* + Ar \tag{6}$$

The use of Ar plasma mainly introduces radicals on the surface, without introducing functionalities or plasma-deposited polymers. In atmospheric air, these radicals interact with the oxygen, only forming peroxide and hydrogen peroxide [36]. They are more likely to reorient toward the interface after the storage of one or two days, decreasing the interfacial free energy. This phenomenon is the ageing effect of plasma treatment.

When TEOS works as the procedure, owning to the electron impact dissociation, SiOx film is introduced to the Al_2O_3-ER surface. In atmospheic air, the excited oxygen also can remove the impurity groups in TEOS and the ligands in the surface adsorbed fragments, thus promoting the formation of Si-O-Si bonds [29].

$$Si(OC_2H_5)_n(OH)_{4-n} + e \rightarrow Si(OC_2H_5)_{n-1}(OH)_{4-n+1} + C_2H_4 + e \tag{7}$$

$$Si(OC_2H_5)(OH)_{4-n} + O \rightarrow S(OC_2H_5)_{n-1} + C_2H_4O + e \tag{8}$$

Al_2O_3-ER surface with deposited SiOx film is stable during the storage. Therefore, the ageing effect slightly affects the treatment effect.

3.4 *Flashover Voltage Measurement Before and After the Treatments*

When testing the flashover voltage, three samples were tested for each treatment condition after a storage of 5 days. Each sample was tested 10 times. Figure 8 shows the flashover voltage before and after the treatments. The average flashover voltage value of untreated samples is about –7.8 kV. The flashover voltage of the plasma etched samples is slightly improved. Because the effect of plasma etching on the surface charge dissipation is not stable, as shown in Fig. 7. Surface chemical components after 5 days also return to the original state. After plasma deposition, average flashover voltage reached –9.3 kV, which is 15% higher than that of the untreated sample. It is mainly due to the appropriate increase in the charge dissipation rate on the Al_2O_3-ER surface. Fast surface charge dissipation can eliminate the surface charge accumulation phenomenon. After direct fluorination, the average flashover voltage is increased by no more than 10% compared to the untreated sample. Although the charge migration rate is accelerated, the high spikes may still exist on the Al_2O_3-ER surface. It will cause the accumulated charges, providing development for the flashover along the surface.

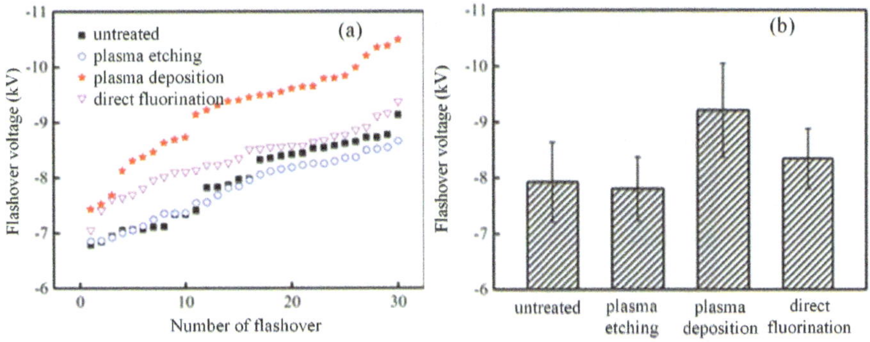

Fig. 8 Flashover voltage of Al_2O_3-ER before and after the treatments. (**a**. flashover voltage **b**. average flashover voltage)

4 Surface Property Analysis

4.1 Surface Morphology

Surface insulating properties is strongly influenced by the surface charge distribution on the Al_2O_3-ER insultator. The change of the surface morphology before and after plasma deposition is shown in Fig. 9. Seen from Fig. 9a, defects in manufacturing

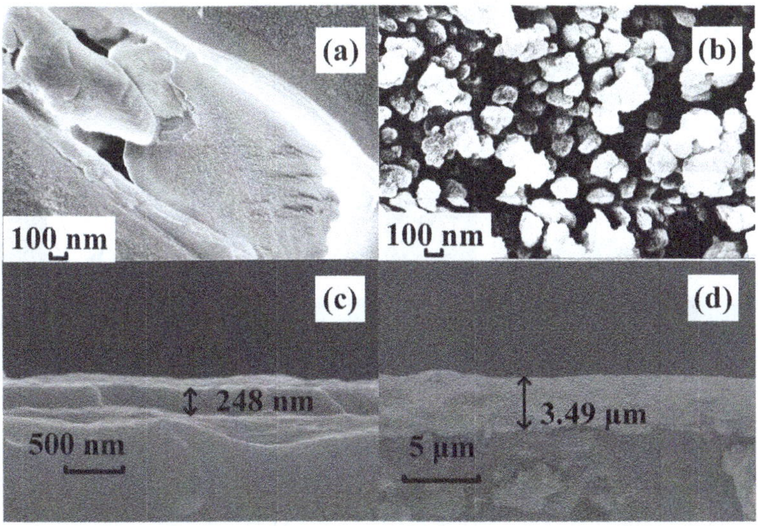

Fig. 9 SEM micrographs of the Al_2O_3-ER before and after the treatments. (**a**. untreated sample **b**. plasma deposited sample **c**. SEM section of plasma deposited sample and **d**. SEM section of direct fluorinated sample)

process could be found on the Al_2O_3-ER surface before the treatment. However, these defects are replaced by some nanoparicles with diameter of 50–100 nm, indicating dense SiO_x nanoparticles form on the Al_2O_3-ER surface after plasma deposition, as seen in Fig. 9b.

The thickness of the film can be obtained by cutting the sample into two parts and observing the cross section with SEM. The cross sections of the plasma deposited and direct fluorinated samples are shown in Fig. 9c, d. Note that the thicknesses of the SiO_x layer and fluorinated layers are 248 nm and 3.49 μm, respectively. Plasma deposition has a thickness of several hundreds of nanometers, it will not affect the properties of the Al_2O_3-ER bulk.

Surface roughness may affect the flashover path and the surface electric field. In the experiment, the surface morphology of Al_2O_3-ER is presented by AFM images. Figure 10 shows the AFM image of the Al_2O_3-ER sample before and after the treatments. Before the treatment, the Al_2O_3-ER surface is uneven with several spikes (Fig. 10a). After plasma etching, the Al_2O_3-ER surface becomes a bit uniform, and the spikes has been cut off due to the impact of high-energy electrons and heavy ions in the plasma (Fig. 10b) [37]. After plasma deopsition, the Al_2O_3-ER surface is covered by a flat SiO_x layer [26]. Moreover, no spike appears on the suface, indicating the decrease of the surface roughness (Fig. 10c). After direct fluorination,

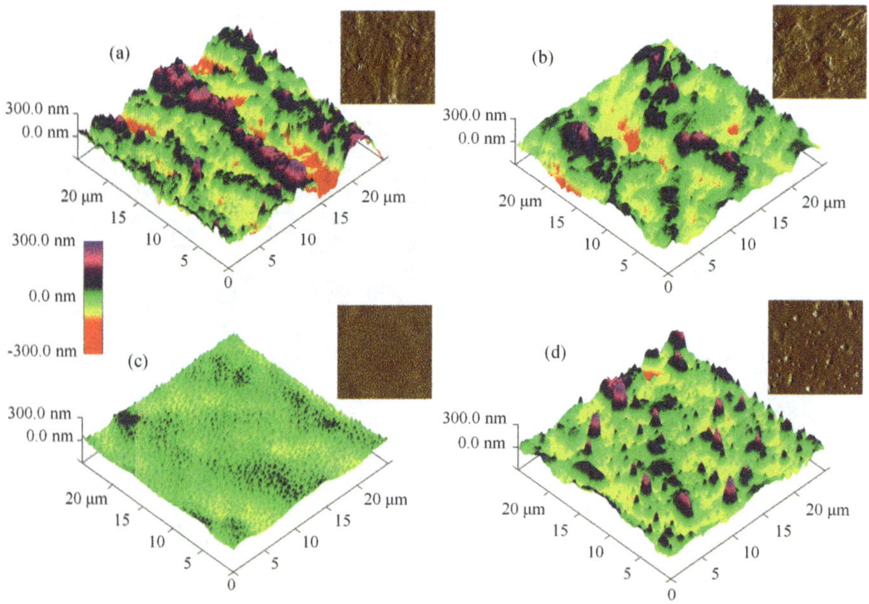

Fig. 10 Three-dimensional view of AFM images on the Al_2O_3-ER surface before and after the treatments (**a.** untreated **b.** plasma etching **c.** plasma deposition **d.** direct fluorination) [28]. Copyright Clearance Center's RightsLink® License Number: 4864701191116

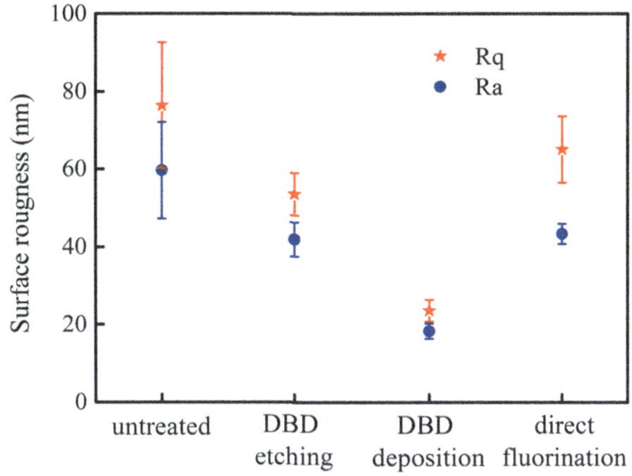

Fig. 11 Surface roughness of the Al_2O_3-ER samples before and after the treatments. Copyright Clearance Center's RightsLink® License Number: 4864701191116

a fluorinated layer covers on the Al_2O_3-ER surface, however, the spikes still exist on the surface (Fig. 10d).

The correspongding mean surface roughness (R_a) and the root-mean-square surface roughness (R_q) are calculated from the AFM image, as shown in Fig. 11. Note that, the surface roughness decreases for all the treated samples. The surface roughness of the plasma deposited sample is the lowest, followed by the plasma etched and direct fluorinated samples. A smooth and flat surface can reduce the distortion of the electric field on the Al_2O_3-ER surface [38]. It is beneficial to the suppression of the accumulated surface charges.

4.2 Chemical Component

In order to deepen the understanding of the chemical element composition before and after the treatments, FTIR and XPS analysis are conducted. The FTIR spectra is taken by ATR mode, the resolution of the wavelength is 2 cm^{-1}. Figure 12 shows the FTIR waveforms on all the samples. Before the treatment, The absorption peaks includes C-H (2962 cm^{-1}, 2927 cm^{-1}, 2872 cm^{-1}, 1460 cm^{-1}, 1180 cm^{-1} and 827 cm^{-1}), C = O (1732 cm^{-1}), C = C (1606 cm^{-1} and 1506 cm^{-1}). All these peaks change a little after plasma etching, because the affected depth is small in such treatment. After plasma deposition, absorption peak at 1046 cm^{-1} increases, and new silicon-containing absorption peaks, Si-O-Si at 1046 cm^{-1} and Si-OH at 960 cm^{-1}, appear on the deposited sample surface. The appearance of these peaks is due to the SiO_x layer on the Al_2O_3-ER surface. After direct fluorination, wide absorption peaks ranged from 900 cm^{-1} to 1350 cm^{-1} is obvious, because the C-F species are introduced on the Al_2O_3-ER surface [30, 39].

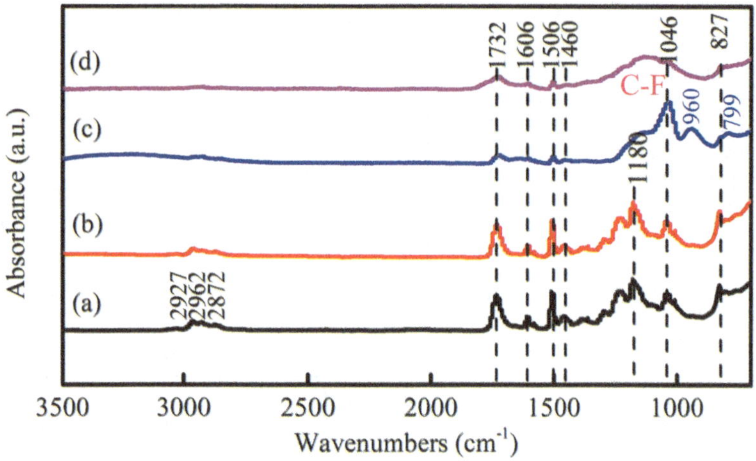

Fig. 12 FTIR spectrum of the Al_2O_3-ER samples before and after the treatments. Copyright Clearance Center's RightsLink® License Number: 4864701191116

Figure 13 shows the XPS spectrum of all the samples. The corresponding element ratios of the chemical components were listed in Table 1. Before the treatment, atomic percent (AT%) of carbon and oxygen species are 58.22% and 23.29%, respectively. Small amount of silicon species comes from the curing agent in the Al_2O_3-ER sample. After plasma etching, oxygen species significantly increases while the carbon species decreases. It is because many C-C bonds are replaced by the C-O and C = O bonds. After plasma deposition, both oxygen and silicon species increase due to the appearance of Si-O-Si and Si-OH functional groups. After direct fluorination,

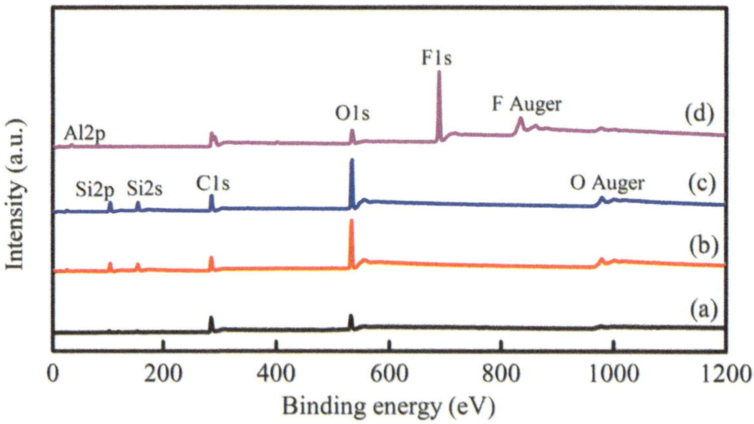

Fig. 13 XPS spectrum of the Al_2O_3-ER samples before and after the treatments. (**a**. untreated **b**. plasma etching **c**. plasma deposition **d**. direct fluorination) Copyright Clearance Center's RightsLink® License Number: 4864701191116

Table 1 Atomic percent of chemical composition before and after the treatments [29]. Copyright Clearance Center's RightsLink® License Number: 4864701191116

AT(%)	C	O	Si	F	C/O	C/F
Untreated sample	58.22	23.29	17.26	–	250	–
Plasma etching	31.28	48.75	17.98	–	66.9	–
Plasma deposition	34.26	42.22	22.47	–	81.1	–
Direct fluorination	46.44	11.69	–	40.93	397	113

fluorine-containing groups appear on the Al_2O_3-ER surface due to the fluorinated layer.

Figure 14 presents the reconstructed C1s core-level photoemission spectra of Al_2O_3-ER samples before and after the treatments. Principal chemical functional groups introduced into the Al_2O_3-ER surface are separated. The concentration of each chemical component with C1s can be calculated by deconvolution using a Gaussian–Lorentzian fit [21]. Seen from Fig. 14a, only C-C/C-H (284.7 eV) and C-O (286.2 eV) exist on the untreated sample. After plasma etching, C = O/O-C = O (288.8–289.5 eV) functional groups appear on the Al_2O_3-ER surface (Fig. 14b).

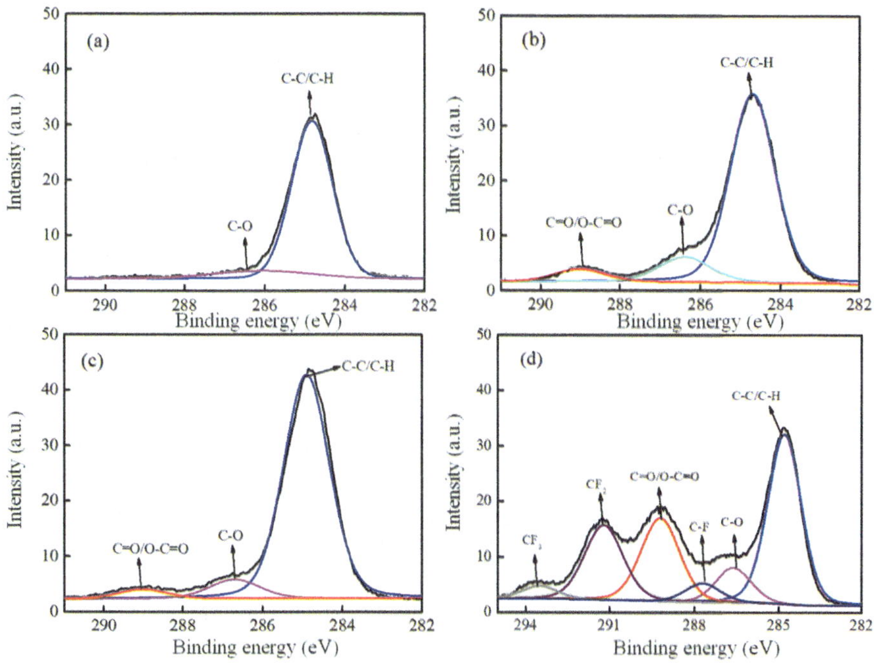

Fig. 14 C1s XPS spectra of the Al_2O_3-ER samples before and after the treatments (**a.** untreated **b.** plasma etching **c.** plasma deposition **d.** direct fluorination). Copyright Clearance Center's RightsLink® License Number: 4864701191116

These groups are typical polar groups, and they will enhance the hydrophilicity on the Al_2O_3-ER sample. After plasma deposition, the same carbon-containing groups can be found. The intensity of the C-C/C-H group is the largest (Fig. 14c). This is because the precursor TEOS contains unbroken C-H and C-C groups, which will also be deposited on the surface. After direct fluorination, three new functional groups, CF_3 (293.5 eV), CF_2 (291.2 eV) and C-F (287.8 eV), are observed in Fig. 14d. It confirms a fluorinated layer appear on the Al_2O_3-ER surface.

4.3 *Surface and Bulk Resistivity*

According to the dissipation path of surface charges, they are more likely to dissipate along the bulk and surface of Al_2O_3-ER sample. Figure 15 shows the surface and bulk resistivities of the Al_2O_3-ER samples before and after the treatments. Compared with Al_2O_3-ER sample before the treatment, the bulk resistivity of the all the treated samples decrease under various treatment conditions, among which the fluorinated sample drops by an order of magnitude. After plasma etching, plasma deposition and direct fluorination treatments, the surface resistivity of the Al_2O_3-ER samples are reduced by about one order of magnitude, three orders of magnitude and two orders of magnitude compared to the untreated sample, respectively. Plasma deposition shows the best performance on the improvement of the surface resistivity among the three treatments.

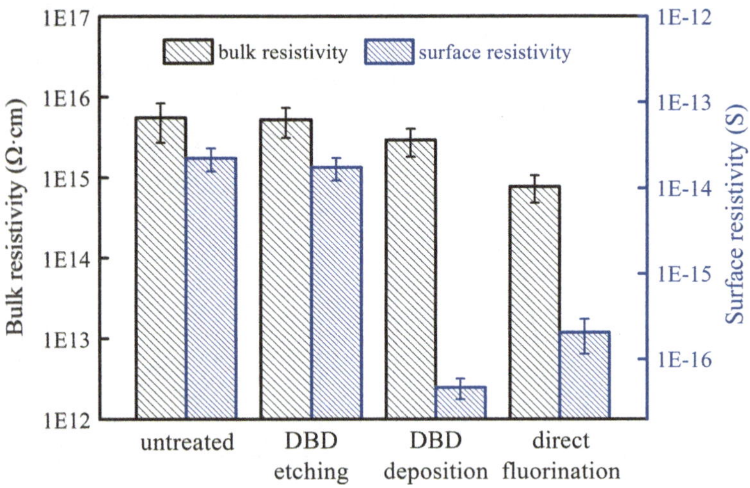

Fig. 15 Surface and bulk resistivities of the Al_2O_3-ER samples before and after the treatments

4.4 Trap Density Distribution

Based on the isothermal current theory, attenuation of the surface potential is assumed to only determined by the relaxation process of free carriers [40]. Equation (9) shows the relationship between the trap density energy distribution of the trap levels and the band gap ($N(E_m)$):

$$I_{d(t)} \propto \frac{dV}{dt} \propto N(E_m)\frac{dE_d}{dt} \tag{9}$$

where $I_{d(t)}$ is the emission current induced from the electrons emitted out of the traps to the conduction band. dV/dt is the decay rate of the surface potential, which is proportional to the $I_{d(t)}$. Ed is the corresponding electron-emitting trap level of the free carriers. It should be pointed out that dV/dt is close related to the emission to the valence band of holes from the trap levels in the lower half of the band gap. In other words, dV/dt is proportional to the trap density energy distribution of the trap levels throughout the band gap ($N(E_m)$).

The energy level depth can be calculated according to the decay time as follows [32]

$$\Delta E = E_0 - E_m = kT \times \ln(vt) \tag{10}$$

where k is Boltzmann's constant, T is the absolute temperature, v is the attempt-to-escape frequency of electrons in the traps. Therefore, the relationship between the energy level depth and the trap density energy distribution for the Al_2O_3-ER sample can be obtained.

Figure 16 shows density of trap energy level of the Al_2O_3-ER samples before and after the treatments. Because of the slow dissipation of the surface charges, the density plots of the untreated and plasma etched samples are incomplete. However, peaks of the trap energy of all the samples are presented. After plasma etching, two peaks at 0.91 and 1.01 eV can be observed. They are a bit smaller than that of the untreated sample (1.03 eV). Peaks of the trap energy are even smaller when plasma deposition and direct fluorination are conducted. The former has two peaks at 0.89 and 0.96 eV, and the latter has one peak at 0.92 eV. Therefore, it could be concluded that both plasma deposition and direct fluorination introduce shallow traps on the Al_2O_3-ER surface. These shallow traps exist in the SiO_x and fluorinated layers. They are beneficial to the acceleration of the surface charges, which are coherent with the decrease of the surface resistivity shown in Fig. 15.

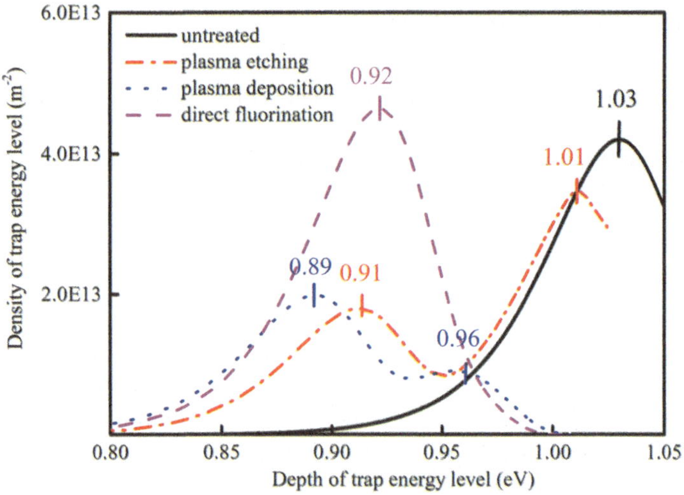

Fig. 16 Density of trap energy level of the Al_2O_3-ER samples before and after the treatments. Copyright Clearance Center's RightsLink® License Number: 4864701191116

5 Discussion

In our experiment, plasma deposition is used to improve the surface dissipation of the Al_2O_3-ER insulator. Figure 17 presents the schematic picture of charge generation, injection and dissipation before and after plasma deposition. Corona discharges occurring around the tip of the needle electrode provide free charges, which are the sources of surface charges. In the case of negative polarity, most of the free charges are CO^{3-} ions [41]. When the DC electric field presents, these ions are forced to move to the surface of the Al_2O_3-ER, and then they are trapped by the local surface state to form surface charges. With the continuous excitation of DC voltage, the surface charges continue to accumulate on the Al_2O_3-ER surface. Due to the effect of local electric field and external electric field, these charges will conduct along the sample surface and transfer to the bulk. When the high voltage is removed, corona

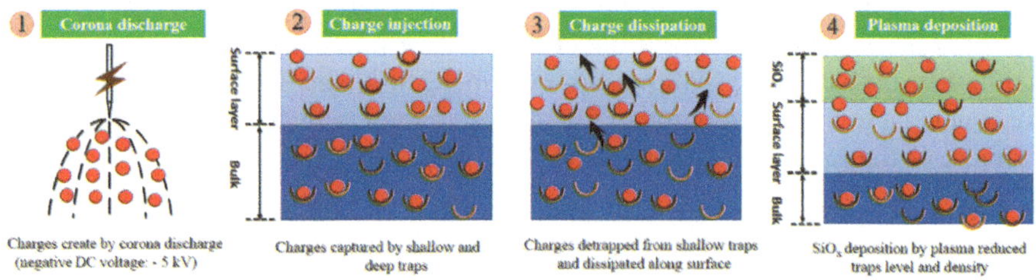

Fig. 17 Schematic picture of the processes for charge accumulation and dissipation [26]. Copyright Clearance Center's RightsLink® License Number: 4864710579632

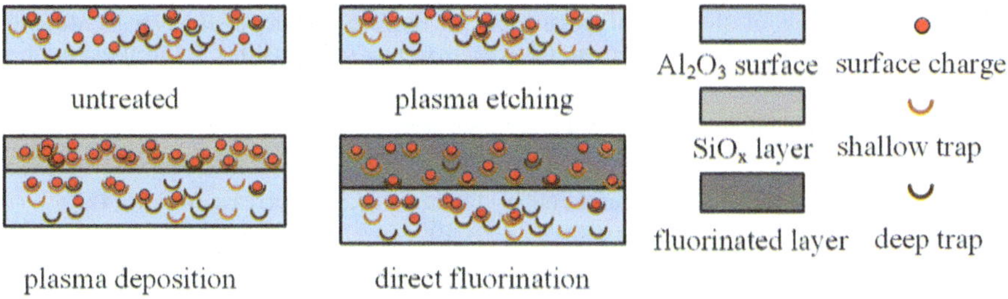

Fig. 18 Schematic picture of Al_2O_3-ER surface before and after the treatments

discharge will be extinguished, and the surface charges begin to dissipate, resulting in the surface potential distribution on the Al_2O_3-ER surface, which is shown in Fig. 4.

After plasma deposition, SiO_x layer covers the Al_2O_3-ER surface, introducing some polar groups (OH, Si-OH and Si-C) on it. Seen from the density of trap energy level, many shallow traps appear in the SiO_x layer. Surface charges generated by the negative corona discharge are captured by these shallow traps. The charges that are located in shallow traps decay at a faster speed than the charges in deep traps, which in turn accelerates surface charge dissipation. The corresponding surface resistivity decreases. Furthermore, the SiO_x layer covers the defects of the Al_2O_3-ER surface. The electric field at the triple junction becomes uniform, suppressing the generation of the surface charges on the surface.

Figure 18 presents schematic picture of Al_2O_3-ER surface before and after the treatments. Different from the plasma deposition, plasma etching and direct fluorination achieve treatment effect by other means. Although some oxygen-containing groups are introduced on the surface, plasma etching mainly change the morphology of the Al_2O_3-ER surface by electron impacting. Note that the depth of the trap energy level for the main part of its density only decreases 0.02 eV compared to the untreated sample. Other depth of the trap energy level decreases from 1.03 to 0.91 eV (Fig. 16). It means the amount of the shallow traps will only increases a little after plasma etching [26, 28]. Direct fluorination also forms a fluorinated layer with the thickness of several micrometers on the Al_2O_3-ER surface. However, according to AFM image, surface roughness of the fluorinated sample is larger than that of the plasma deposited sample. It means the electric field distribution on the plasma deposition is more uniform. Moreover, compared with the plasma deposited sample, fluorinated sample has large surface resistivity and small bulk resistivity, as shown in Fig. 15. Because both shallow and deep traps are introduced on the Al_2O_3-ER surface. Thus, surface charges of fluorinated samples can be dissipated along the surface, but the probability of the surface charge decay along the bulk is also large.

Table 2 listed the typical parameters of the surface charge dissipation compared to other works. Note that the accumulated surface charges usually decrease 50% after direct fluorination. The decay rate exceeds 98% for both direct fluorination and plasma deposition. Therefore, it could be concluded that both direct fluorination

Table 2 Comparison of the typical parameters of surface charge dissipation in different surface treatment approaches. Copyright Clearance Center's RightsLink® License Number: 4864701191116

Treatment method	Treatment time	Surface charges	Surface resistivity	Trap level	Reference
Direct fluorination	15, 30 and 60 min	Decreases 60%	Decreases 3.8×10^5	Decrease from 0.93 to 0.75 eV	[42]
Direct fluorination	30 min	Decreases no more the 5%; Decay rate reaches 90%	Decreases 1.4×10^4	–	[13]
Gliding discharge	10 s	Decreases 50%; Decay rate increases from 41 to 90%	Decreases 40 times	Decrease from 0.83 to 0.81 eV	[24]
Plasma etching	10, 60, and 180 s	Decay rate increases from 2 to 31%	Decreases 8 times	Decrease from 0.80 to 0.79 eV	[25]
Electron beam irradiation	irradiation dose: 100, 500 and 700 kGy	Decreases no more the 15%; Decay rate increases from 3 to 20%	Decreases 5 times	Decrease from 0.896 to 0.87 eV	[43]
Plasma etching	5 min	Decreases 40%; Decay rate increases from 1.3 to 58.12%	Decreases 4.6 times	Decrease from 1.03 to 0.91 eV	This work
Plasma deposition	10 min	Decreases 77%; Decay rate increases from 1.3 to 99%	Decreases 3.4×10^2	Decrease from 1.03 to 0.89 eV	
Direct fluorination	30 min	Decreases no more the 58%; Decay rate increases from 1.3 to 99%	Decreases 1×10^2	Decrease from 1.03 to 0.92 eV	

and plasma deposition could decrease the surface charge accumulation and increase the surface charge decay rate. In addition, it should be pointed out that, compared to direct fluorination, plasma deposition required only one-third of the treatment time, and the treatment process was non-toxic. Therefore, it could be considered as a promising application prospects for suppressing the accumulated surface charges on the Al_2O_3-ER surface.

6 Conclusion

This chapter presents the surface modification of Al_2O_3-ER for improving the surface charge dissipation by atmospheric plasma deposition, plasma etching and direct fluorination, respectively. Accumulation and dissipation characteristics of surface charges on the Al_2O_3-ER surface among different treatments are compared. Surface morphology, chemical component and surface dielectric parameters are used to analyze change of the surface properties before and after the treatments. The following remarks could be proposed:

(1) The maximum surface potentials decrease to 20.49%, 43.75% and 49.18% of original value of the untreated sample after plasma deposition, direct fluorination and plasma etching, respectively. The decay rates of the surface charges exceed 98% after plasma deposition and direct fluorination.
(2) The average flashover voltage increases 10% after plasma deposition.
(3) Surface morphology illustrates that the surface roughness of the plasma deposited sample is the smallest, followed by the plasma etched and fluorinated samples. A flat surface improves the uniformity of electric field, thereby reducing the surface charge accumulation.
(4) SiO_x layer and fluorine layer are formed on the Al_2O_3-ER surface after plasma deposition and direct fluorination due to the introduction of silicon-containing and fluorine-containing groups, respectively. These layers introduce shallow traps on the Al_2O_3-ER surface, accelerating the surface charge dissipation.
(5) The surface resistivity after plasma deposition and direct fluorination, decreases three and two orders of magnitude higher than that of the untreated sample, which effectively improves the surface charge dissipation on the Al_2O_3-ER surface.

In summary, compared to direct fluorination and plasma etching, plasma deposition provides an effective and clean approach for accelerating surface charge dissipation on the Al_2O_3-ER surface.

Acknowledgements This work was supported by the National Science Fund for Distinguished Young Scholars (Grant No. 51925703), the National Natural Science Foundation of China (Grant Nos. U1830135, 51777204, 11575194). Authors thank Professor Xiangyang Liu and Dr. Xu Wang from the Sichuan University, China for their kindly help in surface fluorination.

References

1. Laghari JR, Qureshi AH (1981) Surface flashover of spacers in compressed gas insulated systems. IEEE Trans. Dielectr. Electr. Insul. **EI-16**: 373–387
2. H.C. Miller, Flashover of insulators in vacuum: the last twenty years. IEEE Trans. Dielectr. Electr. Insul. **22**, 3641–3657 (2015)

3. C. Li, C. Lin, B. Zhang, Q. Li, W. Liu, J. Hu, J. He, Understanding surface charge accumulation and surface flashover on spacers in compressed gas insulation. IEEE Trans. Dielectr. Electr. Insul. **25**, 1152–1166 (2018)
4. T. Shao, F. Kong, H. Lin, Y. Ma, Q. Xie, C. Zhang, Correlation between surface charge and DC surface flashover of plasma treated epoxy resin. IEEE Trans. Dielectr. Electr. Insul. **25**, 1267–1274 (2018)
5. C. Li, C. Lin, J. H, W. Liu, Q. Li, B. Zhang, S. He, Y. Yang, F. Liu, J. He, Novel HVDC spacers by adaptively controlling surface charges–part I: charge transport and control strategy. IEEE Trans. Dielectr. Electr Insul **25**, 1238–1247 (2018)
6. P. Lukes, M. Clupek, V. Babicky, P. Sunka, The role of surface chemistry at ceramic/electrolyte interfaces in the generation of pulsed corona discharges in water using porous ceramic-coated rod electrodes. Plasma Process. Polym. **6**, 719–728 (2009)
7. Z. An, W. Chen, K. Chen, W. Gao, Z. Shen, Y. Yang, F. Liu, X. Liu, Z. Zhang, W. Yang, Z. Xing, Comparative study on direct fluorination and surface properties of alumina-filled and unfilled epoxy insulators. IEEE Trans. Dielectr. Electr. Insul. **27**, 85–93 (2020)
8. I. Kacem, M. Daoudi, W. Dridi, H. Sellemi, K. Harzli, G. De Izzara, B. Geslot, H. Guermazi, P. Blaise, F. Hosni, A.F. Al-Hossainy, Effects of neutron–gamma radiation on the free radical contents in epoxy resin: upconversion luminescence and structural stabilization. Appl. Phys. A **125**, 758 (2019)
9. B. Du, Z. Wang, J. Li, H. Liang, Z. Li, Epoxy insulator with surface graded-permittivity by magnetron sputtering for gas-insulated line. IEEE Trans. Dielectr. Electr. Insul. **27**, 197–205 (2020)
10. Y.R. Eker, M. Özcan, A.O. Özkan, H. Kırkıcı, The Influence of Al_2O_3 and TiO_2 additives on the electrical resistivity of epoxy resin-based composites at low temperature. Macromol. Mater. Eng. **304**, 1800670 (2019)
11. B. Du, R. Xu, J. Li, H. Liu, M. Xiao, Z. Li, Z. Hou, Improved charging behaviors and flashover voltage of polypropylene film based on surface molecular structure modification. Thin Solid Films **680**, 12–19 (2019)
12. B. Du, Y. Gao, Gamma-ray irradiation inhibiting surface charge accumulation on polyethylene. IEEE Trans. Dielectr. Electr. Insul. **16**, 876–881 (2009)
13. L. Que, Z. An, Y. Ma, D. Xie, F. Zheng, Y. Zhang, Improved DC flashover performance of epoxy insulators in SF6 gas by direct fluorination. IEEE Trans Dielectr Electr Insul **24**, 1153–1161
14. T. Shao, R. Wang, C. Zhang, P. Yan, Atmospheric-pressure pulsed discharges and plasmas: mechanism, characteristics and applications. High Volt **3**, 14–20 (2018)
15. Y.P. Raizer, *Gas Discharge Physics* (Springer, Berlin, 1991)
16. H.E. Wagner, R. Brandenburg, K.V. Kozlov, A. Sonnenfeld, P. Michel, J.F. Behnke, The barrier discharge: basic properties and applications to surface treatment. Vacuum **71**, 417–436 (2003)
17. C. Zhang, J. Qiu, F. Kong, X. Hou, Z. Fang, Y. Yin, T. Shao, Plasma surface treatment of Cu by nanosecond-pulse diffuse discharges in atmospheric air. Plasma Sci. Technol. **20**, 014011 (2018)
18. T. Shao, K. Long, C. Zhang, P. Yan, S. Zhang, R. Pan, Experimental study on repetitive unipolar nanosecond-pulse dielectric barrier discharge in air at atmospheric pressure. J. Phys. D Appl. Phys. **41**, 215203 (2018)
19. T. Shao, Y. Yu, C. Zhang, D. Zhang, Z. Niu, J. Wang, P. Yan, Y. Zhou, Excitation of atmospheric pressure uniform dielectric barrier discharge using repetitive unipolar nanosecond-pulse generator. IEEE Trans. Dielectr. Electr. Insul. **17**, 1830–1837 (2010)
20. T. Shao, C. Zhang, K. Long, D. Zhang, J. Wang, P. Yan, Y. Zhou, Surface modification of polyimide films using unipolar nanosecond-pulse DBD in atmospheric air. Appl. Surf. Sci. **256**, 3888–3894 (2010)
21. C. Zhang, T. Shao, K. Long, Y. Yu, J. Wang, D. Zhang, P. Yan, Y. Zhou, Surface treatment of polyethylene terephthalate films using DBD excited by repetitive unipolar nanosecond pulses in air at atmospheric pressure. IEEE Trans. Plasma Sci. **38**, 1517–1526 (2010)
22. C. Zhang, Y. Zhou, T. Shao, Q. Xie, J. Xu, W. Yang, Hydrophobic treatment on polymethylmethacrylate surface by nanosecond-pulse DBDs in CF_4 at atmospheric pressure. Appl. Surf. Sci. **311**, 468–477 (2014)

23. T. Shao, W. Yang, C. Zhang, Z. Niu, P. Yan, E. Schamiloglu, Enhanced surface flashover strength in vacuum of polymethylmethacrylate by surface modification using atmospheric-pressure dielectric barrier discharge. Appl. Phys. Lett. **105**, 071607 (2014)
24. C. Zhang, H. Lin, S. Zhang, Q. Xie, C. Ren, T. Shao, Plasma surface treatment to improve surface charge accumulation and dissipation of epoxy resin exposed to DC and nanosecond-pulse voltages. J. Phys. D Appl. Phys. **50**, 405203 (2017)
25. T. Shao, F. Liu, B. Hai, Y. Ma, R. Wang, C. Ren, Surface modification of epoxy using an atmospheric pressure dielectric barrier discharge to accelerate surface charge dissipation. IEEE Trans. Dielectr. Electr. Insul. **24**, 1557–1565 (2017)
26. R. Wang, H. Lin, Y. Gao, C. Ren, K. Ostrikov, T. Shao, Inorganic nanofilms for surface charge control on polymer surfaces by atmospheric-pressure plasma deposition. J. Appl. Phys. **122**, 233302 (2017)
27. F. Kong, S. Zhang, H. Lin, C. Ren, C. Zhang, T. Shao, Effects of nanosecond pulse voltage parameters on characteristics of surface charge for epoxy resin. IEEE Trans. Dielectr. Electr. Insul. **25**, 2058–2066 (2018)
28. C. Zhang, Y. Ma, F. Kong, P. Yan, C. Chang, T. Shao, Atmospheric pressure plasmas and direct fluorination treatment of Al_2O_3-filled epoxy resin: A comparison of surface charge dissipation. Surf. Coat. Technol. **362**, 1–11 (2020)
29. C. Zhang, Y. Ma, F. Kong, R. Wang, C. Ren, T. Shao, Surface charge decay of epoxy resin treated by AP-DBD deposition and direct fluorination. IEEE Trans. Dielectr. Electr. Insul. **26**, 768–775 (2019)
30. B. Du, J. Li, W. Du, Dynamic behavior of surface charge on direct-fluorinated polyimide films. IEEE Trans. Dielectr. Electr. Insul. **20**, 947–954 (2013)
31. A. Crisci, B. Gosse, J.P. Gosse, V. Ollier-Duréault, Surface-potential decay due to surface conduction. Eur. Phys. J. Appl. Phys. **4**, 107–116 (1998)
32. J. Kindersberger, C. Lederle, Surface charge decay on insulators in air and sulfurhexafluorid-part I: simulation. IEEE Trans. Dielectr. Electr. Insul. **15**, 941–948 (2008)
33. J. Kindersberger, C. Lederle, Surface charge decay on insulators in air and sulfurhexafluorid-part II: measurements. IEEE Trans. Dielectr. Electr. Insul. **15**, 949–957 (2008)
34. Y. Jin, C. Ren, L. Yang, D. Wang, Comparative study of the surface cleaning for Ar-/He-based plasma jets at atmospheric pressure. IEEE Trans. Plasma Sci. **43**, 3193–3199 (2015)
35. C. Zhang, T. Shao, Y. Zhou, Z. Fang, P. Yan, W. Yang, Effect of O2 additive on spatial uniformity of atmospheric-pressure helium plasma jet array driven by microsecond-duration pulses. Appl. Phys. Lett. **105**, 044102 (2014)
36. T. Desmet, R. Morent, N. De Geyter, C. Leys, E. Schacht, P. Dubruel, Nonthermal plasma technology as a versatile strategy for polymeric biomaterials surface modification: a review. Biomacromol **10**, 2351–2378 (2009)
37. C. Li, J. He, J. Hu, Surface morphology and electrical characteristics of direct fluorinated epoxy-resin/alumina composite. IEEE Trans. Dielectr. Electr. Insul. **23**, 3071–3077 (2016)
38. C. Li, J. Hu, C. Lin, B. Zhang, G. Zhang, J. He, Surface charge migration and dc surface flashover of surface-modified epoxy-based insulators. J. Phys. D Appl. Phys. **50**, 065301 (2017)
39. Z. An, F. Shan, L. Yang, R. Shen, X. Gu, F. Zheng, Y. Zhang, Unusual effect of temperature on direct fluorination of high temperature vulcanized silicone rubber and properties of the fluorinated surface layers. IEEE Trans. Dielectr. Electr. Insul. **25**, 190–198 (2018)
40. J.G. Simmons, M.C. Tam, Theory of isothermal currents and the direct determination of trap parameters in semiconductors and insulators containing arbitrary trap distributions. Phys. Rev. B **7**, 3706 (1973)
41. S. Hirota, Anisotropy study of electrical conduction in oriented polymers. I. Analysis of bulk electrical conduction parallel to the surface of film. J. Appl. Phys. **50**, 3003–3006 (1979)
42. B. Zhang, G. Zhang, Q. Wang, C. Li, J. He, Z. An, Suppression of surface charge accumulation on Al_2O_3-filled epoxy resin insulator under dc voltage by direct fluorination. AIP Adv. **5**, 127207 (2015)
43. B. Du, X. Kong, Y. Xing, J. Li, Trap distribution of electron beam irradiated epoxy resin under repetitive pulse voltage. IEEE Trans. Dielectr. Electr. Insul. **24**, 3869–3877 (2017)

Promising Functional Graded Materials for Compact Gaseous Insulated Pipelines

Jin Li, Wendong Li, Boxue Du, and Guanjun Zhang

Abstract The electric field distributions along gas–solid interfaces determine the reliability, lifetimes and sizes of gaseous insulated switchgears/pipelines (GIS/GIL), which also affect the reliability of power systems. In this paper, the characteristics of steady and transient electric field distributions were first introduced, followed by the concept of functionally graded materials (FGM). The development histories of FGM applied in electrical engineering and the related optimization methods were described in detail. The field regulation effect and fabrication technology of different FGM insulators were also compared. To overcome the limitations of traditional FGM insulators, the design of surface functional graded materials (SFGM) was proposed, together with their preparation methods and electrical performance. Furthermore, the future development prospects of FGM and SFGM insulators for compact GIS/GIL were summarized.

1 Introduction

Nowadays, large-scale wind farms inland and offshore, as well as solar generation stations, are under planning or construction, which requires reliable and economical power transmissions and conversions. The technology of gaseous insulated switchgears and pipelines (GIS/GIL) invented in the 1960s has been proven feasible since the installation of the first high voltage GIS in 1968 and the first 420 kV GIL in 1974 [1]. GIS and GIL are indispensable for power generation, transmission and distribution, based on their various advantages compared with the traditional methods [2].

J. Li · B. Du (✉)
School of Electrical and Information Engineering, Tianjin University, Tianjin 300072, China
e-mail: duboxue@tju.edu.cn

W. Li · G. Zhang
State Key Laboratory of Electrical Insulation and Power Equipment, Xi'an Jiaotong University, Xi'an 710049, Shaanxi, China

B. Du (ed.), *Polymer Insulation Applied for HVDC Transmission*,
https://doi.org/10.1007/978-981-15-9731-2_20

Spacers are the most critical components in GIS/GIL, playing the role of electric insulation, mechanical support and chamber isolation [3, 4]. However, the mismatch of dielectric parameters at gas–solid interfaces causes electric field (E-field) concentration around the spacers, leading to partial discharges and even flashovers over spacers. For the AC case, the E-field distribution mainly depends on the permittivity of involved dielectrics [5]. For the DC case, the steady-state electric field distribution depends on the conductivity distribution. Since the conductivity of the epoxy insulator is strongly temperature dependent, it is very difficult to design the DC gas–solid insulation system [6].

The electric field optimization of the gas–solid system can help increase the flashover voltage and downsize the devices, thereby saving costs and reducing greenhouse gas emissions [7]. To improve the insulation performance of GIS/GIL, various techniques have been applied to reduce the E-field concentration such as modifying the spacer shape and installing shielding/embedded electrodes [8]. However, these methods have high cost and low efficiency; thus, efficient and simple techniques are highly desired.

New methods to optimize the electric field distribution and improve the withstand voltage of epoxy spacers have been widely researched. Currently, the idea of functionally graded materials (FGM) has become one of the most attractive issues in the field of GIS/GIL insulation design [9–13]. FGM insulators with spatially graded distributions of permittivity/conductivity are designed and fabricated, showing promising performances in suppressing electric field distortions, elevating flashover voltages and simplifying insulation structures [9]. To date, fabrications, multi-physical field simulations and flashover tests of FGM insulators are the research hotspots. However, industry applications will become much more complex, taking the existing technology into consideration [14].

With the continuously increasing demand for electrical power, the status of large-capacity and long-distance power transmissions is particularly prominent. The critical issues in the development of high voltage GIS/GIL are the reliability of the insulation system and the greenhouse effect of SF_6 gas. All the above problems can be effectively solved or relieved by electric field optimizations.

2 Electric Field Distribution in GIS/GIL

2.1 Electric Field Under AC Voltage

In an AC GIS/GIL, for a specific geometry, the electric field distribution is capacitive and mainly depends on the insulation materials' permittivity. Equations for the field calculation are shown below:

$$\nabla \cdot J = 0 \tag{1}$$

$$J = \sigma E + j\omega D \tag{2}$$

$$E = -\nabla V \tag{3}$$

where E is the electric field strength, V/m; V is the electric potential, V; J is the current density, A/m^2; σ is the volume conductivity of the involved dielectrics, S/m; ω is the angular frequency of the power grid, rad/s; and D is the electric flux density, C/m^2.

As the relative permittivity of a material is almost independent of the electric field and temperature, the field distribution in an AC GIS/GIL can be obtained by a steady state calculation, which can then be extrapolated to almost all voltage stresses.

2.2 Electric Field Under DC Voltage

Under DC voltage, the electric field distribution is initially capacitive (similar to AC) and then progressively becomes resistive. The electric field distribution in DC GIS/GIL depends on the electrical conductivities of the insulating materials, which are affected by many factors such as temperature, running time and gas convection, and have a very large uncertainty [15, 16]. When the electric field strength at the gas–solid interface exceeds a certain threshold, processes such as field-electron emission, gas ionization or carrier injection, triggering flashovers along insulator surfaces [17].

2.3 Electric Field Under Transient Voltage

After applying a DC voltage, the transition time of the electric field distribution from capacitive to resistive in GIS/GIL varies from seconds to months depending on the electrical conductivities of the insulating materials (solid and gas) [18, 19]. It is difficult to predict which of the three field distribution states (capacitive, resistive or transient) is the most critical for DC GIS/GIL equipment. Hence, it is necessary to assess the electric field distribution, not only in initial and steady states, but also during the transition.

In the actual operation of DC GIS/GIL, the introduction of lightning impulse voltages subjects the insulator to transient overvoltage superimposed with DC voltages, which easily leads to flashovers on the insulator surface. In addition, when a DC GIS/GIL is subjected to a tripping operation, the residual voltage of the insulator can reach 0.6–1.0 times the standard voltage, remaining there for a long time, and a fast-transient overvoltage generated by the reclosing will be combined with it. The electric field distribution undergoes a process from steady-state to transient-steady state. The overvoltage level and the transition process are closely related to the dielectric and conductance characteristics of epoxy resin insulators [20].

2.4 Electric Field and Long-Term Performance

For gas-insulated equipment, when the local field strength is too high, the electron energy can exceed the electron adhesion of gas molecules, and the dielectric strength will be significantly reduced. At the same time, corona discharges and internal air gap discharges induced by local electric field distortions in actual GIS equipment will lead to deterioration of SF_6 gas and solid insulating materials, thereby reducing the service life of the equipment. The relationship between the breakdown strength and the life time is typically expressed as follows [21].

$$t = t_0 \left(\frac{E}{E_0} \right)^{-n} \tag{4}$$

where E is the breakdown strength; t is the operation time; t_0 and E_0 are values on the life time curve as reference; and n is a voltage endurance coefficient of epoxy insulators. For a typical epoxy spacer, n is obtained in the range from 10 to 20.

Therefore, the electric field distortion at the gas–solid interface is considered the basic cause of flashovers or breakdown of insulators. It is also the development bottleneck of ultra-high voltage gas-insulated equipment, which needs to be solved urgently.

3 Bulk FGM Insulator for Compact GIS/GIL

The current common solutions in response to the above issues can be divided into two categories: The first one is based on structural optimization. The second one is mainly based on the modification of insulating materials, e.g., nanocomposite, surface treatment, nonlinear conductivity materials, etc. [22, 23].

In recent years, researchers have tried to find new technologies to solve the above problems. Some of them have constructed insulating structures with non-uniform distributions of dielectric parameters (dielectric constant/conductivity) to release local electric field [24]. Due to the distinct advantages of functionally graded materials (FGM) in homogenizing electric field distributions and simplifying insulation structures, their great application potential is attracting the attention of scholars worldwide.

3.1 Origins of FGM Insulators

Functionally graded materials (FGM) are advanced engineering materials designed for a specific property or function in which spatial gradations in the structure and/or composition endow them with tailor-made characteristics. Bever et al. first researched the theoretical significance of graded structure composite materials in 1970 [25]. A decade later, Brealey et al. from the General Electric Company first fabricated a capacitance-graded 23 kV post insulator using the lamination method, i.e., laminating and adhering materials with different dielectric properties to compose a graded insulation structure. Compared with ungraded insulators, the withstand voltages of the graded insulators showed an increment of ~10% [26]. Later, Hayashi and Watanabe prepared coaxial disk-type and post-type FGM spacers using the lamination process. For the disk-type FGM spacers, the permittivity distributions decreased from the HV electrode to the GND electrode in the radical direction in different manners. The results showed that the electric field strength at the HV triple junction was greatly weakened by ~44% in a superior manner and the surface breakdown voltage was obviously improved by using FGM spacers compared with the ungraded spacer [27, 28].

Although the effects of lamination method on improving the insulation properties of epoxy insulators have been confirmed by many researchers, there are some limitations on its application. For instance, the lamination method has difficulty achieving a continuous change in dielectric properties, which results in electric field distortions between gradients. Future studies should focus on relieving variations in dielectric properties, reducing defects between different layers and increasing the long-term efficacy to further improve the electric field distribution and the withstand voltage.

3.2 Fabrication of Bulk FGM Insulators

3.2.1 Centrifugation Method

The centrifugation method is one of the classical methods for FGM preparation. Kurimoto et al. first succeeded in fabricating ε-FGM spacers by applying centrifugal force to epoxy/SiO_2/TiO_2 composites [29]. Particles move in the liquid epoxy resin under the centrifugal force, forming gradient distributions of filler contents, i.e. permittivity gradients. The permittivity distribution of an ε-FGM spacer can be controlled by controlling the resin viscosity, particle size, centrifugal force, centrifugal time, etc. In 2006, Kato et al. fabricated FGM spacers with epoxy and TiO_2 fillers (average diameter: 8.7 μm) using the centrifugation method. Simulations and tests showed that the electric field strength at the triple junction of the FGM insulator was weakened by ~20% and the flashover voltage was improved by ~69% in 0.4 MPa SF_6 gas compared with the uniform insulator [30]. In 2012, Hayakawa et al. fabricated GLP-FGM and

U-FGM spacers in disk type [31]. Figure 3 shows the concept and schematic illustration of fabricating GLP-FGM and U-FGM spacers. The maximum electric field strength can be reduced by ~12% with the coaxial-type U-FGM. Compared with the lamination method, the permittivity distributions of GLP-FGM and U-FGM spacers fabricated by the centrifugation method are continuous, and no interface can be observed. However, the controllability of the centrifugation method is weak; it is difficult to fabricate an FGM spacer with an optimal permittivity distribution. To bring this method into practical application, further research on the flow process of epoxy/filler systems is desired.

3.2.2 Printing Technology

The 3D printing technique is considered a novel and promising method for preparing d-FGM insulating spacers [32]. Figure 5 shows 3D printers and the working principles. In 2016, Kurimoto et al. printed a conical insulating spacer layer by layer [33]. The liquid UV-cured-acrylic resin with alumina fillers was poured onto the resin tray and irradiated with a scanning UV light through the transparent bottom of the tray. By elevating the building support plate step by step, the resin was cured layer by layer until the spacer was built up. In 2008, they printed a two-layered permittivity-graded insulator using unfilled and alumina-filled UV-cured resins [34]. In 2019, Zhang et al. printed a non-uniform conductivity spacer by fused deposition modelling (FDM), as shown in Fig. 1 [35]. Acrylonitrile butadiene styrene (ABS), polylactic acid (PLC), ABS/carbon black composites, and PLA/grapheme composites were chosen as printing materials, which were heated, extruded and deposited under 3D control, forming a non-uniform spacer point by point.

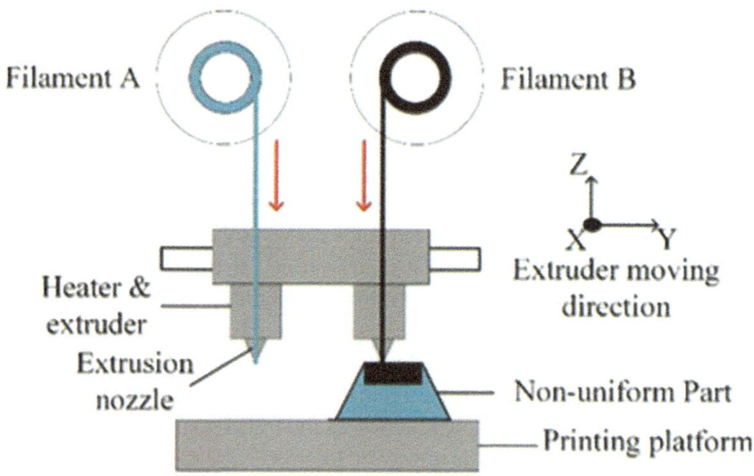

Fig. 1 Schematic illustrations of 3D printers [35]. Copyright Clearance Center's RightsLink® License Number: 4857690254648

The flashover voltage improvement of the non-uniform conductivity insulator was verified both in SF_6 and vacuum under DC voltages. In 0.3 MPa SF_6, the flashover voltage was 23.8% enhanced for the non-uniform ABS insulator, and 16% enhanced for the non-uniform PLA insulator. In vacuum, the flashover voltage was increased by 19.1% for the non-uniform PLA sample, and 20.9% for the non-uniform ABS sample. Since the 3D-printed spacers are built by continuous accumulation of material units (layers or points), this method has the potential to fabricate FGM spacers with any designed permittivity and conductivity distributions inside. However, some technical obstacles must be overcome before industrial application, e.g., limited printing materials, difficult dielectric property adjustment, and uncertain mechanical performances.

3.2.3 Flexible Mixture Casting Method

The flexible mixture casting (FMC) method was proposed by Hayakawa et al. to fabricate ε-FGM spacers [36]. The high-ε resin was prepared by filling the epoxy resin with high-ε $SrTiO_3$ particles, and the low-ε resin was prepared by filling epoxy resin with low-ε SiO_2 particles. Two resins were mixed by a blender, and the mixing ratio was continuously and flexibly controlled by the injection speeds of the high-ε and low-ε resins. The mixture was introduced into a casting mould and accumulated in layers with a controlled mixing ratio. After resin injection, the spacer was put into an oven and cured for 15 h at 80 °C.

Compared with the centrifugation method, the FMC method can be used to fabricate FGM insulators with designed permittivity and conductivity distributions, thereby maximizing the electric field relaxation effect. However, there will be many interfaces composed of neighboring resins, so the mechanical and dielectric properties of the interfaces should be improved before application.

3.2.4 Magnetophoresis and Electrophoresis Methods

In addition to the abovementioned methods, the magnetophoresis and the electrophoresis methods have also been introduced to fabricate FGM. In 2016, Nardi et al. prepared FGM samples by applying an external magnetic field on the suspension system of Fe_3O_4@TiO_2 nanoparticles and epoxy matrix, as shown in Fig. 8 [37]. Numerical simulations clarified that these graded composites can reduce the electric field strength at the triple junction by up to 45% compared with the conventional one. In 2018, Diaham et al. proposed an original method to structure FGM on metalized alumina DBC substrates by local handling of high-permittivity ceramic ($SrTiO_3$) particles in epoxy resin using electrophoresis (i.e., DC voltage application) [38]. By applying FGM, an impressive attenuation of the field reinforcement at the triple point can be obtained, and the breakdown voltages were improved by ~70 and ~30% compared with neat epoxy and the homogeneous composite.

The magnetophoresis and electrophoresis methods are similar to the centrifugation method, but more flexible and simpler; the permittivity distribution can be adjusted by controlling the electric/magnetic field distribution. Before practical application, further research into magnetic/electrostatic preparation and accurate control of magnetophoresis/electrophoresis are desired for fabricating large-scale and complex-shaped insulators.

4 Optimizing Dielectric Parameter Distribution of FGM Insulator

With the development of fabrication techniques for FGM insulators, it becomes feasible to manufacture insulators with designed dielectric parameter distributions to maximize the electric field relaxation effect. Various optimization methods have been developed to modify the distribution of dielectric parameters (conductivity and/or permittivity) in solid insulators while keeping insulators in simple contour form.

4.1 *Iterative Optimization*

In 2006, a computer-aided technique for optimizing FGM solid insulators was proposed by Okubo et al. [39]. The optimal spatial distribution of permittivity was defined as "the distribution which can relax the field stress most effectively inside and around the insulator". The electric field calculation was carried out with the finite element method (FEM), and the electric field optimization in each FEM element was performed by an iterative method. The following two steps were repeated until the termination condition was reached.

(a) Calculate the electric field strength in each FEM element of the insulator, P_i ($1 < i < n$).
(b) Modify the permittivity in each FEM element of the insulator by the following equation [40]:

$$\varepsilon' = \varepsilon_i \cdot \frac{E_i}{E_{obj}} \cdot \frac{\varepsilon_{\max}}{\varepsilon_{\min}} \cdot C \tag{5}$$

where ε_i and ε_i' are the relative permittivity at P_i before and after modification, respectively; E_i is the electric field strength at P_i; and E_{obj} is the objective value of the electric field strength. $\varepsilon_{\max}$ and $\varepsilon_{\min}$ are the limits of the relative permittivity, respectively. $\varepsilon_{\min} = 4$, and $\varepsilon_{\max}$ varies from 8 to 200. C is the modification coefficient, whose initial value $C_0 = E_{\text{obj}}/E_{\max}$ in the first calculation. When

C is <0.01, the variation range of E_{max} is <0.1 kV/mm, and the optimization process stops.

Recently, the authors' team developed a relatively simple method to optimize the dielectric parameter distribution [41]. The non-uniform E-field distribution along the insulator surface is considered the key factor that triggers surface flashovers. The insulation margin of the insulator bulk is so large that there is no need to optimize the E-field distribution inside the insulator, and the optimization objective can be set as the surface electric field distribution. Figure 2 shows the flow chart of the simplified optimization process. The permittivity distribution of the insulator was designed to vary in the z direction, and the surface E-field strength was calculated to evaluate the permittivity distribution. An adjustment coefficient k (z) was defined to modify the permittivity distribution in each iteration, and the E-field non-uniform factor f was defined to check the termination criterion ($f < f_0$).

$$k_i(z) = E_i(z)/E_{i\,\min} \tag{6}$$

$$\varepsilon_{r(i+1)}(z) = k_i(z)\varepsilon_{ri}(z) \tag{7}$$

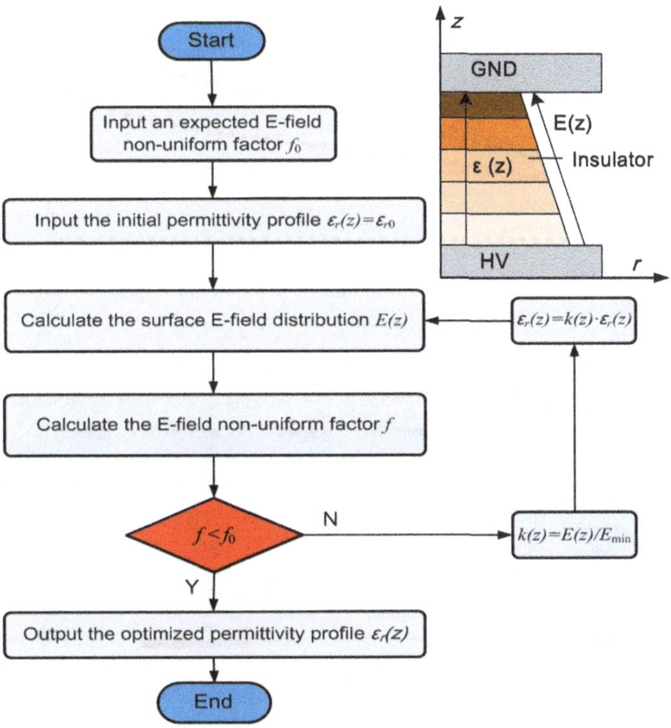

Fig. 2 Flow chart of the optimization process

$$f = E_{i\,\max}/E_{i\,\min} \tag{8}$$

where $E_i(z)$ is the tangential E-field strength along the insulator surface in the ith iteration; $E_{i\max}$ and $E_{i\min}$ are the maximum and minimum E-field strengths in the ith iteration; and $\varepsilon_{ri}(z)$ and $\varepsilon_{r(i+1)}(z)$ are the permittivity distributions in the ith and the $(i+1)$th iterations.

Figure 3 shows the permittivity and electric field distributions during optimization. As the iterative process proceeds, the permittivity distribution converges to the optimal solution; the corresponding electric distribution becomes increasingly uniform with the iteration number. Although the electric field concentration inside the insulator bulk becomes more serious, the electric field relaxation along the gas–solid

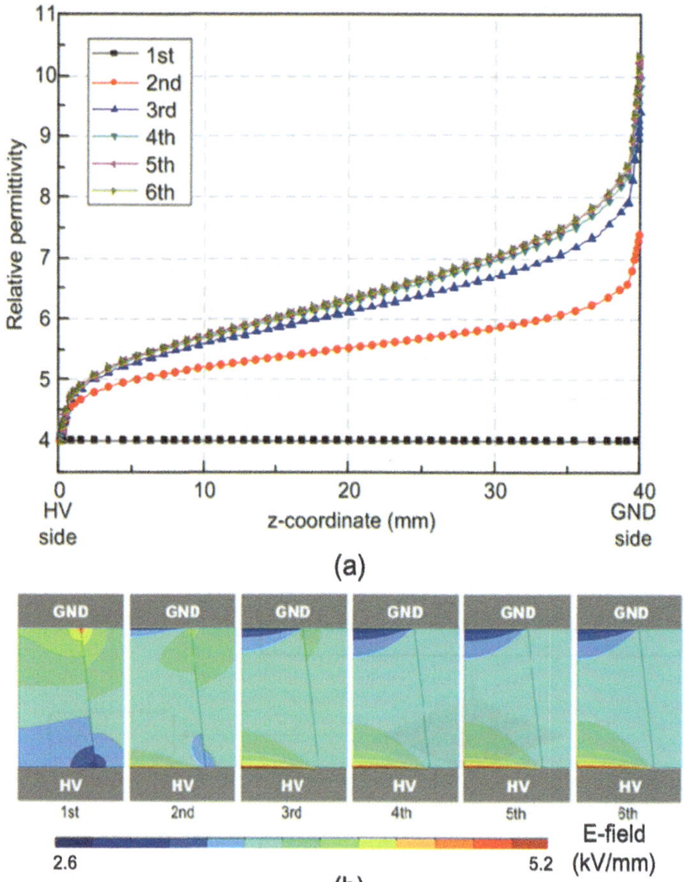

Fig. 3 **a** Permittivity and **b** E-field distributions during optimization

interface can still improve the insulation performance of the gas-sting the permittivity distribution is not suitable for all cases.

4.2 Topology Optimization

Topology optimization is usually used in the field of structural mechanics to improve the mechanical performances by optimizing the material distribution for given design space, loads and other conditions. Recently, Zhang et al. introduced the topology optimization method to optimize the FGM insulator point by point [42]. The spatial distribution of the relative permittivity in an FGM insulator (area I) is represented by the variable density method.

$$\varepsilon(r, z) = (\varepsilon_{\max} - \varepsilon_0)\rho_{design}^{p} + \varepsilon_0, (r, z) \in \Omega_{\mathrm{I}} \tag{9}$$

where ρ_{design} is the control variable, and $\varepsilon_{\max}$ and ε_0 is the maximum and minimum relative permittivity of the FGM insulator. The following multi-objective design is presented to improve the uniformity of the E-field distribution.

$$\begin{cases} \min f = (1-q)f_1 + qf_2 \\ s.t. \begin{cases} 10^{-9} \le \rho^p \le 1 \\ 0 \le \int \rho d\Omega \le \gamma A, \Omega \in \Omega_{\mathrm{I}} \end{cases} \end{cases} \tag{10}$$

where A is the cross-section area of the spacer; q is a weighting parameter balancing f_1 and f_2; and f_1 and f_2 are assigned to control the uniformity and reliability of the permittivity distribution, respectively.

$$\begin{aligned} f_1 &= \frac{1}{C_{ref}} \int |E - E_{mean}|^2 d\Omega, \Omega \in \Omega_{\mathrm{I}} \cup \Omega_{\mathrm{II}} \\ f_2 &= \frac{h_0 h_{\max}}{A} \int |\nabla \rho_{design}|^2 d\Omega, \Omega \in \Omega_{\mathrm{I}} \end{aligned} \tag{11}$$

The global convergent method of moving asymptotes (GCMMA) was used to solve this design problem. Figure 4 shows the permittivity and electric field distributions after optimization. The electric field concentration at the triple junction point is relaxed by the FGM insulator, and a larger permittivity range leads to a better field relaxation effect.

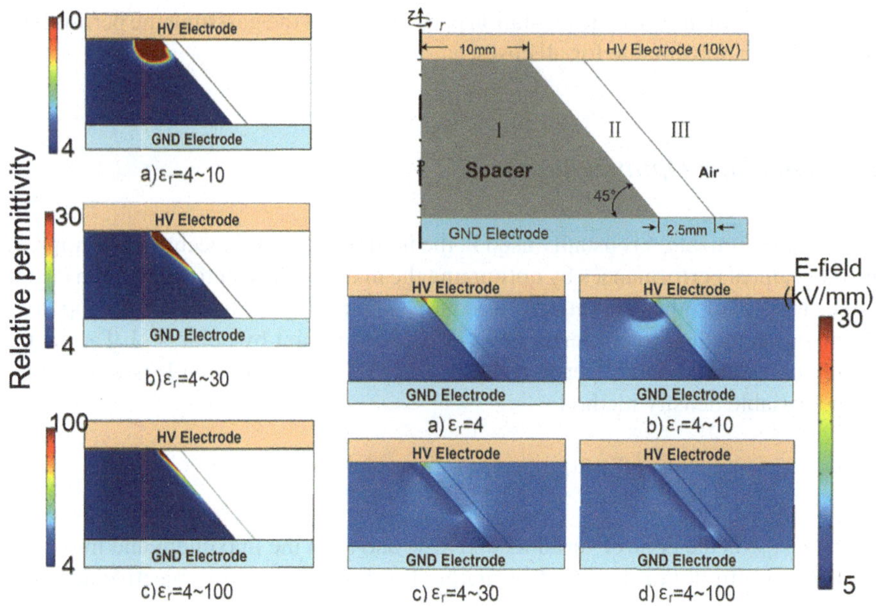

Fig. 4 Permittivity and electric field distributions after optimization [42]. Copyright Clearance Center's RightsLink® License Number: 4857690662310

4.3 Particle Swarm Optimization

Particle swarm optimization (PSO) is a kind of intelligent searching optimization algorithm involving simulating the group cooperation behaviours of birds foraging. The PSO method has the advantages of simple concept, easy realization with only a few tuning parameters and so on. In 2015, Qasim et al. introduced the PSO method to modify the spatial distribution of the permittivity in a spacer [43]. The optimization objective is shown as follows and was solved by the PSO algorithm.

$$\begin{aligned} &\min \ F(\varepsilon) = wgt \cdot \max(E_i) + (1 - wgt) \cdot std\ dev(E_j) \\ &\forall i \in \{\text{the total polygons}\} \\ &\forall j \in \{\text{node points inside insulator}\} \end{aligned} \tag{12}$$

where E_i is the electric field of polygon i in and around the insulator; E_j is the electric field of node point j inside the insulator; ε is a vector of permittivity values of polygons inside the insulator; and wgt is a weighing factor.

After optimization, the maximum electric field in the spacer with a pareto-optimal permittivity distribution is significantly reduced. However, the electric field distribution is not non-monotonic, which may be addressed by a choice of smaller polygons at the cost of greatly increasing the calculation burden.

Recently, the researcher applied this technique to uniform the electric field distribution in a cable termination by using a stress control tube, obtaining an optimized permittivity gradation [44]. The maximum electric field is relaxed by ~62%, and the field utilization factor is improved to ~67%.

To optimize the dielectric parameter distributions, the correct object should first be determined. Currently, most researchers are concerned with the norm of the electric field inside and around the insulator. However, in a gas–solid system, the insulation weakness point is usually the gas–solid interface instead of the insulator bulk. There are tangential and normal electric field components along the gas–solid interface, so their weighting factors during the flashover inducement and development processes are critical for the optimization. Moreover, the mechanical performances of FGM insulators should also be considered in future optimization work.

5 Surface FGM Insulator for Compact GIS/GIL

At present, FGM insulation design mainly involves controlling the spatial gradient distribution of the dielectric constant or conductivity in solid insulation structures, thus effectively suppressing electric field distortions, and ultimately improving the flashover voltage of gas–solid insulation systems [24]. However, FGM insulators are faced with the problems of optimization of the spatial distribution of dielectric parameters, complex fabrication processes and low efficiency in industrial applications. Therefore, novel gradient modification methods are urgently needed [22]. The concept of surface functionally graded materials (SFGM) was proposed to form functionally graded layers on insulator surfaces. Without changing the bulk properties, it it is feasible to optimize the electric field distribution by surface treatment, which means regulating the dielectric parameter distributions on the insulator surface [45].

5.1 *Concept of SFGM*

In our previous paper, equivalent circuits for conventional (Fig. 5a), surface ε-uniform (Fig. 5b) and ε-SFGM insulators (Fig. 5c) were introduced to qualitatively describe the working principle of SFGM by assuming that the equipotential lines were parallel to the top and bottom surfaces of insulators [46]. The conventional insulator has a uniform relative permittivity distribution of ~6.8 inside the bulk. By grading the thickness distribution of the high-permittivity (over 1000) layer in the radial direction, the ε-SFGM insulator can be obtained.

In Fig. 5a, an insulator is divided into n ($n \to \infty$) slices in the height direction. Each slice i has a radius of r_i and a height of dL, forming a capacitor. The conventional insulator can be approximately represented by a capacitor series of C_i, whose value can be described as,

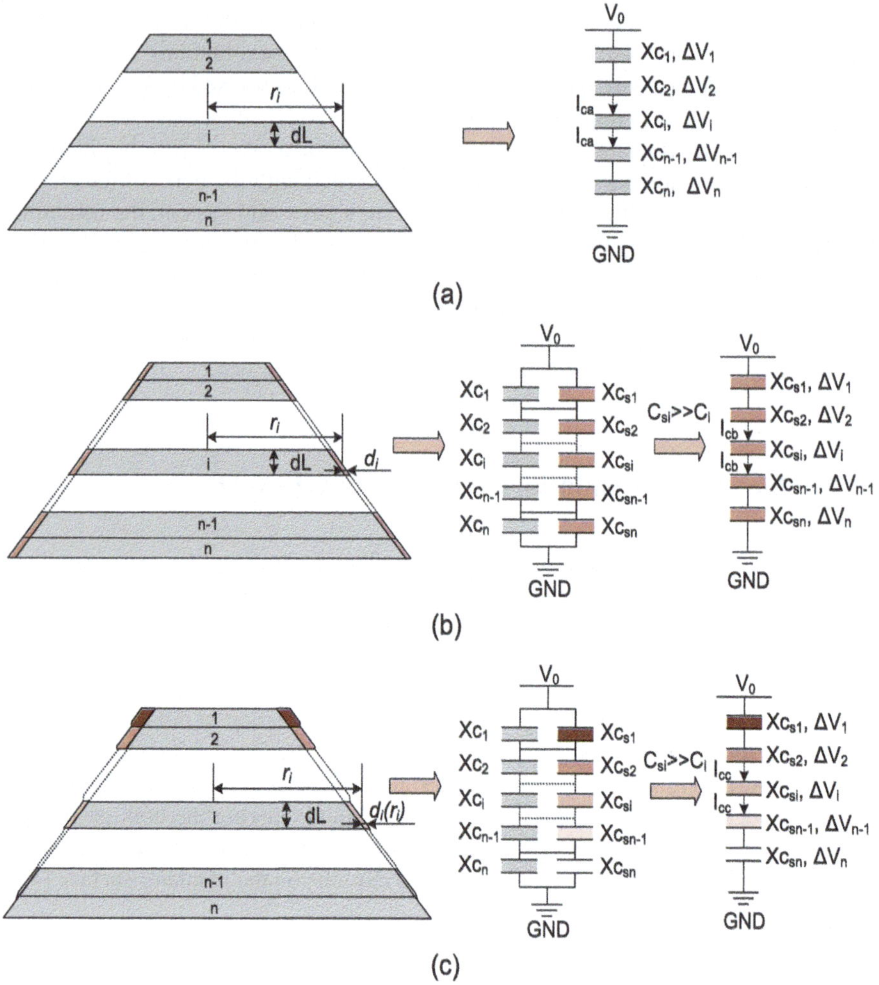

Fig. 5 The equivalent-circuit models for (a) the conventional insulator, (b) the surface ε-uniform insulator and (c) the ε-SFGM insulator [46]. Copyright Clearance Center's RightsLink® License Number: 4857390976179

$$C_i = \frac{\varepsilon_0 \varepsilon_r \pi r_i^2}{dL} \tag{13}$$

According to circuit theory, the electric field strength in each slice can be calculated by the following equation, which indicates that the electric field is inversely proportional to the square radius of a slice and the distribution is extremely uneven.

$$E_i^a = \frac{dV_i}{dL} = \frac{I_{ca}}{2\pi^2 f \varepsilon_0 \varepsilon_r} \cdot \frac{1}{r_i^2} \tag{14}$$

In Fig. 5b, each slice i of the surface layer of the surface ε-uniform insulator can be regarded as a capacitor in a cylindrical ring shape with a radius of r_i, a thickness of d and a height of dL. When the capacitance of the surface capacitor C_{si} is much higher than that of the bulk capacitor C_i, the surface ε-uniform insulator can be approximately represented by a capacitor series of C_{si}. The electric field strength in each slice can be calculated by,

$$E_i^b = \frac{dV_i}{dL} = \frac{I_{cb}}{4\pi^2 f \varepsilon_0 \varepsilon_{rs} d} \cdot \frac{1}{r_i} \quad (15)$$

From the above equation, the electric field of the surface ε-uniform insulator is inversely proportional to the radius of a slice, and the electric field distribution is more uniform than that of the conventional insulator.

$$E_i^c = \frac{I_{cc}}{4\pi^2 f \varepsilon_0 \varepsilon_{rs} d(r_i)} \cdot \frac{1}{r_i} = \frac{const \cdot I_{cc}}{4\pi^2 f \varepsilon_0 \varepsilon_{rs}} \quad (16)$$

By grading the thickness of the high-permittivity surface layer in the form of an inverse proportional function ($d(r_i) = \text{const}/r_i$), a ε-SFGM insulator can be obtained, as shown in Fig. 5c. The electric field can be calculated by,

In this case, the electric field distribution along the ε-SFGM insulator surface is ideally uniform. Although this model does not consider the influence of the stray capacitance, it can help easily understand the working principle of the SFGM insulator.

5.2 *Fabrication of SFGM Insulator*

5.2.1 Dip Coating Method

Miyaji applied the dip coating method to fabricate SFGM insulators with localized conductive layers [47]. They chose σ-SFGM insulators instead of ε-SFGM insulators to relax AC electric field concentrations because it is difficult to prepare huge-permittivity (over 1000) coatings. An alternative approach is to reduce the complex impedances of the surface layer by applying high-conductivity coatings. Although the AC electric field distribution was uniformed, high-conductivity coatings may cause unacceptable conduction loss. Moreover, the conductivity gradient was not continuous, and electric field distortions could be generated between the gradients.

5.2.2 Magnetron Sputtering Method

To fabricate ε-SFGM insulators for AC electric field relaxation, the magnetron sputtering method was proposed to deposit huge-permittivity $BaTiO_3$ layers on insulator

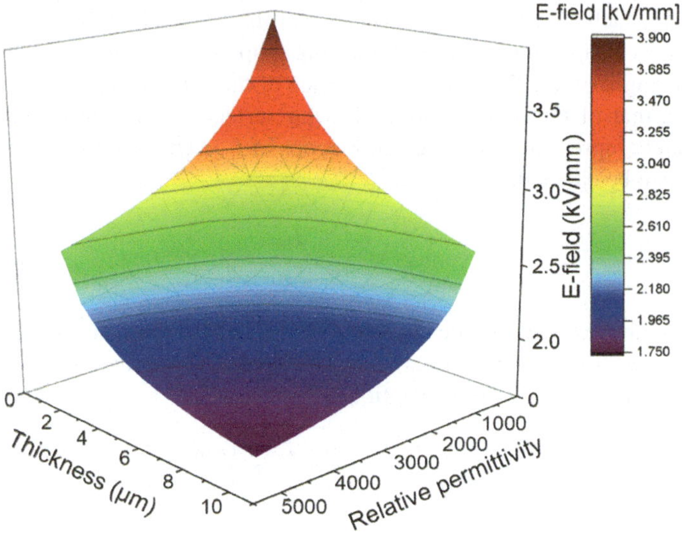

Fig. 6 Combined effects of sputtering layer permittivity and thickness on maximum electric field strength along insulator surface [46]. Copyright Clearance Center's RightsLink® License Number: 4857390976179

surfaces [48]. The electric field relaxation effect of the surface ε-uniform insulator has a "saturation effect". As shown in Fig. 6, the maximum electric field strength declines with increasing the permittivity and thickness of the sputtering layer and finally reaches a stable value when the product of the sputtering layer permittivity and thickness is high enough [46].

To maximize the electric field relaxation effect of the ε-SFGM insulator, the iterative method was used to optimize the thickness distribution of the sputtering layer, as shown in Fig. 7 [49]. After four iterations, the thickness distribution of the sputtering layer converges to the optimal solution. After optimization, the continuous sputtering method was used to fabricate the optimal ε-SFGM insulator. During sputtering, the sputtering time and the layer thickness distributions were controlled by a rotating mask with a designed gap above the insulator. Compared with the conventional insulator, the maximum electric field of the optimal ε-SFGM insulator was reduced by ~70%, and the AC flashover voltage was improved by ~20% in air. Compared with the dip coating method, the ε-SFGM insulator fabricated by the magnetron sputtering method had a much lower conduction loss due to the huge-permittivity layer. In addition, the thickness distribution of the sputtering layer was continuous and optimal, thus exhibiting a better electric field relaxation effect. However, both the dip coating and sputtering methods are faced with the problem of long-term interface stability. Future research should focus on enhancing the binding ability and interfacial compatibility between the sputtering layer and the insulator before practical application.

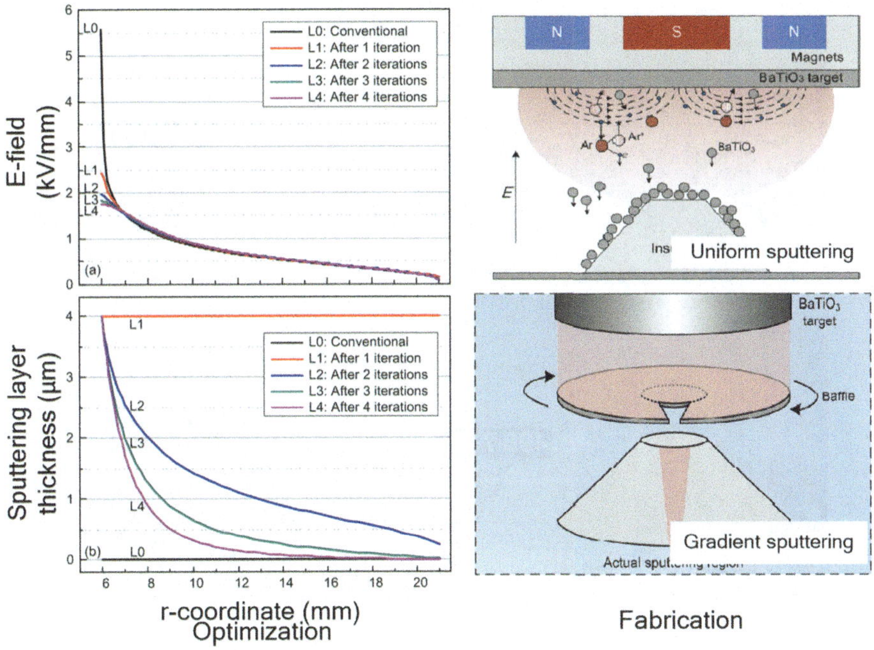

Fig. 7 Optimization and fabrication of ε-SFGM insulator [49]. Copyright Clearance Center's RightsLink® License Number: 4857380962674

5.2.3 Gradient Fluorination Method

Under a DC voltage, the conductivity distribution of the dielectric materials determines the steady-state electric field distribution. Thus, the design of surface conductivity distribution provides an idea for DC electric field control.The fluorination treatment is simple and efficient in improving the surface conductivity and other electrical parameters of polymers [50, 51]. The thickness of the fluorination layer and the surface conductivity of epoxy increase with the fluorination time and the fluorination temperature. The gradient fluorination method was thus proposed by the authors' team to fabricate σ-SFGM insulators with a continuous surface conductivity gradient [52, 53]. During fluorination, a heater was placed on the insulator top to form a continuous temperature gradient, controlling the distribution of fluorination rate and regulating the thickness distribution of the fluorination layer, as shown in Fig. 8a.

By changing the temperature distribution and the fluorination time, continuous σ-SFGM insulators with different surface conductivity gradients can be fabricated, as shown in Fig. 9. An obvious electric field reduction can be observed at the triple junction point of the SFGM insulator, as shown in Fig. 9c. The DC flashover voltage of the σ-SFGM insulator was improved up to ~33% in air compared with the conventional insulator. Compared with the coating and the sputtering methods, SFGM insulators

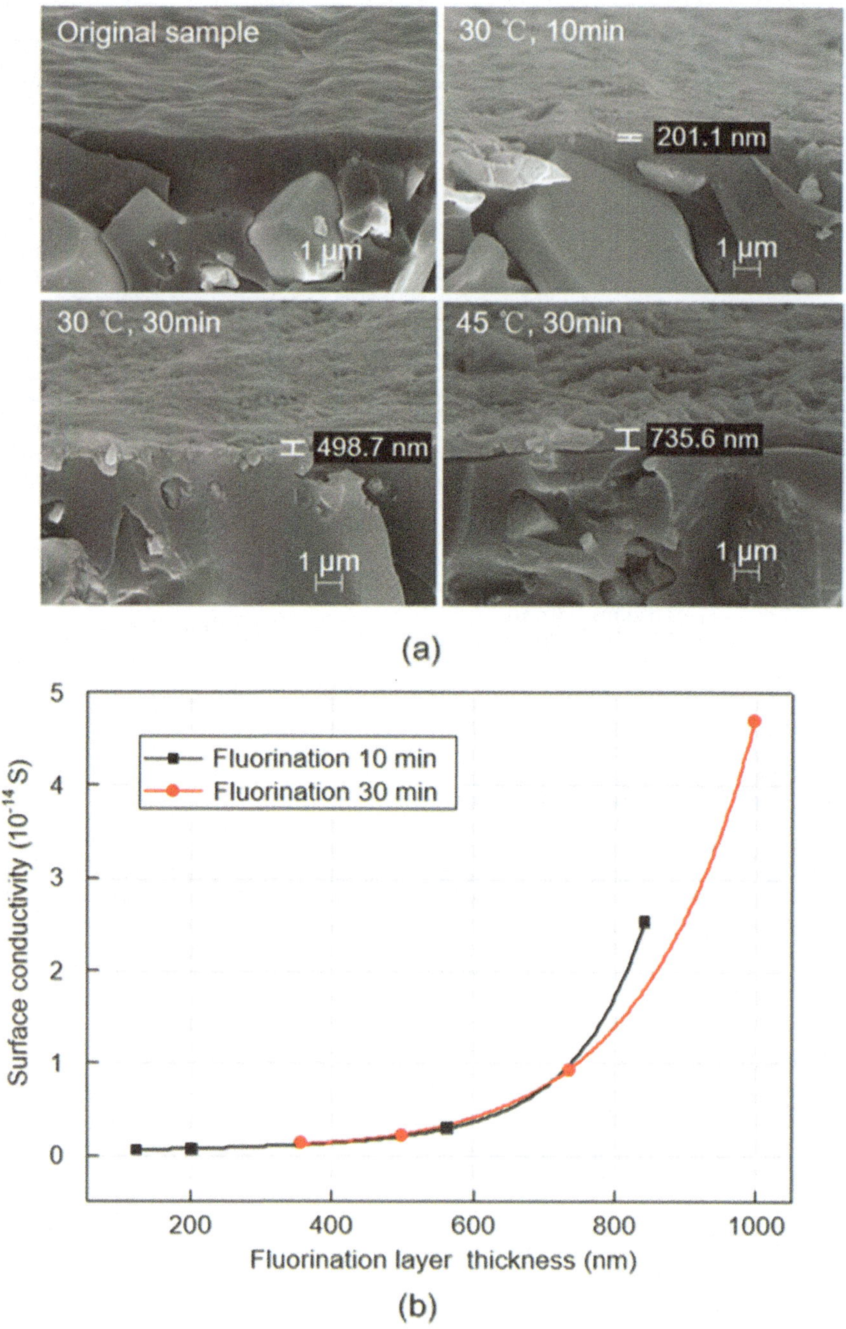

Fig. 8 **a** Cross-sections of the original sample and fluorinated samples of different fluorination times and temperatures, and **b** relationship between the surface conductivity and the fluorination layer thickness [52]

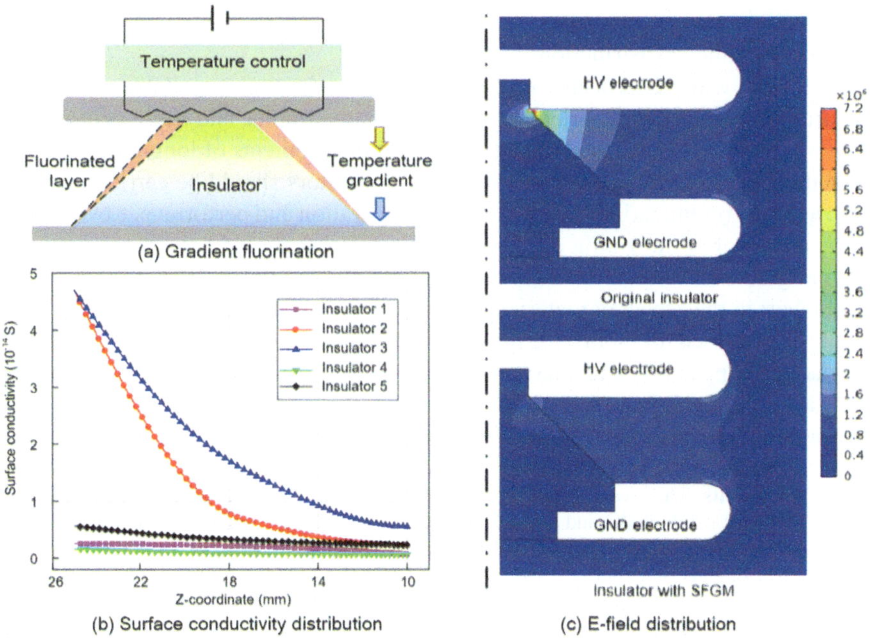

Fig. 9 **a** Diagram of the gradient fluorination method, **b** surface conductivity distributions of SFGM insulators, and **c** electric field distributions [53]

fabricated by the gradient fluorination method have no interface problems. However, the relationship among the surface conductivity, fluorination time and fluorination temperature is complicated and influenced by some external factors. Accurate control of the temperature distribution during fluorination is also critical for obtaining the optimal surface conductivity gradient. After solving the above problems, σ-SFGM insulators fabricated by the gradient fluorination method can be brought into practical application.

6 Conclusion

Suppressing the electric field distortion at gas–solid interfaces is the key of the insulation design to reduce flashovers and further promote the compactness of high-voltage gas insulated equipment. FGM for relaxing electric field concentrations has been developed for decades, and relevant fabrication techniques keep arising. Simulation and experimental test results showed excellent performances of FGM insulators in uniforming electric field distributions and reducing flashovers, but the mechanical performances were not considered. The future development of FGM insulators should be carried out from two aspects. First, the optimization of FGM should consider the mechanical performances of insulators in addition to their electric field regulation effects; second, novel techniques should be proposed to form continuous and optimal dielectric parameter distributions.

Compared with FGM insulators, SFGM insulators have more reliable mechanical performances, and their fabrication techniques are simpler and more economical. However, most SFGM insulators are faced with the problems of long-term interface reliability. The future development of SFGM insulators should be carried out from three aspects: optimization design, material preparation and performance evaluation. In the aspect of optimization design, SFGM insulators should have continuous and optimal gradients to maximize the electric field regulation effect and minimize the loss power; in the aspect of material preparation, novel surface treatment techniques should be proposed for modifying the surface dielectric parameters and enhancing the interface binding ability and interfacial compatibility before practical application; (3) in the aspect of performance evaluation, a performance evaluation system for SFGM insulators should be established and perfected considering actual working conditions.

Acknowledgments This work is supported by the National Natural Science Foundation of China (51807136, 51537008 and U196620069); Natural Science Foundation of Tianjin City (18JCQNJC07300); National Postdoctoral Program for Innovative Talents (BX201700168); Key Laboratory of Engineering Dielectrics and Its Application (Harbin University of Science and Technology), Ministry of Education (KFY1804); Project supported by China State Key Lab. of Power System (SKLD19KM08).

References

1. H. Zhou, G. Ma, Y. Wang, W. Qin, J. Jiang, C. Yan, C. Li, Optical sensing in condition monitoring of gas insulated apparatus: a review. High Volt. **4**(4), 259–270 (2019)
2. J. Sun, L. Sun, W. Chen, Z. Li, X. Yan, Y Xu, Metal particle movement and distribution characteristics under AC voltage and ball-plane electrodes. High Volt. **4**(2), 138–143 (2019)
3. B.X. Du, H.C. Liang, J. Li, C. Zhang, Temperature dependent surface potential decay and flashover characteristics of epoxy/SiC composites. IEEE Trans. Dielectr. Electr. Insul. **25**(2), 631–638 (2018)
4. H.C. Liang, B.X. Du, J. Li, Z.H. Wang, Mechanical stress distribution and risk assessment of 110 KV GIS insulator considering Al_2O_3 settlement. High Volt. **4**(1), 65–71 (2019)
5. M. Talaat, A. El-Zein, M. Amin, Electric field simulation for uniform and FGM cone type spacer with adhering spherical conducting particle in GIS. IEEE Trans. Dielectr. Electr. Insul. **25**(1), 339–351 (2018)
6. J.Y. Xue, H. Wang, J.H. Chen, K.F. Li, Y.Q. Liu, B.P. Song, J.B. Deng, G.J. Zhang, Effects of surface roughness on surface charge accumulation characteristics and surface flashover performance of alumina-filled epoxy resin spacers. J. Appl. Phys. **124**(8), 083302 (2018)
7. B.X. Du, Q. Du, J. Li, H.C. Liang, Carrier mobility and trap distribution dependent flashover characteristics of epoxy resin. IET Gener. Transm. Distrib. **12**(2), 466–471 (2018)
8. W.A. Stygar, J.A. Lott, T.C. Wagoner, V. Anaya, H.C. Harjes, H.C. Ives, Improved design of a high-voltage vacuum-insulator interface. Rev. Mod. Phys. **8**(5), 050401 (2005)
9. K. Kato, M. Kurimoto, H. Shumiya, H. Adachi, S. Sakuma, H. Okubo, Application of functionally graded material for solid insulator in gaseous insulation system. IEEE Trans. Dielectr. Electr. Insul. **13**(2), 362–372 (2006)

10. N. Hayakawa, Y. Miyaji, H. Kojima, K. Kato, Electric field reduction by functionally graded materials (FGM) with permittivity and conductivity distribution, in *IEEE International Conference on Conference on Electrical Insulation and Dielectric Phenomena (CEIDP)*, Toronto, Canada (2016), pp. 32–35
11. Z. Liu, W.D. Li, Y.B. Wang, G.Q. Su, G.J. Zhang, Y. Cao, D.C. Li, Topology optimization and 3D-printing fabrication feasibility of high voltage FGM insulator, in *IEEE International Conference on High Voltage Engineering Applications (ICHVE)*, Chengdu, China (2016), pp. 1–4
12. H.J. Ju, B. Kim and K.C. Ko, Optimal design of an elliptically graded permittivity spacer configuration in gas insulated switchgear. IEEE Trans. Dielectr. Electr. Insul.**18**(4), 1268–1273 (2011)
13. I.A. Metwally, Reduction of electric-field intensification inside GIS by controlling spacer material and design. J. Electrostat. **70**(2), 217–224 (2012)
14. L. Zhang, C. Lin, C. Li, S. V. Suraci, G. Chen, U. Riechert, T. Shahsavarian, M. Hikita, Y. Tu, Z. Zhang, D. Fabiani, J.L. He, Gas–solid interface charge characterisation techniques for HVDC GIS/GIL insulators. High Volt. **5**(2), 95–109 (2020)
15. B.X. Du, H.C. Liang, J. Li, Z.Y. Ran, Electrical field distribution along SF6/N2 filled DC-GIS/GIL epoxy resin spacer. IEEE Trans. Dielectr. Electr. Insul. **25**(4), 1202–1210 (2018)
16. H.Y. Zhou, G.M. Ma, C.R. Li, C. Shi, S.C. Qin, Impact of temperature on surface charges accumulation on insulators in SF6-filled DC-GIL. IEEE Trans. Dielectr. Electr. Insul. **24**(1), 601–610 (2017)
17. Y. Luo, J. Tang, C. Pan, Z. Pan, Y. Li, Z. Cui, Dynamics of surface charge and electric field distributions on basin-type insulator in GIS/GIL due to voltage polarity reversal. High Volt. **5**(2), 151–159 (2020)
18. S. Okabe, Phenomena and mechanism of electric charges on spacers in gas insulated switchgears. IEEE Trans. Dielectr. Electr. Insul. **14**(1), 46–52 (2007)
19. A. Winter, J. Kindersberger, Stationary resistive field distribution along epoxy resin insulators in air under DC voltage. IEEE Trans. Dielectr. Electr. Insul. **19**(5), 1732–1739 (2012)
20. U. Straumann, M. Schuller, C.M. Franck, Theoretical investigation of HVDC disc spacer charging in SF6 gas insulated systems. IEEE Trans. Dielectr. Electr. Insul. **19**(6), 2196–2205 (2012)
21. M. Kurimoto, K. Kato, M. Hanai, M. Takei, H. Okubo, Application of functionally graded material for reducing electric field on electrode and spacer interface. IEEE Trans. Dielectr. Electr. Insul. **17**(1), 256–263 (2010)
22. J. Li, B.X. Du, Progress on electrical field distribution along gas-solid interface in compact gas insulated system and its regulation methods. High Volt. Eng. **45**(8), 2619–2628 (2019). ((in Chinese))
23. A. Winter, J. Kindersberger, Transient field distribution in gas-solid insulation systems under DC voltages. IEEE Trans. Dielectr. Electr. Insul. **21**(1), 116–128 (2014)
24. G.J. Zhang, W.D. Li, Z. Liu, G.Q. Su, B.P. Song, R.D. Xie, J.B. Deng, Research progress on dielectric functionally graded materials for electrical insulation. Proc. CSEE **37**(14), 4232–4245 (2017) (in Chinese)
25. M.B. Bever, P.E. Duwez, Gradients in composite materials. Mater. Sci. Eng. **10**(1), 1–8 (1972)
26. R.H. Brealey, P.W. Juneau, J.E. Zlupko, Fabrication design and electrical evaluation of bulk-graded filled polymer post insulators, in *IEEE International Conference on. Electrical Insulation Conference (EIC)*, Philadelphia, PA, USA (1982), pp. 225–228
27. N. Hayashi, K. Kawahara, M. Sumikura, M. Hara, F. Endo, Electric field control by permittivity functionally graded materials and their lightning impulse withstand voltages for surface breakdown, in *IEEE International Symposium Electrical Insulation (ISEI)*, Boston, MA, USA (2002), pp. 260–263
28. S. Watanabe, N. Hayashi, H. Takeuchi, Y. Uchida, D. Dykes, G. Touchard, M. Ieda, Electrical applications of titanium-based FGMs manufactured by progressive lamination, in *IEEE International Conference on Conduction and Breakdown in Solid Dielectrics*, Vasteras, Sweden (1998), pp. 539–542

29. M. Kurimoto, K. Kato, H. Adachi, S. Sakuma, H. Okubo, Fabrication and experimental verification of permittivity graded solid spacer for GIS, in *IEEE International Conference on Electrical Insulation Dielectric Phenomena (CEIDP)*, Cancun, Mexico (2002), pp. 789–792
30. K. Kato, M. Kurimoto, H. Shumiya, H. Adachi, S. Sakuma, H. Okubo, Application of functionally graded material for solid insulator in gaseous insulation system. IEEE Trans. Dielectr. Electr. Insul. **13**(1), 362–372 (2006)
31. N. Hayakawa, J. Shimomura, T. Nakano, M. Hanai, K. Kato, H. Okubo, Fabrication technique of permittivity graded materials (FGM) for disk-type solid insulator, in *Annual Report Conference on Electrical Insulation Dielectric Phenomena (CEIDP)*, Montreal, Canada (2012), pp. 32–35
32. W.D. Li, X.Y. You, H.B. Mu, J.B. Deng, G.J. Zhang, Numerical optimization and 3D-printing fabrication concept of high voltage FGM insulator, in *IEEE PES Asia-Pacific Power and Energy Engineering Confernece (APPEEC)* (2015), pp. 1–4
33. M. Kurimoto, H. Ozaki, Y. Yamashita, T. Funabashi, T. Kato, Y. Suzuoki, Dielectric properties and 3D printing of UV-cured acrylic composite with alumina microfiller. IEEE Trans. Dielectr. Electr. Insul. **23**(5), 2985–2992 (2016)
34. M. Kurimoto, T. Sawada, T. Kato, Y. Suzuoki, 3D printing of 2 layered permittivity-graded material using UV-cured-resin/alumina composite, in *IEEE International Conference Properties and Applications of Dielectric Materials (ICPADM)*, Xi'an City, China (2018), pp. 1010–1013
35. X.R. Li, Z. Liu, W.D. Li, G.Y. Sun, J.Y. Xue, C. Wag, G.J. Zhang, 3D printing fabrication of conductivity non-uniform insulator for surface flashover mitigation. IEEE Trans. Dielectr. Electr. Insul. **26**(4), 1162–1169 (2019)
36. N. Hayakawa, Y. Miyaji, H. Kojima, K. Kato, Simulation on discharge inception voltage improvement of GIS spacer with permittivity graded materials (ε-FGM) using flexible mixture casting method. IEEE Trans. Dielectr. Electr. Insul. **25**(1), 1318–1323 (2018)
37. T. Nardi, N. Mora, F. Rachidi, Y. Leterrier, Graded-permittivity polymer nanocomposites as superior dielectrics. Compos. Sci. Technol. **129**, 1–9 (2016)
38. S. Diaham, Z. Valdez Nava, L. Lévêque, T.T. Le, L. Laudebat, T. Lebey, An original in-situ way to build field grading materials (FGM) with permittivity gradient using electrophoresis, in *IEEE Conference on Dielectrics (ICD)*, Budapest, Hungary (2018), pp. 1010–1013
39. H. Okubo, H. Shumiya, M. Ito, K. Kato, Optimization techniques on permittivity distribution in permittivity graded solid insulators, in *IEEE International Symposium Electrical Insulation (ISEI)*, Toronto, ON, Canada (2006), pp. 519–522
40. M. Talaat, A. El-Zein, M. Amin, Developed optimization technique used for the distribution of U-shaped permittivity for cone type spacer in GIS. Electr. Power Syst. Res. **163**, 754–766 (2018)
41. Z.Y. Ran, B.X. Du, J. Li, H.C. Liang, Electric field regulation of insulator interface by FGM with conductivity for superconducting-GIL. IEEE Trans. Appl. Supercond. **29**(2), 7701005 (2019)
42. W.D. Li, X.R. Li, B.H. Guo, C. Wang, Z. Liu, G.J. Zhang, Topology optimization of truncated cone insulator with graded permittivity using variable density method. IEEE Trans. Dielectr. Electr. Insul. **26**(1), 1–9 (2019)
43. S.A. Qasim, N. Gupta, Use of particle swarm optimization in the computation of an optimal permittivity distribution in functionally graded material insulators, in *IEEE International Conference Condition Assessment Techniques in Electrical Systems (CATCON)*, Bangalore, India (2015), pp.184–188
44. S.A. Qasim, N. Gupta, 2D spatial optimization of functionally graded materials used as insulators for field stress control. IET Sci. Meas. Technol. **13**(8), 1203–1211 (2018)
45. J. Li, H.C. Liang, B.X. Du, Z.H. Wang, Surface functional graded spacer for compact HVDC gaseous insulated system. IEEE Trans. Dielectr. Electr. Insul. **26**(2), 664–667 (2019)
46. B.X. Du, Z.H. Wang, J. Li, M. Xiao, H.C. Liang, Surface FGM insulator based on BaTiO3 magnetron sputtering for electric field grading of AC gas insulated power apparatus. IEEE Access **7**(1), 62681–62688 (2019)

47. Y. Miyaji, H. Kojima, K. Kato, N. Hayakawa, Fabrication and evaluation of permittivity Graded Materials with $SrTiO_3$ particles and localized conductive layer, in *IEEE International Conference Electrical Insulation Dielectric Phenomena (CEIDP)*, Chengdu, China (2016), pp. 631–634
48. B.X. Du, J.P. Jiang, J. Li, W.B. Zhu, Effects of ZnO Magnetron sputtering on surface charge and flashover voltage of oil-impregnated paper. High Volt. **4**(4), 308–315 (2019)
49. B.X. Du, Z.H. Wang, J. Li, H.C. Liang, Epoxy insulator with surface graded-permittivity by magnetron sputtering for gas-insulated line. IEEE Trans. Dielectr. Electr. Insul. **27**(1), 175–182 (2020)
50. J. Li, B.X. Du, J.G. Su, H.C. Liang, Y. Liu, Surface layer fluorination-modulated space charge behaviors in HVDC cable accessory. Polymers **10**(5), 500 (2018)
51. J. Li, B.X. Du, H. Xu, Suppressing interface charge between LDPE and EPDM for HVDC cable accessory insulation. IEEE Trans. Dielectr. Electr. Insul. **24**(3), 1331–1339 (2017)
52. B.X. Du, Z.Y. Ran, J. Li, H.C. Liang, Novel insulator with interfacial σ-FGM for DC compact gaseous insulated pipeline. IEEE Trans. Dielectr. Electr. Insul. **26**(3), 818–825 (2019)
53. B.X. Du, Z.Y. Ran, J. Li, H.C. Liang, Fluorinated epoxy insulator with interfacial conductivity graded material for HVDC gas insulated pipeline. IEEE Trans. Dielectr. Electr. Insul. (accepted)

Surface Functionally Graded Insulator for High Voltage Gas Insulated Apparatus

Hang Yao, Boxue Du, Jin Li, and Zehua Wang

Abstract Electric field optimization is a key technique to ensure the compact structure and reliable operation for AC gas insulated power apparatus. In order to obtain a more uniform electric field distribution, the concept of surface functionally graded materials (SFGM) materials for polymer insulation was proposed in this chapter. SFGM spacer which applying the surface functionally graded material (SFGM) with a designed thickness (*d*) distribution of high permittivity along the spacer surface. The magnetron sputtering technique was used to fabricate the *d*-uniform spacer and the *d*-SFGM spacer by depositing BaTiO3 layers. Theoretical analysis indicates that the electric field distribution is optimized using the *d*-SFGM spacer, followed by the *d*-uniform spacer. The finite element simulations were used to investigate the dielectric losses and electric field distributions of different spacers. The results show that for the case of the *d*-uniform spacer, a greater thickness or higher permittivity of the sputtering layer results in a more uniform electric field distribution. With the application of the *d*-SFGM spacer, the maximum electric field strength and dielectric loss are both significantly reduced compared to the ordinary spacer. Finally, experimental tests were conducted to further verify the feasibility of SFGM. The AC flashover voltage of the *d*-SFGM spacer is much higher than those of the conventional spacer and the *d*-uniform spacers. Therefore, the *d*-SFGM spacer has great potential for improving the reliability of the AC gas insulated power apparatus.

1 Introduction

The gas insulated power apparatus plays an important role in the generation, transmission and distribution of the power energy, combining reliability with low losses, low electro- magnetic interference, high transmission capacity and safety [1, 2]. Solid spacers made of epoxy/Al_2O_3 composites provide mechanical support and electrical insulation for AC gas insulated power devices [3]. However, flashover along the

H. Yao · B. Du (✉) · J. Li · Z. Wang
School of Electrical and Information Engineering, Tianjin University, Tianjin 300072, China
e-mail: duboxue@tju.edu.cn

B. Du (ed.), *Polymer Insulation Applied for HVDC Transmission*,
https://doi.org/10.1007/978-981-15-9731-2_21

insulating spacer surface can easily occur, which limits the security and reliability of operation and further development of AC GIL [4].

It is generally considered that the highly non-uniform electric field distribution along the spacer surface is an important cause of flashover [5]. Because of the higher relative permittivity of the solid spacer compared to the surrounding gas, the electric field is enhanced along the gas–solid interface, especially at the triple junction [6]. Structural optimization of the solid spacer, including additional electrodes and complex spacer geometries shapes, has been applied to uniform the distribution of electric field [7]. Nevertheless, these traditional techniques complicate the structure of the equipment and increase the manufacturing costs [8, 9]. Therefore, novel methods for regulating the electric field and increasing the flashover voltage of spacers have recently been proposed.

In an AC system, the electric field distribution mainly depends on the permittivity of the dielectrics involved. By designing the permittivity distribution inside the spacer, the electric field concentration can be regulated, which is the principle of functionally graded materials (FGM). To further uniform the electric field distribution along the spacer, the concept of functionally graded materials (FGM) with a spatially inhomogeneous permittivity distribution was originally applied to spacer [10]. The laminating method was used to prepare the graded post-type spacer, whose flashover voltage under lightning impulse voltage was increased by 10–25% compared to that of the conventional spacer [11]. The centrifugation technique was then developed to fabricate permittivity (ε)-FGM conical spacers with a continuous permittivity distribution [12–15]. The results showed that the application of ε-FGM spacers can reduce the maximum electric field near the triple junction and increase the flashover voltage compared to ε-uniform insulators under lightning impulse voltage. Recently, 3D printing technology was used to fabricate FGM spacers; however, more research and development are required for industrial applications of such FGM epoxy spacers [16–18]. Existing fabrication processes such as lamination, centrifugation, and 3D printing show some limitations. First, the control range of permittivity for FGM spacer by bulk doping is small (4–80), which limits the effect of electric field regulation. Second, these methods not only degrade mechanical performances of spacers but also are complex, which makes the FGM spacers hard to be brought into the practical application.

The electric field relaxation of the FGM spacer mentioned above is at the cost of complicated processing process, degraded mechanical properties and increased leakage current. Without affecting the bulk properties, surface treatment is considered a simpler but more effective method than traditional modification methods of suppressing flashover because flashover mainly occurs along the spacer surface. To break the above limitations, the concept of d-SFGM, i.e., applying FGM to the spacer surface, was proposed to form thickness (d) gradient surface layers of ultrahigh permittivity. The radio-frequency (RF) magnetron sputtering technique was used to deposit $BaTiO_3$ layers on cone-type spacers to obtain the d-uniform and d-SFGM spacers. Subsequently, the fracture section was observed using a scanning electron microscope (SEM) to investigate the influence of sputtering time on the layer thickness. Then the dielectric parameters of sputtering layer were calculated from the capacitance and resistance measurements. Next, the effects of layer permittivity and thickness on electric field distribution and the dielectric loss of spacer were investigated via finite element simulations. Finally, the flashover voltages of the conventional, *d*-uniform and *d*-SFGM spacers were measured to verify the feasibility of SFGM.

2 Concept of SFGM

Figure 1 shows equivalent-circuit models for the conventional spacer (Fig. 1a), *d*-uniform spacer (Fig. 1b) and *d*-SFGM spacer (Fig. 1c). The conventional spacer consists of epoxy/Al_2O_3 composites, whose volume relative permittivity is about

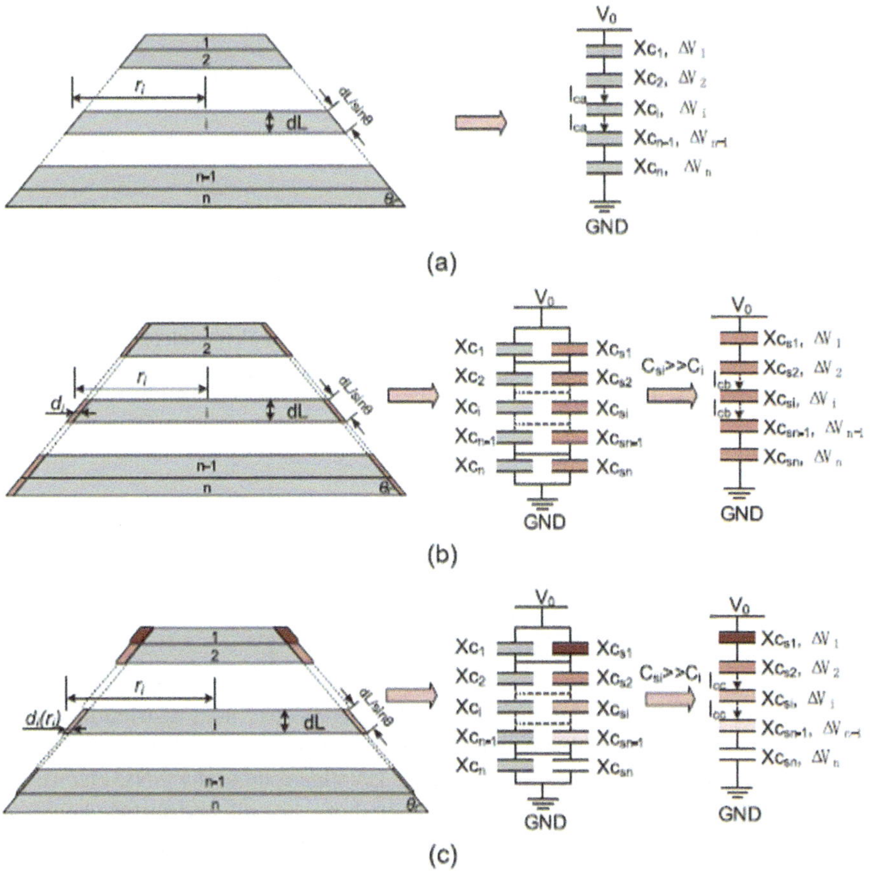

Fig. 1 Equivalent-circuit models for cone-type spacers. **a** The conventional spacer, **b** the *d*-uniform spacer, and **c** the *d*-SFGM spacer

6.8. The *d*-uniform spacer consists of two parts: the bulk consisting of epoxy/Al2O3 composite with a low relative permittivity and the surface layer of $BaTiO_3$ with an ultrahigh relative permittivity (over 1000). When the thickness of the surface layer is designed to be functionally graded along the spacer surface, a *d*-SFGM spacer can be obtained. To simplify the model for easy understanding, the stray capacitance of air is ignored because the permittivity of air is quite low compared with the spacer and the $BaTiO_3$ layer.

As shown in Fig. 1a, a cone-type spacer can be divided into n elements. Each element i can be regarded as a cylinder with a radius of r_i and a height of dL. Denoting the capacitance value of each element i as C_i, the conventional spacer can be approximately considered a capacitor series of C_i. According to the capacitance formula, the capacitance value of element i can be calculated as,

$$C_i = \frac{\varepsilon_0 \varepsilon_r \pi r_i^2}{dL} \tag{1}$$

where ε_0 and ε_r are the vacuum permittivity and the relative permittivity of the spacer, respectively. The capacitive reactance of element i can be calculated according to the following equation:

$$X_{Ci} = \frac{1}{2\pi f C_i} \tag{2}$$

where f is the power frequency, which is set to 50 Hz in this chapter. Under AC voltage, the leakage current primarily consists of the capacitive current Ic. Without considering the conductive current, the electric potential drop dVi across element i can be calculated as follows:

$$dV_i = I_c X c_i \tag{3}$$

Based on the above analysis, the electric field strength along element i surface of the conventional spacer can be described as,

$$E_i^a = \frac{dV_i}{dL/\sin\theta} = \frac{I_{ca}\sin\theta}{2\pi^2 f \varepsilon_0 \varepsilon_r} \cdot \frac{1}{r_i^2} \tag{4}$$

According to Eq. (4), it can be inferred that the electric field distribution along the spacer surface is extremely uneven. The maximum electric field strength appears near the HV electrode and decreases from the HV electrode to the GND electrode.

For the *d*-uniform spacer as shown in Fig. 1b, both the bulk and surface layer can be divided into n elements. Each element i of the surface layer can be regarded as a cylindrical ring with a radius of r_i, a thickness of d and a height of dL. According to the capacitance formula, the capacitance value of each surface layer element$i$ can be approximately formulated as,

$$C_{si} = \frac{\varepsilon_0 \varepsilon_{rs} 2\pi r_i d}{dL} \tag{5}$$

where ε_{rs} is the relative permittivity of the surface layer. In the equivalent-circuit model for the d-uniform spacer, the bulk capacitor series and the surface layer capacitor series are connected in parallel. When $C_{si} >> C_i$, the equivalent circuit can be simplified to only the surface layer capacitor series. The electric field strength along element i surface of the d-uniform spacer can be expressed as,

$$E_i^b = \frac{dV_i}{dL/\sin\theta} = \frac{I_{cb}\sin\theta}{4\pi^2 f \varepsilon_0 \varepsilon_{rs} d} \cdot \frac{1}{r_i} \tag{6}$$

Compared with the conventional spacer, the electric field distribution along the d-uniform spacer is more uniform because the electric field strength is proportional to r^{-1} instead of r^{-2}, as shown by Eqs. (4) and (6), respectively.

If the surface layer thickness is designed to be functionally graded along the spacer surface ($d\ (r_i) = \text{const}/r_i$), as shown in Fig. 1c, then the electric field strength along surface of element i of the d-SFGM spacer can be described as,

$$E_i^c = \frac{I_{cc}\sin\theta}{4\pi^2 f \varepsilon_0 \varepsilon_{rs} d(r_i)} \cdot \frac{1}{r_i} \tag{7}$$

Equation (7) indicates that the electric field strength along the d-SFGM spacer is constant in the ideal case.

The ideal electric field distributions along the conventional spacer, d-uniform spacer and d-SFGM spacer are shown in Fig. 2. Considering some idealized hypotheses have been made to simplify the model, the electric field distribution calculated by using the equivalent-circuit models have some deviations. Thus, the finite element simulation is necessary to obtain the accurate electric field distributions of spacers.

3 Fabrication of SFGM Spacer

3.1 Magnetron Sputtering

The cone-type spacers used in the experiments were comprised of epoxy/Al_2O_3 composites, produced following the industrial manufacturing process. In this chapter, three d-uniform spacers and one d-SFGM spacer with $BaTiO_3$ as the deposition material were fabricated using an RF magnetron sputtering apparatus, including a JPGF-400 magnetron sputterer cooled with cold-water circulation, which has been widely used in the semiconductor processing field [19].

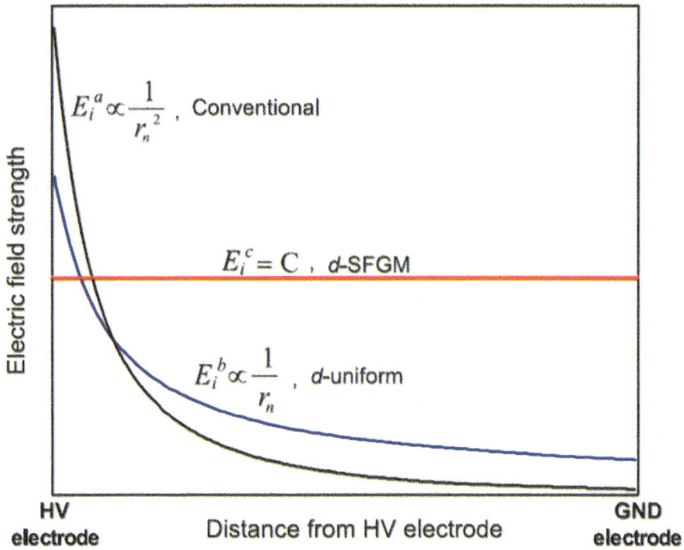

Fig. 2 Ideal electric field distributions for cone-type spacers

The experimental setup for magnetron sputtering of spacers is shown in Fig. 3a. The $BaTiO_3$ target with the diameter of 60 mm and the thickness of 3 mm was provided by CHINO New Material Technology Co. Ltd. The cone-type spacer was placed below the $BaTiO_3$ target and remained intact during the sputtering process. By adjusting both the position and hole size of the baffle, the sputtering area of spacer can be controlled. Before sputtering, the chamber was evacuated to achieve the required pressure of less than 510–3 Pa. The input RF power was 90 W, and the gas flow rate of $Ar:O_2$ was set to 4:1 (SCCM). In the magnetron sputtering process, the target is bombarded by energetic gaseous ions generated by stable RF glow discharge plasma to cause the removal of target atoms, which then condense on spacers as a thin film [20]. Although secondary electrons move to the substrate, the magnetic field can restrict them to a nearby target, thus increasing the possibility of collision ionization of electrons and atoms. The increased ionization efficiency results in the increased ion bombardment on the target, giving higher sputtering rates and higher deposition rates on the substrate [20], i.e., the spacer (Fig. 4).

For a constant sputtering power, the sputtering layer thickness is dependent on the sputtering time. To obtain *d*-uniform spacers with different layer thicknesses, the sputtering times of three *d*-uniform spacers were set to 30, 60 and 90 min. Figure 3b shows the surface gradient arrangement for the *d*-SFGM spacer. The cone-type spacer was divided into four gradients with equal height, and the sputtering times of each gradient were 90, 60, 30 and 0 min from top to bottom.

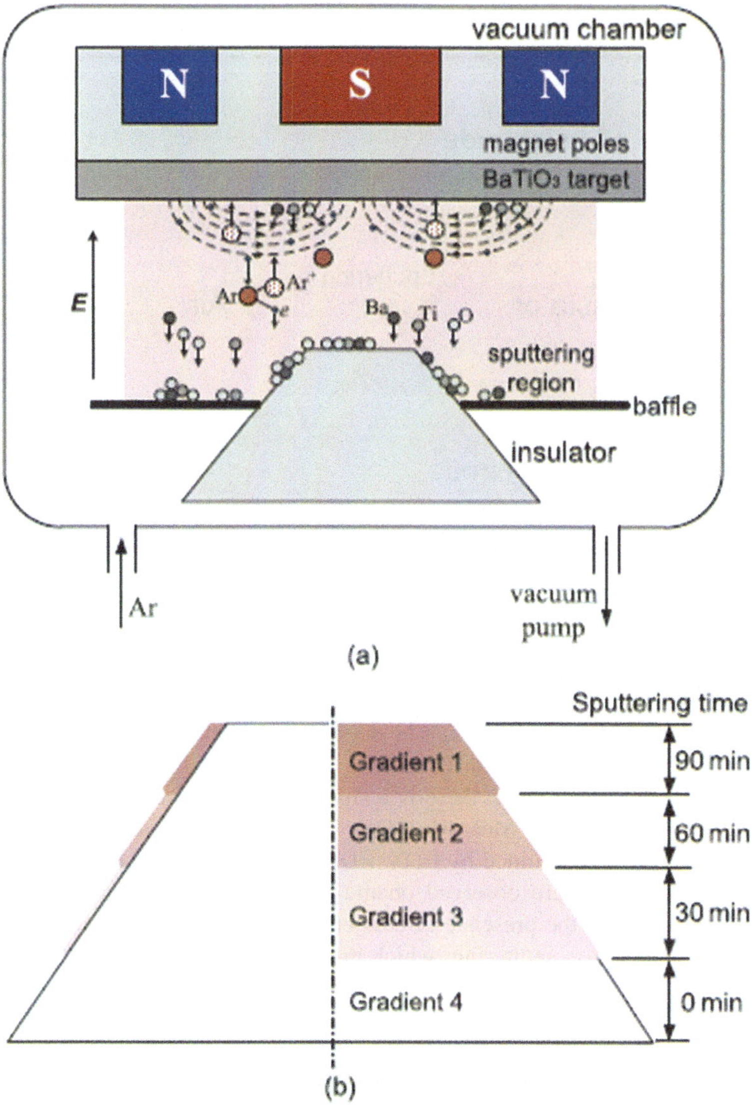

Fig. 3 Preparation of sputtered spacers. **a** Experimental setup for $BaTiO_3$ magnetron sputtering of spacers. **b** Gradient arrangement along the spacer surface

3.2 *Characterization of $BaTiO_3$Layer*

The characterization of the $BaTiO_3$ layer primarily focused on its microstructural properties, dielectric properties and the thickness-time relationship. In order to investigate the microstructural changes of the spacer surface before and after sputtering in

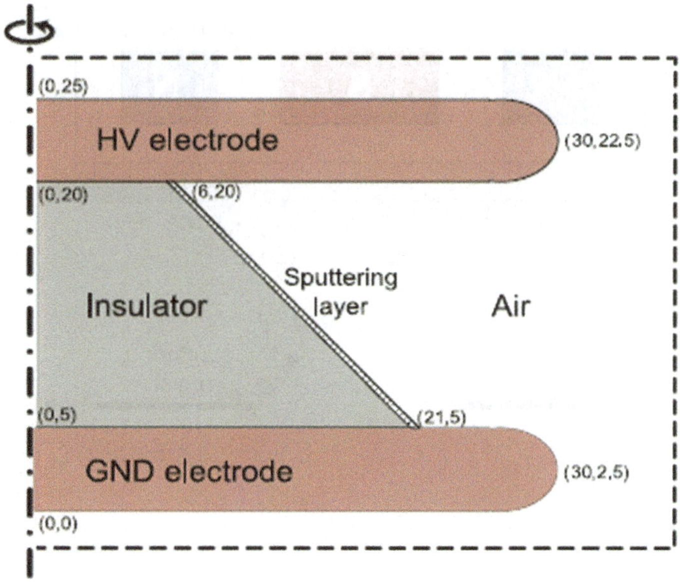

Fig. 4 Simulation model for the cone-type spacer

detail, the relative composition and surface morphology were respectively analyzed by SEM. As shown in Fig. 5, it can be seen that the Al_2O_3 fillers are uniformly dispersed in the epoxy matrix at the micron level in the SEM images. Due to the formation of a deposited layer, there is a significant change on the spacer surface after magnetron sputtering. Meanwhile, the chemical composition of the $BaTiO_3$ deposited layer was determined by EDS. EDS spectrum results show that only the elements C, O and Al are observed on the surface of the spacer surface before magnetron sputtering. the presence of elements C, O, Al, Ti and Ba on the spacer surface after magnetron sputtering, which proves that the $BaTiO_3$ sputtering layer was successfully formed on the surface of the spacer.

In order to study the relationship between sputtering time and $BaTiO_3$ sputtering layer thickness, the $BaTiO_3$ layer thicknesses of different sputtering times, including 30, 60 and 90 min, were measured, and a curve fitting was carried out, as shown in Fig. 5. As the sputtering time increases, the sputtering layer thickness increases according to the following exponential relationship [21]:

$$d = 1.058e^{\frac{t}{62.7}} - 1.058 \tag{8}$$

where d is the thickness of the sputtering layer, μm, and t is the sputtering time, min. This exponential function is very suitable for the measurements and will be used in the rotatable baffle gap design.

To obtain the relative permittivity of the $BaTiO_3$ sputtering layer in the frequency range from 50–105 Hz at room temperature, the $BaTiO_3$ layer was deposited on a

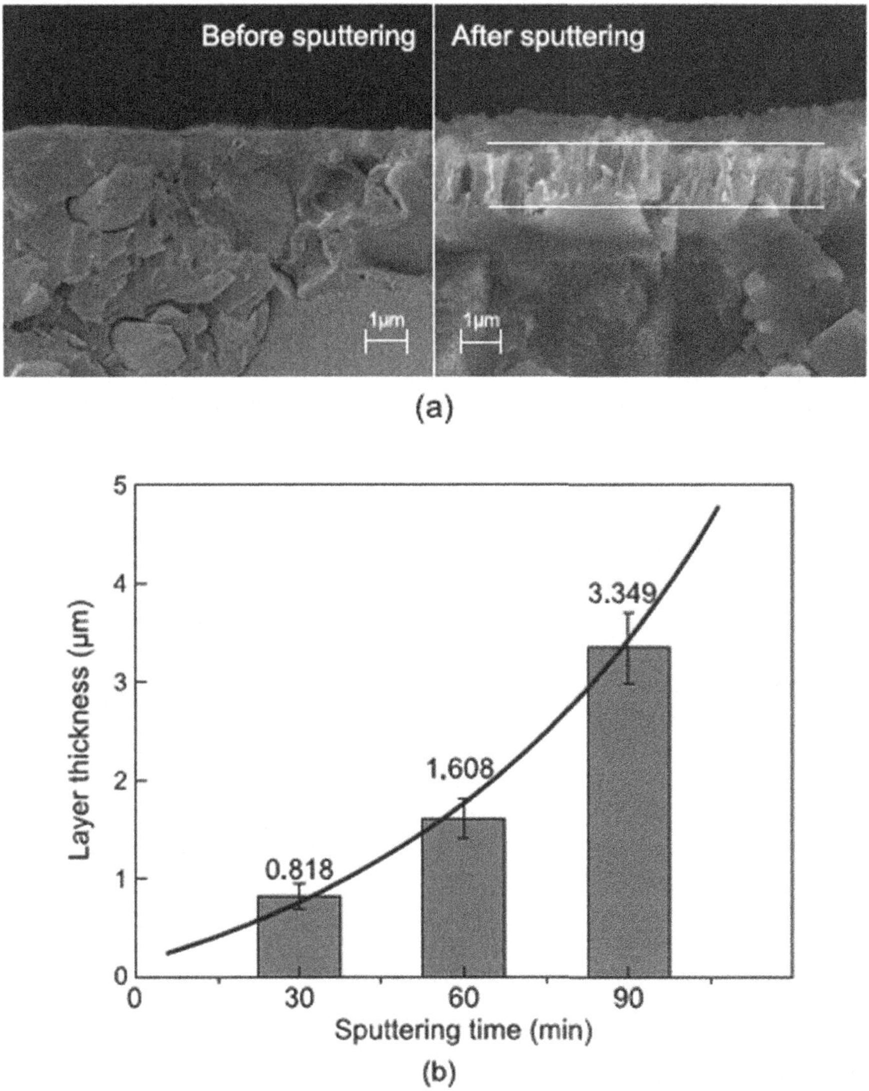

Fig. 5 Effects of sputtering time on the sputtering layer thickness. **a** Cross-section SEM images of the insulators. **b** Relationship between the layer thickness and the sputtering time

smooth copper plate used as a substrate. A small electrode with the diameter of 50 mm was pressed tightly on the $BaTiO_3$ sputtering layer. The LCR meter was connected to two electrodes, and the capacitance of the $BaTiO_3$ layer was measured at an alternating voltage of 10 mV AC voltage at different frequencies; from the obtained data, the relative permittivity can be calculated according to the capacitance formula. For different sputtering times, the relative dielectric constant values are slightly different. Overall, the relative permittivity drops rapidly in the range of 50–104 Hz until it finally reaches its minimum value of ~20 above 104 Hz.

3.3 SFGM Spacer Fabrication Method

The spacer with a uniform thickness $BaTiO_3$ sputtering layer can be fabricated by the method shown in Fig. 3a, while the fabrication of an SFGM spacer with a graded $BaTiO_3$ distribution requires additional setup. In order to obtain a continuously graded layer thickness distribution along the spacer surface, a rotatable baffle with a specific shaped gap is placed above the spacer. Therefore, the actual sputtering area on the spacer surface can be controlled by the gap shape. At position r on spacer surface, the ratio of the arc length of the actual sputtering area to the circumference represents the effective sputtering time per unit time. This ratio decreases continuously from top to bottom along the spacer surface, which means that the thickness of the sputtered layer continuously decreases when the baffle is slowly rotated. Thus, the shape of the baffle gap can be used to control the thickness distribution of the sputtered layer.

To achieve the optimal layer thickness distribution along the SFGM spacer surface, a specific shape of baffle gap should be designed. Figure 5b shows the relationships among optimal sputtering layer thickness, position r along the spacer surface and sputtering time, which could guide the design of the gap shape. The curve in the xOy coordinate plane, which is obtained by iterative calculation, represents the optimal layer thickness distribution along the spacer surface. The curve in the xOz coordinate plane represents the relationship between the layer thickness and the sputtering time. By combining the above two curves, the optimal sputtering time distribution along the spacer surface can be obtained, as shown in the yOz coordinate plane. This curve also represents the effective sputtering time along the spacer surface. Therefore, the optimal sputtering layer thickness distribution can be transformed into the optimal effective sputtering time along the epoxy spacer surface.

3.4 Dielectric Parameters Measurement

To obtain the permittivity and conductivity of the sputtering layer, the capacitance and resistance were measured using the (LCR) meter under 1 V AC voltage of 50 Hz. Under the same conditions as above, $BaTiO_3$ was deposited on a sheet copper to form an even layer for better accuracy of measurement results. Then another smaller electrode with diameter of 50mmwas pressed tightly on the sputtering layer. The $BaTiO_3$ permittivity value can be calculated according to the capacitance formula, which is about 1600 in this chapter. Similarly, the volume conductivity of sputtering layer can be obtained from the measured resistance value, which is about 1×10^{-7} S/m.

4 Electric Field Simulation

4.1 Simulation Model

Based on the Maxwell–Wagner theory, the finite element simulations were conducted to investigate the effects of the sputtering layer on electric field distribution of the solid spacer. Figure 5 shows the calculation model which is based on a rotationally symmetric system. In the figure, the physical dimensions of spacer and electrodes are presented in the form of coordinates in millimeters. An AC voltage of 15 kV was applied at the HV electrode. The governing equations for the electric calculation are as follows:

$$\nabla \cdot J = 0 \tag{9}$$

$$J = \sigma E + j\omega D \tag{10}$$

$$E = -\nabla V \tag{11}$$

where $\boldsymbol{E}$ is the electric field strength, V/m; V is the electric potential, V; $\boldsymbol{J}$ is the current density, A/m^2; σ is the volume conductivity of the involved dielectrics, S/m; ω is the angular frequency of the power grid, rad/s; $\boldsymbol{D}$ is the electric flux density, C/m^2.

4.2 Effect of Sputtering Layer Permittivity

Since the layer permittivity value can vary with preparation parameters such as sputtering gas, sputtering power and target purity, it is necessary to explore the effects of the sputtering layer permittivity on electric field distribution along the spacer surface, as shown in Fig. 6. In this simulation, the thickness of the sputtering layer on the spacer surface is uniform, with a value of 4 μm. The relative permittivity is set to 500–5000. The creepage distance $l = 0$ mm is the location of the HV electrode, and the creepage distance $l = 21$ mm is the location of the GND electrode. The electric field strength declines from the HV triple junction to the GND triple junction. Therefore, the flashover is primarily triggered by the region surrounding the HV triple junction. To illustrate the electric field distribution near the HV triple junction, the logarithm of the creepage distance is presented. The maximum electric field strength at the triple junction clearly decreases with increasing permittivity of the sputtering layer. However, when the permittivity reaches a value of 3500, the electric field distribution is only slightly affected by the permittivity of the sputtering layer. The "saturation effect" of the relaxation function on electric field of the sputtering layer with high

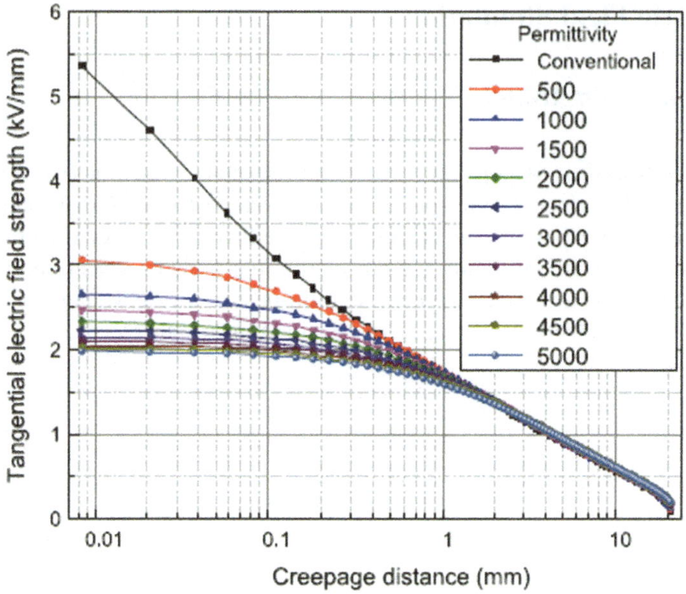

Fig. 6 Electric field distributions along the conventional spacer and $BaTiO_3$-sputtered spacers with different layer permittivities

permittivity can be explained as follows. According to the equivalent-circuit model shown in Fig. 1b, the capacitance value C_{si} of each surface layer element i increases with increasing permittivity of the sputtering layer. For a higher permittivity of the layer, the assumption of $C_{si} >> C_i$ is more realistic, and the electric field distribution is more consistent with Eq. (6). When the permittivity reaches a certain value, the assumption of $C_{si} >> C_i$ becomes true. Therefore, the continued growth of the layer permittivity has only a slight influence on the electric field distribution along the surface of the $BaTiO_3$-sputtered spacer.

Because the permittivity and conductivity of sputtering layer are both higher than those of epoxy/Al_2O_3 composites, it is meaningful to analyze the leakage current and dielectric loss of spacer. By integrating the current density on the HV or GND electrode, the leakage current of the spacer can be obtained. Figure 7 shows the real component (I_R) and imaginary component (I_C) of leakage currents as well as dielectric losses of the $BaTiO_3$-sputtered spacers with different levels of layer permittivity. As the relative permittivity of the sputtering layer varies from 500 to 5000, the imaginary component of spacer leakage current continues to increase from 14.0 to 16.9 μA, because of the increasingly lower capacitive reactance of sputtering layer; but the real component of leakage current almost keeps constant. As for the dielectric loss, the ration of I_R to I_C, has inevitable decline trend.

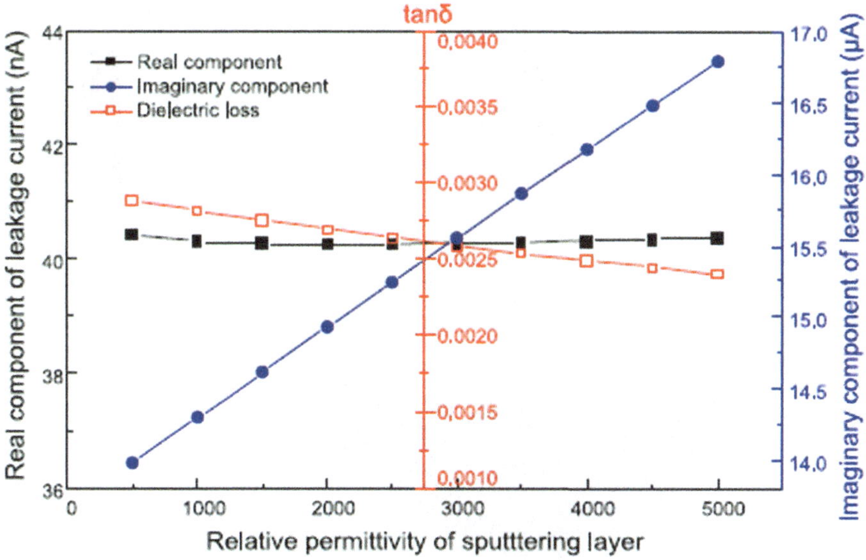

Fig. 7 Leakage currents and dielectric losses of the $BaTiO_3$-sputtered spacer with different layer permittivities

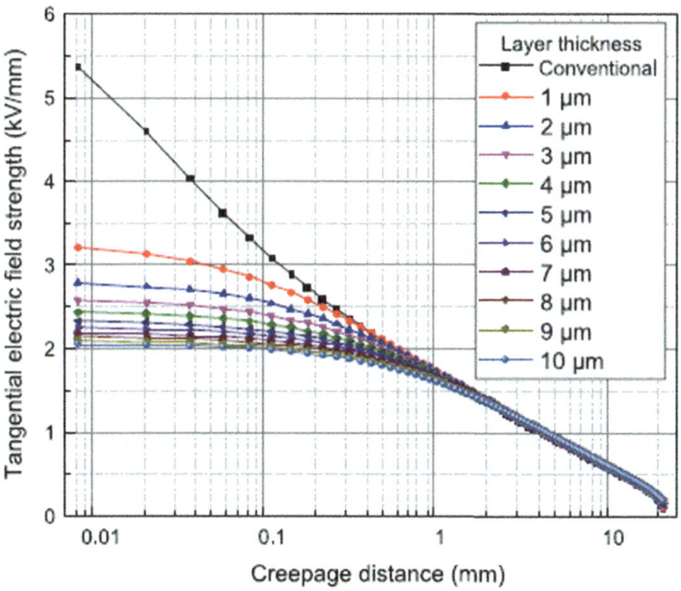

Fig. 8 Electric field distributions along the conventional insulaton and $BaTiO_3$-sputtered spacers with different layer thicknesses

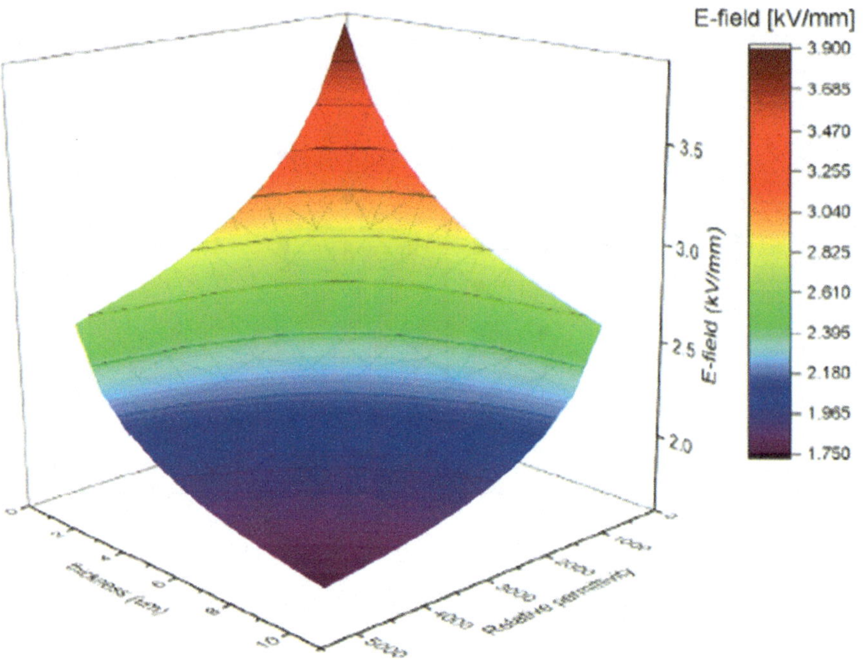

Fig. 9 Combined effects of sputtering layer permittivity and thickness on maximum electric field strength along spacer surface

4.3 Effects of Sputtering Layer Thickness

Figure 6 shows the electric field distributions along the conventional spacer and $BaTiO_3$-sputtered spacers with different sputtering layer thicknesses. The permittivity value of sputtering layer is set to 1600 in this section, and layer thickness ranges from 1 to 10 μm. By introducing the $BaTiO_3$ layer, the electric field strength is significantly reduced at the triple junction. With increasing layer thickness, the effect of the electric field relaxation function grows substantially.

And when the layer thickness reaches 8 μm, the electric field distribution becomes stable. Similarly, the capacitance value C_{si} of element i increases with increasing thickness of the sputtering layer. For a thicker layer, the assumption of $C_{si} >> C_i$ is more realistic, and the electric field distribution is more likely to obey Eq. (6). It is clear that when the layer thickness reaches a certain value, the assumption of $C_{si} >> C_i$ will hold true. The electric field changes slightly with the layer thickness, and the "saturation effect" of the electric field relaxation function appears again. As shown in Fig. 7, the maximum electric field strength declines with increasing both the sputtering layer permittivity and thickness, and finally reaches a stable value when the product of sputtering layer permittivity and thickness is high enough (Figs. 8 and 9).

Figure 10 shows the leakage currents and dielectric losses of the $BaTiO_3$-sputtered spacers with different layer thicknesses. The thicker sputtering layer, the increasingly reduced capacitive reactance and impedance of the sputtering layer causes the $BaTiO_3$-sputtered spacer to have a larger I_C and I_R. Meanwhile, the rising rate of I_R is higher than that of $I_{C,}$ which causes increased loss of spacer. In practical applications, an optimal sputtering layer thickness should be chosen by considering both the electric field distribution and the dielectric loss.

4.4 Application of Prepared Spacers

As described, the "saturation effect" of the electric field relaxation function exists in the *d*-uniform spacer with increasing permittivity and/or thickness of the sputtering layer. For superior electric field distribution, the *d*-SFGM spacer is necessary to be presented. The prepared spacers, three *d*-uniform spacers and a *d*-SFGM spacer, were applied to compare the electric field relaxation function and dielectric losses of spacers. Figure 11 shows the electric field distributions along the conventional spacer, *d*-uniform spacers and *d*-SFGM spacer. The permittivity of sputtering layer is 1600, which has been measured in front section. According to the theoretical analysis, to further improve the electric field distribution compared to the *d*-uniform spacer, the *d*-SFGM spacer was fabricated by grading the thickness of sputtering layer, as shown in Fig. 3b. It is obvious that the maximum electric field strength at the HV triple junction is further reduced by the application of *d*-SFGM spacer compared with *d*-uniform spacers. Because of the thickness variations of the different gradients of sputtering layers, three electric field distortions can be observed. The electric field distortions can be suppressed by increasing the division of the gradient. Especially, the continuously graded distribution of sputtering layers could completely eliminate these distortions, because there is no mutation in thickness.

In addition to the enhanced effect of relaxation on electric field strength, the *d*-SFGM spacer is also found to possess a lower leakage current and dielectric loss than the 3.349 μm sputtered spacer, as shown in Fig. 12. Therefore, the *d*-SFGM spacer has a greater potential for application in AC gas insulated power apparatus.

5 Flashover Test Results

In the previous sections, the feasibility of the *d*-funiform spacer and the *d*-SFGM spacer with $BaTiO_3$-sputtered layers to achieve a uniform electric field distribution has been validated via theoretical analysis and electrical simulation. In this section, flashover tests were conducted to evaluate the insulation performance of spacers.

The test spacer was placed between parallel plane electrodes with a diameter of 60 mm. One of the electrodes was connected with a standard AC voltage through a limiting resistance, and the other with the ground. The AC flashover test was

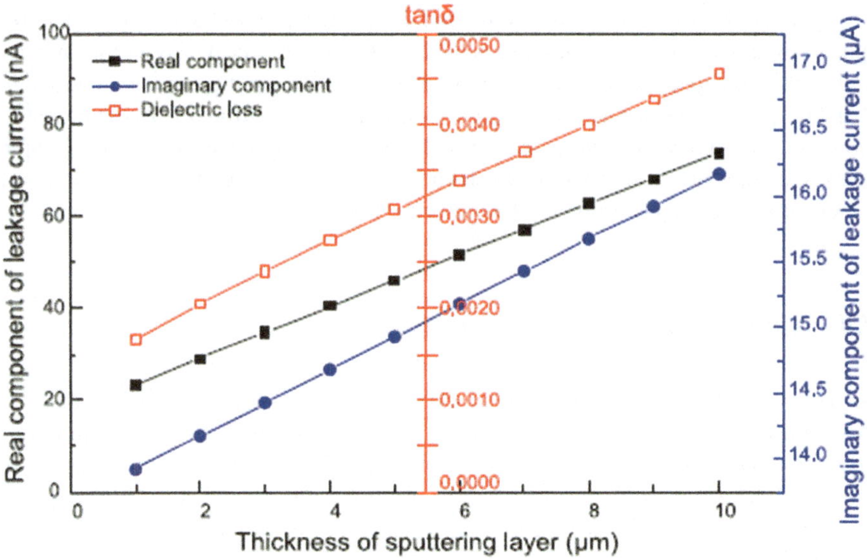

Fig. 10 Leakage currents and dielectric losses of the $BaTiO_3$-sputtered spacers with different layer thicknesses

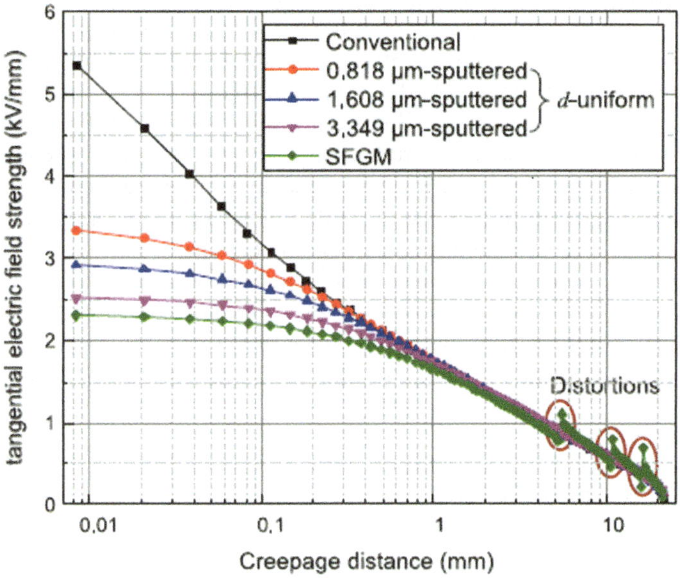

Fig. 11 Electric field distributions along the conventional, *d*-uniform and *d*-SFGM spacers

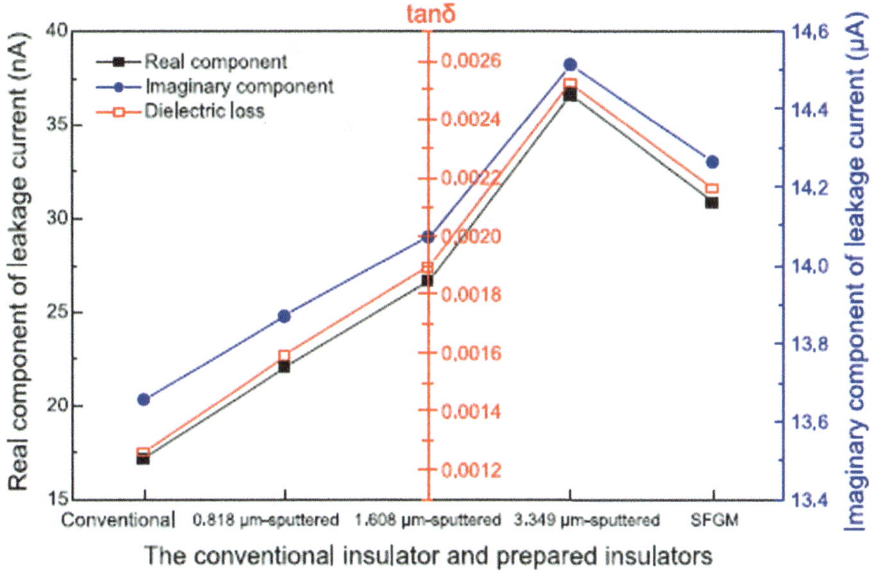

Fig. 12 Leakage currents and dielectric losses of four kinds of prepared spacers

conducted in a chamber at the temperature of 20 °C. At least fifteen flashover tests were performed for each kind of spacer. Then the Weibull plots were used to analyze the flashover voltages of the conventional spacer, *d*-uniform spacer and *d*-SFGM spacer.

Figure 13 shows the flashover voltages of the conventional spacer, *d*-uniform spacers (0.818-, 1.608-, and 3.349-μm-sputtered spacers) and the *d*-SFGM spacer. The observed phenomena are consistent with the electric field distribution. For the conventional spacer, the flashover voltage is ~15.2 kV. The flashover voltage of the *d*-uniform spacer is higher than that of the conventional spacer and increases with increasing layer thickness. Among the *d*-uniform spacers, the 3.349 μm-sputtered spacer has the highest flashover voltage of ~17.1 kV. By applying the *d*-SFGM spacer, the lower electric field strength near the HV triple junction causes the flashover voltage to reach as high as ~18.1 kV. Such promotion of 20% on spacer flashover voltage is of great significance to prolong service life, save space and reduce GHG emission for AC gas insulated power apparatus. If the d-distribution is refined, according to Eq. (7), the flashover voltage of *d*-SFGM spacer can be significantly improved.

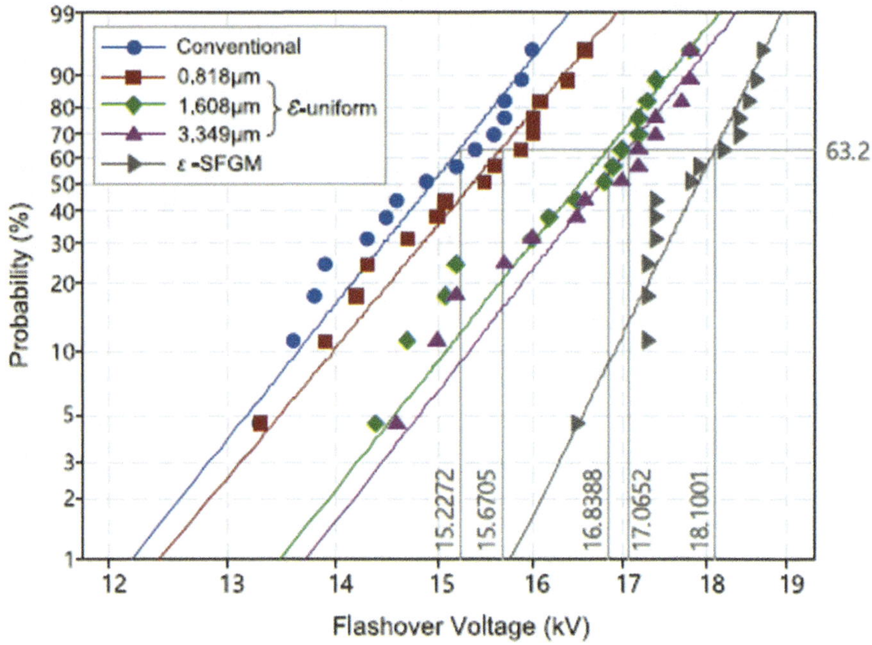

Fig. 13 Weibull distributions of flashover voltages of the conventional spacer, ε-uniform spacers and ε-SFGM spacer

6 Conclusion

The concept of SFGM was proposed and applied to cone-type spacers in this chapter. The electric field distributions, leakage currents, dielectric losses and flashover voltages of the conventional spacer, *d*-uniform spacer and *d*-SFGM spacer were investigated. The main conclusions of this chapter are as follows:

(1) The equivalent-circuit models indicate that the *d*-SFGM spacer has the most uniform electric field distribution, followed by the d-uniform spacer and the conventional spacer.
(2) For the *d*-uniform spacer, as the permittivity of the sputtering layer increases, the electric field distribution becomes increasingly uniform, and the dielectric loss of spacer decreases. As the sputtering layer thickness increases, the electric field distribution is increasingly improved. Considering the "saturation effect" of layer thickness on electric field relaxation, an optimal thickness should be designed to achieve the better electric field distribution and lower dielectric loss.
(3) Compared with the *d*-uniform spacers, by the introduction of the *d*-SFGM spacer, the maximum electric field strength at the HV triple junction is further reduced, with much lower dielectric loss.
(4) Flashover tests confirm that the *d*-SFGM spacer has the highest AC flashover voltage compared with the conventional spacer and the *d*-uniform spacer.

Therefore, the *d*-SFGM spacer has great potential for practical application in AC gas insulated power apparatus.

References

1. H. Koch, F. Goll, T. Magier, K. Juhre, Technical aspects of gas insulated transmission lines and application of new insulating gases. IEEE Trans. Dielectr. Electr. Insul. **25**(4), 1448–1453 (2018)
2. K.D. Srivastava, M.M. Morcos, A review of some critical aspects of insulation design of GIS/GIL systems, in *Proceedings of IEEE/PES Transmission and Distribution Conference and Exposition*, Atlanta, GA, USA, November 2001, pp. 787–792
3. B.X. Du et al., Temperature dependent surface potential decay and flashover characteristics of epoxy/SiC composites. IEEE Trans. Dielectr. Electr. Insul. **25**(2), 631–638 (2018)
4. B. Du, Q. Du, J. Li, H. Liang, Carrier mobility and trap distribution dependent flashover characteristics of epoxy resin. IET Gener. Transmiss. Distrib. **12**(2), 466–471 (2018)
5. B.X. Du, H.C. Liang, J. Li, Z.Y. Ran, Electrical field distribution along SG6/N2 filled DC-GIS/GIL epoxy spacer. IEEE Trans. Dielectr. Electr. Insul. **25**(4), 1202–1210 (2018)
6. M. Talaat, A. El-Zein, M. Amin, Electric field simulation for uniform and FGM cone type spacer with adhering spherical conducting particle in GIS. IEEE Trans. Dielectr. Electr. Insul. **25**(1), 339–351 (2018)
7. W.A. Stygar et al., Improved design of a high-voltage vacuum-spacer interface. Phys. Rev. Accel. Beams **8**(5), Article no. 050401 (2005)
8. K. Kato, M. Kurimoto, H. Shumiya, H. Adachi, S. Sakuma, H. Okubo, Application of functionally graded material for solid spacer in gaseous insulation system. IEEE Trans. Dielectr. Electr. Insul. **13**(2), 362–372 (2006)
9. M. Kurimoto, K. Kato, M. Hanai, Y. Hoshina, M. Takei, H. Okubo, Application of functionally graded material for reducing electric field on electrode and spacer interface. IEEE Trans. Dielectr. Electr. Insul. **17**(1), 256–263 (2010)
10. S. Watanabe et al., Electrical applications of titanium-based FGMs manufactured by progressive lamination, in *Proceedings of IEEE 6th International Conference Conduction Breakdown Solid Dielectr. (ICCBSD)*, Vasteras, Sweden, June 1998, pp. 539–542
11. N. Hayashi, K. Kawahara, M. Sumikura, M. Hara, F. Endo, Electric field control by permittivity functionally graded materials and their lightning impulse withstand voltages for surface breakdown, in *Proceedings of IEEE International Symposium Electrical Insulation (ISEI)*, Boston, MA, USA, April 2002, pp. 260–263
12. N. Hayakawa, J. Shimomura, T. Nakano, M. Hanai, K. Kato, H. Okubo, Fabrication technique of permittivity graded materials (FGM) for disk-type solid spacer, in *Proceedings of Annual Report Conference Electrical Insulation Dielectric Phenomena (CEIDP)*, Montreal, QC, Canada, October 2012, pp. 32–35.
13. N. Hayakawa, Y. Miyaji, H. Kojima, K. Kato, Electric field reduction by functionally graded materials (FGM) with permittivity and conductivity distribution, in *Proceedings of IEEE Conference Electrical Insulation Dielectric Phenomena (CEIDP)*, Toronto, ON, Canada, October 2016, pp. 627–630
14. J. Ishiguro, M. Kurimoto, H. Kojima, K. Kato, H. Okubo, N. Hayakawa, Electric field control in coaxial disk-type solid spacer by functionally graded materials (FGM), in *Proceedings of IEEE Conference Electrical Insulation Dielectric Phenomena (CEIDP)*, Des Moines, IA, USA, October 2014, pp. 663–666
15. N. Hayakawa, J. Ishiguro, H. Kojima, K. Kato, H. Okubo, Fabrication and simulation of permittivity graded materials for electric field grading of gas insulated power apparatus. IEEE Trans. Dielectr. Electr. Insul. **23**(1), 547–554 (2016)

16. Z. Liu et al., Topology optimization and 3D-printing fabrication feasibility of high voltage FGM spacer, in *Proceedings of IEEE International Conference High Voltage Engineering Application (ICHVE)*, Chengdu, China, September 2016, pp. 1–4
17. Z. Liu, W.-D. Li, L.-Y. Zhang, X.-R. Li, J.-B. Deng, G.-J. Zhang, Simulation design and 3D-printing fabrication of conductivity graded insulator, in *Proceedings of IEEE 1st International Conference Electrical Materials Power Equipment (ICEMPE)*, Xi'an, China, May 2017, pp. 171–174
18. M. Kurimoto, H. Ozaki, Y. Yamashita, T. Funabashi, T. Kato, Y. Suzuoki, Dielectric properties and 3D printing of UV-cured acrylic composite with alumina microfiller. IEEE Trans. Dielectr. Electr. Insul. **23**(5), 2985–2992 (2016)
19. Y. Imanishi, M. Taguchi, K.-I. Onisawa, Effect of sublayer surface treatments on ZnO transparent conductive oxides using dc magnetron sputtering. Thin Solid Films **518**(11), 2945–2948 (2010)
20. P.J. Kelly, R.D. Arnell, Magnetron sputtering: a review of recent developments and applications. Vacuum **56**(3), 159–172 (2000)
21. F. Gao, J. An, BST thin films deposited by RF-magnetron sputtering and dielectric property research, in *Proceedings of IEEE Advanced Information Management, Communicates, Electronic and Automation Control Conference (IMCEC)*, Xi'an, China, October 2016, pp. 1808–1811

Mechanical Reliability of Large-Scaled Epoxy Insulator in GIS

Boxue Du, Ying Zhang, Jin Li, and Hucheng Liang

Abstract Large insulators produce internal defects or internal micro-cracks caused by internal and external mechanical stress during production and operation, and insulation breakdown under the combined action of mechanical stress and strong electric field is also one of the bottleneck problems that limit the reliability of large-capacity power systems. The settlement of alumina (Al_2O_3) during curing and the stress distribution of insulators under different laying methods were studied. The risk coefficient is defined to assess the operational dependability of insulators with different laying methods. The conclusion is as follows: The settlement of alumina causes the insulator density to vary from 2.13 to 2.32 g/cm^3. And t The amount of alumina obviously affects the mechanical properties of the composite system. The first principal stress is concentrated on the upper interface, threatening the mechanical reliability of this insulator. The laying method V has the lowest risk coefficient of 31.5, which should be the optimal laying method. In practical applications, the insulator should be laid in way V, and laid down to ensure its safe operation.

B. Du (✉) · Y. Zhang · J. Li · H. Liang
School of Electrical and Information Engineering, Tianjin University, No. 92, Weijin Road, Nankai District, Tianjin City, China
e-mail: duboxue@tju.edu.cn

Y. Zhang
e-mail: zhangying2019@tju.edu.cn

J. Li
e-mail: lijin@tju.edu.cn

H. Liang
e-mail: hcliang@tju.edu.cn

B. Du (ed.), *Polymer Insulation Applied for HVDC Transmission*,
https://doi.org/10.1007/978-981-15-9731-2_22

1 Introduction

Large insulators provide insulation and support for Gas insulated switchgear and gas insulated line. Insulators withstand the dual effects of electrical stress and mechanical stress during operation, requiring them to have good mechanical and electrical properties [1–3]. At present, the research on insulators is mainly focused on electrical insulation [4], but less on the mechanical stress of insulators. But a research showed that mechanical failures accounted for 39.3% of all GIS failures, even higher than insulation failures of 38.1% [5]. Therefore, it is necessary to study and improve the mechanical reliability of large insulators.

The mechanical stress causes the epoxy insulator to fail due to internal and external reasons: internally, during the casting of the epoxy insulator, internal stress will be generated during the curing and cooling stages: first, the insulator is not uniformly radiated and solidifies. The process has poor controllability, and uneven curing is easy to occur, resulting in uneven stress distribution; second, the insulator shrinks greatly during the casting process, and uneven shrinkage is easy to occur, resulting in uneven stress distribution; third, due to the insulator's The expansion coefficient is $3.0 \pm 0.4 \times 0^{-5}/°C$, which has a large difference from the thermal expansion coefficient of the metal insert. When the Joule heat generated by the through-current conductor causes a temperature gradient distribution, the difference in mechanical properties between the two causes the metal conductor There is a stress concentration area at the interface with the epoxy resin, which leads to the partial peeling between the epoxy insulator and the center conductor to form a micro air gap. In terms of external factors, epoxy insulators are subject to external stress loads during assembly and operation, and there will be obvious stress concentration inside, causing internal defects or even cracking of the insulator: First, if it appears during the assembly of the insulator The three-phase misalignment caused by the uncorrected position will cause the insulator to bear additional stress during the operation of the equipment; second, the alternating electromagnetic force generated during operation causes the pipe shell to vibrate, and the long-term vibration is in the form of mechanical stress Acting on the insulator eventually leads to insulation failure; third, the fast transient overvoltage generated by the reclosing is more than several times the operating voltage. The transient overvoltage causes the casing to undergo abnormal mechanical vibration and subject the insulator to short-term strong mechanical stress effect.

Most large insulators are cast from EP resin, which becomes uneven during the curing process due to the sedimentation of additives. Tests have shown that the difference between the upper and lower density of the 110 kV insulator is about 0.12 $\mu g/cm^3$. The content of aluminium oxide (Al_2O_3) and the way of laying will also have a important impact on the mechanical properties. The study found that the content of additives is strongly related to the mechanical properties of the composite materials (elastic modulus, compressive strength, tensile strength, etc.) [6, 7]. Large density changes will inevitably significantly affect the mechanical properties of composite insulators.

This chapter studies the Al_2O_3 settlement and mechanical stress distribution during the vulcanization of insulators. Because the mechanical stress of the insulator in different directions is different, the way it is laid will affect whether the insulator is prone to failure. To determine whether the insulator is vulnerable to mechanical damage, this article defines the risk factor. The laying method V has the lowest risk coefficient, which is 31.5, and it is judged that the laying method V is the best. In practical applications, keeping the mold gate down is an effective and easy way to avoid mechanical damage to the insulator.

2 Experimental Setup

2.1 Density Measurement

In practice, the materials of large insulators are generated by the physical blending of multiple materials. As shown in Fig. 1a, the mass ratio of epoxy resin, curing agent and Al_2O_3 particles is 100:38:330. Mix the epoxy resin, curing agent and Al_2O_3 particles uniformly Into an upright mold in a vacuum. It is then cured at high

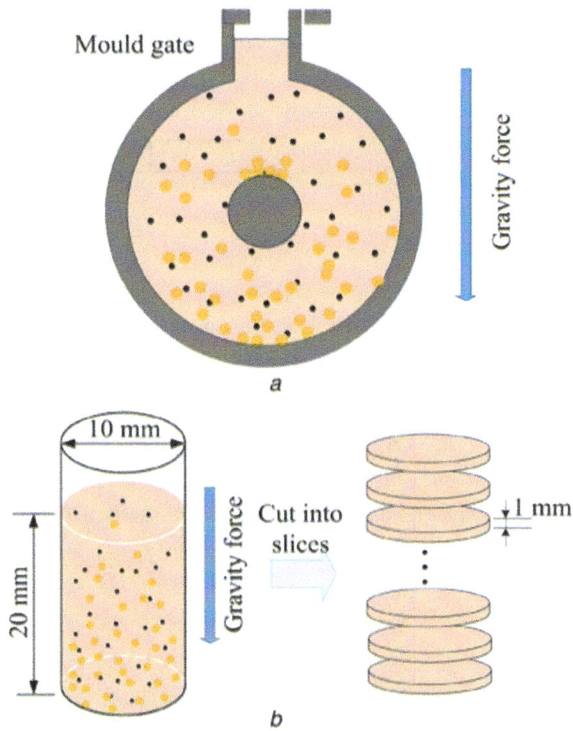

Fig. 1 **a** Casting process of large insulators. **b** Cutting process of cylindrical samples

temperature for more than 8 h, where the temperature exceeds 130°. Due to the low viscosity of the composite at the beginning, the micro-sized Al_2O_3 particles can easily settle under the action of gravity. Therefore, the density of the insulator is not uniform, and its mechanical properties will also be affected. By manufacturing some cylindrical samples, as shown in Fig. 1b, the settlement of Al_2O_3 particles during the curing process of the insulator was studied. The sample is 10 mm in diameter and 20 mm in length. The sample was then cut into 1 mm thick slices with a microtome. By measuring the density of the sheet through the displacement method, can get how the density of the entire sample is distributed.

2.2 *Mechanical Property Measurement*

On account of the sedimentation of Al_2O_3 particles, the density of the insulator material changes with the direction of gravity. Samples with different Al_2O_3 contents and densities of 2.14, 2.19, 2.24, 2.29 and 2.34 g/cm3 were prepared, and their mechanical properties were tested. Figure 2 shows how to apply tensile and compressive stress to the sample. Tensile and compressed specimens have different shapes, and the compression test sample is cylindrical. Simulation parameters are provided for the mechanical stress distribution, and the mechanical stress performance parameters of different samples are obtained.

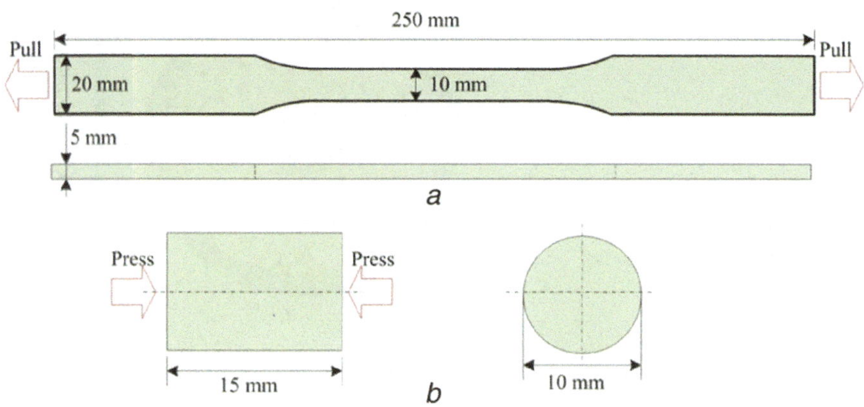

Fig. 2 Samples for different mechanical stress tests. **a** Tensile sample. **b** Compressive sample

3 Simulation Model

3.1 Al_2O_3Settlement

The settled particles of micro-sized alumina are controlled by the following equation [8]:

$$m\frac{dv_t}{dt} = mg - F_V - F_b \tag{1}$$

$$m = \frac{\pi}{6}D_f^3\rho_f \tag{2}$$

$$F_b = \frac{\pi}{6}D_f^3(\rho_f - \rho_{ep})g \tag{3}$$

$$F_v = 3\pi\eta D_f v_t \tag{4}$$

where m is the mass of Al_2O_3 particles, kg; g is the acceleration of gravity, m/s^2; F_b is the buoyancy force, N; F_v is the viscous force N; D_f is the particle size of Al_2O_3 particles, m; *ep* resin density and Al_2O_3, respectively Particle density, kg/m^3; characterizes the viscosity of the EP/Al_2O_3 system, Pa s; v_t is the velocity of Al_2O_3 particles at time t.

The velocity of Al_2O_3 particles impacting the terminal after a short acceleration, set dv/dt = 0 to get the free settling velocity v_0.

$$v_0 = \frac{gD_f^2(\rho_f - \rho_{ep})}{18\eta} \tag{5}$$

Because Al_2O_3 particles also affect each other, the hindered settling velocity v can be expressed as:

$$v = v_0(1 - C)^n \tag{6}$$

where n is the measured coefficient, which is set to 2.22. C is the Al_2O_3 concentration, whose starting value is set to 7.95×10^{11} m^{-3}.

Figure 3 is the distribution diagram of alumina and the change of viscosity with the content of additives in the composite system. Figure 3a shows that the diameter of Al_2O_3 particles varies from 1 to 60 μ, mainly concentrated at 10–30 μ. From Formula (5), it can be seen that the larger the particles, the faster the sedimentation speed. It can be seen from Fig. 3b that as the curing time increases, the viscosity of the composite system increases. It can be seen from (5) that the sedimentation rate of Al_2O_3 particles is inversely proportional to the viscosity of the composite system. From the figure, after 70 min, the viscosity of epoxy resin is very high,, and

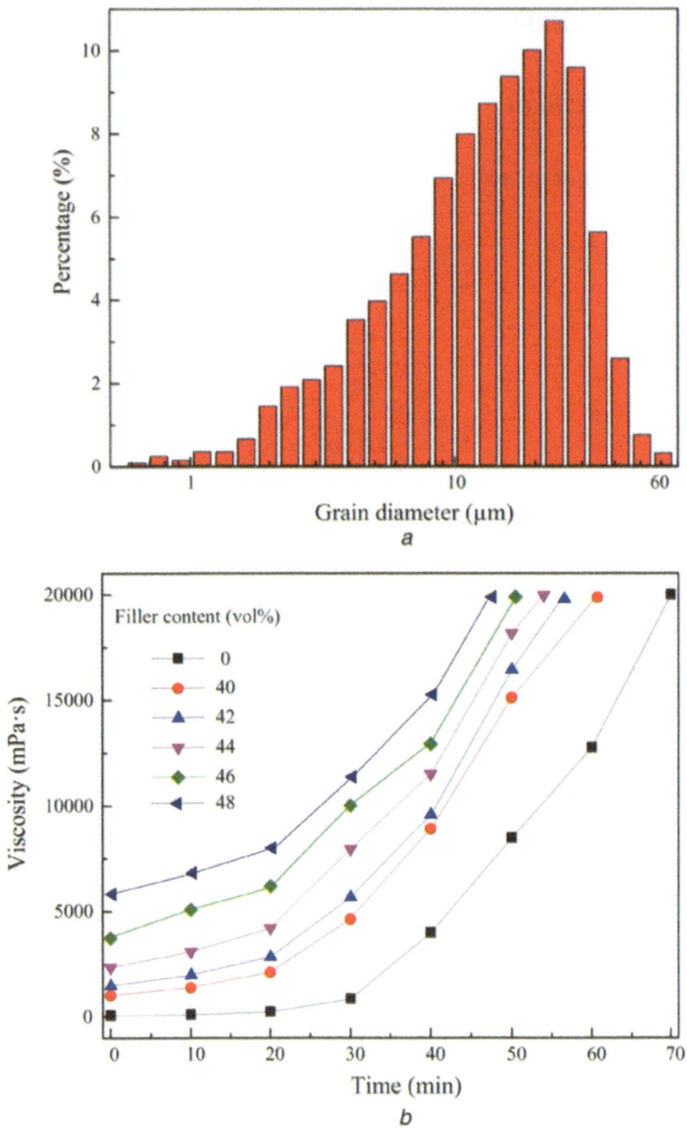

Fig. 3 **a** Al_2O_3 diameter distribution and measured viscosity of EP/Al_2O_3 system. **b** Viscosity change of composite system

the sedimentation rate of Al_2O_3 was so slow that it was negligible. Therefore, this paper sets the gravity settlement simulation time to 70 min. In addition, the viscosity of the composite system increases with increasing Al_2O_3 content. The viscosity of the composite system can be described as:

$$\eta = \eta_0(1 - C/C_{\max})^{-m} \quad (7)$$

In the formula, η is the viscosity of the pure EP resin, Pa s; C_{max} is the maximum Al_2O_3 concentration, m^3; m is the correction coefficient, obtained by fitting the data in Fig. 3b.

Figure 4 simulates the Al_2O_3 settlement process in the composite system The mathematical model of the settlement of the cylindrical specimen is shown in Fig. 4a

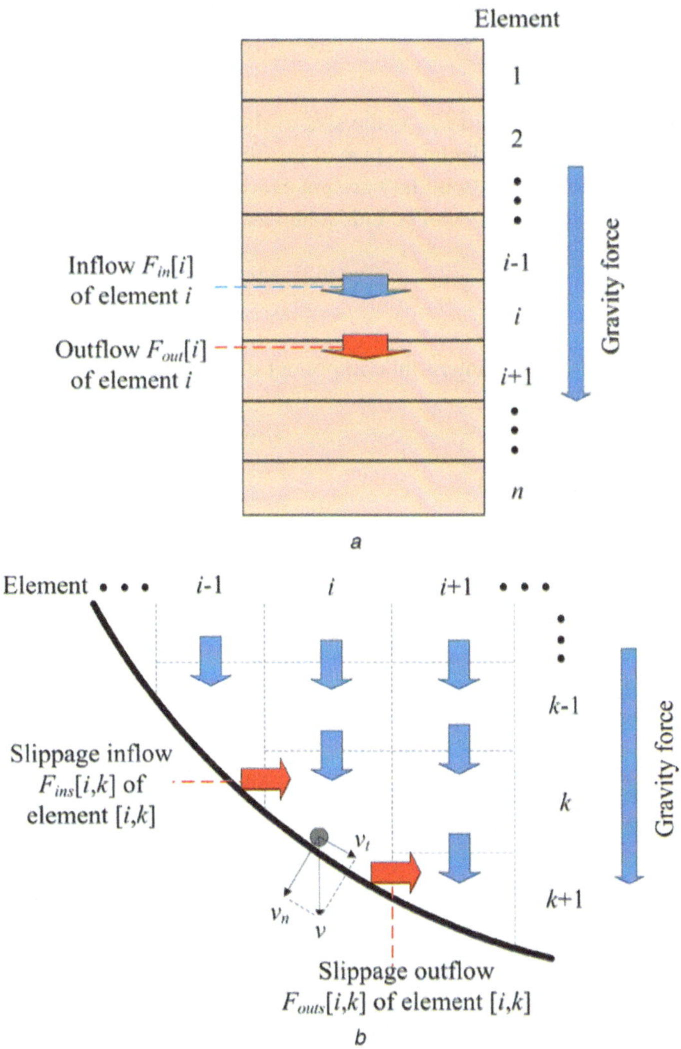

Fig. 4 Columnar sample and large insulator Al_2O_3 settlement simulation model. **a** Columnar sample. **b** Large insulator

[9], which is one-dimensional. It can be seen that the cylinder is divided into n parts, and the Al_2O_3 content of each part can be calculated by the following formula:

$$C[i]_t = C[i]_{t-dt} + F_{in}[i] - F_{out}[i] \tag{8}$$

where $C[i]$ is the Al_2O_3 concentration of element i, m^{-3}, and $F_{in}[i]$ and $F_{out}[i]$ are the inflow and outflow of part i, m^{-3}, respectively.

$$F_{in}[i+1] = F_{out}[i] = \frac{C[i]vdt}{L_{ele}} \tag{9}$$

where dt is the time step s and L_{ele} is the size of the element, set to 0.001 m.

Because the particles inside the surface of the large insulator have a slip behavior, the inflow and outflow at the boundary are not exactly the same, as shown in Fig. 4b. For element $[i, k]$, the outflow can be calculated as:

$$F_{outs}[i,k] = \frac{C[i,k]v_t dt}{L_{ele}} \tag{10}$$

In the formula, v_t is the tangential component of the final velocity, m/s.

3.2 Mechanical Stress

In the horizontally-laid GIS, the two insulators are 6 m apart and have a diameter of 315 mm. The diameter of the hollow conductor is 80 mm. The specific structure is shown in Fig. 5. The material of the large insulator is EP/Al_2O_3 composite, the groove and conductor are made of cast aluminum, and the groove is filled with 0.4 MPa sulfur hexafluoride to improve the strength of the medium. This article studies how static stress is distributed, which follows Hooke's law [10, 11]

$$\varepsilon_x = \frac{1}{E}[\sigma_x - \mu(\sigma_y + \sigma_z)] \tag{11}$$

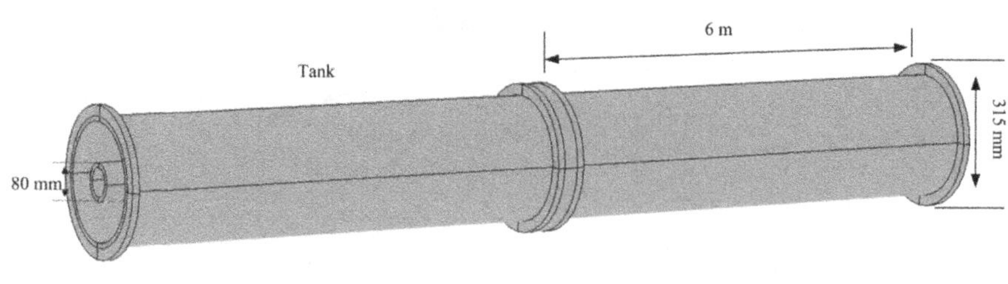

Fig. 5 Local structure of a horizontally laid 110 kV GIS

$$\varepsilon_y = \frac{1}{E}[\sigma_y - \mu(\sigma_z + \sigma_x)] \tag{12}$$

$$\varepsilon_z = \frac{1}{E}[\sigma_x - \mu(\sigma_x + \sigma_y)] \tag{13}$$

$$\gamma_{xy} = \frac{2(1+\mu)}{E}\tau_{xy} \tag{14}$$

$$\gamma_{yz} = \frac{2(1+\mu)}{E}\tau_{yz} \tag{15}$$

$$\gamma_{xz} = \frac{2(1+\mu)}{E}\tau_{xz} \tag{16}$$

In the formula, the normal stress is σ and the tangential stress is τ, N/m^2; ε and γ are normal and shear strains, respectively; E is the elastic modulus and N/m^2; μ is the Poisson's ratio.

Figure 6a shows the geometry of the 110 kV GIS insulator, and Fig. 6b shows the simplified bolt model. Set 6 pairs of symmetrically distributed screw holes on the edge of the large insulator. Its function is to fix the insulator to the water tank with bolts. The specific bolt type and tightening torque are shown in Fig. 6a. GIS tank and large insulator are relatively stationary, so the pretension F_B is used instead of the bolt. After calculation, F_B is set to 4.6 105 N.

Because all electrical devices must bear weight, all objects have the effect of gravity. Therefore, on the inner and outer surfaces of the tank body, the surface and edges of the insulator. Different surface pressures are applied to the conductor surface. Since only the stress distribution on the insulator is considered, fixed constraints are imposed on both ends of the tank.

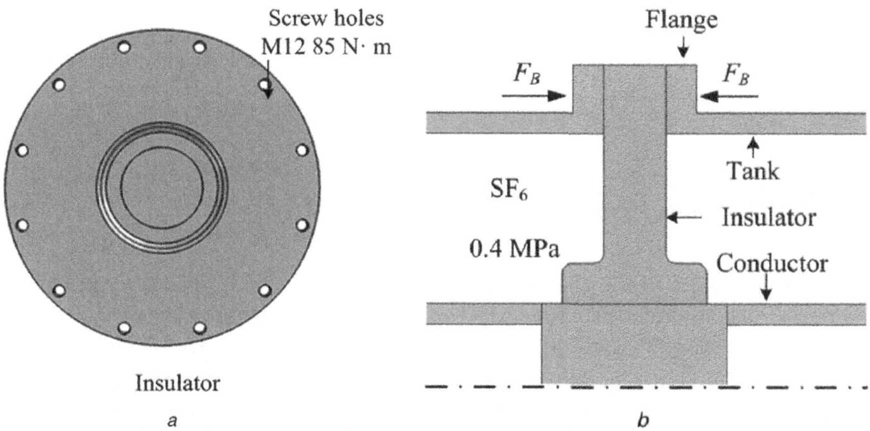

Fig. 6 **a** Structural drawing of large insulator. **b** Simplified bolt model

Table 1 Mechanical parameters in the mechanical stress simulation

Parameter	Poisson's ratio	Density/kg · m^{-3}	Elastic modulus/Pa
Casting aluminum	0.34	2.8×10^3	7×10^{10}
EP/Al_2O_3	$\mu(\rho)$	$\rho(x, y, z)$	$E(\rho)$
EP/Al_2O_3	$\mu(\rho)$	$\rho(x, y, z)$	$E(\rho)$

Table 1 lists the mechanical parameters of insulator composite materials when simulating mechanical stress.

4 Results and Discussion

4.1 Density Distribution

Figure 7a is a comparison diagram of measured density distribution and simulated density distribution of a cylindrical sample. In order to ensure the uniformity of the density of the cylindrical specimen in the compression test, the specimen must be taken from the middle area. It can be seen that the difference between the simulated data and the measured data is very small, and the trend is consistent. Therefore, the mathematical model of the simulated insulator is speculated to be credible. The maximum value of the sample density is obtained at the bottom of the sample, while the minimum value is obtained at the top of the sample. In the central region, the sample density basically does not change with location, it is 2.24 g/cm^3. Figure 7b shows the density distribution inside the simulated GIS insulator. From Fig. 7b, the maximum density of the insulator is at the edge of the lower half of the insulator and the upper interface with the conductor, and the minimum value of the density of the insulator is at the edge of the upper half of the insulator and the lower interface with the conductor. Studies have shown that alumina affects the mechanical properties of composite materials. Therefore, the 110 kV GIS insulator density that changes greatly affects its mechanical properties.

4.2 Mechanical Property

Figure 8a is a graph of the change of the elastic modulus and Poisson's ratio of the composite system under different Al_2O_3 contents. When the density of the sample increased, the Poisson's ratio of the composite system did not change significantly, and the elastic modulus increased significantly, from 8000 to 12,400 MPa. Figure 8b shows the tensile and compressive strength of the composite system under different

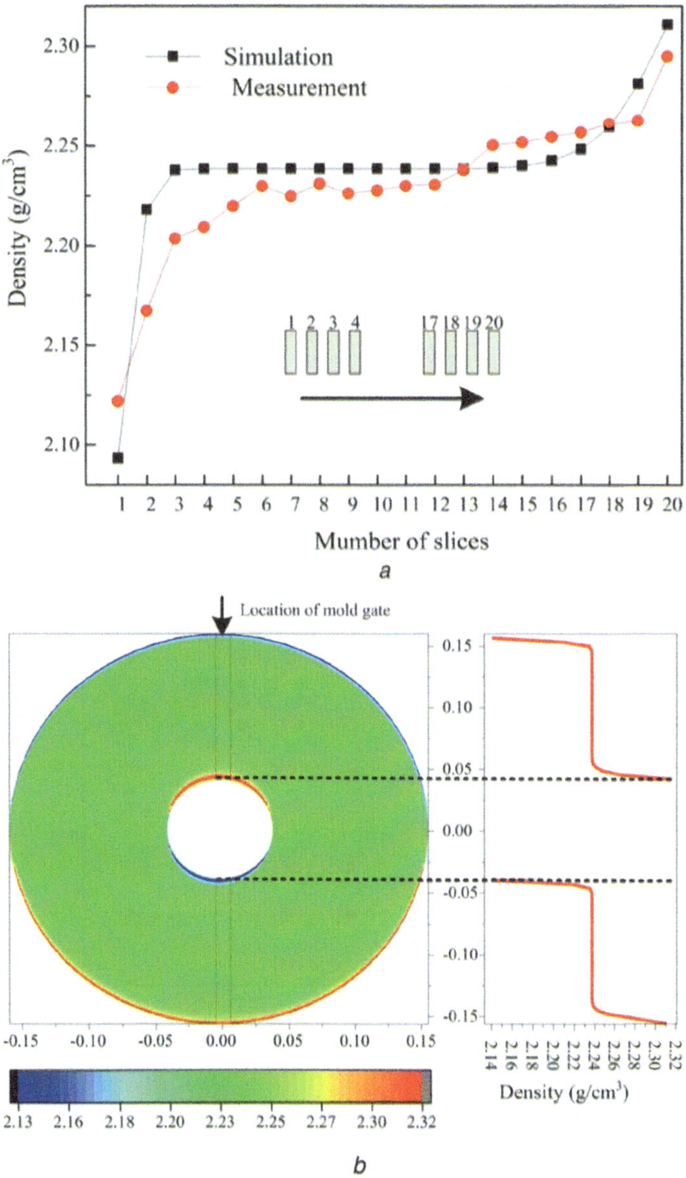

Fig. 7 **a** Density distribution of cylindrical samples; **b** Density distribution of large insulators

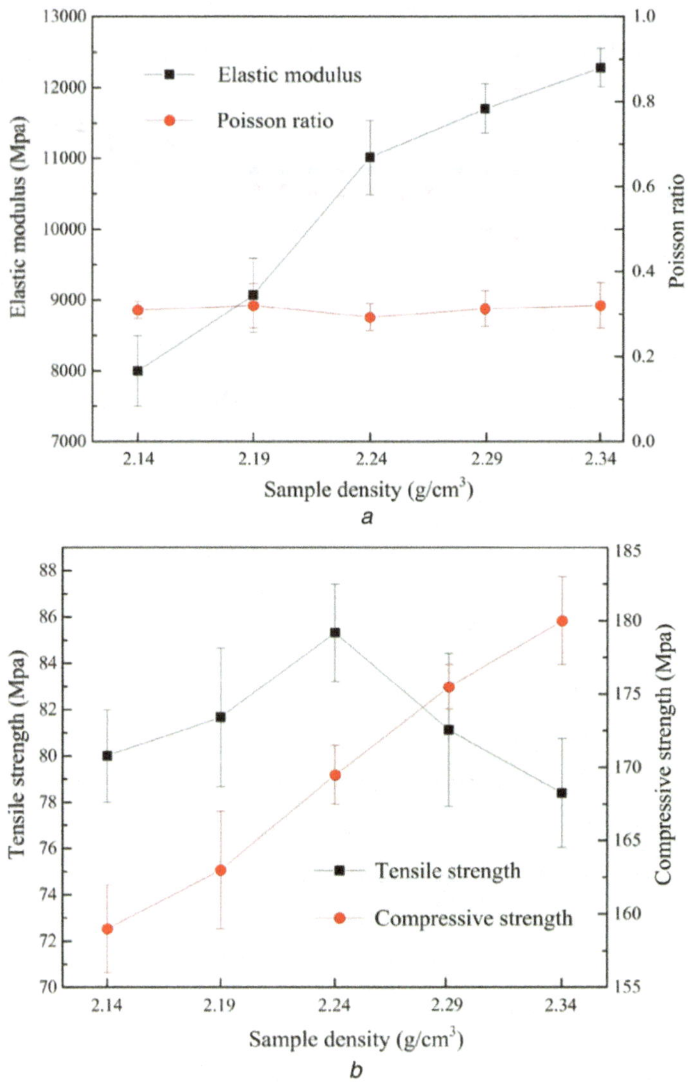

Fig. 8 **a** Elastic modulus and Poisson's ratio of EP/Al_2O_3 system. **b** Tensile and compressive strength of EP/Al_2O_3 system

Al_2O_3 contents. As the density of the sample increases, the tensile strength of the composite system increases first and then decreases, and the compressive strength increases continuously. It can be seen from the figure that the maximum tensile strength of the sample is 85 MPa.

4.3 Stress Distribution and Laying Method

Set the insulator density to 2.24 g/cm^3. If the density change is not taken into consideration, the stress of large insulators in the 110 kV GIS under the horizontal laying method is simulated. Figure 9 is a simulation diagram of the stress of a large insulator. When decomposing the mechanical stress, there needs to be a diagonal section where the shear stress is zero. The stress can be decomposed into the first principal stress, the second principal stress and the third principal stress in numerical order. The positive and negative stress values represent tensile stress or compressive stress, and the positive value of the principal stress is tension, and vice versa. Figure 9a is a simulation diagram of the first principal stress. It can be seen from the figure that the tensile force is locally distributed on the upper interface in contact with the

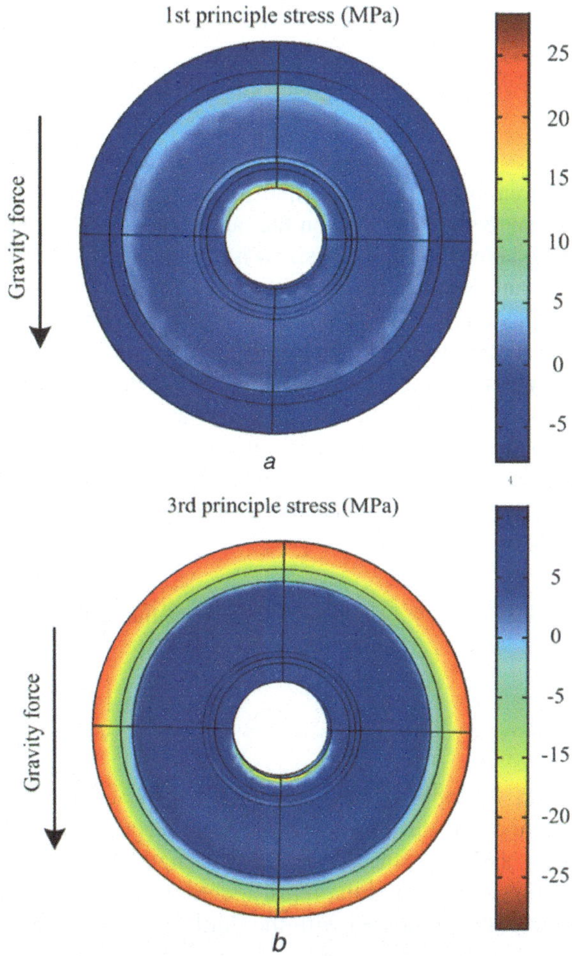

Fig. 9 Large-scale insulator stress simulation diagram. **a** First principal stress simulation diagram. **b** Third principal stress simulation diagram

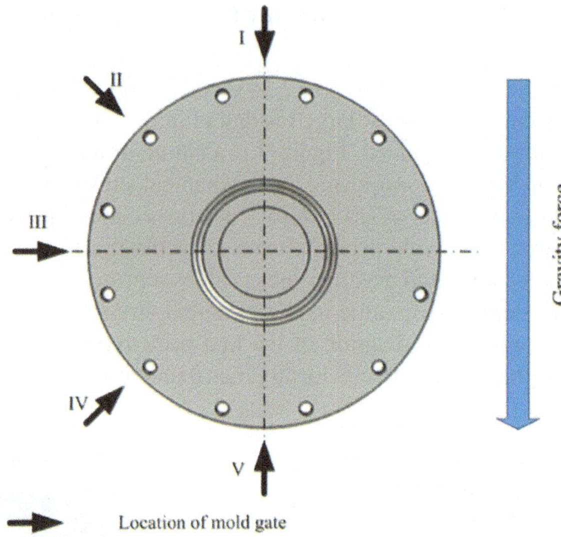

Fig. 10 Diagram of different insulator laying methods

conductor, and the maximum tensile force is about 27 MPa. Figure 9b is a simulation diagram of the third principal stress. It can be seen from the figure that the pressure is mainly concentrated on the outer and upper and lower interfaces in contact with the conductor. The edge stress is caused by the action of the bolt, and the maximum pressure is about 29 MPa.

Obviously, the mechanical properties and stress distribution of GIS insulators are directional, not symmetrical. Therefore, in the case of such placement of GIS, the laying method obviously affects whether the insulator is prone to mechanical failure. As shown in Fig. 10, there are five ways of laying insulators. In the first method, the insulator keeps the mold gate up. Then rotate the insulator counterclockwise at four angles, respectively for the remaining four placement methods. In order to compare whether the insulator is easy to find mechanical damage, define the risk factors as follows:

$$\beta_1 = \max\left\{\frac{\sigma_1}{\sigma_t}\right\} \tag{17}$$

$$\beta_3 = \max\left\{\frac{\sigma_3}{\sigma_{bc}}\right\} \tag{18}$$

where β_1 and β_3 are the first and third risk coefficients, respectively; σ_1 and σ_3 are the first and third principle stresses, respectively, MPa; σ_t is the tensile strength, MPa; and σ_{bc} is the compressive strength, MPa.

Figure 11 shows the hazard factor of insulators under different laying methods. Obviously, compared with the third risk factor, the first risk factor is much higher.

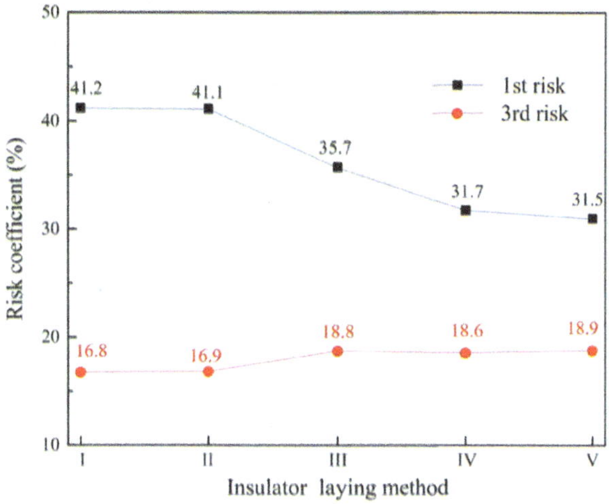

Fig. 11 Risk coefficients of different insulator laying methods

This shows that insulators are more prone to tensile damage, so most of the mechanical damage is due to tensile forces, and most manufacturers only pay attention to their ability to withstand tensile stress. As the insulator moves from mode I counterclockwise to mode V, the first risk factor continues to decrease. Method V seems to be the best with only 31.5 risk factors.

Because the maximum value of the first risk factor is closely related to whether it is susceptible to tensile stress damage, the size and distribution of tensile stress of large insulators is very important. Figure 12 shows the distribution of the first principal stress and tensile strength at the interface of the insulator.

The Fig. 12 shows that, no matter what way the insulator is laid, the maximum value of the first principal stress appears at the upper interface. It can be seen from Fig. 12 that the maximum value of the first principal stress of the insulator laying method I is the highest, the laying method is the third, and the laying method V is the smallest. Therefore, it is verified that the laying method of the insulator can change the stress concentration of the first interface. The study also found that changes in the laying method affect the tensile strength of large insulators. As shown in Fig. 12a, in the laying method I, the maximum value of the first principal stress and the minimum value of the tensile strength appear simultaneously and the maximum value of the first principal stress is the highest. Therefore, this laying method is most prone to mechanical stress damage, and the first risk factor is the highest, reaching 41.2. As shown in Fig. 12b, in the laying method III, although the maximum value of the first principal stress is almost unchanged and also appears at the upper interface, the tensile strength of the upper interface increases. Therefore, the first risk factor has a significant decrease compared with Method I, reaching 35.7. As shown in Fig. 12c, in the laying method V, although the maximum value of the first principal stress and the minimum value of the tensile strength appear simultaneously, and the tensile strength is only 80 MPa, the maximum value of the first principal stress The value has dropped significantly compared to method I, so the first risk factor has dropped

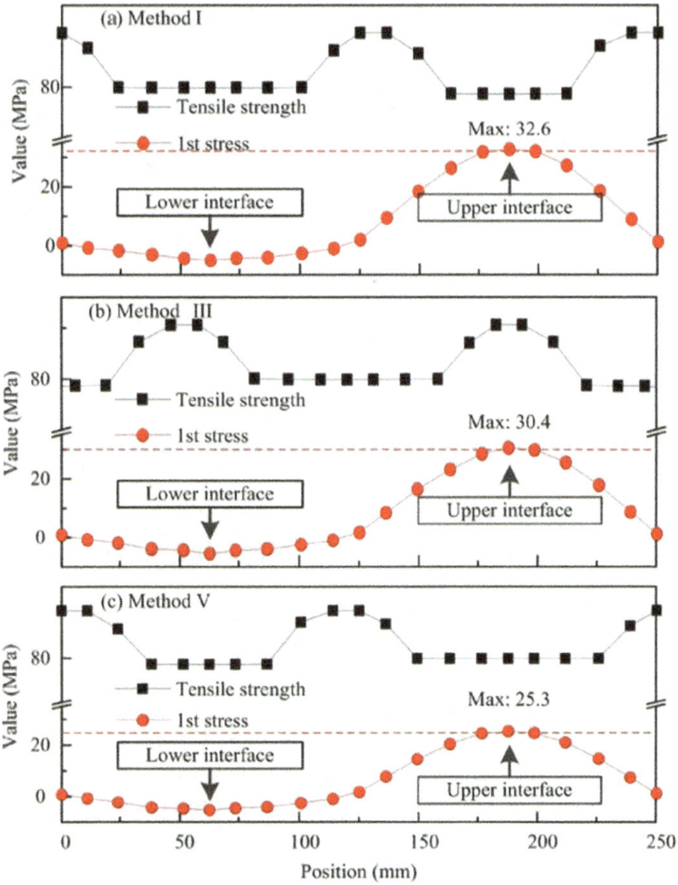

Fig. 12 First principle stress and the tensile strength distributions on the interface between insulator and conductor by different laying methods. **a** Method I. **b** Method III. **c** Method V

to as low as 31.5. It can be seen that in practice, in order to reduce the possibility of mechanical damage to the insulator and ensure the long-term safe operation of the system, the insulator should be laid down on the mold door.

5 Conclusion

The density distribution and mechanical performance parameters of large insulators containing alumina were studied. The main summary is as follows:

1. Due to the sedimentation of Al_2O_3 in the insulator, its density is not uniform. The maximum density appears in the upper interface between the conductor and its

bottom edge area, the minimum density appears in the lower interface between the conductor and its top edge area.

2. The content of alumina affects the mechanical properties of insulator composites. As the alumina content increases, the elastic modulus of the composite system increases, but the Poisson's ratio hardly changes. The compressive strength of the composite system increased linearly with the Al_2O_3 content, and the tensile strength showed a nonlinear trend of increasing first and then decreasing.
3. The first principal stress is located at the upper interface of the conductor, and the third principal stress is located at the edge of the conductor and the lower interface of the conductor. The first principal stress concentration makes large insulators susceptible to mechanical stress damage.
4. The laying method V is the best, and its risk factor is 31.5. In practice, the insulator should be laid down to improve its mechanical reliability.

References

1. M. Ehsani, M.E. Zeynali, M. Abtahi et al., LDPE/EPDM blends as electrical insulators with unique surface electrical and mechanical properties. Iranian Polymer J. **18**(1), 37–47 (2009)
2. M. Ehsani, H. Borsi, E. Gockenbach, et al., Improvement of electrical, mechanical and surface properties of silicone insulators, in *The 17th Annual Meeting of the IEEE Lasers and Electro-Optics Society, LEOS 2004*, Boulder, CO, USA, (2004), pp. 623–626
3. J.T. Burnham, R.J. Waidelich, Gunshot damage to ceramic and nonceramic insulators. IEEE Trans. Power Delivery **12**(4), 1651–1656 (1997)
4. L. Liu, et al., The influence of electric field distribution on insulator surface flashover, in *2018 IEEE Conference on Electrical Insulation and Dielectric Phenomena (CEIDP)* (2018), pp. 255–258
5. Y.L. Cui, H.Z. Ma, T.Y. Wang, et al., Fault analysis of GIS based on fault tree theory. High Volt. Appar. **51**(7), 125–129 (2015)
6. A. Omrani, L.C. Simon, A.A. Rostami, The effects of alumina nanoparticle on the properties of an epoxy resin system. Mater. Chem. Phys. **114**(4), 145–150 (2009)
7. S. Zhao, L.S. Schadler, R. Duncan et al., Mechanisms leading to improved mechanical performance in nanoscale alumina filled epoxy. Compos. Sci. Technol. **68**(14), 2965–2975 (2008)
8. T.N. Smith, A model of settling velocity. Chem. Eng. Sci. **53**(2), 315–323 (1998)
9. H. Hayakawa, J. Ishiguro, H. Kojima et al., Fabrication and simulation of permittivity graded materials for electric field grading of gas insulated power apparatus. IEEE Trans. Dielectr. Electr. Insul. **23**(1), 547–554 (2016)
10. G. Kadkhodaei, K. Sheshyekani, M. Hamzeh, Coupled electric–magnetic–thermal–mechanical modelling of busbars under short-circuit conditions. IET Gener. Transm. Distrib. **10**(4), 955–963 (2016)
11. H. Wang, Z. Guo, H. Feng et al., Simulation and experimental study on mechanical performance of UHV GIS spacer in hydrostatic test. IEEE Trans. Dielectr. Electr. Insul. **24**(5), 3122–3131 (2017)

Insulation Design of Superconducting Gas Insulated Transmission Line

Boxue Du, Jianan Dong, Jin Li, Mingyang Wang, and Zhaoyu Ran

Abstract To improve the insulation performance of superconducting gas insulated transmission line (S-GIL), this chapter puts forward the two E-field regulating methods: E-field adaptive insulator and surface functional graded material (SFGM) insulator. The two simulation models of insulator in S-GIL are completed for optimize the E-field distribution of insulator. For the method of E-field adaptive insulator, the optimal E-field non-uniformity factor can be reduced to 1.1 considering the E-field distribution and leakage current. This method can be realized by means of 3D print technology; For surface functional graded material (SFGM) insulator, a proper value of parameter k can be selected to realize the optimal E-field distribution. The insulator with nonlinear surface conductivity has superior properties. The leakage current can be restricted an acceptable value of 2.6×10^{-16} A. This method can be realized by means of magnetron sputtering technology or fluorination treatment. It is hoped that this chapter can contribute a useful reference guide for the design and application of S-GIL.

B. Du (✉) · J. Dong · J. Li · M. Wang · Z. Ran
School of Electrical and Information Engineering, Tianjin University, Weijin Road 92, Tianjin, People's Republic of China
e-mail: duboxue@tju.edu.cn

J. Dong
e-mail: djn058@tju.edu.cn

J. Li
e-mail: lijin@tju.edu.cn

M. Wang
e-mail: tyutwzh@163.com

Z. Ran
e-mail: ranzhaoyu@tju.edu.cn

B. Du (ed.), *Polymer Insulation Applied for HVDC Transmission*,
https://doi.org/10.1007/978-981-15-9731-2_23

1 Introduction

The modem society dependence on electricity is rapidly increasing, and the electric grid will be expected to satisfy with this growing demand. The social requirement to use sustainable and environmentally friendly methods for power generation and distribution is also increasing, which requires the adoption of new technologies [1]. In this respect, high temperature superconducting (HTS) devices are one of novel technologies to meet the demands the growing power demands of the future and can be applied in future power transport. HTS has some certain benefits over the conventional copper transmission lines, such as higher current density, and low line losses as well as reducing impact on the surrounding environment [2–4]. HTS technology has shown the application potential in some conditions including naval vessels, electric aircraft, high-energy physics, data center power distribution and so on [5–7]. However, liquid nitrogen as a refrigerant of cryogenic superconducting equipment cannot work normally under some operating conditions, such as below the temperature range of liquid nitrogen. Therefore, gaseous helium (GHe) has been regarded as substitute for liquid nitrogen (LN_2) due to the wider operating temperature range [8]. But, one of the key weaknesses of GHe is that partial discharge is to be triggered as the traditional design of HTS cables are applied in GHe environment [9, 10]. Therefore, it is very urgent to find an alternative insulation design to realize the dielectric properties in HTS cables.

In recent years, the gas insulation transmission line has been significantly developed in practical application. Based on it, the novel HTS cable insulation design, Superconducting gas-insulated transmission line (S-GIL), is proposed in [11]. In S-GIL, the gas medium as refrigeration and dielectric insulation are applied for the cable [12]. According to previous studies and the practical experience of the conventional gas insulated equipment, the unreasonable dielectric parameters along the insulator may result in E-field (E-field) concentration in the vicinity of the triple junction (gas phase, insulators and the electrode) causing practical discharges and eventually initiating surface flashovers inside GIL [13, 14]. Consequently, it is very necessary to improve the E-field distribution on the insulator surface of S-GIL for the stable operation of power system. In order to improve the insulation performance of S-GIL, kinds of methods have been investigated to homogenize the E-field distribution of insulators for instant optimizing insulator shape, installing shielding electrode and so on [15, 16]. However, these methods are complicated and costly but the effects are limited, and thus it is vital to discover the simple and highly efficient method used for the E-field regulation.

In order to break the constrain, there are two feasible E-field regulating methods proposed in this chapter, the E-field adaptive insulator and surface functional graded material (SFGM) insulator. The model of insulator was completed to discuss the effects of optimization methods on the E-field distribution along the insulator. The electric-field non-uniformity factor of the S-GIL insulator was introduced to assess reasonability and validity of the optimization method. Besides, the feasibility analysis of two optimization methods for improving insulation performance of insulator in

S-GIL system was carried out by means of calculating leakage current. It is hoped that this chapter can contribute a useful reference guide for the design and application of S-GIL.

2 E-Field Adaptive Insulator

In this section, the E-field adaptive insulator as a novel method is puts forward to optimize the E-field distribution and insulation performance of S-GIL insulator. The iterative method is adopted to adjust the conductivity profile of S-GIL insulator. It is worth mentioning that 3D printing technology can be used to realize the conductivity gradient insulator [17].

2.1 The Model of the S-GIL Design

The schematics of S-GIL design is showed in Fig. 1. The S-GIL is consisting of cable, cable tray, tri-post insulators and cryostat. The cable as HV (High Voltage)electrode is supported by the cable tray. The cryostat is enclosure as the ground electrode. Post insulator plays the role of mechanical support and electric insulation. The flexible connection between the wall of the cryostat and post insulators is adopted to reduce mechanical stress damage caused by contraction, expansion or vibration the connection.

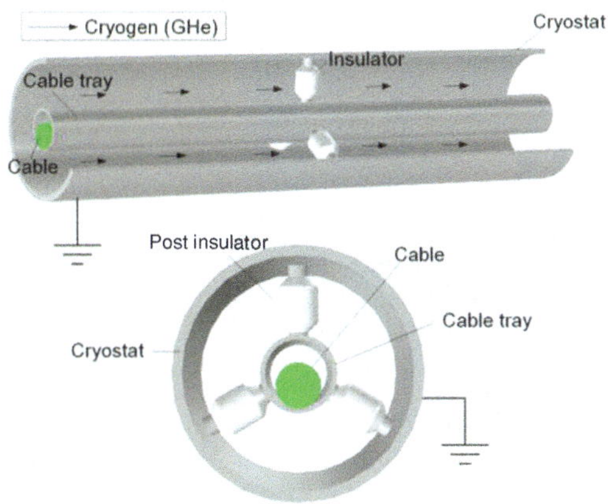

Fig. 1 The schematics of the S-GIL design [18]. Copyright Clearance Center's RightsLink ® License Number: 4854490238660

2.2 Simulation Design of E-Filed Adaptive Insulator

DC-GIL is confronted with more severe serious problems than AC-GIL. On the basis of the electromagnetic field theory, the E-field distribution under an ac voltage is capacitive and mainly determined by the permittivity of the insulating materials, the E-field under dc voltage is determined by conductivity of the insulating materials, resulting in a so-called resistive E-field distribution [19]. In order to improve the E-field distribution of insulator, it is necessary to design a reasonable conductivity profile of insulator. In this section, the iterative method is adopted to realize the E-field adaptive insulator in S-GIL.

In this section, the simulation model of S-GIL insulator is used to investigate E-field distribution. Figure 2 presents the simplified geometric model profile of S-GIL insulator. The volume conductivity of the post insulator is designed as a gradient distribution along the z direction to optimize the E-field distribution $E(z)$ of the S-GIL insulator.

The flow chart of conductivity profile optimization of S-GIL insulator is displayed in Fig. 3. A constant volume conductivity of the post insulator $(\sigma(z) = \sigma_0)$ is set, and then the simulation calculation is done to obtain surface E-field distribution $E(z)$ in the first iterative calculation.

Under the DC electric stress, if local E-field is too high, it can be reduced by increasing the local conductivity; on the contrary, if local E-field is too low, it can be increased by decreasing the local conductivity [20, 21]. Therefore, the conductivity profile of post insulator is updated in each iteration by defining the adjustment coefficient k_i as showed below,

$$\sigma_{i+1}(z) = k_i \sigma_i(z) \tag{1}$$

$$k_i = \left(\frac{E_i(z)}{E^i_{\min}} \right)^\alpha \tag{2}$$

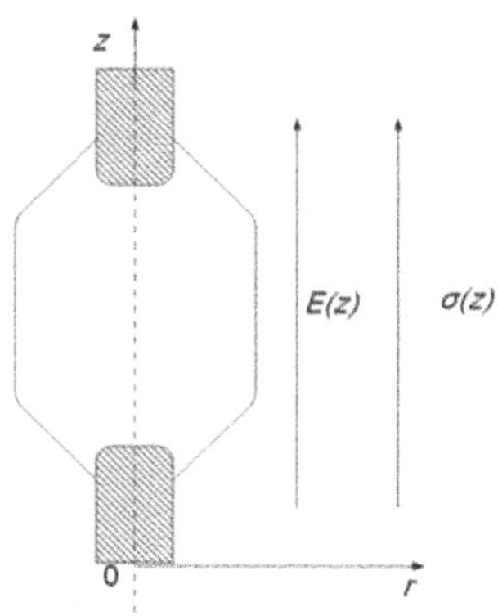

Fig. 2 The simplified geometric model profile of S-GIL insulator [18]. Copyright Clearance Center's RightsLink ® License Number: 4854490238660

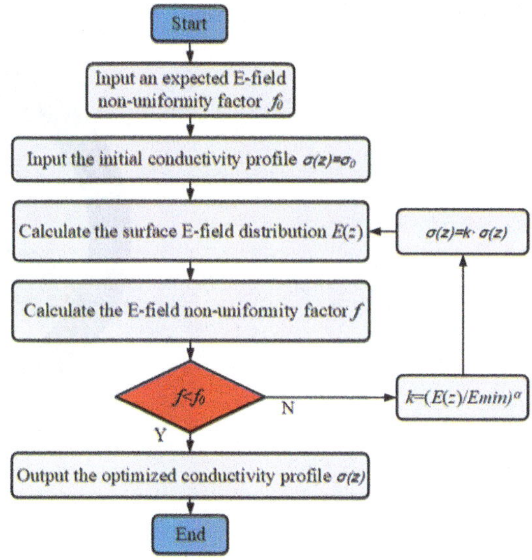

Fig. 3 Flow chart of conductivity profile optimization of S-GIL insulator [18]. Copyright Clearance Center's RightsLink ® License Number: 4854490238660

Among them, σ_{i+1} is the $(i + 1)$-th conductivity profiles, and σ_i is i-th conductivity profiles; $E_i(z)$ and E^i_{min} are the E-field distribution and the minimum E-field strength along the insulator in the i-th iteration, respectively; The adjustment coefficient is defined as k_i in the i-th iteration; In order to adjust the convergence speed, the coefficient α is defined, and is set to 1 in this section.

When the iteration reaches a certain number of times, the adjustment coefficient k becomes extremely close to 1. Meanwhile, the E-field distribution will become approximated to the uniform distribution. The optimal solution of the conductivity profile is obtained. At the beginning of iteration, f_0 is set as the expected E-field non-uniformity factor. It is used to terminate iteration when the E-field distribution reaches to our expectation. The non-uniformity factor f will be calculated in each iteration step. The iteration process will stop if the termination criterion ($f < f_0$) is evaluated to true.

2.3 Results and Discussion

In this section, the multi-physics coupling analysis software COMOSL Multiphysics is used to calculate the DC E-field distribution along the S-GIL insulator. The E-field distribution of S-GIL in steady state before optimization is showed in Fig. 4. There is non-uniform E-field distribution along the insulator surface. The maximum E-field strength is up to 1.5 kV/mm, while the minimum E-field strength as low as 0.1 kV/mm. To evaluate the degree of E-field inhomogeneity, the E-field non-uniformity factor f is defined as,

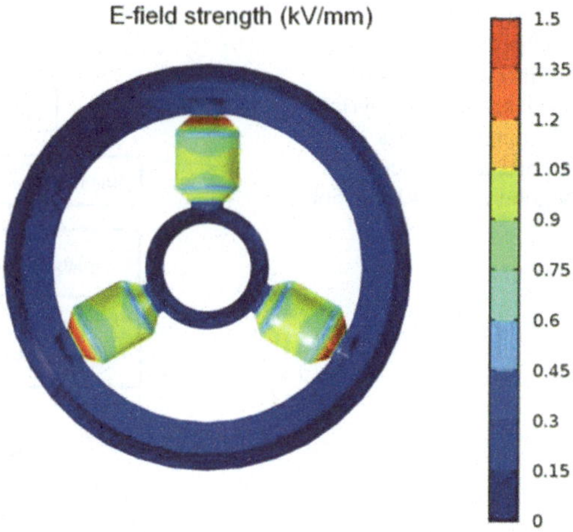

Fig. 4 E-field distribution of S-GIL before optimization [18]. Copyright Clearance Center's RightsLink ® License Number: 4854490238660

$$f = \frac{E_{max}}{E_{av}} \tag{3}$$

Among them, E_{max} and E_{av} are the maximum and the average E-field strength along the insulator is defined, kV/mm; Noted that the increase of f indicates the enhancement of the local E-field and the deterioration of the overall E-field distribution along the insulator surface. Therefore, as small as possible f is expected, which means there is more uniform E-field distribution along the insulator surface. Before optimization, the E-field non-uniformity f is obtained as ~1.8 by calculation using Eq. (3). It means that partial discharge is more likely to be initiated at the place where the local E-field is concentrated, even resulting in flashover fault.

The novel concept of E-field adaptive insulator is proposed in order to relax the local E-field distortion and improve the E-field distribution as well as improve the voltage rating of S-GIL. The iteration method, which has been introduced before is adopted to optimize the conductivity profile of insulator. When the non-uniformity factor f decreases to the expected value f_0, the iterative calculation will end. This section set the expected E-field non-uniformity factor f_0 as 1.05.

The distribution of the E-field along the insulator of the S-GIL is showed in Fig. 5 at the different iteration calculations. The results show that the E-field distribution is more and more even with the increase of the iteration steps. The maximum E-field strength is greatly weakened after 14 iteration calculations. The maximum E-field strength is 1.5 kV/mm and 0.82 kV/mm before and after optimization, respectively. As we know, the higher the E-field strength, the more likely to discharge. At the position of maximum E-strength, partial discharge (PD) will be initiated when the dielectric strength of GHe is lower than the maximum E-field strength in the insulated pipeline of GHe. It can be known that the maximum E-field strength before optimization is about 1.45 times higher than that after optimization. Therefore, it is

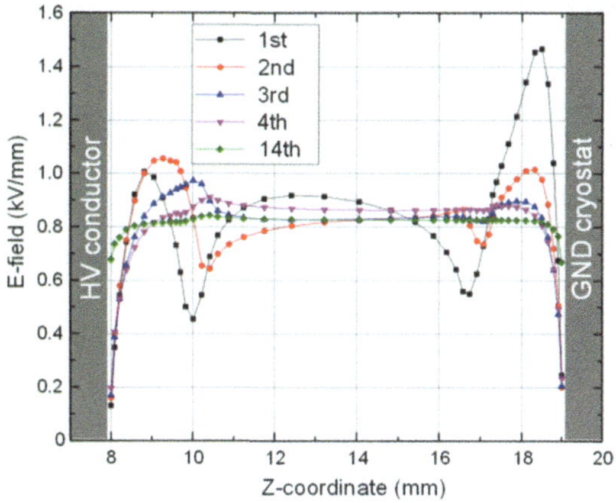

Fig. 5 The E-field distribution on the S-GIL insulator surface in each iteration [18]. Copyright Clearance Center's RightsLink ® License Number: 4854490238660

estimated that the voltage of partial discharge of S-GIL will be increased by about 45%, and the voltage rating will also be increased.

The volume conductivity profile of S-GIL insulator at the different iteration calculations is shown in Fig. 6. The results show that the conductivity of insulator is higher in the middle and lower on both sides. Consequently, the volume conductivity on the both sides of insulator has to be reduced for relaxing the E-field strength. On the contrary, the volume conductivity in the middle part is increased. In particular, in the middle of the insulator, the volume conductivity of the first iteration is eighter orders of magnitude higher than that of the last iteration. It should not be ignored that the leakage current significantly increases due to the increase of volume conductivity of insulator even if the E-field distribution is improved and become more even than before. Consequently, the balance between the E-field distribution and leakage current has to be discovered for the practical application of S-GIL.

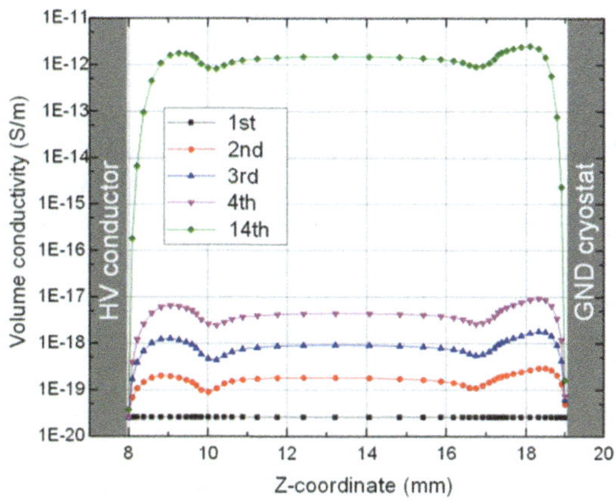

Fig. 6 The conductivity profile of S-GIL insulator in each iteration [18]. Copyright Clearance Center's RightsLink ® License Number: 4854490238660

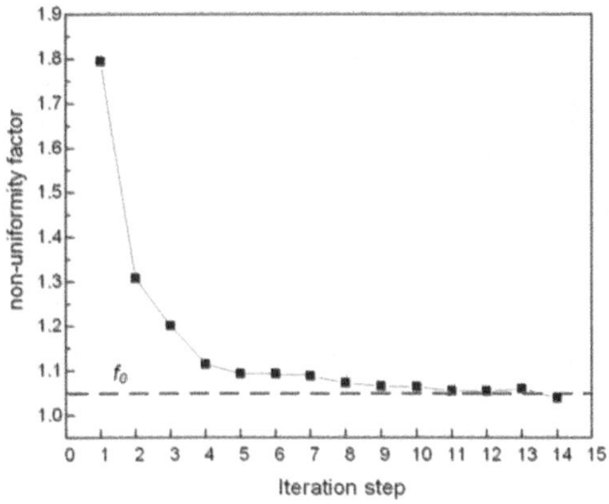

Fig. 7 The process in which the nonuniformity factor f decreases with increasing of iterations [18]. Copyright Clearance Center's RightsLink ® License Number: 4854490238660

The process in which the nonuniformity factor *f* decreases with increasing of iterations is showed in Fig. 7. As can be seen that the non-uniformity factor *f* is significantly reduced by about 63.8% from 1.8 to 1.1 in the first four iterative calculations. It means that the leakage current in the fourth iterative calculation is approximately hundreds of times that in the first iterative calculation.

In the next ten iterative calculations, the E-field non-uniformity factor f is only slightly reduced by 0.05, but the volume conductivity i.e. leakage current has increased by approximately six orders of magnitude. There is a less improvement in E-field distribution but a huge increase in dielectric loss of insulator. Therefore, the demand for E-field uniformity has to be properly reduced so as to lower the increase of dielectric loss. The best setting of E-field non-uniformity factor f of the S-GIL insulator should be 1.1 with the consideration of the E-field distribution and leakage current.

2.4 Summary

This section puts forward a new concept with application of the E-field adaptive insulator in S-GIL to optimize the insulation design of S-GIL for the future integrated energy system. The iterative method is adopted to design novel S-GIL insulation with the gradient volume conductivity in the z direction. With the increase of the iterative step, the E-field non-uniformity factor f becomes less and the leakage current gradually raises. Consequently, the balance between the E-field distribution and leakage current has to be discovered for the practical application of S-GIL. For the S-GIL insulator, the best setting of E-field non-uniformity factor f of the S-GIL insulator should be 1.1 with the consideration of the E-field distribution and leakage current.

3 Insulator with Surface Graded Material

This section puts forward a novel concept with surface functionally graded material (SFGM) to optimize the E-field distribution and improve insulation performance of insulators in S-GIL. The influence of SFGM on the E-field distribution along the insulator was investigated by means of establishing the cone-type insulator model with surface graded conductivity. To discover an optimal parameter of surface conductivity, the different surface conductivity profiles of cone-type insulator were set up. To evaluate the effectiveness of surface graded conductivity, this section defined the E-field non-uniformity factor *f* of the S-GIL insulator. In addition, the electric conductive losses were calculated to analyzed the feasibility of practical application of the SFGM insulator in S-GIL system. It is worth mentioning that such kind of surface conductivity gradient insulator can be produced by magnetron sputtering technology or fluorination treatment [15, 16]. Thus, the feasibility of actual application of the SFGM insulator in S-GIL is greatly increased.

3.1 The Model of the S-GIL Insulator

To evaluate the effectiveness of SFGM, the simplified model of S-GIL insulator is built based on the structure of insulator in conventional gas insulated line, which is showed in Fig. 8. The E-field distribution along the cone-type insulator will be similar to that of the actual irregular insulator when the HV electrode is applied DC voltage. If the conical insulator is separated into finite conical elements, each element can be regarded as a cylinder. The conductivity of each cylinder is proportional to the square of the radius according to the law of conductivity. The conductivity near the HV electrode is much bigger than that near the GND electrode. Consequently, the E-field will be dramatically non-uniform in the vicinity of ground electrode when the conductor is applied direct current voltage.

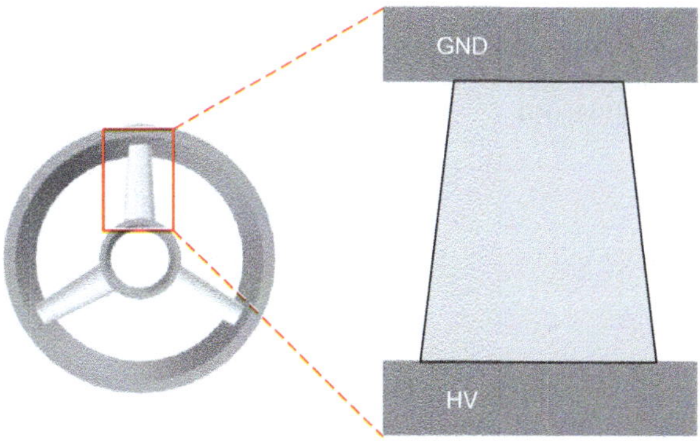

Fig. 8 The model of S-GIL insulator [22]. Copyright Clearance Center's RightsLink ® License Number: 4854481446040

In this section, the insulator surface conductivity can be regard as the Ohmic conductivity of the skin layer of insulator. As a result, outer skin layer and internal insulator are combined to form the entire insulator. In the other word, the conductivity of insulator can be regard as the parallel connection of that of volume and surface. Noted that the conductivity of insulator surface is set to be greater than that of the internal insulator. According to Ohm's law, the surface current should be much larger than the volume current of insulator under the DC voltage. The E-field distribution can be homogenized to avoid E-field concentration by reasonably adjusting the surface conductivity profile of insulator.

3.2 Simulation of Surface Conductivity Gradient Insulator

The surface conductivity of the post insulator is designed as a gradient distribution along the z direction in order to optimize the E-field distribution *E(z)* of the S-GIL insulator. The surface conductivity has to be increased from HV electrode to ground electrode for raising the low E-field strength and reducing the intensity of distorted E-field. Figure 9 presents the simplified geometric model profile of S-GIL insulator.

Equation (3) and (4) are used to express the relationship between insulator surface conductivity $\sigma(z)$ and z as follows. Different trends of the surface conductivity are shown in Eq. (1) and (2). The former represents that the surface conductivity $\sigma(z)$ is increased at constant rate with the increase of z, while the latter represents that the surface conductivity $\sigma(z)$ is increased at non-uniform rate with the increase of z.

$$\sigma(z) = k(z - z_0) + \sigma_0 \tag{4}$$

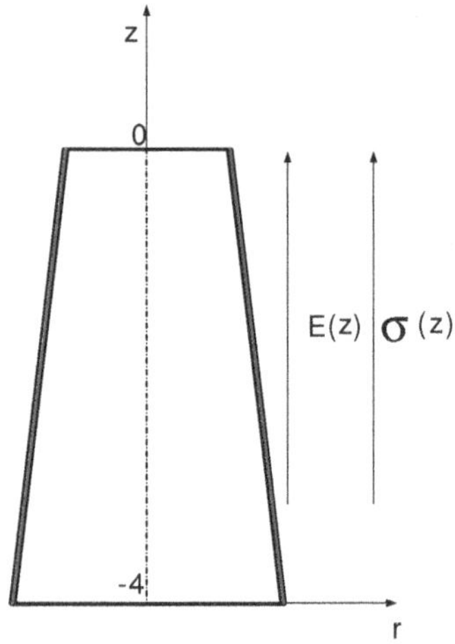

Fig. 9 The simplified geometric model profile of S-GIL insulator [22]. Copyright Clearance Center's RightsLink ® License Number: 4854481446040

Fig. 10 2D axisymmetric simulation model of insulator [22]. Copyright Clearance Center's RightsLink ® License Number: 4854481446040

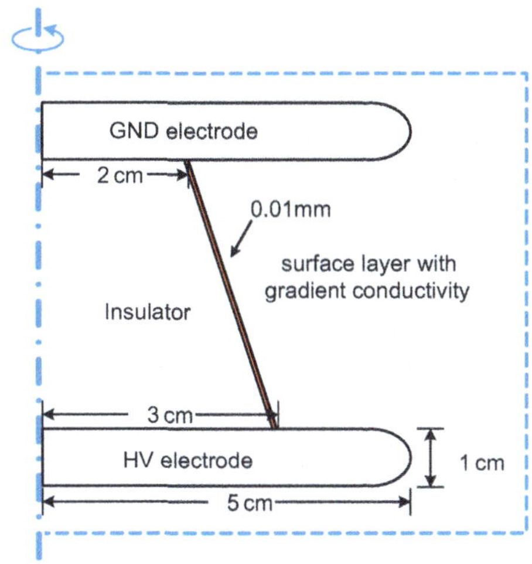

$$\sigma(z) = k(z - z_0)^2 + \sigma_0 \tag{5}$$

Among them, k is the increase rate which has to be regulated during calculating, S/m^2. z_0 is the coordinate at the bottom of the insulator in the z direction, m. σ_0 is the insulator surface conductivity when z equals z_o, which keeps constant, S/m.

The finite element method is adopted to obtain the E-field distribution on the insulator surface in the steady state. The E-field calculation is carried on the 2D axisymmetric simulation model of insulator which is showed in Fig. 10. The voltage applied on the HV electrode is direct current voltage 150 kV. The control equations are showed as follows,

$$\nabla \cdot J = 0 \tag{6}$$

$$J = E \cdot \sigma \tag{7}$$

$$E = -\nabla V \tag{8}$$

Among them, J is the current density flowing through insulator, A/m^2; E is the E-field strength inside the S-GIL, V/m; σ is the conductivity of the involved dielectrics, S/m; V is the electric potential, V.

The finite element simulation software COMSOL Multiphysics was used to computed the E-field distribution of S-GIL. The E-field non-uniformity factor f is adopted to estimate the effectiveness of surface conductivity gradient insulator, which is mentioned in Eq. (3).

3.3 Results and Discussion

By calculating the Eq. (4), the linear surface conductivity profile on the insulator surface with parameter k is obtained as showed in Fig. 11. When k equals to 0, the insulator surface conductivity is a constant. Based on the surface conductivity, the E-field distribution along the insulator can be calculated with parameter k. The E-field distribution is showed in Fig. 12. The results show that the E-field distortion is so serious near the triple junction of the grounding electrode. The minimum E-field strength is 1.23 kV/mm. However, the maximum E-field strength is up to 9.22 kV/mm. It means that E-field distribution is the extremely non-uniform. The surface conductivity along the insulator plays an important role in E-field distribution. As k increases, the uniformity of E-field distribution increases, but at a certain threshold of k, the uniformity of E-field distribution decreases again. The E-field is enhanced around the triple junction of the HV electrode. It can be explained that

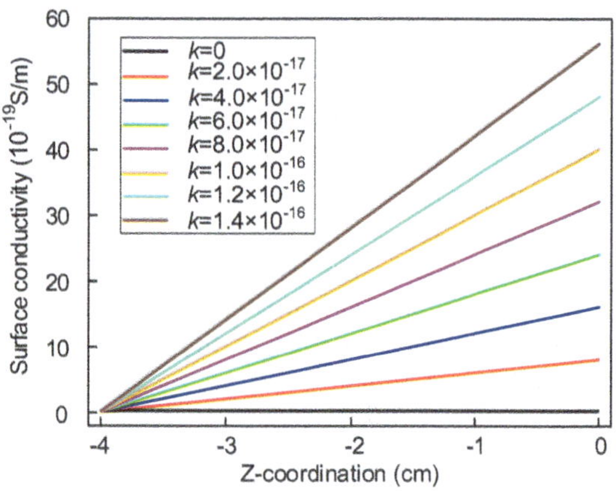

Fig. 11 Linear surface conductivity profile along the insulator [22]. Copyright Clearance Center's RightsLink ® License Number: 4854481446040

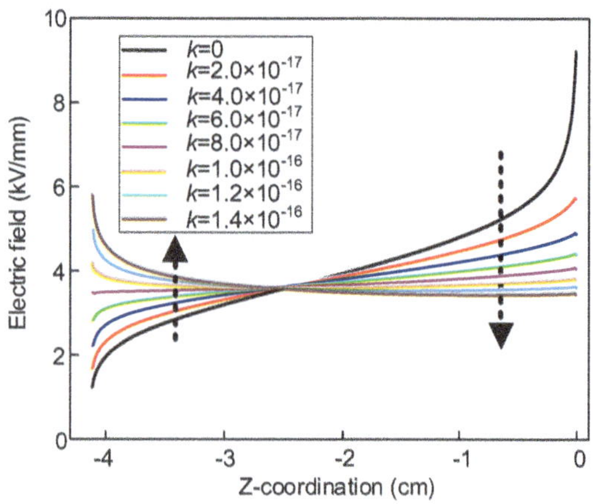

Fig. 12 E-field distribution on insulator surface in S-GIL [22]. Copyright Clearance Center's RightsLink ® License Number: 4854481446040

the over-high surface conductivity of insulator near the HV electrode. Noted that the double arrows in contrary direction express as the reverse tendency of E-field with the increase of k at the position near the two electrodes.

Figure 13 shows the change of E-field non-uniformity factor and parameter k. As the increase of k, the E-field non-uniformity factor firstly decreases. Nevertheless, when k increases further and reaches a critical value, the E-field non-uniformity factor increases. It means that the linear surface conductivity of insulator brings a contrary influence on the E-field distribution. It's obvious that the lowest non-uniformity factor can be obtained when k is set to 8×10^{-17} S/m^2, which is approximately 1.1.

Figure 14. the variation of leakage current with parameter k. It should not be ignored that the leakage current increase with k, this is due to the raising surface conductivity of insulator. To realize the most uniform state of the E-field distribution, k is set to be 8×10^{-17} S/m^2 and the leakage current is up to 3.1×10^{-16} A

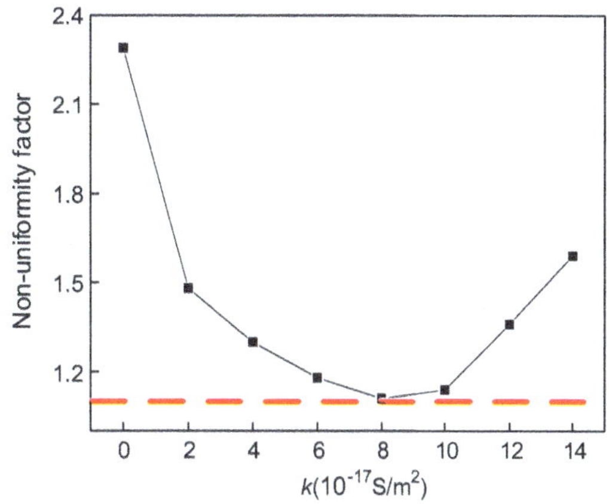

Fig. 13 The variation of E-field non-uniformity factor with parameter k [22]. Copyright Clearance Center's RightsLink ® License Number: 4854481446040

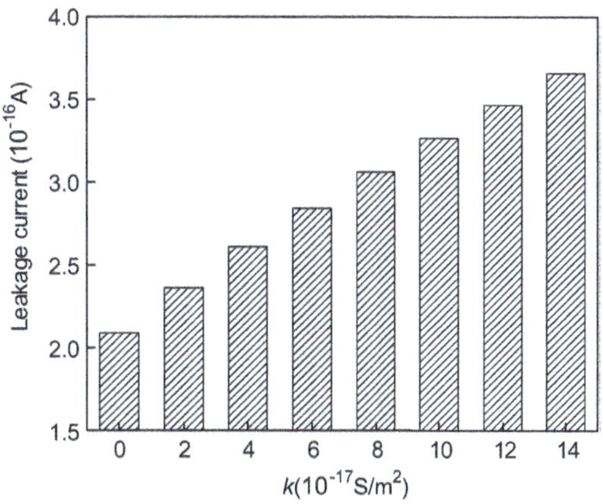

Fig. 14 The variation of leakage current with parameter k [22]. Copyright Clearance Center's RightsLink ® License Number: 4854481446040

approximately. The result shows that surface conductive design has the advantage of lower electric conductive loss.

To further investigate the effect of insulator surface conductivity on the E-field distribution, the variation of nonlinear conductivity profiles with parameter *k* is obtained in Fig. 15. It can be found that the conductivity of insulator surface near earth electrode are not different from those in Fig. 11. However, the conductivity of insulator surface in Fig. 15 increases more rapidly with z-axis.

The E-field distribution on the insulator surface with nonlinear surface conductivity is showed in Fig. 16. When k is set 0 to 3.5×10^{-19} S/m^2, the most distorted E-field at the triple junction of ground electrode is weaken by 69.4%. It indicates that the insulator with nonlinear surface conductivity possesses a better result on the relaxation of E-field. However, it should be noticed that the E-field around the HV electrode is only slightly enhanced as the increasing of k. Therefore, the choice of different *k* should be careful.

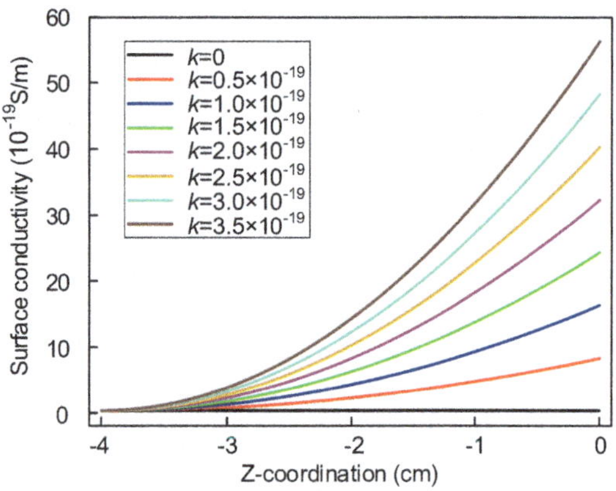

Fig. 15 The variation of nonlinear conductivity profiles with different *k* [22]. Copyright Clearance Center's RightsLink ® License Number: 4854481446040

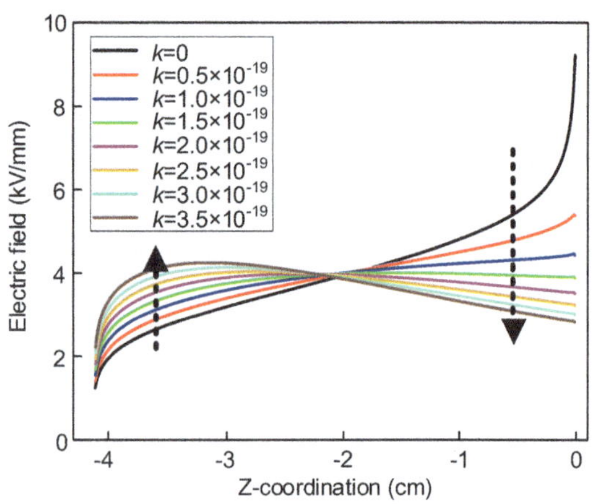

Fig. 16 The variation of E-field with different *k* [22]. Copyright Clearance Center's RightsLink ® License Number: 4854481446040

Figure 17 shows the variation of E-field distribution with different k. As the increase of k, the E-field non-uniformity factor firstly decreases. Nevertheless, when k increases further and reaches a critical value of 1.5×10^{-19} S/m^2, the non-uniformity factor is as low as 1.1, which is the most homogeneous E-field distribution. At this time, the approximately uniform E-field distribution is obtained with the best linear surface conductivity.

Figure 18 shows the change trend of leakage current (electric conductive loss) and k. The leakage current increase with k due to the raising surface conductivity of insulator. It is important to take the leakage current into consideration for choosing the best parameter. Under the premise of lower E-field non-uniformity factor, the electric conductive loss should be lower as much as possible. Therefore, the best value of k is 1.5×10^{-19} S/m^2. At this time, the leakage current is only $2.6 \times$

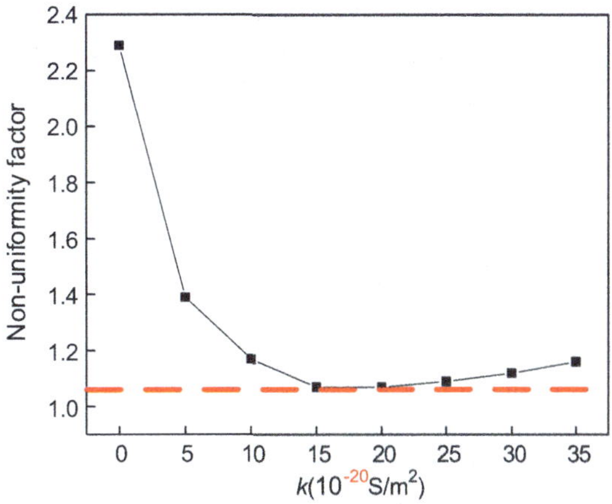

Fig. 17 The variation of E-field distribution with different k [22]. Copyright Clearance Center's RightsLink ® License Number: 4854481446040

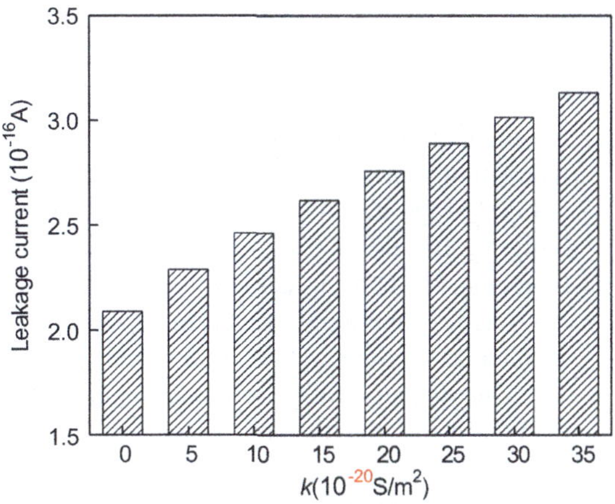

Fig. 18 The variation of leakage current with different k [22]. Copyright Clearance Center's RightsLink ® License Number: 4854481446040

10^{-16} A. It can be found that insulator with nonlinear conductivity possesses a lower leakage current compared with the insulator with the linear surface conductivity.

3.4 Summary

This section puts forward a novel concept with surface functionally graded material to optimize the insulation design of S-GIL system. The surface graded conductivity of insulator is realized by means of simulation method. The influence of surface graded conductivity on the E-field distribution along the cone-type insulator is investigated. The results show that the E-field distribution on the insulator surface can be improved by adjusting the linear or nonlinear profiles of surface conductivity when the parameter is in a suitable range. When parameter k is selected as an appropriate value, the E-field distribution can be optimized as much as possible, while the leakage current can be accepted. Taking both the E-field optimization and electric conductive losses due to leakage current into consideration, the insulator with nonlinear surface conductivity has better properties.

3.5 Conclusion

In this chapter, the two feasible E-field regulating methods are proposed: E-field adaptive insulator and surface functional graded material (SFGM) insulator. The models of insulator are established by the finite element simulation software to discuss the effects of the two methods on the E-field regulation along the insulator. The E-field non-uniformity factor of insulator in S-GIL is introduced to assess reasonability and validity of the optimization method. Besides, the leakage current is considered to analyze the feasibility of two optimization methods for improving insulation performance of insulator in S-GIL system. The main study two results obtained are as follows:

1. For the method of E-field adaptive insulator, the optimal E-field non-uniformity factor firstly increases and then decrease. In order to obtain insulator with the optimal properties, the E-field non-uniformity factor should be set to 1.1 considering the E-field distribution and leakage current. This method can be realized by means of 3D print technology.
2. For surface functional graded material (SFGM) insulator, when parameter k is selected as a proper value, the optimal E-field non-uniformity factor is as low as 1.1. The insulator with nonlinear conductivity possesses a lower leakage current compared with the insulator with the linear surface conductivity. The current can be limited to an acceptable value of 2.6×10^{-16} A. This method can be realized by means of magnetron sputtering technology or fluorination treatment.

It is hoped that this chapter can contribute a useful reference guide for the design and application of S-GIL.

References

1. A.P. Malozemoff, The power grid and the impact of hightemperature superconductor technology: An overview. *Superconductors in the Power Grid: Materials and Applications,* pp. 1–28 (2015)
2. J.X. Jin, Y.J. Tang, X.Y. Xiao, B.X. Du, Q.L. Wang, J.H. Wang, S.H. Wang, Y.F. Bi, J.G. Zhu, HTS power devices and systems: principles, characteristics, performance, and efficiency. IEEE Trans Appl. IEEE Trans. Appl. Supercond. **26**(7), 1–26 (2016)
3. W. Kim, H. Kim, J. Cho, Y.S. Choi, S. Kim, Electrical and mechanical characteristics of insulating materials for HTS DC Cable and cable joint. IEEE Trans. Appl. Supercond. **25**(3), 1–4 (2015)
4. B.X. Du, M. Wang, J. Li, W. Zhu, J. Jiang and X. Kong, Partial discharge behaviors of low dielectric Loss PTFE/PP composite insulation in LN2, *IEEE Int. Conf. Applied Superconductivity and Electromagnetic Devices,* pp. 1–2 (2018)
5. S. Pamidi, C.H. Kim, and L. Graber, High-temperature superconducting (HTS) power cables cooled by helium gas. in *Superconductors in the Power Grid: Materials and Applications,* pp. 225–260 (2015)
6. P. Cheetham, et al., Exploration of additive manufacturing for HTS cable components for electric aircrafts. *AIAA/IEEE Electric Aircraft Technologies Symposium.,* pp. 1–13 (2018)
7. P. Cheetham, C. Kim, L. Graber, S. Pamidi, Feasibility of superconducting gas-insulated transmission lines for electric aviation applications. IEEE Int. Conf. ESARS-ITEC., pp. 1–6 (2018)
8. N.G. Suttell, C.H. Kim, S.V. Pamidi, J.C. Ordonez, Transient thermal finite element modeling of hts cable systems cooled with gaseous helium. IEEE Trans. Appl. Supercond. **27**(4), 1–5 (2017)
9. L. Graber, W.J. Kim, P. Cheetham, C.H. Kim, H. Rodrigo, S.V. Pamidi, Dielectric properties of cryogenic gas mixtures containing helium, neon, and hydrogen. *IOP Conf. Ser., Mater. Sci. Eng.*, vol. 102, pp. 12018 (2015)
10. P. Cheetham, W. Kim, C.H. Kim, S.V. Pamidi, L. Graber, H. Rodrigo, Use of partial discharge inception voltage measurements to design a gaseous helium cooled high temperature superconducting power cable. IEEE Trans. Dielectr. Electr. Insul. **24**(1), 191–199 (2017)
11. P. Cheetham, J. Wagner, T. Stamm, C. Kim, L. Graber, S. Pamidi, (2018). Electric field analysis of insulating spacers for a superconducting gas-insulated transmission line. *In 2018 12th International Conference on the Properties and Applications of Dielectric Materials (ICPADM),* pp. 185–189 (2018)
12. P. Cheetham, C.H. Kim, S. Pamidi, L. Graber, Optimization of a superconducting gas-insulated transmission line. IEEE Trans. Dielectr. Electr. Insul. **26**(3), 930–938 (2019)
13. B.X. Du, Q. Du, J. Li, H. Liang, Carrier mobility and trap distribution dependent flashover characteristics of epoxy resin. IET Gener. Transm. Distrib. **12**(2), 466–471 (2018). https://doi.org/10.1049/iet-gtd.2017.0984
14. B.X. Du, Z.H. Wang, J. Li, H.C. Liang, Z.H. Li, Epoxy insulator with surface graded-permittivity by magnetron sputtering for gas-insulated line. IEEE Trans. Dielectr. Electr. Insul. **27**(1), 197–205 (2020)
15. B.X. Du, H.C. Liang, J. Li, Novel spacer coated with functionally graded ZnO film for HVDC gas insulated line. IEEE Trans. Dielectr. Electr. Insul. **27**(1), 231–239 (2020)
16. B.X. Du, Z.Y. Ran, J. Li, H.C. Liang, Novel insulator with interfacial σ-FGM for DC compact gaseous insulated pipeline. IEEE Trans. Dielectr. Electr. Insul. **26**(3), 818–825 (2019)

17. Z. Liu et al., "Simulation design and 3D-printing fabrication of conductivity graded insulator," in *Proc. IEEE Int. Conf. Electr. Mat. Power Equipment*, 2017, pp. 171–174
18. Z. Wang, B. Du, J. Li, H. Liang, H. Liu, Application of E-Field Adaptive Insulator for DC GHe-Insulated Pipeline. IEEE Trans. Appl. Supercond. **29**, 1–3 (2019)
19. M.L. Fu. et al., Simulation Research on the Application of Epoxy Resin with Nonlinear Conductivity in the ±600 kV Converter Transformer Bushing. High Voltage Eng. **45**(05), 1476–1482 (2019). (in Chinese)
20. R. Oishi, H. Kojima, K. Kato and N. Hayakawa, E-field relaxation effect of functionally graded materials (FGM) with conductive multilayer coating. in Proc. IEEE Conf. Electr. Insul. Dielectr. Phenom., pp. 46–49 (2017)
21. Y. Miyaji et al., Fabrication and evaluation of permittivity Graded Materials with $SrTiO_3$ particles and localized conductive layer. in *Proc. IEEE Conf. Electr. Insul. Dielectr. Phenom.*, 2016, pp. 631–634
22. Z.Y. R, B.X. Du, J. Li, Electric field regulation of insulator interface by FGM with conductivity for superconducting-GIL. *IEEE Trans. Appl. Supercond.*, vol. 29 (2019)

Polymer Insulation for HVDC Capacitors

Polymer Dielectrics for Film Capacitors Applied in HVDC Transmission

Sang Cheng and Qi Li

Abstract Capacitors are fundamental components of DC-AC inverters and AC-DC converters used for HVDC transmission. Among various types of capacitors, polymer film capacitors stand out for their operational reliability, high charge-discharge speed, low cost and light weight. This chapter provides an overview of recent progress in advanced polymer dielectric materials towards film capacitor applications, including linear dielectric polymers, ferroelectric dielectric polymers, multilayer polymers and polymer nanocomposites. The basic characteristics of these materials, along with their advantages and disadvantages for capacitor applications are reviewed. The fundamental requirements for capacitor dielectric polymers are discussed firstly. Challenges and future opportunities are highlighted at the end of this chapter.

1 Introduction

Electrostatic capacitors are fundamental passive components in electrical circuits [1]. Parallel-plate capacitors, a classical model of electrostatic capacitors, consisting of a pair of metal electrodes separated by the dielectric material can store energy in the form of electrical fields established in the dielectric material by holding opposite charges on the electrodes [2]. The capacitance (C) is the critical parameter reflecting the energy storage capacity of capacitors, i.e., capacitors having higher capacitance tend to store a larger amount of electrical energy. The capacitance of parallel-plate capacitors is expressed by

$$C=\varepsilon_0\varepsilon_r A/d \tag{1}$$

where ε_0 is the permittivity of vacuum, ε_r is the relative permittivity of dielectric material, A is the area of metal electrodes and d is the thickness of the dielectric material. Among all the energy storage devices including electrostatic capacitors,

S. Cheng · Q. Li (✉)
State Key Lab of Power Systems, Department of Electrical Engineering, Tsinghua University, Beijing 100084, China
e-mail: qili1020@tsinghua.edu.cn

B. Du (ed.), *Polymer Insulation Applied for HVDC Transmission*,
https://doi.org/10.1007/978-981-15-9731-2_24

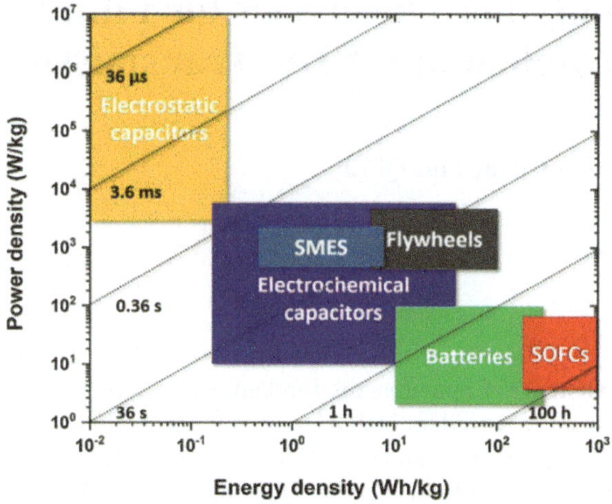

Fig. 1 Ragone plot of different energy storage devices. Copyright Clearance Center's RightsLink® License Number: 4854090183130

electrochemical capacitors (i.e., supercapacitors), superconducting magnetic energy storage (SMES) systems, batteries, etc., electrostatic capacitors stand out for the ultrahigh charge-discharge speed, meaning that they can release electrical energy in a short period of time and provide ultrahigh power density (Fig. 1) [3]. Moreover, electrostatic capacitors take the advantages of less loss and better ability to withstand various voltages compared to other kinds of capacitors, e.g., electrolytic capacitors [4]. As a result, electrostatic capacitors are widely employed in many pulsed power systems including electromagnetic weapons, medical defibrillators, high speed trains and so on [5].

Electrostatic capacitors can be mainly divided into polymer types and ceramic types according to the different dielectric materials employed in capacitors. The ceramic types include multi-layer ceramic capacitors (MLCCs), mica capacitors and thin film capacitors, while the polymer types mainly take the structure of polymetric film capacitors [4]. Ceramic capacitors can operate at high temperature but suffer from high loss, low operating fields and poor reliability. On the contrary, polymer film capacitors possess the advantages of high operating electrical fields, great stability, low cost and light weight, while the maximum operating temperature is limited to a relatively low level [6]. In this chapter, we mainly focus on polymer film capacitors. First, we introduce the basic structure and the applications in the HVDC transmission system of practical polymer film capacitors. Second, fundamental requirements for capacitor polymer dielectric materials are discussed. Third, we provide an overview of present polymer dielectric materials as well as emphasize the advantages and disadvantages of them. Finally, we discuss the prospects of advanced polymer dielectric materials.

2 Polymer Film Capacitors and Their Applications in HVDC Transmission

Metallized polymer films are the mainstream dielectrics of present polymer film capacitors, where a thin layer (20–100 nm) of metals (aluminum, zinc, or alloy) is vacuum-deposited onto the dielectric material as electrodes [7, 8]. Metallized polymer film capacitors have excellent operational reliability for the graceful failure characteristic known as the "self-clearing" capability (Fig. 2) [9, 10]. When localized breakdown occurs in a spot of the capacitor, the current flowing through the capacitor is restricted by the metallization resistivity and the heat generated by energy dissipation can "clear" the breakdown site. As a result, an open circuit rather than a short circuit forms at the breakdown site and the capacitor can continuously operate at the expense of slightly decreased capacitance. In other kinds of capacitors with metal foils as electrodes, localized breakdown will lead to catastrophic dielectric failure of capacitors and harm the safe operation of the whole system. A practical polymer film capacitor is typically wound from two layers of single-side-metallized polymer film (Fig. 3) [11].

Capacitors behave differently under voltages with different frequencies. In direct-current (DC) circuits, capacitors are considered open circuits where DC currents are completely blocked by capacitors. In alternating-current (AC) circuits, capacitors charge and discharge cyclically with the alternation of applied voltages where capacitors act as "resistance" to AC currents. The value of "resistance" is expressed by the capacitive reactance [12]

$$X_c = -\frac{1}{2\pi f C} \tag{2}$$

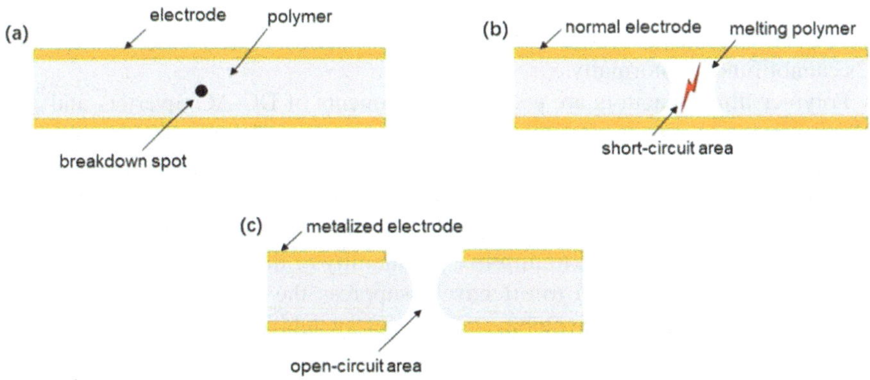

Fig. 2 Schematic of a self-clearing process. **a** Localized breakdown occurs at a spot of defect inner the polymer. **b** short-circuit area forms due to melting polymer in a normal capacitor; **c** open-circuit area forms in a metalized capacitor. Copyright Clearance Center's RightsLink® License Number: 4854100719544

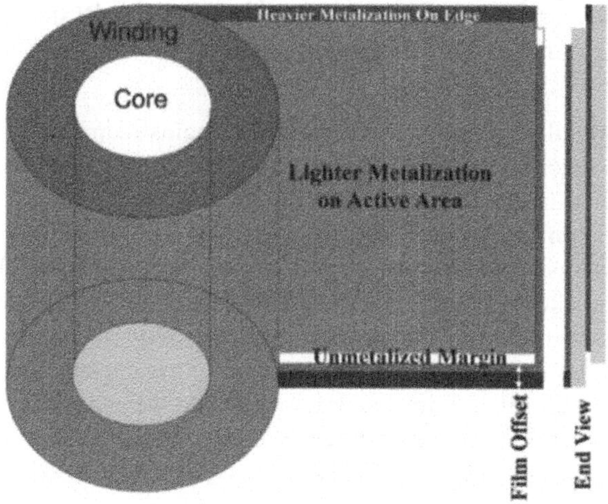

Fig. 3 Schematic of the construction of a metallized polymer film capacitor winding without the end connections. Reprinted with permission from reference [7]. Copyright (2018) American Chemical Society

where f is the frequency of applied voltages. Since the capacitive reactance is inversely proportional to the frequency, capacitors can be considered short circuits in high-frequency circuits.

Apart from capacitive reactance, practical capacitors have additional equivalent series resistance (ESR) and equivalent series inductance (ESL) due to the imperfection within the dielectric material [12]. In AC circuits, currents flowing through ESR will generate heat continuously, which is harmful to the device. The ESR also determines the self-clearing capability of capacitors since the clear energy is associated with the metallization resistivity. The ESR can neither too high that the clear energy is insufficient to "clear" the breakdown site, nor too low that the clear energy damages adjacent layers of the dielectric material. The ESL is an important parameter for capacitors applied in high-speed logic circuits (ICs). Because of the ESL, there exists a resonant frequency where capacitors behave like pure resistances and ICs cannot function normally.

Polymer film capacitors are essential components of DC-AC inverters and AC-DC converters employed in HVDC transmission systems. The function of capacitors includes DC filter, DC link, snubber and output filter [13]. Figure 4 shows one of the typical inverter circuits where C_1 is the DC filter capacitor, C_2 is the DC link capacitor, C_3 is the snubber capacitor and C_4 is the output filter capacitor. The DC filter capacitors are designed to improve the stability of the DC input of the circuit. The DC-link capacitors aim to effectively suppress the voltage ripple appearing on the DC link during the switching process of switching elements. The snubber capacitors are designed to absorb energy as well as to limit the rate of change in voltage or current during turn-on and turn-off process. The output filter capacitors filter the high harmonic within the output voltage and ensure that the output voltage is approximately sinusoidal. In the 800 kV HVDC transmission system, capacitors occupy ~60% of the converter tower weight and ~50% of the volume [14]. In the

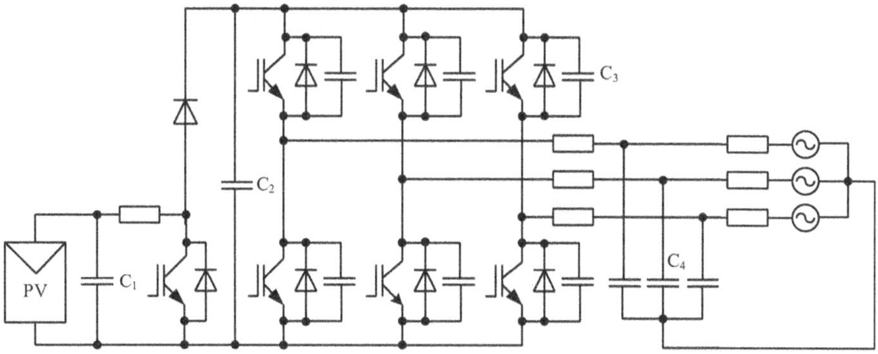

Fig. 4 A typical circuit of AC-DC inverters

long-distance HVDC transmission system, it is a challenge to keep the voltage stable. Power capacitors are also needed to absorb the voltage fluctuations in power grid.

3 Requirements for Capacitor Polymer Dielectric Materials

Advanced HVDC transmission systems have urgent demands for high-performance capacitors possessing wide capacitance range, high breakdown strength, high energy density, long service time, etc. Actually, not only the intrinsic characteristics of polymer dielectric materials but also the manufacture technologies determine the electrical performance of polymer film capacitors. For example, developing novel deposition technologies for preparing metallized polymer film capacitors may decrease the contact resistance between electrodes and polymer material, and therefore reduce the ESR of capacitors. In this section, we focus on the basic requirements for high-performance polymer dielectric materials.

3.1 Energy Storage Performance

Energy density (U_e), defined as the energy stored in per volume or weight, is one of the foremost parameters for the evaluation of polymer dielectric materials. High energy density contributes to reducing the weight and volume of capacitors for a given energy output, which is of great significance to reduce the size of electrical devices where capacitors occupy a large amount of volume and weight. The energy density of polymer dielectric capacitors is generally calculated by the equation [15].

$$U_e = \int E dD \tag{3}$$

where E is the applied electrical field and D is the electrical displacement induced by applied electrical field. For linear dielectric materials in which D is proportional to E, Eq. (3) can be simplified as

$$U_e = \frac{1}{2}\varepsilon_0\varepsilon_r E^2 \tag{4}$$

where ε_0 is the vacuum permittivity ($= 8.85 \times 10^{-12}$ F/m) and ε_r is the relative dielectric permittivity of the dielectric materials. In practical polymer dielectric materials, the discharged energy density is principally lower than charged energy density per charge-discharge cycle due to the dielectric loss caused by dissipation factor and conduction. Notably, the energy loss will not only reduce the energy density that polymer dielectrics can deliver, but also convert to Joule heat and induce temperature rise within the polymer dielectrics. The temperature rise may lead to thermal breakdown of the dielectric materials [16].

The energy storage performance of dielectric materials can be assessed by *D-E* hysteresis loop approach (Fig. 5) (usually obtained by using a Sawyer-Tower circuit). The integrated area between the charging curve and the electrical displacement ordinate reflects the charged energy density U_c, while the integrated area between the discharging curve and the electrical displacement ordinate stands for the dischargeable energy density U_d. The area enclosed by the charging curve and the discharging curve represents the energy loss (U_l) during one charge-discharge cycle. The U_d/U_c ratio is known as charge-discharge efficiency. Dischargeable energy density and charge-discharge efficiency are two of the most widely used parameters to assess the energy storage capability of polymer dielectrics.

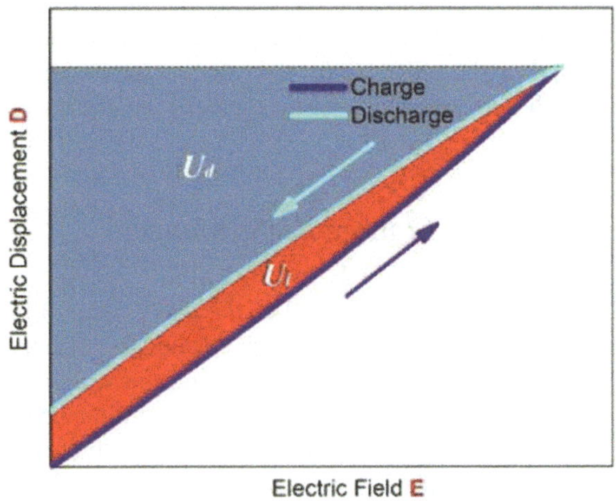

Fig. 5 Illustration of the *D-E* loop. Copyright Clearance Center's RightsLink® License Number: 4854090911138

3.2 Relative Permittivity and Dissipation Factor

Relative permittivity (ε_r) (also called dielectric constant) of dielectric materials, defined as the factor by which the electric field between the electrodes is decreased relative to vacuum, reflects the dielectric response to applied electric fields. Namely, it is the ratio of the capacitance of a capacitor using the dielectric material as a dielectric to a capacitor using vacuum as a dielectric. Equation (4) shows that the energy density is proportional to ε_r, indicating that the energy density of capacitors can be promoted by raising the relative permittivity of polymer dielectric materials. Dissipation factor, which contributes to the ESR of capacitors, reflects the loss-rate of energy during the polarization-depolarization process. Under relatively low electric fields, dissipation factor is the main source of energy loss in dielectric materials. Therefore, it is desirable to obtain polymer dielectric material with high relative permittivity and low dissipation factor.

Relative permittivity and dissipation factor of dielectric materials are usually variable under different temperature and electrical-field frequency owing to the fact that the polarization of dipoles and space charges relies on the mobility and relaxation time. In a high-quality capacitor, it is necessary for the dielectric material to have stable relative permittivity and dissipation factor over a certain temperature and frequency range. Moreover, the dissipation factor of polymer dielectric materials as a function of frequency is commonly used to analyze the polarization mechanisms and the correlation of dielectric properties with structural development [17, 18].

The measurement of relative permittivity and dissipation factor is usually conducted using LCR meter or impedance analyzer under low electric fields (<1 V). Although the measurement does not necessarily reflect the high-field performance of polymer dielectric materials considering that the electrical performance becomes even worse under high electric fields, it provides a quick overview of the dielectric materials.

3.3 Breakdown Strength and Leakage Current

Breakdown strength (E_b) is the maximum electric field dielectric materials can sustain without an abrupt reduction in insulating properties. Since the energy density is quadratically proportional to applied electric fields, dielectric materials with low breakdown strength cannot operate under high electric fields and thus the maximum energy density is largely limited. Various breakdown mechanisms, e.g., free volume breakdown, electromechanical breakdown and thermal breakdown, have been proposed to understand the dielectric failure occurring in dielectric materials [19]. Although none of them can comprehensively predict the breakdown process, they point out critical factors affecting the dielectric strength of dielectric materials including temperature, Young's modulus, relative permittivity, thermal conductivity and so forth. In practical capacitors, the breakdown strength is largely dependent on the manufacture technologies, e.g., defects in polymer films may largely reduce the breakdown strength of metallized polymer film capacitors since breakdown tends to occur at the weakest regions in dielectric materials. The breakdown strength of polymer dielectric films is also dependent on the film thickness [20].

DC breakdown strength, which is obtained by recording several breakdown voltages of dielectric materials under a constant ramped voltage and analyzing the data using Weibull statistic, is most commonly used to represent the dielectric strength. In practical applications, a certain margin between the rated operating field and the breakdown strength of dielectric materials needs to be reserved because the dielectric performance of polymer dielectrics declines rapidly as the applied electric field approaches the breakdown strength.

Leakage current, caused by the conduction of charge carriers, is another source of energy loss in dielectric materials. Leakage current increases with elevated temperature and electric fields due to the temperature- and electric-field-dependent conduction mechanisms such as Schottky emission, thermionic-field emission, Poole-Frenkel emission, hopping conduction and ohmic conduction [21]. At relatively low temperature, energy loss caused by leakage current is negligible relative to that caused by metallization resistance in metallized polymer film capacitors. As temperature rises, the increased leakage current turns into a considerable issue since polymers usually have low thermal conductivity and the heat generated by conduction loss cannot dissipate fast enough to keep the thermal equilibrium within polymers, which results in the thermal runaway of polymer dielectric materials [22].

3.4 Self-clearing Capability, Processability and Thermal Stability

Apart from electrical properties, structural, mechanical and thermal properties of polymer dielectrics are also important for capacitor applications. For example, the self-clearing capability of metalized polymer film capacitors, which ensures the operational reliability of the device containing capacitors, relies on the chemical composition of polymer dielectrics. It is proven that the polymers with lower ratios of (hydrogen + oxygen) to carbon exhibit poorer self-clearing capability than those with higher (hydrogen + oxygen)/carbon ratios since the ungraceful failure is closely related to the free carbon produced during the breakdown process [8]. The hydrogen and oxygen can react with the free carbon, which reduces the conductivity of breakdown path.

As the capacitance of polymer films is inversely proportional to the thickness, reducing the thickness of polymer dielectrics can effectively improve the energy density of capacitors for a given voltage. Unfortunately, not all polymer dielectric materials can be processed into films thin enough (usually <12 μm) to obtain sufficient energy density due to the limitations of manufacture technologies as well as the reduction in dielectric performance, e.g., the breakdown strength of polymer films decreases as the film thickness is lower than a certain level [23]. A practical strategy to produce thin polymer films is to extrude molten polymers to films with moderate thickness followed by stretching them to the desirable thickness. The tensile strength is a critical parameter concerned with the manufacture of thin polymer films [24]. The tensile strength can neither be too high that the stretching method is inapplicable, nor be too low that the polymer films cannot be wound into capacitors.

The demands for capacitors operating under high temperature have become more urgent in advanced electrical systems. Polymers, however, usually have poor thermal stability due to the low glass-transition temperature (T_g) associated with the weak molecular interactions [25]. As the temperature approaches T_g, the free volume and the propagation of charge carriers in polymers increase, which leads to an abrupt reduction in dielectric properties. Incorporation of rigid structures, e.g., π conjunction, aromatic/heteroaromatic molecular skeletons and high-strength chemical bonds into skeletons of polymers is an effective strategy to achieve high thermal stability [26].

4 Advanced Polymer Dielectric Materials

To develop high-performance capacitors, many commercial and noncommercial polymers have been explored as dielectric materials. In this section, we will introduce basic characteristics of them and point out their advantages as well as disadvantages for practical capacitor applications.

4.1 Classification of Polymer Dielectrics

Polymer dielectric materials are clarified into five categories based on different dipole and domain structures (Fig. 6) [27]. The ferroelectric (FE) domain containing spontaneously aligned dipoles plays an important role in determining the dielectric responses of dielectric materials. Linear dielectric materials, which exhibit a linear *D-E* behavior, have no net dipole or FE domain within them, leading to a low orientational polarization under applied electric fields. As a result, linear dielectric polymers generally exhibit low relative permittivity and low dielectric loss. By adding isolated dipoles into linear polymers, the dipolar glass forms in which dipole rotations are independent due to the weak interactions between dipoles and matrix. The dipolar glass exhibits slim *D-E* loops without any hysteresis loss. Incorporation of more dipoles may turn the dipolar glass into paraelectric polymers where the dipole-dipole and dipole-matrix interactions are strengthened. Compared to linear dielectric polymers, both dipolar glass and paraelectric polymers have higher relative permittivity at the expense of increased dielectric loss. Ferroelectric polymers refer to a class of polymers containing a large amount of FE domains [28]. The FE domains provide large orientational polarization under applied electric field, while they are

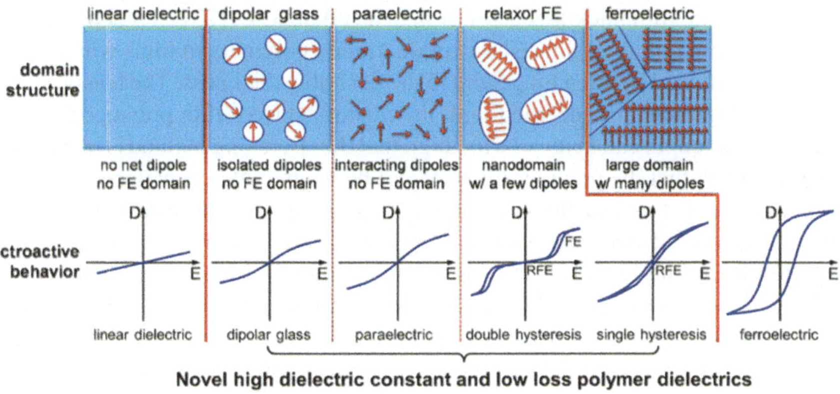

Fig. 6 Different dipole and FE domain structures with increasing dipole-dipole or domain-domain interactions and corresponding electroactive responses in D-E loops. Reprinted with permission from reference [27]. Copyright (2014) American Chemical Society

difficult to reverse back when the electric field is removed. Consequently, ferroelectric polymers show high relatively permittivity and broad *D-E* loops indicating high hysteresis loss. Relaxor ferroelectric polymers, which have slimmer *D-E* loops indicating lower hysteresis loss compared to FE polymers, are a class of special ferroelectric polymers modified by electron irradiating or incorporation of proper monomers [29]. Linear polymers and ferroelectric polymers are most widely studied as capacitor dielectric materials.

4.2 Linear Dielectric Polymers

Linear dielectric polymers have no high-dipole-moment polar groups in their chemical structures, so typical features of them are the low dielectric constant accompanied by low dielectric loss. Commercially available capacitors mainly employ biaxial oriented linear dielectric polymers as dielectric materials including biaxially oriented polypropylene (BOPP), poly(ethylene terephthalate) (PET), poly(phenylene sulfide) (PPS), poly(ethylene 2,6-naphthalate) (PEN) and so on. The orientation process can not only reduce the film thickness, but also improve the dielectric properties. Since these polymers are semi-crystalline, the spherulites within these polymers are aligned during the orientation process, serving as barriers to restrict charge transport and suppress the propagation of electrical trees [30]. Table 1 summarizes the linear dielectric polymers.

BOPP is the mainstream dielectric polymer used in commercial polymer film capacitors currently [31, 32]. As nonpolar polymers, BOPP has the dielectric loss as low as 0.02% but suffers from relatively low dielectric constant of ~2.2 at 1 kHz, which limits the energy density to 4 J/cm^3. BOPP-based polymer film capacitors are

Table 1 Overview of linear dielectric polymers

Material	ε_r (1 kHz)	tanδ (1 kHz)	E_b (MV/m)	Reference
BOPP	2.2	0.002	700 (1.6 μm)	[32]
PTFE	2.1	0.0005	443 (7 μm)	[33]
PC	2.8	0.001	600 (3 μm)	[60]
PET	3.3	0.003	580 (0.7 μm)	[60]
PPS	2.8	0.0005	500 (1.2 μm)	[60]
PEN	3.0	0.003	600 (1 μm)	[60]
PI	2.7–3.5	0.0013–0.0026	154–303 (7.6–127 μm)	[5]
PEI	3.15	0.0012	200 (25 μm)	[5]
PBI	3.3	0.001	402 (55 μm)	[5]
FPE	3.2–3.5	0.0031–0.007	220–320 (100–120 μm)	[5]

applicable in electrical devices operating on various conditions since the dielectric constant and dielectric loss of BOPP are quite stable over a broad range of frequency. BOPP also exhibits excellent self-clearing behavior due to the high ratio of (hydrogen + oxygen) to carbon of 2 in chemical composition. Defect-free BOPP films can be processed into ultrathin thickness (<1.6 μm), ensuring a high breakdown strength of BOPP of over 700 MV/m. One major drawback of BOPP is the low thermal stability and the maximum operating temperature of BOPP is only 105 °C.

Polytetrafluoroethylene (PTFE), developed by replacing all the hydrogen atoms in PP with fluorine atoms, is considered a potential candidate for BOPP under relatively high temperature. PTFE has similar dielectric constant and dissipation factor to BOPP but can sustain higher temperature of <200 °C [33]. However, it is difficult to obtain thin PTFE films with high quality using the traditional melt-extrusion technology and an effective strategy is to disperse PTFE powder into solution followed by sintering. Such films tend to have pin holes and exhibit low breakdown strength.

Polycarbonate (PC), prepared from carbonic acids and dihydric alcohols, has a dielectric constant of 2.8 and a dielectric loss of <0.02% at 1 kHz. PC capacitors are widely manufactured during 1950s and 2000s, which stand out for their temperature-independent capacitance [22], i.e., the capacitance varies only 3% from 25 °C to 125 °C. Solvent-cast PC films possess breakdown strength comparable to that of BOPP films.

Poly(ethylene terephthalate) (PET) is a kind of polar polymer which has ester functional groups. PET has similar dielectric properties to PC at room temperature and 1 kHz, while the dielectric constant and dielectric loss vary a lot with temperature and frequency owing to the presence of polar groups [34]. Ultrathin PET films (<5 μm) with high breakdown strength (>500 MV/m) can be obtained using biaxial stretching techniques and the self-clearing capability of PET is good.

Poly(phenylene sulfides) (PPS) and poly(ethylene 2,6- naphthalates) (PEN) are considered substitutes for PC and PET, respectively, due to the similar chemical structure. Differently, PPS and PEN have higher thermal stability induced by either high crystallinity or rigid chemical structure.

To meet the increasing demands for high-temperature capacitors, much effort has been made to develop dielectric materials based on commercial heat-resistant polymer dielectrics (HTPDs). One common obstacle to the capacitor applications of the HTPDs is the cost, which is much higher than that of polymers mentioned above. These HTPDs cannot be widely used in capacitor industry until a cost-effective manufacturing technique is developed.

Imide polymers, which contain imide ring in conjunction with the aromatic ring in the main chains, are a class of HTPDs investigated mostly for capacitor applications. Polyimide (PI), prepared from the condensation of dianhydride and diamine (or diisocyanate) monomers followed by imidization reaction, has been widely used in insulation industry for over 50 years but rarely in capacitor industry [35, 36]. Serving as polymer dielectrics for capacitor applications, PI has a decent dielectric constant of 3.2 and dielectric loss of 0.1% at 25 °C and 1 kHz accompanied by acceptable breakdown strength. However, the conduction loss in PI increases sharply under elevated temperature attributed to multiple temperature-dependent conduction mechanisms and the charge-discharge efficiency of PI under 200 MV/m at 150 °C is only 76%. In addition, it is difficult to obtain PI films with thickness below 12 μm because most PIs are insoluble and infusible. Poly(ether imides) (PEI), synthesized by introducing flexible ether linkages into the skeleton of PI, has better processability than PI at the cost of reduced thermal stability. Defect-free PEI films below 5 μm can be manufactured using the melt-extrusion technique [37]. PEI exhibits relatively stable dielectric constant of 3.1 and dielectric loss of 0.2% over a broad temperature and frequency range. At 150 °C, the charge-discharge efficiency of PEI reaches 90%, surpassing all commercially available polymers.

Polybenzimidazoles (PBI) with ultrahigh glass-transition temperature is usually used in textile industry [38]. PBI has good chemical and thermal stability as well as excellent insulating properties, while the relatively high dissipation factor and temperature-dependent dielectric performance limit its applications in capacitors.

Fluorene polyester (FPE) is one of the state-of-the-art dielectric polymers synthesized from the reaction of fluorene bisphenol and phthaloyl chloride [7]. FPE has relatively stable dielectric constant of 3.2 and dielectric loss of <0.3% at 1 kHz from 25 °C to 300 °C, indicating that it is suitable for high-temperature applications. FPE films below 5 μm have been successfully produced in an industrial scale.

To address the issue that the energy density of linear dielectric polymers is limited to a low level by the low dielectric constant, polar groups with high dipole moment, e.g., -CN [~4.0 D], $-NO_2$ [~3.6 D] and $-SO_2$ [~4.3D], are introduced into the main chains or the side chains of linear dielectric polymers [39, 40]. It is confirmed that the strategy can significantly improve the dielectric constant and slightly increase the dielectric loss of dielectric polymers, leading to an overall increase in the maximum energy density of dielectric polymers. Moreover, the polar groups attached to the side chains are more desirable since they flip more easily than polar groups attached to the main chains.

4.3 Ferroelectric Polymers

Ferroelectric polymers have FE domains which can induce large orientational polarization under applied electric fields. When the electric fields are removed, however, the FE domains are difficult to reverse back, resulting in large hysteresis loss. Therefore, typical features of ferroelectric dielectric polymers are high dielectric constant accompanied by high dielectric loss.

Ferroelectric polymers include poly(vinylidene fluoride) (PVDF) and its copolymers and terpolymers. PVDF, whose chemical formula is $(CH_2–CF_2)_n$, is featured by replacing half of the hydrogen atoms in PP with fluorine atoms. The breakdown strength of PVDF is as high as over 700 MV/m [41]. PVDF with proper molecular structures exhibits high dielectric constant (~12) for two main reasons [15]. On one hand, the high electronegativity of fluorine atoms leads to the high dipole moment [1.92 D] of C–F bonds. On the other hand, the fluorine atom is only slightly larger than the hydrogen atom, ensuring the high packing density of C–F bonds. PVDF has two types of torsional bond arrangements including trans (T) where the adjacent CH_2 and CF_2 are at 180° to each other and gauche$^\pm$ ($G^\pm$) where the adjacent CH_2 and CF_2 are at ±60° to each other [42]. The two types of torsional bond arrangements lead to three main chain conformations of PVDF, known as TG^+TG^-, $TTTG^+TTTG^-$, and *all-trans*. In TG^+TG^- and $TTTG^+TTTG^-$ conformations, the dipole moment of the chain is low since the CF_2 units are aligned in different direction. In *all-trans* conformation, the chain has large dipole moment due to the same orientation of CF_2 units. Different chain conformations result in four major crystalline phases of PVDF (α, δ, β, γ) (Fig. 7). In the α phase, two chains with TG^+TG^- conformation are antiparallel to each other and the crystal exhibits no net dipole moment. In the δ phase, two chains with TG^+TG^- are oriented in the same direction and the crystal is macroscopically polar. In the β phase, two chains with *all-trans* conformation are oriented in the identical direction, providing a large net dipole moment. The γ phase contains chains with $TTTG^+TTTG^-$ conformation and shows lower net dipole moment than the β phase. The crystalline phase of PVDF is largely dependent on the processing conditions. The α phases are more desired when lower hysteresis loss is required and the β phases are more desired when higher polarization is required. Although PVDF with appropriate molecular structures can deliver a large energy density, it is unsuitable for capacitor applications attributable to the high hysteresis loss.

To further improve the dielectric performance of PVDF, various monomers are employed to copolymerize with VDF, obtaining PVDF-based copolymers (i.e., those polymerized from two different types of monomers) and terpolymers (i.e., those polymerized from three different types of monomers). The introduction of trifluoroethylene (TrFE) stabilizes the β-phase in PVDF and the P(VDF-TrFE) exhibits high ferroelectricity with high electric displacement accompanied by high hysteresis loss [43]. As a result, the dischargeable energy density of P(VDF-TrFE) is low because a large amount of stored energy cannot be released. It is believed that the hysteresis loss of ferroelectric polymers is associated with the energy barrier between the FE

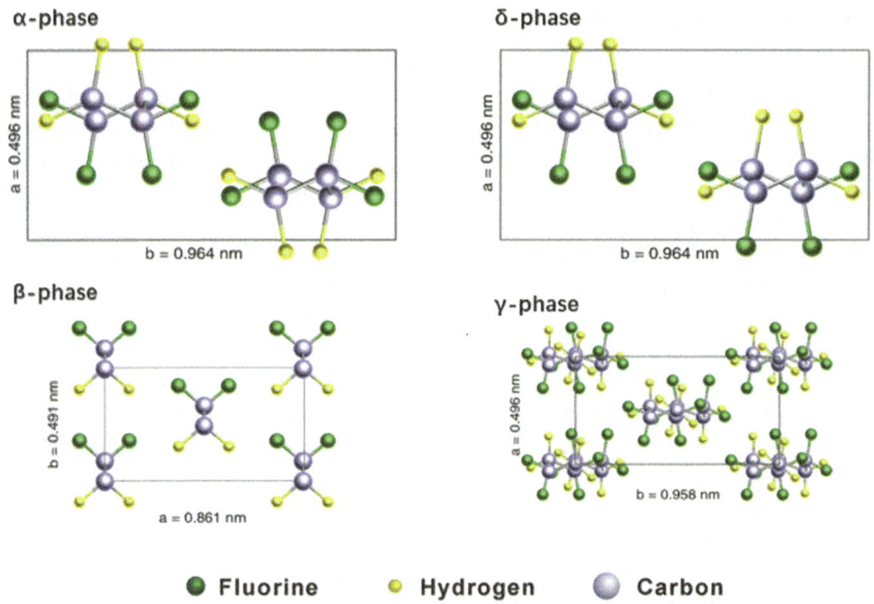

Fig. 7 Schematic view of the unit cells of the four crystalline phases of PVDF. Copyright Clearance Center's RightsLink® License Number: 4854101120832

domains and reducing the domain size can reduce the energy barrier for orientational polarization of FE domains, thereby reducing hysteresis loss [44]. Capitalizing on this idea, the third bulky monomer such as chlorotrifluoroethylene (CTFE) and 1,1-chlorofluoroethylene (CFE) are introduced into P(VDF-TrFE) to break the FE domains into nano-sized ferroelectric domains [45]. Consequently, the relaxor ferroelectric polymer P(VDF-TrFE-CFE) has a dielectric constant as high as 50 at 1 kHz and a high energy density of ~10 J/cm^3 as well as a slim *D-E* loop suggesting low hysteresis loss [46].

For relaxor ferroelectric polymers, the saturation of polarization induced by the high dielectric constant under relatively low electric fields (< 100 MV/m) prevents them from obtaining higher energy density. To solve this problem, poly(vinylidene fluoride–hexafluoropropylene) [P(VDF-HFP)] copolymer and P(VDF-CTFE) copolymer with high concentration of bulky HFP or CTFE monomers are developed [47, 48]. Relative to relaxor ferroelectric polymers, these copolymers have decreased dielectric constant but maintain the low hysteresis loss. As a result, the discharged energy density of P(VDF-CTFE) reaches 25 J/cm^3 under an electric field of 600 MV/m, and the maximum discharged energy density of P(VDF-HFP) even exceeds 30 J/cm^3 [48].

4.4 Multilayer Dielectric Polymers

Due to the intrinsic limitations of single-layer polymers, further improvement on the dielectric and energy storage properties is difficult to realize by just modifying the molecular structure of polymers. For example, it is difficult to synthesize a kind of linear polymers possessing dielectric constant as high as ferroelectric dielectric polymers. Therefore, multilayer polymer films consisting of ferroelectric polymers which have high dielectric constant and linear polymers which have high breakdown strength and low loss are studied to achieve higher energy density and higher charge-discharge efficiency. These multilayer films are usually prepared by coextruding the two components using the state-of-the-art coextrusion system developed by Case Western Reserve University (Fig. 8A). The individual layer can be processed into as thin as 10 nm [49].

In multilayer polymer films, many effects absent in single-layer polymers have a large influence on the electrical properties [50]. For example, the Maxwell–Wagner–Sillars (MWS) interfacial polarization resulting from the difference of relative permittivity between adjacent layers may lead to the accumulation of charges and ions at the interface (Fig. 8B). The accumulated charges and ions may lower the local electric field and serve as traps to capture injected electrons, effectively suppressing the movement of charge carriers and improving the breakdown strength. Nanoconfinement effect, which denotes a phenomenon that the hysteresis loss of multilayer films declines with increased layer numbers when the thickness of individual layer decreases to nanoscale, is another unique phenomenon occurring in multilayer films. The nanoconfinement effect is related to restricted ion migration in thinner high-conductivity layers. It is noteworthy that the nanoconfinement effect is absent in ultrathin (<100 nm) individual layers since they cannot provide sufficient ions to form interfacial polarization [51]. The dielectric performance of multilayer polymer films is largely dependent on the interfacial adhesion i.e., local breakdown can cause

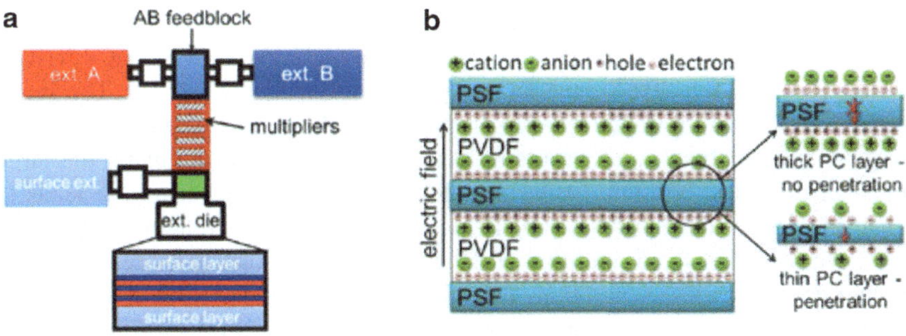

Fig. 8 **A** Schematic representation of the multilayer coextrusion process. **B** Schematic of the interfacial polarization from both impurity ions and free electrons. Reprinted with permission from reference [50]. Copyright (2017) American Chemical Society

layer lamination close to breakdown sites, where further breakdown process is easy to occur considering the low breakdown strength of air.

The multilayer film composed of polysulfone (PSF), which is a kind of linear dielectric polymers having low dielectric constant ($\varepsilon_r = 3$) and high breakdown strength, and poly(vinylidene fluoride) (PVDF), which is a kind of ferroelectric dielectric polymers having high dielectric constant ($\varepsilon_r = 12$) and high hysteresis loss, is the state-of-art multilayer dielectric polymer showing decent dielectric and energy storage properties over a broad temperature range [52]. According to the bilayer model, the electric field applied in PSF layers is 4 times that in PSF layers, so the hysteresis loss of PVDF layers is suppressed due to the weak applied electric field and the high breakdown strength of PSF prevents the overall multilayer films from breaking down. Influenced by complicated effects mentioned above, the PSF/PVDF multilayer film maintained a high Weibull breakdown strength of over 500 MV/m from 25 °C to 150 °C and showed a low hysteresis loss at room temperature.

4.5 Polymer Nanocomposites

The dielectric constant and the breakdown strength are two major parameters determining the energy density of dielectric materials. Generally, polymers have high breakdown strength but suffer from low dielectric constant, yet ceramics possess high dielectric constant but break down at relatively low electric field. In addition, there exist remarkable difference in other performance aspects between dielectric polymers and dielectric ceramics such as thermal conductivity, Young's modulus and electrical conductivity. To combine the advantages of dielectric polymers and dielectric ceramics as well as to overcome their shortcomings, nanocomposites composed of polymer matrix and nano-sized ceramic dopants are developed.

Most literature works in dielectric nanocomposites are concerned with the improvement of dielectric constant while maintaining a high breakdown strength and low dielectric loss. One of the most practiced strategy is to incorporate high-permittivity ceramic nanoparticles, e.g., $BaTiO_3$, $Ba_xSr_{1-x}TiO_3$, $CaCu_3Ti_4O_{12}$ (CCTO), Pb $(Zr,Ti)O_3$ (PZT) and TiO_2, into the polymer matrix [53]. The introduction of these nanofillers effectively promotes the dielectric constant of polymer matrix but brings about some problems in the meantime. On one hand, the large contrast of dielectric constant between nanoparticles and polymer matrix leads to the distortion of local electric field and reduces the breakdown strength of nanocomposites. On the other hand, the high surface energy of nanoparticles results in the agglomeration and phase separation, also reducing the breakdown strength. Coating nanoparticles with appropriate organic surface modifiers, e.g., organosilanes, phosphonates, dopamine, small-molecule carboxylic acids and organic macromolecules can improve the compatibility between polymer matrix and nanoparticle, thereby promoting the dielectric properties of nanocomposites [54]. For example, the P(VDF-HFP)-based nanocomposite containing 50 vol.% of phosphonate-modified $BaTiO_3$

nanoparticles had a maximum energy density of 6.1 J/cm^3, along with high permittivity ($\approx$37), low dielectric loss (<0.007) and high breakdown strength ($\approx$210 MV/m) [55]. Coating nanoparticles with inorganic materials possessing moderate dielectric constant, e.g., Al_2O_3, can ease the inhomogeneity of local electric field and improve the breakdown strength.

The shape of nanofillers is confirmed to have an impact on the dielectric properties of nanocomposites. 1-D nanofillers with high aspect ratios such as nanorods, nanowires and nanofibers, which have high lattice constant and high electrical displacement, are more effective in improving the dielectric constant. The dielectric constant of P(VDF-TrFE-CFE)-based nanocomposites containing 10% volume fraction of barium strontium titanates (BSTs) nanowires is 20% higher than that of nanocomposites containing 10% volume fraction of BST nanoparticles [56]. 2-D nanofillers such as nanosheets serve as topological barriers in the polymer matrix, restricting the movement of charge carriers and suppressing the propagation of electrical trees, thereby improving the breakdown strength of nanocomposites.

Apart from high-permittivity nanofillers, high-insulation ceramic nanofillers such as BN and Al_2O_3 are also added into polymers to improve the insulating properties [22]. For example, the incorporation of 10 vol.% boron nitride nanosheets (BNNSs) into cross-linked divinyltetramethyldisiloxane-bis(benzocyclobutene) (c-BCB) improved the breakdown strength by 50% at 250 °C and reduced the conduction loss from 18% for neat polymers to 3% for nanocomposites at 150 °C and 200 MV/m. As a result, the nanocomposite had a high discharged energy density of 2.2 J/cm^3 at 150 °C. The introduction of BNNSs also improved the thermal conductivity of polymer matrix due to the high thermal conductivity of BNNSs, which led to a better high-temperature performance of the nanocomposite.

Ternary nanocomposites containing both high-permittivity nanofillers and high-insulation nanofillers are also developed to realize concurrently improved dielectric constant and breakdown strength [57]. Furthermore, Luo et al. developed the ternary nanocomposite containing BNNSs with $BaTiO_3$ nanoparticles tightly encapsulated in it, which suppressed the negative influence of $BaTiO_3$ nanoparticles on the insulation properties of nanocomposites [58]. As a result, the nanocomposites exhibited higher breakdown strength than those containing randomly dispersed BNNSs and $BaTiO_3$. A unique design of ternary nanocomposites is the sandwich structured nanocomposites where the outer layers having low ε_r and high E_b are doped with high-insulation nanofillers and the inner layer having high ε_r, low E_b is doped with high-permittivity nanofillers (Fig. 9) [59]. The inner layer can provide large polarization and the outer layers can prevent the film from breakdown, and therefore high discharged energy density of nanocomposites can be achieved.

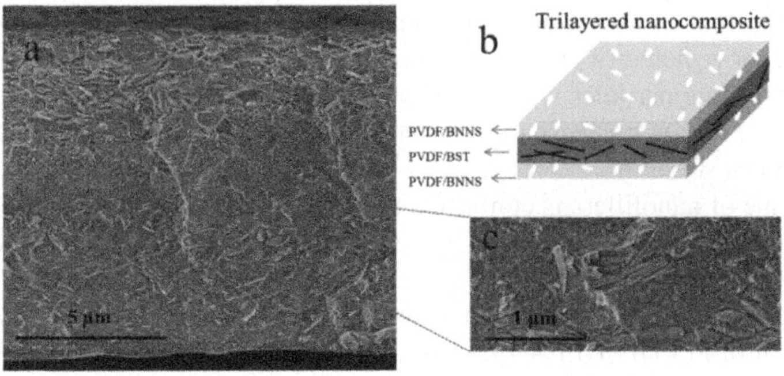

Fig. 9 **a** Cross-sectional SEM image of trilayer-structured nanocomposites with 8 vol.% BST NW in the central layer. **b** Schematic of the trilayered film composed of PVDF/BNNS as outer layers and PVDF/BST as the middle layer BST. **c** SEM image of the interfacial area between PVDF/BNNS and PVDF/BST layers. Copyright Clearance Center's RightsLink® License Number: 4854111192145

5 Conclusion

Capacitors are fundamental components of DC-AC inverters and AC-DC converters employed in HVDC transmission. Metallized polymer film capacitors are preferred over other types of capacitors for the self-clearing capability, high charge-discharge speed, low cost and light weight. The electrical performance of polymer film capacitors is largely dependent on the characteristics of polymer dielectric materials. Basic requirements for capacitor dielectric polymers include high energy density, high charge-discharge efficiency, high dielectric constant, low dissipation factor, high breakdown strength, low leakage current, good self-clearing capability, decent film processability and thermal stability. Various polymers have been explored as dielectric materials. Linear dielectric polymers, which typically have low dielectric loss and low dielectric constant, are most commercially used in capacitors. Ferroelectric polymers, which typically have high dielectric constant and high dielectric loss, can deliver a large energy density. Multilayer dielectric polymers can combine the advantages of linear dielectric polymers and ferroelectric dielectric polymers. Polymer nanocomposites can overcome some intrinsic limitations of pure polymers and achieve high energy density. However, there remain some problems for the development of advanced polymer dielectric materials. First, due to the lack of polar groups, the energy density of linear dielectric polymers is still limited to a low level by the low dielectric constant. How to improve the dielectric constant of linear dielectric polymers without increasing the dielectric loss remains challenging. Second, the dielectric loss of currently available ferroelectric dielectric polymers is still much higher than acceptable for practical applications. Third, more effective film processing techniques need to be developed for the scalable production of uniform and defect-free nanocomposite dielectric films.

References

1. W.J. Sarjeant, I.W. Clelland, R.A. Price, Proc. IEEE **89**(6), 846 (2001)
2. B.J. Chu, X. Zhou, K. Ren, B. Neese, M. Lin et al., Science **313**, 334 (2006)
3. L. Yang, X. Kong, F. Li, H. Hao, Z. Cheng, H. Liu, J.F. Li, S. Zhang, Prog. Mater Sci. **102**, 72 (2019)
4. Q. Tan, P. Irwin, Y. Cao, IEEE Trans. FM **126**, 1209 (2006)
5. Q. Li, F.Z. Yao, Y. Liu, G. Zhang, H. Wang, Q. Wang, Annu. Rev. Mater. Res. **48**, 219 (2018)
6. W.J. Sarjeant, J. Zirnheld, F.W. Mcdougall, IEEE Trans. Plasma Sci. **26**(5), 1368 (1998)
7. J. Ho, S.G. Greenbaum, A.C.S. Appl, Mater. Interfaces. **10**(35), 29189 (2018)
8. C.W. Reed, S.W. Cichanowski, IEEE Trans. Dielectr. Electr. Insul. **1**(5), 904 (1994)
9. D.A. McLean, Proc. IRE. **38**(9), 1010 (1950)
10. H.G. Wehe, Bell Lab. Rec. **27**(9), 317 (1949)
11. S.A. Boggs, J. Ho, T.R. Jow, I.E.E.E. Electr, Insul. Mag. **26**(2), 7 (2010)
12. 'Capacitor', https://en.wikipedia.org/wiki/Capacitor, Accessed Dec. 15, 2019
13. https://www.compotec-electronics.com/wordpress/wp-content/uploads/Capacitors-in-Power-Inverters-Roland-Gallay.pdf
14. Q. Li, S. Cheng, I.E.E.E. Electr, Insul. Mag. **36**(2), 7 (2020)
15. Q. Chen, Y. Shen, S. Zhang, Q. M. Zhang., Annu. Rev. Mater. Res. **45**, 433 (2015)
16. N. Zebouchi, M. Bendaoud, R. Essolbi, D. Malec, B. Ai, H. Giam, J. Appl. Phys. J. **79,** 2497 (1996)
17. J. Mijovic, J.-W. Sy, T.K. Kwei, Macromolecules **30**(10), 3042 (1997)
18. L. Y. Yang, X. Y. Li, E. Allahyarov, P. L. Taylor, Q. M. Zhang and L. Zhu, Polymer **54**(7), 1709 (2013)
19. M. Ieda, IEEE Trans. Dielectr. Electr. Insul. **EI-15**, 206 (1980)
20. D.Q. Tan, J. Appl. Polym. Sci. **137**(33), 49379 (2020)
21. M. Ieda, IEEE Trans. Electr. Insul. **19**, 162 (1984)
22. Q. Li, L. Chen, M.R. Gadinski, S. Zhang, G. Zhang et al., Nature **523**, 576 (2015)
23. J. Cozens, I.R.E. Trans, Compon. Parts **6**(2), 114 (1959)
24. 'Steinerfilm Dielectrics for Film Capacitors', http://www.steinerfilm.de/en/our-products/, Accessed Dec. 15, 2016
25. R.J. Young, P.A. Lovell, *Introduction to Polymers*. CRC Press (2011)
26. X.M. Zhang, J.G. Liu, S.Y. Yang, Rev. Adv. Mater. Sci. **46**(1), 22 (2016)
27. L. Zhu, J. Phys. Chem. Lett. **5**(21), 3677 (2014)
28. W. Xia, Z. Zhang, IET Nanodielectrics **1**(1), 17 (2018)
29. L.E. Cross, Ferroelectrics **76**(1), 241 (1987)
30. B. Li, F. Salcedo-Galan, P.I. Xidas, E. Manias, A.C.S. Appl, Nano Mater. **1**(9), 4401 (2018)
31. S.J. Laihonen, U. Gafvert, T. Schutte, U.W. Gedde, IEEE Trans. Dielectr. Electr. Insul. **14**(2), 275 (2007)
32. J. Ho, R. Ramprasad, S. Boggs, IEEE Trans. Dielectr. Electr. Insul. **14**(5), 1295 (2007)
33. Q. Tan, P. Irwin, Y. Cao, IEEJ Transactions on Fundamentals and Materials **126**(11), 1153 (2006)
34. W. Reddish, Trans. Faraday Soc. **46**, 459 (1950)
35. D.J. Liaw, K.L. Wang, Y.C. Huang, K.R. Lee, J.Y. Lai, C.S. Ha, Prog. Polym. Sci. **37**(7), 907 (2012)
36. S. Diaham, S. Zelmat, M.L. Locatelli, S. Dinculescu, M. Decup, T. Lebey, IEEE Trans. Dielectr. Electr. Insul. **17**(1), 18 (2010)
37. D. Tan, L. Zhang, Q. Chen, P. Irwin, J. Electron. Mater. **43**(12), 4569 (2014)
38. S. Bourbigot, X. Flambard, *Fire Mater.* **26**(4–5), 155 (2002)
39. J.T. Bendler, C.A. Edmondson, M.C. Wintersgill, D.A. Boyles, T.S. Filipova, J.J. Fontanella, Eur. Polym. J. **48**, 830 (2012)
40. J.T. Bendler, D.A. Boyles, C.A. Edmondson, T.S. Filipova, J.J. Fontanella, M.A. Westgate, M.C. Wintersgill, Macromolecules **46**, 4024 (2013)

41. M. Hikita, M. Nagao, G. Sawa, M. Ieda, J. Phys. D **13**(4), 661 (1980)
42. Q. Li, Q. Wang, Macromol. Chem. Phys. **217**(11), 1228 (2016)
43. T. Furukawa, Phase Trans: Multinat J **18**(3–4), 143 (1989)
44. X. Zhou, B. Chu, Q.M. Zhang, IEEE Trans. Ultrason. Ferroelectr. Freq. Control **53**(8), 1540 (2006)
45. R.J. Klein, J. Runt, Q.M. Zhang, Macromolecules **36**(19), 7220 (2003)
46. S. Wu, M. Shao, Q. Burlingame, X. Chen, M. Lin, K. Xiao, Q.M. Zhang, Appl. Phys. Lett. **102**(1), 1 (2013)
47. X. Zhou, B. Chu, B. Neese, M. Lin, Q.M. Zhang, IEEE Trans. Dielectr. Electr. Insul. **14**(5), 1133 (2007)
48. X. Zhou, X.H. Zhao, Z. Suo, C. Zou, J. Runt, S. Liu, Appl. Phys. Lett. **94**(16), 162901 (2009)
49. M. Ponting, A. Hiltner, E. Baer, Macromolecules **294**(1), 19 (2010)
50. E. Baer, L. Zhu, Macromolecules **50**(6), 2239 (2017)
51. J.K. Tseng, S. Tang, Z. Zhou, M. Mackey, J.M. Carr, R. Mu, Polymer **55**(1), 8 (2014)
52. K. Yin, J. Zhang, Z. Li, J. Feng, C. Zhang, X. Chen, J. Appl. Polym. Sci. **136**(20), 47535 (2019)
53. Z.M. Dang, J.K. Yuan, S.H. Yao, R.J. Liao, Adv. Mater. **25**(44), 6334 (2013)
54. S. Ogitani, S.A. Bidstrup-Allen, P.A. Kohl, IEEE Trans. Adv. Packag. **23**(2), 313 (2000)
55. P. Kim, N.M. Doss, J.P. Tillotson, P.J. Hotchkiss, M.J. Pan, S.R. Marder, J. Li, J.P. Calame, J.W. Perry, ACS Nano **3**(9), 2581 (2009)
56. G. Zhang, X. Zhang, T. Yang, Q. Li, L.Q. Chen, S. Jiang, Q. Wang, ACS Nano **9**(7), 7164 (2015)
57. Q. Li, K. Han, M.R. Gadinski, G. Zhang, Q. Wang, Adv. Mater. **26**(36), 6244 (2014)
58. S. Luo, J. Yu, S. Yu, R. Sun, L. Cao, W.H. Liao, C.P. Wong, Adv. Energ. Mater. **9**(5), 1803204 (2019)
59. F. Liu, Q. Li, J. Cui, Z. Li, G. Yang, Y. Liu, L. Dong, C. Xiong, H. Wang, Q. Wang, Adv. Funct. Mater. **27**(20), 1606292 (2019)

Improvement of Dielectric Properties of Polypropylene Film for HVDC Metallized Film Capacitors

Meng Xiao, Boxue Du, Ranran Xu, Zhaoyu Ran, Haoliang Liu, Jiwen Xing, and Kailun Fan

Abstract In this chapter, the improvement of electrical properties of PP films based on crystallization regulation was studied. PP films with different content of β-nucleating agent (β-NA) and film samples doped with an organic phosphorus nucleating agent under three cooling processes were prepared. The crystalline morphology and dielectric properties was researched by experimental methods. The effect of crystallization control on the transportation process of charge carriers was analyzed and the mechanism of breakdown improvement was discussed. Based on the results, the relationship between the crystallization parameters and the electrical properties is established. A comprehensive control method for the conduction loss and breakdown performance of the PP film is proposed, showing the potential for application in metallized film capacitors. It was concluded that improving the insulation performance through crystallization control is feasible, and this method shows great potential for the modification of PP films.

M. Xiao (✉) · B. Du · R. Xu · Z. Ran · H. Liu · J. Xing · K. Fan
Tianjin University, No. 92 Weijin Road, Nankai District, Tianjin City, China
e-mail: tjuxiaomeng@tju.edu.cn

B. Du
e-mail: duboxue@tju.edu.cn

R. Xu
e-mail: xuranran0512@126.com

Z. Ran
e-mail: 2991673951@qq.com

H. Liu
e-mail: TJU_liuhaoliang@tju.edu.cn

J. Xing
e-mail: xingjiwen@163.com

K. Fan
e-mail: 1921500766@qq.com

B. Du (ed.), *Polymer Insulation Applied for HVDC Transmission*,
https://doi.org/10.1007/978-981-15-9731-2_25

1 Introduction

The development of high-voltage direct-current (HVDC) transmission has progressed dramatically because of its advantages, such as a long conveying distance, large capacity and high efficiency [1, 2]. As the core component in converter stations, metallized film capacitors play a significant role in suppressing the commutation overshoot voltage and avoiding severe damage to thyristors [3, 4]. As a result of an increase in the voltage level and transmission capacity, the requirements for the performance of metalized film capacitors have become more stringent.

Polypropylene is a kind of semicrystalline polymer that acts as the main insulator in capacitors [5]. A metallized film capacitor consists of PP films and metallized electrodes, and the films have a dominant impact on the safety and stability of the whole system [6]. During long-term operation, the PP films experience the combined effects of the electrothermal field, during which the molecular chains break and degrade, and microdefects form inside the PP films. Local breakdown occurs because of the increased free path of electrons and impact ionization, which greatly shorten the lifetime of the films even though there is a self-healing capability. In addition, the decrease in the breakdown strength due to the temperature rise caused by the dielectric loss cannot be ignored [7]. Therefore, methods for improving the breakdown strength without sacrificing the loss characteristics of PP films have attracted widespread attention in recent years [8, 9].

The modification of the PP mainly includes physical and chemical methods, of which the former generally possesses lower investment costs and less time input. Since the thickness of the thin PP film is at the μm level, the inorganic filling methods are restricted due to the poor dispersion of the filler particles. There is a strong relationship between the microscopic characteristics and the macroscopic properties of the PP insulation. The control of the crystallization process was evaluated and showed the potential for application to the structure optimization of PP films [10]. The low crystallinity of PP induces an increase in electron transport in the amorphous region, which results in an increase in impact ionization. Furthermore, the low density of the spherulite polymer causes a decrease in the dielectric breakdown strength of the PP material [11, 12]. The results showed an increase in the crystallization peak temperature of the composites and a preferred orientation of the polymer crystallites as the concentration of the nucleating agent increased [13]. When the cooling rate is 10 °C/min and the concentration of 1,3:24-bis(3,4-dimethylobenzylideno)sorbitol (DMDBS) is 0.2 wt%, the crystallization peak temperature of the isotatic PP can be increased from 118.7 °C for the original PP to 130.1 °C [14]. The doping of the organic phosphorus nucleating agent content ranged from 0 to 0.8% in [15]. The crystallization peak temperature increased by 10 °C, and the nucleation density was highly elevated. The tensile and flexural strengths of the nucleated composite increased by 15%, indicating improved mechanical properties. In addition, PP with 0.15% 1,2,3-tridesoxy-4,6:5,7-bis-O-[(4-propylphenyl) methylene]nonitol (TBPMN), in a certain cooling rate range, crystallized at a temperature lower than the crystallization temperature of nonnuclear PP [16]. Quenched PP samples have electrical properties that are superior to isothermally crystallized or naturally cooled samples [17]. Although regulating the crystallization is promising for improving the properties of PP, the focus on electrical characteristics is scarce, and the relationship between the electrical characteristics and the microstructure is unknown. Moreover, the application of such a modification method on PP films has yet to be verified by experiments.

In this chapter, effects of crystalline morphology on dielectric properties of PP film will be the studied. PP film with different content of β-nucleating agent (β-NA) was prepared. The crystallization property and the morphology of spherulites of samples were tested. The dielectric loss, DC conductivity and breakdown strength were evaluated. The trap energy level distribution was analyzed. PP film samples doped with an organic phosphorus nucleating agent under three cooling processes were prepared to evaluate the effects of regulating the crystallization. The micromorphology and crystal structure were observed. In addition, the conductivity of the film sample was tested at different temperatures to compare the conduction loss. The DC breakdown strength was investigated, and the mechanism was discussed.

2 Dielectric Properties Dependent on Crystalline Morphology of PP Film for HVDC Capacitors Application

2.1 Experimental Arrangement

2.1.1 Sample Preparation

The PP resin was supplied by Sinopec Hainan Refining & Chemical Limited Company, and the type of the PP resin is PPH-T03. The β-NA, N,N'-dicyclohexylterephthalamide (TMB-5), is a kind of rod-like micro molecule with a certain length to diameter ratio, which was purchased from Shanxi Provincial Institute of Chemical Industry. The β-NA was dispersed into PP by a banbury mixer with the filler contents of 0.3, 0.5 and 0.7 wt%. The mixing temperature was set at 190 °C, and the mixing process was continued 20 minutes for homogeneous mixing. The mixture was made into film by a stainless-steel mold. The filming process was continued for 10 min and the pressure was set at 25 MPa. The films were cooled by wind from 190 to 120 °C and the cooling rate was approximately 10 °C per min. Samples with the NA content of 0.3 wt%, 0.5 wt% and 0.7 wt%, were denoted by PP-0.03%, PP-0.05% and PP-0.07%, respectively. Before experiments, all samples were wiped by anhydrous alcohol to eliminate impurities on the surface.

2.1.2 Sample Characterization

Crystalline forms of samples were characterized by wide-angle X-ray diffraction (XRD, Rigaku D/MAX 2500). The target material of Cu was employed and the excitation voltage is 40 kV. The scanning rate was set at 6 °/min. The scanning range of 2θ diffraction peak was 5°–40°.

The crystalline morphology was observed by polarized optical microscopy (POM) (59XF-PC). Samples were melted at 190 °C and keep the molten state for 3 min. Then, samples were cooled with the rate of 2°C/min. During the cooling process, the disordered molecular chains fold into crystals and the crystalline morphological was recorded by a digital camera.

2.1.3 Dielectric Loss and DC Conductivity Experiments

The dielectric loss can be represented by the current integration. Three kinds of different type currents flow though the tested samples, containing transient current, absorption current, and the conduction current. The transient current belongs to the polarization current without loss, while the absorption current and the conduction current lead loss. The current integration of samples was measured using the Q(t) method. An integrating capacitor is inserted between a high voltage terminal and the specimen to integrate all the currents [20, 21]. The Q(t) is the integration of all kinds current over time. The Q(t) can be obtained as follow equation.

$$Q(t) = \int_0^t I_{tr}(t)dt + \int_0^t I_a(t)dt + \int_0^t I_c(t)dt \tag{1}$$

where $I_{tr}(t)$ is the transient current, $I_a(t)$ is the absorption current, $I_c(t)$ is the conduction current, t is the time.

The behavior of dielectric loss can be also represented by the charging ratio λ. The charging ratio λ is defined as follow.

$$\lambda = \frac{Q(t)}{Q_i} = \frac{\int_0^t I_{tr}(t)dt + \int_0^t I_a(t)dt + \int_0^t Ic(t)dt}{Q_i} \tag{2}$$

where the Qi is the initial current integration.

The Q(t) of all samples were measured at 85 °C. The electric field varied from 5 to 100 kV/mm. The time of applying voltage was 120 s.

DC conductivity was tested by a three-electrode system, with cylindrical-plate stainless-steel electrodes (the cylindrical electrode is 25 mm in diameter as the measuring electrode and the plate electrode is 50 mm in diameter as the high—voltage electrode) and a guard ring electrode (the guard ring electrode is 30 mm in inner diameter and is grounding for eliminating surface leakage currents). The conduction current was measured by an electrometer (Keithley 6517B) at temperature from 25 to 85 °C. The current was recorded every 1 s for 30 mins. The average value of last 100 current values was regard as the steady-state conduction current. Then the DC conductivity could be calculated based on the steady-state conduction current.

2.1.4 Breakdown Strength Tests

DC breakdown strength test was performed using a dielectric strength tester with a high-voltage DC power supply, a plate electrode (60 mm in diameter) and a ball electrode (20 mm in diameter). The plate electrode was grounded and the ball electrode was connected to the high-voltage DC power supply. The sample with the thickness of 30 μm and the dielectric strength tester were immersed in insulating oil. The breakdown strength tests were conducted at 85 °C and the voltage rising rate was set at 0.5 kV/s. The two-parameter Weibull distribution was employed to analyze the DC breakdown strength of samples from the experimental data.

2.2 Results

2.2.1 Crystalline Type and Crystalline Morphology

The crystalline types of samples with different contents of β-NA were tested. The XRD patterns are shown in Fig. 1. For PP film, there are five diffraction peaks, 29 = 14.1°, 16.9°, 18.5°, 21.0° and 21.8°, corresponding to the crystal face of (110), (040), (130), (111) and (131), respectively, which belong to the characteristic diffraction

Fig. 1 The X-ray diffraction patterns of PP, PP-0.03%, PP-0.05% and PP-0.007%

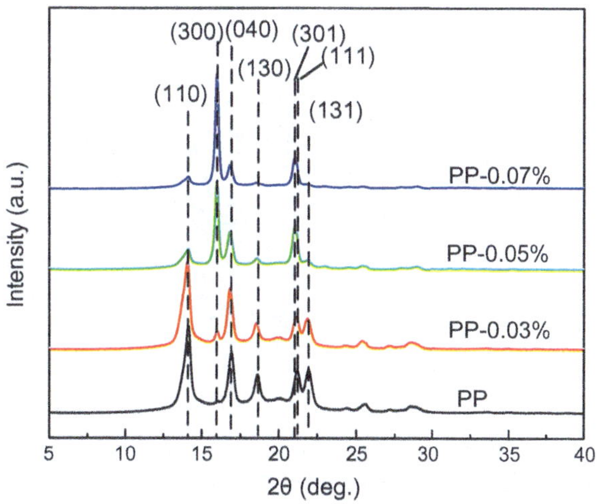

peaks of α crystalline type. For PP-0.01% and PP-0.03%, there are six diffraction peaks, 2θ = 14.1°, 16.2, 16.9°, 18.5°, 21.2° and 21.8°, corresponding to the crystal face of (110), (300), (040), (130), (301) and (131), respectively. The crystal face of (110), (040), (130), (301) and (131) belong to the characteristic diffraction peaks of α crystalline type, while the crystal face of (300) belongs to the characteristic diffraction peaks of β crystalline type. For PP-0.05% and PP-0.10%, there are five diffraction peaks, 2θ = 14.1°, 16.2, 16.9°, 21.2° and 21.8°. It can be observed that with the increasing of the content of β-NA, the content of β crystal increases, while the content of α crystal decreases. When the content of β-NA is 0.05%, the content of β crystal in PP film is highest. The XRD pattern of PP-0.10% is same with the XRD pattern of PP-0.05%. The results of XRD patterns indicate that the β crystal of PP-0.05% is highest.

Figure 2 shows the POM images of samples with different β-NA content during the crystallization process. For PP film in Fig. 2a, there are few lamellae fold by disordered molecular chains first. Then, lamellae grow up into spherulites. The spherulite present radioactive growth from the crystal nucleus, which is the typical morphology of α crystalline type. Next, spherulites grow up further and contact with other spherulites. At last, all spherulites contact with other spherulites around and the spherulites no longer grow up. It can be observed that α crystals are relative big and there are clear boundaries among spherulites.

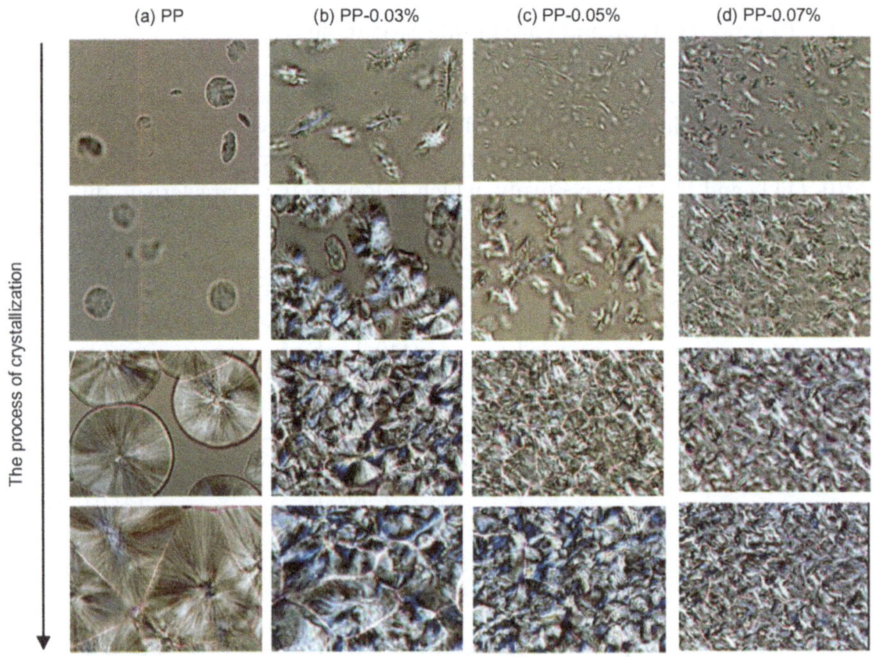

Fig. 2 The crystallization morphology of different sample during the process of crystallization **a** PP, **b** PP-0.03%, **c** PP-0.05%, **d** PP-0.07%

As shown in Fig. 2b–d, the crystallization processes of samples with different content of β-NA present different characteristics, comparing with the crystallization processes of PP film. For PP-0.03%, the number of lamellae in Fig. 2b is more than the number of lamellae in Fig. 2a. The lamellae in Fig. 2b grow around NAs. The lamellae clusters are parallel to each other, the morphology of which is very different with the morphology of α crystal. Considering with the results of XRD, it can be inferred that the lamellae clusters in Fig. 2b are β crystals. Then, with the growth of crystals, it can be clearly observed that there are both α crystals and β crystals in PP-0.03%. At last, crystals no long grow and the size of spherulites of PP-0.03% is much smaller than the size of spherulites of PP. Besides, the boundaries among spherulites in PP-0.03% are blurred.

For PP-0.05%, the number of lamellae in Fig. 2c is much more than the number of lamellae in Fig. 2b. The size of lamellae is smaller than the size of spherulites of PP film and PP-0.03% in Fig. 2a, b. Besides, boundaries among spherulites in PP-0.03% are much more blurred than the boundaries in PP film and PP-0.03%. For PP-0.07%, the number of lamellae in Fig. 2d is the most and the size of lamellae is minimum. It can be inferred that the existence of excess β-NA limits the growth of lamellae, leading to the imperfection of the spherulites morphology.

2.2.2 Dielectric Loss and DC Conductivity

The current integration measured by the method of Q(t) reflects all currents, containing transient current, absorption current and conduction current, through the dielectric during the testing time. The results of Q(t) reflect the dielectric loss, containing polarization loss and conduction loss. The current integration of samples under different electric field at 85 °C were measured, as shown in Fig. 3. The current integration Q(t) is closely related with the applied electric field. The current integration of PP film increases with the increasing electric field from 10 to 100 kV/mm. The current integration is about 41 nC at the electric field of 10 kV/mm, while the current integration is about 1349 nC at the electric field of 100 kV/mm. The current integration of samples with different β-NA content in Fig. 3b–d are lower than that of PP film in Fig. 3a. The results indicate that crystalline morphology affects the current integration in samples. Especially, the current integration in PP-0.05% is least, indicating that the morphology of PP-0.05% presents the best effect of suppressing dielectric loss.

The charging ratio λ of samples under different electric field are presented in Table 1. The charging ratio λ can reflect conduction loss more intuitive, because the transient current and absorption current decrease quickly. The charging ratio λ is mainly contributed by conduction current. The electric field greatly affects the charging ratio λ. The charging ratio λ of all samples are low when the applied electric field is low. The conduction current increases under a higher electric field, leading to a larger charging ratio λ. The charging ratio λ of PP increase from 1.23 to 3.14 with the increasing electric field from 10 to 100 kV/mm. While, the charging ratio of PP-0.05% increase from 1.05 to 3.02 with the increasing electric field from 10

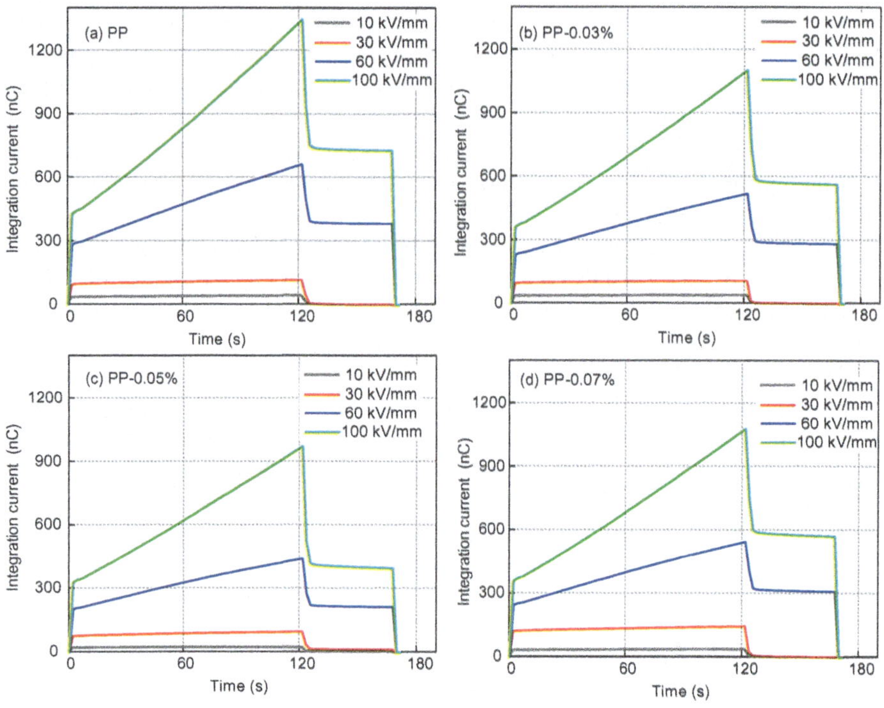

Fig. 3 The current integration of different sample at 85 °C, **a** PP, **b** PP-0.03%, **c** PP-0.05%, **d** PP-0.07%

Table 1. The charging ratio of samples at 85 °C

Electric field		Charging ratio λ (%)		
(kV/mm)	PP	PP-0.03%	PP-0.05%	PP-0.07%
10	1.23	1.13	1.05	1.12
30	1.38	1.14	1.11	1.15
60	2.32	2.24	2.20	2.22
100	3.14	3.05	3.02	3.09

to 100 kV/mm. Comparing the charging ratio λ of all samples, it's inferred that the existence of β-spherulites can effectively decrease the conduction current.

Further, the DC conductivity was measured to represent the DC conduction loss. The DC conductivities of samples with different contents of β-NA are presented in Fig. 4. In Fig. 4a, the conductivities of all samples increase with the increasing electric field. The higher electric field enhances the directed migration of charge and the charge injection from electrodes to the sample based the theory of Poole-Frenkel effect and Schottky effect [22–24]. The conductivities of samples with β-NA are lower than the conductivity of PP film at different electric field. It can be found

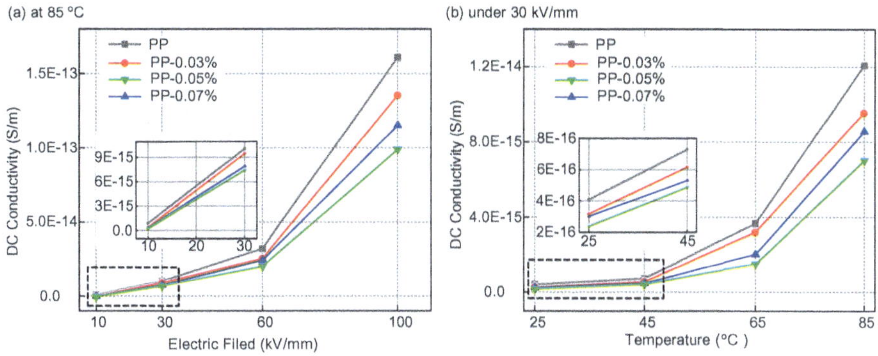

Fig. 4 DC conductivity of samples with different NA content, **a** under different electric field at 85 °C, **b** at different temperature under 30 kV/mm

that the conductivity of PP-0.05% is lowest. In Fig. 4b, the DC conductivities of all samples increase with the increasing temperature. The DC conductivity of all samples are relative low at temperature from 25 to 45 °C. When the temperature increases from 45 to 85 °C, the DC conductivities of all samples increase significantly. The charge carriers can obtain more kinetic energy and the transport process of carriers will be promoted at higher temperature. Besides, the higher temperature will enhance the Poole-Frenkel effect and Schottky effect, which will increase the injection and proliferation of charges in dielectric film. The samples with β-NA present a lower DC conductivity at different temperature, comparing with PP film. The results show that the existence of β crystals can effectively decrease the DC conductivity, which is consistent with the results of current integration.

The DC conductivity of dielectric film increases at high temperature and high electric field, which will produce more heat and increase the temperature rise in capacitors. The increased temperature in capacitors will further increase the DC conductivity of dielectric film. When the heat production is more than heat dissipation, the thermal breakdown will take place. Therefore, decreasing the conductivity is beneficial. In Fig. 4a, b, the conductivities of samples with β-NA are lower than the conductivity of PP film, which indicates that the crystalline morphology can affect the charge injection and charge carrier transport.

2.2.3 DC Breakdown Strength

The Weibull distribution of DC breakdown strength of samples at 85 °C is shown in Fig. 5. The breakdown strength at 63.2 percentile is defined as the characteristic breakdown strength, which can reflect the breakdown strength of samples. The breakdown strength under positive voltage is shown in Fig. 5a. The characteristic breakdown strength of PP film, PP-0.03%, PP-0.05% and PP-0.07% are 304.1, 347.6, 376.8 and 369.3 kV/mm, respectively. It can be seen that the breakdown strength

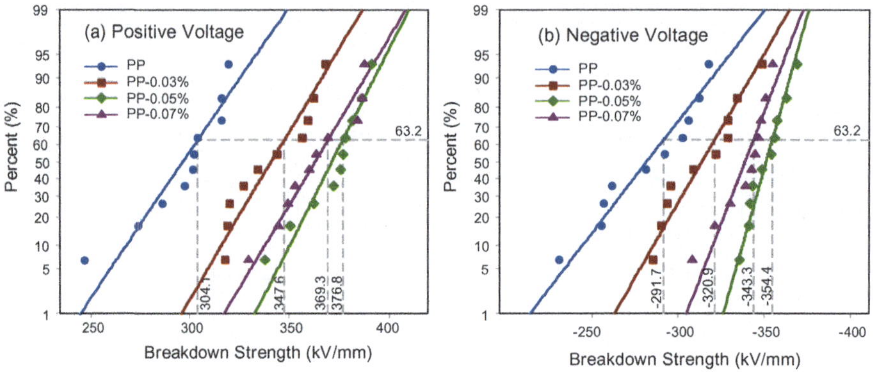

Fig. 5 Weibull distribution of DC breakdown strength at 85 °C: **a** under positive voltage and **b** under negative voltage

of samples with β-NA under positive voltage is strengthened to varying degrees comparing with PP film. The breakdown strength of samples under negative voltage is shown in Fig. 5b, the characteristic breakdown strength of PP, PP-0.03%, PP-0.05% and PP-0.07% are –291.7, –320.9, –354.4 and –343.3 kV/mm, respectively. The breakdown strength of all samples under negative voltage is lower than that under positive voltage, because the injection of electrons is easier than the injection of holes. The results indicate that the crystalline morphology affects the DC breakdown strength, and the formation of β spherulites can increase the DC breakdown strength effectively.

2.3 Discussion

2.3.1 Crystal Model

Based on the results of POM images the structure models of spherulite and interface region are summarized, and presented in Fig. 6. As shown in Fig. 6a1, the lamellae of α spherulite grow radially outward from the centre of the spheru-lite, which consists of short, tightly crossed lamellae [25]. At the same time, bifurcations appeared in the radial-growth lamellae and the bifurcation angle is approximately 90 ° [26]. The space between the radial lamellae is also filled with short cross-hatching lamellae, the thickness of which is less than the thickness of the dominant radial lamellae, as shown in Fig. 6a2. In the centre of β spherulite, the lamellae are assembled into a bundle structure in parallel. The bundle structure grows continuously at both ends, and the lamellae bend and grow into a complete spherulite, as shown in Fig. 6b1. The β spherulites have longer curved lamellae. The lamellae magnification model of β spherulite is shown in Fig. 6b2, and lamellae are approximately parallel to each other [26]. In the theory of Flory [27], the lamellar structure conforms to the plug-board

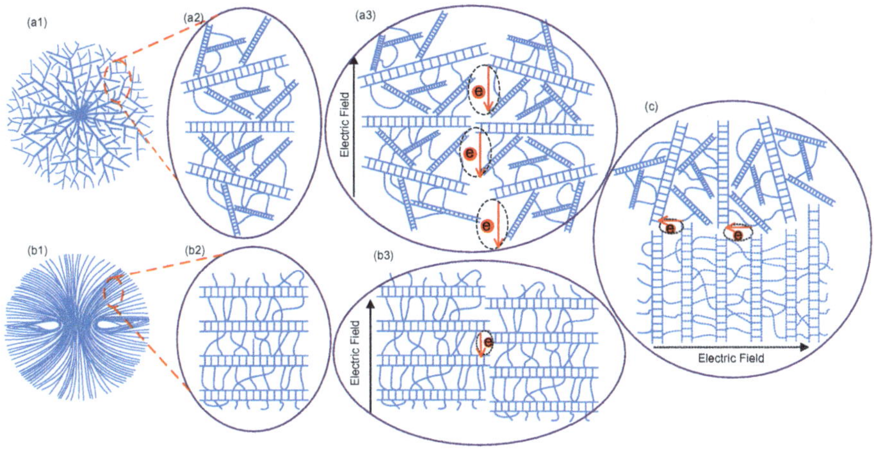

Fig. 6 Model of crystal and crystal boundary: **a1** α crystal, **a2** lamellar crystals of α crystal, **a3** boundary between α crystals, β crystal, **b2** lamellar crystals of β crystal, **b3** boundary between β crystals, **c** boundary between α crystal and β crystal

model, that is, when molecular chains are completely random into lamellae, two adjacent segments of a lamella are non-adjacent segments and segments belonging to different molecules. Molecular chains on the surface of lamellae, like the plug wires, are irregular and form amorphous region.

The lamellae radial growing in α spherulite branch and grow at an angle of approximately 90 °, a large free volume may be formed at the location of branch. At the same time, there are lamellae intersections at the interface between α spherulites, as shown in Fig. 6a3, and larger free volumes may be generated at the lamellae intersections. The radial growth of β lamellae results in smaller lamellar branching angles and smaller probability of large free volumes. The inter- facial molecular chains between β spherulites grow interactively, so the boundaries of β-spherulites are indistinct [28, 29], as shown in Fig. 6b3. There are more molecular chains at the interface region of β spherulites. There are also lamellae intersections and intergrowth molecular chains at the interface region between α spherulite and β spherulite, as shown in Fig. 6c. The free volume in Fig. 6c is also smaller than that in Fig. 6a3.

Based on the microstructure of α spherulites, β spherulites and the interface regions, the effects of microstructure on dielectric properties can be speculated. Injected electrons will be accelerated to collide with molecular chains at the edge of the free volume, leading to the breakdown of molecular chains, as shown in Fig. 6a3, b3 and c. The breakdown of molecular chains will form low density area, where is the origin region of electric breakdown. Besides, the molecular chains in amorphous region may also affect the traps energy level distribution and affect the transport of charge carriers, which is closely related with the conduction loss of PP [30]. Results of breakdown strength in Fig. 5 confirmed the inference of the relationship between microstructure and breakdown. The crystalline type in PP film is α spherulite and there are bigger free volume in α spherulites and boundaries among α spherulites.

The injected electron will be accelerated in free volume to collide with the molecule chain at the edge of the free volume, as shown in Fig. 6a3, b3 and c. The bigger free volumes leading to more collapsing force of electrons. The breakdown of molecule chains collided by high speed electrons will form low density area, and then leads to electric breakdown. For samples with β-NA, the formation of β spherulites decreased the free volume which will limit the speed up of electrons. Therefore, the breakdown strength of samples modified by β-NA are increased.

2.3.2 Trap Energy Level Distribution

The conduction mechanism of PP film is considered to be the hopping conduction. The process that charge carriers transport from one localized state to another localized state is referred to as "hopping transport". The conduction current is related with temperature, electric field. From the microscopic view, the conduction current is related with the charge carrier density and carrier mobility [23]. The carrier mobility is related with the hopping distance and the energy of carriers. The carrier density is related with the trap energy level distribution.

In this paper, the trap energy level distribution was analysed using themethod of isothermal discharge current (ICD) [30]. The trap energy level distribution of samples at temperature of 25–85 °C were shown in Fig. 7. In Fig. 7a, the deep trap energy level and the deep trap density first increase with the increasing β-NA content, and then decrease with the increasing β-NA content. The deep trap energy level of PP-0.05% is highest. There are more molecular chains at the interface region of β spherulites, comparing with α spherulites. The more molecular chains in are the main reason of the increased deep trap energy level. Charges captured by deep traps can hardly escape from deep traps. Charges injected from electrodes will trap in deep traps, which form an opposite electric field with the applied electric field and suppress the injection of charges into film. Besides, the introduced deep trap energy levels by β spherulites can limit the number of charge carriers and the charge carrier mobility. The limitation of charge carrier transportation is beneficial to reduce the DC conductivity and dielectric loss. The results of trap energy level distribution are consistent with the results of conductivity in Fig. 4b. Besides, the deep trap energy levels introduced by β spherulites can also benefit the breakdown strength for the limitation of carrier mobility.

The deep trap energy levels of all sample were deepened with the increasing temperature, as shown in Fig. 7a–d. Charges obtain more energy at high temperature promoting the escape probability from the deeper traps. The promoted transportation of charges will increase the conductivity, as shown in Fig. 4b. The deep trap energy level of samples with β-NA is deeper than PP at high temperature, as shown in Fig. 7c–d. Therefore, the conductivity of samples with β-NA is lower than PP at high temperature. The deep trap energy level of PP-0.05% is highest in all samples, as shown in Fig. 7c–d. Based on the analysis above, the effects and mechanism of different crystalline morphology on conductivity and breakdown strength are detailed.

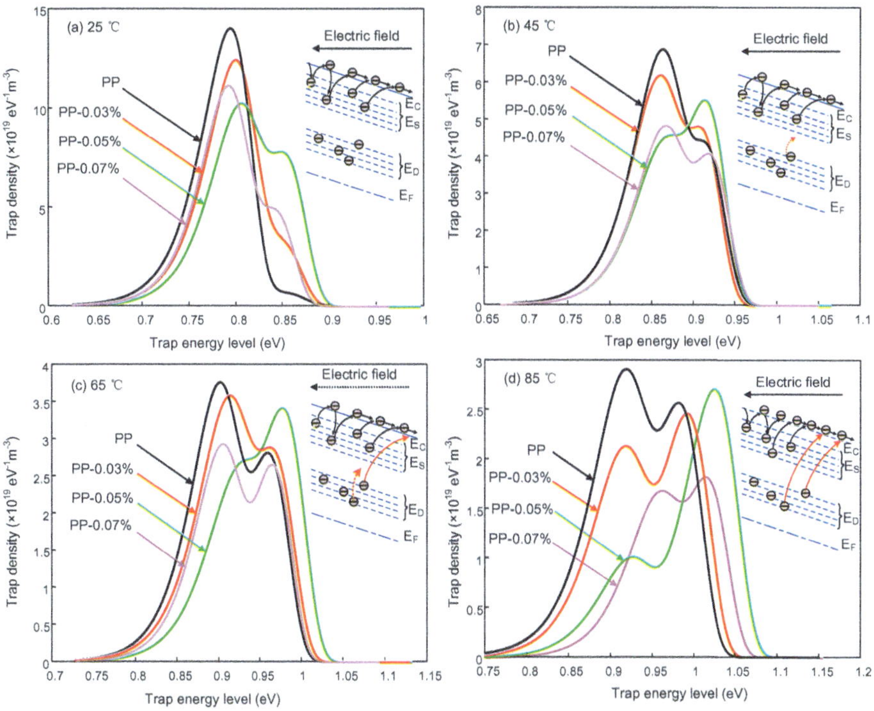

Fig. 7 Trap energy level distribution of samples at different temperature **a** 25 °C, **b** 45 °C, **c** 65 °C and **d** 85 °C. (E_c is the bottom level of the conduction band; E_S is the shallow trap energy levels; E_D is the deep trap energy levels; E_f is the Fermi level)

3 Effect of Crystallization Regulation on the Breakdown Strength of Metallized Polypropylene Film Capacitors

3.1 Experimental Preparation

3.1.1 Sample Preparation

The PP film samples consisted of PP particles and a nucleating agent with a thickness of ~25 μm. The PP particles were provided by Hainan Petrochemical Company, and the organic phosphorus nucleating agent (sodium 2,2'-methylenebis (4,6-di-tert-butylphenyl) phosphate) was manufactured by Kangjin Technological Company, with a high nucleation efficiency and low cost. The structure of the nucleating agent is shown in Fig. 8, and the preparation steps are described below.

(1) After the PP particles were melted and the nucleating agent was ground, both were quickly mixed in a double roll machine at 190 °C for 10 min to

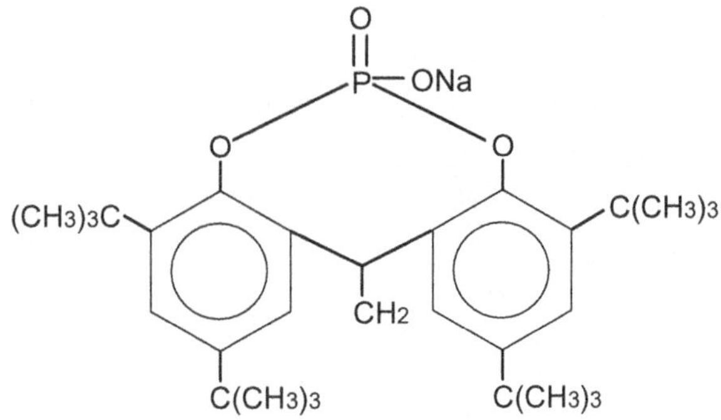

Fig. 8 Structure of the nucleating agent

Table 2 Test samples with different doping contents under three cooling methods

Sample	Content					
	0%	0.005%	0.01%	0.03%	0.05%	0.07%
A	A-0	A-0.5	A-1	A-3	A-5	A-7
B	B-0	B-0.5	B-1	B-3	B-5	B-7
C	C-0	C-0.5	C-1	C-3	C-5	C-7

avoid damage to the PP and obtain a homogenous mixture. The content of the nucleating agent ranged from 0 to 0.07 wt%.

(2) The film samples were stereotyped by hot pressing at 190 °C and a pressure of 35 MPa for 5 min. This process produced samples with a thickness of ~25 μm and a side length of 9 cm.

Then, the samples were crystallized by three cooling methods.

Sample A: The samples were annealed in cold water at a temperature of 15 ± 2 °C for 5 min. The temperature of the sample was decreased from 190 to 15 °C and then the samples were cleaned and dried.

Sample B: The samples were cooled under a pressure of 35 MPa from 190 to 120 °C and then cooled to a room temperature of 25 °C in air.

Sample C: The samples were cooled under a pressure of 35 MPa from 190 to 150 °C and then cooled to a room temperature of 25 °C in air. The order of the average cooling rate from fast to slow is sample A > sample C > sample B. The film samples were named as shown in Table 2.

3.1.2 Experimental Methods

Polarizing optical microscopy (POM) was used to observe the crystallization process and the microstructure of the PP composite samples. The thermodynamic properties of the samples were tested by differential scanning calorimetry (DSC). In addition, the X-ray photoelectron spectroscopy (XPS) results of the different samples were compared and analyzed. Based on these results, the crystallinity was calculated.

The conduction loss is a significant part of the whole loss of PP materials and changes substantially with the temperature. To study the relationship between the crystallization and the dielectric properties, the conductivity of the film samples at different temperatures was measured using a three-electrode system [18]. The electric field was kept at a constant value of 27.28 kV/mm. The humidity was kept at 28%, and the temperatures were 25, 55 and 85 °C. The DC breakdown strength test was conducted by a ball-plate electrode system in oil, and the circuit is shown in [19]. The form of the applied voltage was directly increased at a rate of 0.1 kV/s. Each sample was tested 15 times, and the test points were dispersed. The edge of the samples was avoided to ensure the consistency of the thickness. The breakdown strength was calculated as the ratio of the breakdown voltage to the thickness of the test point of the sample. The experimental data were analyzed with a Weibull plot, through which the characteristics of the original data distribution can be effectively reflected.

3.2 *Results*

3.2.1 Crystallization Regulation

Figure 9 shows the crystallization processes of the pure PP and the PP composite samples, from which it can be seen that the doping of the nucleating agent does not induce a new crystal form. After doping with the nucleating agent, the number of nucleation sites are obviously increased at the start of crystallization, as shown in Figure a1 and b1. The red circles in Figure a4 and b4 represent the spherulites with a relatively complete structure when the crystallization ends. The average diameter of the six spherulites is calculated, and for the PP composite samples with 0.05% nucleating agent, the average diameter is approximately 48.61 μm, which represents

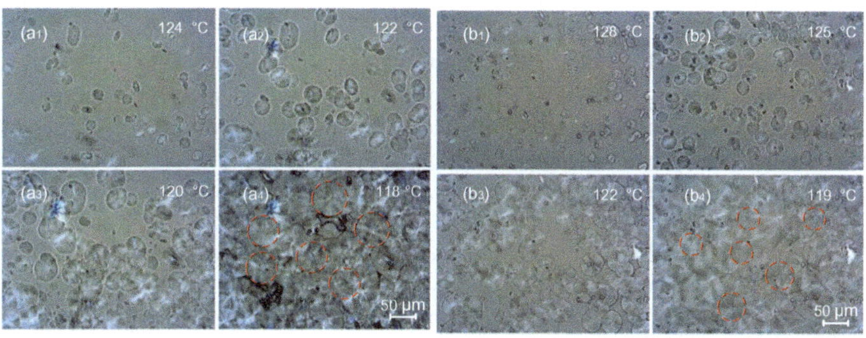

Fig. 9. The whole crystallization processes of the (**a**) and the **b** PP composite samples with 0.05 wt% nucleating agent

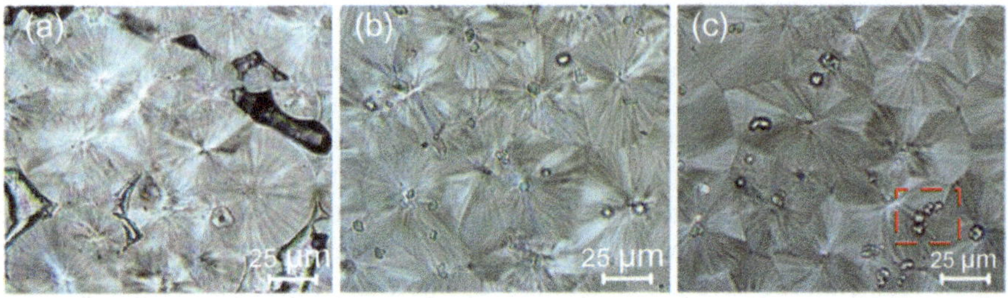

Fig. 10 Microstructures of the **a** pure PP, **b** PP composite samples with 0.01 wt% nucleating agent and **c** PP composite samples with 0.05 wt% nucleating agent

a reduction of 30.20% compared with that of the pure PP. In addition, the crystallization of the sample with a nucleating agent occurs earlier than that of the pure PP. The crystallization temperature is elevated from 124 °C to approximately 128 °C, showing the impact of the nucleating agent. The two benzene ring structures in the nucleating agent form a V-shaped structure, which confines the helical structure of the PP molecules, thereby effectively stabilizing the PP molecules. During the crystallization process, the nucleating agent, as a heterogeneous crystal nucleus, crystallizes earlier than the PP melt. The PP molecules are epitax-ially crystallized on the surface of the nucleating agent crystal until a complete spherulite is formed.

The microstructures of the PP composite samples that contain 0 wt%, 0.01 wt% and 0.05 wt% of the nucleating agent are shown in Fig. 10. The spherulite distribution in the original PP film is scattered. With increasing nucleating agent content, the crystal area increases, and the density of the spherulite distribution increases. However, when the content is high, the doping of the nucleating agent causes a reuniting phenomenon, as shown by the red dotted line in Figure Ira, which should be considered because it may break the structure of the spherulite.

3.2.2 Crystallization Characteristics

The results of DSC testing are shown in Fig. 11. Figure 11a shows that the crystallization temperature (Tc) of the PP composite sample increases slightly from 113 to 116 °C after doping with the nucleating agent. From the perspective of thermodynamics, crystallization is impacted by molecular diffusion and nucleation processes. After the nucleating agent is added, the nucleation free energy of the system is greatly reduced, which is conducive to nucleation. The heterogeneous nucleation caused by the nucleating agent causes the molecular chains to be attached to it and arranged in an orderly manner, which requires less energy than homogeneous nucleation. Thus, PP that contains the nucleating agent crystallizes at an increased temperature. In addition, as shown in Figure b, the melting peak of the PP sample is shifted to the right, changing from 160 to 167 °C. The increase in the crystallinity and the close arrangement of the molecules lead to an increase in intermolecular forces and the melting temperature (Tm). The melting enthalpy (ΔHm) is elevated significantly, as

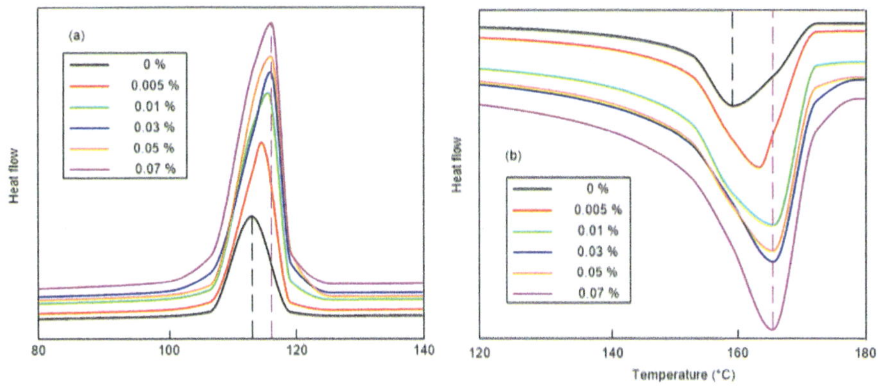

Fig. 11 DSC curves of the PP samples

shown by the peak value, which can be attributed to the improved microstructure that resulted from the addition of the nucleating agent.

Figure 12 shows the XRD curves of the different PP samples. Figure a shows that after adding the nucleating agent, the diffraction peak angles for the thin film sample do not change significantly compared with those for the pure PP film, which further proves that they all comprised the α crystal form. Samples A and C have the same trend as sample B. Figure b shows that the cooling method has an obvious effect on the internal structure of PP film samples A, B and C. The crys-tallinity degree (*Xc*) of the PP film is shown in Table 3 and is calculated using the formula below:

$$Xc = \frac{Ic}{Ic + Ia} \tag{3}$$

$$Ic = I\alpha(110) + I\alpha(040) + I\alpha(130) + I\alpha(111) + I\alpha(131) \tag{4}$$

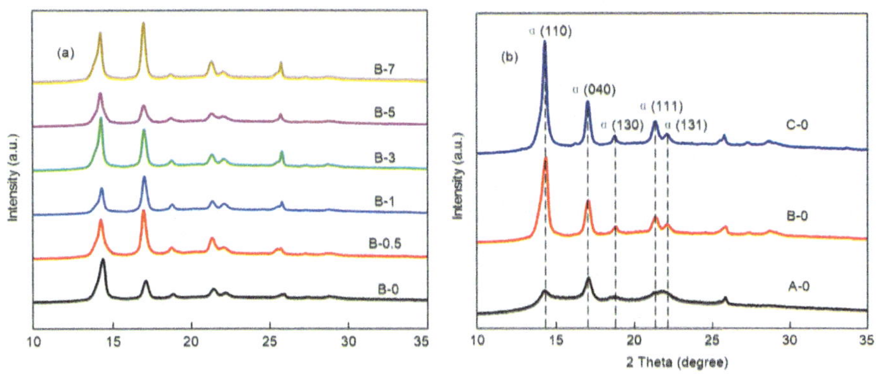

Fig. 12 XPS curves of the PP samples

Table 3 Crystallinities of the PP films with different doping contents and cooling methods

Sample	Doping contents and *Xc*					
A	A-0	A-0.5	A-1	A-3	A-5	A-7
	26.74	28.80	32.32	35.84	36.27	40.27
B	B-0	B-0.5	B-1	B-3	B-5	B-7
	32.43	41.32	46.61	51.41	52.75	54.35
C	C-0	C-0.5	C-1	C-3	C-5	C-7
	34.14	44.39	48.44	51.69	52.67	53.85

where *Ic* is the whole diffraction intensity of the crystal area and *Ia* is the whole scattering intensity of the amorphous region. $I\alpha(110)$, $I\alpha$ (040), $I\alpha$ (130), $I\alpha(111)$, and $I\alpha(131)$ are the intensities of the a-crystal peaks attributed to the (110), (040), (130), (111), and (131) planes of monoclinic cell, respectively. The 20 values for the peaks are 14.4°, 17.1°, 18.8°, 21.4°, and 22.1°.

Sample B-7 possesses the highest *Xc* of 54.35%, which is increased by 67.59% compared with *Xc* for B-0 and by 34.96% compared with *Xc* for A-7. It can be concluded that the value of *Xc* has a positive correlation with the content of the nucleating agent. Moreover, the relatively slow cooling rate promotes the crystallization process. After adding the nucleating agent, additional molecular chains and segments tend to be arranged in an orderly manner, which increases the density and crystallinity. For samples A-0~A-7, crystallization occurs under extreme nonisothermal conditions. The macromolecular segments do not have enough time to fold and form a wafer. At this time, the polymer relaxation process cannot keep up with the temperature change. Thus, the molecular chains form a bulky amorphous region because they could not crystallize during the rapid cooling process (Table 3).

3.2.3 Conductivity Characteristics

Figure 13 shows the conductivities of the PP samples at different temperatures . At 25 °C, there is a slight decrease and then an increase with the nucleating agent doping content. For the PP film with different contents of the nucleating agent at relatively high temperatures of 55 °C or 85 °C, the conductivity is basically un-changed and has a minor reduction. The densification of certain structures, especially the space between supramolecular structures, reduces the mobility of the ions and thereby decreases the conductivity, which demonstrates how the nucleating agent restricts the conduction loss. When the temperature is elevated from 25 to 55 and 85 °C, the conductivity of the film sample increases greatly from 10^{-14} to 10^{-12} S/m. The movement of ions in a polymer solid is limited by the barrier between chains. Increased segment motion at high temperatures provides a greater free volume for ion migration; thus, the conductivity increases.

It is worth noting that the conductivity of PP film sample A may be the highest among the other two series of samples B and C. The conductivity of the crystalline

Fig. 13 Conductivities of the PP film samples at different temperatures

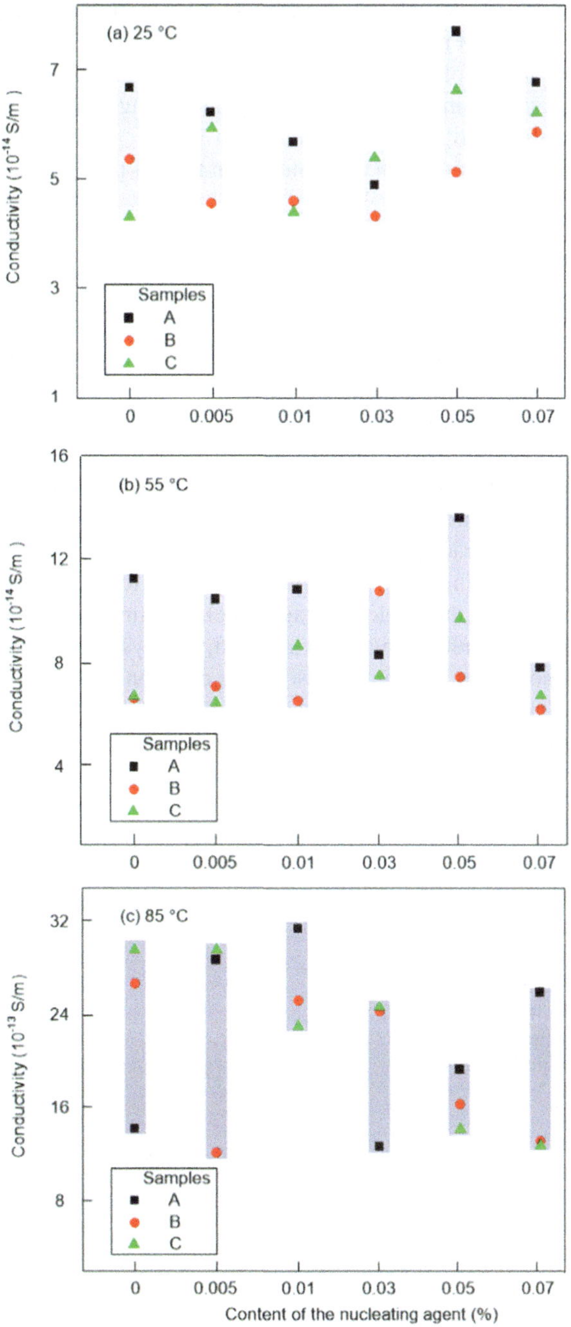

region in the polymer is much smaller than that of the amorphous region; thus, the conduction process mainly occurs in the amorphous region. The comparatively low *Xc* of sample A is responsible for the high conductivity. Based on the discussion above, the introduction of the nucleating agent does not increase the conduction loss in the PP film. In addition, it can be speculated that the fast rate of cooling has a negative impact on the conduction loss of the PP film.

3.2.4 Breakdown Strength

Figure 14 shows the breakdown strengths of PP film samples B-0~B-7 with different contents of nucleating agent at 25 °C, analyzed by Weibull plots. The effect of the nucleating agent content is significant, and with the addition of the nucleating agent, the breakdown strength of the film sample first increases and then decreases. Sample B-1 possesses the highest scale parameter of 610.0 kV/mm as well as the highest shape parameter of 9.22 compared with the scale parameter of 497.7 kV/mm and the shape parameter of 6.87 for B-0, indicating a reliable improvement in the DC breakdown characteristics. When the content reaches 0.07 wt%, the breakdown value of the film sample is basically equivalent to that of the pure PP sample. An appropriate doping of the nucleating agent helps to improve the microstructure of the PP film. After the nucleating agent is added, the crystalline region in the film increases and the amorphous region decreases. The electron acceleration process is limited; thus, the breakdown strength increases. Although the crystallinity increases with increasing content, the agglomeration caused by the poor dispersion of the excessive nucleating agent cannot be ignored because it decreases the electrical characteristics to an extent.

The effects of different temperatures on the breakdown strength distributions of the PP film samples are investigated, as shown in Figure 15. By comparing the results of the original PP film sample and the sample containing the nucleating agent with the optimal content of 0.01 wt%, it can be seen that as the temperature rises from 25

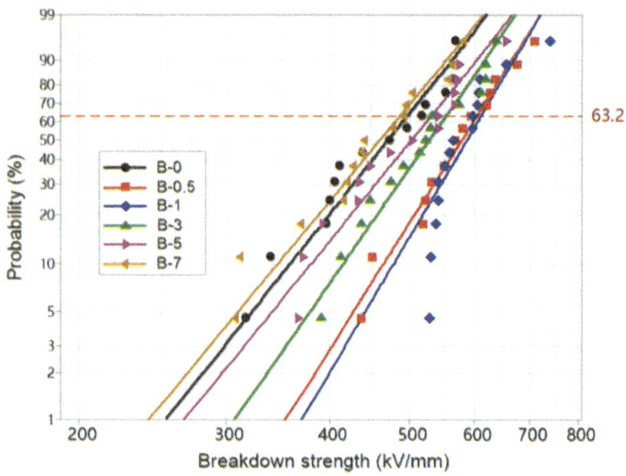

Fig. 14 Breakdown strength distributions of the PP film samples with different contents at 25 °C

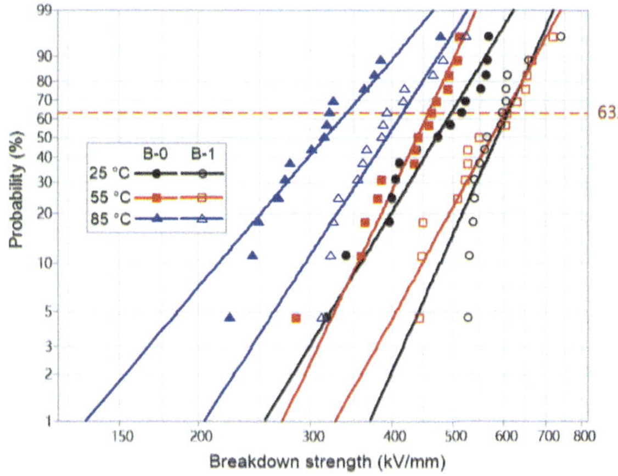

Fig. 15 Breakdown strength distributions of the PP film samples at the testing temperature of 25 °C, 55 °C and 85 °C

to 85 °C, and both kinds of samples have a deteriorated performance, particularly at the high temperature of 85 °C. The breakdown strength reduction of B-1 is less than that of B-0. The reason can be explained as follows: the high temperature causes the number of hot electrons to increase. The molecular chains are easily impacted to cause fracture, which rapidly increases the low-density region and promotes carrier ionization. Finally, the overall breakdown strength of all film samples is reduced. For sample B-1, the existence of the nucleating agent helps to produce structures with an increased integrity and density. Compared with the amorphous region, the molecular chain relaxation in the crystalline region is relatively weak at high temperatures, and the energy required to destroy the molecular chains is large. The higher the melting temperature of the PP film is, the slower the rate of the breakdown strength decreases with the temperature.

To further analyze the effects of the cooling rate, the increase in the breakdown value with different treatments is calculated and compared with the goal of avoiding the influence of the experimental error. Figure 16 shows the percent improvement in the average strength of the film sample with 0.01 wt% nucleating agent compared to that of the pure film under the same experimental conditions. It can be seen clearly that the breakdown strength improvement of sample B-1 is the highest among the three series of data. The percent improvements are 25.7%, 31.2% and 23.9% at 25 °C, 55 °C and 85 °C, respectively, showing a stable trend under different conditions. Based on the results above, the cooling treatment method plays a substantial role in the regulation of the breakdown characteristics. On the one hand, the slow rate of the cooling process is conducive to the formation of crystal regions, which may have the potential to improve the breakdown performance of the PP film sample. On the other hand, when the cooling rate is too fast, the quality of the film formation is affected; for example, the uniformity of the film thickness is reduced, and wrinkles are likely to occur.

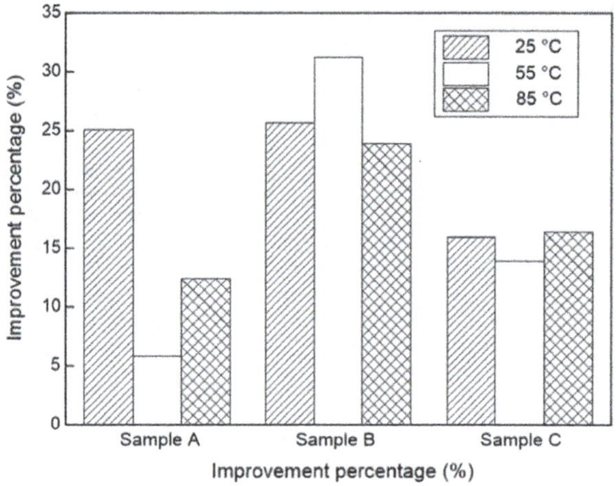

Fig. 16 Percentage improvements of the average breakdown strength of the film sample with different cooling methods

3.2.5 Discussion

Figure 17 shows the model for the electron transport in the PP film. It can be seen in area 1 that the spherulites consist of radially grown wafers and epitaxial tangential wafers. Some of the molecular chains in the amorphous region are bound between the wafers; thus, the arrangement is relatively tight. In area 2, which is the interface between the spherulites, the wafers are not tightly connected due to the high packing density of the wafers in single spherulites. Molecular chains at the boundaries of the

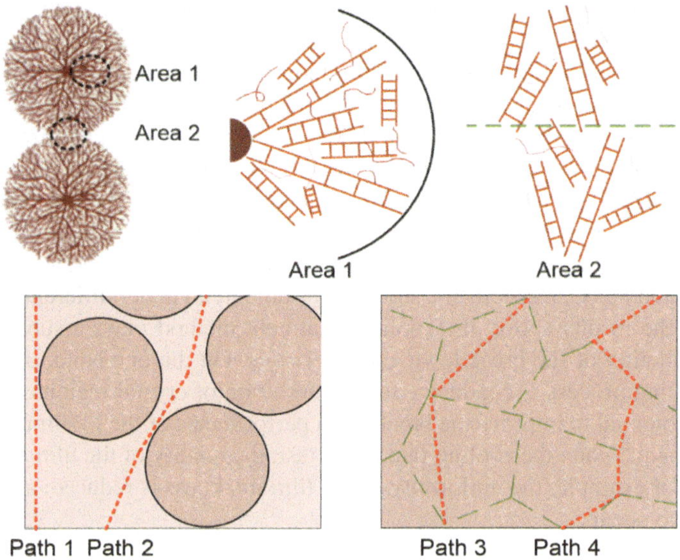

Fig. 17 Model for the electron transport in the PP film

spherulites are tightly entangled, increasing the difficulty of electron transport. The closer the amorphous region is to the crystal region, the greater the molecular chains are constrained by the wafers, and the more ordered the distribution is.

When a DC electric field is applied to pure PP film sample B-0, the electrons are preferentially transported in the amorphous regions, where the molecular chains are loosely arranged in a disordered manner, such as path 1 and path 2. After the crystallization is improved by doping with the nucleating agent and cooling at a slow rate, the microstructure of B-1 is improved, and the density of the spherulites increases and they slightly change their shape. The decreased number of amorphous regions in the film likely force the electrons to move around the spherulites. If the electrons choose to transport in the amorphous region of the crystal, the route is simplified as path 3 consisting of area 1. If the electrons avoid moving in the spherulites as much as possible, they can only choose to transport along the grain boundaries as path 4 consisting of area 2. Whether the electrons are transported from inside the spherulites or from the interface between the spherulites, their movement is greatly restricted, which not only reduces the conduction loss but also inhibits the destruction of the molecular chains and increases the breakdown strength of the PP film. In view of the above analysis, the crystalline struc-ture determines the transport of the electrons, which affects the electrical properties of the PP film. The control of the electronic transport process connects the mi- crostructure and macroelectrical characteristics, making it possible to improve the breakdown strength by adjusting the crystallization characteristics.

4 Conclusion

The effects of crystalline morphology and its mechanism on the conduction loss and breakdown strength of PP film were studied.

(1) The results of XRD indicate that the β spherulites were introduced by the β-NA. The POM results show that the β spherulites are relative smaller than α spherulites. The interfacial molecular chains between β spherulites grow interactively and the boundaries of β-spherulites are indistinct.
(2) The results of Q(t) and DC conductivity show that the formation of β spherulites can suppress the charge injection and decrease the conduction loss of PP film. The DC breakdown strength at 85 °C was improved by the formation of β spherulites, and the DC breakdown strength of PP-0.05% is highest.
(3) The crystal model explains the effects mechanism of crystalline morphology on DC breakdown strength. The results of IDC indicate that morphology characteristics of β spherulites introduce more deep trap energy levels, which inhibits the charge injection and migration, decreases the dielectric loss and DC conductivity.

(4) With the doping of the nucleating agent, the crystallization temperature, the melting temperature and the melting enthalpy are elevated. In addition, the nucleating agent and a slow cooling rate improve the crystallinity of the PP.
(5) Both the content of the nucleating agent and the cooling rate do not obviously affect the conductivity of the sample. The temperature plays a dominant role in the conduction loss of the PP film. The reduced amorphous regions and free volume result in a lower conductivity.
(6) An appropriate nucleating agent content (0.01 wt%) and slow cooling rate improve the average breakdown strength of the PP film by 25.7%, 31.2% and 23.9% compared with that of the pure PP sample at temperatures of 25 °C, 55 °C and 85 °C, respectively.

Reference

1. S.H. Kung, G.J. Kish, Multiport modular multilevel converter for DC systems. IEEE Trans. Power Del. **34**(1), 73–83 (2019)
2. H.Y. Ye et al., Review on HVDC cable terminations. High Vol. 3(2), 79–89 (2018)
3. P. Guido, R. Maurizio, Pulse handling capability of energy storage metallized film capacitors. IEEE Trans. Plasma **28**(5), 1603–1606 (2000)
4. M.H. El-Husseini et al., Thermal simulation for geometric optimization of metallized polypropylene film capacitors. IEEE Trans. Industr. Appli. **38**(3), 713–718 (2002)
5. E.J. Barshaw et al., High Energy Density (HED) Biaxially-Oriented Poly-Propylene (BOPP) capacitors for pulse power applications. IEEE Trans. Magnet. **43**(1), 223–225 (2007)
6. H. Li et al., The capacitance loss mechanism of metallized film capacitor under pulsed discharge condition. IEEE Trans. Electr. Insul. **18**(6), 2089–2094 (2011)
7. M. Horak, P. Mach, Study of electrical ageing of polypropylene film capacitors using third harmonics measurement. IEEE 20th Inter. Symp. Desi. Tech. Etectr. Pack. (SIITME), 55–59 (2014)
8. H. Li et al., Electric field and temperature dependence of electrical conductivity in Biax-ially oriented polypropylene films. IEEE Trans. Plasma Sci. **42**(11), 3585–3591 (2014)
9. W. Sima et al., Accumulative failure characteristic of polypropylene film in surge capacitor under repeated impulse voltage. High. Vol. Eng. **44**(2), 358–365 (2018)
10. Y.H. Wu et al., A remarkable suppression on space charge in isotatic polypropylene by inducing the β-crystal formation. Appl. Phys. Lett. **107**(11), 112901 (2015)
11. B.X. Du, L.W. Zhu, T. Han, Effect of ambient temperature on electrical treeing and breakdown phenomenon of polypropylene with repetitive pulse voltage. IEEE Trans. Electr. Insul. **24**(4), 2216–2224 (2017)
12. K. Matsui et al., Space charge behavior in low-density polyethylene at Pre-breakdown. IEEE Trans. Electr. Insul. **12**(3), 406–415 (2005)
13. C. Marco et al., Comparative study of the nucleation activity of third-generation sorbitol-based nucleating agents for isotactic polypropylene. J. Appl. Polym. Sci. **84**(13), 2440–2450 (2002)
14. Y.F. Zhang, T. Zhou, Non-Isothermal crystallization behaviors of isotactic polypropylene with and without 1,3:2,4-Bis(3,4-Dimethylbenzylidene) Sorbitol as a Nucleating Agent. Int. J. Polym. Anal. Charact. **15**(7), 450–459 (2010)
15. Q. Gui et al., Effects of an organic phosphorus nucleating agent on crystallization behaviors and mechanical properties of poly(propylene). J. Appl. Polymer Sci. **88**(2), 297–301 (2003)
16. E.K. Jürgen, F. chawe, I. Budde, Alig, "Non-isothermal crystallization of polypropylene with sorbitol-type nucleating agents at cooling rates used in processing". Polymer Int. **68**(2), 240–247 (2019)

17. W. Zhang et al., Effect of the β-crystal formation in isotactic polypropylene used for eco-friendly insulating material. 1st International Conference on Electrical Materials and Power Equipment.(ICEMPE) (2017), pp. 419–423
18. B.X. Du, Z.L. Li, Z.R. Yang, Field-dependent conductivity and space charge behavior of silicone rubber/SiC composites. IEEE Trans. Electr. Insul. **23**(5), 3108–3116 (2016)
19. B.X. Du et al., Effect of graphene nanoplatelets on space charge and breakdown strength of PP/ULDPE blends for HVDC cable insulation. IEEE Trans. Electr. Insul. **25**(6), 2405–2412 (2018)
20. T. Takada et al., New diagnostic method of electrical insulation properties based on current integration. IEEE Trans. Dielectr. Electr. Insul. **24**(4), 2549–2558 (2017)
21. T. Takada et al., Space charge accumulation in double-Layer dielectric systems-measurement methods and quantum chemical calculations. IEEE Electr. Insul. M. **35**(5), 36–46 (2019)
22. J.G. Simmons, M.C. Tam, Theory of isothermal currents and the direct determination of trap parameters in semiconductors and insulators containing arbitrary trap distributions. Phys. Rev. B **7**(8), 3706–3713 (1973)
23. F.C. Chiu, A review on conduction mechanisms in dielectric films. Adv. Mater. Sci. Eng. **7**, 1–18 (2014)
24. J. Ho, T.R. Jow, High field conduction in heat resistant polymers at elevated temperature for Metallized Film Capacitors. *IEEE Power Modulator and High Voltage Conference (IPMHVC)*, pp. 399–402 (2012)
25. M. Ieda, Electrical conduction and carrier traps in polymeric materials. IEEE Trans. Dielectr. Electr. Insul. **19**(3), 163–178 (1984)
26. M.R. Huang, X.G. Li, B.R. Fang, β-nucleators and β-crystalline form of isotactic polypropylene. J. Appl. Polym. Sci. **56**(10), 1323–1337 (1995)
27. F. Rybnikar, Transition of β to α phase in isotactic polypropylene. J. Macromol. Sci. Part B-Phys. **30**, 201–223 (1991)
28. P. Flory, On the morphology of the crystalline state in polymers. J. Am. Chem. Soc. **84**(15), 2857–2867 (1962)
29. W.C. Xiao et al., Influence of a novel β-nucleating agent on the structure, morphology, and nonisothermal crystallization behavior of isotactic polypropylene. J. Appl. Polym. Sci. **111**(2), 1076–1085 (2009)
30. W. Zhang et al., Effect of β-crystals on the mechanical and electrica properties of β-nucleated isotactic polypropylene. IEEE Trans. Dielectr. Electr. Insul. **26**(3), 714–721 (2019)

High Temperature Dielectric Materials for Electrical Energy Storage

Tiandong Zhang and Qingguo Chi

Abstract Dielectric materials for electrical energy storage at elevated temperature have attracted much attention in recent years. Comparing to inorganic dielectrics, polymer-based organic dielectrics possess excellent flexibility, low cost, lightweight and higher electric breakdown strength and so on, which are ubiquitous in the fields of electrical and electronic engineering. However, the poor thermal stability and unsatisfied capacitive performance at elevated temperature severely limit their further application under extreme conditions. Seeking effective ways to reduce the electrical conduction loss are the most important key to improve the capacitive performances of dielectrics at high-temperature and high electric field. This chapter is mainly focused on the high-temperature energy storage performances of polymer-based dielectrics. The key parameters for obtaining an excellent capacitive performances should be generally considered and briefly introduced in Sect. 2. The measurements techniques for electrical energy storage are introduced in Sect. 3, including in-direct *D-E* hysteresis loop test and direct discharge test. Most importantly, an overview of recent progress on improving the energy storage performances of polymer-based dielectrics at elevated temperature have been summarized in Sect. 4. The latest meaningful findings may inspire a new wave of research and further broader applications of polymer-based dielectrics in the electrical and electronic engineering under extreme conditions.

1 Introduction

Dielectric materials have been widely used in the field of the electrical and electronic engineering, one of the most common applications is used as the core of capacitors [1–3]. Dielectric capacitors are different from that of supercapacitors and batteries due to their rapid charge and discharge rate, high open-circuit voltage, excellent temperature stability and good safety [4, 5]. They can be used as power generator,

T. Zhang · Q. Chi (✉)
Key Laboratory of Engineering Dielectrics and Its Application, Ministry of Education, Harbin University of Science and Technology, Harbin 150080, People's Republic of China
e-mail: qgchi@hotmail.com

B. Du (ed.), *Polymer Insulation Applied for HVDC Transmission*,
https://doi.org/10.1007/978-981-15-9731-2_26

conversion devices, filter capacitors, tuning capacitors, coupling capacitors and so on, which are widely applied in the field of new energy system, power grid, wind power and photovoltaic energy, electronics [6, 7].

In recent years, with the increasing demand of energy storage capacitors worked at extreme high-temperature condition, the dielectric materials, such as the polymer films, with excellent high-temperature energy storage performances are in urgent need of explorations [8]. For examples, the electronic control system of the hybrid electric vehicle (HEV) always run over a temperature of 140 °C, but the biaxially oriented polypropylenes (BOPP) as the main part of the capacitor in power inverters must be operating at a temperature lower than 100 °C. To keep the capacitor operating steadily, an additional cooling system should be involved to refrigerate the capacitor, resulting in the increased cost and volume of devices [9]. Especially, in some cases, it is really difficult to attach cooling system in oil & gas exploration equipment or in aerospace craft (maximum operation temperature >200 °C) due to the limitation of operating conditions, the only way is to develop high-temperature dielectric materials which possess good thermal stability and outstanding energy storage performances at elevated temperature [8, 10].

High-temperature dielectric materials for energy storage should possess some qualifications, such as high thermal stability, low dielectric loss and conductivity at high-temperature, excellent insulation. With the increase of temperature and applied electric field, the significant increasing conductivity of dielectric materials leads to the serious energy loss and joule heat, which remarkably reduces the energy storage performance and lifetime of capacitors. Various kinds of dielectric materials have been investigated for high-temperature energy storage capacitors, including organic polymer films and inorganic ceramics or inorganic thin films. Such as polyetherimide (PEI) [11], cross-linked divinyltetramethyldisiloxane–bis(benzocyclobutene) (c-BCB) [12], mica [13], regenerated cellulose (RC) [14], poly(4-methyl-1-pentene) (P4MP) [15], polyimide (PI) [16], poly(methyl methacrylate) (PMMA) [17], poly(arylene ether urea) (PEEU) [18], polyaramid (PA) [19], sulfonylated poly(2,6-dimethyl-1,4-phenylene oxide) (SO_2–PPO) [20], $(Pb_{0.9}La_{0.1})(Zr_{0.65}Ti_{0.35})O_3$ ceramic [21], $BaZr_{0.2}Ti_{0.8}O_3$ film [22] and so on. Many strategies have been put forward to restrain the conduction loss and improve the energy storage performance at high-temperature, such as doping inorganic nanofillers in the polymer matrix, constructing sandwich structure or multilayer thin films, growing barrier layer between electrodes and polymer films. Although the high-temperature energy storage has been optimized constantly by developing good dielectric materials, some major challenges are still existed. To meet the requirements of much higher temperature and much better energy storage performance, great efforts should be paid on the way to develop novel dielectric structures and fabrication techniques.

2 Key Parameters of High Temperature Dielectric Materials

2.1 Dielectric Constant and Dielectric Loss

It is well known that the positive relation exists between the dielectric constant (ε_r) and energy storage density (U_e) for dielectric capacitors. For example, the U_e of the linear dielectric materials can be calculated by Eq. (1)

$$U_e = 1/2\varepsilon_0\varepsilon_r E^2 \tag{1}$$

where E is the applied electric field to dielectric materials, ε_0 is the dielectric constant of vacuum. Obviously, the larger ε_r is favorable to obtain the higher U_e. About a decades ago, researchers were focused on improving the ε_r of polymer films by utilizing the percolation theory [23, 24], where composites were prepared by adding a conductive filler into an insulating polymer matrix. In recent year, another effective method to increase the ε_r polymer films is to fill high dielectric inorganic fillers to polymer matrix, a significant increase of ε_r can be induced by the interfacial polarization effect [25–27]. However, the increased dielectric loss are also easily induced by both ways, especially with a high content of inorganic fillers. The large dielectric constant mismatch of polymer and inorganic ceramic fillers, poor interfacial compatibility and the intrinsic properties of inorganic fillers attribute to the undesirable dielectric loss. High dielectric loss not only leads to reduced U_e and lowered energy storage efficiency but also generates Ohmic heat to endanger the reliability and lifetime of dielectric polymers [9].

In fact, according to the previous reports, the dielectric materials used for high-temperature energy storage have been paid much attention to entitle the dielectric constant to have a good temperature stability, rather than to improve the ε_r value. It was reported that at the frequency of interest for common power conditioning (104 Hz), c-BCB/BNNS (boron nitride nanosheets, abbreviated as BNNS) exhibits excellent dielectric stability from room temperature to 300 °C, and the temperature coefficient of ε_r for c-BCB/BNNS (65 p.p.m $°C^{-1}$) is much lower than that of FPE (308 p.p.m $°C^{-1}$) and Kapton (498 p.p.m $°C^{-1}$). More importantly, the dielectric loss of c-BCB/BNNS just increases from 0.09% to 0.13% within the measured temperature regions, however, obvious increase of dielectric loss can be found in the other polymer dielectrics [8]. Besides, comparing to pure PI films, $BaTiO_3$/PI nanocomposites possess weak dependence of dielectric loss temperature, which can be attributed to the restriction on movement of polymer molecular chains [16]. Excellent dielectric constant temperature stability and low dielectric loss at elevated temperature are very crucial to obtain the desirable high-temperature energy storage performance of dielectric materials.

2.2 Conduction and Breakdown Strength

High-temperature dielectric materials for energy storage are always subjected to high electric field and elevated temperature which easily cause the increased electric conduction loss and large leakage current. Severe conduction loss results in the unexpected energy loss and joule heat in capacitors, which has being the main bottleneck for high-temperature dielectric materials. The conduction mechanisms of dielectric materials include electrode-limited conduction mechanism (Schottky emission, Fowler-Nordheim tunneling, direct tunneling, thermionic-field emission) and bulk-limited conduction mechanism (Poole-Frenkel emission, hopping conduction, ohmic conduction, space-charge-limited conduction, ionic conduction and grain-boundary-limited conduction). Schottky emission and Poole-Frenkel emission are the most common conduction mechanisms for dielectric materials [9]. Schottky emission describes that once the electrons in metal electrodes obtain enough energy supplied by thermal excitation at elevated temperature, which will overcome the energy barrier at the interfacial regions of electrodes and dielectric materials and inject to dielectric materials. It is one of the inducements to result in degraded energy storage property at elevated temperature. Besides, Poole-Frenkel emission involves that the electrons induced by thermal excitation will emit from traps into the conduction band of the dielectric materials. Of course, even if the obtained thermal activation energy of carriers is lower than the barrier between the adjacent trap sites, the carriers can still transport in dielectric materials due to the hopping conduction mechanism, and hopping conduction mechanism have been well explained the experimental results in previous studies [8, 28]. Herein, the representative conduction mechanism, such as Schottky emission and the P-F emission, can be expressed by the following Eqs. (2) and (3), respectively [29].

$$J = \frac{4\pi q k^2 m^*}{h^3} \exp\left[\frac{-q\left(\varphi_B - \sqrt{qE/4\pi\varepsilon_r\varepsilon_0}\right)}{kT}\right] \tag{2}$$

$$J = q\mu N_C E \exp\left[\frac{-q\left(\varphi_T - \sqrt{qE/4\pi\varepsilon_r\varepsilon_0}\right)}{kT}\right] \tag{3}$$

where J is the current density, m^* is the effective electron mass in dielectric, T is the absolute temperature, q is the electronic charge, $q\varphi_B$ is the Schottky barrier height, E is the applied electric field, k is the Boltzmann's constant, h is the Planck's constant, ε_0 and ε_r are the same as defined before. μ is the electronic drift mobility, N_C is the density of states in the conduction band, $q\varphi_T$ is the trapping energy level.

It can be found that the current density increases with the rising temperature, the leakage current increases exponentially with increasing electric fields Either injection or de-trapping of electrons may induce carriers transportation and increase the conductivity, so it is necessary to reduce the leakage current (or conductivity) at elevated temperature for the energy storage dielectric materials. Some effective ways have been put forward to restrain the electrons injection from electrodes and prevent the trapping charges de-trapping from the traps, such as growing a barrier layer between electrodes and dielectric films, introducing deep traps by doping with inorganic nanoparticles, the corresponding works will be further discussed later.

The conductivity of dielectric materials is much more significant at elevated temperature, leading to the increased probability of thermal runaway and reduced the electric breakdown strength E_b of dielectric materials. The failure mechanisms for polymer dielectrics at elevated temperature are mainly thermal breakdown and electromechanical breakdown [30]. Once the accumulated heat induced by dielectric loss and joule heat is much higher than the heat dissipation, the thermal breakdown may occur. Equation (4) shows the thermal instability, the first term dT/dt corresponds to the steady state thermal breakdown, and the second item $div(\lambda \cdot gradT)$ corresponds to the impulse state thermal breakdown, remarkably, the endurable electric breakdown strength decreases significantly with the increasing temperature [31].

$$C_v = \frac{dT}{dt} - div(\lambda \cdot gradT) = \sigma E^2 \tag{4}$$

where C_v is the specific heat per unit volume and temperature, respectively. λ and σ are the thermal and electrical conductivity, respectively.

Electromechanical breakdown is usually induced by the deformation of polymer-based dielectrics, and the large mismatch of elasticity modules between inorganic additives and polymer matrix can also lead to mechanical force and accumulated charges at interfacial regions. The E_b can be described quantitatively by Eq. (5) [9].

$$E_b = 0.606\left(\frac{Y}{\varepsilon_0 \varepsilon_r}\right)^{0.5} \tag{5}$$

where Y the Young's modulus. Generally, Y of the polymer dielectrics gradually decreases with the increasing temperature, especially near the temperature of glass transition temperature (T_g) or melting point (T_m), resulting in the occurrence of the electromechanical breakdown. It is necessary to point out that the electric breakdown behavior may also contribute to free volume breakdown strength, the thermal excursions near T_g dramatically increase the free volume which accelerates the mobility of carriers in the polymer dielectrics. Although many methods have been proposed to improve the breakdown strength of polymer-based dielectrics, such as tailoring the interface of inorganic fillers and organic matrix, modulating the shape, distribution, and orientation of the inorganic fillers, or introducing hierarchical structure, and so on. Besides, to enhance the structure stability of polymer which possesses high T_g or T_m may be the most effective way to remain the desirable E_b at elevated temperature.

2.3 Thermal Parameter

T_g is one of the most key parameters to evaluate the thermal stability of polymer dielectrics, and T_m can be also used to assess the temperature capability of the polymer with high crystalline. As aforementioned, dielectric property of polymer-based dielectrics is sensitive to the structure stability at elevated temperature. Once the polymer is exposed to harsh thermal condition, the large dielectric loss may be induced by thermal degradation. Even worse, the thermal degradation can enable carriers to move more easily due to the structure instability, such as the scission of molecular chains and the plasticizing of deliberated small molecules. The mobile carriers further accelerate the degradation of polymer and their dielectric property, leading to the serious deteriorate of long-term working stability of capacitors. To enhance the thermal stability (high T_g or T_m) of polymer dielectrics is very important to guarantee the stable operation of capacitors. Several approaches have been used to improve the thermal stability. Adding antioxidants, such as radical scavengers or peroxide scavengers, can improve the oxidative reliability. Incorporating strong intermolecular interactions in polymer chains which are good for keeping the thermal stability of structure at high-temperature [7]. Crosslinking can limit the mobility of molecular chains and improve the thermal stability, further reduce the dielectric loss due to the steric effect for carriers [32].

It is inevitable to produce joule heat in dielectrics at high electric field and elevated temperature, and the energy storage property will be degraded when the joule heat accumulates seriously. Usually, the polymer dielectrics always display a ultra-low thermal conductivity, leading to the accumulation of joule heat. However, if the dielectric possesses outstanding thermal conductivity, the joule heat can be dissipated efficiently from polymer dielectrics to external heat-transfer medium, this is in favor of keeping the steady operation of capacitors. After incorporating BNNS which has a high thermal conductivity (300–2000 $Wm^{-1}K^{-1}$), resulting in a significant increase of thermal conductivity of polymer-based composites and an uniform distribution of the internal temperature for the polymer capacitors [8, 33].

3 Measurement of Energy Storage Property

3.1 D-E Hysteresis Loop Test

To quantitatively evaluate the energy storage property, such as energy density and efficiency, the mainstream way is based on the ferroelectric test, where the electrical displacement induced by an electric field (*D-E* hysteresis loop) should be tested and the energy storage property can be calculated by Eqs. (6)–(7). η is the energy storage efficiency, U_e is the same as defined in Eq. (1), U_{total} is the total charged energy. As shown in Fig. 1a, the recoverable energy density is determined from the area in yellow during the discharging process, and the red area represents the energy loss.

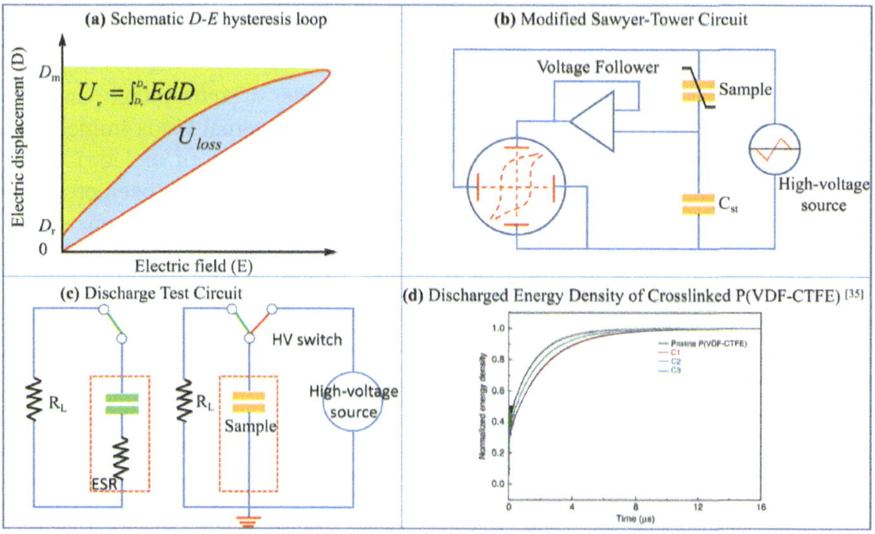

Fig. 1 Measurement methods for energy storage performance of dielectrics

$$U_e = \int_{D_r}^{D_m} EdD \tag{6}$$

$$\eta = \frac{U_e}{U_{\text{total}}} \times 100\% \tag{7}$$

It is a mature technology for that ferroelectric hysteresis loops, the circuit used for *D-E* loop measurement was the traditional Sawyer-Tower circuit or modified Sawyer-Tower circuit [34]. The schematic diagram of the modified Sawyer-Tower circuit is given in Fig. 1b, when the dielectric is subjected to a triangular unipolar electric field, an unipolar *D-E* loop can be obtained. U_e and η can be calculated according to Eqs. (6) and (7). Of course, a heating furnace is essential to measure the *D-E* loop at different temperatures and to analyze the energy storage property of dielectric materials at high-temperature. According to our experience, due to the existence of polarization relaxation and de-polarization hysteresis effect of some kinds of dielectrics, the frequency of applied electric field exhibits a deep influence on the polarization behaviors. It should be also noted that the cyclic loading electric field and the increase of measured temperature are easy to degrade the fatigue resistance of dielectrics, leading to further increased dielectric loss.

3.2 Discharge Test

In actual applications, the capacitors are charged under a high voltage, and then the charged electrical energy will be discharged to an external loads immediately [35, 36]. In order to characterize the discharged performance of dielectric materials or capacitors under real-world applications, a direct method has been proposed. The circuit is given in Fig. 1c, containing a high voltage generator, a high voltage MOSFET switch, a resistor R_L as load, electrical interconnections, sample as capacitor. Figure 1d shows the discharged performance of cross-linked P(VDF-CTFE), where C1, C2 and C3 represent the films containing 1, 3 and 5 wt% 1,4-bis(t-butyl peroxy) diisopropylbenzene initiator and 3 wt% triallyl isocyanurate cross-linking agent. Herein, a resistor R_L of 20 kΩ was used as a load, and the discharge time was defined as the time for the discharge energy in a load resistor to reach 95% of the final value, for example, the discharge time for C2 sample was 4.8 μs [35].

Herein, R_L can be flexibly changed so that the discharging behavior can be measured over a broad time domain. It is inevitable for the existence of dielectric loss which be equivalent to series resistance (ESR), so that the total charged energy can not be released in full to R_L. But if the value of $R_L >> ESR$, U_e is similar to U_{total}, this is an ideal situation. Actually, a portion of U_{total} will be dissipated at *ESR*, and U_e and *ESR* can be expressed by the following Eqs. (8) and (9).

$$U_e = U_{\text{total}} \frac{R_L}{R_L + ESR} \quad (8)$$

$$\text{ESR} = \frac{\tan\delta}{\omega C} \quad (9)$$

where C is the capacitance of sample, ω is the angular frequency, and tanδ is the dielectric loss of the sample. It can be found that (1) The larger R_L is, the more electrical energy can be discharged, resulting in high charge-discharge efficiency. (2) At the initial time (high frequency) of discharging process, the effect of *ESR* on the discharging behavior can be ignored, a large amount of U_{total} will be released immediately. But at the end of discharge process (lower frequency), *ESR* has a negative effect on discharging electrical energy. (3) R_L also has a deep influence on the discharge time of capacitors, generally, the lower R_L shortens the discharge time.

4 Energy Storage of Polymer-Based Dielectrics

4.1 Polymer-Based Nanocomposites with 0D Fillers

Polymer-based nanocomposites have attracted so much attention in recent year due to amount of inorganic nanofillers (described as 0D) can induce large interfacial regions and substantial effects. Even though many kinds of inorganic nanofillers, such as conductive nanofillers, semi-conductive nanofillers and insulating ferroelectric nanofillers with high dielectric constant, are choose to improve the energy storage performance of polymer films, to enhance the polarization and breakdown strength simultaneously is still a bottleneck problem. Taking ferroelectric nanofillers as an example, large mismatch of dielectric constant between fillers and polymer always lead to serious local electric field concentration at interfacial regions, reducing the breakdown strength of composites. Besides, high filling content of inorganic fillers easily introduces interfacial structure defects due to the poor compatibility of nanofillers and polymer. For the sake of improving the energy storage performance at elevated temperature, it may be more important to reduce conduction loss than that to blindly pursue high dielectric constant of dielectric materials.

Energy storage performances of representative polymer-based nanocomposites with 0D nanofibers at elevated temperature are given in Table 1. In 2019, Zhang et al. reported that the semi-crystalline polymer poly(arylene ether urea) (PEEU) filled with 0.21 vol.% Al_2O_3 nanoparticles (diameter~20 nm) exhibited a high U_e of 5 J/cm^3 and η of 90% at 150 °C. Such ultra-low content of Al_2O_3 weakens the hydrogen bonding and expands the interchain spacing, simultaneously, the deep traps are introduced to reduce the conduction loss at high-temperature, both the breakdown strength and dielectric constant are improved [18]. It was also reported that by filling 0.5 vol.% Al_2O_3 nanoparticles into the semi-crystalline polymer TFE/HFP/VDF

Table 1 Energy storage performances of representative polymer-based nanocomposites with 0D nanofibers

Polymer-based nanocomposites with 0D fillers	Room temperature 25 °C			High temperature 150 °C			References
	E MV/m	U_e J/cm^3	η %	E MV/m	U_e J/cm^3	η %	
PEEU	600	8.2	90	200	0.83	90	[18]
0.21vol.% Al_2O_3/PEEU	900	27	90	400	5	90	
PAEK	400	2.5	~95	150	0.36	-	
0.35vol.% Al_2O_3/PAEK	500	4.4	~85	250	1.3	~85	
PEI	400	2.2	~95	275	1.01	~90	
0.32 vol.% Al_2O_3/PEI	400	3.2	>95	250	1.25	~85	
PI	450	~5.2	86.7	300	~1.5	~40	[16]
3 vol.% $BaTiO_3$/PI	275	1.7	~52	200	~0.6	~42	
PI	-	-	-	250	0.82	55.7	[39]
7 vol.% Al_2O_3/PI	-	-	-	250	1.12	93.7	
5 vol.% HfO_2/PI	-	-	-	250	1.21	91	
1 vol.% TiO_2/PI	-	-	-	250	1.11	81.8	
7.5vol.%Al_2O_3/c-BCB	-	-	-	250	0.89	78.9	[12]

(THV), a significant reduced conduction loss at high-temperature was obtained, which can be attributed to the increase of traps depth caused by filling Al_2O_3 nanoparticles [37]. Similarly, the discharged density of amorphous polyetherimide (PEI) filled with 0.32 vol.% Al_2O_3 nanoparticles is 1.12 times larger than that of pure PEI films at 150 °C [38]. PI-based composites filled with Al_2O_3, HfO_2 and TiO_2 nanoparticles with different bandgap exhibited substantial improvement in energy storage performances at high-temperature [39]. However, when ferroelectric $BaTiO_3$ nanoparticles (100 nm) were incorporated into polyimide (PI), ε_r gradually increased with rising filling content, but the nanocomposites possessed a sever degradation of energy density due to the compromised E_b at elevated temperature [16]. According to the previous reports, three key factors should be considered in improving the energy storage performance of polymer-based nanocomposites at elevated temperature, the first one is to choose the inorganic nanofillers with a moderate ε_r which is close to that of polymer matrix; the second one is that the nanofillers entitled with large bandgap may be better; the last one is to load nanofillers with a lower content. ε_r can be enhanced due to the interfacial polarization, more importantly, the reduced conduction loss and enhanced E_b can be also achieved by introducing deep traps or increasing traps depth.

4.2 *Polymer-Based Nanocomposites with 1D Fillers*

One dimensional (1D) inorganic nanofibers possess high aspect ratio and anisotropy characteristics, which have been widely employed to improve the energy storage performance of polymer dielectrics. After adjusting the distribution pattern (parallel or perpendicular to the direction of applied electric field) or changing the filling content of 1D fillers in polymer films, the energy storage can be optimized. However, most of the studies are focused on improving energy storage performances at room temperature. For example, the $PbZrTiO_3$ (PZT), TiO_2 wrapped PZT (TiO_2@PZT) or $BaTiO_3$ nanowires were tuned to align in the direction of applied electric field and enabled the nanocomposites to possess a large polarization and excellent energy storage density at a low electric field [40–42]. Zhang reported that 1D nanofibers aligned in the direction which is perpendicular to electric field, the breakdown strength of nanocomposites can be enhanced remarkably due to the hindered the growth of the breakdown path [43]. Furthermore, the orthogonal configurations by controlling the orientation of 1D nanofibers was proposed to resolve the paradox between ε_r and E_b, $U_e = 25.5$ J/cm^3 and $\eta = 76.3\%$ were obtained for $BaTiO_3$PVDF-HFP matrix at room temperature under an electric field of 690 kV/mm [44, 45]. The above results show that by filling or adjusting the distribution of 1D nanofibers can be used to significantly optimize the energy storage performance of composites at room temperature, contributing to high aspect ratio of 1D nanofillers, but few studies are focused on improving the capacity at elevated temperature. It was reported that 1 vol.% BZT-BCT nanofibers incorporated into PI matrix, U_e decreased from 2.2 J/cm^3 to 1.3 J/cm^3 from 25 to 150 °C, as shown in Fig. 2 [46].

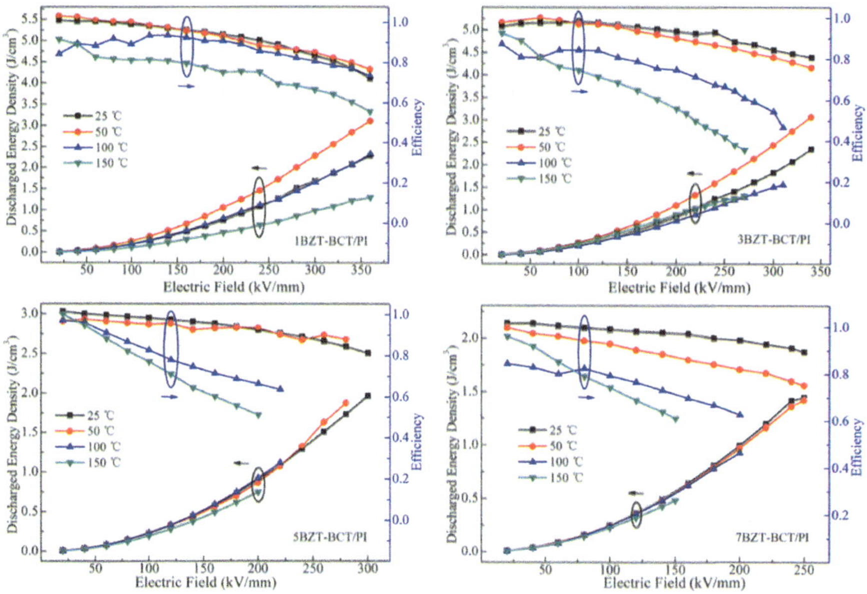

Fig. 2 Energy storage performances of PI-based nanocomposites with 1D nanofibers. Reprinted from permission from Ref. [46]. Copyright ® 2019, American Chemical Society

Similarly, 1 vol.% $BaTiO_3$ nanofibers were filled into PI matrix, the nanocomposites exhibited a decreased leakage current density and increased thermal conductivity, U_e of 2.1 J/cm^3 and 4 J/cm^3 were obtained at 150 °C and 100 °C with η larger than 90%, respectively [47]. The phase-field simulation indicated that comparing with that of random 0D nanoparticles and 2D nanosheet, E_b of nanocomposites with 1D nanofibers in In-plane distribution displayed a much more serious decrease [48, 49]. Although the polarization can be effectively improved by utilizing high aspect ratio, the undesirable local electric field concentration may be detrimental to E_b, this may be one of the most key reasons for that few studies focused on improving the high-temperature energy storage performance of dielectrics by using 1D nanofibers. Table 2 summarizes the energy storage performances of representative polymer-based nanocomposites with 1D nanofibers, it can also clearly see that the capacitive performances at elevated temperature of nanocomposites with 1D fillers are few reported in comparison to that at room temperature.

4.3 Polymer-Based Nanocomposites with 2D Fillers

Two dimensional (2D) inorganic nanosheets, such as BN or Al_2O_3 nanosheets have been used to improve the capacitive performances of dielectrics. BN nanosheets with

Table 2 Energy storage performances of representative polymer-based nanocomposites with 1D nanofibers

Polymer-based nanocomposites with 1D fillers	Room temperature 25 °C			High temperature 150 °C			References
	E MV/m	U_e J/cm^3	η %	E MV/m	U_e J/cm^3	η %	
TiO_2@PZT/ P(VDF-TrFE-CTFE)	143	6.9	~50	-	-	-	[41]
3 vol.% $BaTiO_3$/ (PVDF-CTFE)	240	10.8	61.4	-	-	-	[42]
3 vol.% aligned BZCT@SiO_2/PVDF	560	18.9	53.3	-	-	-	[43]
3 vol.% BZCT @Al_2O_3@SiO_2/PVDF	440	20.1	58.6	-	-	-	[44]
$BaTiO_3$/PVDF-HFP	690	25.5	76.3	-	-	-	[45]
7.5 vol.% Al_2O_3/c-BCB	-	-	-	250	0.98	88.73	[12]
BCZT/PI	360	2.2	71.7	360	1.3	56	[46]
$BaTiO_3$/PI	400	~4.1	>90	300	2.1	90	[47]

a moderate ε_r(3~5), high thermal conductivity (in-plane~600 W·m^{-1}·K^{-1} and out of-plane~30 W·m^{-1}·K^{-1}) and excellent insulation (bandgap~5.2 eV) [50]. The excellent insulation and thermal conductivity of inorganic 2D nanosheets not only enhance the breakdown strength of nanocomposites, but also improve the heat dissipation, and the electrical conductivity can be significantly suppressed due to the formation of deep traps and the charge-blocking barrier, the phase-field model also proves that the breakdown path and breakdown strength are significantly dependent on the microstructure of the nanocomposite, by incorporating with 2D nanosheets is beneficial to markedly enhance E_b and capacitive performance [48].

Although the preparation for BN nanosheets, such as liquid-phase exfoliation method, is time assuming and low-yeild, BN nanosheets is still the most popular and widely used. The electromechanical breakdown behavior has been regarded as the main electric breakdown mechanism in PVDF-based dielectrics, after filling with 12 wt% BN nanosheets into P(VDF-TrFECFE), both of the Young's modulus and thermal conductivity were enhanced, leading to a significant improvements on E_b(610 MV/m) which was 1.7 times higher than pristine terpolymer at room temperature [33]. Similarly, Liu et al. reported that by filling 12 wt% BN nanosheets endowed BN/PMMA nanocomposites to possess a U_e of 0.866 J/cm^3 and η of 97% at 70 °C, however, both of capacitive performances and served temperature can not meet the high-temperature requirements [17]. The thermal stability (T_g >350 °C) of thermal crosslinked c-BCB outperforms many other polymer dielectrics. The values of U_e(2.2 J/cm^3) and η(>90%) under 400 MV/m and 150 °C had obtained in c-BCB/BNNS nanocomposites, the capacitive performances measured at 200 MV/m and 150 °C were much superior to BOPP at 70 °C. Even at 200 °C, c-BCB/BNNS still delivered 2 J/cm^3, which was two times higher than that of pristine PEI, and leakage current was one order of magnitude lower than PEI. These results demonstrated that by incorporating BN nanosheets into the polymer matrix can significantly hinder conduction loss and thus improve energy storage performance [8].

In the consideration of the limited production of BN nanosheets by liquid-phase exfoliation method, Li et al. successfully prepared Al_2O_3 nanosheets through hydrothermal synthesis method. Comparing to 0D and 1D Al_2O_3, by filling with 2D Al_2O_3 nanosheets into c-BCB enabled nanocomposites to have a homogeneous distribution of local electric fields, and the local electric fields were serious concentration at interfacial regions for 0D Al_2O_3/c-BCB and 1D Al_2O_3/c-BCB nanocomposites. E_b of 7.5 vol.% 2D Al_2O_3/c-BCB was 489 MV/m, which was even greater than that of the high-temperature c-BCB/BNNS nanocomposites (421 MV/m). Most importantly, the reported excellent capacitive performance, U_e of 3.31 J/cm^3 and η of >90% measured at 450 MV/m and 150 °C, that exceeds all the capacitive performance of the current polymer dielectrics and nanocomposites at elevated temperatures [12]. Recently, a novel method was proposed to dramatically enhance the breakdown strength of PVDF-based nanocomposites. Different from 2D BN or 2D Al_2O_3 nanosheets, $Ca_2Nb_3O_{10}$ (CNO) nanosheets were considerable negative charges on the surface due to their intrinsic stoichiometry, which generated a local electric field to slow down or even block the energetic electrons moving towards 2D CNO nanosheets and significantly increase E_b(792 MV/m) of nanocomposite and enhance U_e(36.2 J/cm^3) accordingly [51]. But the high-temperature energy storage performances were not mentioned in this work, this method has great potential to be used for improving energy storage at elevated temperature in future. Table 3 shows energy storage performances for representative polymer-based nanocomposites with 2D nanofibers.

Table 3 Energy storage performances of representative polymer-based nanocomposites with 2D nanofibers

Polymer-based nanocomposites with 2D fillers	Room temperature			High temperature			References
	E MV/m	U_e J/cm^3	η %	E MV/m	U_e J/cm^3	η %	
	25 °C			150 °C			
7.5 vol.% Al_2O_3/c-BCB	-	-	-	450	3.31	>90	[12]
10 vol.% BNNS/C-BCB	-	-	-	400	2.2	>90	[8]
5 vol.% BNNS/PI	-	-	-	250	1.08	91.8	[39]
12 wt% BNNS/PMMA	-	-	-	70 °C			
				200	0.866	97	[17]
12 wt% BNNS/ P(VDF-TrFE-CFE)	650	20.3	>75	-	-	-	[33]
0.9 wt% $Ca_2Nb_3O_{10}$/ PVDF	792	36.2	~60	-	-	-	[51]

4.4 Polymer-Based Inhomogeneous Dielectrics

Previous reports have shown that E_b and ε_r are difficult to improve simultaneously in a single-phase dielectrics, especially at elevated temperature. To construct the artificial inhomegeneous dielectrics has been an effective way to improve the electrical energy storage performances, which aims to adopt the advantages and avoid the disadvantages of individual materials. Herein, we mainly focus on polymer-based inhomogeneous dielectrics, including three types: (1) compositional gradient layer-structured polymer-based nanocomposites; (2) sandwiched-structured polymer-based nanocomposites with high ε_r layer and high E_b layer; (3) inorganic barrier-layer or block-layer in polymer dielectrics.

The compositional gradient layer-structured nanocomposites exhibited ultra-high E_b. Shen et al. reported that the gradient distribution of $BaTiO_3$ nanoparticles in P(VDF-HFP) matrix has been successfully prepared by combing electrospinning technique with hot pressing, and the content of fillers gradually increased from the outer surface layer to the center core layer. The simulation results indicated that the developments of breakdown paths of nanocomposite with randomly distributed $BaTiO_3$ nanoparticles was much faster than that of gradient structured nanocomposites. Of particular, E_b continuously increased with the increase of grade number, the maximum E_b of 676.3 kV/mm was obtained for the gradient structured nanocomposites with 7 grades, which was ascribed to the improved mechanical properties along the out-of-plane direction.U_e of 18 J/cm^3 and η of ~65% were obtained at room temperature [52]. The same research group reported that another mechanism to enhance the capcacitive performance is that the ions trapped at the interfaces in gradient structured nanocomposites lead to form a strong depolarization electric field, which can hamper the mobility of carriers along the direction of the applied electric field, meanwhile, the outer layers with high E_b layer as can restrained charge injection from electrodes, suppresses electric conduction. The synergistically effects contribute to enhanced apparent breakdown strength [53, 54]. However, although the artificial gradient nanocomposites display outstanding energy storage performances, the capacitive performances at elevated temperature have not been mentioned for the compositional gradient structured nanocomposites.

The sandwiched-structured nanocomposites have attracted much attention in recent years. In contrast to compositional gradient structured nanocomposites, the fabrication process and configuration seem to be simplified for the sandwiched-structured nanocomposites which mainly contains a middle layer with high ε_r and outer layers with high E_b, in the majority situation, two kinds of fillers are employed in different layers, respectively [55, 56]. Li et al. prepared the sandwiched-structured nanocomposites (abbreviated as SSN) by sequentially casting methods, including a c-BCB/BNNS outer layers, a c-BCB/$BaTiO_3$ middle layer. After optimizing the filling content of BN nanosheets (10 vol.%) and $BaTiO_3$ nanoparticles (25 vol.%), the sandwiched-structured nanocomposites described as SSN-25 displayed a U_e of 1.1 J/cm^3 and η of 93% at 150 °C and 200 MV/m, the capacitive performance of SSN-25 at elevated temperature was superior to all of the single-layer nanocomposites as well as other representative high-temperature dielectrics. More importantly, the simulation results indicated that the leakage current density of SSN-25 at 50 MV/m and 150 °C was similar to that of tetralayered polymer nanocomposites, meaning that more sophisticated multilayered structures may not be beneficial to further reduce the leakage current for dielectrics operating at elevated temperature [55]. The similar sandwiched-structured PI-based nanocomposites, BN/PI served as outer layers and BZT-BCT/PI acted as central layer, was constructed and U_e of 1.83 J/cm^3 and η of ~68% were achieved at 150 °C, as shown in Fig. 3 [46].

In order to reduce the charge injection from electrodes and prevent the de-trapping charge carriers from moving in the polymer at elevated temperature, the inorganic barrier-layer or block-layer with excellent insulation, such as SiO_2 or BN, has been introduced into the polymer nanaocomposites to enhance the capacitive performances at high-temperature. Hexagonal boron nitride (h-BN) films was prepared by chemical vapor deposition method on the surface of Cu foil, and then transferred onto PEI surfaces, leading to an obvious reduction of dielectric loss from 74% of pristine PEI to 7% of 19-layer h-BN/PEI composites at 200 °Cand 300 MV/m. A significant enhancement of U_e(1.2 J/cm^3) and η(>90%) were obtained in PEI coated with h-BN composites even at 200 °C. The interfacial potential barrier increased from 2.93 eV of PEI/Au to 5.1 eV of PEI/h-BN/Au, meaning that the charges were more difficult to

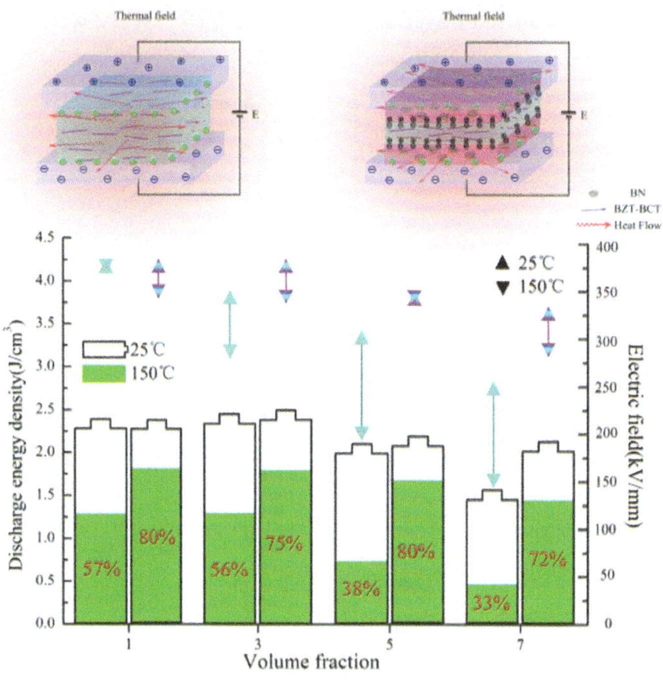

Fig. 3 Energy storage performances of PI-based nanocomposites with 1D nanofibers. Reprinted from permission from Ref. [46]. Copyright ® 2019 American Chemical Society

be injected from metal electrodes [11]. Liu et al. proposed that the inorganic barrier layer of h-BN was prepared by electrospinning technique, polycarbonate (PC) films coated with h-BN possessed a U_e of 5.52 J/cm^3 and η(87.25%) at 100 °C and 500 MV/m, and U_e of 3.39 J/cm^3 and η(78.59%) at 150 °C and 400 MV/m, respectively. The optimized mechanism was that synergistic effects of the reduced charge injection and the enhanced thermal conductivity in the in-plane direction, preventing thermal runaway and improving the capacitive performance stability at elevated temperature, as shown in Fig. 4 [56].

In the consideration of the low controllability of transfer process and low yield of h-BN by electrospinning, a scalable and high-throughput approach, plasma-enhanced chemical vapor deposition (PECVD), was proposed to fabricate SiO_2 barrier layer on the surfaces of polymer films. The optimized coating thickness of SiO_2 was180 nm, which endowed the polymer to have the optimal electrical resistivity and breakdown strength. Taking BOPP as example, η of BOPP fell below 90% at 110 MV/m, conversely, η of BOPP coated with SiO_2 barrier layer kept above 90% even at 400 MV/m. In order to verify the generality of coating SiO_2 barrier layer on improving the capacitive performance at elevated temperature, the SiO_2-coated PEI, PC and other representative dielectric films were also investigated, showing that the energy storage performances at elevated temperatures were significantly enhanced in comparison to that of pristine polymer. The outstanding performance should be attributed to the increased potential barrier at the dielectric/electrode interfaces [57]. More interestingly, BNNSs as an interlayer between two PVDF layers were fabricated by

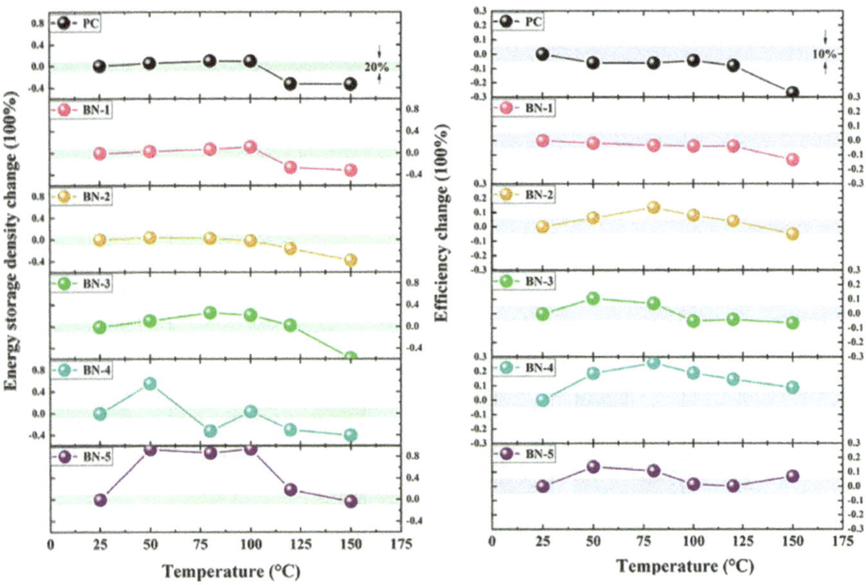

Fig. 4 Energy storage performances of PC-based inhomogeneous dielectrics with h-BN layer. Reprinted from permission from Ref. [56]. Copyright ® 2020 Elsevier

simple casting method. Different from the mechanism of limited charge injection from electrodes, the improved capacitive performance by introducing BNNSs interlayer can be attributed to the formation of topological barrier layer and scatting centers inside of polymer, which can impede the transportation of charge carriers and suppress the overall leakage current. A superior U_e of 14.3 J/cm^3 and η of 73% were obtained at 612 MV/m and room temperature [58]. Although the capacitive performances at elevated temperature have not been mentioned in this work, we still highly look forward to confirm the generality of this configuration on enhancing the energy storage performance at elevated temperature in near future. Energy storage performances of representative polymer-based inhomogeneous dielectrics are listed in Table 4.

5 Conclusion

In summary, high-temperature dielectric materials for electrical energy storage should be endowed with good thermal stability, low electrical conduction loss, excellent electrical insulation. According to the previous studies, it can be found that the degraded energy storage performances of polymer-based dielectrics are mainly induced by insufficient heat tolerance and serious conduction loss at high-temperature and high electric field. Especially, the key to improve the capacitive performances at elevated temperature is to reduce conduction loss rather than increase permittivity. As yet many strategies have been proposed to improve high-temperature energy storage performances of polymer-based dielectrics, such as filling with inorganic nano-fillers with different dimension, adjusting the distribution patterns of nano-fillers in polymer matrix, or designing the inhomogeneous structure, leading to a significantly reduced electrical conduction loss under high-temperature and high electric field due to introducing deep traps, increasing potential barrier between electrodes and polymer interfaces or blocking transmission path of charge carriers. Linear dielectrics may be more appropriate to be high-temperature energy storage materials than non-linear ferroelectric dielectrics. In the future, much more effort should be made to develop more efficient manufacturing techniques with high productivity, low cost, and to put forward advanced measurement methods for evaluating the capacitive performances of dielectrics. It is also highly expected to synthesize new polymer dielectrics with outstanding electrical, mechanical and thermal performances for high-temperature practical applications.

Table 4 Energy storage performances of representative polymer-based inhomogeneous dielectrics

Polymer-based nanocomposites with multilayers	Room temperature 25 °C			High temperature 150 °C			References
	E MV/m	U_e J/cm^3	η %	E MV/m	U_e J/cm^3	η %	
5 vol.% of $BaTiO_3$/ P(VDF-HFP) with 7 grades	670	18	65	–	–	–	[52]
P(VDF-HFP)/P(VDF-HFP)-$BaTiO_3$ with 16 layers	800	30.15	78	–	–	–	[53]
BZT/BNNS P(VDF-HFP) with gradient structure	678	23.4	83	–	–	–	[54]
PVDF/BNNS/PVDF with sandwiched structure	612	14.3	73	–	–	–	[58]
c-BCB/BNNS/BT with sandwiched structure (SSN-25)	–	–	–	200	1.1	93	[55]
h-BN/PC/h-BN with sandwiched structure	550	4.96	90.66	400	3.39	78.59	[56]
h-BN coated PEI	–	–	–	–	~2.1	~90	[11]
SiO_2 coated PEI	–	–	–	400	2.12	>90	[57]
SiO_2 coated PEN	–	–	–	350	1.75	>90	
SiO_2 coated PI	–	–	–	300	1.24	>90	
SiO_2 coated PC	–	–	–	400	1.79	>90	
SiO_2 coated FPE	–	–	–	400	2.06	>90	

References

1. Prateek, V.K. Thakur, R.K. Gupta. recent progress on ferroelectric polymer-based nanocomposites for high energy density capacitors: synthesis, dielectric properties, and future aspects. Chem. Rev. **116**(7), 4260–4317 (2016)
2. A. Chauhan, S. Patel, R. Vaish, C.R. Bowen, Anti-ferroelectric ceramics for high energy density capacitors. Materials **8**(12), 8009–8031 (2015)
3. J.L. Pan, K. Li, J.J. Li, T. Hsu, Q. Wang, Dielectric characteristics of poly(ether ketone ketone) for high temperature capacitive energy storage. Appl. Phys. Lett. **95**(2), 022902 (2009)
4. M.O. Aboelfotoh, C. Feger, Frequency dependence of dielectric loss in thin aromatic polyimide films. Phys. Rev. B **47**(20), 13395–13400 (1993)
5. L. Cheng, W.F. Liu, X.W. Liu, C.M. Liu, S.T. Li, Z.L. Xing, Online degradation of biaxial-orientated polypropylene film from HVDC filter capacitors. IEEE Trans. Dielectr. Electr. Insul. **26**(1), 26–33 (2019)
6. M. Ritamäki, I. Rytöluoto, K. Lahti, Performance metrics for a Modern BOPP capacitor film. IEEE Trans. Dielectr. Electr. Insul. **26**(4), 1229–1237 (2019)
7. B.Y. Fan, F.H. Liu, G. Yang, H. Li, G.Z. Zhang, S.L. Jiang, Q. Wang, Dielectric materials for high-temperature capacitors. IET Nanodielectrics **1**(1), 32–40 (2018)
8. Q. Li, L. Chen, M.R. Gadinski, S.H. Zhang, G.Z. Zhang, H.Y. Li, A. Haque, L.Q. Chen, T.N. Q. Jackson. Wang. Flexible High-Temperature Dielectric Materials from Polymer Nanocomposites. Nature **523**(7562), 576–579 (2015)
9. Q. Li, F.Z. Yao, Y. Liu, G.Z. Zhang, H. Wang, Q. Wang, High-temperature dielectric materials for electrical energy storage. Annu. Rev. Mater. Sci. **48**(1), 219–243 (2018)
10. Z.B. Zhang, M.H. Litt, L. Zhu, Understanding the paraelectric double hysteresis loop behavior in mesomorphic even-numbered nylons at high temperatures. Macromolecules **50**(15), 5816–5829 (2017)
11. A. Azizi, M.R. Gadinski, Q. Li, M.A. Alsaud, J.J. Wang, Y. Wang, B. Wang, F.H. Liu, L.Q. Chen, N. Alem, Q. Wang, High-performance polymers sandwiched with chemical vapor deposited hexagonal boron nitrides as scalable high-temperature dielectric materials. Adv. Mater. **29**(35), 1701864 (2017)
12. H. Li, D. Ai, L.L. Ren, B. Yao, Z.B. Han, Z.H. Shen, J.J. Wang, L.Q. Chen, Q. Wang, Scalable polymer nanocomposites with record high-temperature capacitive performance enabled by rationally designed nanostructured inorganic fillers. Adv. Mater. **31**(23), 1900875 (2019)
13. X.W. Xu, W.L. Liu, Y. Li, Y.F. Wang, Q.B. Yuan, J. Chen, R. Ma, F. Xiang, H. Wang, Flexible mica films for high-temperature energy storage. J. Materiomics **4**(3), 173–178 (2018)
14. J.P. Lao, H.A. Xie, Z.Q. Shi, G. Li, B. Li, G.H. Hu, Q.L. Yang, C.X. Xiong, Flexible regenerated cellulose/boron nitride nanosheet high-temperature dielectric nanocomposite films with high energy density and breakdown strength. ACS Sustain. Chem. Eng. **6**(5), 7151–7158 (2018)
15. S. Gupta, I. Offenbach, J. Ronzello, Y. Cao, S.A. Boggs, R.A. Weiss, M. Cakmak. Evaluation of Poly(4-methyl-1-pentene) as a Dielectric Capacitor Film for High-Temperature Energy Storage Applications. J. Polym. Sci., Part B: Polym. Phys. **55**(20), 1497–1515 (2017)
16. W.D. Sun, X.J. Lu, J.Y. Jiang, X. Zhang, P.H. Hu, M. Li, Y.H. Lin, C.W. Nan, Y. Shen, Dielectric and energy storage performances of Polyimide/$BaTiO_3$ nanocomposites at elevated temperatures. J. Appl. Phys. **121**(24), 244101 (2017)
17. F.H. Liu, Q. Li, Z.Y. Li, Y. Liu, L.J. Dong, C.X. Xiong, Q. Wang, Poly(methylmethacrylate)/Boron nitride nanocomposites with enhanced energy density as high temperature dielectrics. Compos. Sci. Technol. **142**, 139–144 (2017)
18. T. Zhang, X. Chen, Y. Thakur, B. Lu, Q.M. Zhang, J. Runt. A highly Scalable Dielectric Metamaterial with Superior Capacitor Performance Over a Broad Temperature. Sci. Adv. **6**(4), eaax6622 (2020)
19. D. Xu, W.H. Xu, T. Seery, H.B. Zhang, C.Y. Zhou, J.H. Pang, Y.H. Zhang, Z.H. Jiang, Rational design of soluble polyaramid for high efficiency energy storage dielectric materials at elevated temperatures. Macromol. Mater. Eng. **305**(3), 1900820 (2019)

20. Z.B. Zhang, D.H. Wang, M.H. Litt, L.S. Tan, L. Zhu, High temperature and high energy density dipolar glass polymers based on Sulfonylated Poly(2,6-dimethyl-1,4-phenylene oxide). Angew. Chem. Int. Ed. **57**(6), 1528–1531 (2018)
21. J.H. Gao, Y.B. Liu, Y. Wang, D. Wang, L.H. Zhong, X.B. Ren. High Temperature-Stability of $(Pb_{0.9}La_{0.1})(Zr_{0.65}Ti_{0.35})O_3$ Ceramic for Energy-Storage Applications at Finite Electric Field Strength. Scr. Mater. **137**, 114–118 (2017)
22. Z.X. Sun, C.R. Ma, X. Wang, M. Liu, L. Lu, M. Wu, X.J. Lou, H. Wang, C.L. Jia. Large Energy Density, Excellent Thermal Stability, and High Cycling Endurance of Lead-Free $BaZr_{0.2}Ti_{0.8}O_3$ Film Capacitors. ACS Appl. Mater. Interfaces **9**(20), 17096–17101 (2017)
23. Y. Shen, Y.H. Lin, M. Li, C.W. Nan, High dielectric performance of polymer composite films induced by a percolating interparticle barrier layer. Adv. Mater. **19**(10), 1418–1422 (2007)
24. Z.M. Dang, J.K. Yuan, J.W. Zha, T. Zhou, S.T. Li, G.H. Hu, Fundamentals, processes and applications of high-permittivity polymer-matrix composites. Prog. Mater Sci. **57**(4), 660–723 (2012)
25. Z.B. Pan, L.M. Yao, J.W. Zhai, X. Yao, H. Chen, Interfacial coupling effect in organic/inorganic nanocomposites with high energy density. Adv. Mater. **30**(17), 1705662 (2018)
26. H. Luo, X.F. Zhou, C. Ellingford, Y. Zhang, S. Chen, K.C. Zhou, D. Zhang, C.R. Bowen, C.Y. Wan, Interface design for high energy density polymer nanocomposites. Chem. Soc. Rev. **48**(16), 4424–4465 (2019)
27. M.F. Guo, J.Y. Jiang, Z.H. Shen, Y.H. Lin, C.W. Nan, Y. Shen, High-energy-density ferroelectric polymer nanocomposites for capacitive energy storage: enhanced breakdown strength and improved discharge efficiency. Mater. Today **29**, 49–67 (2019)
28. M.O. Aboelfotoh, C. Feger, A. Castellano, R. Kaufman, S. Molis, Dielectric loss in thin films of an aromatic polyimide. Appl. Phys. Lett. **62**(18), 2286–2288 (1993)
29. F. Chiu, A review on conduction mechanisms in dielectric films. Adv. Mater. Sci. Eng. **2014**, 1–18 (2014)
30. J.R. Laghari, W.J. Sarjeant, Energy-storage pulsed-power capacitor technology. IEEE Trans. Power Electron. **7**(1), 251–257 (1992)
31. X.Y. Huang, B. Sun, Y.K. Zhu, S.T. Li, P.K. Jiang, High-k polymer nanocomposites with 1D filler for dielectric and energy storage applications. Prog. Mater Sci. **100**, 187–22 (2019)
32. M.X. Yuan, G. zhang, B. Li, M. Chung, R. Rajagopalan, M.T. Lanagan. Thermally-Stable Low-Loss Polymer Dielectrics Enabled by Attaching Crosslinkable Antioxidant to Polypropylene. ACS Appl. Mater. Interfaces **12**(12), 14154–14164 (2020)
33. Q. Li, G.Z. Zhang, F.H. Liu, K. Han, M.R. Gadinski, C.X. Xiong, Q. Wang, Solution-Processed ferroelectric terpolymer nanocomposites with high breakdown strength and energy density utilizing boron nitride nanosheets. Energy Environ. Sci. **8**(3), 922–931 (2015)
34. T. Yoshimura, N. Fujimura, Polarization hysteresis loops of ferroelectric gate capacitors measured by sawyer-tower circuit. Jpn. J. Appl. Phys. **42**(9), 6011–6014 (2003)
35. P. Khanchaitit, K. Han, M.R. Gadinski, Q. Li, Q. Wang, Ferroelectric polymer networks with high energy density and improved discharged efficiency for dielectric energy storage. Nat. Commun. **4**(1), 2845 (2013)
36. B.J. Chu, X. Zhou, K.L. Ren, B. Neese, M.R. Lin, Q. Wang, F. Bauer, Q.M. Zhang, A dielectric polymer with high electric energy density and fast discharge speed. Science **313**(5785), 334–336 (2006)
37. Y. Thakur, M.H. Lean, Q.M. Zhang, Reducing conduction losses in high energy density polymer using nanocomposites. Appl. Phys. Lett. **110**(12), 122905 (2017)
38. Y. Thakur, T. Zhang, C. Iacob, T.N. Yang, J. Bernholc, L.Q. Chen, J. Runt, Q.M. Zhang, Enhancement of the dielectric response in polymer nanocomposites with low dielectric constant fillers. Nanoscale **9**(31), 10992–10997 (2017)
39. D. Ai, H. Li, Y. Zhou, L.L. Ren, Z.B. Han, B. Yao, W. Zhou, L. Zhao, J.M. Xu, Q. Wang, Tuning nanofillers in in situ prepared polyimide nanocomposites for high-temperature capacitive energy storage. Adv. Energy Mater. **10**(16), 1903881 (2020)
40. H.X. Tang, Y.R. Lin, H.A. Sodano, Enhanced energy storage in nanocomposite capacitors through aligned PZT nanowires by uniaxial strain assembly. Adv. Energy Mater. **2**(4), 469–476 (2012)

41. D. Zhang, W.W. Liu, R. Guo, K.C. Zhou, H. Luo, High discharge energy density at low electric field using an aligned titanium dioxide/lead zirconate titanate nanowire array. Adv. Sci. **5**(2), 1700512 (2018)
42. B. Xie, H.B. Zhang, Q. Zhang, J.D. Zang, C. Yang, Q.P. Wang, M.Y. Li, S.L. Jiang, Enhanced energy density of polymer nanocomposites at a low electric field through aligned $BaTiO_3$ nanowires. J. Mater. Chem. **5**(13), 6070–6078 (2017)
43. Y. Zhang, C.H. Zhang, Y. Feng, T.D. Zhang, Q.G. Chen, Q.G. Chi, L.Z. Liu, G.F. Li, Y. Cui, X. Wang, Z.M. Dang, Q.Q. Lei, Excellent energy storage performance and thermal property of polymerbased composite induced by multifunctional one-dimensional nanofibers oriented in-plane direction. Nano Energy **56**, 138–150 (2019)
44. Y. Zhang, C.H. Zhang, Y. Feng, T.D. Zhang, Q.G. Chen, Q.G. Chi, L.Z. Liu, X. Wang, Q.Q. Lei, Energy storage enhancement of P(VDF-TrFE-CFE)-based composites with double-shell structured BZCT nanofibers of parallel and orthogonal configurations. Nano Energy **66**, 104195 (2019)
45. X. Zhang, J.Y. Jiang, Z.H. Shen, Z.K. Dan, M. Li, Y.H. Lin, C.W. Nan, L.Q. Chen, Y. Shen, Polymer nanocomposites with ultrahigh energy density and high discharge efficiency by modulating their nanostructures in three dimensions. Adv. Mater. **30**(16), 1707269 (2018)
46. Q.G. Chi, Z.Y. Gao, T.D. Zhang, C.H. Zhang, Y. Zhang, Q.G. Chen, X. Wang, Q.Q. Lei, Excellent energy storage properties with high-temperature stability in sandwich-structured polyimide-based composite films. ACS Sustain. Chem. Eng. **7**(1), 748–757 (2019)
47. P.H. Hu, W.D. Sun, M.Z. Fan, J.F. Qian, J.Y. Jiang, Z.K. Dan, Y.H. Lin, C.W. Nan, M. Li, Y. Shen, Large energy density at high-temperature and excellent thermal stability in polyimide nanocomposite contained with small loading of $BaTiO_3$ nanofibers. Appl. Surf. Sci. **458**, 743–750 (2018)
48. Z.H. Shen, J.J. Wang, J.Y. Jiang, Y.H. Lin, C.W. Nan, L.Q. Chen, Y. Shen, Phase-field model of electrothermal breakdown in flexible high-temperature nanocomposites under extreme conditions. Adv. Energy Mater. **8**(20), 1800509 (2018)
49. Z.H. Shen, J.J. Wang, Y.H. Lin, C.W. Nan, L.Q. Chen, Y. Shen, High-throughput phase-field design of high-energy-density polymer nanocomposites. Adv. Mater. **30**(2), 1704380 (2018)
50. S.B. Luo, J.Y. Yu, S.H. Yu, R. Sun, L.Q. Cao, W. Liao, C.P. Wong, Significantly enhanced electrostatic energy storage performance of flexible polymer composites by introducing highly insulating-ferroelectric microhybrids as fillers. Adv. Energy Mater. **9**(5), 1803204 (2019)
51. Z.W. Bao, C.M. Hou, Z.H. Shen, H.Y. Sun, G.Q. Zhang, Z. Luo, Z.Z. Dai, C.M. Wang, X.W. Chen, L.B. Li, Y.W. Yin, Y. Shen, X.G. Li. Negatively Charged Nanosheets Significantly Enhance the Energy-Storage Capability of Polymer-Based Nanocomposites. Adv. Mater. 1907227 (2020)
52. Y.D. Jiang, X. Zhang, Z.H. Shen, X.H. Li, J.J. Yan, B.W. Li, C.W. Nan, Ultrahigh breakdown strength and improved energy density of polymer nanocomposites with gradient distribution of ceramic nanoparticles. Adv. Funct. Mater. **30**(4), 1906112 (2020)
53. J.Y. Jiang, Z.H. Shen, J.F. Qian, Z.K. Dan, M.F. Guo, Y. He, Y.H. Lin, C.W. Nan, L.Q. Chen, Y. Shen, Synergy of Micro-/Mesoscopic interfaces in multilayered polymer nanocomposites induces ultrahigh energy density for capacitive energy storage. Nano Energy **62**, 220–229 (2019)
54. J.Y. Jiang, Z.H. Shen, X.K. Cai, J.F. Qian, Z.K. Dan, Y.H. Lin, B.L. Liu, C.W. Nan, L.Q. Chen, Y. Shen, Polymer nanocomposites with interpenetrating gradient structure exhibiting ultrahigh discharge efficiency and energy density. Adv. Energy Mater. **9**(15), 1803411 (2019)
55. Q. Li, F.H. Liu, T.N. Yang, M.R. Gadinski, G.Z. Zhang, L.Q. Chen, Q. Wang, Sandwich-structured polymer nanocomposites with high energy density and great charge-discharge efficiency at elevated temperatures. PNAS **113**(36), 9995–10000 (2016)
56. G. Liu, T.D. Zhang, Y. Feng, Y.Q. Zhang, C.H. Zhang, Y. Zhang, X. Wang, Q.G. Chi, Q.G. Chen, Q.Q. Lei, Sandwich-structured polymers with electrospun boron nitrides layers as high-temperature energy storage dielectrics. Chem. Eng. J. **389**, 124443 (2020)

57. Y. Zhou, Q. Li, B. Dang, Ya. Yang, T. Shao, H. Li, J. Hu, R. Zeng, J.L. He, Q. Wang, A scalable, high-throughput, and environmentally benign approach to polymer dielectrics exhibiting significantly improved capacitive performance at high temperatures. Adv. Mater. **30**(49), 1805672 (2018)
58. Y.K. Zhu, Y.J. Zhu, X.Y. Huang, J. Chen, Q. Li, J.L. He, P.K. Jiang, High energy density polymer dielectrics interlayered by assembled boron nitride nanosheets. Adv. Energy Mater. **9**(40), 1901826 (2019)

Correction to: Polypropylene Insulation Materials for HVDC Cables

Jun-Wei Zha, Ming-Sheng Zheng, Wei-Kang Li, George Chen, and Zhi-Min Dang

Correction to: Chapter "Polypropylene Insulation Materials for HVDC Cables" in: B. Du (ed.), *Polymer Insulation Applied for HVDC Transmission*, https://doi.org/10.1007/978-981-15-9731-2_4

In the original version of the book, the following corrections have been incorporated: In the chapter "Polypropylene Insulation Materials for HVDC Cables", the affiliations of the authors have been updated as follows:

George Chen
University of Southampton, Southampton, UK
Zhi-Min Dang
Tsinghua University, Beijing, China
Wei-Kang Li
Global Energy Interconnection Research Institute Corporation, Beijing, China
The chapter and the book have been updated with the changes.

The updated version of this chapter can be found at
https://doi.org/10.1007/978-981-15-9731-2_4

B. Du (ed.), *Polymer Insulation Applied for HVDC Transmission*,
https://doi.org/10.1007/978-981-15-9731-2_27

The manufacturer's authorised representative in the EU is Springer Nature Customer Service Centre GmbH, Europaplatz 3, 69115 Heidelberg, Germany. If you have any concerns regarding our products, please contact ProductSafety@springernature.com

Printed and bound by CPI Group (UK) Ltd, Croydon, CR0 4YY

30/09/2026

02225288-0019